FOUNDATIONS OF INTEGER PROGRAMMING

FOUNDATIONS OF INTEGER PROGRAMMING

Harvey M. Salkin
Kamlesh Mathur

Department of Operations Research
Weatherhead School of Management
Case Western Reserve University
Cleveland, Ohio

with contributions by
Robert Haas

North-Holland

New York • Amsterdam • London

Elsevier Science Publishing Co., Inc.
655 Avenue of the Americas, New York, New York 10010

Sole distributors outside the United States and Canada:

Elsevier Science Publishers B.V.
P.O. Box 211, 1000 AE Amsterdam , the Netherlands

©1989 by Elsevier Science Publishing Co., Inc.

Library of Congress Cataloging-in-Publication Data
Salkin, Harvey M.
Foundations of integer programming / Harvey M. Salkin, Kamlesh
Mathur; with contributions by Robert Haas.
p.   cm.
Includes bibliographical references.
ISBN 0-444-01231-1
1. Integer programming.   I. Mathur, Kamlesh.   II. Haas, Robert,
Dr.   III. Title.
T57.74.S255 1989
519.7'7--dc20   89-16919
CIP

ISBN 0-444-01231-1

Current printing (last digit):
10 9 8 7 6 5 4 3 2 1

Manufactured in the United States of America

To all those we love ...
HMS, KM

Contents

Preface   **xvii**

1   Introduction to Integer Programming   1

  1.1   Linear Programs with Integer Variables   1

  1.2   Uses and Applications   3

        1.2.1   Formulations That Allow Integer Variables   3

        1.2.2   Classical Applications and Case Studies   8

        Problems   20

2   Review of Linear Programming   27

  2.1   The Linear Programming Problem   27

  2.2   Graphical Solution and Geometric Concepts   28

  2.3   The Simplex Algorithm   31

        2.3.1   Definitions   31

        2.3.2   Fundamental Theorems   34

        2.3.3   The Simplex Algorithm   35

        2.3.4   The Two-Phase Simplex Method   42

  2.4   The Revised Simplex Algorithm   47

  2.5   Duality in Linear Programming   53

  2.6   The Dual Simplex Algorithm   57

  2.7   The Beale Tableau   62

        Problems   67

3   Using Linear Programming to Solve Integer Programs   70

  3.1   Graphical Solutions to Mixed Integer or Programs   70

  3.2   Solving an Integer Programming Problem as a Linear

        Program   73

        3.2.1   Unimodularity   74

  3.3   Obtaining Integer Programming Solutions by Rounding

        Linear Programming Solutions   76

3.4　An Overview of Approaches for Solving Integer (or Mixed-
Integer) Problems 78

3.4.1　Cutting Plane Techniques (Chapters 4, 5, 6, 7) . . 79

3.4.2　Enumerative Methods (Chapters 8, 9) 79

3.4.3　Partitioning Algorithms (Chapter 10) 79

3.4.4　Group Theoretic Algorithms (Chapter 11) 79

Problems 80

4　Dual Fractional Integer Programming　84

4.1　The Basic Approach 84

4.2　Notation: The Beale Tableau 87

4.3　The Form of the (Gomory) Cut 89

4.4　Illustrations 90

4.5　The Derivation of the Cut 99

4.5.1　Congruence 99

4.5.2　Derivation 100

4.6　Some Properties of Added Inequalities 102

4.7　Algorithm Strategies 116

4.7.1　Number of Possible Inequalities 116

4.7.2　Choosing the Source Row 116

4.7.3　Dropping Inequalities 117

4.8　A Geometric Derivation 118

4.9　Finiteness 121

Problems 123

Appendix A: Convergence using the Dantzig Cut 132

Appendix B: A Variation of the Basic Approach: The
Accelerated Euclidean Algorithm 135

Appendix C: Geometrically Derived Cuts 142

5　Dual Fractional Mixed Integer Programming　150

5.1　The Basic Approach 150

5.2　The Form of the Cut 151

5.3　An Illustration 152

5.4　The Derivation of the Cut 155

5.5　Applying the Mixed Cut to the Integer Program 159

5.6　Finiteness 163

Problems 166

6 Dual All-Integer Integer Programming **170**

6.1 The Basic Approach 170
6.2 The Form of the Cut 172
    6.2.1 The Rules for Finding λ 173
6.3 Illustrations 173
6.4 Derivations: The Form of the Cut, the Pivot Column,
    and the λ Selection Rules 179
    6.4.1 The Form of the Cut 179
    6.4.2 Choosing the Pivot Column 181
    6.4.3 λ Selection Rules 182
6.5 Some Properties of the Added Inequalities 184
    6.5.1 The Relative Strength of the Added Inequalities . 184
    6.5.2 A Second Derivation of the Fractional Cut, and
    Relation to the All-Integer Cut 186
6.6 Algorithm Strategies 187
    6.6.1 Choosing the Source Row 187
    6.6.2 Dropping Inequalities 188
6.7 Finiteness 188
    6.7.1 If Only the Cost Row Is All-Integer 189
    Problems 190

7 Primal All-Integer Integer Programming **197**

7.1 Introduction 197
7.2 The Tableau, the Rudimentary Primal Algorithm 198
    7.2.1 The Rudimentary Algorithm 199
    7.2.2 The Rudimentary Primary Algorithm 200
7.3 A Convergent Algorithm: The Simplified Algorithm (SPA) 204
    7.3.1 Modifications 1 and 2: Introducing a Reference
    Row and Selecting the Pivot Column 205
    7.3.2 Modification 3: Acceptable Source Row Selection
    Rules 206
    7.3.3 The Simplified Primal Algorithm (SPA) 211
7.4 A Second Reference Equation: Using the Dual Variables 213
7.5 Illustrations (SPA) 216
7.6 Optimality without Dual Feasibility 226
7.7 Convergence 228
    7.7.1 Finiteness under Rule 1 231
    7.7.2 Finiteness under Rule 2 232
    7.7.3 Finiteness under Rule 3 233

Problems 234

Appendix A: Cycling in Rudimentary Promal Algorithm 240

8 Branch and Bound Enumeration   245

8.1 Introduction 245
8.2 The Problem, Notation, and the Basic Result 247
    8.2.1 The Problem, Notation 247
    8.2.2 The Basic Result and Geometric Interpretations . 248
8.3 The Enumeration Tree, Algorithm Formulation, An Example 255
    8.3.1 The Tree, Algorithm Formulation 255
    8.3.2 An Example 257
8.4 The Basic Approach, A Second Example 261
    8.4.1 The Basic Approach 261
    8.4.2 A Second Example 263
8.5 A Variation of the Basic Approach 264
8.6 Specialization for the Zero-One Problem 267
8.7 Node Selection, Branching Rules, and Penalties 269
    8.7.1 Node Selection 269
    8.7.2 Branching Rules and Penalties 272
    Problems 277
    Appendix A: Computational Details of Examples 8.3 and 8.4 287

9 Search Enumeration   298

9.1 Introduction 298
9.2 The Basic Approach, the Tree 299
    9.2.1 The Basic Approach 300
    9.2.2 The Tree 300
9.3 The Point Algorithm: Implicit Enumeration Criteria . . 304
    9.3.1 Ceiling Tests 305
    9.3.2 Infeasibility Test 306
    9.3.3 Cancellation Zero Test 306
    9.3.4 Cancellation One Test 307
    9.3.5 Linear Programming 308
    9.3.6 Post-Optimization, Penalties 309
    9.3.7 Surrogate Constraints 310
9.4 The Point Algorithm: Branching Strategies 314
    9.4.1 Preferred Sets 314

9.4.2 The Balas Test 317
9.4.3 Other Branching Rules 318
9.5 The Generalized Origin, Restarts 318
9.6 Search Enumeration in $0 - 1$ Mixed Integer
Programming 322
Problems 325
Appendix A: A Sample Search Algorithm and Its Imple-
mentation 339
A.1 The Basic Approach; The Point Algorithm 340
A.2 The Bookkeeping Scheme 341
Appendix B: Computational Experience 343
B.1 Program Description 343

10 Partitioning in Mixed Integer Programming **346**

10.1 Introduction 346
10.2 Posing the Mixed Integer Program as an Integer
Program 347
10.3 The Partitioning Algorithm 352
10.4 Properties of the Partitioning Algorithm 356
10.5 Finiteness 360
Problems 363
Appendix A: Application of the Partitioning Algorithm
to the Uncapacitated Plant Location Problem 370
A.1 Problem [P1]'s Equivalent Integer Program [I] 371
A.2 Solving the Linear Program [DL] 372
A.3 Summary of the Partitioning Algorithm 374
A.4 An Illustrative Example 375

11 Group Theory in Integer Programming **380**

11.1 Introduction 380
11.2 The Group Minimization Problem 383
11.2.1 The Group $G(\bar{\alpha})$ 387
11.2.2 Solving the GMP 393
11.3 Solving the Group Minimization Problem 394
11.3.1 Dynamic Programming Algorithms 395
11.3.2 A Sufficient Condition for $x_B \geq 0$ 405
11.3.3 An Enumeration Algorithm 410
11.4 The Group Minimization Problem Viewed as a Network
Problem 415

    11.4.1 Formulation 415

    11.4.2 Solving the Integer Program by Solving the Net-
           work Problem 420

11.5 An Equivalent Group Minimization Problem 422

11.6 The Isomorphic Factor Group $G(\mathbf{A})/G(\mathbf{B})$ 430

    11.6.1 Definitions, Results 430

    11.6.2 The Isomorphic Groups $G(\bar{\alpha})$ and $G(\lambda)$ 434

    11.6.3   The Subgroup Decomposition of $G(\mathbf{A})/G(\mathbf{B})$ . . 439

    11.6.4 The Order of $G(\bar{\alpha})$ and $G(\lambda)$ 442

11.7 The Geometry 443

    11.7.1 The Corner Polyhedron ($\mathbf{x}'$ Space) 443

    11.7.2 The Corner Polyhedron ($\mathbf{x_N}$ Space) 446

    11.7.3 Relating the Corner Polyhedra 448

    11.7.4 The Master (Corner) Polyhedron 453

    11.7.5 Generating Valid Inequalities from the Faces of
           Master Polyhedra 461

    Problems 467

    Appendix A: A Shortest Route Algorithm 485

A.1 Specialization of the Shortest Path Algorithm for the
Group Minimization Problem 491

    Appendix B: Diagonalizing the Basis – Smith's Normal
Form 492

    Appendix C: Computational Experience 499

C.1 Computer Programmed Group Minimization Algorithms 499

    C.1.1 A Dynamic Programming Algorithm 499

    C.1.2 Branch and Bound Algorithm 501

    C.1.3 Shortest Route Algorithms 502

C.2 Reducing The Order Of The Group 503

    C.2.1 Scaling 503

    C.2.2 Relaxation 505

    C.2.3 Decomposition 506

    Appendix D: Implementation of a Group Theoretic Al-
gorithm 508

D.1 Linear Programming Module 508

D.2 Construction of FGMP 508

D.3 Solution of the FGMP 509

D.4 Branch and Bound Algorithm 509

D.4.1 Node Selection Rule 509

D.4.2 Implicit Branching Rule 509

D.4.3 Node Omission Rule 510

D.4.4 Combining the Implicit Branching Rule and the
Node Omission Rule 510

12 The Knapsack Problem   **513**

12.1 Introduction 513

12.2 Applications and Uses of Knapsack and Related
Problems 516

     12.2.1 Capital Budgeting 516

     12.2.2 The Cutting Stock Problem 519

     12.2.3 Loading Problems 524

     12.2.4 Change Making Problem 525

     12.2.5 Other Uses 525

12.3 Reducing Integer Programs to Knapsack Problems: Aggregating Constraints 526

     12.3.1 An Aggregation Process 527

     12.3.2 An Improved Aggregation Process 533

12.4 Algorithms 538

     12.4.1 Dynamic Programming Techniques 538

     12.4.2 A Periodic Property 546

     12.4.3 Branch and Bound Algorithms: General Knap-
sack Problem 554

     12.4.4 Lagrangian Multiplier Methods 559

     12.4.5 Network Approaches 564

12.5 Branch and Bound Algorithms: 0-1 Knapsack Problem . 567

     12.5.1 The Linear Programming Solution 567

     12.5.2 The Upper Bound Solution 568

     12.5.3 The Lower Bound Solution 569

     12.5.4 Reduction Algorithm 569

     12.5.5 The Branch and Bound Algorithm 571

     Problems 578

**13 The Set Covering Problem, the Set Partitioning Prob-
lem**   **590**

13.1 Introduction 590

13.2 Set Covering and Networks 593

     13.2.1 The Node Covering Problem 593

13.2.2 The Matching Problem 595
13.2.3 Disconnecting Paths 597
13.2.4 The Maximum Flow Problem 599
13.3 Applications 600
13.3.1 Airline Crew Scheduling 600
13.3.2 Truck Scheduling 602
13.3.3 Political Redistricting 602
13.3.4 Information Retrieval 603
13.4 Relevant Results 603
13.5 Algorithms 614
13.5.1 A Search Algorithm for the Set Partitioning Problem 615
13.5.2 A Search Algorithm for the Set Covering Problem 618
Problems 625
Appendix A: Computational Experience 633
A.1 Cutting Plane Algorithms 633
A.2 Enumerative Algorithms 635
A.3 Summary 638

**14 The Fixed Charge Problems: The Plant Location Prob-
lem and Fixed Charge Transportation Problem   639**

14.1 Introduction 639
14.1.1 The Plant Location Problem 639
14.1.2 The Fixed Charge Transportation Problem 641
14.2 Algorithms for the Plant Location Problem 643
14.2.1 A Branch and Bound Algorithm for the Fixed
Charge Problem 644
14.2.2 A Branch and Bound Algorithm for the Plant
Location Problem 646
14.3 Branch and Bound Algorithm for the Fixed Charge Transportation Problem 652
Problems 656
Appendix A: Computational Experience 664
A.1 The General Fixed Charge Problem 664
A.2 The Fixed Charge (or Capacitated Plant Location) Prob-
lem 666
A.3 The (Uncapacitated) Plant Location Problem 666
A.4 Summary 667

15 The Traveling Salesman Problem **669**

15.1 Introduction 669
15.2 Mathematical Formulation 670
15.3 Algorithms 673
    15.3.1 Bounding Rules 675
    15.3.2 Branching Rules 677
15.4 Approximate Algorithms 682
    15.4.1 Tour Construction Procedures 682
    15.4.2 Tour Improvement Algorithms 689
    Problems 692
    Appendix A: The Assignment Algorithm 695
    Appendix B: Some Counterexamples to Heuristic Algo-
    rithms for the Traveling Salesman Problem 700
B.1 Cheapest Insertion Algorithm (nonsymmetric case) . . . 700
B.2 Cheapest Insertion Algorithm (symmetric case) 701
B.3 Convex Hull Heuristic 702
B.4 Two Edge Interchange (2-opt) Algorithm 703

Bibliography **705**

Author Index **745**

Subject Index **751**

Preface

We have found the literature devoted to integer programming very difficult to read, principally because it assumes substantial prerequisite knowledge and (or) basic concepts are often confused by voluminous mathematical notation. This book overcomes these difficulties and presents a unifying, easy to read exposition of integer linear programming techniques and applications. Wherever possible, concepts are presented through motivating passages, examples, and graphical illustrations.

The book assumes understanding of matrix arithmetic and some background in linear programming. However, Chapter 2 reviews the fundamental results of the simplex and dual simplex algorithms used to develop integer programming results and algorithms in the subsequent chapters.

The book can naturally be used for reference purposes, but is designed as a text with emphasis on building a strong foundation of each integer programming technique. Each chapter is structured so that introductory and background material are presented first. This is usually followed by the chapter's principal content in logical order, and then by more detailed points and mathematical proofs. Problems are at the end of each chapter and are arranged chronologically with the chapter's sections. Appendices appear wherever appropriate and usually describe computational results, algorithm design for computer implementation, or to provide details of numerical examples.

Chapter 1 contains an introduction to the field and includes several well known applications. Chapter 2 reviews fundamental linear programming results and develops the simplex and dual simplex algorithms. Chapter 3 discusses classes of integer programs which may be solved by the simplex algorithm. For problems which may not be solved by the simplex method, the pitfalls of rounding the LP fractional values to produce an integer solution are also described and illustrated. The next nine chapters are algorithm oriented. In particular, Chapters 4, 5, 6, and 7 present cutting plane techniques, and Chapters 8 and 9 describe enumerative methods. Chapter 10 relates to Benders' partitioning algorithm specialized for the mixed integer programming problem. Chapter 11 is devoted to the use of group theory in integer programming and discusses the pioneering work of Ralph E. Gomory. Chapters 12, 13, 14, and 15 describe integer programming models with particular structure and corresponding special purpose algorithms, results, and applications.

The narrative style we adopted coupled with the use of detailed illustrations and the many topics discussed naturally produced a large manuscript. As a result, it became necessary to omit some topics such as integer nonlinear programming. However, some of these appear in the problems as appropriate extensions to the material discussed in the chapters.

We feel that it is not possible to cover the entire text in a one semester course. However, in a (fourteen week) one semester graduate course at Case Western Reserve University, we get through most of the material in the first ten Chapters and sometimes parts of Chapters 11 and 12. Other possible course plans depend on the number of weeks available and emphasis desired. Sample teaching sequences are listed in the table below.

COURSE EMPHASIS		(COURSE DURATION)		
		Quarter		Semester
Techniques	1	Chapters 1,2,3	1	Chapters 1,2,3
	2	Early Parts of Chapters 4,5,6,7,8,9	2	Most of Chapters 4,5,6,7,8,9
	3	Chapter 10	3	Most of Chapter 10
			4	Part of Chapter 11
Techniques and applications	1	Chapters 1,2,3	1	Chapters 1,2,3
	2	Early Parts of Chapters 4,6,8,9	2	Early Parts of Chapters 4,5,6,7,8,9
	3	Early Parts of Chapter 12	3	Early Parts of Chapter 10
	4	Chapters 13,14	4	Most of Chapters 12,13,14,15

During the development of this book, we have received contributions and help from numerous graduate students and other individuals. We are especially indebted to Dr. Robert Haas for his contributions and suggestions which have produced a much more readable text, and also to Gregory Pollock for his constructive criticism and for proofreading the manuscript many times. Appreciation is also extended to Bardia Kamrad, Gadi Rabinowitz, and Leah Chu. The Department of Operations Research and our colleagues deserve our thanks for providing a conducive environment and general support. Last, but not least, we are indebted to Elaine Iannacelli for providing her typing and artistic skills.

H. M. S.
K. M.
Cleveland, Ohio
July, 1989

Chapter 1

Introduction to Integer Programming

1.1 Linear Programs with Integer Variables

A *linear program* (LP) is a mathematical model where we wish to find a set of nonnegative values for the unknowns or variables which maximizes or minimizes a linear equation or objective function while satisfying a system of linear constraints. In many situations a linear programming model may not be sufficient as the variables must take on integer values. For example, we cannot build 1.37 schools, manufacture 11.74 aircraft, or award 10.48 research contracts. A linear program with some, but not all, of the variables required to be integer, is a *mixed integer (linear) program* (MIP). If all the variables are required to be integer, we have an *integer (linear) program* (IP).

Using matrix notation, a mixed integer program[1] may be written as

$$(\text{MIP}) \quad \text{Maximize} \quad \mathbf{cx} + \mathbf{dy} = z \qquad (1)$$
$$\text{subject to} \quad \mathbf{Ax} + \mathbf{Dy} \le \mathbf{b}, \qquad (2)$$
$$\mathbf{x} \ge \mathbf{0}, \mathbf{y} \ge \mathbf{0}, \qquad (3)$$
$$\text{and} \quad \mathbf{x} \text{ integer}$$

where $\mathbf{c} = (c_j)$ is an n row (or cost) vector,
$\mathbf{d} = (d_j)$ is an n' row (or cost) vector,
$\mathbf{A} = (a_{ij})$ is an m by n (constraint) matrix,

---

[1] We shall interchangeably use MIP, mixed problem, mixed program, mixed integer program, mixed integer problem, and mixed integer linear program. Corresponding equivalences will be used when referring to the integer linear program.

$$\mathbf{D} = (d_{ij}) \quad \text{is an } m \text{ by } n' \text{ (constraint) matrix,}$$
$$\mathbf{b} = (b_i) \quad \text{is an } m \text{ column vector of constants}$$
$$\text{(or, simply, the right-hand side),}$$
$$\mathbf{x} = (x_j) \quad \text{is an } n \text{ vector of integer variables,}$$
$$\text{and} \quad \mathbf{y} = (y_j) \quad \text{is an } n' \text{ vector of continuous variables.}$$

When $n' = 0$, the continuous variables $\mathbf{y}$ vanish and we have an integer program. If $n = 0$, there are no integer variables $\mathbf{x}$ and the problem reduces to a linear program. Mixed integer (or integer) programs in which the integer variables are constrained to be 0 or 1 are called *zero-one mixed integer (or integer) programs* . The following result shows that every mixed integer (or integer) program in which the integer variables are bounded above can be posed as a zero-one mixed integer (or integer) problem.

Theorem 1.1 *Suppose in problem MIP (or IP) each $x_j \leq u_j$ (a positive integer); then the mixed integer (or integer) program is equivalent to a zero-one mixed integer (or integer) program.*

Proof: Replace each x_j by either (i) $\sum_{k=1}^{u_j} t_{kj}$ where the t_{kj}'s are 0-1 variables, and omit the $x_j \leq u_j$ constraints (since $\sum_{k=1}^{u_j} t_{kj}$ can at most be u_j), or (ii) $\sum_{k=0}^{l_j} 2^k t_{kj}$ where the t_{kj}'s are 0-1 variables and l_j is the smallest integer such that[2] $\sum_{k=0}^{l_j} 2^k = 2^{l_j+1} - 1 \geq u_j$ and retain the $x_j = \sum_{k=0}^{l_j} 2^k t_{kj} \leq u_j$ constraints. The result follows since either substitution allows x_j to take on any integer value between 0 and u_j. (This is evident in (i) and also in (ii), since every integer can be written in the base two.) ∎

Example 1.1 If $u_j = 107$, then in (ii), $l_j = 6$, since

$$\sum_{k=0}^{6} 2^k = 2^7 - 1 \geq 107$$

and

$$\sum_{k=0}^{5} 2^6 - 1 < 107;$$

---

[2]The sum of the finite geometric series $a_0 + a_1 + a_2 + \ldots + a_N$ is $(a_0 - a_0 r^{N+1})/(1-r)$, where $r = a_{i+1}/a_i$ $(i = 0, 1, \ldots N-1)$; and thus the equality.

therefore, replace x_j by

$$\sum_{k=0}^{6} 2^k t_{kj} = t_{0j} + 2t_{1j} + 4t_{2j} + 8t_{3j} + 16t_{4j} + 32t_{5j} + 64t_{6j},$$

and retain the constraint

$$\sum_{k=0}^{6} 2^k t_{kj} \leq 107.$$

Observe that whenever a general integer variable is written as the sum of 0-1 variables, the problem size increases rapidly with u_j. Although it is sometimes easier to solve a zero-one mixed integer or integer problem than a general one, the growth in the number of integer variables on such a transformation usually makes it computationally unattractive except perhaps for problems with small values of u_j. Nevertheless, the theorem is of some value since certain results are easier to show for zero-one problems.

1.2 Uses and Applications

The importance of mixed integer and integer programming is well established. Many mathematical programs can be converted to problems with integer variables (Section 1.2.1). Also, as already mentioned, many situations yield linear programming formulations where some or all of the variables are required to be integer. Included in these are scheduling, location, network, radio frequency management, and selection problems which appear in industry, military, education, health, and other environments. Some classical models are given below (Section 1.2.2), while a more complete discussion may be found in Balinski [50], Balinski and Spielberg [51], and Dantzig [158]. Other applications (mainly of the scheduling and sequencing type which are not described in later chapters) are listed in references.

1.2.1 Formulations That Allow Integer Variables

Restrictions on the Solution Values

Suppose a variable z_j is allowed to take on only one of several values, say $v_{11}, v_{12}, \ldots, v_{1N}$. This is equivalent to setting

$$z_j = x_1 v_{11} + x_2 v_{12} + \ldots + x_N v_{1N}$$

with

$$x_1 + x_2 + \ldots + x_N = 1,$$

and,

$$x_j = 0 \text{ or } 1 \quad (j = 1, \ldots, N).$$

We may generalize this to the case where the vector $\mathbf{z}$ is allowed to be either the vector $\mathbf{v}_1, \mathbf{v}_2, \ldots,$ or $\mathbf{v}_N$, by setting

$$\mathbf{z} = x_1\mathbf{v}_1 + x_2\mathbf{v}_2 + \ldots + x_N\mathbf{v}_N,$$

with

$$x_1 + x_2 + \ldots + x_N = 1,$$

and

$$x_j = 0 \text{ or } 1 \quad (j = 1, \ldots, N).$$

Piecewise Convex Constraint Sets

Consider a model whose constraint set consists of several convex subsets. The union of the subsets is not necessarily convex. (They may, for example, be disjoint.) To be a solution, a point must satisfy some of the subsets. Specifically, we wish to

$$\begin{array}{ll} \text{maximize} & \mathbf{cy} \\ \text{subject to} & \mathbf{b}_i - \mathbf{A}_i\mathbf{y} \geq \mathbf{0} \quad (i = 1, \ldots, p) \end{array}$$

for at least t of the p constraint subsets[3]. Here $\mathbf{A}_i$ is a matrix and $\mathbf{b}_i$ is a vector. This is equivalent to the mixed integer program

$$\begin{array}{lll} \text{maximize} & \mathbf{cy} \\ \text{subject to} & (\mathbf{b}_i - \mathbf{A}_i\mathbf{y}) - \mathbf{L}_i x_i \geq \mathbf{0} & (i = 1, \ldots, p), \\ & \displaystyle\sum_{i=1}^{p} x_i = p - t \\ \text{and} & x_i = 0 \text{ or } 1 & (i = 1, \ldots, p), \end{array}$$

where $\mathbf{L}_i$ is a lower bound on the vector function $\mathbf{b}_i - \mathbf{A}_i\mathbf{y}$. When $x_i = 1$ the constraints reduce to $\mathbf{b}_i - \mathbf{A}_i\mathbf{y} \geq \mathbf{L}_i$ which are automatically satisfied and may be ignored; if $x_i = 0$ the inequality becomes $\mathbf{b}_i -$

---

[3] A linear programming algorithm for the $t = 1$ case is described by Fricks[209].

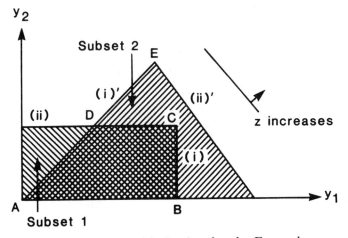

Figure 1.1: Feasible Region for the Example

$\mathbf{A}_i\mathbf{y} \geq \mathbf{0}$. So, if $p-t$ of the x_i's are 1, t of the x_i's are 0, and a solution $\mathbf{y}$ will satisfy $\mathbf{b}_i - \mathbf{A}_i\mathbf{y} \geq \mathbf{0}$ for at least t of the p constraint sets. As an example consider,

$$
\begin{array}{llrcll}
\text{maximize} & y_1 & + & y_2 & = & z \\
\end{array}
$$

$$
\begin{array}{llrrll}
\text{subject to} & y_1 & & \leq & 4 & (i) \\
& & y_2 & \leq & 2 & (ii)
\end{array} \left.\vphantom{\begin{array}{l}1\\1\end{array}}\right\} \text{subset 1}
$$

$$
\begin{array}{rrrll}
-y_1 & + & y_2 & \leq & 0 & (i)' \\
4y_1 & + & 3y_2 & \leq & 24 & (ii)'
\end{array} \left.\vphantom{\begin{array}{l}1\\1\end{array}}\right\} \text{subset 2}
$$

$$
\text{and} \qquad y_1, \qquad y_2 \;\geq\; 0
$$

for at least one of the two sets (i.e., $t = 1$). The (nonconvex) set of feasible solutions appears in Fig. 1.1. Note that the four constraints taken together define the region ABCD, but the optimal solution to this problem occurs at the point E.

To obtain the equivalent mixed integer program we first take the lower bound vector for

$$
\begin{pmatrix} 4 - y_1 \\ 2 - y_2 \end{pmatrix} \text{and} \begin{pmatrix} y_1 - y_2 \\ 24 - 4y_1 - 3y_2 \end{pmatrix} \text{to be} \begin{pmatrix} -M \\ -M \end{pmatrix},
$$

where M is some large positive number. Then the equivalent problem is

$$
\begin{aligned}
\text{maximize} \quad & y_1 + y_2 && = z \\
\text{subject to} \quad 4 - \; & y_1 && + M x_1 && \geq 0, \\
2 \quad & - \; y_2 + M x_1 && \geq 0, \\
& y_1 - y_2 && + M x_2 \geq 0, \\
24 - \; & 4y_1 - 3y_2 && + M x_2 \geq 0, \\
& && x_1 + x_2 = 1, \\
\text{and} \quad & y_1,\, y_2 \geq 0; && x_1,\, x_2 = 0 \text{ or } 1.
\end{aligned}
$$

Observe that constraints of the form $(\mathbf{b}_i - \mathbf{A}_i \mathbf{y}) - x_i \mathbf{v}_i \leq \mathbf{0}$ where $\mathbf{v}_i$ is an upper bound on the vector function $\mathbf{b}_i - \mathbf{A}_i \mathbf{y}$, appear when the constraint subsets are $\mathbf{b}_i - \mathbf{A}_i \mathbf{y} \leq \mathbf{0}$.

Piecewise Linear Objective Functions

Suppose that the objective function z is the sum of piecewise linear functions in one variable; that is, $z = \sum_j f_j(y_j)$, with each $f_j(y_j)$ as in Fig. 1.2. (The subscript j is omitted in the figure.)

Then, since any point y in the closed interval $[\bar{y}_i, \bar{y}_{i+1}]$ may be written as $a_i \bar{y}_i + a_{i+1} \bar{y}_{i+1}$, where $a_i + a_{i+1} = 1$ and $a_i, a_{i+1} \geq 0$, we have, by the linearity of f in the interval, that $f(y) = a_i f(\bar{y}_i) + a_{i+1} f(\bar{y}_{i+1})$. This suggests setting

$$
f(y) = a_1 f(\bar{y}_1) + a_2 f(\bar{y}_2) + \ldots + a_N f(\bar{y}_N),
$$

where

$$
a_1 \bar{y}_1 + a_2 \bar{y}_2 + \ldots + a_N \bar{y}_N = y
$$

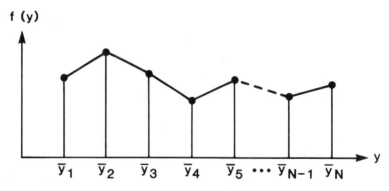

Figure 1.2: Piecewise Linear Function

$$
\begin{array}{ccccccccl}
a_1 & + & a_2 & + & \cdots & + & a_N & = & 1 \\
a_1 & & & & & & & \leq & x_1 \\
& & a_2 & & & & & \leq & x_1 + x_2 \\
& & & \ddots & & & & & \vdots \\
& & & & a_{i+1} & & & \leq & x_i + x_{i+1} \\
& & & & & \ddots & & & \vdots \\
& & & & & & a_N & \leq & x_{N-1}
\end{array}
$$

and $x_1 + x_2 + \ldots + x_{N-1} = 1$, $x_i = 0$ or 1 $(i = 1, \ldots, N-1)$,

where $y \geq 0$ and $a_i \geq 0$ $(i = 1, \ldots, N)$. When an x_i is 1 (with the rest 0), we have $a_i + a_{i+1} = 1$ and the other a_i's must be 0, or the program then considers points y in the interval $[\bar{y}_i, \bar{y}_{i+1}]$ and $f(y)$ reduces to $a_i f(\bar{y}_i) + a_{i+1} f(\bar{y}_{i+1})$. As an example consider

$$
\begin{array}{lll}
\text{maximize} & f_1(y_1) & + \; y_2 \\
\text{subject to} & y_1 & + \; y_2 \leq 3, \\
\text{and} & y_1, & \quad y_2 \geq 0,
\end{array}
$$

where

$$
f_1(y_1) = \begin{cases} y_1 & \text{for } 0 \leq y_1 \leq 2 \\ 4 - y_1 & \text{for } 2 \leq y_1 \leq 3 \\ 0 & \text{otherwise.} \end{cases}
$$

Then (see Fig. 1.3) $f_1(y_1) = a_1 f_1(0) + a_2 f_1(2) + a_3 f_1(3) = 2a_2 + a_3$,

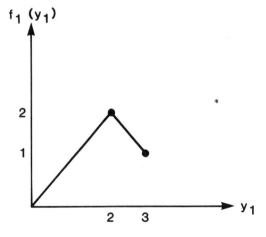

$f_1 (y_1)$ Figure 1.3: Function $f_1(y_1)$

and $a_1 \bar{y}_1 + a_2 \bar{y}_2 + a_3 \bar{y}_3 = a_1(0) + a_2(2) + a_3(3) = y_1$, or the equivalent problem is

$$
\begin{array}{lllllll}
\text{maximize} & & & 2a_2 & + & a_3 & + & y_2 & \\
\text{subject to} & & & 2a_2 & + & 3a_3 & + & y_2 & \leq 3, \\
& a_1 & & & & & & & \leq x_1, \\
& & & a_2 & & & & & \leq x_1 + x_2, \\
& & & & & a_3 & & & \leq x_2, \\
& a_1 & + & a_2 & + & a_3 & & & = 1, \\
& y_1 & = & 2a_2 & + & 3a_3 & & & \geq 0, \\
& x_1 & + & x_2 & = & 1, & x_1, & x_2 & = 0 \text{ or } 1, \\
\text{and} & a_1, & & a_2, & & a_3, & & y_2 & \geq 0.
\end{array}
$$

The transformation given here can be used to approximate a nonlinear function. A nonlinear function can also be linearized by first transforming it into a polynomial function with zero-one variables (Balas [18]), Hammer and Rudeanu [311]) and then transforming the polynomial function into a linear function with zero-one variables (Balas [18], Watters [593], Zangwill [617]). However, as suggested by the example these transformations tend to dramatically increase the number of variables and constraints. Methods to achieve more economical linear representations of zero-one polynomial programming problems are discussed by Glover [254], and Glover and Woolsey [268], [269]. A computational study by Taha [559] indicates that the benefit (in terms of ease of solution) gained by using these transformations is data dependent, and conversion may not be worthwhile. Related work includes the relationship between a zero-one integer program and a quadratic programming problem (Problem 1.4) as discussed in Bowman and Glover [89], Kennington and Fyffe [363], and Raghavachari [470], [471]; other transformations are in Garfinkel and Nemhauser [217], Granot and Hammer [290], and Petersen [459].

1.2.2 Classical Applications and Case Studies

In practice a few integer programming models represent a vast number of real world problems, and as a result they have been titled and considerable work has been done on developing algorithms and computer packages for them (a listing is in Salkin [507]). We now describe these models. Applications and algorithms are left to subsequent chapters.

Facility Location Problems

In the simplest case m sources (or facility locations) produce a single commodity for n customers each with a demand for d_j units ($j = 1, \ldots, n$). If a particular source i is operating (or facility is built), it has a fixed cost $f_i \geq 0$ and a production capacity $M_i > 0$ associated with it. There is also a positive cost g_{ij} for shipping a unit from source i to customer j. The problem is to determine the location of the operating sources so that capacities are not exceeded and demands are met, all at a minimal total cost. All data are assumed to be integral.

To model this problem, we let z_{ij} be the amount shipped from source i to customer j, and define x_i to be 1 if source i is operating and 0 if it is not. Then the integer programming model is

$$\text{minimize} \quad \sum_{i=1}^{m} \sum_{j=1}^{n} g_{ij} z_{ij} + \sum_{i=1}^{m} f_i x_i \tag{4}$$

$$\text{subject to} \quad \sum_{i=1}^{m} z_{ij} = d_j \quad (j = 1, \ldots, n), \tag{5}$$

$$\sum_{j=1}^{n} z_{ij} \leq M_i x_i \quad (i = 1, \ldots, m), \tag{6}$$

$$z_{ij} \geq 0 \quad (\text{all } i, j), \tag{7}$$

$$\text{and} \quad x_i = 0 \text{ or } 1 \quad (i = 1, \ldots, m). \tag{8}$$

The objective function (4) is the total shipping cost, i.e., $\sum_i \sum_j g_{ij} z_{ij}$, plus the total fixed cost, i.e., $\sum_i f_i x_i$; note that f_i contributes to this sum only when $x_i = 1$ or source i is operating. Constraints (5) guarantee that each customer's demand is met. Inequality (6) ensures that we do not ship from a source which is not operating[4] (M_i is an upper bound on the amount that may be shipped from source i) and it also restricts production from exceeding capacity. Even though by definition z_{ij} is discrete, we need only the nonnegativity conditions (7) because it can be shown that constraints (8) with (5) and (6) ensure that $z_{ij} \geq 0$ will mean that z_{ij} is integer in the optimal solution (see Chapter 3).

In certain situations, because of available equipment or expected demand, it may be reasonable to assume that the amount produced at each facility i will never exceed the capacity M_i. The resulting problem

---

[4]This can also be guaranteed by using constraints of the form $z_{ij} \leq M_i x_i$ for all i, j

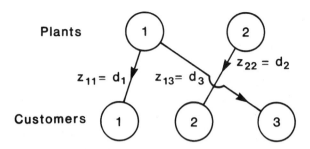

Figure 1.4: A 2 Plants 3 Customers Example

is sometimes referred to as the *uncapacitated plant location problem.* Under this assumption, for any fixed zero-one value of $\mathbf{x}$ the shipping cost is minimized if each customer's demand d_j is satisfied from the open plant i (i.e., $x_i = 1$) with minimal cost g_{ij}. To illustrate, consider Fig. 1.4, where plants 1 and 2 are open to satisfy the demands of three customers. Also, suppose $g_{11} < g_{21}, g_{13} < g_{23}$, and $g_{12} > g_{22}$. Then, to minimize the total shipping cost, we meet the demands of customers 1 and 3 from plant 1 and customer 2 from plant 2. (Alternate optima exist when there are equal shipping costs. For example, if $g_{11} = g_{21}$, minimal solutions contain $z_{11} = K$ and $z_{21} = d_1 - K$ where $0 \le K \le d_1$. However, there is always an optimal solution with d_j units obtained from one plant.)

The above discussion suggests a reformulation of the problem. Suppose we denote the fraction of demand d_j satisfied by plant i as y_{ij}, i.e., $y_{ij} = z_{ij}/d_j$. Then for any $\mathbf{x}$ there is a minimal solution with $y_{ij} = 0$ or 1 for all i, j. The set of inequalities (6) which has the sole function of prohibiting shipments from closed plants and allowing them otherwise, may be replaced by $\sum_{j=1}^{n} y_{ij} \le nx_i$. Letting $c_{ij} = g_{ij}d_j$ yields the reformulated problem (where the constraints (5)′ and (7)′ ensure that $y_{ij} \le 1$)

$$\text{minimize} \quad \sum_{i=1}^{m}\sum_{j=1}^{n} c_{ij}y_{ij} + \sum_{i=1}^{m} f_i x_i \tag{4}'$$

$$\text{subject to} \quad \sum_{i=1}^{m} y_{ij} = 1 \quad (j = 1, \ldots, n), \tag{5}'$$

$$\sum_{j=1}^{n} y_{ij} \le nx_i \quad (i = 1, \ldots, m), \tag{6}'$$

$$y_{ij} \ge 0 \quad (\text{all } i, j), \tag{7}'$$

and $\qquad x_i = 0 \text{ or } 1 \qquad (i = 1, \ldots, m).$ $\qquad\qquad$ (8)′

Multicommodity Distribution System Design

This problem is a generalization of a facility location problem where we have several commodities, and shipment from a plant to a customer occurs through a distribution center. The particular formulation given here is from Geoffrion and Graves [228] and can be described using the network in Figure 1.5. In particular, we have plants j $(j = 1, 2, \ldots, n)$ each having the capacity of producing S_{ij} units of commodity i $(i = 1, 2, \ldots, q)$. The demand for commodity i by customer l $(l = 1, 2, \ldots, m)$ is D_{il} and may be satisfied by shipping via a distribution center $(k = 1, \ldots, p;$ c.f. Figure 1.5). The p possible locations for the distribution centers are known, but the particular sites to be used are to be determined. The problem is to determine which distribution centers to use so that all customer demands are satisfied, production capacities are not exceeded, and the total distribution cost, that is, the fixed cost of opening the distribution centers and the transportation cost, is minimized.

To model this problem, we let

$\qquad i:$　index for commodities, $i = 1, \ldots, q$,

$\qquad j:$　index for plants, $j = 1, 2, \ldots, n$,

$\qquad k:$　index for distribution centers, $k = 1, 2, \ldots, p$,

$\qquad l:$　index for customers, $l = 1, 2, \ldots, m$,

$\quad S_{ij}:$　production capacity for commodity i at plant j,

$\quad D_{il}:$　demand for commodity i by customer l,

$L_k, U_k:$　minimum and maximum allowable total annual flow through the distribution center at site k,

$\quad f_k:$　fixed annual operating cost of the distribution center at site k,

$\quad v_k:$　per unit cost of flow through the distribution center at site k,

$\ c_{ijkl}:$　per unit cost of producing and shipping commodity i from plant j to customer l via distribution center at site k,

$\ x_{ijkl}:$　the amount of commodity i shipped from plant j to customer l via distribution center at site k,

$\quad y_{kl}:$　a 0-1 variable, where $y_{kl} = 1$ if distribution center k serves customer l, and $y_{kl} = 0$ otherwise, and

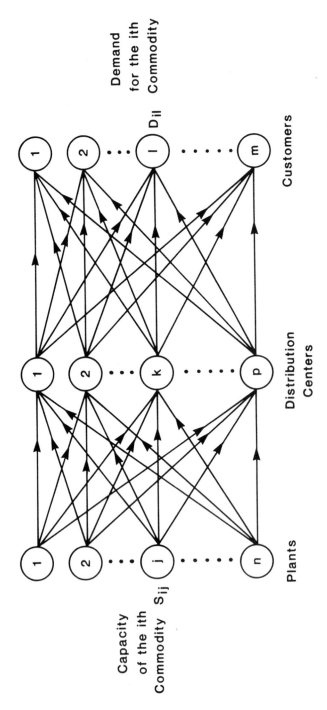

Figure 1.5: A Multicommodity Distribution Network

z_k: a 0-1 variable, where $z_k = 1$ if a distribution center is opened at site k, and 0 otherwise.

The mixed integer programming model is

$$\text{minimize} \quad \sum_{i=1}^{q}\sum_{j=1}^{n}\sum_{k=1}^{p}\sum_{l=1}^{m} c_{ijkl}x_{ijkl} + \sum_{k=1}^{p} f_k z_k + \sum_{k=1}^{p}\sum_{i=1}^{q}\sum_{l=1}^{m} v_k D_{il}y_{kl} \quad (9)$$

$$\text{subject to} \quad \sum_{k=1}^{p}\sum_{l=1}^{m} x_{ijkl} \leq S_{ij} \qquad \text{(all } i,j\text{)}, \qquad (10)$$

$$\sum_{j=1}^{n} x_{ijkl} = D_{il}y_{kl} \qquad \text{(all } i,k,l\text{)}, \qquad (11)$$

$$\sum_{k=1}^{p} y_{kl} = 1 \qquad (l = 1,\ldots,m), \qquad (12)$$

$$L_k z_k \leq \sum_{i=1}^{q}\sum_{l=1}^{m} D_{il}y_{kl} \leq U_k z_k \quad (k = 1,\ldots,p), \qquad (13)$$

$$x_{ijkl} \geq 0 \qquad \text{(all } i,j,k,l\text{)}, \qquad (14)$$

$$y_{kl} = 0 \text{ or } 1 \qquad \text{(all } k,l\text{)}, \qquad (15)$$

$$\text{and} \quad z_k = 0 \text{ or } 1 \qquad (k = 1,\ldots,p). \qquad (16)$$

The objective function (9) is the total distribution cost which is the total transportation cost plus the total fixed cost for the distribution centers and the variable cost of flow through the distribution centers; constraint (10) guarantees that the capacity of each plant for each commodity is not exceeded. Constraint (11) guarantees that each customer's demand is met. Note that constraint (11) forces x_{ijkl} (for all i and j) to take value 0, if $y_{kl} = 0$. That is, whenever $y_{kl} = 0$, customer l is not served by distribution center k. Constraint (12) ensures that each customer l is served by only one distribution center. Constraint (13) requires that the total flow through distribution center k, if open (that is, $z_k = 1$) be between the lower bound L_k and the upper bound U_k. Constraint (13) also ensures the logical relationship between the y and z variables; that is, $z_k = 1$ if and only if $y_{kl} = 1$ for some l.

Resource-Task Scheduling

In this situation we have n resources which must perform m tasks where each resource can perform some or all of the tasks. For each resource j $(j = 1,\ldots,n)$, we may list, subject to time, location, and other

Variables	$(x_{11}$	x_{21}	x_{31}	$\ldots$	x_{12}	x_{22}	$\ldots$	$\ldots$	x_{1n}	$\ldots)$	$=\mathbf{x}$
Costs	$(c_{11}$	c_{21}	c_{31}	$\ldots$	c_{12}	c_{22}	$\ldots$	$\ldots$	c_{1n}	$\ldots)$	$=\mathbf{c}$
Resource			1			2			n		
Combination	1	2	3	$\ldots$	1	2	$\ldots$		1	$\ldots$	
Tasks											
$(\mathbf{e_1})$ 1	1	0	0	$\ldots$	0	0	$\ldots$		1	$\ldots$	≥ 1
$(\mathbf{e_2})$ 2	0	1	0	$\ldots$	1	0	$\ldots$		1	$\ldots$	≥ 1
$\vdots$	$\vdots$	$\vdots$	$\vdots$		$\vdots$	$\vdots$			$\vdots$		$\vdots$
$(\mathbf{e_m})$ m	1	1	1	$\ldots$	0	1	$\ldots$		1	$\ldots$	≥ 1

$$\underbrace{\qquad\qquad\qquad\qquad\qquad\qquad\qquad}_{E}$$

Figure 1.6: Resource Scheduling Problem

constraints, all possible ways a resource can do the tasks. In particular, for each resource j, a zero-one matrix can be developed where the rows correspond to the tasks, the columns to the possible combinations, and an entry is $1(0)$ if in that combination resource j can (cannot) do the task corresponding to the row. Suppose also that a cost c_{ij} of using the ith possible combination of resource j can be computed (Figure 1.6). Then we may model the problem by defining the binary variable x_{ij} to be 1 if the ith combination of resource j is used, and $x_{ij} = 0$ otherwise. To ensure that each task is done at least once, we have

$$\mathbf{e_t x} \geq 1 \quad \text{for each task } (t = 1, \ldots, m) \tag{17}$$

where $\mathbf{e_t}$ is the row of the binary matrix corresponding to task t and $\mathbf{x}$ is a vector whose components are the corresponding binary variables x_{ij} (Figure 1.6). Note that the m inequalities (17) are equivalent to

$$\mathbf{E x} \geq \mathbf{e}, \tag{18}$$

where E is a zero-one matrix whose rows are $\mathbf{e_t}$ and $\mathbf{e}$ is an m column of ones. As we wish to minimize the total cost, the objective function is

$$\text{minimize } \mathbf{cx}, \tag{19}$$

where $\mathbf{c}$ is a vector whose elements are the corresponding costs c_{ij} (Figure 1.6). Combining (18) and (19) we have the integer program

$$\begin{array}{ll}
\text{minimize} & \mathbf{cx} \\
\text{subject to} & \mathbf{Ex} \geq \mathbf{e} \\
\text{and} & \mathbf{x} \quad \text{binary.}
\end{array}$$

This problem is often referred to as a *set covering problem*. It is often necessary that each task be done exactly once. In this case, the constraints (18) are $\mathbf{Ex} = \mathbf{e}$, and the model is the *set partitioning problem*

$$\begin{array}{ll}
\text{minimize} & \mathbf{cx} \\
\text{subject to} & \mathbf{Ex} = \mathbf{e} \\
\text{and} & \mathbf{x} \quad \text{binary.}
\end{array}$$

A Loading Problem

Suppose that a relief plane has cargo weight capacity b and is to be loaded with items each with weight a_j and relative (nonnegative) value c_j. The problem is to load the plane so as to maximize its total relative value. The situation is described by the integer program with a single constraint

$$\begin{array}{ll}
\text{maximize} & \displaystyle\sum_{j=1}^{n} c_j x_j \\
\text{subject to} & \displaystyle\sum_{j=1}^{n} a_j x_j \leq b \\
\text{and} & x_j = 0 \text{ or } 1 \quad (j = 1, \ldots, n),
\end{array}$$

where $x_j = 1$ if item j is loaded on the plane and $x_j = 0$ if not. The problem is usually called a *knapsack problem*. It relates to a hiker who can carry at most b pounds and is to select items of weight a_j and relative value c_j so as to maximize the value of the knapsack.

The Traveling Salesman Problem

A traveling salesman must visit n cities, each exactly once. The distance between every pair of cities (i, j), denoted by d_{ij} $(i \neq j)$, is known and may depend on the direction travelled (i.e., d_{ij} does not necessarily equal d_{ji}). The problem is to find a tour which commences and terminates at the salesman's home city and minimizes the total distance travelled.

Suppose we label the home city as city 0 and as city $n + 1$. (Then we may think of the salesman's initial location as city 0 and the desired final location as city $n + 1$.) Also, introduce the zero-one variable x_{ij} $(i = 0, 1, \ldots, n, \ j = 1, \ldots, n + 1, \ i \neq j)$, where $x_{ij} = 1$ if the salesman travels from city i to j, and $x_{ij} = 0$ otherwise. To guarantee that each city (except city 0) is entered exactly once, we have (Figure 1.7(a))

$$\sum_{\substack{i=0 \\ i \neq j}}^{n} x_{ij} = 1 \qquad (j = 1, \ldots, n + 1).$$

Similarly, to ensure that each city (except city $n + 1$) is left exactly once, we have (Figure 1.7(b))

$$\sum_{\substack{j=1 \\ j \neq i}}^{n+1} x_{ij} = 1 \qquad (i = 0, \ldots, n).$$

These constraints, however, do not eliminate the possibility of sub-

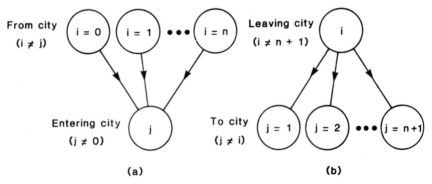

Figure 1.7: Constraints of The Traveling Salesman Problem

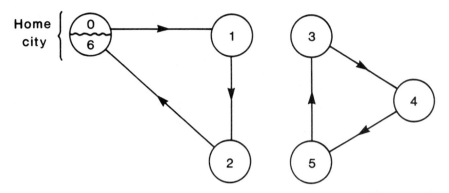

Figure 1.8: Example of Subtours

tours or "loops" as indicated by Figure 1.8, where the solution is $x_{01} = x_{12} = x_{26} = 1, x_{34} = x_{45} = x_{53} = 1$, and $x_{ij} = 0$ otherwise (with $n = 5$). One way of eliminating the subtour possibility is to add the constraints

$$a_i - a_j + (n+1)x_{ij} \leq n \quad (i = 0, 1, \ldots, n, \; j = 1, \ldots, n+1, \; i \neq j),$$

where a_i is a real number associated with city i.

To show that a solution containing loops cannot satisfy these constraints, consider any subtour except the one containing the home city. Then, if we sum the inequalities corresponding to the $x_{ij} = 1$ around the loop, the $a_i - a_j$ cancel each other and we are left with $(n+1)N \leq nN$, where N is the number of arcs in the subtour, which is a contradiction. (Relating to Figure 1.8 and the loops $3 \rightarrow 4 \rightarrow 5$, the constraints are $a_3 - a_4 + 6 \leq 5, a_4 - a_5 + 6 \leq 5$, and $a_5 - a_3 + 6 \leq 5$; or, after adding, $18 \leq 15$.)

On the other hand, to see that these constraints may be satisfied when there are no subtours, define $a_0 = 0, a_{n+1} = n+1$, and let $a_i = K$ if city i is the Kth city visited on the tour. Then, when $x_{ij} = 1$, we have $a_i - a_j + (n+1) = K - (K+1) + (n+1) = n$. Also, since $1 \leq a_i \leq n+1$ $(i = 1, \ldots, n+1)$, the difference $a_i - a_j$ is always $\leq n$ (all i, j), and thus the constraints are satisfied when $x_{ij} = 0$.

To complete the model we wish to minimize the total distance $\sum_{i=0}^{n} \sum_{\substack{j=1 \\ j \neq i}}^{n+1} d_{ij} x_{ij}$. Or an integer programming formulation of the traveling salesman problem is to find variables x_{ij} and arbitrary real numbers a_i which

$$\text{minimize} \quad \sum_{\substack{i=0}}^{n} \sum_{\substack{j=1 \\ j \neq i}}^{n+1} d_{ij} x_{ij}$$

$$\text{subject to} \quad \sum_{\substack{i=0 \\ i \neq j}}^{n} x_{ij} = 1 \qquad (j = 1, \ldots, n+1),$$

$$\sum_{\substack{j=1 \\ j \neq i}}^{n+1} x_{ij} = 1 \qquad (i = 0, 1, \ldots, n),$$

$$a_i - a_j + (n+1)x_{ij} \leq n \quad \begin{array}{l}(i = 0, 1, \ldots, n, \\ j = 1, \ldots, n+1, i \neq j),\end{array}$$

$$\text{and} \qquad x_{ij} = 0 \text{ or } 1 \qquad \begin{array}{l}(i = 0, 1, \ldots, n, \\ j = 1, \ldots, n+1, i \neq j),\end{array}$$

where $x_{0,n+1} = 0$ (since $x_{ij} = 0$ for $i = j$). This formulation originally appeared in Tucker [578].

Radio Frequency Management

Two or more electromagnetic signals transmitted from and/or received on a small platform (e.g., a ship, satellite, or airplane), tend to produce what is called intermodulation interference (see [422] or [423]). Consider, for example, the ship portrayed in Figure 1.9 which is simultaneously transmitting N signals at the known frequencies $F_1, F_2, \ldots, F_N$. If at the same time, it is receiving a signal at a frequency F_R, this signal will be interfered with if, either (i) F_R is a harmonic of one of the transmitting frequencies (say) F_j or (ii) F_R is an intermodulation frequency of $F_1, F_2, \ldots, F_N$. F_R is a *harmonic* of the transmitting frequency F_j if $F_R = QF_j$, where Q is a positive integer. Also, if F_R is an *intermodulation frequency* of the transmitting frequencies $F_1, F_2, \ldots, F_N$, by definition, it means that

$$F_R = \sum_{i=1}^{N} F_i x_i,$$

where x_i are integers. In either case, Q (defined by $Q = \sum_{i=1}^{N} |x_i|$ in case

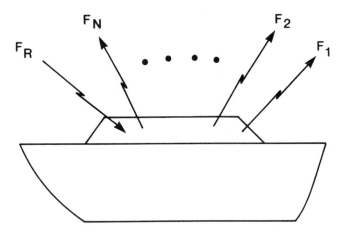

Figure 1.9: Transmission To and From a Ship

(ii)), is called the *order of intermodulation*. In practice the magnitude of the interference increases as the order Q decreases. Hence, to evaluate the clarity of a given frequency assignment $F_1, F_2,$ $\ldots, F_N$, we wish to find the intermodulation or harmonic interference with lowest order Q. If this order is too small, then we must either change the receiving frequency F_R or one (or more) of the transmitting frequencies $F_1, F_2, \ldots, F_N$. To find the smallest order Q, we solve the integer programming problem.

$$\text{minimize} \quad Q = \sum_{i=1}^{N} |x_i|$$

$$\text{subject to} \quad \sum_{i=1}^{N} F_i x_i = F_R$$

$$\text{and} \quad x_i \quad \text{integer} \quad (i = 1, \ldots, N).$$

Algorithms for the model described above are in [433] and [422]. In larger, more complex fleet assignment problems, where we wish to assign both receiving and transmitting frequencies to many ships with a known radio communication plan, we often solve the above model as a subproblem in an iterative radio frequency assignment algorithm (see, e.g., [422]).

Problems

Problem 1.1 (Piecewise convex constraint sets) Suppose that some of the sets $\mathbf{b}_i - \mathbf{A}_i\mathbf{y} \leq \mathbf{0}$ in section 1.2.1 are unbounded above. Should the reformulation be modified? Illustrate your answer by solving

$$
\begin{array}{llll}
\text{maximize} & y_1 & + & y_2 \\
\text{subject to} & y_1 & & \leq 2 \quad \text{and/or} \\
& & y_2 & \leq 3 \\
\text{and} & y_1, & y_2 & \geq 0.
\end{array}
$$

Problem 1.2 Suppose we formulate the piecewise convex constraint set problem appearing in section 1.2.1 as the mixed integer nonlinear program

$$
\begin{array}{lll}
\text{maximize} & \mathbf{cy} & \\
\text{subject to} & x_i(\mathbf{b}_i - \mathbf{A}_i\mathbf{y}) \geq 0 & (i = 1, \ldots, p), \\
& \displaystyle\sum_{i=1}^{p} x_i = t, & \\
\text{and} & x_i = 0 \text{ or } 1 & (i = 1, \ldots, p).
\end{array}
$$

Is this formulation correct? If so, discuss a solution technique which is computationally the same when applied to the mixed integer *linear* formulation appearing in section 1.2.1. Illustrate.

Problem 1.3 Consider the piecewise linear function $f(y)$ in Fig. 1.10, which is linear in two intervals,
Then an equivalent function (see section 1.2.1) is:

$$
f(y) = \alpha_0 f(\bar{y}_0) + \alpha_1 f(\bar{y}_1) + \alpha_2 f(\bar{y}_2),
$$

where

$$
\begin{array}{rcl}
y & = & \alpha_0 \bar{y}_0 \; + \; \alpha_1 \bar{y}_2 \; + \; \alpha_2 \bar{y}_3, \\[4pt]
1 & = & \alpha_0 \quad + \; \alpha_1 \quad + \; \alpha_2, \\[4pt]
x_0 & & \geq \alpha_0, \\
x_0 \; + \; x_1 & & \geq \alpha_1, \\
x_1 & & \geq \alpha_2, \\
x_0 \; + \; x_1 & = & 1, \\
\alpha_0, \; \alpha_1, \; \alpha_2 & \geq & 0,
\end{array}
$$

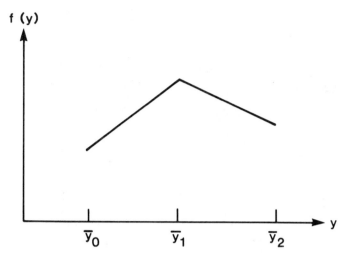

Figure 1.10: Function $f(y)$ for Problem 1.3

and

$$x_0, \ x_1 = 0 \text{ or } 1.$$

Are there redundant constraints on the x_i's and α_j's? Explain. Is your answer correct when there are n linear intervals?

Problem 1.4 (Raghavachari [470],[471], Kennington and Fyffe [363]) Consider the zero-one integer program

(IP) maximize **cx**,
     subject to **Ax** $\le$ **b**,
               $0 \le$ **x** $\le$ **e**,
     and **x** integer ,

and the quadratic program

(QP) maximize **cx** $- M$**x**^{t}(**e** $-$ **x**),
      subject to **Ax** $\le$ **b**,
      and $0 \le$ **x** $\le$ **e**,

where **e** is a vector of ones, M is a large positive number, and the row vector **x**^{t} is the transpose of the vector **x**. By proving the results below, show that the integer program may be solved by solving the quadratic program. Intuitively, why do these results seem true?

a) If x^* is an optimal solution to **QP** and x^* is integer, then it is optimal for **IP**.

b) If x^* is an optimal solution to **QP** and x^* does not have all integer components, then **IP** has no integer feasible solution.

c) If **QP** has no feasible solution then **IP** has no integer feasible solution.

Problem 1.5 Given the facility location problem with capacity constraints which appear in section 1.2.2, show that the inequalities in the model imply that the shipping variables will be integer in an optimal solution. Is this result true for every optimal solution? Illustrate your answer using the data appearing in the next problem (part (d)) with the additional capacity restrictions $M_1 = 300$ and $M_2 = 350$. (This capacitated problem can be solved by inspection.)

Problem 1.6 (Facility location problem) Starting with the reformulated facility location problem appearing section 1.2.2:

a) Prove that there exists an optimal solution with all $y_{ij} = 0$ or 1.

b) Discuss when fractional solutions are possible.

c) Suggest what you consider to be an efficient solution procedure. Why does your scheme work?

d) Using the method obtained in (c), solve the two-plant, three-customer problem whose data appears below.

g_{ij}				d_j			f_i	
	1	2	3	1	2	3	1	2
1	5	7	6	150	170	260	90	100
2	5	6	7					

Problem 1.7 (Plant location problem with side constraints) Remodel the plant location problem (section 1.2.2) to allow for each of the following additional conditions. In each case can the model be simplified?

a) Only a subset of the plants may be open.

b) The cost of opening a plant is equal to a fixed cost K_i plus the amount produced at the plant.

c) The facilities can be listed according to regions, and because of customer convenience each region k must have at least l_k (a positive number) plants open.

d) A total amount of F dollars is available for construction.

Problem 1.8 (Airline crew scheduling with crew base constraints) Formulate the following problem as an integer (linear) program: An airline with n crews must make m flight legs. Each crew is currently in one of N locations (or cities) and the possible combination of flights each crew may take can be enumerated. Because of union rules, however, at most u_k and at least l_k crews at city k ($k = 1, \ldots, N$) may be assigned flight duty. The cost of using crew j on the ith combination of flights is c_{ij} ($j = 1, \ldots, n$). The problem is to obtain an assignment which covers all flights and satisfies the union rules at the minimum total cost.

Problem 1.9 (Multi-item knapsack problem) Consider the knapsack problem

$$\text{maximize} \quad \sum_{j=1}^{n} c_j x_j = z,$$

$$\text{subject to} \quad \sum_{j=1}^{n} a_j x_j \leq b,$$

$$x_j \geq 0 \text{ and integer},$$

where c_j is the relative (positive) value of item j, a_j its (positive) weight, b is the (knapsack's) weight capacity, and x_j is the number of items j to put in the knapsack.

a) If x_j is not required to be integer, solve the eight variable problem whose data is

c_j	8	10	9	8	6	2	7	5
a_j	4	2	7	3	5	1	3	5

and $b = 23$. (Hint: Let $z_j = a_j x_j$ for $j = 1, \ldots, n$ and transform the problem to the variables z_j. The linear programming solution is then obvious.)

b) Obtain an integer feasible solution by changing the value of one variable in the solution from (a).

c) Using (b), what is the range of values for an optimal integer solution? Why?

Problem 1.10 (Symmetric traveling salesman problem) Consider the traveling salesman problem with $d_{ij} = d_{ji}$ for all i, j. Can the integer programming model which appears in section 1.2.2 be simplified?

Problem 1.11 (The four color problem) A classical unsolved problem in mathematics is the four color problem. The question is: given any map which divides a plane into regions, can we use no more than four colors to color it so that no two regions with a common boundary have the same color.

a) Formulate the requirements as linear constraints with integer variables. (Hint: Let the variable in region i be denoted by x_i, where $x_i = 0, 1, 2, 3$, depending on which color is used for that region.)

b) Why is it not possible to develop an objective function and solve the integer program (where the constraints are from (a)) so that the four color problem is answered?

Problem 1.12 (Capital budgeting) The ABC Investment Company can invest capital funds in any five probjects during the next three years. Each project if funded has different annual capital requirements and a known present worth of future returns as listed in the table below. The expected amount of capital available each year is also known and listed in the table.

Project	Capital Requirements (in thousands of dollars)			Present worth of future returns (in thousands of dollars)
	Year 1	Year 2	Year 3	
1	25	15	15	100
2	35	25	25	120
3	20	10	5	80
4	60	0	0	170
5	40	10	10	150
Capital available	100	50	50	

Formulate an integer programming model to determine which projects should be funded so as to maximize the total present worth of future returns while the budget constraints are satisfied. List any assumptions.

Problem 1.13 (Satellite constellation design [516]) A satellite constellation is a series of satellites in predetermined orbits circling the globe. Typically a satellite constellation is used for navigational purposes from known points j $(j = l, \ldots, q)$ on the globe. At any segment of time a particular satellite may not be visible from a point j. Thus, let a_{ijt} be 1 (0) if satellite i is visible (is not visible) from point j during time segment t. Now consider a constellation with M satellites, each with a known orbit around the earth which requires T time segments. That is, we know the 0-1 values for all a_{ijt} $(i = 1, \ldots, M, j = 1, \ldots, q,$ and $t = 1, 2, \ldots, T)$.

Suppose budget considerations require that only K satellites $(K < M)$ be launched. Formulate an integer programming model to decide which K of the M satellites in the constellation should be launched under each of the following objectives.

a) Maximize the average number of satellites visible at the q points during the day consisting of T equal time segments.

b) Maximize the number of pairs (j, t), $(j = 1, \ldots, q;\ t = 1, \ldots, T)$, for which at least k $(k \le K)$ satellites are visible at point j during time segment t.

Explain why each objective (a) and (b) makes sense in context of the satellite constellation design scenario.

Problem 1.14 Consider a situation where to make an item i $(i = 1, \ldots, K)$, it first must be processed on machine 1, then machine 2, then machine 3, $\ldots$, and finally on machine M. Although the sequence of processing each item on the machines is fixed, namely, $1, 2, \ldots, M$, the order of processing an item i at each machine is to be determined and may vary from machine to machine. Let t_{ij} be the processing time required by item i on machine j. Formulate the problem as an integer program where we wish to determine the order of processing the items at each machine so as to minimize the total time necessary to process all the items. List any assumptions.

Problem 1.15 Reformulate problem 1.14, where the order of processing the items at each machine is required to be the same. List any assumptions.

Problem 1.16 (Cutting stock problem) Suppose we are manufacturing paper in large rolls with a fixed length and width of L (inches). Each roll is then cut into pieces of width l_j (inches), $j = 1, 2, \ldots, K$. Let $\mathbf{a}_i = (a_{i1}, a_{i2}, \ldots, a_{iK})$, $i = 1, 2, \ldots, I$, be a vector of nonnegative integers representing a particular cutting pattern. Specifically, $\mathbf{a}_i$ is the cutting pattern where the roll is cut up into a_{i1} pieces of width l_1, a_{i2} pieces of width $l_2, \ldots$, and a_{iK} pieces of width l_K. Then corresponding to each cutting pattern $\mathbf{a}_i$, $w_i = L - \sum_{j=1}^{K} a_{ij} l_j$ is the waste. During the current planning horizon, we need n_j pieces of width l_j ($j = 1, \ldots, K$). Formulate an integer programming model to determine how many times each cutting pattern $\mathbf{a}_i$ ($i = 1, 2, \ldots, I$), should be used so as to minimize the total waste in all the rolls used.

Chapter 2

Review of Linear Programming

2.1 The Linear Programming Problem

As mentioned in Chapter 1, a linear programming problem is a problem in which we wish to maximize or minimize a linear objective function subject to linear equality or inequality constraints and nonnegativity of the variables. In general, a linear programming problem can be stated as the model

$$
\begin{array}{lrcl}
\text{minimize} & \mathbf{cx} & = & z \\
\text{subject to} & \mathbf{Ax} & = & \mathbf{b} \\
\text{and} & \mathbf{x} & \geq & \mathbf{0},
\end{array}
\qquad
\begin{array}{r}
(1) \\
(2) \\
(3)
\end{array}
$$

$$
\begin{array}{llll}
\text{where} & \mathbf{c} & = & (c_j) \quad \text{is an } n \text{ row (or cost) vector,} \\
& \mathbf{A} & = & (a_{ij}) \quad \text{is an } m \text{ by } n \text{ constraint matrix,} \\
& \mathbf{b} & = & (b_i) \quad \text{is an } m \text{ column vector,} \\
\text{and} & \mathbf{x} & = & (x_j) \quad \text{is an } n \text{ vector of variables.}
\end{array}
$$

Note that if a linear programming problem has inequality constraints, $\mathbf{Ax'} \leq \mathbf{b}$ (or $\mathbf{Ax'} \geq \mathbf{b}$), $\mathbf{x'} \geq \mathbf{0}$, they can be converted to the equality constraints $\mathbf{Ax'} + \mathbf{Ix_s} = \mathbf{b}$ (or $\mathbf{Ax'} - \mathbf{Ix_s} = \mathbf{b}$), $\mathbf{x} = (\mathbf{x'}, \mathbf{x_s}) \geq \mathbf{0}$, where $\mathbf{I}$ is an m by m identity matrix, by adding an m column of slack (or subtracting an m column of surplus) variables $\mathbf{x_s}$. Converting a system of constraints to equation form with a nonnegative right hand side through the use of slack or surplus variables is often referred to as putting the constraints in standard form.

2.2   Graphical Solution and Geometric Concepts

In this section we show that linear programs with two or three variables may be solved graphically. The procedure yields several intuitive results which, when formalized, suggest an algorithm for the linear programming problem.

Example 2.1 Consider the following problem

$$
\begin{array}{rlrll}
\text{Maximize} & 2x_1 & + & 3x_2 & = & z \\
\text{subject to} & x_1 & + & x_2 & \leq & 10 \\
& x_1 & + & 2x_2 & \leq & 12 \\
\text{and} & x_1 & \geq & 0, \; x_2 & \geq & 0.
\end{array}
$$

Each inequality constraint restricts the feasible solution to a half-space in a two dimensional graph. The feasible region is the intersection of the half spaces satisfying the linear inequalities and the nonnegativity constraints and is the shaded region in Figure 2.1. The linear pro-

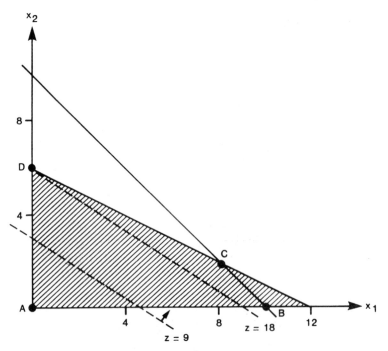

Figure 2.1: Graphical Method

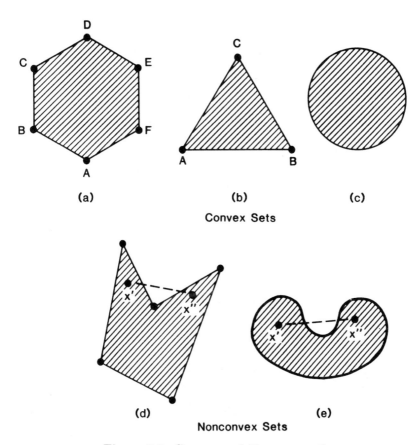

(a)   (b)   (c)

Convex Sets

(d)   (e)

Nonconvex Sets

Figure 2.2: Convex and Nonconvex Sets

gramming problem is thus equivalent to finding a point (x_1^*, x_2^*) in the shaded area which gives a maximum value to the objective function $z = 2x_1 + 3x_2$. The search for such a point is simplified due to the results below. Before giving these, we require certain definitions.

A set of points S is a *convex set* if, given any two points $\mathbf{x}', \mathbf{x}''$ in S, all points $\alpha\mathbf{x}' + (1 - \alpha)\mathbf{x}''$ belong to the set S, where $0 \leq \alpha \leq 1$. Geometrically, S is a convex set if for any two points in the set, the line segment joining these two points lies entirely in the set. Note that Figures 2.2(a), (b), and (c) display convex sets, while the sets in Figures 2.2(d), and (e) are not convex.

An *extreme point* of a convex set is a point which can be removed from the set and the set still remains convex. For example, in Figures

2.2(a) and (b) the extreme points are the labeled corner points while in Figure 2.2(c) the extreme points comprise the boundary of the circle.

Theorem 2.1 *The feasible region of a linear programming problem is a convex set.*

To prove the theorem we take two points $\mathbf{x}'$ and $\mathbf{x}''$ which satisfy the constraints of the linear programming problem and then show that any point $\mathbf{x} = \alpha\mathbf{x}' + (1 - \alpha)\mathbf{x}''$, $0 \leq \alpha \leq 1$, also satisfies the constraints. The proof is left to the problems.

Theorem 2.2 *If the feasible region of a linear programming problem is bounded, then at least one optimal solution occurs at one of its extreme points.*

The result is illustrated in Figure 2.1. In that diagram the broken lines represent lines of constant value of the objective function. As this constant value changes the objective function line moves parallel to itself. Therefore, to find the maximum value of the objective function, geometrically we move the line of constant objective function value as far as possible in the direction in which z increases. That is, a further increase beyond this value would produce an objective function line which has no points in common with the feasible region. In this problem, the line $2x_1 + 3x_2 = 22$ is the limiting line at the extreme point labeled C . Increasing z further moves the line outside the feasible region. Note that this problem has a unique optimal solution and it occurs at an extreme point. However, had the objective function been: $z = x_1 + 2x_2$, then any point on the line segment joining the extreme points labeled C and D would be optimal. In this case, we have *alternate optimal solutions*. Note that this does not violate Theorem 2.2. The proof of Theorem 2.2 is left to Problem 2.3.

Finally, as an extreme point is defined by the intersection of constraints, and there is a finite number of such constraints, we have another important result. The proof is left to Problem 2.4.

Theorem 2.3 *The feasible region of a linear programming problem has a finite number of extreme points.*

Thus to solve a bounded linear programming problem, we need to evaluate only the extreme points. For problems with several variables the graphical method does not work as the graph cannot be drawn. To overcome this, we develop an iterative method, called the *simplex method*, which solves any linear program by nonredundantly searching over the extreme points for an optimal one. To develop the simplex method, we need to characterize an extreme point algebraically using the constraint equations. We must then show, given a particular extreme point, whether it is an optimal one, and if not, how to find another extreme point so that the objective function improves. The method also indicates when the linear program is infeasible or has an unbounded solution (cf. Figure 2.3).

2.3 The Simplex Algorithm

The linear programming problem (1)-(3) can be written as

$$\text{minimize} \quad z = \sum_{j=1}^{n} c_j x_j \tag{4}$$

$$\text{subject to} \quad \sum_{j=1}^{n} a_{ij} x_j = b_i \quad (i = 1, 2, \ldots, m) \tag{5}$$

$$\text{and} \quad x_j \geq 0 \quad (j = 1, \ldots, n). \tag{6}$$

Without loss of generality, we assume that $\mathbf{A} = (a_{ij})$ is m by $n, n > m$, and the rank of $\mathbf{A}$ is m. We require the definitions and results listed below.

2.3.1 Definitions

Feasible Solution   A set of values for the vector $\mathbf{x}$ which satisfies constraints (5) and (6) is a feasible solution. A feasible solution which minimizes the objective function (4) is an *optimal solution* .

Extreme Point Let S be a convex set. A point $\mathbf{x} \in S$ is said to be an extreme point of S if it cannot be expressed as a convex combination of two other distinct points in S . That is, $\mathbf{x}$ is an extreme point of S if there do not exist $\mathbf{x}' \in S$, and $\mathbf{x}'' \in S$ $(\mathbf{x}' \neq \mathbf{x}'')$ such that

$$\mathbf{x} = \alpha \mathbf{x}' + (1 - \alpha)\mathbf{x}'' \text{ for some } 0 < \alpha < 1.$$

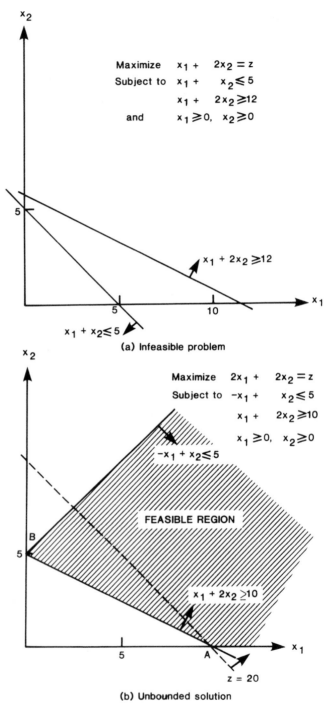

Figure 2.3:
Infeasible and
Unbounded
Linear Programs

Note that this is an algebraic characterization of the definition given in Section 2.2.

Basic Feasible Solution   Under the assumption that rank $(\mathbf{A}) = m$, a nonsingular submatrix $\mathbf{B}$ from $\mathbf{A}$, formed by any m linearly independent columns from $\mathbf{A}$, is a *basis* for m dimensional space or simply a basis. Corresponding to a basis $\mathbf{B}$ from $\mathbf{A}$, we may rearrange columns so that $\mathbf{A} = (\mathbf{B}, \mathbf{N})$ and $\mathbf{x} = (\mathbf{x_B}, \mathbf{x_N})$, or $\mathbf{Ax} = \mathbf{b}$ becomes $\mathbf{Bx_B} + \mathbf{Nx_N} = \mathbf{b}$; here $\mathbf{x_B}$ are the m basic variables corresponding to the columns in $\mathbf{B}$ and $\mathbf{x_N}$ are the $n - m$ nonbasic variables corresponding to the remaining $n - m$ columns in $\mathbf{N}$. The solution found by setting the nonbasic variables $\mathbf{x_N} = \mathbf{0}$ and $\mathbf{x_B} = \mathbf{B}^{-1}\mathbf{b}$ is a *basic solution*. A basic solution which is also nonnegative is a *basic feasible solution*. A basic feasible solution is *nondegenerate* if the values for its m basic variables are all nonzero.

Homogeneous and Extreme Homogeneous Solution   Corresponding to the system $\mathbf{Ax} = \mathbf{b}$, $\mathbf{x} \geq \mathbf{0}$, a *homogeneous solution* is a vector $\mathbf{y}$ satisfying

$$\mathbf{Ay} = \mathbf{0} \quad \text{and} \quad \mathbf{y} \geq \mathbf{0}.$$

A homogeneous solution is called an *extreme homogeneous solution* if and only if it is a basic feasible solution to the system

$$\mathbf{Ay} = \mathbf{0}, \mathbf{y} \geq \mathbf{0}$$

and $\quad \displaystyle\sum_{j=1}^{n} y_j = 1.$

Notice that $\mathbf{Ax} = \mathbf{b}$, $\mathbf{x} \geq \mathbf{0}$ has a nonzero homogeneous solution if and only if the feasible region is unbounded. (Why?)

Extreme Half-Line   Let S be the set of feasible solutions to the linear program (4)-(6). Suppose $\bar{\mathbf{x}}$ is a basic feasible solution and $\bar{\mathbf{y}}$ is an extreme homogeneous solution of the system (5)-(6). Then every point on the half line

$$\{\mathbf{x} : \bar{\mathbf{x}} + \theta\bar{\mathbf{y}}, \ \theta \geq 0\}$$

is a feasible solution. (Why?) This half-line is an *extreme half-line* of S if every point $\hat{x}$ on it can be represented as $\hat{x} = \alpha x' + (1 - \alpha) x''$, for some $0 < \alpha < 1$, and $x' \in S$, $x'' \in S$, where x' and x'' both lie on the half line.

2.3.2 Fundamental Theorems

In this section we list without proof some basic results which are used in the derivation of the simplex algorithm. The proofs may be found in Bazaraa and Jarvis [59], Dantzig [159], Gass [211], Hadley [303], Murty [439], or Solow [541].

Theorem 2.4 *The set S of feasible solutions to the constraints (5)–(6) is a convex set. A point $\bar{x} \in S$ is an extreme point of S if and only if it is a basic feasible solution (BFS) to the system of equations.*

Theorem 2.5 *If the system (5)–(6) has a feasible solution, then it has a basic feasible solution.*

Theorem 2.6 *If the linear program (4)–(6) has an optimal solution, then it has an optimal basic feasible solution.*

Theorem 2.7 *Let S be the set of feasible solutions to the constraints (5)–(6). Then any feasible solution x can be expressed as either*

    i) a convex combination of basic feasible solutions (if S is bounded), or

    ii) the sum of a convex combination of basic feasible solutions and a nonnegative linear combination of extreme homogeneous solutions (if S is not bounded).

In general, if $x^1, x^2, \ldots, x^p$ are the basic feasible solutions and $y^1, y^2, \ldots, y^r$ are the extreme half lines of S ($r = 0$, if there are no extreme half lines), then any feasible solution $\bar{x}$ can be represented as

$$\bar{x} = \sum_{j=1}^{p} \lambda_j x^j + \sum_{k=1}^{r} \alpha_k y^k$$

where
$$\sum_{j=1}^{p} \lambda_j = 1$$

and $\quad \lambda_j \geq 0 \quad (j = 1, \ldots, p) \quad$ and $\quad \alpha_k \geq 0 \quad (k = 1, \ldots, r)$.

Illustration of Theorem 2.7 is left to Problem 2.10.

2.3.3 The Simplex Algorithm

The previous discussion and results suggest that an optimal solution to a linear programming problem can be obtained by somehow evaluating all its basic feasible solutions and finding one with a minimum value for the objective function. However, the problem with applying this approach directly is that the feasible region may have an enormous number of extreme points (hence basic feasible solutions) and thus each one cannot be found explicitly. To overcome this difficulty, we develop a procedure, referred to as the simplex algorithm, which moves from a known (or current) extreme point along an edge of the feasible region to a neighboring extreme point which has an improved value of the objective function. We thus search nonredundantly over a subset of the extreme points, or equivalently, we need to evaluate only some of the basic feasible solutions. The steps are listed below and are subsequently illustrated.

Step 0 (Finding a starting basic feasible solution)

If the constraints (5)–(6) contain a full set of slack variables and the right hand side (b_i) is nonnegative, then the slack variables may serve as an initial set of basic variables, with the remaining variables nonbasic. Otherwise, as illustrated in Section 2.3.4, we will use the simplex method itself to find an initial basic feasible solution.

Step 1 (The current basic feasible solution)

At the beginning of each simplex iteration we have a current basic feasible solution to the system (5)–(6). The basic variables are $\mathbf{x_B} = (x_1, x_2, \ldots, x_m)$ and the nonbasic variables are $\mathbf{x_N} = (x_{m+1}, \ldots, x_n)$. Then (5) is the same as $\mathbf{Ax} = \mathbf{b}$, or $\mathbf{Bx_B} + \mathbf{Nx_N} = \mathbf{b}$, where $\mathbf{x} =$

$$x_1 \qquad + \bar{a}_{1,m+1}x_{m+1} + \bar{a}_{1,m+2}x_{m+2} + \ldots + \bar{a}_{1,n}x_n = \bar{b}_1$$

$$x_2 \qquad + \bar{a}_{2,m+1}x_{m+1} + \bar{a}_{2,m+2}x_{m+2} + \ldots + \bar{a}_{2,n}x_n = \bar{b}_2$$

$$\ddots \qquad \vdots \qquad \vdots \qquad \qquad \vdots \qquad \vdots$$

$$x_m + \bar{a}_{m,m+1}x_{m+1} + \bar{a}_{m,m+2}x_{m+2} + \ldots + \bar{a}_{m,n}x_n = \bar{b}_m$$

Table 2.1: System of Equations in Canonical Form

$(\mathbf{x_B}, \mathbf{x_N})$. Multiplying $\mathbf{Bx_B} + \mathbf{Nx_N} = \mathbf{b}$ by $\mathbf{B}^{-1}$ gives

$$\mathbf{x_B} + \mathbf{B}^{-1}\mathbf{Nx_N} = \mathbf{B}^{-1}\mathbf{b}. \tag{7}$$

The system (7) is said to be in *canonical form* and can be written as in Table 2.1. In the table $\mathbf{B}^{-1}\mathbf{N} = (\bar{a}_{i,m+j})$, $\mathbf{B}^{-1}\mathbf{b} = (\bar{b}_i)$, and $\bar{b}_i \geq 0$ ($i = 1, \ldots, m$) as we are assuming that $(\mathbf{x_B} = \mathbf{B}^{-1}\mathbf{b}, \ \mathbf{x_N} = \mathbf{0})$ is a feasible solution.

Step 2 (Test for optimality)

The objective function can be written as

$$z = \mathbf{cx} = \mathbf{c_B}\mathbf{x_B} + \mathbf{c_N}\mathbf{x_N},$$

where $\mathbf{c} = (\mathbf{c_B}, \mathbf{c_N})$. Solving for $\mathbf{x_B}$ in (7), and substituting for $\mathbf{x_B}$ in the objective function gives $z = \mathbf{c_B}(\mathbf{B}^{-1}\mathbf{b} - \mathbf{B}_N^{-1}\mathbf{x_N}) + \mathbf{c_N}\mathbf{x_N} = \mathbf{c_B}\mathbf{B}^{-1}\mathbf{b} + (\mathbf{c_N} - \mathbf{c_B}\mathbf{B}^{-1}\mathbf{N})\mathbf{x_N}$, or

$$z = c_1\bar{b}_1 + c_2\bar{b}_2 + \ldots + c_m\bar{b}_m + \bar{c}_{m+1}x_{m+1} + \ldots + \bar{c}_n x_n, \tag{8}$$

where

$$\bar{c}_{m+j} = c_{m+j} - \sum_{i=1}^{m} c_i \bar{a}_{i,m+j}. \tag{9}$$

Denoting $\sum_{i=1}^{m} c_i \bar{b}_i$ as z_c, the value of the objective function corresponding to the current basic feasible solution can be written as

$$z = z_c + \bar{c}_{m+1} x_{m+1} + \ldots + \bar{c}_n x_n. \tag{10}$$

In (10), the coefficients $\bar{c}_{m+j}$ are called the *reduced costs*. As the current basic feasible solution $(\mathbf{x_B}, \mathbf{x_N})$ is found by setting $\mathbf{x_N} = \mathbf{0}$, it is clear from (10) that $(\mathbf{x_B} = \mathbf{B^{-1}b}, \mathbf{x_N} = \mathbf{0})$ is optimal with $z = z_c$ whenever all the reduced costs $\bar{\mathbf{c}} = (\bar{c}_{m+1}, \ldots, \bar{c}_n)$ are nonnegative. The discussion for the minimization problem (4) – (6) is summarized in Theorem 2.8.

Theorem 2.8 *(Test for Optimality) A basic feasible solution* $\mathbf{x_B}$ *is optimal with an objective function value* $z = z_c = \mathbf{c_B B^{-1} b}$ *if all the reduced costs* $\bar{c}_j$ $(j = m + 1, \ldots, n)$ *are nonnegative.*

If the condition of Theorem 2.8 is satisfied, then the algorithm terminates with the current basic feasible solution as optimal. Otherwise, there exists a nonbasic variable x_j such that $\bar{c}_j < 0$, and we proceed to Step 3 to seek an improved basic feasible solution.

Step 3 (Improving the current basic feasible solution)

By (10), $\bar{c}_j$ is the change in the value of the objective function for a per unit increase in the nonbasic variable x_j. Hence, if $\bar{c}_j < 0$ for one or more j, it appears reasonable to select a nonbasic variable with the most negative $\bar{c}_j$ and try to increase it as much as possible from its current value of 0. Thus, defining

$$\bar{c}_s = \min_{j | \bar{c}_j < 0} \bar{c}_j,$$

we want to increase x_s as much as possible from its current value of zero, while keeping all other nonbasic variables nonbasic and thus at value 0. It turns out (see, e.g., Hadley [303]) that by doing this we geometrically move along an edge of the feasible region. If there is a limiting value $\theta > 0$ for x_s that maintains feasibility, then when $x_s = \theta$ we arrive at a neighboring extreme point corresponding to an improved basic feasible solution. To find the neighboring extreme point we use the computations explained below.

From Table 2.1, we may solve for the current basic variables (x_i) in terms of x_s; note that all other nonbasic variables (x_{m+i}) are at value 0. Then

$$x_i = \bar{b}_i - \bar{a}_{is} x_s \quad (i = 1, \ldots, m). \tag{11}$$

From equation (10) we have

$$z = z_c + \bar{c}_s x_s.$$

As $\bar{c}_s < 0$ and $x_s \geq 0$, to make z as small as possible we wish to make x_s as large as possible. To maintain feasibility we must have $x_i \geq 0$ for $(i = 1, \ldots, m)$, and by (11), the largest value for x_s is

$$\theta = \underset{i|\bar{a}_{is} > 0}{\text{minimum}} \; \bar{b}_i / \bar{a}_{is}. \tag{12}$$

Note that if $\bar{a}_{is} \leq 0$ $(i = 1, \ldots, m)$, by (11), x_s can be made arbitrarily large and each x_i $(i = 1, \ldots, m)$ remains nonnegative. As $\bar{c}_s < 0$, this means z becomes arbitrarily small, or the problem is unbounded. This is summarized in the theorem below.

Theorem 2.9 *(Unboundedness)   If for some nonbasic variable x_s, $\bar{c}_s < 0$ and $\bar{a}_{is} \leq 0$ for $i = 1, \ldots, m$, then feasible solutions exist for which the value of the objective function can be arbitrarily small; that is, the linear program is unbounded.*

If one or more $\bar{a}_{is} > 0$, then we compute θ using (12), denoting by r the row in which the minimum occurs. Ties are broken arbitrarily. In certain rare instances when degeneracy is present, and $\theta = 0$, additional rules to break ties may be necessary to eliminate the possibility, called cycling, of never leaving an extreme point (see, e.g. Hadley [303] or Solow [541]). When x_s becomes a basic variable and increases to value θ, the basic variable x_r decreases to 0 and becomes nonbasic. In this case, provided $x_s = \theta > 0$, we have found a new basic feasible solution, which has a lower value for the objective function.

Using this new set of basic variables we can create a new system of equations in canonical form. Step 4 provides a computationally efficient way of doing this. The process is called a pivot step, with row r labeled the *pivot row* and the column corresponding to x_s the *pivot column*.

	$x_1\ x_2\ \ldots\ x_r\ \ldots\ x_m$	$x_{m+1}\ \ldots\ x_s\ \ldots\ x_n$	
x_1	1	$\bar{a}_{1,m+1}\ \ldots\ \bar{a}_{1s}\ \ldots\ \bar{a}_{1n}$	b_1
x_2	$\ 1$	$\bar{a}_{2,m+1}\ \ldots\ \bar{a}_{2s}\ \ldots\ \bar{a}_{2n}$	b_2
$\vdots$	$\ddots$	$\vdots\qquad\vdots\qquad\vdots$	$\vdots$
OUT $\rightarrow$ x_r	$\ 1$	$\bar{a}_{r,m+1}\ \ldots\ \boxed{\bar{a}_{rs}}\ \ldots\ \bar{a}_{rn}$	b_r
$\vdots$	$\ddots$	$\vdots\qquad\vdots\qquad\vdots$	$\vdots$
x_m	$\ 1$	$\bar{a}_{m,m+1}\ \ldots\ \bar{a}_{ms}\ \ldots\ \bar{a}_{mn}$	$\bar{b}_m$
Reduced costs	$0\ \ 0\ \ldots\ 0\ \ldots\ 0$	$\bar{c}_{m+1}\ \ldots\ \bar{c}_s\ \ldots\ \bar{c}_n$	z
		$\uparrow$ IN	

Table 2.2: Simplex Tableau

The element $\bar{a}_{rs}$ is the *pivot element* (cf. Step 4). Sometimes x_s is referred to as the "IN" variable and x_r the "OUT" variable.

Step 4 (Pivot step)

The system of equations in canonical form (cf. Table 2.1) is written in a detached coefficient form, termed a *simplex tableau*, as in Table 2.2.

To do the pivot step we use elementary row operations which transform a system of linear equations into an equivalent system with the same set of feasible solutions. In particular, we may, without altering the set of feasible solutions:

1. Multiply any equation by a *nonzero* constant, and/or

2. Add a (positive or negative) constant multiple of one equaiton to any other.

As we wish to have x_s serve as the basic variable in the rth row of the tableau (cf. Table 2.2) it must have a +1 coefficient in this row and a

0 coefficient in every other row of the tableau. The pivot step is the following:

(i) Divide row r, the pivot row, by the pivot element $\bar{a}_{rs}$.

(ii) For each row i ($i = 1, \ldots, m;\ i \neq r$), multiply the new row r obtained in (i), by $-\bar{a}_{is}$ and add it to row i. This means in the new row i, $\bar{a}_{is} = 0$ ($i \neq r$) and, in general, $\bar{a}_{ij}$ ($i = 1, \ldots, m,\ i \neq r$) becomes $\bar{a}_{ij} - \bar{a}_{is}\bar{a}_{rj}/\bar{a}_{rs}$.

As the bottom row of the tableau really corresponds to equation (10), with basic variable z, we may also use (ii) to update this row (see also [303]). In particular, it is easy to show that

(iii) $\bar{c}_j$ becomes $\bar{c}_j - \bar{c}_s\,\bar{a}_{rj}/\bar{a}_{rs}$

Subsequent to a pivot step, the algorithm returns to Step 1.

Steps 1 through 4 are used iteratively until we either find an optimal solution (Step 2) or show that the problem is unbounded (Step 3). We illustrate the computations.

Example 2.2   Consider the following linear programming problem

$$
\begin{array}{rrrrrcl}
\text{minimize} & -x_1 & - & 2x_2 & - & 3x_3 & = & z \\
\text{subject to} & x_1 & + & x_2 & + & x_3 & \leq & 5 \\
& 2x_1 & + & 2x_2 & + & 3x_3 & \leq & 9 \\
& & & & & x_3 & \leq & 2 \\
\text{and} & & & \multicolumn{4}{l}{x_1 \geq, x_2 \geq 0, x_3 \geq 0.}
\end{array}
$$

We first convert the problem to one with equality constraints (or constraints in standard form) by adding slack variables s_1, s_2, and s_3. The problem is then

$$
\begin{array}{rl}
\text{Minimize} & -x_1 - 2x_2 - 3x_3 + 0s_1 + 0s_2 + 0s_3 = z \\
\text{subject to} & x_1 + x_2 + x_3 + s_1 \qquad\qquad\quad = 5 \\
& 2x_1 + 2x_2 + 3x_3 \qquad + s_2 \qquad = 9 \\
& x_3 \qquad\qquad\qquad + s_3 = 2 \\
\text{and} & x_1 \geq 0, x_2 \geq 0, x_3 \geq 0, s_1 \geq 0, s_2 \geq 0, s_3 \geq 0.
\end{array}
$$

Note that the constraint equations above are also in canonical form with the slack variables serving as the basic variables. That is, $s_1 = 5, s_2 = 9, s_3 = 2$ and $x_1 = x_2 = x_3 = 0$ is a basic feasible solution. The associated basis $\mathbf{B} = \mathbf{I}$ corresponds to the identity columns associated with s_1, s_2 and s_3.

Iteration 0 The first simplex tableau is below.

Basic variables (BV)	s_1	s_2	s_3	x_1	x_2	x_3	Basic feasible solution (BFS)	Ratio $(\bar{b}_i/\bar{a}_{is})$ for $a_{is} > 0$
s_1	1	0	0	1	1	1	5	5
s_2	0	1	0	2	2	3	9	3
OUT $\rightarrow s_3$	0	0	1	0	0	①	2	2
Reduced Costs $\bar{c}_j$	0	0	0	-1	-2	-3 ↑ IN	$z = 0$	

The IN variable here is $x_s = x_3$ as it has the most negative reduced cost $\bar{c}_j$ (cf. equation (10)) and the OUT variable is $x_r = s_3$ as it has the minimum $\bar{b}_i/\bar{a}_{is}$ ratio among all the rows with $\bar{a}_{is} > 0$ (cf. equation (12)). Performing a pivot step gives the tableau below.

Iteration 1

Basic variables (BV)	s_1	s_2	s_3	x_1	x_2	x_3	Basic feasible solution (BFS)	Ratio $(\bar{b}_i/\bar{a}_{is})$ for $a_{is} > 0$
s_1	1	0	-1	1	1	0	3	3
OUT $\rightarrow s_2$	0	1	-3	2	②	0	3	3/2
x_3	0	0	1	0	0	1	2	—
Reduced Costs $\bar{c}_j$	0	0	3	-1	-2 ↑ IN	0	$z = -6$	

Note that had we listed the basic columns (s_1, s_3, x_s) first, it is easy to see that the constraint equations in the tableau are in canonical form. Performing another pivot step, using x_2 as the IN variable and s_2 as the OUT variable, produces the tableau below.

Iteration 2

Basic variables (BV)	s_1	s_2	s_3	x_1	x_2	x_3	Basic feasible solution (BFS)
s_1	1	$-1/2$	$1/2$	0	0	0	$3/2$
x_2	0	$1/2$	$-3/2$	1	1	0	$3/2$
x_3	0	0	1	0	0	1	2
Reduced Costs $\bar{c}_j$	0	1	0	1	0	0	$z = -9$

Since all the reduced costs are nonnegative, the basic feasible solution $s_1 = 3/2$, $x_2 = 3/2$, $x_3 = 2$, and $x_1 = s_2 = s_3 = 0$, with $z = -9$, is optimal.

2.3.4 The Two-Phase Simplex Method

Consider the linear program

$$\begin{array}{rlll}
\text{minimize} & 2x_1 + 3x_2 + x_3 & = z \\
\text{subject to} & x_1 + x_2 + x_3 & = 5 \\
& 2x_1 + x_2 + 3x_3 & \geq 9 \\
\text{and} & x_1 \geq 0, x_2 \geq 0, x_3 \geq 0.
\end{array}$$

In standard form, this problem may be written as

$$\begin{array}{rlll}
\text{minimize} & 2x_1 + 3x_2 + x_3 + 0s_1 \\
\text{subject to} & x_1 + x_2 + x_3 & = 5 \\
& 2x_1 + x_2 + 3x_3 - s_1 & = 9 \\
\text{and} & x_1 \geq 0, x_2 \geq 0, x_3 \geq 0, s_1 \geq 0.
\end{array}$$

To initiate the simplex computations, we need to find a starting basic feasible solution. If the linear program when placed in standard form has a full set of slack variables, resulting from a full set of less than or equal to inequalities with a nonnegative right-hand side, then the slack variables may immediately serve as the basic variables. In this case, we have an obvious starting basic feasible solution (cf. Example 2.2). However, as in the example above, this is usually not the case and we need to somehow find a starting basic feasible solution. One way of doing this is to use a "two phase simplex algorithm". In Phase 1 we either find an initial basic feasible solution or show that none exists. If Phase 1 terminates with a basic feasible solution, this is used to start the computations of Phase 2, which proceed by the simplex algorithm discussed in the previous section. If Phase 1 ends without finding a basic feasible solution to the original problem then it is infeasible and the computations terminate. The Phase 1 algorithm is discussed below.

Phase 1 Corresponding to the problem (4)–(6), Phase 1 considers the linear programming problem below, sometimes referred to as the Phase 1 problem.

$$\text{Minimize} \quad w = x_{n+1} + x_{n+2} + \ldots x_{n+m}$$

$$\text{subject to} \quad \sum_{j=1}^{n} a_{ij} x_j + x_{n+i} = b_i \qquad (i = 1, 2, \ldots, m)$$

$$\text{and} \quad x_j \geq 0. \qquad (j = 1, \ldots, n+m)$$

We assume here that the right-hand side (b_i) is nonnegative. (If not, multiply the corresponding equations by -1 before adding the new variable x_{n+i}.) The nonnegative variables $x_{n+1}, x_{n+2}, \ldots, x_{n+m}$ are called *artificial variables*. Note that setting $x_{n+i} = b_i$ $(i = 1, \ldots, m)$, and $x_j = 0$ $(j = 1, \ldots n)$, provides a starting basic feasible solution. Also note that $w \geq 0$, and $w = 0$ if and only if $x_{n+i} = 0$ for $i = 1, \ldots, m$. The simplex computations applied to the Phase 1 problem terminate in one of the following ways.

(i) The minimum $w > 0$. This indicates that the original problem (5)-(6) has no feasible solution. For if there were a feasible solution, it would be a feasible solution to the enlarged problem with $x_{n+i} = 0$ $(i = 1, \ldots, m)$ and $w = 0$, contradicting that the minimum $w > 0$.

(ii) The minimum $w = 0$. This means that all artificial variables are at value zero. If at the optimal solution, all artificial variables are nonbasic, then the optimal basis to the Phase 1 problem is a basis for the original problem and serves as the starting basic feasible solution for the Phase 2 computations. On the other hand, if at the optimal solution, one or more of the artificial variables is still basic and at value zero, we still proceed to Phase 2. However, in this case, we must ensure that during the Phase 2 computations any artificial variable which is basic and at value 0 never becomes positive. One way of doing this is to simply eliminate each artificial variable which is basic at value 0 from the tableau. This is accomplished by pivoting on any nonzero element in the row of the simplex tableau correspondig to an artificial variable which is basic at value 0 prior to starting the Phase 2 computations. If all enteries in the tableau happen to be 0 it means that the corresponding constraint is redundant and the row of zeroes in the tableau should be dropped.

We illustrate the two-phase simplex method using the example given in the beginning of this section. Before presenting the computations, notice that an artificial variable is introduced solely to serve as a basic variable. Thus, once it leaves the set of basic variables it may be dropped. In the example below the columns corresponding to x_4 and x_5 are therefore not being kept once they become nonbasic.

Example 2.3
Phase 1 computations

The Phase 1 problem is

$$\text{minimize} \qquad\qquad\qquad x_4 + x_5 = w$$

$$\text{subject to} \quad x_1 + x_2 + \ x_3 + \qquad x_4 \qquad = 5$$

$$2x_1 + x_2 + 3x_3 - s_1 + \qquad x_5 = 9$$

$$x_1 \geq 0, x_2 \geq 0, x_3 \geq 0, s_1 \geq 0,$$

and $\qquad x_4 \geq 0, x_5 \geq 0,$

where x_4 and x_5 are artificial variables and provide us with the initial basic feasible solution, $x_4 = 5, x_5 = 9$, and $x_1 = x_2 = x_3 = s_1 = 0$.

Iteration 0 The initial simplex tableau is

Basic variables (BV)	x_1	x_2	x_3	s_1	x_4	x_5	Basic feasible solution (BFS)	Ratio $(\bar{b}_i/\bar{a}_{is})$ for $a_{is} > 0$
x_4	1	1	1	0	1	0	5	5
OUT $\to x_5$	2	1	③	-1	0	1	9	3
$\bar{c}_j$	-3	-2	-4	1	0	0	$w = 14$	

↑
IN

The reduced costs for the basic variables are 0 and for nonbasic variables they are computed using (9). As before (cf. Section 2.3.3), $x_s = x_3$, and $x_r = x_5$. A pivot step provides the updated tableau and after a second pivot operation the Phase 1 computations terminate.

Iteration 1

Basic variables (BV)	x_1	x_2	x_3	s_1	x_4	Basic feasible solution (BFS)	Ratio $(\bar{b}_i/\bar{a}_{is})$ for $a_{is} > 0$
OUT $\to x_4$	1/3	②/③	0	1/3	1	2	3
x_3	2/3	1/3	1	$-1/3$	0	3	9
$\bar{c}_j$	$-1/3$	$-2/3$	0	$-1/3$	0	$w = 2$	

↑
IN

Iteration 2

Basic variables (BV)	x_1	x_2	x_3	s_1	Basic feasible solution (BFS)
x_2	1/2	1	0	1/2	3
x_3	1/2	0	1	−1/2	2
$\bar{c}_j$	0	0	0	0	$w = 0$

Since $w = 0$ and all reduced costs are nonnegative, Phase 1 terminates. The basic variables x_2 and x_3 at the optimal solution provide a basic feasible solution to the original (Phase 2) problem. This means that the x_2 and x_3 rows of the simplex tableau remain the same and to start Phase 2 we need only find the values of the reduced costs.

Phase 2 computations

To start Phase 2 we need to compute, using (9), the reduced cost based on the original objective function. This completes the initial tableau for Phase 2 given below. Subsequent to a pivot step, we obtain the optimal solution.

Iteration 0

Basic variables (BV)	x_1	x_2	x_3	s_1	Basic feasible solution (BFS)	Ratio $(\bar{b}_i/\bar{a}_{is})$ for $a_{is} > 0$
OUT → x_2	1/2	1	0	(1/2)	3	6
x_3	1/2	0	1	−1/2	2	−
$\bar{c}_j$	0	0	0	−1	$z = 11$	

$$\uparrow$$
$$\text{IN}$$

Iteration 1

Basic variables (BV)	x_1	x_2	x_3	s_1	Basic feasible solution (BFS)
s_1	1	2	0	1	6
x_3	1	1	1	0	5
$\bar{c}_j$	1	2	0	0	$z = 5$

2.4   The Revised Simplex Algorithm

The simplex algorithm presented in the previous section requires that all the constraints be kept in canonical form. As the problem size grows, storage and computational requirements limit the applicability of this approach. The revised simplex method which we present here alleviates these problems by requiring only the inverse of the current basis which is updated at each iteration. All other values are computed as needed from their mathematical relationship to the basis inverse. The computations are explained below.

The linear programming problem (4)-(6) may be written as

$$\text{minimize} \quad \sum_{j=1}^{n} c_j x_j \tag{13}$$

$$\text{subject to} \quad \sum_{j=1}^{n} \mathbf{a}_j x_j = \mathbf{b} \tag{14}$$

$$\text{and} \quad x_j \geq 0 \quad (j = 1, \ldots, n), \tag{15}$$

where $\mathbf{a}_j = (a_{1j}, a_{2j}, \ldots, a_{mj})^t$ and $\mathbf{b} = (b_1, b_2, \ldots, b_m)^t$.

Suppose we have a basic feasible solution to (14) with basic variables $\mathbf{x_B} = (x_{j_1}, x_{j_2}, \ldots, x_{j_m})$, corresponding costs $\mathbf{c_B} = (c_{j_1}, c_{j_2}, \ldots, c_{j_m})$ and basis $\mathbf{B} = (\mathbf{a}_{j_1}, \mathbf{a}_{j_2}, \ldots, \mathbf{a}_{j_m})$, with inverse $\mathbf{B}^{-1}$.

Since all nonbasic variables are at value zero, equation (14) may be written as $\mathbf{Bx_B} = \mathbf{b}$, or,

$$\mathbf{x_B} = \mathbf{B}^{-1}\mathbf{b} = \bar{\mathbf{b}}. \tag{16}$$

To determine if (16) gives an optimal basic feasible solution, we need to compute the reduced cost $\bar{c}_j$ for all nonbasic variables x_j. From (9), we have $\bar{c}_j = c_j - \mathbf{c_B}\bar{\mathbf{a}}_j$, where $\bar{\mathbf{a}}_j = \mathbf{B}^{-1}\mathbf{a}_j$. Hence

$$\bar{c}_j = c_j - \mathbf{c_B}\mathbf{B}^{-1}\mathbf{a}_j, \tag{17}$$

Or

$$\bar{c}_j = c_j - \boldsymbol{\pi}\mathbf{a}_j. \tag{18}$$

where $\boldsymbol{\pi}$ denotes the *simplex multipliers* $\mathbf{c_B}\mathbf{B}^{-1}$. Using (18), we may compute the reduced cost $\bar{c}_j$ for each nonbasic variable. If all the reduced costs are nonnegative, the current basic feasible solution is optimal; if not, we find the IN variable x_s using

$$\bar{c}_s = \underset{j|\bar{c}_j < 0}{\text{minimum}}\ \bar{c}_j. \tag{19}$$

To compute the OUT variable x_{j_r} we compute θ using the minimum ratio test given by (12). To do this, we first find the *updated column* $\bar{\mathbf{a}}_s$ using

$$\bar{\mathbf{a}}_s = \mathbf{B}^{-1}\mathbf{a}_s. \tag{20}$$

If $\bar{\mathbf{a}}_s \leq \mathbf{0}$, the linear programming problem is unbounded and the computations terminate; if not, the maximum value θ that x_s can have while maintaining a basic feasible solution, is given by

$$\theta = \frac{\bar{b}_r}{\bar{a}_{rs}} = \underset{i|\bar{a}_{is} > 0}{\text{minimum}}\ \frac{\bar{b}_i}{\bar{a}_{is}}. \tag{21}$$

				x_s
x_{j_1}			$\bar{b}_1$	$\bar{a}_{1s}$
x_{j_2}			$\bar{b}_2$	$\bar{a}_{2s}$
$\vdots$		$\mathbf{B}^{-1}$	$\vdots$	$\vdots$
x_{j_r}			$\bar{b}_r$	$\left(\bar{a}_{rs}\right)$
$\vdots$			$\vdots$	$\vdots$
x_{j_m}			$\bar{b}_m$	$\bar{a}_{ms}$
	$-\pi_1$ $\quad$ $-\pi_2$ $\quad$ $\cdots$ $\quad$ $-\pi_m$			$\bar{c}_s$

Table 2.3: The Revised Simplex Tableau

Naturally, when $x_s = \theta$, $x_{j_r} = 0$ and is thus the OUT variable.

The computations may be kept in the tableau displayed in Table 2.3. For ease of computation, the simplex multipliers $\boldsymbol{\pi} = (\pi_i)$ and the right hand side $\mathbf{x_B} = \bar{\mathbf{b}} = (\bar{b}_i)$, along with $\mathbf{B}^{-1}$, are kept and updated.

When x_s is increased to value θ, we move to an improved basic feasible solution corresponding to the basis $\hat{\mathbf{B}}$, where the basic variable x_{j_r} is replaced by x_s. To update $\mathbf{B}^{-1}$, $\boldsymbol{\pi}$, and $\mathbf{x_B}$ to yield $\hat{\mathbf{B}}^{-1}$, $\hat{\boldsymbol{\pi}}$, and $\hat{\mathbf{x}}_{\mathbf{B}}$ respectively we use the computations given below.

Pivot Computations:

(i) *Updating* $\mathbf{B}^{-1}$: To find the new basis inverse $\hat{\mathbf{B}}^{-1}$ we note that

$$\mathbf{B} = (\mathbf{a}_{j_1}, \ldots, \mathbf{a}_{j_{r-1}}, \mathbf{a}_{j_r}, \mathbf{a}_{j_{r+1}}, \ldots, \mathbf{a}_{j_m}) \text{ and }$$

$$\hat{\mathbf{B}} = (\mathbf{a}_{j_1}, \ldots, \mathbf{a}_{j_{r-1}}, \mathbf{a}_s, \mathbf{a}_{j_{r+1}}, \ldots, \mathbf{a}_{j_m})$$

Using (20) we have that

$$\mathbf{a}_s = \sum_{\substack{i=1 \\ i \neq r}}^{m} \bar{a}_{is}\mathbf{a}_{j_i} + \bar{a}_{rs}\mathbf{a}_{j_r}$$

Solving for $\mathbf{a}_{j_r}$ gives

$$\mathbf{a}_{j_r} = \eta_1 \mathbf{a}_{j_1} + \ldots + \eta_{r-1} \mathbf{a}_{j_{r-1}} + \eta_r \mathbf{a}_s + \ldots + \eta_m \mathbf{a}_{j_m}; \qquad (22)$$

where, $\eta_i = (-\bar{a}_{is}/\bar{a}_{rs})$ $(i = 1, \ldots, m; \ i \neq r)$ and $\eta_r = (1/\bar{a}_{rs})$. Or, defining the column vector $\boldsymbol{\eta} = (\eta_1, \ldots, \eta_m)^t$, we have

$$\mathbf{a}_{j_r} = \hat{\mathbf{B}}\boldsymbol{\eta}. \qquad (23)$$

By defining the m by m elementary matrix $\mathbf{E}$ as

$$\mathbf{E} = \begin{bmatrix} 1 & 0 & \ldots & 0 & \eta_1 & \ldots & 0 \\ 0 & 1 & \ldots & 0 & \eta_2 & \ldots & 0 \\ \vdots & \vdots & & \vdots & \vdots & & \vdots \\ 0 & 0 & & 1 & \eta_{r-1} & \ldots & 0 \\ 0 & 0 & & 0 & \eta_r & \ldots & 0 \\ \vdots & \vdots & & \vdots & \vdots & & \vdots \\ 0 & 0 & & 0 & \eta_m & \ldots & 1 \end{bmatrix}$$

where $\boldsymbol{\eta}$ is rth column of $\mathbf{E}$, we have that $\hat{\mathbf{B}}\mathbf{E} = \mathbf{B}$; or

$$\hat{\mathbf{B}}^{-1} = \mathbf{E}\mathbf{B}^{-1} \qquad (24)$$

Expression (24) allows us to find the new values of the inverse $\hat{\mathbf{B}}^{-1}$ from the old values in $\mathbf{B}^{-1}$.

(ii) *Updating* $\boldsymbol{\pi}$:   By definition, the new simplex multipliers are

$$\hat{\boldsymbol{\pi}} = (c_{j_1}, c_{j_2} \ldots c_{j_{r-1}}, c_s, c_{j_{r+1}}, \ldots, c_{j_m})\hat{\mathbf{B}}^{-1}.$$

Using (24) with the last expression gives

$$\hat{\boldsymbol{\pi}} = (c_{j_1}, c_{j_2} \ldots c_{j_{r-1}}, \lambda, c_{j_{r+1}}, \ldots, c_{j_m})\mathbf{B}^{-1}, \qquad (25)$$

where,

$$\lambda = \sum_{\substack{i=1 \\ i \neq r}}^{m} \eta_i c_{j_i} + \eta_r c_s = c_{j_r} + \frac{\bar{c}_s}{\bar{a}_{rs}}.$$

Substituting for λ in (25), gives

$$\hat{\boldsymbol{\pi}} = \mathbf{c_B B}^{-1} + (0, 0, \ldots, 0, \lambda - c_{j_r}, 0, \ldots, 0)\, \mathbf{B}^{-1};$$

or equivalently,

$$\hat{\boldsymbol{\pi}} = \boldsymbol{\pi} + (\lambda - c_{j_r})\, \mathbf{B_r}^{-1} = \boldsymbol{\pi} + \frac{\bar{c}_s}{\bar{a}_{rs}} \mathbf{B_r}^{-1}, \qquad (26)$$

where $\mathbf{B_r}^{-1}$ is the rth row of $\mathbf{B}^{-1}$. Using (26), we may find the new values of $\hat{\boldsymbol{\pi}}$ from the old values $\boldsymbol{\pi}$.

(iii) *Updating* $\bar{\mathbf{b}}$: The new basic feasible solution $\mathbf{x}_{\hat{B}}$ is given by $(x_1, \ldots, x_{r-1}, x_s, x_{r+1} \ldots, x_m)^t = \hat{\mathbf{B}}^{-1}\mathbf{b}$. Using (24), this means that $\mathbf{x}_{\hat{B}} = \mathbf{E}\mathbf{B}^{-1}\mathbf{b} = \mathbf{E}\mathbf{x_b}$. Or, using $\mathbf{x_B} = \bar{\mathbf{b}} = (\bar{b}_i)$, gives

$$\mathbf{x}_{\hat{B}} = \begin{bmatrix} \bar{b}_1 - \dfrac{\bar{a}_{1s}}{\bar{a}_{rs}} \bar{a}_r \\ \vdots \\ \dfrac{\bar{b}_r}{\bar{a}_{rs}} \\ \vdots \\ \bar{b}_m - \dfrac{\bar{a}_{ms}}{\bar{a}_{rs}} \bar{b}_r \end{bmatrix} \qquad (27)$$

Using (27) we may find the new values of the basic variables $\mathbf{x}_{\hat{B}}$ from the old ones $\mathbf{x_B} = \bar{\mathbf{b}}$.

The revised simplex tableau uses equations (24), (26), and (27) to update the tableau. A closer look at these expressions reveals that updating the revised simplex tableau (that is, updating $\mathbf{B}$, $\boldsymbol{\pi}$, and $\mathbf{x_B}$) is equivalent to performing the pivot step (i), (ii), and (iii), where x_{j_r}

corresponds to the pivot row, x_s corresponds to the pivot column, and $\bar{a}_{rs}$ is the pivot element (cf. Section 2.3). We summarize the revised simplex algorithm below.

Revised Simplex Algorithm

Step 1. Find an initial basic feasible solution $\mathbf{x_B} = \mathbf{B}^{-1}\mathbf{b} = \bar{\mathbf{b}}$; compute $\boldsymbol{\pi} = \mathbf{c_B}\mathbf{B}^{-1}$.

Step 2. For each nonbasic variable, compute $\bar{c}_j = c_j - \boldsymbol{\pi}\mathbf{a}_j$. If all $\bar{c}_j$ are non-negative, the current basic feasible solution is optimal, stop; if not, find the IN variable indexed by s, using $\bar{c}_s = \text{minimum}_j\ \bar{c}_j$. Go to Step 3.

Step 3. Compute $\bar{\mathbf{a}}_s = \mathbf{B}^{-1}\mathbf{a}_s$. If $\bar{\mathbf{a}}_s \leq \mathbf{0}$, the problem is unbounded, stop. If not, using (21) compute the minimum ratio θ and determine the OUT variable x_{j_r}.

Step 4. Using (24), (26), and (27) perform a pivot step. This updates $\mathbf{B}^{-1}$, $\boldsymbol{\pi}$, and $\mathbf{x_B}$ to yield $\hat{\mathbf{B}}^{-1}$, $\hat{\boldsymbol{\pi}}$, and $\hat{\mathbf{x}}_\mathbf{B}$, respectively. Go to Step 2.

Example 2.4 (from Example 2.2) The slack variables s_1, s_2, and s_3 correspond to the initial basic variables; the first tableau is below, with $\mathbf{c_B} = (0,0,0)$ or $\boldsymbol{\pi} = \mathbf{c_B}\mathbf{B}^{-1} = (0,0,0)$, and $\bar{c}_j = c_j - \boldsymbol{\pi}\mathbf{a}_j = c_j$. As $c_1 = -1, c_2 = -2$ and $c_3 = -3$; x_3 is the IN variable. Using $\bar{\mathbf{a}}_3 = \mathbf{B}^{-1}\mathbf{a}_3 = \mathbf{a}_3$, (21) gives $\theta = \text{minimum}\ (5/1, 9/3, 2/1) = 2$ and s_3 is the OUT variable. We perform a pivot operation yielding the next tableau. Subsequent iterations are computationally similar and are in the tableaux below.

#1				$\bar{b}$	$\bar{a}_3$
s_1	1	0	0	5	1
s_2	0	1	0	9	3
s_3	0	0	1	2	①
$-\pi$	0	0	0		-3

#2				$\bar{b}$	$\bar{a}_2$
s_1	1	0	-1	3	1
s_2	0	1	-3	3	(2)
x_3	0	0	1	2	0
$-\pi$	0	0	3		-2

$$\bar{c}_1 = c_1 - \pi a_1 \quad = -1$$
$$\bar{c}_2 = c_2 - \pi a_2 \quad = -2$$
$$- - - - - - - - - -$$
$$\theta = \text{minimum } \{3/1, 3/2\}$$
$$= 3/2$$

#3				$\bar{b}$
s_1	1	$-1/2$	$1/2$	$3/2$
x_2	0	$1/2$	$-3/2$	$3/2$
x_3	0	0	1	2
$-\pi$	0	1	0	

In Tableau #3, the reduced costs are $-1 - \pi \mathbf{a}_1 = 1$ (for x_1), $0 - \pi(0,1,0)^t = 1$ (for s_2), and $0 - \pi(0,0,1)^t = 0$ (for s_3). As all the reduced costs are nonnegative, we have the optimal solution $x_1 = 0$, $x_2 = 3/2$, and $x_3 = 2$.

2.5   Duality in Linear Programming

In this section, we introduce a new linear program defined from the original one. This is done because it turns out that the original problem can be solved by solving the new problem which may be easier. We give the details.

Consider the following linear programming problem

minimize	$\mathbf{cx}$ +	$\mathbf{dy}$	= z	(28)
subject to	$\mathbf{Ax}$ +	$\mathbf{Fy}$	= $\mathbf{b}$	(29)
	$\mathbf{Ex}$ +	$\mathbf{Gy}$	$\geq \mathbf{h}$	(30)
and		$\mathbf{x} \geq \mathbf{0}$, and $\mathbf{y}$ unrestricted,		(31)

where $\mathbf{x}$ is an n vector, $\mathbf{y}$ is an r vector, $\mathbf{A}$ is an m by n matrix and $\mathbf{E}$

Primal Problem			Dual Problem		
Minimize	$\mathbf{cx}$	$= z$	Maximize	$\boldsymbol{\pi}\mathbf{b}$	$= v$
subject to	$\mathbf{Ax}$	$= \mathbf{b}$	subject to	$\boldsymbol{\pi}\mathbf{A}$	$\leq \mathbf{c}$
and	$\mathbf{x}$	$\geq \mathbf{0}$	and $\boldsymbol{\pi}$ unrestricted		

Table 2.4: Primal-Dual Pair with Primal Problem in Standard Form

is a p by n matrix. All other vectors and matrices are of appropriate sizes.

Corresponding to this problem, called the *primal problem*, consider the following linear program

$$\text{maximize} \quad \boldsymbol{\pi}\mathbf{b} + \boldsymbol{\mu}\mathbf{h} = v \tag{32}$$
$$\text{subject to} \quad \boldsymbol{\pi}\mathbf{A} + \boldsymbol{\mu}\mathbf{E} \leq \mathbf{c} \tag{33}$$
$$\boldsymbol{\pi}\mathbf{F} + \boldsymbol{\mu}\mathbf{G} = \mathbf{d} \tag{34}$$
$$\text{and} \quad \boldsymbol{\pi} \text{ unrestricted, and } \boldsymbol{\mu} \geq \mathbf{0}, \tag{35}$$

where $\boldsymbol{\pi}$ and $\boldsymbol{\mu}$ are row vectors of size m and p, respectively. This problem is called the *dual* of the primal problem (28)-(31). A particular case of this primal-dual pair is the situation where the primal problem has only equality constraints (that is, constraint (30) is not present) and all variables are nonnegative (that is, the primal problem does not include the variables $\mathbf{y}$). This primal-dual pair can be written as in Table 2.4.

Note that the primal problem in the general form (28)-(31) can be changed to standard form (as in Table 2.4) by using the following transformations

(i) Replace an unconstrained variable by the difference of two non-negative variables.

(ii) Convert an inequality constraint to an equality constraint by adding a slack or subtracting a surplus variable.

Hence, in this section all results are stated for the primal-dual pair in Table 4. We have the following useful results.

Theorem 2.10 *(Weak Duality Theorem) Let* $\mathbf{x}$ *be a primal feasible solution and* $z = \mathbf{cx}$ *be the corresponding value of the primal objective* *function. Also, let $\boldsymbol{\pi}$ be a dual solution and $v = \boldsymbol{\pi}\mathbf{b}$ be the corresponding value for the dual objective function. Then,*

$$z \geq v$$

Proof: Since $\boldsymbol{\pi}$ is a solution to the dual,

$$\boldsymbol{\pi}\mathbf{A} \leq \mathbf{c}. \tag{36}$$

As $\mathbf{x}$ is feasible, $\mathbf{x} \geq \mathbf{0}$, and multiplying inequality (36) by $\mathbf{x}$ yields,

$$\boldsymbol{\pi}\mathbf{A}\mathbf{x} \leq \mathbf{c}\mathbf{x} = z \tag{37}$$

but $\boldsymbol{\pi}\mathbf{A}\mathbf{x} = \boldsymbol{\pi}\mathbf{b} = v$, hence $v \leq z$. ∎

Theorem 2.11 *(Fundamental Duality Theorem) If the primal and the dual problems both have feasible solutions, then both have optimal solutions, and minimum z = maximum v.*

Proof: By Theorem 2.10, the value of the primal objective function is bounded below by v corresponding to the dual solution $\boldsymbol{\pi}$; similarly the value of the dual objective function is bounded above by z corresponding to the primal feasible solution $\mathbf{x}$. Hence both primal and dual problems have optimal solutions.

Now let $\mathbf{B}$ be the optimal primal basis corresponding to the basic variables $\mathbf{x_B}$. We can partition the primal problem as

$$\begin{aligned} \text{minimize} \quad & \mathbf{c_B}\mathbf{x_B} + \mathbf{c_N}\mathbf{x_N} = z \\ \text{subject to} \quad & \mathbf{B}\mathbf{x_B} + \mathbf{N}\mathbf{x_N} = \mathbf{b} \\ \text{and} \quad & \mathbf{x_B} \geq \mathbf{0},\ \mathbf{x_N} \geq \mathbf{0}. \end{aligned}$$

The simplex multipliers (cf. Section 2.4) associated with the optimal basis $\mathbf{B}$ are $\boldsymbol{\pi}^* = \mathbf{c_B}\mathbf{B}^{-1}$, and the value of the primal objective function is $z^* = \mathbf{c_B}\mathbf{B}^{-1}\mathbf{b}$.

Since $\mathbf{B}$ is an optimal basis, the reduced costs $\bar{\mathbf{c}} = \mathbf{c} - \boldsymbol{\pi}^*\mathbf{A}$ are nonnegative. Hence, $\mathbf{c} - \boldsymbol{\pi}^*\mathbf{A} \geq \mathbf{0}$, or $\boldsymbol{\pi}^*\mathbf{A} \leq \mathbf{c}$.

	Dual Problem	
	Feasible	Infeasible
Feasible	Min z = Max v (optimal solution)	Primal unbounded
Infeasible	Dual unbounded	Both problems are infeasible

Primal Problem (row label)

Table 2.5: Summary of Duality Results

Thus, $\boldsymbol{\pi}^*$ is a solution to the dual with objective function value $v = \boldsymbol{\pi}^*\mathbf{b} = \mathbf{c_B}\mathbf{B}^{-1}\mathbf{b} = z^*$. As z^* is an upper bound on v, $\boldsymbol{\pi}^*$ is optimal for the dual problem. But by definition $\mathbf{x_B}$ is optimal for the primal problem and $z^* = $ minimum $z = $ maximum v; hence the desired result. ∎

Theorem 2.12 *If either the primal or the dual problem has an unbounded solution, then the other problem has no feasible solution (i.e., is infeasible).*

Proof: Let us assume that the primal is unbounded, then by Theorem 2.10, $\boldsymbol{\pi}\mathbf{b} \leq -\infty$ for all solutions $\boldsymbol{\pi}$ to the dual. If the dual has a solution, then $\boldsymbol{\pi}\mathbf{b}$ is finite, contradicting the above condition. Hence, the dual has no feasible solution. A similar argument holds in case the dual is unbounded. ∎

The results of the last three theorems are summarized in the Table 2.5.

Theorem 2.13 *(Complementary Slackness Condition) If $\mathbf{x}^*$ is a primal feasible solution and $\boldsymbol{\pi}^*$ is a solution to the dual, then $\mathbf{x}^*$ is optimal for the primal and $\boldsymbol{\pi}^*$ is optimal for the dual, if and only if*

$$(\mathbf{c} - \boldsymbol{\pi}^*\mathbf{A})\mathbf{x}^* = \mathbf{0}.$$

Proof:

(i) If $\mathbf{x}^*$ and $\boldsymbol{\pi}^*$ are primal and dual optimal solutions, respectively, then by Theorem 2.11, $\mathbf{c}\mathbf{x}^* = \boldsymbol{\pi}^*\mathbf{b}$. Since $\mathbf{x}^*$ is a primal feasible solution,

$$\mathbf{A}\mathbf{x}^* = \mathbf{b}.$$

Multiplying both sides by $\boldsymbol{\pi}^*$ yields,

$$\boldsymbol{\pi}^*\mathbf{A}\mathbf{x}^* = \boldsymbol{\pi}^*\mathbf{b}.$$

Or, $\mathbf{c}\mathbf{x}^* - \boldsymbol{\pi}^*\mathbf{A}\mathbf{x}^* = \mathbf{c}\mathbf{x}^* - \boldsymbol{\pi}^*\mathbf{b} = 0$, or $(\mathbf{c} - \boldsymbol{\pi}^*\mathbf{A})\mathbf{x}^* = 0$.

(ii) If $(\mathbf{c} - \boldsymbol{\pi}^*\mathbf{A})\mathbf{x}^* = 0$. Then $\mathbf{c}\mathbf{x}^* - \boldsymbol{\pi}^*\mathbf{b} = \mathbf{c}\mathbf{x}^* - \boldsymbol{\pi}^*\mathbf{A}\mathbf{x}^* = 0$. By Theorem 2.10, this means that $\mathbf{x}^*$ is optimal for the primal problem and $\boldsymbol{\pi}^*$ is optimal for the dual problem. ∎

The condition $(\mathbf{c} - \boldsymbol{\pi}^*\mathbf{A})\mathbf{x}^* = 0$ is referred to as *complementary slackness*. Note that in the simplex algorithm, corresponding to a basis $\mathbf{B}$, we have $\mathbf{x} = (\mathbf{x_B}, \mathbf{x_N}) = (\mathbf{B}^{-1}\mathbf{b}, 0)$, and the simplex multipliers $\boldsymbol{\pi} = \mathbf{c_B}\mathbf{B}^{-1}$ always satisfy the complementary slackness condition since

$$(\mathbf{c} - \boldsymbol{\pi}\mathbf{A})\mathbf{x} = (\mathbf{c_B} - \boldsymbol{\pi}\mathbf{B})\mathbf{x_B} + (\mathbf{c_N} - \boldsymbol{\pi}\mathbf{N})\mathbf{x_N} = 0.$$

At optimality $\mathbf{c} - \boldsymbol{\pi}\mathbf{A} \geq 0$ implies that the simplex multipliers $\boldsymbol{\pi}$ is a solution to the dual. Thus, at each iteration the simplex algorithm generates the vectors $\mathbf{x}$ and $\boldsymbol{\pi}$, where $\mathbf{x}$ is a primal feasible solution, and $\mathbf{x}$ and $\boldsymbol{\pi}$ satisfy the complementary slackness condition. However, $\boldsymbol{\pi}$ does *not* satisfy the constraints to the dual problem until optimality, when it is also a solution to the dual (i.e., the corresponding dual slack variables become nonnegative) and by Theorem 2.13, $\boldsymbol{\pi}$ is thus optimal for the dual and $\mathbf{x}$ is optimal for the primal.

Using duality we can now describe another simplex method, called the *dual simplex algorithm* to solve linear programming problems. At each iteration this algorithm generates a pair of vectors $\mathbf{x}$ and $\boldsymbol{\pi}$ satisfying complementary slackness; $\boldsymbol{\pi}$ is always dual feasible, but $\mathbf{x}$ becomes primal feasible only at optimality. If the primal problem is infeasible, the algorithm terminates when the dual problem exhibits unboundedness.

2.6   The Dual Simplex Algorithm

Consider the following primal-dual pair, with the primal in standard form.

Primal Problem	Dual Problem
Minimize $\quad$ $\mathbf{cx}$	Maximize $\quad$ $\boldsymbol{\pi}\mathbf{b}$
subject to $\quad$ $\mathbf{Ax} = \mathbf{b}$	subject to $\quad$ $\boldsymbol{\pi}\mathbf{A} \leq \mathbf{c}$
and $\quad$ $\mathbf{x} \geq \mathbf{0}$	and $\boldsymbol{\pi}$ unrestricted

The dual simplex algorithm requires a basis $\mathbf{B}$ such that the simplex multipliers $\boldsymbol{\pi} = \mathbf{c_B B}^{-1}$ are a solution to the dual (that is, $\boldsymbol{\pi}\mathbf{A} \leq \mathbf{c}$, or $\mathbf{c} - \boldsymbol{\pi}\mathbf{A} \geq \mathbf{0}$), and $\mathbf{B}^{-1}\mathbf{b}$ is not necessarily nonnegative. Equivalently, the basis $\mathbf{B}$ need not be primal feasible, but all the reduced costs $\bar{\mathbf{c}} = \mathbf{c} - \boldsymbol{\pi}\mathbf{A}$ are nonnegative. Such a basis may not always be available, limiting the usefulness of this algorithm. However, if in the primal problem, the coefficient matrix $\mathbf{A}$ contains a minus identity matrix $-\mathbf{I}$, corresponding to a full set of surplus variables, and $\mathbf{b} \geq \mathbf{0}$ and $\mathbf{c} \geq \mathbf{0}$, then setting $\mathbf{B} = -\mathbf{I}$, with $\mathbf{x_B} = \mathbf{B}^{-1}\mathbf{b} = -\mathbf{b} \leq \mathbf{0}$ and $\bar{\mathbf{c}} = \mathbf{c} - \mathbf{c_B}\mathbf{B}^{-1}\mathbf{A} = \mathbf{c} \geq \mathbf{0}$, provides a primal not feasible, basic solution and a solution $\boldsymbol{\pi} = \mathbf{c_B}\mathbf{B}^{-1} = \mathbf{0}$ to the dual (cf. Example 2.4).

Corresponding to the basis $\mathbf{B}$, let $\mathbf{x_B} = (x_1, \ldots x_m)$ be the associated vector of basic variables and $\mathbf{x_N} = (x_{m+1}, \ldots, x_n)$ be the vector of nonbasic variables. The tableau corresponding to this basis (cf. Table 2, Section 2.3.3) can be displayed as in Table 2.6. We list the steps of the dual simplex algorithm below.

Step 1 (Optimality Criterion)
If corresponding to the current basis $\mathbf{B}$, $\bar{\mathbf{b}} = \mathbf{B}^{-1}\mathbf{b} \geq \mathbf{0}$, then $\mathbf{x} = (\mathbf{x_B}, \mathbf{x_N}) = (\bar{\mathbf{b}}, \mathbf{0})$, is a primal feasible solution. As $\boldsymbol{\pi} = \mathbf{c_B}\mathbf{B}^{-1}$ is

Table 2.6: Tableau in Canonical Form

a dual solution, with $\mathbf{x}$ and $\boldsymbol{\pi}$ satisfying complementary slackness, we have that $\mathbf{x}$ is an optimal solution to the primal and we terminate. If not (i.e., one or more basic variables is negative), go to Step 2.

Step 2 (Find the Pivot Row and Pivot Column)
Pick a row r such that $\bar{b}_r < 0$. Row r is the pivot row and x_r is the OUT variable. Now if we pivot on a *negative* element in this row, then x_r leaves the basis to become nonbasic and the new basic variable becomes positive (since we divide the pivot row by the pivot element.) However, the pivot element should also be selected so that we maintain a solution to the dual problem. Equivalently, the reduced costs must remain nonnegative. Thus, if the pivot element is $\bar{a}_{rs}$, we must have,

$$\bar{c}_j - \frac{\bar{c}_s \bar{a}_{rj}}{\bar{a}_{rs}} \geq 0 \ (j = 1, \ldots, n)$$

To satisfy the above, the pivot column indexed by s, or the IN variable x_s, is selected so that

$$\frac{\bar{c}_s}{-\bar{a}_{rs}} = \operatorname*{minimum}_{j \mid \bar{a}_{rj} < 0} \frac{\bar{c}_j}{-\bar{a}_{rj}}.$$

Ties for the minimum ratio are broken arbitrarily. As mentioned before, when degeneracy is present, additional rules to prevent the dual simplex method from cycling may be necessary (see, e.g., Solow [541]).

If $\bar{a}_{rj} \geq 0$ for all j, then the primal problem is infeasible, stop. (This follows because the rth constraint

$$x_r + \sum_{j=m+1}^{n} \bar{a}_{rj} x_j = \bar{b}_r$$

can never be satisfied as $\bar{b}_r < 0$ and all terms in the left-hand-side of the equality are nonnegative.)

Step 3 (Pivot Step) In the new basis, the basic variable x_r is replaced by the nonbasic variable x_s. The new tableau in canonical form can be obtained by using the pivot operations described in Section 2.3.3.
    Note that as described in (26), the new simplex multiplier $\hat{\boldsymbol{\pi}}$ can be written as

$$\hat{\boldsymbol{\pi}} = \boldsymbol{\pi} + \frac{\bar{c}_s}{\bar{a}_{rs}} \mathbf{B_r^{-1}}, \tag{38}$$

where $\mathbf{B_r^{-1}}$ is the rth row of $\mathbf{B^{-1}}$. Multiplying both sides of (38) by $\mathbf{b}$ yields,

$$\hat{\pi}\mathbf{b} = \pi\mathbf{b} + \frac{\bar{c}_s}{\bar{a}_{rs}}\mathbf{B_r^{-1}}\mathbf{b}$$

or

$$\hat{\pi}\mathbf{b} = \pi\mathbf{b} + \frac{\bar{c}_s}{\bar{a}_{rs}}\bar{b}_r. \tag{39}$$

Since $\bar{b}_r < 0$ and $\bar{a}_{rs} < 0$, $\hat{\pi}\mathbf{b} \geq \pi\mathbf{b}$; or, the value of the dual objective function can never get worse. Moreover, if the dual is nondegenerate, that is, if $\bar{c}_j > 0$, $j = m + 1, \ldots, n$, then $\hat{\pi}\mathbf{b} > \pi\mathbf{b}$. In this case, since there is a finite number of basis, and at each iteration the dual objective function increases, it follows that a basis is never repeated; so the algorithm converges in a finite number of steps.

Finite convergence even with degeneracy present in the dual problem can be guaranteed by using an anticycling procedure such as the lexicographic rule (see, e.g., Hadley [303], Solow [541] or Murty [439]). A dual simplex algorithm which incorporates lexicographic rules to avoid cycling is sometimes referred to as a *dual lexicographic simplex method*.

Example 2.5 Consider following linear programming problem

$$
\begin{array}{llrcrcl}
\text{minimize} & 4x_1 & + & 5x_2 & & \\
\text{subject to} & -x_1 & - & 4x_2 & \leq & -5 \\
& -3x_1 & - & 2x_2 & \leq & -7 \\
\text{and} & x_1, & & x_2 & \geq & 0.
\end{array}
$$

After adding the slack variables s_1 and s_2, the problem in standard form is

$$
\begin{array}{llrcrcrcrcl}
\text{minimize} & 4x_1 & + & 5x_2 & & & & & & \\
\text{subject to} & -x_1 & - & 4x_2 & + & s_1 & & & = & -5 \\
& -3x_1 & - & 2x_2 & & & + & s_2 & = & -7 \\
\text{and} & x_1, & & x_2, & & s_1, & & s_2 & \geq & 0.
\end{array}
$$

The basis corresponding to the variables (s_1, s_2) provides an immediate solution to the dual. Hence, it makes sense to use the dual simplex algorithm.

Iteration 0
The tableau in canonical form is given below. Note that the dual so-

lution need never explicitly appear and we solve the problem using strictly primal notation. In the tableau we have $x_r = s_2$ and $x_s = x_1$. After two pivot steps, we obtain an optimal solution.

Basic variables	s_1	s_2	x_1	x_2	Basic solution
s_1	1	0	-1	-4	-5
OUT $\rightarrow$ s_2	0	1	$\left(-3\right)$	-2	-7
$\bar{c}_j$	0	0	4	5	$z = 0$
$\dfrac{\bar{c}_j}{-\bar{a}_{rj}}$ for $j \mid \bar{a}_{rj} < 0$	—	—	4/3	5/2	
				$\uparrow$ IN	

Iteration 1

Basic variables	s_1	s_2	x_1	x_2	Basic solution
OUT $\rightarrow$ s_1	1	$-1/3$	0	$-10/3$	$-8/3$
x_1	0	$-1/3$	1	$2/3$	$7/3$
$\bar{c}_j$	0	$4/3$	0	$7/3$	$z = 28/3$
$\dfrac{\bar{c}_j}{-\bar{a}_{rj}}$ for $j \mid \bar{a}_{rj} < 0$	—	4	—	$7/10$	
				$\uparrow$ IN	

Iteration 2

Basic variables	s_1	s_2	x_1	x_2	Basic solution
x_2	$-3/10$	$1/10$	0	1	$8/10$
x_1	$1/5$	$-2/5$	1	0	$9/5$
$\bar{c}_j$	$7/10$	$11/10$	0	0	$z = 56/5$

Since $x_1 = 9/5$ and $x_2 = 4/5$ is a primal feasible solution, it is optimal.

2.7   The Beale Tableau

In this section, we describe another tableau which may be used to represent linear programming computations. This tableau is often used in integer programming algorithms. For the basis $\mathbf{B}$ corresponding to the basic variables $\mathbf{x_B} = (x_1, x_2, \ldots, x_m)$, the tableau in canonical form as discussed in Section 2.3 can be displayed as below.

BV	x_1 x_2 $\ldots$ x_m	x_{m+1}	$\ldots$ x_j $\ldots$	x_n	$-z$	(BFS)
x_1	1	$\bar{a}_{1,m+1}$	$\ldots$	$\bar{a}_{1n}$	0	$\bar{b}_1$
x_2	$\quad$ 1	$\bar{a}_{2,m+1}$	$\ldots$	$\bar{a}_{2n}$	0	$\bar{b}_2$
$\vdots$	$\ddots$	$\vdots$		$\vdots$		$\vdots$
x_m	$\qquad$ 1	$\bar{a}_{m,m+1}$	$\ldots$	$\bar{a}_{mn}$	0	$\bar{b}_m$
$\bar{c}_j$	0 0 $\ldots$ 0	$\bar{c}_{m+1}$	$\ldots$	$\bar{c}_n$	1	$-\bar{z} = -\mathbf{c_B}\mathbf{B}^{-1}\mathbf{b}$

Note that in order to represent the relationship of the objective function z to the nonbasic variables, we have augmented the tableau by adding a column corresponding to the variable $(-z)$. The tableau represents the following system of constraints:

$$x_i \;=\; \bar{b}_i \;+\; \sum_{j=m+1}^{n} \bar{a}_{ij}(-x_j) \quad (i = 1, 2, \ldots, m)$$

$$(-z) \;=\; -\bar{z} \;+\; \sum_{j=m+1}^{n} \bar{c}_j(-x_j).$$

The Beale tableau given in Table 2.7 displays these and the identity relationships

$$x_i = -(-x_i) \quad (i = m+1, \ldots, n).$$

From the table, we note that row 0 corresponds to the objective func-

All variables	Constant values	Nonbasic variables				
		$(-x_{m+1})$	$\cdots$	$(-x_{m+j})$	$\cdots$	$(-x_n)$
$(-z)$	$(-\bar{z})$	$\bar{c}_{m+1}$	$\cdots$	$\bar{c}_{m+j}$	$\cdots$	$\bar{c}_n$
x_{m+1}	0	-1	$\cdots$	0	$\cdots$	0
$\vdots$	$\vdots$	$\vdots$		$\vdots$		$\vdots$
x_{m+j}	0	0	$\cdots$	-1	$\cdots$	0
$\vdots$	$\vdots$	$\vdots$		$\vdots$		$\vdots$
x_n	0	0	$\cdots$	0	$\cdots$	-1
x_1	$\bar{b}_1$	$\bar{a}_{1,m+1}$	$\cdots$	$\bar{a}_{1,m+j}$	$\cdots$	$\bar{a}_{1n}$
x_2	$\bar{b}_2$	$\bar{a}_{2,m+1}$	$\cdots$	$\bar{a}_{2,m+j}$	$\cdots$	$\bar{a}_{2n}$
$\vdots$	$\vdots$	$\vdots$		$\vdots$		$\vdots$
x_r	$\bar{b}_r$	$\bar{a}_{r,m+1}$	$\cdots$	$\bar{a}_{r,m+j}$	$\cdots$	$\bar{a}_{rn}$
$\vdots$	$\vdots$	$\vdots$		$\vdots$		$\vdots$
x_m	$\bar{b}_m$	$\bar{a}_{m,m+1}$	$\cdots$	$\bar{a}_{m,m+j}$	$\cdots$	$\bar{a}_{mn}$

Table 2.7: The Beale Tableau

tion, and each row, 1 to n, is associated with one of the original variables x_j $(j = 1, \ldots, n)$. Also, for each $j = 1, \ldots, n - m$, let

$$
\boldsymbol{\alpha}_j = \begin{bmatrix} \bar{c}_{m+j} \\ 0 \\ \vdots \\ -1 \\ \vdots \\ 0 \\ \bar{a}_{1,m+j} \\ \vdots \\ \bar{a}_{m,m+j} \end{bmatrix} \begin{array}{l} \left.\rule{0pt}{5em}\right\} \begin{array}{l}(n-m)\text{ elements with} \\ -1\text{ in the }j\text{th row}\end{array} \\ \left.\rule{0pt}{2.5em}\right\} m\text{ elements} \end{array}
$$

and

$$\alpha_0 = \begin{bmatrix} z \\ 0 \\ \vdots \\ 0 \\ \bar{b}_1 \\ \vdots \\ \bar{b}_m \end{bmatrix} \begin{array}{l} \left.\rule{0pt}{2.5em}\right\} (n-m) \text{ elements} \\[1em] \left.\rule{0pt}{2.5em}\right\} m \text{ elements} \end{array}$$

Then using vector notation, the Beale tableau corresponds to the following equations

$$\mathbf{x} = \alpha_0 + \sum_{j=1}^{n-m} \alpha_j(-x_{m+j}), \tag{40}$$

where $\mathbf{x} = (x_{m+1}, x_{m+2}, \ldots, x_n, x_1, x_2, \ldots, x_m)^t$.

If we apply the dual lexicographic simplex algorithm to the Beale tableau, the pivot row, indexed by r, satisfies

$$\bar{b}_r = \operatorname*{minimum}_{\bar{b}_i < 0} \bar{b}_i$$

The selection of the pivot column in the dual simplex method was explained in the previous section. Special rules may also be added if necessary to prevent cycling (see, e.g., Solow [541]). The selection of this column may be expressed by using the concept of lexicographic ordering; namely, a vector $\mathbf{x}$ is lexicographically smaller than a vector $\mathbf{y}$ (written $\mathbf{x} \overset{\mathcal{L}}{\prec} \mathbf{y}$) if the first nonzero term of $\mathbf{x} - \mathbf{y}$ is negative. Then the pivot column indexed by k, is found using

$$\frac{\alpha_k}{|\bar{a}_{rk}|} \overset{\mathcal{L}}{\prec} \frac{\alpha_j}{|\bar{a}_{rj}|} \quad \text{for all } j \text{ such that } \bar{a}_{rj} < 0.$$

The last expression follows from the standard rules to select a pivot column when using the dual simplex method (as explained in the previous section) expanded to include anticycling rules (see, e.g., [541] or [439]).

A vector $\mathbf{x}$ is lexicographic smaller than a vector $\mathbf{y}$, written $\mathbf{x} \overset{\mathcal{L}}{\prec} \mathbf{y}$, if and only if the first nonzeroterm of $\mathbf{x} - \mathbf{y}$ is negative.

During a simplex iteration the variable x_{m+k} replaces x_r as the basic variable. To update the Beale tableau, note that

$$x_r = \bar{b}_r + \sum_{j=1}^{n-m} \bar{a}_{r,m+j} \left(-x_{m+j}\right)$$

or

$$\left(-x_{m+k}\right) = -\frac{\bar{b}_r}{\bar{a}_{r,m+k}} + \frac{1}{\bar{a}_{r,m+k}} x_r - \sum_{\substack{j=1 \\ j \neq k}}^{n-m} \frac{\bar{a}_{r,m+j}}{\bar{a}_{r,m+k}} (-x_{m+j}). \quad (41)$$

Substituting (41) in (40) yields

$$\mathbf{x} = \left(\boldsymbol{\alpha}_0 - \frac{\bar{b}_r}{\bar{a}_{r,m+k}} \boldsymbol{\alpha}_k\right) + \left(\frac{-1}{\bar{a}_{r,m+k}} \boldsymbol{\alpha}_k\right)(-x_r) +$$
$$\sum_{j=1}^{n-m} \left(\boldsymbol{\alpha}_j - \frac{\bar{a}_{r,m+j}}{\bar{a}_{r,m+k}} \boldsymbol{\alpha}_k\right)(-x_{m+j}). \quad (42)$$

Equation (42) suggests following pivot rules:

(i) In the list of nonbasic variables, replace $-x_{m+k}$ by $-x_r$.

(ii) Update all columns $\boldsymbol{\alpha}_j$ to $\hat{\boldsymbol{\alpha}}_j$ using

$$\hat{\boldsymbol{\alpha}}_0 = \boldsymbol{\alpha}_0 - \frac{\bar{b}_r}{\bar{a}_{r,m+k}} \boldsymbol{\alpha}_k \qquad \hat{\boldsymbol{\alpha}}_k = \frac{-1}{\bar{a}_{r,m+k}} \boldsymbol{\alpha}_k$$

and $\quad \hat{\boldsymbol{\alpha}}_j = \boldsymbol{\alpha}_j - \frac{\bar{a}_{r,m+j}}{\bar{a}_{r,m+k}} \boldsymbol{\alpha}_k \quad (j = 1, \dots m; \; j \neq k)$

Note that updating the tableau in Section 2.3.3 involved elementary row operations using the pivot row, whereas updating the Beale tableau involves elementary column operations using the pivot column. This follows since the pivot operation transforms the pivot element to -1 and all other elements in the pivot row to 0. We illustrate the pivot operations using the example given in Section 2.6 which was solved by the dual simplex method.

Example 2.6 (From Example 2.5) The first Beale tableau containing the example is below.

OUT

↓

#1	α_0	α_1	α_2
		$-x_1$	$-x_2$
$(-z)$	0	4	5
x_1	0	-1	0
x_2	0	0	-1
s_1	-5	-1	-4
s_2	-7	-3	-2 ← IN

Pivot column selection:

minimum$\{4/3, 5/2\} = 4/3$,

thus x_1 becomes basic.

To update the tableau, we perform elementary column operations using α_1. The objective is to transform the pivot element to -1 and all other elements in the pivot row to 0. This involves the following steps:

(i)   Divide α_1 by 3 ($\hat{\alpha}_1 = \frac{1}{3}\alpha_1$).

(ii) Subtract $(7/3)\alpha_1$ from α_0 ($\hat{\alpha}_0 = \alpha_0 - \frac{7}{3}\alpha_1$).

(iii) Subtract $(2/3)\alpha_1$ from α_2 ($\hat{\alpha}_2 = \alpha_2 - \frac{2}{3}\alpha_1$).

These steps result in tableau #2.

OUT

↓

#2	α_0	α_1	α_2
		$-s_2$	$-x_2$
$(-z)$	$-28/3$	$4/3$	$7/3$
x_1	$7/3$	$-1/3$	$2/3$
x_2	0	0	-1
s_1	$-8/3$	$-1/3$	$-10/3$ ← IN
s_2	0	-1	0

Pivot column selection:

minimum$\{4, 7/10\} = 7/10$.

Using the last column in tableau #2 and the s_1 row to perform elementary column operations yields the tableau below. As the primal solution, exhibited as the α_0 column, is feasible, it is optimal.

	α_0	α_1	α_2
#3		$-s_2$	$-s_1$
$(-z)$	$-112/10$	$11/10$	$7/10$
x_1	$9/5$	$-2/5$	$1/5$
x_2	$4/5$	$1/10$	$-3/10$
s_1	0	0	-1
s_2	0	-1	0

Optimal solution:

$$x_1 = 9/5$$
$$x_2 = 4/5$$
$$z = 112/10$$

Problems

Problem 2.1   Solve the following problems using the graphical method.

(i) Maximize $3x_1 + 5x_2$
     subject to $0.5x_1 + 0.5x_2 \leq 100$
                  $x_1 + 2x_2 \leq 300$
     and $x_1, \quad x_2 \geq 0.$

(ii) Maximize $2x_1 + x_2$
      subject to $x_1 - 2x_2 \leq 0$
                  $-x_1 + x_2 \leq 1$
      and $x_1 \geq 1, \; x_2 \geq 0.$

(iii) Maximize $x_1 + 2x_2 + x_3$
       subject to $x_1 + x_2 + x_3 \leq 5$
                  $x_3 \leq 3$
       and $x_1 \geq 0, x_2 \geq 0, x_3 \geq 0.$

Problem 2.2   Prove Theorem 2.1.

Problem 2.3   Prove Theorem 2.2.

Problem 2.4   Prove Theorem 2.3.

Problem 2.5   Prove Theorem 2.4.

Problem 2.6   Prove Theorem 2.5.

Problem 2.7   Prove Theorem 2.6.

Problem 2.8   Prove Theorem 2.7.

Problem 2.9   Consider the following linear program

$$
\begin{array}{ll}
\text{minimize} & \mathbf{cx} \\
\text{subject to} & \mathbf{Ax} = \mathbf{b} \\
\text{and} & \mathbf{x} \geq \mathbf{0}.
\end{array}
$$

Prove that $\mathbf{Ax} = \mathbf{b}$, $\mathbf{x} \geq \mathbf{0}$ has a nonzero homogeneous solution if and only if the feasible region is unbounded.

Problem 2.10   Consider the following linear program

$$
\begin{array}{lrcrcl}
\text{maximize} & 2x_1 & + & 2x_2 & = & z \\
\text{subject to} & -x_1 & + & x_2 & \leq & 5 \\
 & x_1 & + & 2x_2 & \leq & 10 \\
\text{and} & \multicolumn{4}{l}{x_1 \geq 0, \quad x_2 \geq 0.}
\end{array}
$$

(a) Find all the basic feasible solutions.

(b) Find all extreme half lines for the set of feasible solutions.

(c) Interpret the results of (a) and (b) graphically, and illustrate the results of Theorem 2.7.

Problem 2.11   Solve the linear program in Problem 2.1 using the simplex algorithm. Relate each iteration to the graph.

Problem 2.12   Let $S = \{\mathbf{x} \mid \mathbf{Ax} = \mathbf{b}, \ \mathbf{0} \leq \mathbf{x} \leq \mathbf{u}\}$, where $\mathbf{x} = (x_j)$ is an n vector, $\mathbf{u} = (u_j)$ is a vector of upper bounds, and $\mathbf{A}$ is an m by n matrix. Show that a vector $\mathbf{x} \in S$ is an extreme point of S if and only if $n - m$ values of x_j are either 0 or u_j, the remaining m values for x_j are between 0 and u_j, and the columns corresponding to these m variables form a basis for $\mathbf{Ax} = \mathbf{b}$.

Problem 2.13   Using the result given in Problem 2.12, explain how the steps of the simplex algorithm may be modified to take into account variables with upper bounds while using a basis for $\mathbf{Ax = b}$.

Problem 2.14   Generalize the result given in Problem 2.12 for $S = \{x | \mathbf{Ax = b}, \; \mathbf{l \leq x \leq u}\}$. Modify the simplex algorithm to enforce the bounds on variables while only using a basis for $\mathbf{Ax = b}$.

Problem 2.15   Solve the problem below using the dual simplex algorithm. At each iteration:

(i) give and compare the standard simplex tableau and the Beale tableau.

(ii) list the value of the primal variables and the dual variables and show that complementary slackness is maintained.

$$
\begin{array}{lrcrcrcr}
\text{Minimize} & 2x_1 & + & 3x_2 & + & 4x_3 \\
\text{subject to} & -x_1 & - & x_2 & - & x_3 & \leq & -5 \\
& -x_1 & + & x_2 & & & \leq & -2 \\
& & & & & x_3 & \leq & 2 \\
\text{and} & \multicolumn{7}{l}{x_1 \geq 0, x_2 \geq 0, x_3 \geq 0.}
\end{array}
$$

Problem 2.16   Solve the linear program given in Problem 2.15 using the two-phase simplex algorithm.

Chapter 3

Using Linear Programming to Solve Integer Programs

3.1 Graphical Solutions to Mixed Integer or Integer Programs

For ease of exposition, we rewrite the mixed integer program

$$
\begin{array}{llll}
\text{(MIP)} & \text{maximize} & \mathbf{cx} + \mathbf{dy} = z \\
& \text{subject to} & \mathbf{Ax} + \mathbf{Dy} \leq \mathbf{b}, \\
& & \mathbf{x} \geq \mathbf{0}, \ \mathbf{y} \geq \mathbf{0}, \\
& \text{and} & \mathbf{x} \text{ integer},
\end{array}
$$

where $\mathbf{A}$ is m by n, $\mathbf{D}$ is m by n', and all other vectors are of appropriate size.

From linear programming (cf. Chapter 2), we know that the intersection of the hyperplanes corresponding to the linear constraints $\mathbf{Ax} + \mathbf{Dy} \leq \mathbf{b}$, $\mathbf{x} \geq \mathbf{0}$, and $\mathbf{y} \geq \mathbf{0}$ is a convex polyhedron and that a maximum value of the linear objective function $\mathbf{cx} + \mathbf{dy} = z$ occurs at one of its extreme points. In general, however, the values of the variables $\mathbf{x}$ at an optimal extreme point will not be integer. However, since the polyhedron contains all solutions with $\mathbf{x}$ integer, we may graphically solve mixed integer or integer problems by first finding an optimal extreme point of its associated linear program (sometimes referred to as the linear programming relaxation of the mixed integer problem) and then by parallel shifting the hyperplane $\mathbf{cx} + \mathbf{dy} = z$ back into the polyhedron until it intersects the first point $(\mathbf{x}, \mathbf{y})$ with $\mathbf{x}$ integer. Such a solution is optimal. This procedure allows us to solve problems with up to a total of three variables graphically.

Example 3.1 (Graphical solution)

$$
\begin{array}{rrrrl}
\text{Maximize} & 2x & - & y & = z \\
\text{subject to} & 5x & + & 7y & \leq 45, \qquad (1) \\
 & -2x & + & y & \leq 1, \qquad (2) \\
 & 2x & - & 5y & \leq 5, \qquad (3) \\
 & x, & & y & \geq 0, \qquad (4) \\
\text{and} & \multicolumn{4}{l}{x \text{ integer.}}
\end{array}
$$

The linear programming feasible region, defined by the constraints (1), (2), (3), and (4), is the convex polyhedron spanned by the extreme points ABCDE (Fig. 3.1). The maximal value of z is 35/3 and occurs at the extreme point $C = (y, x) = (5/3,\ 20/3)$. The set of feasible

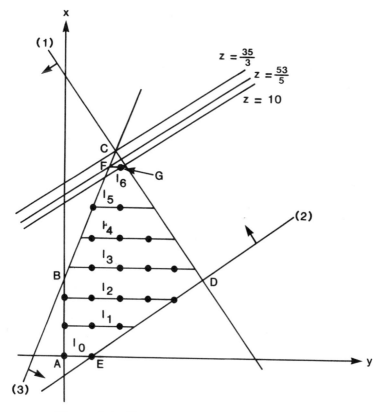

Figure 3.1: Graphical Solution to Example 3.1

solutions with x integer is made up of the points on the line segments $\ell_0, \ell_1, \ell_2, \ell_3, \ell_4, \ell_5,$ and, ℓ_6. If the z hyperplane is pushed back into the polyhedron the first point reached with x integer is $F = (7/5, 6)$ when $z = 53/5$ (Fig. 3.1). If y is also required to be integer, solutions to the integer program occur at the intersection of the line segments ℓ_i ($i = 0, 1, \ldots, 6$) with the hyperplanes $y = K$, where K is a nonnegative integer. The results are the 20 dotted points (Figure. 3.1). A parallel shift of the z hyperplane further into the polyhedron yields the optimal integer solution $G = (2,6)$, with $z = 10$. Note that G is an interior point.

Example 3.2 (No solution to the integer program)

$$\begin{aligned}
\text{Maximize} \quad & x_1 + x_2 = z \\
\text{subject to} \quad & -4x_1 + x_2 \leq -1, \quad &(5)\\
& 4x_1 + x_2 \leq 3, \quad &(6)\\
\text{and} \quad & x_1, \quad x_2 \geq 0 \quad \text{and integer.}
\end{aligned}$$

The linear programming feasible region is spanned by the extreme points ABC (Figure. 3.2) and the optimal solution to the LP occurs at

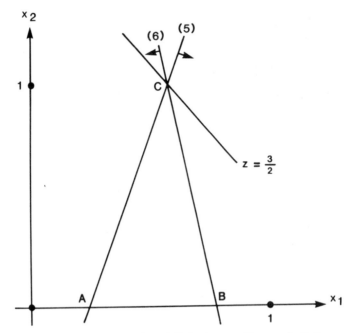

Figure 3.2: Graphical Solution to Example 3.2

$C = (1/2, 1)$. However, there are no integer points (x_1, x_2) inside the polyhedron; therefore, the integer program has no feasible solution.

The previous discussion suggests that useful information can be extracted from the associated linear program. In particular, since the solutions to the MIP or IP are contained in the linear programming feasible region, we have

Theorem 3.1 *The maximal value of the objective function to the mixed integer (or integer) program solved as a linear program is an upper bound on the value of any mixed integer (or integer) feasible solution.*

Corollary 3.1.1 *If the optimal solution to the mixed (or integer) problem solved as a linear one is integer in its integer constrained variables it solves the mixed (or integer) program.*

Corollary 3.1.2 *If the mixed (or integer) problem solved as a linear one is infeasible, then so is the mixed (or integer) program.*

3.2 Solving an Integer Programming Problem as a Linear Program

Consider the following integer programming problem

$$
\begin{array}{lrcll}
\text{maximize} & \mathbf{cx} & = & z & \quad (7) \\
\text{subject to} & \mathbf{Ax} & = & \mathbf{b} & \quad (8) \\
\text{and} & \mathbf{x} & \geq & \mathbf{0} \quad \text{and integer,} & \quad (9)
\end{array}
$$

where $\mathbf{A}$ is m by n with rank m, all vectors are of appropriate size, and $\mathbf{A}$ and $\mathbf{b}$ are assumed to be integer. Corollary 3.1.1 indicates that if the linear programming relaxation of the above problem is solved using the simplex algorithm and the linear programming solution is integer, then it solves the integer program.

In this section we define a class of integer programming problems for which the linear programming optimal solution will always satisfy the integrality constraints. Such problems can therefore be solved using the simplex algorithm. Before classifying these problems, we need to discuss unimodularity.

3.2.1 Unimodularity

The matrix $\mathbf{A}$ is said to be unimodular if the determinant of every square submatrix of $\mathbf{A}$ is either $0, +1$, or $, -1$. This means that $\mathbf{A}$ is unimodular, if and only if every submatrix is unimodular. (The properties of unimodular matrices may be found in [521].)

Now consider the linear programming solution of (7)-(9). If the coefficient matrix $\mathbf{A}$ in (8) is unimodular, then any basis $\mathbf{B}$ from $\mathbf{A}$ is also unimodular. Also, if $\mathbf{x_B}$ are the basic variables, then from $\mathbf{Bx_B} = \mathbf{b}$ we have

$$\mathbf{x_B} = \mathbf{B}^{-1}\mathbf{b}.$$

Now from Cramer's rule, $\mathbf{B}^{-1} = \mathbf{B}^{*}/|\mathbf{B}|$, where $\mathbf{B}^{*}$ is the adjoint of $\mathbf{B}$. As $\mathbf{B}$ is unimodular, all elements of $\mathbf{B}^{*}$ are either 0, $+1$, or -1 and $|\mathbf{B}|$ is either $+1$ or -1. Thus all elements of $\mathbf{B}^{-1}$ are either 0, $+1$, or -1, and from $\mathbf{x_B} = \mathbf{B}^{-1}\mathbf{b}$, we have $\mathbf{x_B}$ is integer as $\mathbf{b}$ is assumed to be integer. Hence, every basic feasible solution, and therefore the optimal linear programming solution to (7)-(9), satisfies the integrality requirements. Solutions to such integer programs may thus be obtained by omitting the integer constraints and solving the relaxed problem.

Classical transportation problems (Problem 3.4), assignment problems (Example 3.3 below), and minimum cost network flow problems (Problem 3.5) belong to the class of integer programs which have a unimodular coefficient matrix and thus can be solved using the simplex method. We illustrate with the example below

Example 3.3 (Assignment problem)

We wish to assign three jobs to three machines so that each job i $(i = 1, 2, 3)$ is assigned to one machine and each machine j $(j = 1, 2, 3)$ is to receive exactly one job. If c_{ij} is the processing cost of job i on machine j, then the assignment problem is to assign the jobs to the machines in a way that minimizes the total processing cost. Let x_{ij} be a $0-1$ variable which is 1 if job i is assigned to machine j, and 0 otherwise. Then it is easy to verify that the assignment problem is the $0-1$ integer program below.

Minimize $\displaystyle\sum_{i=1}^{3}\sum_{j=1}^{3} c_{ij}x_{ij}$

subject to $\displaystyle\sum_{i=1}^{3} x_{ij} \quad = \quad 1 \quad (j = 1, 2, 3)$

$\displaystyle\sum_{j=1}^{3} x_{ij} \quad = \quad 1 \quad (i = 1, 2, 3)$

and $\qquad x_{ij} \quad = \quad 0 \text{ or } 1 \quad (\text{all } i, j).$

For this problem, the coefficient matrix $\mathbf{A}$ can be written as below

$$
\begin{array}{c}
\\
\text{Machine}\\
\text{Constraints}\\
\\
\text{Job}\\
\text{Constraints}
\end{array}
\left\{
\begin{array}{c}
x_{11}\ x_{12}\ x_{13}\ x_{21}\ x_{22}\ x_{23}\ x_{31}\ x_{32}\ x_{33}\\
\left[
\begin{array}{ccccccccc}
1 & 0 & 0 & 1 & 0 & 0 & 1 & 0 & 0\\
0 & 1 & 0 & 0 & 1 & 0 & 0 & 1 & 0\\
0 & 0 & 1 & 0 & 0 & 1 & 0 & 0 & 1\\
1 & 1 & 1 & 0 & 0 & 0 & 0 & 0 & 0\\
0 & 0 & 0 & 1 & 1 & 1 & 0 & 0 & 0\\
0 & 0 & 0 & 0 & 0 & 0 & 1 & 1 & 1
\end{array}
\right]
\end{array}
\right.
$$

Note that the matrix $\mathbf{A}$ has the following properties:

(i) Each element is either 0 or 1, hence the determinant of every square submatrix of size 1 is either 0 or 1.

(ii) Every column has exactly two 1's, one corresponding to the machine constraints, and the other associated with the job constraints.

(iii) The sum of the rows corresponding to the machine constraints is the same as the sum of the job constraints (a vector of all 1's). Hence, the rows of $\mathbf{A}$ are not linearly independent so the rank($\mathbf{A}$) < 6, and the determinant of every square submatrix of size 6 is 0.

Using these remarks, we prove by induction that the matrix $\mathbf{A}$ is unimodular. In particular, suppose the determinant of all submatrices of $\mathbf{A}$ of size up to $(m - 1)$ is either $0, +1$, or -1 and let $\mathbf{B}$ be a square submatrix of size m. There are three cases:

1) If $\mathbf{B}$ has a column of 0's, $|\mathbf{B}| = 0$.

2) If $\mathbf{B}$ has a column containing only one 1, expand the determinant of $\mathbf{B}$ by the cofactors of that column. Thus $|\mathbf{B}| = \pm|\mathbf{B}'|$, where $\mathbf{B}'$ is the submatrix of $\mathbf{B}$ obtained by deleting a column which has one 1 and the corresponding row. By hypothesis $|\mathbf{B}'| = 0, 1$, or -1; hence $|\mathbf{B}| = 0, 1$, or -1 as well.

3) If every column of $\mathbf{B}$ has two 1's, then one 1 in each column is from the machine constraints, the other from the job constraints. As in remark (iii) above, it follows that the rows of $\mathbf{B}$ are dependent and thus $|\mathbf{B}| = 0$.

3.3   Obtaining Integer Programming Solutions by Rounding Linear Programming Solutions

Most integer programming problems cannot be solved as linear programs. For these problems, the geometry discussed in Section 3.1 suggests that we might solve the mixed (or integer) problem as a linear one and then round the integer constrained variables which are not integer either up or down. This procedure does in many instances produce a solution which may only slightly violate the constraints and is "good enough." This is especially true for problems where the accuracy of the data is questionable, or when the integer variables are expected to take on large values. (For example, the cost of producing 732,333 parts rather than 732,334–or for that matter, 732,000—is negligible.) On the other hand, the optimal mixed (or integer) solution is often required when the model data are believed to be accurate and integer variables take on smaller values, since the rounding may result in a point whose value is relatively far from an optimal solution, or it may violate crucial constraints. The examples that follow demonstrate the usefulness, as well as the shortcomings, of rounding techniques.

Example 3.4 (Knapsack problem)
Consider the following knapsack problem (cf. Section 1.2)

$$
\begin{array}{lllllll}
\text{maximize} & 5x_1 & + & 3x_2 & + & 4x_3 & & & (10) \\
\text{subject to} & 3x_1 & + & 2x_2 & + & 4x_3 & \leq & 6 & (11) \\
\text{and} & x_1, & & x_2, & & x_3 & = & 0 \text{ or } 1. & (12)
\end{array}
$$

It is easy to verify (Problem 3.7) that $(1, 1, 1/4)$ is the optimal linear programming solution. Furthermore, it is easy to show (cf. Problem 3.6) that the optimal linear programming solution to the knapsack problem will have at most one variable with a fractional value. The integer solution found by rounding the fractional value to 0 is called the round-down solution. In this example, the round-down solution is $(1,1,0)$ and it turns out that this is also the optimal integer solution. In general, the round-down linear programming solution for a knapsack problem cannot be guaranteed to be an optimal integer one. However, as the knapsack problem has a single constraint with positive coefficients and a positive right-hand side, rounding down the single variable with a fractional part to the next lowest integer, will always produce a feasible integer solution. Notice that if z_{IP} is the value of the objective function for the round-down solution and z_{LP} is the value for the optimal linear programming solution, then as z_{LP} is an upper bound on the integer solution, the difference $z_{LP} - z_{IP}$ is the maximum error when using the round-down solution. In many cases, this error is expected to be small so that the round-down solution is good enough.

Example 3.5 (Glover and Sommer [266])
Consider the following facility location problem (cf. Chapter 1).

		Shipping costs g_{ij} Customers					Production capacity
		1	2	3	4	5=n	M_i
	1	93	70	48	68	81	2
	2	45	89	97	85	96	3
Sources	3	92	93	58	37	99	2
	4	55	103	55	57	38	3
	m=5	74	60	78	54	52	2
Demands d_j		1	1	1	1	1	

In this problem we are assuming that the fixed cost of building a facility f_i is 0 $(i = 1,\ldots,5)$ and can be ignored; also, the situation requires that if a facility is operating it must produce (and ship) its capacity. (In relation to the model given in Section 1.2.2, this means that the inequalities (3) become equalities.) As the total demand is 5, the latter assumption implies that there are just six combinations of sources which can supply the customers; namely, (1,2), (1,4), (2,3), (2,5), (3,4), and (4,5). The problem with the corresponding pairs of variables $(x_1, x_2), (x_1, x_4)$, etc. at 1 and the remaining x_i variables at 0 is a linear program. The optimal solutions to the six linear programs are listed below, where unlisted variables have value 0.

Facility variables	Shipping variables	Total cost
$x_1 = x_2 = 1$	$z_{12} = z_{13} = z_{21} = z_{24} = z_{25} = 1$	344
$x_1 = x_4 = 1$	$z_{12} = z_{13} = z_{41} = z_{44} = z_{45} = 1$	268
$x_2 = x_3 = 1$	$z_{21} = z_{22} = z_{25} = z_{33} = z_{34} = 1$	325
$x_3 = x_4 = 1$	$z_{32} = z_{34} = z_{41} = z_{43} = z_{45} = 1$	278
$x_2 = x_5 = 1$	$z_{21} = z_{22} = z_{23} = z_{54} = z_{55} = 1$	337
$x_4 = x_5 = 1$	$z_{41} = z_{43} = z_{45} = z_{52} = z_{54} = 1$	262 (optimal)

When the integer restrictions on the facility variables are dropped, the optimal linear programming solution is

$$x_1 = x_3 = x_5 = 1/2, x_2 = x_4 = 1/3,$$

$$z_{13} = z_{21} = z_{34} = z_{45} = z_{52} = 1 \tag{13}$$

with total cost 228. It is evident that this solution cannot be rounded to produce any of the six integer solutions since three of the x_i in (13) would have to be rounded down to 0. But then the solution becomes infeasible because the capacity constraints ((3) in Section 1.2.2)

$$\sum_{j=1}^{5} z_{ij} = M_i x_i \qquad (i = 1, \ldots, 5)$$

are violated, since $x_i = 0$ implies $z_{ij} = 0$ for all j and in the solution (13) $z_{ij} = 1$ for each i.

3.4   An Overview of Approaches for Solving Integer (or Mixed-Integer) Programming Problems

In Section 3.2, we showed that certain integer programs may be solved as linear ones. Clearly, however, most integer programs will not exhibit this integrality property, and the simplex method, *per se*, will generally not solve mixed integer (or integer) programs.

The principal approaches for solving such problems are (a) cutting plane techniques, (b) enumerative methods, (c) partitioning algorithms, and (d) group theoretic approaches. We briefly describe these leaving the details to subsequent chapters.

3.4.1   Cutting Plane Techniques (Chapters 4, 5, 6, 7)

The general intent of cutting plane algorithms is to deduce supplementary inequalities or "cuts" from the integrality and constraint requirements which when added to the existing constraints eventually produce a linear program whose optimal solution is integer in the integer constrained variables.

3.4.2   Enumerative Methods (Chapters 8, 9)

The intent here is to enumerate, either explicitly or implicitly, all possible solution candidates to the mixed integer or integer program. The feasible solution which maximizes the objective function is optimal. An enumerative algorithm can usually be characterized as a bookkeeping scheme which keeps track of the enumeration, coupled with a "point or node algorithm" whose objective is to show via criteria which are contrived from the integrality and constraint requirements that certain related integer points cannot yield improved solutions; that is, to implicitly enumerate large numbers of points. The efficiency of the enumerative technique usually depends largely on the effectiveness of its point algorithm.

3.4.3   Partitioning Algorithms (Chapter 10)

Benders [79] in 1962 showed that a mixed integer program can be posed as an integer problem. The difficulty in solving the integer formulation arises out of the large number of constraints which must be generated. To overcome this obstacle Benders proposed a "partitioning algorithm" which solves the integer programming equivalent of the mixed problem by solving a series of related integer and linear programs.

3.4.4   Group Theoretic Algorithms (Chapter 11)

In a classical paper on the cutting plane algorithm, Gomory [276] discussed the relationships between the integer program and particular group structures.[1] Based on this work, Gomory [278] in 1965 showed

---

[1] A group is a set which is closed under an associative binary (arithmetic) operation and has a zero element and inverses.

that by relaxing the nonnegativity, but not integrality, constraints on certain variables, any integer program can be represented as a minimization problem defined on a group. Algorithms for solving this group problem and hence the integer program have appeared in the literature. The use of these problems in generating valid inequalities for any integer program, originally discussed by Gomory [279], has also been extensively treated by Gomory and Johnson [284,285].

Although a very substantial effort has been (and is being) made to develop efficient algorithms for linear programs with discrete variables, we cannot claim that every mixed integer or integer program can be solved in practice. There are known problems with less than 50 integer variables that cannot be solved in a reasonable length of time using existing techniques and computer facilities. The algorithms simply refuse to converge in some instances. It is not even clear which method is best for a particular problem. Thus composite approaches are often tried.

Fortunately, most real world formulations yield highly structured constraint matrices. Often, special purpose algorithms that explicitly take advantage of the structure, and hence are usually considerably more efficient than the classical techniques, may be developed. These, however, do not preclude the use or study of the well-known methods. In fact, special purpose algorithms are usually variations of the classical techniques whose basic properties often support the convergence of the specialized scheme.

At present, general algorithms can usually solve problems with 50 to 100 variables and as many constraints. As the size of the problem grows beyond these limits, the probability of solving it in a reasonable amount of computer time decreases dramatically. On the other hand, special purpose techniques may be able to solve problems with several thousand variables and several hundred rows (as in the case of set covering problems). Additionally, very large but highly structured problems may be attacked with specially devised heuristic methods; these methods may incorporate optimization procedures to solve integer or linear subproblems that may occur during the heuristic algorithm. Such heuristic methods are devised on an as-needed basis and are mentioned as appropriate in subsequent chapters (see, e.g., the Multi- dimensional Knapsack Problem, Chapter 12).

Problems

Problem 3.1 Solve the following integer programs graphically.

(i) Maximize $2x_1 + 3x_2$
     subject to $x_1 + x_2 \leq 5$
     $\quad\qquad\qquad 2x_1 + 3x_2 \leq 12$
     and $x_1 \geq 0, \quad x_2 \geq 0$ and integer.

(ii) Minimize $3x_1 + x_2$
     subject to $10x_1 + 10x_2 \leq 22$
     $\quad\qquad\qquad x_1 + 2x_2 \leq 2.8$
     and $x_1 \geq 0, \quad x_2 \geq 0$ and integer.

(iii) Maximize $3x_1 + 5x_2$
     subject to $0.5x_1 + 0.5x_2 \leq 100$
     $\quad\qquad\qquad x_1 + 2x_2 \leq 300$
     and $x_1 \geq 0, \quad x_2 \geq 0$ and integer.

Problem 3.2 Show that the matrix A given below is unimodular.

$$A = \begin{pmatrix} 1 & 1 & 0 & 0 \\ 0 & 0 & 1 & 1 \\ 1 & 0 & 1 & 0 \\ 0 & 1 & 0 & 1 \end{pmatrix}$$

Problem 3.3 Solve the linear programming relaxation of the following problem using the Simplex Algorithm.

Minimize $2x_1 + 3x_2 + 7x_3 + 9x_4$
subject to $\mathbf{Ax} = \mathbf{b}$
and $\mathbf{x} \geq \mathbf{0},$

where $\mathbf{x} = (x_1, x_2, x_3, x_4)$, $\mathbf{b} = (200, 300, 250, 250)$, and the matrix $\mathbf{A}$ is given in Problem 3.2. Verify that the linear programming solution satisfies the integrality constraints.

Problem 3.4 (*Transportation Problem*) Suppose we have m plants which produce a single product to service n customers. Suppose in a given time period there are a_i units available at plant i ($i = 1, \ldots, m$) and b_j units required by customer j ($j = 1, \ldots, n$). Further suppose the total supply equals the total demand; that is, $\sum_{i=1}^{m} a_i = \sum_{j=1}^{n} b_j$. The per unit shipping cost from plant i to customer j is c_{ij}. Define the mn shipping variables x_{ij} to be the number of units to send from plant i to customer j.

(a) Explain why the integer programming model representing this situation is

$$\text{minimize} \quad \sum_{i=1}^{m}\sum_{j=1}^{n} c_{ij}x_{ij}$$

$$\text{subject to} \quad \sum_{j=1}^{n} x_{ij} = a_i \quad (i = 1,\ldots,m)$$

$$\sum_{i=1}^{m} x_{ij} = b_j \quad (j = 1,\ldots,n)$$

$$\text{and} \qquad\qquad x_{ij} \geq 0 \text{ and integer (all } i,j).$$

(b) Show that the coefficient matrix for this problem is unimodular.

(c) What is the relationship between the transportation model given here and the assignment model given in Example 3.3?

Problem 3.5 (*Transhipment Problem*) Consider a network $G = (N, E)$, where N is a set of nodes and E is a set of directed arcs. We wish to ship v units of a product from a source node s to a sink node t at the minimum total cost. The per unit shipping cost of a unit along the arc (i,j) is c_{ij} and each arc (i,j) can have at most u_{ij} units of flow (i.e., shipped across it). Suppose x_{ij} is the amount of shipment or flow from node i to node j along the arc (i,j).

(a) Explain why the integer programming model representing this situation is

$$\text{Minimize} \quad \sum_{(i,j)\in E} c_{ij}x_{ij}$$

$$\text{subject to} \quad \sum_{j:(i,j)\in E} x_{ij} - \sum_{j:(j,i)\in E} x_{ji} = \begin{cases} v & i = s \\ 0 & i \neq s,t \\ -v & i = t \end{cases}$$

$$0 \leq x_{ij} \leq u_{ij} \text{ and integer (all } i,j).$$

(*Hint*: The constraints in this formulation represent the equilibrium conditions which require that, for all nodes except s and t, the total incoming flow should equal the total outflow. Also, the net flow out of node s and into node t is v.)

(b) Show that the coefficient matrix of this problem is unimodular.

Problem 3.6 Consider the knapsack problem

$$\text{maximize} \quad \sum_{i=1}^{n} c_i x_i$$

$$\text{subject to} \quad \sum_{i=1}^{n} a_i x_i \leq b$$

$$\text{and} \quad 0 \leq x_i \leq u_i \text{ and integer.}$$

Show that if the variables are indexed so that $c_i/a_i \geq c_{i+1}/a_{i+1}$, the optimal linear programming solution to this problem is

$$x_i = \begin{cases} u_i & \text{for } i < i^* \\ (b - \sum_{i=1}^{i^*-1} a_i u_i)/a_{i^*} & \text{for } i = i^* \\ 0 & \text{for } i > i^*, \end{cases}$$

where i^* is the smallest index such that $\sum_{i=1}^{i^*} a_i u_i \geq b$.

Problem 3.7 Consider the following knapsack problem

$$\begin{array}{llll} \text{Maximize} & 5x_1 + 3x_2 + 4x_3 \\ \text{subject to} & 3x_1 + 2x_2 + 4x_3 \leq 6 \\ \text{and} & x_1, \quad x_2, \quad x_3 = 0 \text{ or } 1. \end{array}$$

(a) Solve it as a linear program using the result given in Problem 3.6.

(b) Using the optimal linear programming solution from (a), what is the integer programming solution found by rounding the fractional variable to 0? What is the maximum error?

Chapter 4

Dual Fractional Integer Programming

4.1 The Basic Approach

In Chapter 3 we briefly discussed cutting plane algorithms. Recall that in these techniques new constraints are derived which are satisfied by every integer (or mixed integer) solution[1] and are appended to the constraint set so as to eventually produce a linear program whose optimal solution is integer in the integer constrained variables. This chapter concerns itself with a cutting plane algorithm for the integer program which utilizes the dual simplex method and allows fractional numbers in computation–hence the designation "dual fractional". We outline the basic approach for the integer program; extension to the mixed case appears in the next chapter.

Step 1 Starting with an all-integer tableau,[2] solve the integer program as a linear one.[3] If it is infeasible, so is the integer problem–terminate. If the optimal solution is all-integer, the integer program is solved–terminate. Otherwise, go to
Step 2.

---

[1] In this chapter and the next, unless otherwise specified, an integer solution means an integer feasible solution.

[2] This requires rational data. When irrational numbers are present, convergence of the algorithm cannot usually be guaranteed.

[3] The dual problem can always be made feasible by bounding the feasible region. For example, adding the constraint $\sum_{j=1}^{n} x_j \leq M$, where the x_j's are the nonbasic variables and M is a large positive number, to the bottom of the simplex tableau, and pivoting on the element in the new row which is in the lexicographically smallest column produces a dual feasible solution (see Example 4.2 and Problem 4.1).

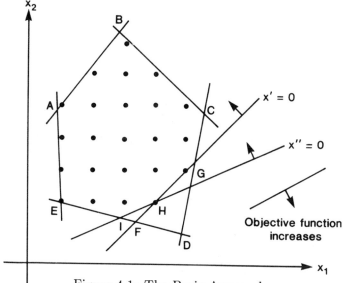

Figure 4.1: The Basic Approach

Step 2 Derive a new inequality constraint (or "cut") from the integrality and other current constraint requirements which "cuts off" the (current) optimal point (i.e., the linear programming problem displays infeasibility) but does not eliminate any integer solution. Add the new inequality to the bottom of the simplex tableau which then exhibits primal infeasibility. (The value of the slack variable associated with the new row will be negative.) Go to Step 3.

Step 3 Reoptimize the new linear program using the dual (lexicographic) simplex method. If the new linear program is infeasible,[4] the integer problem has no solution–terminate. If the new optimum is in integers, the integer program is solved–terminate. Otherwise, go to Step 2.

In Figure 4.1, for example, suppose that $ABCDE$ is the feasible region associated with the original constraints, where the dots correspond to the integer solutions and point D is the (first) optimal linear pro-

---

[4]From the dual theorem of linear programming, primal infeasibility will be indicated by the dual problem being unbounded. In computations, this means that a tableau will be reached where a pivot column cannot be found.

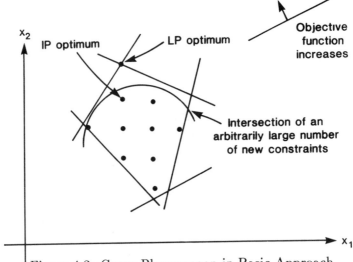

Figure 4.2: Curve Phenomenon in Basic Approach

gramming (LP) solution. A new constraint $x' \geq 0$ is added which cuts off D, and upon reoptimization the new optimal LP solution is point F. (The feasible region is $ABCGFE$.) A second constraint $x'' \geq 0$ is appended which eliminates the noninteger point F. (The feasible region is $ABCGHIE$.) On resolving the linear program, the optimal integer solution, i.e., vertex H, is found.

It is often relatively simple to obtain inequalities that are satisfied by every integer solution and that also cut off the current continuous (or LP) optimum. The difficulty is to prove that a *finite* number of such constraints will produce a linear program whose optimal solution is integer. It is possible, for example, to obtain the "curve phenomenon" displayed by Figure 4.2. The curved boundary is the intersection of an arbitrarily large number of cuts. (The general difficulties of cutting plane algorithms are described by Halfin [308], Jeroslow [347],[348], Jeroslow and Kortanek [349], [350], and Rubin [490].)

An example of a cut which did not lead to a convergent algorithm was proposed by Dantzig [156] in 1959. The valid constraint is $\sum_{j \in J} x_j \geq 1$, where J is the set of indices corresponding to the nonbasic variables appearing in the optimal LP solution. (The inequality is allowed since the current solution with nonbasic variables equal to zero is assumed not all integer, hence, in any integer solution at least one of the nonbasic variables must be positive; but since these variables are integer constrained, at least one of them must be one.) However, it has been shown by Gomory and Hoffman [283] that the basic approach with this cut may not converge for large classes of problems. An exposition of their results appears in Appendix A. Other cutting planes, derived from geometric arguments, which also do not lead directly to a convergent algorithm, are described in Appendix C.

What has become the classical convergent dual fractional algorithm for the integer program was developed by Ralph Gomory [276] in 1958. The algorithm uses the basic approach (cf. Section 4.1) where the cut is inequality (1) in Section 4.3. Before presenting this inequality we introduce the notation.

4.2   Notation: The Beale Tableau

For ease of exposition and to conform to standard notation, we write the integer program in the following enlarged mode (cf. Section 2.7)

Maximize

$$x_0 \;=\; a_{00}+ \quad a_{01}(-x_1)+ \quad a_{02}(-x_2)+\ldots+ \quad a_{0n}(-x_n) \quad \text{objective function}$$

subject to

$$
\left.
\begin{aligned}
x_1 &= & -1(-x_1) & & & \\
x_2 &= & & -1(-x_2) & & \\
\vdots & & & & \ddots & \\
x_n & & & & & -1(-x_n)
\end{aligned}
\right\} \text{original variables}
$$

$$
\left.
\begin{aligned}
x_{n+1} &= a_{n+1,0}+ a_{n+1,1}(-x_1)+ a_{n+1,2}(-x_2)+\ldots+ a_{n+1,n}(-x_n) \\
\vdots & \qquad \vdots \\
x_{n+m} &= a_{n+m,0}+a_{n+m,1}(-x_1)+a_{n+m,2}(-x_2)+\ldots+a_{n+m,n}(-x_n)
\end{aligned}
\right\} \text{original constraints}
$$

$$x_j \;\geq 0 \qquad (j = 0, 1, \ldots, n + m)$$

and

$$x_j \quad \text{integer} \ (j = 0, 1, \ldots, n + m).$$

The following comments are obvious.

Remarks

1. In relation to the integer program,

Variables	Constant values	Non-basic variables				
	1	$(-x_1)$	$\cdots$	$(-x_j)$	$\cdots$	$(-x_n)$
$x_0 =$	a_{00}	a_{01}	$\cdots$	a_{0j}	$\cdots$	a_{0n}
$x_1 =$	0	-1					≥ 0
$\vdots$	$\vdots$		$\ddots$				$\vdots$
$x_j =$	0			-1			≥ 0
$\vdots$	$\vdots$				$\ddots$		$\vdots$
$x_n =$	0					-1	≥ 0
$x_{n+1} =$	$a_{n+1,0}$	$a_{n+1,1}$	$\cdots$	$a_{n+1,j}$	$\cdots$	$a_{n+1,n}$	≥ 0
$\vdots$	$\vdots$	$\vdots$		$\vdots$		$\vdots$	$\vdots$
$x_{n+m} =$	$a_{n+m,0}$	$a_{n+m,1}$	$\cdots$	$a_{n+m,j}$	$\cdots$	$a_{n+m,n}$	≥ 0

Table 4.1: Beale Tableau

$$
\begin{aligned}
\text{maximize} \quad & \mathbf{cx} = z, \\
\text{subject to} \quad & \mathbf{Ax} \leq \mathbf{b},\ \mathbf{x} \geq \mathbf{0} \\
\text{and} \quad & \mathbf{x} \ \text{integer},
\end{aligned}
$$

where $\mathbf{A} = (a_{ij})$ is m by n; we have $x_0 = z, a_{00} = 0, \mathbf{c} = (c_j) = (-a_{0j})$, $\mathbf{b} = (b_i) = (a_{n+i,0})$ $(i = 1,\ldots,m)$, $a_{ij} = a_{n+i,j}$ $(i = 1,\ldots,m; j = 1,\ldots,n)$, $\mathbf{x} = (x_j)$ $(j = 1,\ldots,n)$, and x_{n+i} $(i = 1,\ldots,m)$ are the m slack variables corresponding to the constraints $\mathbf{Ax} \leq \mathbf{b}$.

2. The requirements that the x_j $(j = 1,\ldots,n)$ be integer and the data be integral (constraints are cleared of fractions) ensure that the value of the objective function (x_0) and the slack variables $(x_{n+1},\ldots,x_{n+m})$ are integral for any integer solution.

Table 4.1 exhibits the enlarged problem in the convenient Beale tableau (cf. Chapter 2), where each variable x_i $(i = 0,1,\ldots,n+m)$ appears as a linear combination of the nonbasic variables $x_1,\ldots,x_n$. In addition, we use the following notation:

J	contains the n indices of the current non-basic variables as they appear from left to right on the topmost line of the tableau. Initially, $J = 1, 2, \ldots, n$.
$J(j)\ (j = 1, \ldots, n)$	is the jth element in J. For example, if $n = 3$ and $J = 2, 5, 1$, then $J(1) = 2, J(2) = 5$, and $J(3) = 1$.
$\overset{\mathcal{L}}{\succ} \mathbf{0}\ (\overset{\mathcal{L}}{\prec} \mathbf{0})$	denotes a lexicographically positive (negative) column, equivalently, one whose first nonzero element is positive (negative).
$[y]$	means the largest integer less than or equal to y.

$\alpha_j\ (j = 0, 1, \ldots, n)$ is the jth column of the tableau; that is,

$$\alpha_j = \begin{pmatrix} a_{0j} \\ \vdots \\ a_{n+m,j} \end{pmatrix}$$

4.3   The Form of the (Gomory) Cut

At the optimal simplex tableau $a_{0j} \geq 0\ (j = 1, \ldots, n)$, $a_{i0} \geq 0\ (i = 1, \ldots, n + m)$ and $\alpha_j \overset{\mathcal{L}}{\succ} 0\ (j = 1, \ldots, n)$.[5] (Since a_{00} may be negative, we cannot say that $\alpha_0 \overset{\mathcal{L}}{\succ} 0$.) Further, if $a_{i0}\ (i = 0, 1, \ldots, n + m)$ is all-integer, the integer program is solved with $x_i = a_{i0}\ (i = 0, \ldots, m + n)$. Otherwise there is at least one "generating" or "source" row v with $x_v = a_{v0} + \sum_{j=1}^{n} a_{vj}(-x_{J(j)})$ and a_{v0} fractional. Then the (kth) "Gomory cut" has the form

$$x_{n+m+k} = -f_{v0} + \sum_{j=1}^{n} (-f_{vj})(-x_{J(j)}) \geq 0, \tag{1}$$

where x_{n+m+k} is the "Gomory slack variable" associated with the (kth)

---

[5]At the risk of confusion, we are omitting notation specifying iteration or tableau number; $\alpha_j \overset{\mathcal{L}}{\succ} 0\ (j = 1, \ldots, n)$ is guaranteed by the *lexicographic* dual simplex method.

added inequality and $f_{vj} = a_{vj} - [a_{vj}]$ $(j = 0, 1, \ldots, n)$. Observe that $0 < f_{v0} < 1$, $0 \leq f_{vj} < 1$ $(j = 1, \ldots, n)$, and when the inequality is appended to the tableau, primal infeasibility is introduced, since $x_{n+m+k} = -f_{v0} < 0$. We will show that (1) is satisfied by every integer solution and that it has properties sufficient to support a finite algorithm. Before going into the details we present an example.

4.4   Illustrations

The circled entry in the tableau corresponds to the pivot and the arrow ($\rightarrow$) designates the source row. The latter is the constraint with the largest fractional a_{i0} component. In case of ties the largest number is selected. (The motivation for this and other source row selection rules may be found in Section 4.7).

Example 4.1 (Balinksi [50])

$$
\begin{array}{lrrcl}
\text{Maximize} & -4x_1 & - & 5x_2 & = & x_0 \\
\text{subject to} & -x_1 & - & 4x_2 & \leq & -5, \\
& -3x_1 & - & 2x_2 & \leq & -7, \\
\text{and} & x_1, & & x_2 & \geq & 0, \quad \text{integer.}
\end{array}
$$

After introducing nonnegative slack variables x_3 and x_4 we obtain Tableau #1. It displays primal infeasibility and dual feasibility. Hence, the standard lexicographic dual simplex method is used[6] (cf. Chapter 2).

---

[6]The dual of the associated linear program is

$$
\begin{array}{lrrcll}
\text{minimize} & -5w_1 & - & 7w_2 & = & w_0 \\
\text{subject to} & -w_1 & - & 3w_2 & \geq & -4 \quad (w_{s1}), \\
& -4w_1 & - & 2w_2 & \geq & -5 \quad (w_{s2}), \\
\text{and} & w_1, & & w_2 & \geq & 0;
\end{array}
$$

multiplying each constraint inequality by -1 and adding nonnegative slack variables w_{s1} and w_{s2} yields

$$
\begin{array}{rcrcrcrcl}
w_1 & + & 3w_2 & + w_{s1} & & & = & 4, \\
4w_1 & + & 2w_2 & & + & w_{s2} & = & 5,
\end{array}
$$

Thus, setting nonbasic variables $w_1 = w_2 = 0$ yields a basic feasible solution with basic variables $w_{s1} = 4$ and $w_{s2} = 5$. These values appear in the x_0 row of the tableau.

#1	1	$(-x_1)$	$(-x_2)$	$(x_1 = x_2 = 0)$
x_0	0	4	5	Pivot row selection:
x_1	0	-1	0	Minimum $\{-5, -7\} = -7$,
x_2	0	0	-1	or x_4 becomes nonbasic,
x_3	-5	-1	-4	Pivot column selection:
x_4	-7	$\boxed{-3}$	-2	Minimum $\{4/|-3|, 5/|-2|\} = 4/3$

or x_1 becomes basic.

#2	1	$(-x_4)$	$(-x_2)$	$(x_1 = 7/3, x_2 = 0)$
x_0	$-28/3$	$4/3$	$7/3$	
x_1	$7/3$	$-1/3$	$2/3$	
x_2	0	0	-1	
x_3	$-8/3$	$-1/3$	$\boxed{-10/3}$	
x_4	0	-1	0	

#3	1	$(-x_4)$	$(-x_3)$	$(x_1 = 18/10, x_2 = 8/10)$
x_0	$-112/10$	$11/10$	$7/10$	$a_{10} = 18/10$
$\rightarrow x_1$	$18/10$	$-4/10$	$2/10$	$f_{10} = 18/10 - [18/10] = 8/10$,
x_2	$8/10$	$1/10$	$-3/10$	$a_{11} = -4/10$,
x_3	0	0	-1	$f_{11} = -4/10 - [-4/10] = 6/10$
x_4	0	-1	0	$a_{12} = 2/10$
x_5	$-8/10$	$\boxed{-6/10}$	$-2/10$	$f_{12} = 2/10 - [2/10] = 2/10$

Cut: $x_5 = -8/10 - [6/10](-x_4)$
$$-[2/10](-x_3)$$

Tableau #3 exhibits the optimal linear programming solution $x_0 = -112/10$ with $x_1 = 18/10, x_2 = 8/10$, and $x_3 = x_4 = 0$. Row 1 is selected as the source row. On pivoting, we obtain Tableau #4. A second cut is added and after another pivot the optimal integer solution $x_1 = 2$ and $x_2 = x_3 = x_4 = 1$ with $x_0 = -13$ is found.

#4	1	$(-x_5)$	$(-x_3)$	
x_0	-76/6	11/6	2/6	
x_1	14/6	-4/6	2/6	
$\rightarrow x_2$	4/6	1/6	-2/6	$(x_1 = 14/6, x_2 = 4/6)$
x_3	0	0	-1	
x_4	8/6	-10/6	2/6	
x_5	0	-1	0	
x_6	-4/6	-1/6	$\boxed{-4/6}$	

#5	1	$(-x_5)$	$(-x_6)$	
$\rightarrow x_0$	-13	7/4	2/4	
x_1	2	-3/4	2/4	
x_2	1	1/4	-2/4	
x_3	1	1/4	-6/4	$(x_1 = 2, x_2 = 1)$
x_4	1	-7/4	2/4	Optimal tableau
x_5	0	-1	0	
x_6	0	0	-1	
x_7	0	-3/4	$\boxed{-2/4}$	

Suppose we express the new constraints in terms of the original nonbasic variables (i.e., x_1 and x_2). Then we have $x_5 = -8/10 + (6/10)x_4 + (2/10)x_3 \geq 0$, but $x_3 = -5 + x_1 + 4x_2$ and $x_4 = -7 + 3x_1 + 2x_2$; substituting, $x_5 = -6+2x_1+2x_2 \geq 0$ or $x_1+x_2 \geq 3$. Also, $x_6 = -4/6+ (1/6)x_5+(4/6)x_3 \geq 0$; but $x_5 = -6+2x_1+2x_2$ and $x_3 = -5+x_1+4x_2$; substituting yields $x_6 = -5+x_1+3x_2 \geq 0$ or $x_1+3x_2 \geq 5$. Observe that in each case the Gomory slack variable can be expressed as an all-integer linear combination of the original nonbasic variables. In Theorem 4.1 we prove that this is always the case.

Another interesting observation is that an all-integer integer–optimal tableau may be obtained by continuing to add cuts. Suppose we choose row 0 as the source row, then we add the cut $x_7 = 0 - (3/4)(-x_5) - (2/4)(-x_6)$ to the bottom of Tableau #5 and pivot on $-2/4$. The result is Tableau #6. Another cut derived from either the x_1, x_2, x_4 or x_6 row, and pivot produces the all-integer Tableau #7. We show that this post-optimality phase will always yield an all-integer tableau in Theorem 4.2.

#6	1	$(-x_5)$	$(-x_7)$
x_0	-13	1	1
$\rightarrow x_1$	2	$-3/2$	1
x_2	1	1	-1
x_3	1	$5/2$	-3
x_4	1	$-5/2$	1
x_5	0	-1	0
x_6	0	$3/2$	-2
x_7	0	0	-1
x_8	0	$(-1/2)$	0

$(x_1 = 2, x_2 = 1)$

#7	1	$(-x_8)$	$(-x_7)$
x_0	-13	2	1
x_1	2	-3	1
x_2	1	2	-1
x_3	1	5	-3
x_4	1	-5	1
x_5	0	-2	0
x_6	0	3	-2
x_7	0	0	-1
x_8	0	-1	0

$(x_1 = 2, x_2 = 1)$

Optimal all-integer Tableau

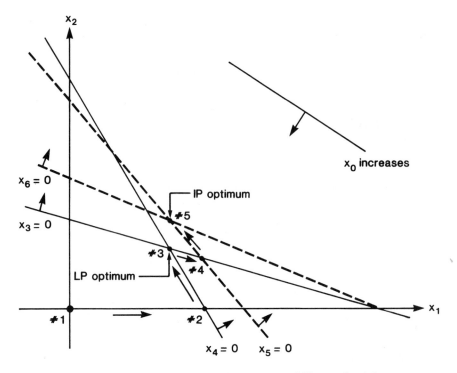

Figure 4.3: Graphical Illustration of Example 4.1

The correspondence between the Tableaux and (x_1, x_2) space is portrayed in Figure 4.3. Of course, since the constant column α_0 does not change, Tableaux #5, #6, and #7 correspond to the same point. The inequalities $x_7 \geq 0$ and $x_8 \geq 0$, when written in terms of x_1 and x_2, are $2x_1 + 3x_2 \geq 7$ and $x_1 + x_2 \geq 3$, respectively. Observe that these inequalities must be redundant since, when generated, they are expressed as linear combinations of existing constraints. In particular, $x_7 = (3/4)x_5 + (2/4)x_6$ and $x_8 = (1/2)x_5$.

The second illustration has no integer solutions but does contain continuous ones. It takes three cuts to eliminate the entire (linear programming) feasible region.

Example 4.2 (Salkin [504])

$$\text{Maximize} \quad x_1 \; + \; x_2 \; = \; x_0$$

$$\text{subject to} \quad -4x_1 \ + \ x_2 \ \leq \ -1,$$
$$4x_1 \ + \ x_2 \ \leq \ 3,$$
$$\text{and} \quad x_1, \quad x_2 \ \geq \ 0, \quad \text{integer.}$$

After adding nonnegative slack variables x_3 and x_4 we have Tableau #1. However, it is neither primal nor dual feasible. So to obtain dual feasibility we bound the feasible region by adding the redundant constraint $x_1 + x_2 \leq 10$ or $x_1 + x_2 + s = 10$, where s is a nonnegative slack variable, to the bottom of the Tableau. (The second inequality and the nonnegativity of x_1 and x_2 guarantee that the added constraint is always satisfied.) After pivoting on the element in the adjoined row and column 1, the lexicographically smallest column, we obtain the dual feasible Tableau #2. Computations now proceed as in the previous example.

#1	1	$(-x_1)$	$(-x_2)$
x_0	0	-1	-1
x_1	0	-1	0
x_2	0	0	-1
x_3	-1	-4	1
x_4	3	4	1
s	10	(1)	1

#2	1	$(-s)$	$(-x_2)$	(dual feasible)
x_0	10	1	0	
x_1	10	1	1	
x_2	0	0	-1	
x_3	39	4	5	
x_4	-37	-4	(−3)	
s	0	-1	0	

#3	1	$(-s)$	$(-x_4)$
x_0	10	1	0
x_1	$-7/3$	$-1/3$	$1/3$
x_2	$37/3$	$4/3$	$-1/3$
x_3	$-68/3$	$\boxed{-8/3}$	$5/3$
x_4	0	0	-1
s	0	-1	0

#4	1	$(-x_3)$	$(-x_4)$
$\rightarrow x_0$	$12/8$	$3/8$	$5/8$
x_1	$4/8$	$-1/8$	$1/8$
x_2	1	$4/8$	$4/8$
x_3	0	-1	0
x_4	0	0	-1
x_5	$-4/8$	$\boxed{-3/8}$	$-5/8$

$(x_1 = 4/8, x_2 = 1)$

Minimum $\left\{ \frac{3/8}{|-3/8|}, \frac{5/8}{|-5/8|} \right\}$ tie;

to maintain lexicographically positive

columns α_1 and α_2, compute

minimum $\left\{ \frac{-1/8}{|-3/8|}, \frac{1/8}{|-5/8|} \right\} = -\frac{1}{3}$;

or x_3 becomes basic.

Tableau #4 exhibits the (first) optimum continuous solution $x_1 = 4/8, x_2 = 1$, and $x_3 = x_4 = 0$ with $x_0 = 12/8$. Note that the s row is omitted. This is allowed since the corresponding constraint was introduced solely to achieve dual feasibility. Hence, once its slack variable becomes basic, it and its defining row may be dropped.

Also observe that at Tableau #4 the minimal ratio rule to determine the pivot column yields a tie, i.e.,

$$\frac{3/8}{|-3/8|} = \frac{5/8}{|-5/8|}.$$

So we compare

$$\frac{-1/8}{|-3/8|} \text{ with } \frac{1/8}{|-5/8|}.$$

Or, for columns 1 and 2 to remain lexicographically positive we use column 1 as the pivot column. After a total of three cuts and five more Tableaux we demonstrate that there is no integer solution.

#5	1	$(-x_5)$	$(-x_4)$	
x_0	1	1	0	
$\rightarrow x_1$	2/3	−1/3	1/3	
x_2	1/3	4/3	−1/3	
x_3	4/3	−8/3	5/3	$(x_1 = 2/3, x_2 = 1/3)$
x_4	0	0	−1	
x_5	0	−1	0	
x_6	−2/3	−2/3	$\boxed{−1/3}$	

#6	1	$(-x_5)$	$(-x_6)$	
x_0	1	1	0	
x_1	0	−1	1	
x_2	1	2	−1	
x_3	−2	$\boxed{−6}$	5	$(x_1 = 0, x_2 = 1)$
x_4	2	2	−3	
x_5	0	−1	0	
x_6	0	0	−1	

#7	1	$(-x_3)$	$(-x_6)$	
$\rightarrow x_0$	4/6	1/6	5/6	
x_1	2/6	−1/6	1/6	
x_2	2/6	2/6	4/6	
x_3	0	−1	0	$(x_1 = x_2 = 2/6)$
x_4	8/6	2/6	−8/6	
x_5	2/6	−1/6	−5/6	
x_6	0	0	−1	
x_7	−4/6	$\boxed{−1/6}$	−5/6	

#8	1	$(-x_7)$	$(-x_6)$	
x_0	0	1	0	
x_1	1	−1	1	
x_2	−1	2	−1	
x_3	4	−6	5	$(x_1 = 1, x_2 = -1)$
x_4	0	2	−3	
x_5	1	−1	0	
x_6	0	0	−1	
x_7	0	−1	0	

#9	1	$(-x_7)$	$(-x_2)$	
x_0	0	1	0	
x_1	0	1	1	
x_2	0	0	−1	
x_3	−1	4	5	$(x_1 = x_2 = 0)$
x_4	3	−4	−3	
x_5	1	−1	0	
x_6	1	−2	−1	
x_7	0	−1	0	

The x_3 row of Tableau #9 indicates that the dual is unbounded, or the primal is infeasible, Thus, the integer program has no solution. Writing the Gomory slacks $x_5, x_6,$ and x_7 in terms of x_1 and x_2 yields

$$
\begin{aligned}
x_5 &= 1 - x_1 - x_2 \geq 0 \text{ or } x_1 + x_2 \leq 1, \\
x_6 &= 1 - 2x_1 - x_2 \geq 0 \text{ or } 2x_1 + x_2 \leq 1, \text{ and} \\
x_7 &= - x_1 - x_2 \geq 0 \text{ or } x_1 + x_2 \leq 0.
\end{aligned}
$$

Figure 4.4 relates Tableaux #4 through #9 to the computations in (x_1, x_2) space.

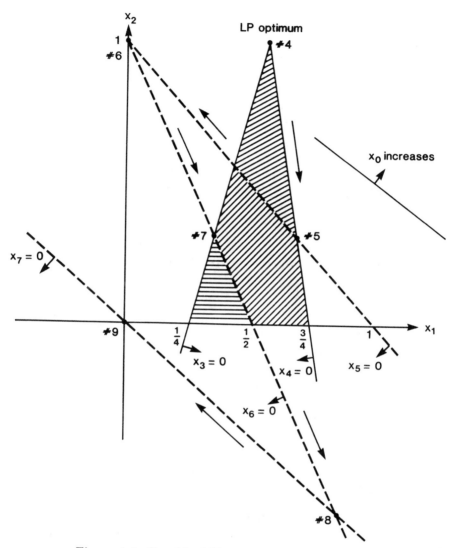

Figure 4.4: Graphical Illustration of Example 4.2

4.5   The Derivation of the Cut

4.5.1   Congruence

In order that the reader may fully understand the derivation and remarks that follow we introduce the notion of congruence. A complete
discussion may be found in Lang [384].

Definition 4.1 The number x is *congruent* to y modulo K if there exists an integer I such that $x - y = IK$. The congruence relationship is written as $x \equiv y \pmod{K}$.

Examples

$$
\begin{aligned}
2 &\equiv 7 & (mod\ 5), \\
3 &\equiv -1 & (mod\ 2), \\
1 &\equiv 0 & (mod\ 1).
\end{aligned}
$$

Observe that $x \equiv y \pmod{1}$ if and only if $x - y$ is an integer. In particular, $x \equiv 0 \pmod{1}$ means that x is an integer. The definition justifies the following properties. The proofs are obvious and left to Problem 4.10.

Properties of the Congruence Relation

   i) $x \equiv x \pmod{K}$.

   ii) If $x \equiv y \pmod{K}$ and $y \equiv z \pmod{K}$, then $x \equiv z \pmod{K}$.

   iii) If $x \equiv y \pmod{K}$, then $y \equiv x \pmod{K}$.

   iv) If $x \equiv y \pmod{K}$ and z is an integer, then $xz \equiv yz \pmod{K}$.

   v) If $x \equiv y \pmod{K}$ and $x' \equiv y' \pmod{K}$, then $x + x' \equiv y + y' \pmod{K}$. Further, if x and y' are integers, then $xx' \equiv yy' \pmod{K}$.

4.5.2 Derivation

Consider the derivation of the first cut. In the optimal simplex tableau there is a row v with a_{v0} not integral. (Otherwise, the integer program is solved.) The corresponding equation is

$$
x_v = a_{v0} + \sum_{j=1}^{n} a_{vj}(-x_{J(j)}),
$$

where $a_{v0} > 0$. Since x_v is required to be integer, every integer solution must satisfy

$$
x_v \equiv 0 \pmod{1}
$$

or

$$0 \equiv a_{v0} + \sum_{j=1}^{n} a_{vj}(-x_{J(j)}) \pmod{1}. \tag{2}$$

Now, we may subtract integer amounts from (2) without destroying the congruence relationship. In particular, since $0 \equiv [a_{v0}] \pmod{1}$ we have (by property v) that

$$0 \equiv f_{v0} + \sum_{j=1}^{n} a_{vj}(-x_{J(j)}) \pmod{1}, \tag{3}$$

where $0 < f_{v0} < 1$. (Recall that $f_{ij} = a_{ij} - [a_{ij}]$.) Also, since in every integer solution $x_{J(j)}$ is integer, we may subtract from (3) the congruences $0 \equiv [a_{vj}](-x_{J(j)}) \pmod{1}$. This reduces each coefficient in (3) to its fractional part. That is, we obtain

$$0 \equiv f_{v0} + \sum_{j=1}^{n} f_{vj}(-x_{J(j)}) \pmod{1}, \tag{4}$$

which (by properties i and v) is equivalent to

$$f_{v0} \equiv \sum_{j=1}^{n} f_{vj}x_{J(j)} \pmod{1}. \tag{5}$$

But, for any nonnegative solution, the right-hand side in (5) is positive, and it differs from the left-hand side by an integer. Hence, the value of $\sum_{j=1}^{n} f_{vj}x_{J(j)}$ must be either $f_{v0}, 1 + f_{v0}, 2 + f_{v0}$, etc. Consequently,

$$f_{v0} \leq \sum_{j=1}^{n} f_{vj}x_{J(j)}.$$

After adding the nonnegative slack variable x_{m+n+1}, we have the Gomory cut

$$x_{m+n+1} = -f_{v0} + \sum_{j=1}^{n} (-f_{vj})(-x_{J(j)}) \geq 0. \tag{6}$$

It remains to be shown that the kth cut is in the form of inequality (6). To prove this it suffices to show that the value of a Gomory slack variable is necessarily integer in every integer solution. This follows since, by (6), the Gomory slack variable is defined as $-(f_{v0} + \sum_{j=1}^{n} f_{vj}(-x_{J(j)}))$, which means by (4), that it is congruent to 0 (mod 1) and must therefore be an integer. (This result also follows directly from Theorem 4.1 below). Hence, every nonbasic variable in (2) is integer constrained and the derivation goes through unchanged.

Finally, by the derivation, every integer solution must satisfy the generated inequalities.

Before presenting several results we give the derivation of the first cut which appears in Example 4.1. All congruence relations are understood to be modulo 1.

Example (from Example 4.1)

From Tableau #3 the source row is

$$x_1 = 18/10 - (4/10)(-x_4) + (2/10)(-x_3).$$

Since $x_1 \equiv 0$, we have $0 \equiv 18/10 - (4/10)(-x_4) + (2/10)(-x_3).$
Subtracting $0 \equiv 1$
yields $0 \equiv 8/10 - (4/10)(-x_4) + (2/10)(-x_3);$
adding $0 \equiv 1(-x_4)$
gives $0 \equiv 8/10 + (6/10)(-x_4) + (2/10)(-x_3).$
Or, $8/10 \equiv (6/10)x_4 + (2/10)x_3,$
which means that $8/10 \leq (6/10)x_4 + (2/10)x_3.$

Hence, we may add the nonnegative slack variable x_5 to obtain

$$x_5 = -8/10 - (6/10)(-x_4) - (2/10)(-x_3) \geq 0.$$

This equation was adjoined to Tableau #3.

4.6   Some Properties of Added Inequalities

In this section we give several interesting properties of the generated inequalities. The first shows that an added inequality is equivalent to an all-integer inequality in the original nonbasic variables. The second proves that, if there is an integer solution, we can always obtain an all-integer integer optimal tableau. These two observations have been noted previously. The initial tableau is assumed to be all-integer.

Theorem 4.1 *Each added inequality (the Gomory cut) becomes an all integer inequality when expressed in terms of the original nonbasic variables.*

Proof: To see this, let the generating row for the first inequality be

$$x_v = a_{v0} + \sum_{j=1}^{n} a_{vj} \left(-x_{J(j)} \right).$$

Then the generated constraint is

$$x_{m+n+1} = -f_{v0} + \sum_{j=1}^{n} (-f_{vj})(-x_{J(j)})$$

$$= ([a_{v0}] - a_{v0}) + \sum_{j=1}^{n} ([a_{vj}] - a_{vj})(-x_{J(j)})$$

$$= \left\{ [a_{v0}] + \sum_{j=1}^{n} [a_{vj}](-x_{J(j)}) \right\} - \left\{ a_{v0} + \sum_{j=1}^{n} a_{vj}(-x_{J(j)}) \right\}.$$

Now, each nonbasic variable $x_{J(j)}$ is either an original nonbasic variable or an original basic variable. In the former case, we have $x_{J(j)} = x_g$ for some $g = 1, \ldots, n$. And in the latter case we have $x_{J(j)} = a_{n+i,0} + \sum_{j=1}^{n} a_{n+i,j}(-x_j)$ for some $i = 1, \ldots, m$. But the coefficients in the last equality are all-integer since the original data is assumed to be integral. Hence each $x_{J(j)}$ can be written as an all-integer linear combination of the original nonbasic variables. Thus, the first bracketed right-hand side term gives an all-integer expression when written in terms of $x_1, \ldots, x_n$.

The second bracketed right-hand side term is equal to x_v. Again, this variable is either an original nonbasic variable or it appeared as an integer linear combination of the original nonbasic variables. In any case, the second bracketed term gives an all-integer expression when written in terms of $x_1, \ldots, x_n$. Hence, $x_{m+n+1} \geq 0$ becomes an all-integer inequality when expressed in terms of the original nonbasic variables.

By thinking of the new inequality as part of the original program we may reapply the same line of reasoning to show that the second cut becomes an all-integer inequality. The proof for the third cut is the same as the first two, and so forth. ∎

Theorem 4.2 *After integer optimality is exhibited we may continue to add cuts (derived from rows containing fractions) to obtain an all-integer integer optimal tableau.*

Proof: First observe that after the integer optimum is found, subsequent cuts, derived from any row i in the tableau that still contains fractions (if none exist the tableau is all-integer), have an $f_{i0} = 0$ term since α_o is integer. Consequently, successive pivot steps will leave the values in the 0 column unchanged; or these tableaux maintain the optimal integer point.

The determinant $|\mathbf{B}|$ of the on-hand ($n+1$ by $n+1$) dual basis $\mathbf{B}$ is the product of minus the value of all preceding pivot elements.[7] But at the post-optimality phase the negative of the pivot element is a proper fraction, thus $|\mathbf{B}|$ is positive and decreases after each pivot step. Also, every basis $\mathbf{B}$ is an integer matrix since each of its rows is the negative of some row of $\mathbf{A}^1$, the initial integer matrix; hence, $|\mathbf{B}|$ is integer. Thus the post optimality pivoting will terminate if the tableau becomes all-integer or if $|\mathbf{B}|$ becomes 1. Actually, the two cases coincide; that is, if the tableau is all-integer then $|\mathbf{B}|$ is necessarily 1. To see this, denote the elements in the kth tableau as the matrix $\mathbf{A}^k$, and write the current dual basis as $\mathbf{B}$, with inverse $\mathbf{B}^{-1}$. Then, from linear programming, we have $\mathbf{A}^k = \mathbf{A}^1 \mathbf{B}^{-1}$ or $\mathbf{A}^k \mathbf{B} = \mathbf{A}^1$. Now, since $\mathbf{A}^1$ contains a negative identity matrix $-\mathbf{I}$, there must exist rows of $\mathbf{A}^k$ which comprise the square matrix $\mathbf{C}$ such that $\mathbf{C}\mathbf{B} = -\mathbf{I}$ or $-\mathbf{C} = \mathbf{B}^{-1}$.[8] Now, if $\mathbf{A}^k$ is all-integer it follows that $-\mathbf{C}$ and hence $\mathbf{B}^{-1}$ is all-integer and therefore $|-\mathbf{C}| = |\mathbf{B}^{-1}| = 1/|\mathbf{B}|$ is an integer. But since $|\mathbf{B}|$ itself is an integer, $|\mathbf{B}| = 1$. On the other hand, if $|\mathbf{B}| = 1$ we have (using $\mathbf{B}^{-1} = \mathbf{B}^{+}/|\mathbf{B}|$) that

---

[7]A well-known result from linear programming ([159],[303]) is that the determinant of the on-hand (dual) basis, corresponding to the current basic feasible solution, is the product of (minus the value of) all preceding pivot elements.

At each tableau we have a dual basic feasible solution. Thus, we consider the on-hand dual basis which can be directly associated with the negative of rows in the first simplex tableau if the dual objective function w_0 is written as a constraint and if the coefficients of the dual basic variables are written row-wise in the basis. Its first row is then the $n + 1$ identity vector $(1, 0, \ldots, 0)$ corresponding to w_0; see the next example.

[8]All existing cuts are expressed in terms of the original nonbasic variables and adjoined to the original program. Our doing this does not cause the argument to differ when Gomory slack variables become nonbasic or after their columns enter the dual basis. Also, to be precise we add the trivial relationship $-1 = -1(1)$ to the top of the tableau.

$$\mathbf{A}^k = \mathbf{A}^1\mathbf{B}^{-1} = \mathbf{A}^1\frac{\mathbf{B}^+}{|\mathbf{B}|} = \mathbf{A}^1\mathbf{B}^+$$

an integer matrix. This follows because, since $\mathbf{B}$ is integer, so is its adjoint $\mathbf{B}^+$. ∎

To clarify the preceding proof we discuss the computations for the last three tableaux in Example 4.1.

Example (from Example 4.1)

The optimal integer solution appears in Tableau #5 and is $x_0 = -13, x_1 = 2, x_2 = x_3 = x_4 = 1$, and $x_5 = x_6 = x_7 = 0$. The initial matrix $\mathbf{A}^1$ from Tableau #1 is below, where we have included the Gomory slack variables x_5, x_6, x_7, and x_8 in terms of the original nonbasic variables, and the trivial equation $-1 = -1(1)$ to take the dual objective function into account.

#1	1	$(-x_1)$	$(-x_2)$
-1	-1	0	0
x_0	0	4	5
x_1	0	-1	0
x_2	0	0	-1
x_3	-5	-1	-4
x_4	-7	-3	-2
x_5	-6	-2	-2
x_6	-5	-1	-3
x_7	-7	-2	-3
x_8	-3	-1	-1

$\mathbf{A}^1$

The dual of the associated linear program may be written as

$$\text{minimize} \quad w_0$$
$$\text{subject to} \quad w_0 + 5w_1 + 7w_2 + 6w_3 + 5w_4 + 7w_5 + 3w_6 \qquad\qquad = 0$$
$$w_1 + 3w_2 + 2w_3 + w_4 + 2w_5 + w_6 + w_7 \qquad = 4$$
$$4w_1 + 2w_2 + 2w_3 + 3w_4 + 3w_5 + w_6 \qquad + w_8 = 5$$
$$\text{and} \qquad w_1, w_2, \ldots, w_8 \geq 0.$$

At Tableau #1, w_0, w_7, and w_8 are basic variables, so their coefficients row-wise yields the dual basis,

$$\mathbf{B} = \begin{pmatrix} 1 & 0 & 0 \\ 0 & 1 & 0 \\ 0 & 0 & 1 \end{pmatrix} \quad \text{and} \quad |\mathbf{B}| = 1.$$

At Tableau #5, x_5 and x_6 are nonbasic (that is, w_0, w_3, and w_4 are the basic variables for the dual), again writing their coefficients row-wise yields, [9]

$$\mathbf{B} = \begin{pmatrix} 1 & 0 & 0 \\ 6 & 2 & 2 \\ 5 & 1 & 3 \end{pmatrix}$$

and $|\mathbf{B}| = 4$, which is the product of the negative value of the preceding pivot elements, i.e.,

$$4 = (4/6)(6/10)(10/3)(3).$$

At Tableau #6, x_5 and x_7 are nonbasic variables, so

---

[9]To explicitly see that the dual basis has these values, note that at Tableau #5, w_3 and w_4 are dual basic variables. This follows because the dual simplex method maintains complementary slackness. That is $x_j w_{sj} = 0$ for all j ($j \neq 0$), and $w_i x_{si} = 0$ for all i ($i \neq 0$); where w_i (w_{sj}) is the ith original (jth slack) dual variable and x_j (x_{si}) is the jth original (ith slack) primal variable. To maintain these equalities, the dual simplex algorithm pairs the variables (in our example) as below.

	original		slack					
Primal	x_1	x_2	x_3	x_4	x_5	x_6	x_7	x_8
Dual	w_7	w_8	w_1	w_2	w_3	w_4	w_5	w_6
	slack				original			

That is, any simplex tableau contains a pair of primal and dual basic solutions, where if a primal variable is nonbasic (basic) the corresponding dual variable is basic (nonbasic). This enforces complementary slackness. In our example, since x_5 and x_6 are nonbasic, the dual simplex method will ensure that w_3 and w_4 are basic. Including the dual objective function w_0 as the first basic variable, the coefficients of (w_0, w_3, w_4) written row-wise produce the basis

$$\mathbf{B} = \begin{pmatrix} 1 & 0 & 0 \\ 6 & 2 & 2 \\ 5 & 1 & 3 \end{pmatrix}.$$

$$\mathbf{B} = \begin{pmatrix} 1 & 0 & 0 \\ 6 & 2 & 2 \\ 7 & 2 & 3 \end{pmatrix} \quad \text{and} \quad |\mathbf{B}| = 2 = 4(2/4).$$

At the all-integer Tableau #7, x_8 and x_7 are nonbasic variables, or

$$\mathbf{B} = \begin{pmatrix} 1 & 0 & 0 \\ 3 & 1 & 1 \\ 7 & 2 & 3 \end{pmatrix} \quad \text{and} \quad |\mathbf{B}| = 1 = 2(1/2).$$

Note that

$$
\begin{array}{ccc}
\mathbf{A}^7 & \mathbf{B} & = \mathbf{A}^1
\end{array}
$$

$$
\begin{bmatrix}
-1 & 0 & 0 \\
-13 & 2 & 1 \\
2 & -3 & 1 \\
1 & 2 & -1 \\
1 & 5 & -3 \\
1 & -5 & 1 \\
0 & -2 & 0 \\
0 & 3 & -2 \\
0 & 0 & -1 \\
0 & -1 & 0
\end{bmatrix}
\begin{pmatrix}
1 & 0 & 0 \\
3 & 1 & 1 \\
7 & 2 & 3
\end{pmatrix}
=
\begin{bmatrix}
-1 & 0 & 0 \\
0 & 4 & 5 \\
0 & -1 & 0 \\
0 & 0 & -1 \\
-5 & -1 & -4 \\
-7 & -3 & -2 \\
-6 & -2 & -2 \\
-5 & -1 & -3 \\
-7 & -2 & -3 \\
-3 & -1 & -1
\end{bmatrix}
$$

Previously, we used the fact that the array $\mathbf{A}^k$ corresponding to the kth tableau is obtained by post-multiplying the all-integer matrix $\mathbf{A}^1$, the original data, by the dual basis inverse $\mathbf{B}^{-1}$ associated with tableau k. We may also use this to show that every element in $\mathbf{A}^k$ can be expressed as $L/|\mathbf{B}|$, where L is an integer. To see this, we have

$$\mathbf{A}^k = \mathbf{A}^1\mathbf{B}^{-1} = \mathbf{A}^1\frac{\mathbf{B}^+}{|\mathbf{B}|},$$

where $\mathbf{B}^+$ is the adjoint of $\mathbf{B}$. But $\mathbf{B}$ is integer since it is made up of negative rows of $\mathbf{A}^1$. Thus, $\mathbf{B}^+$ is integer; or $\mathbf{A}^1\mathbf{B}^+$ is an integer matrix. Hence, each element in the current tableau may be written as a fraction whose numerator is an integer and whose denominator is the determinant of the current basis. This fact can be extended to the matrix $\mathbf{F}(\mathbf{A}^k)$, whose rows are the fractional parts of the corresponding rows of $\mathbf{A}^k$. That is, an element of $\mathbf{F}(\mathbf{A}^k)$, f_{ij}, equals $a_{ij} - [a_{ij}]$. Thus, since

$$f_{ij} = a_{ij} - [a_{ij}] = \frac{L}{|\mathbf{B}|} - \frac{[a_{ij}]|\mathbf{B}|}{|\mathbf{B}|} = \frac{L - [a_{ij}]|\mathbf{B}|}{|\mathbf{B}|}$$

we have that each f_{ij} may be written as an integer divided by $|\mathbf{B}|$. Moreover, since $|\mathbf{B}|$ is a positive integer and f_{ij} is a proper fraction, it follows that $f_{ij} = I_j/|\mathbf{B}|$, where I_j is a nonnegative integer smaller than $|\mathbf{B}|$. Hence, the Gomory cut $\sum_{j=1}^{n} f_{ij} x_{J(j)} \geq f_{i0}$ may be expressed as

$$\sum_{j=1}^{n} \frac{I_j}{|\mathbf{B}|} x_{J(j)} \geq \frac{I_0}{|\mathbf{B}|},$$

where I_j $(j = 1, \ldots, n)$ is a nonnegative integer less than $|\mathbf{B}|$ and I_0 is a positive integer since $0 < f_{i0} < |\mathbf{B}|$. We use the rewritten version of the added inequality in the next theorem, which appeared in Salkin and Breining [510].

Theorem 4.3 *Consider the hyperplane*

$$\frac{I_0}{|\mathbf{B}|} = \sum_{j=1}^{n} \frac{I_j}{|\mathbf{B}|} x_{J(j)},$$

corresponding to an added inequality. Then the hyperplane passes through at least one integer point (not necessarily inside the feasible region) if and only if the greatest common divisor (gcd) of $I_1, \ldots, I_n$ divides I_0.[10]

Proof: Suppose $(\lambda_1, \ldots, \lambda_n)$ is an integer point. Then for it to be on the hyperplane we must have

$$\frac{I_0}{|\mathbf{B}|} = \frac{I_1}{|\mathbf{B}|}\lambda_1 + \ldots + \frac{I_n}{|\mathbf{B}|}\lambda_n,$$

which means $I_0 = I_1\lambda_1 + \ldots + I_n\lambda_n$. Now, an elementary result from number theory says that integers $\lambda_1, \ldots, \lambda_n$ that satisfy the previous equality exist if and only if the gcd of $I_1, \ldots, I_n$ divides I_0 (Saaty [500]). This, in effect, proves the assertion. ∎

Example 4.3 (Gomory [276]) The example below illustrates the theorem.

$$\text{Maximize} \quad 3x_1 \quad - \quad x_2 \quad = \quad x_0$$

---

[10]Recall that if a and b are integers, not both zero, their greatest common divisor (gcd) is defined to be the largest integer which divides both a and b.

$$\text{subject to} \quad 3x_1 \quad - \quad 2x_2 \quad \leq \quad 3,$$
$$-5x_1 \quad - \quad 4x_2 \quad \leq \quad -10,$$
$$2x_1 \quad + \quad x_2 \quad \leq \quad 5,$$
$$\text{and} \quad x_1, \quad x_2 \quad \geq \quad 0, \quad \text{integer.}$$

With nonnegative slack variables x_3, x_4, and x_5 the first simplex tableau appears below and is neither primal nor dual feasible. Adding the redundant constraint $x_1 + x_2 \leq 5$, with slack variable s, to the bottom of the first tableau and pivoting on the element in this row and the α_1 column (the lexicographically smallest column) yields Tableau #2. The optimal simplex Tableau #4 is obtained after two more pivots.

#1	1	$(-x_1)$	$(-x_2)$
x_0	0	-3	1
x_1	0	-1	0
x_2	0	0	-1
x_3	3	3	-2
x_4	-10	-5	-4
x_5	5	2	1
→ s	5	(1)	1

#2	1	$(-s)$	$(-x_2)$
x_0	15	3	4
x_1	5	1	1
x_2	0	0	-1
→ x_3	-12	-3	(−5)
x_4	15	5	1
x_5	-5	-2	-1
s	0	-1	0

#3	1	$(-s)$	$(-x_3)$
x_0	27/5	3/5	4/5
x_1	13/5	2/5	1/5
x_2	12/5	3/5	-1/5
x_3	0	0	-1
x_4	63/5	22/5	1/5
$\rightarrow x_5$	-13/5	$\boxed{-7/5}$	-1/5
s	0	-1	0

#4	1	$(-x_5)$	$(-x_3)$
x_0	30/7	3/7	5/7
x_1	13/7	2/7	1/7
x_2	9/7	3/7	-2/7
x_3	0	0	-1
x_4	31/7	22/7	-3/7
x_5	0	-1	0

$$B = \begin{pmatrix} 1 & 0 & 0 \\ -3 & -3 & 2 \\ -5 & -2 & -1 \end{pmatrix}$$

$$|B| = 7$$

In Tableau #4, suppose x_1 is used as the source row. Then the generated inequality is

$$x_6' = -6/7 - (1/7)(-x_3) - (2/7)(-x_5) \geq 0.$$

Or, the hyperplane associated with this constraint is

$$6/7 = (1/7)x_3 + (2/7)x_5. \tag{7}$$

Now, the gcd of 1 and 2 is 1, which divides 6. Thus, by the theorem, the hyperplane (7) intersects integer points. To actually see this, write the equality in terms of x_1 and x_2. This yields the hyperplane $2 - 2x_1 = 0$ in (x_1, x_2) space. It contains an infinite number of integer points (Figure 4.5).

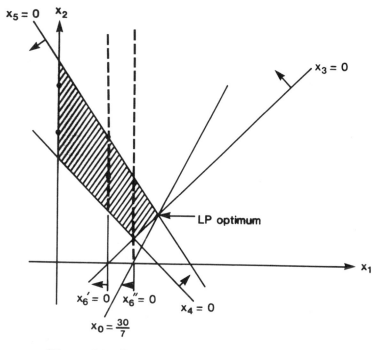

Figure 4.5: Graphical Illustration of Example 4.3

Now, suppose $2x_1$ is used as the generating row. (In Section 4.7 we will see that any integer multiple of rows may be used as a source row.) That is, the cut is generated from

$$2x_1 = 26/7 + (2/7)(-x_3) + (4/7)(-x_5).$$

Then the generated inequality is

$$x_6'' = -5/7 - (2/7)(-x_3) - (4/7)(-x_5) \geq 0.$$

Or, the hyperplane associated with this constraint is

$$5/7 = (2/7)x_3 + (4/7)x_5. \tag{8}$$

Now, however, the gcd of 2 and 4 is 2, which does *not* divide 5. Hence there cannot exist an integer point satisfying (8). To explicitly see this, transform the hyperplane (8) to (x_1, x_2) space. The result is the hyperplane $3 - 2x_1 = 0$, which, of course, does not pass through any integer point. Figure 4.5 portrays the discussion in (x_1, x_2) space.

Observe that in this example the hyperplane which passed through at least one integer point passed, in fact, through an infinite number of them. We prove that this is always the case.

Theorem 4.4 *If the hyperplane corresponding to an added inequality goes through one integer point, then it goes through an infinite number of integer points.*

Proof:[11] Write the hyperplane as

$$\sum_{j=1}^{n} \frac{I_j}{|\mathbf{B}|} x_{J(j)} = \frac{I_0}{|\mathbf{B}|}.$$

Then for an integer point $(\lambda_1, \ldots, \lambda_n)$ to be on the hyperplane we must have $\sum_{j=1}^{n} I_j \lambda_j = I_0$.

Now, if $I_1 = 0$ we have that all integer points of the form $(K\lambda_1, \lambda_2, \ldots, \lambda_n)$, where K is any integer, are on the hyperplane. Similarly, if $I_2 = 0$ all-integer points of the form $(\lambda_1, K\lambda_2, \ldots, \lambda_n)$ lie on the hyperplane. Thus, suppose I_1 and I_2 are different from 0. Then all points of the form $(\lambda_1 + KI_2, \lambda_2 - KI_1, \ldots, \lambda_n)$ are on the hyperplane. In any case the assertion is proved. ∎

To complete the discussion consider the case where the hyperplane

$$x' = \frac{-I_0}{|\mathbf{B}|} + \sum_{j=1}^{n} \frac{I_j}{|\mathbf{B}|} x_{J(j)} = 0, \tag{9}$$

associated with the cut (row indices are omitted)

$$x' = -f_0 + \sum_{j=1}^{n} f_j x_{J(j)} \geq 0, \tag{10}$$

does not pass through an integer point. Then the inequality (10) can be improved if we push its hyperplane into the feasible region until it intersects the first integer point. (With regard to Figure 4.6 the inequality $x'' \geq 0$ is stronger than $x' \geq 0$). We explain the computations.

If there is no integer point satisfying (9) we have, by Theorem 4.3, that the gcd of $I_1, \ldots, I_n$ which we call d $(d > 0)$, does not divide I_0.

---

[11]A more complete proof, which gives the general form of the points on the hyperplane, appears in Saaty [500].

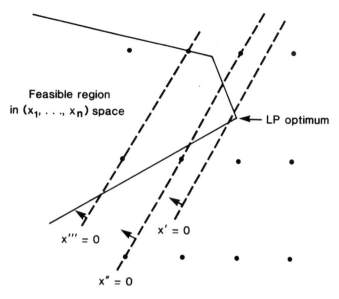

Figure 4.6: Derivation of Stronger Cuts

Suppose, we write the equality (9) in terms of the original nonbasic variables. The result is the all-integer inequality (11).

$$x' = n_0 + \sum_{j=1}^{n} n_j x_j \geq 0. \tag{11}$$

Now, since the hyperplanes (inequalities) (10) and (11) are equivalent it follows that (11) is void of integer solutions, or the gcd of $n_1, \ldots, n_n$, which we label d' ($d' > 0$), does not divide n_0. (Why?) Moreover, the hyperplane (9), or equivalently (10), can pass through an integer point only if d divides I_0 and d' divides n_0. Thus, in (11) we may change the value of n_0 by 1 and not eliminate any integer solutions. But the other coefficients retain their values, or changing n_0 by 1 in (11) amounts to changing $f_0 = I_0/|\mathbf{B}|$ by 1 (or I_0 by $|\mathbf{B}|$) in (9).[12] Also, since the feasible region in $x_{J(j)}$ space is in the first quadrant and the

---

[12]Observe that the new all-integer inequality

$$x'' = (n_0 - 1) + \sum_{j=1}^{n} n_j x_j \geq 0$$

may be generated from the source row used to obtain $x' \geq 0$. Or, changing I_0 by $|\mathbf{B}|$ yields a valid cut. This is not necessarily true if I_0 is altered by arbitrary integer amounts.

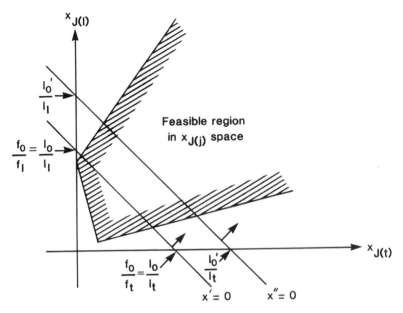

Figure 4.7: Choosing the Source Row

inequality points away from the origin, the further away the intercepts of the hyperplane (9) with each $x_{J(j)}$ axis the stronger the cut. Thus, if the coefficients $I_j/|\mathbf{B}|$ $(j = 1, \ldots, n)$ remain the same, strengthening the inequality $x' \geq 0$ is equivalent to *increasing* the value of $I_0/|\mathbf{B}|$. (This follows since the intersection of the hyperplane (9) with the $x_{J(t)}$ axis is the point $x_{J(t)} = I_0/I_t$, $x_{J(j)} = 0$ $(j = 1, \ldots, n, j \neq t)$; see Figure 4.7.) Notice that increasing $I_0/|\mathbf{B}|$ is equivalent to decreasing $-I_0/|\mathbf{B}|$, and thus n_0 decreases.

Now, the same line of reasoning can be applied to the new inequality (hyperplane) (12)

$$\frac{-(I_0 + |\mathbf{B}|)}{|\mathbf{B}|} + \sum_{j=1}^{n} \frac{I_j}{|\mathbf{B}|} x_{J(j)} \geq 0. \tag{12}$$

That is, if d (which has not changed) divides $I_0 + |\mathbf{B}|$ we have the desired constraint. Otherwise we increase $I_0 + |\mathbf{B}|$ by $|\mathbf{B}|$ and retest d. In effect the process is repeated until we find the smallest positive integer K such that d divides $I_0 + K|\mathbf{B}|$.[13] The resulting inequality

---

[13]In general, it may be difficult to find the gcd of several integers. This is usually done by employing the Euclidean Algorithm (see Saaty [500]). An estimate on the number of computations when there are two integers appears in Dixon [172]).

is $x'' \geq 0$ (see Figure 4.6). To clarify the discussion we improve the second derived inequality in the previous example.

Example (from Example 4.3)

The source row is

$$2x_1 = 26/7 + (2/7)(-x_3) + (4/7)(-x_5).$$

The generated inequality (hyperplane) is ($|\mathbf{B}| = 7$)

$$x_6' = -5/7 - (2/7)(-x_3) - (4/7)(-x_5) \geq 0. \tag{10'}$$

Or, in terms of the original nonbasic variables x_1 and x_2,

$$x_6' = 3 - 2x_1 \geq 0. \tag{11'}$$

Now, the gcd of 2 and 4 is 2, which does not divide 5. Hence, add 1 to $5/7$ in $(10)'$ and the new cut is

$$x_6'' = -12/7 - (2/7)(-x_3) - (4/7)(-x_5) \geq 0, \tag{12'}$$

which passes through integer points since 2 divides 12. As previously noted, increasing $5/7$ by 1 in (10)' amounts to reducing 3 by 1 in (11)'. Or, in terms of x_1 and x_2, (12)' is

$$x_6'' = 2 - 2x_1 \geq 0, \tag{12''}$$

which clearly intersects an infinite number of integer points (Figure 4.5).

Up to this time we have not mentioned integer points *inside* the feasible region. A generated inequality (hyperplane) such as (10) can be *most* improved if we could replace it by the one whose hyperplane passes through the first integer solution. (In Figure 4.6 the hyperplane $x''' = 0$ defines the strongest cut from $x' = 0$.) The difficulty is, however, to determine whether the hyperplane passes through an integer solution and, if not, to obtain one that does. Precise conditions, as before, are not available. Nevertheless, a procedure might be to first obtain the hyperplane which passes through an integer point. Then test all-integer points on it "near" the feasible region for solutions. If an integer solution cannot be found, increase f_0 by 1 and repeat the search. The problem with this approach is that there are usually many integer points near the feasible region on the hyperplane. This is especially true when $n - 1$, its dimension, is large. Thus, even if a systematic enumeration scheme could be developed it would almost surely be computationally unworthwhile.

4.7   Algorithm Strategies

4.7.1   Number of Possible Inequalities

If we take a second look at the derivation of the cut it becomes apparent that any integer linear combination of rows that yields a fractional constant element may be used as a source row. The resulting left-hand side does not have to be nonnegative. Thus it is possible to generate an arbitrarily large number of inequalities from each tableau.

4.7.2   Choosing the Source Row

We have already mentioned that the first quadrant contains the feasible region in $(x_{J(1)}, \ldots, x_{J(n)})$ space. Also, if we write the hyperplane defining the cut as $\sum_{j=1}^{n} f_j x_{J(j)} = f_0$, then its intersection with each $x_{J(j)}$ axis $(j = 1, \ldots, n)$ occurs at the point $(0, \ldots, 0, f_0/f_j, 0, \ldots, 0)$ which is above the origin. (Row indices are being omitted.) Thus, the hyperplane must appear as in Figure 4.7. Or we may say that the larger the value of f_0/f_j, the stronger the cut. Hence, the objective when selecting the source row is to produce an inequality or hyperplane in which these ratios are as large as possible.

Since source rows will usually yield inequalities in which the f_0/f_j ratios are large for some variables and small for others, it is not clear which row or combination of rows to select. A classical rule is to generate the inequality from the row with the largest fractional component. Hopefully, by choosing f_0 large the ratios f_0/f_j will be large. This, of course, is not always true. Probably the criterion is used more because of its ease of implementation, and a uniformly better one has yet to be found. Another simple-to-use rule is to generate the cut from the first row with a fractional constant component. This may seem more arbitrary than the first choice. However, a rule which selects the first "fractional row" every finite number of tableaux is, in fact, necessary to support finiteness of the algorithm (Section 4.9).

A third rule might be to add several different cuts and let the simplex method choose one. Or, perhaps, select the one that gives the largest decrease in the objective function. Appended inequalties that are not used may be dropped.

4.7.3 Dropping Inequalities

A standard practice is to drop each added inequality immediately after its slack re-enters the set of basic variables. That is, when a new inequality is introduced it is used as the pivot row so its slack becomes a nonbasic variable. When (and if) this slack variable returns to the basic set, it and its defining row are omitted. Since there are n nonbasic variables this strategy puts an upper limit of n on the number of Gomory slack variables present in the tableau at any one time. However, observe that constraints which could later be used as pivot rows may be omitted. That is, an optimal LP solution may subsequently correspond to a portion of the feasible region previously cut off (see Figure 4.8). (Of course, it cannot then be all-integer.) Nevertheless,

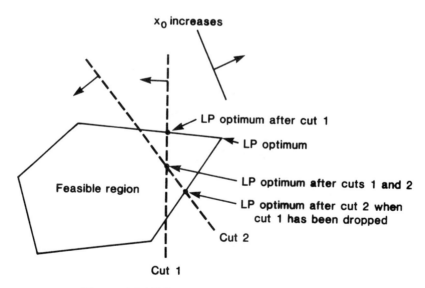

Figure 4.8: Effect of Dropping Inequalities

the contention is that it is more beneficial to generate new cuts than reuse "old" ones. When this strategy is adopted, we do not need the identity rows $x_{m+n+k} = -1(-x_{m+n+k})$ $(k = 1, 2, \ldots)$.

4.8   A Geometric Derivation

Recall that the negative identity rows are always present in the tableaux and any integer linear combination of rows (variables) yielding a fractional constant component may be used to generate the cut. Hence, it is always possible to obtain a source row (row indices are being omitted)

$$L = a_0 + \sum_{j=1}^{n} a_j(-x_{J(j)}), \tag{13}$$

with $a_j \geq 0$ for $j = 1, \ldots, n$. Here, L is some integer linear combination of variables x_i (that is, rows i), and each a_j $(j = 1, \ldots, n)$ is the corresponding sum of coefficients a_{ij}. Now, in any integer solution L must be an integer (though not necessarily nonnegative). Consequently, we may add or subtract integer multiples of $[a_0]$ to (13) and, as in section 4.5 derive the valid inequality

$$x' = -f_0 + \sum_{j=1}^{n} (-a_j)(-x_{J(j)}) \geq 0. \tag{14}$$

Observe that inequality (14) is weaker than the constraint obtained by also reducing each a_j $(j = 1, \ldots, n)$ to its fractional part. Moreover, the two cuts are the same only when each coefficient a_j $(j = 1, \ldots, n)$ in the source row (13) is less than one. Interestingly, we can derive the cut (14) from (13) via a geometrical argument.

Suppose for the moment, that we are about to derive the first inequality. Then L is an integer combination of the original basic and (or) nonbasic variables. Since each original basic variable initially apeared as an integer equality in terms of the nonbasic ones, we may rewrite L as

$$L = n_0 + \sum_{j=1}^{n} n_j x_j,$$

where n_j $(j = 0, 1, \ldots, n)$ is an integer and the x_j's $(j = 1, \ldots, n)$ are, of course, the original nonbasic variables.

Now, the current optimal values for $x_1, \ldots, x_n$ and the optimal extreme point solution are found by setting the $x_{J(j)}$ variables to 0. Then (13) is

$$n_0 + \sum_{j=1}^{n} n_j x_j = a_0. \tag{15}$$

Since (13) is derived from the rows of the current tableau, it follows that the optimum values of $x_1, \ldots, x_n$ satisfy (15); or, equivalently, the hyperplane $L = a_0$ passes through the extreme point P in the original nonbasic variable space. Furthermore, since in (13) the $a_j (j = 1, \ldots, n)$ are nonnegative, we have that $\sum_{j=1}^{n} a_j(-x_{J(j)})$ is nonpositive for any feasible solution $(x_{J(1)}, \ldots, x_{J(n)})$; or, equivalently, for any feasible solution $(x_1, \ldots, x_n)$. Hence,

$$L = a_0 + \sum_{j=1}^{n} a_j(-x_{J(j)})$$

takes on its maximum at the vertex P with no part of the feasible region in $(x_1, \ldots, x_n)$ space above it. This is portrayed in Figure 4.9.

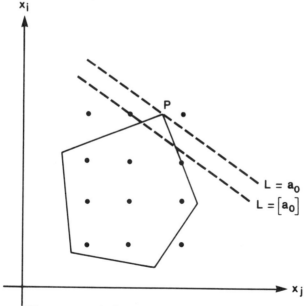

Figure 4.9: A Geometric Derivation of a Cut

If a_0 is not an integer, i.e. $f_0 > 0$, the hyperplane $L = a_0$ can be pushed parallel into the feasible region until $L = [a_0]$. We claim that there cannot be an integer point between $L = a_0$ and $L = [a_0]$. For if $L = a_0'$ took on an integer value for some a_0', $[a_0] < a_0' < a_0$, we would have $L = n_0 + \sum_{j=1}^n n_j x_j = a_0'$ for some choice of integers $\{ x_j \}$. This is impossible because a_0' is not an integer while the $n_j's$ and $x_j's$ are. Hence $L \leq [a_0]$ is a legitimate cut. Finally, by using equality (13) and noting that $f_0 = a_0 - [a_0]$ we see that $L \leq [a_0]$ is inequality (14).

Since any Gomory slack variable becomes an all-integer equality when written in terms of the original nonbasic variables, the geometric derivation is unchanged for subsequent cuts.

Recall that each a_j in (13) may be written as $I_j'/|\mathbf{B}|$, where I_j' is a nonnegative integer and $\mathbf{B}$ is the current basis. Also, $f_0 = I_0/|\mathbf{B}|$, where I_0 is a positive integer smaller than $|\mathbf{B}|$. Hence, the inequality (14) may be written as

$$x' = \frac{-I_0}{|\mathbf{B}|} + \sum_{j=1}^n \frac{I_j'}{|\mathbf{B}|} x_{J(j)} \geq 0. \tag{16}$$

We then have the analogue of Theorems 4.3 and 4.4. That is,

Theorem 4.5 *The hyperplane $x' = 0$ either is void of integer points or it passes through an infinite number of them. Moreover, the latter case occurs if and only if the gcd of $I_j', \ldots, I_n'$ divides I_0.*

Proof: To prove the "moreover" part we replace I_j by I_j' in Theorem 4.3, and proceed as in the proof of that theorem. To show the first assertion we reprove Theorem 4.4. ∎

Now, if the hyperplane $x' = 0$ does not contain integer points, it can be pushed parallel to itself in the feasible region until it intersects one (not necessarily inside the feasible region). The computations, as explained in the closing paragraphs of Section 4.6, amount to finding the smallest positive integer K such that gcd of $I_1', \ldots, I_n'$ divides $I_0 + K|\mathbf{B}|$. Then the improved inequality is found by adding K to $I_0/|\mathbf{B}|$ in (14). The result is the stronger inequality $x'' \geq 0$ as depicted in Figure 4.6. To clarify this section, we give the geometric derivation of the first cut appearing in Example 4.1.

Example (from Example 4.1) From Tableau #3, the source row is

$$x_1 = 18/10 - (4/10)(-x_4) + (2/10)(-x_3).$$

To obtain the source row (13) we subtract the $x_4 = -1(-x_4)$ row from x_1, or

$$L = x_1 - x_4 = 18/10 + (6/10)(-x_4) + (2/10)(-x_3). \qquad (13)'$$

The inequality is therefore

$$x_1 - x_4 \leq [18/10] = 1.$$

But $x_4 = -7 + 3x_1 + 2x_2$; substituting yields

$$7 - 2x_1 - 2x_2 \leq 1,$$

or the cut is

$$2x_1 + 2x_2 \geq 6.$$

Observe that this is the inequality which was derived analytically from the x_1 row. Of course, this must be so since in (13)'

$$a_1 = 6/10 < 1 \text{ and } a_2 = 2/10 < 1.$$

4.9   Finiteness

To show that the algorithm converges to the optimal integer solution after a finite number of tableaux (iterations), it is necessary to assume that some lower bound M is known for the value of x_0. (One can always be found when the feasible region is bounded.) We also suppose that the *first* row with a noninteger constant component is selected as the source row. (Actually this requirement may be weakened to one which ensures that the first fractional row is selected as the source row every finite number of iterations. However, to simplify the discussion, we use the first fractional row rule.)

Now assume that the algorithm is *not* finite. Then, because we are using the lexicographic dual simplex method, there exists an infinite sequence of tableaux (indexed by k) whose 0 column is lexicographically decreasing. That is, $\alpha_o^k \overset{\mathcal{L}}{\succ} \alpha_o^{k+1} \overset{\mathcal{L}}{\succ} \alpha_o^{k+2} \ldots$ etc. Since α_o decreases lexicographically, its first component a_{00} must be nonincreasing. Then it must eventually remain fixed at some integer value. For if not, then row 0 generates the cut (17), which is used as the pivot row.

$$x' = -f_{00} + \sum_{j=1}^{n} (-f_{0j}) \left(-x_{J(j)} \right) \geq 0. \tag{17}$$

If p is the pivot column index, then, as derived in Section 2.7,

$$a'_{00} = a_{00} - \frac{a_{0p} f_{00}}{f_{0p}}, \tag{18}$$

where a prime denotes an element after the pivot. Now, the dual simplex method guarantees that $a_{0p} \geq 0$. But as f_{0p}, the negative value of the pivot element and fractional part of a_{0p}, is different from 0, it follows that $a_{0p} \neq 0$ or

$$a_{0p} > 0 \tag{19}$$

and therefore

$$a_{0p} = [a_{0p}] + f_{0p} \geq f_{0p}; \quad \text{or} \quad \frac{a_{0p}}{f_{0p}} \geq 1. \tag{20}$$

Hence,

$$a'_{00} \leq a_{00} - f_{00} = [a_{00}]. \tag{21}$$

Or, each time row 0 is used as the source row a_{00} is decreased at least to the next integer. Equivalently, a_{00} can contain a fractional part for only a finite number of tableaux; otherwise, it would eventually fall below its lower bound. Let k' be the iteration at which a_{00} is fixed for all iterations $k \geq k'$.

Now that the first component in α_o is fixed, we have, since the infinite sequence is lexicographically decreasing, that a_{10} is nonincreasing for all $k \geq k' + 1$. Further, a_{10} is bounded below by 0 for all $k \geq k' + 1$. For if not, then row 1 becomes the pivot row, and if p is the pivot column index, we have

$$a'_{00} = a_{00} - \frac{a_{0p}a_{10}}{a_{1p}}.$$

Now, $a_{10} < 0$ and the pivot element $a_{1p} < 0$; hence $a_{10}/a_{1p} > 0$. Then a_{0p} must be 0 since, by assumption $a'_{00} = a_{00}$. This contradicts the lexicographic positivity of α_p. (If $a_{1j} \geq 0$ for $j = 1, \ldots, n$ the integer program has no solution.)

To continue the proof we can show that there is an iteration $k'' \geq k' + 1$ such that a_{10} remains fixed at a nonnegative integer value for all iterations $k \geq k''$. To do this we simply replace the first subscript in equations (17), (18), (19), (20), and (21) by 1, and repeat the same argument (with $M = 0$). The only point that requires an explanation is the reason why $a_{1p} > 0$ for all $k \geq k''$. This is evident since α_p is lexicographically positive and a_{0p} must then be 0 in order for a_{00} to remain its fixed integer value; also a_{1p} cannot be 0 because $-f_{1p}$, the pivot element, is different from 0.

To complete the proof we repeat the argument for each of the $m+n+1$ original variables. In this way we obtain an all-integer 0 column. This integer solution is optimal, since all the inequalities which are added cut off the current optimum vertex but never eliminate any integer solutions. Hence, the feasible region is eventually reduced to one with an all-integer extreme point that gives x_0 (which is never changed) its maximal value.

Problems

Unless otherwise specified, the notation, the simplex tableau, and the Basic Approach (Section 4.1) are the same as in the chapter. Also, "dual" will mean "lexicographic dual."

Problem 4.1 Consider a simplex tableau which is neither primal nor dual feasible. Suppose we add the redundant constraint

$$s = M + \sum_{j=1}^{n} 1(-x_{J(j)}) \geq 0$$

to the bottom of the tableau (s is a nonnegative slack variable and M is a large positive number). Show that after a pivot on the 1 element in the new row which is in the lexicographically smallest column, the tableau exhibits dual feasibility. Give an example.

Problem 4.2 Suppose a simplex tableau is as in Problem 4.1. Give another simple way of obtaining dual feasibility. Illustrate your method.

Problem 4.3 Suppose, corresponding to some tableau, we have the row (row indices are omitted)

$$x' = -a_0 + \sum_{j=1}^{n} a_j(-x_{J(j)}), \tag{1}$$

where $a_j > 0$ for $j = 1, \ldots, n$. With the current primal solution not integer and (1) used as a generating row, prove that

a) $\sum_{j=1}^{n} a_j x_{J(j)} \geq 1$ is a constraint which must be satisfied by every integer solution.

b) $\sum_{j=1}^{n} a_j^+ x_{J(j)} \geq 1$, where $a_j^+ = 1$, if a_j is integer and $a_j^+ = f_j = a_j - [a_j]$ otherwise, is also a constraint which must be satisfied by every integer solution. Also, show that it is stronger than the previous inequality.

Problem 4.4 Show that the cuts (inequalities) appearing in Problem 4.3 (a) and (b) do not have the properties required in the convergence proof of the fractional algorithm (Section 4.9). Also, prove that such inequalities cannot necessarily convert an integer optimal tableau to an all-integer integer optimal tableau (see Section 4.6).

Problem 4.5 Given the integer program

$$
\begin{array}{lrcrcl}
\text{minimize} & x_1 & + & x_2 & = & x_0 \\
\text{subject to} & 4x_1 & + & 10x_2 & \geq & 12, \\
& 10x_1 & + & 4x_2 & \geq & 12, \\
\text{and} & x_1, & & x_2 & \geq & 0, \text{integer.}
\end{array}
$$

a) Show that the Basic Approach with the valid inequality

$$s = -1 + \sum_{j=1}^{n} (-1)(-x_{J(j)}) \geq 0$$

where s is a nonnegative slack variable, cannot solve the above problem (see Appendix A).

b) Solve the problem utilizing the Basic Approach with the classical Gomory fractional cut. Relate the computations to (x_1, x_2) space.

Problem 4.6 Given the integer program

$$
\begin{array}{rlrlrlrl}
\text{maximize} & x_1 & + & 3x_2 & + & 10x_3 & = & x_0 \\
\text{subject to} & 2x_1 & + & 4x_2 & + & 8x_3 & \le & 15, \\
& x_1 & & & & & \le & 4, \\
& & & x_2 & & & \le & 2, \\
& & & & & x_3 & \le & 3/2, \\
\text{and} & x_1, & & x_2, & & x_3 & \ge & 0, \text{ integer.}
\end{array}
$$

Show (a) and (b) of the preceding problem. In (b), do not relate the computations to (x_1, x_2, x_3) space.

Problem 4.7 Using the classical (Gomory) cut and the Basic Approach, solve the knapsack problem

$$
\begin{array}{rl}
\text{maximize} & 3x_1 + 4x_2 + 8x_3 + 2x_4 + 7x_5 + x_0 \\
\text{subject to} & 2x_1 + x_2 + 3x_3 + 5x_4 + 6x_5 \le 14 \\
\text{and} & x_j = 0 \text{ or } 1 \quad (j = 1, \ldots, 5).
\end{array}
$$

Problem 4.8 Solve the following integer program using the Basic Approach with the classical cut.

$$
\begin{array}{rlrlrlrl}
\text{Maximize} & 2x_1 & + & 3x_2 & + & 4x_3 & + & 7x_4 & = & x_0 \\
\text{subject to} & 4x_1 & + & 6x_2 & + & 2x_3 & + & 8x_4 & = & 16, \\
& x_1 & + & 2x_2 & - & 6x_3 & + & 7x_4 & = & 3, \\
\text{and} & x_j \ge 0, \text{ integer} & (j = 1, \ldots, 4).
\end{array}
$$

Problem 4.9 For each tableau in Example 4.1 (Section 4.4) find:

a) The value of all primal and dual variables.

b) The primal basis and the dual basis.

c) The determinant of the basis.

Give a geometric explanation of what each iteration does to the dual problem. Can you relate each tableau to some dual space? Explain.

Problem 4.10 Prove the congruence relations (i) through (v) appearing in the initial part of Section 4.5. Show that property (v) is not necessarily true if x and y' (or x' and y) are not integer.

Problem 4.11   Consider the inequality (row indices are omitted)

$$x' = -f_0 + \sum_{j \in J^1} (-f_j)\left(-x_{J(j)}\right) +$$

$$\sum_{j \in J^2} -\left(\frac{f_0}{1-f_0}\right)(1-f_j)\left(-x_{J(j)}\right) \geq 0, \qquad (2)$$

where $J^1 = \{j | f_j \leq f_0\}$ and $J^2 = \{j | f_j > f_0\}$, derived from some row of a simplex tableau. Of course, the f_j's are the fractional parts, $f_0 > 0$, and the $x_{J(j)}$'s are integer constrained variables.

a) Show that inequality (2) does not eliminate any integer solution. That is, it is a valid cut.

b) Explain why the nonnegative slack variable x' need not necessarily be an integer in every integer solution. What problems are thus created when the cut (2) is used with the Basic Approach?

c) Using the Basic Approach with inequality (2), resolve Example 4.2. (*Hint:* Select x_1 as the source row.)

Problem 4.12   Consider the Basic Approach with inequality (2), appearing in Problem 4.11. (The first $m + n + 1$ row are eligible to generate the inequality.) Suppose, for a class of integer programs, that whenever a nonbasic variable is a slack variable corresponding to the inequality (2) its coefficient in the cut or pivot row is 0. Prove that such an algorithm will converge for the class of problems described above. (In the next chapter we show that the algorithm can converge for all problems when (2) is modified to allow for continuous variables.)

Problem 4.13   Prove that if $I_0, I_1, \ldots, I_n$ are integers and $I_0(I_1+\ldots+I_n) \neq 0$, then there exist integers $\lambda_1, \ldots, \lambda_n$ such that $\sum_{j=1}^n I_j\lambda_j = I_0$ if and only if the greatest common divisor of $I_1, \ldots, I_n$ divides I_0. (Recall that this result is used to prove Theorem 4.3.)

Problem 4.14   Consider the source row (from Example 4.4)

$$2x_1 = 26/7 + 2/7(-x_3) + 4/7(-x_5).$$

Then the Gomory cut is

$$x' = -5/7 - 2/7(-x_3) - 4/7(-x_5) \geq 0. \tag{3}$$

Suppose that

$$|\mathbf{B}| = 7, \ x_3 = 3 - 3x_1 + 2x_2, \ x_5 = 5 - 2x_1 - x_2,$$

and x_1 and x_2 are the original nonbasic variables. We know, since the greatest common divisor (gcd) of 2 and 4 (i.e., 2) does not divide 5, that the hyperplane $x' = 0$ does not intersect any integer points. Is it possible to improve the inequality $x' \geq 0$ by increasing 5/7 to 14/7 in (3)? (Then the gcd of 2 and 4 divides 14.) Explain.

Problem 4.15 The (first) optimal simplex tableau in Example 4.4 is

	1	$-x_3$	$-x_5$	
x_0	30/7	5/7	3/7	
x_1	13/7	1/7	2/7	
x_2	9/7	-2/7	3/7	$|\mathbf{B}| = 7$
x_3	0	-1	0	
x_4	31/7	-3/7	22/7	
x_5	0	0	-1	

Show that at most seven different Gomory cuts can be derived from the above tableau.

Problem 4.16 Show that the geometrically derived cut (Section 4.8) is never stronger than the analytically derived cut (Section 4.5) when both are generated from the same row.

Problem 4.17

a) Is the lexicographic simplex method necessary in the finiteness proof (Section 4.8)?

b) Does the algorithm converge if the fractional cuts are not kept?

Problem 4.18 Show that when a derived inequality is added to the bottom of the simplex tableau it cannot affect the lexicographic positivity of any column.

Problem 4.19   Prove that the Basic Approach with the fractional cut is finite when the cut is derived from the first fractional row every finite number of iterations. Does the maximum fractional rule yield a finite algorithm? Explain.

Problem 4.20   What would be involved in developing an efficient algorithm which always obtains a Gomory cut whose hyperplane passes through an integer solution? An integer feasible solution?

Problem 4.21 Does the Basic Approach with the classical inequality necessarily converge if the initial tableau is not all integer? Explain.

Problem 4.22 Consider the following algorithm proposed by Gomory [276]. (**B** is the current basis.)

Step 1 Obtain a simplex tableau which exhibits dual feasibility.
Step 2

   i) If the tableau is both nonoptimal and contains fractions, we either perform a dual simplex iteration or derive a classical cut and pivot on an element in its row. (The element is determined by the dual simplex rules.)

   ii) If the tableau is all integer but not optimal, a dual simplex iteration is made.

   iii) If the tableau is optimal but not all integer a cut is added and a dual simplex pivot on an element in its row is made.

   iv) If the tableau is optimal and all integer the integer optimal solution has been found–terminate.

Step 3 If at any time during Step 2 $|\mathbf{B}|$ exceeds N (a fixed positive number at least 1), apply a technique which maintains dual feasibility and makes $|\mathbf{B}| \leq N$. Then return to Step 2.

   a) What would be a "technique" for Step 3?

   b) Resolve Example 4.1, using this algorithm with $N = 4$.

   c) Repeat (b) with $N = 1$.

d) Prove that the algorithm is finite when applied to integer programs that have integer (feasible) solutions. (*Hint:* Assume the algorithm is not finite. Then there must exist an infinite sequence of tableaux with α_o decreasing lexicographically and with $|\mathbf{B}|$ constant. But an integer solution exists, hence α_o is bounded below. So each of its components must eventually remain fixed. If the final α_o is not all integer, repeat the argument.)

e) How might the algorithm be modified to accomodate integer programs that do not have integer solutions?

Problems 4.23 through 4.29 relate to Appendix B.

Problem 4.23 Solve the integer program given in Problem 4.5, using the Accelerated Euclidean algorithm.

Problem 4.24 For each tableau appearing in Example 4.4, find the dual basis. Also, relate the tableaux to (x_1, x_2) space. Include *all* composite and Gomory cuts.

Problem 4.25 Prove or disprove the following: "The composite cut is a Gomory cut which may be generated from some integer linear combinations of rows in the current optimal tableau."

Problem 4.26 In the Accelerated Euclidean algorithm, is it possible to select the same source row more than once in Step 2? If so, can the algorithm cycle in this step?

Problem 4.27 Prove that the composite cut used in the Accelerated Euclidean algorithm always cuts off the current linear programming optimum.

Problem 4.28 Does the Accelerated Euclidean algorithm necessarily converge? If not, what modifications would ensure convergence?

Problem 4.29 Suppose Step 2 of the Accelerated Euclidean algorithm is replaced by

Step 2 Update row p by pivoting on the element in the (current) Gomory cut in column K. If row p becomes all integer, go to Step 4. Otherwise, go to Step 3.

Show that the resulting algorithm cannot, in general, work.

The remaining problems relate to Appendix C.

Problem 4.30   Using the technique illustrated, derive an intersection cut from Example 4.2. Is it stronger than the first Gomory cut used?

Problem 4.31   Show that the Basic Approach (Section 4.1) with the intersection cut (1) (Appendix C) produces a convergent algorithm if Gomory fractional cuts are used every finite number of iterations. Does the proof go through when the Gomory cuts are not used?

Problem 4.32   Let x^* be a nondegenerate optimal (basic) linear programming solution in n space, and let a_j $(j = 1,\ldots,n)$ be the direction in which an extreme point neighbor of x^* lies. Show that there are n linearly independent points $u_j = x^* + a_j\lambda_j$, where $\lambda_j > 0$ $(j = 1,\ldots,n)$. Is this result true when x^* is degenerate?

Problem 4.33   Show that the intersection cut (1) (Appendix C) does not eliminate any integer solutions. (*Hint*: First prove that $\beta x^* \neq \beta_0$, where x^* is an extreme point and thus cannot be on the line segment joining two other points in the feasible region. Then show that the region cut off by $\beta x \geq \beta_0$ lies entirely in, e.g., the hypersphere, which means that it is void of integer solutions.) Generalize the proof so that any convex region which intersects the unit hypercube in only its vertices can yield an intersection cut.

Problem 4.34   Let x^* be an optimal linear programming extreme point solution in n space. Prove the following:

a) The hypercube containing x^* is defined by $[x^*] \leq x \leq [x^*] + e$, where $[y]$ is the largest integer $\leq y$ and e is an n column of ones.

b) The hypersphere circumscribing the hypercube is defined by

$$\|x - ([x^*] + \frac{1}{2}e)\| \leq \sqrt{n}/2,$$

where $\|y\| = \sqrt{\sum_i y_i^2}$, and $y = (y_i)$.

c) The hyperdiamond circumscribing the hypercube is defined by

$$\|x - ([x_*] + \frac{1}{2}e)\|_1 \leq n/2,$$

where $\|y\|_1 = \sum_i |y_i|$.

d) Use the hyperdiamond (c) to find the intersection cut for Example 4.1. Is it stronger than the one obtained using the hypersphere (b)?

Problem 4.35 In terms of relative strength, ease of implementation, etc., do intersection cuts seem preferable to Gomory cuts? Can the derivation of intersection cuts lead to substantial computer round-off errors?

Problem 4.36 Why is it necessary to drop constraints corresponding to basic variables which are at value 0 in x^* (an optimal extreme point) prior to generating an intersection cut? What if these equations are kept?

Problem 4.37 (Balas [26]) Consider the integer program

$$
\begin{array}{lrcrcll}
\text{maximize} & -x_1 + 4x_2 & = & x_0 & & & \\
\text{subject to} & 2x_1 & + & 4x_2 & \leq & 7 & (x_3), \\
& 10x_1 & + & 3x_2 & \leq & 15 & (x_4), \\
\text{and} & x_1, & & x_2 & \leq & 0 & \text{and integer.}
\end{array}
$$

a) Solve it as a linear program.

b) Using a hypersphere, derive an intersection cut.

c) Append the cut found in (b) to the optimal linear programming tableau. Perform one pivot step. Does the new tableau exhibit the optimal integer solution? Check your results graphically.

Problem 4.38 When x^*, an optimal extreme point. has k integer components there are 2^k possible unit hypercubes. Show that the strongest cut will be obtained from the one with the largest intersection with the feasible region.

Appendix A
Convergence Using the Dantzig Cut

(Dantzig [156], Gomory and Hoffman [283])

Consider the optimal simplex tableau associated with an integer program. If the primal solution is not all integer, then as indicated by Dantzig [156], at least one of the nonbasic variables $x_{J(j)}$ must be positive. But they must also be integer in every integer solution. Hence $x_{J(j)} \geq 1$ for some $j = 1, \ldots, n$. Thus a constraint which must be satisfied by every integer solution is

$$\sum_{j=1}^{n} x_{J(j)} \geq 1 \tag{1}$$

After introducing the nonnegative slack variable x', (1) becomes

$$x' = -1 + \sum_{j=1}^{n} (-1)(-x_{J(j)}) \geq 0. \tag{2}$$

Dantzig suggested using the cut (2) with the Basic Approach as outlined in Section 4.1. (Observe that x' is integer in very integer solution.) However, Gomory and Hoffman [283] have shown that such an algorithm will not converge for certain classes of integer programs. In particular, for problems which do not meet the conditions described in Theorems 4.6 and 4.7, we present their results.

Write the integer program as

$$
\begin{aligned}
\text{maximize } x_0 &= a_{00} &&+ \sum_{j=1}^{n} a_{0j}(-x_j) \\
\text{subject to } x_j &= &&-1(-x_j) \geq 0 \ (j = 1, \ldots, n) \\
x_{n+i} &= a_{n+i,0} &&+ \sum_{j=1}^{n} a_{n+i,j}(-x_j) \geq 0 \ (i = 1, \ldots, m) \\
\text{and} \quad x_j &\quad \text{integer} && (j = 1, \ldots, n+m).
\end{aligned}
\tag{3}
$$

Let $\mathbf{x} = (x_1, \ldots, x_{n+m})$, not necessarily integral, satisfy (3) and let $\bar{x}_1, \ldots, \bar{x}_{n-1}$ be the $n-1$ smallest values of $\mathbf{x}$ ordered so that $\bar{x}_j \leq \bar{x}_{j+1}$.

Also, denote the slack of the kth added inequality (2) by x_{m+n+k}. Then we have the following lemma.

Lemma 4.1 *If* $\sum_{j=1}^{n-1} \bar{x}_j \geq 1$ *then* $x_{m+n+k} \geq \bar{x}_{n-1}$ *for all* k.

Proof: Consider the first added inequality (2). With the regard to that inequality let $x_{J(M)}$ be the largest among the $x_{J(j)}$ $(j = 1, \ldots, n)$ in the feasible solution $\mathbf{x}$.
Then,

$$x_{m+n+1} = \sum_{j=1}^{n} x_{J(j)} - 1 = x_{J(M)} + \sum_{\substack{j=1 \\ j \neq M}}^{n} x_{J(j)} - 1 \geq x_{J(M)} \geq \bar{x}_{n-1}.$$

The first inequality sign follows, since

$$\sum_{\substack{j=1 \\ j \neq M}}^{n} x_{J(j)} - 1 \geq \sum_{j=1}^{n-1} \bar{x}_j - 1,$$

which is nonnegative by hypothesis. The second inequality sign is allowed because $x_{J(M)}$ is the largest among n values of the variables, hence it is at least as large as $\bar{x}_{n-1}$, the largest among $n - 1$ values.

Now, since $x_{m+n+1} \geq \bar{x}_{n-1}$, the set $\bar{x}_1, \ldots, \bar{x}_{n-1}$ does not change after the introduction of the first inequality; hence we can repeat the same argument to show that $x_{m+n+2} \geq \bar{x}_{n-1}$. The proof may be repeated any number of times. ∎

We use the lemma to prove the following theorem.

Theorem 4.6 *If $\mathbf{x}^0$ is an optimal integer solution to (3), then the algorithm, suggested by Dantzig [156], can converge only if the $n - 1$ smallest components of $\mathbf{x}^0$, i.e.,$\bar{x}_{n-1}^0$, are all zero. (Geometrically, $\mathbf{x}^0$ must lie on an edge of the feasible region in $(x_1, \ldots, x_n)$ space.)*

Proof: If the algorithm is to converge we must eventually obtain a feasible region whose optimal extreme point is all integer. But if $\mathbf{x}^0$ is an extreme point (vertex), at least n of its components are 0. (Since an extreme point is defined by the intersection of at least n hyperplanes of the form $x_j^0 = 0$.) Suppose, $\bar{x}_{n-1}^0 > 0$; this means, since $\mathbf{x}^0$ is integer, that $\bar{x}_{n-1}^0 > 0$ for all k. So the extreme point solution $\mathbf{x}^0$ can never be found since n of its components cannot be made 0. Thus, for convergence $\bar{x}_{n-1}^0 = 0$, or $\bar{x}_1^0 = \ldots = \bar{x}_{n-1}^0 = 0$. ∎

To illustrate Theorem 4.6 consider

$$
\begin{array}{rrcl}
\text{maximize} & x_1 \; + \; x_2 & = & x_0 \\
\text{subject to} & 0 \; \le \; x_1 & \le & 3/2 \\
& 0 \; \le \; x_2 & \le & 3/2 \\
\text{and} & x_1, x_2 \text{ integer.}
\end{array}
$$

Clearly, the optimal integer solution is $x_1 = x_2 = 1$, which is in the interior of the feasible region in (x_1, x_2) space. Thus the variables $x_1, x_2, x_3 = 3/2 - x_1$, and $x_4 = 3/2 - x_2$ are all positive at the optimal integer point. Or, the algorithm cannot converge for this problem..

Observe that the condition of Theorem 4.6 is always satisfied by integer programs with zero-one variables, since in these programs we have the additional constraints

$$x_j \le 1 \quad \text{or} \quad x_j' = 1 - x_j \ge 0 \text{ for } j = 1, \ldots, n.$$

The x_j''s are, of course, nonnegative slack variables. Then for *every* zero-one vector $\mathbf{x} = (x_1, \ldots, x_n, x_{n+1}, \ldots, x_{n+m}, x_1', \ldots, x_n')$ at least n of the variables $(x_1, \ldots, x_n, x_1', \ldots, x_n')$ are zero. (Geometrically each possible zero-one solution is a vertex of the unit hypercube. Also, each vertex is defined by the intersection of at least n hyperplanes.) However, the algorithm does not necessarily converge for these problems as we have a second necessary condition.

Theorem 4.7 *Let $\mathbf{x}^0$ be an optimal integer solution and x_0^0 be the corresponding value of x_0. Let $\mathbf{x}$ be any solution satisfying (3) with its corresponding value of x_0 greater than x_0^0. Then for the algorithm to converge we must have $\sum_{j=1}^{n-1} \bar{x}_j < 1$.*

Proof: Suppose that $\sum_{j=1}^{n-1} \bar{x}_j \ge 1$. Then $\bar{x}_{n-1} > 0$ and by the lemma $x_{m+n+k} \ge \bar{x}_{n-1} > 0$ for all k. Or, $\mathbf{x}$ satisfies every added inequality. Thus it can never be cut off and the (worse) solution $\mathbf{x}^0$ can never be found. ∎

To illustrate Theorem 4.7 consider

$$
\begin{aligned}
\text{maximize} \quad & -3x_1 \; - \; 3x_2 \; - \; 4x_3 \; = \; x_0 \\
\text{subject to} \quad & x_1 \qquad\qquad + \quad x_3 \; \leq \; 3, \\
& x_1 \; + \; x_2 \qquad\qquad \geq \; 3/2, \\
& 0 \leq x_j \leq 1 \;\; (j = 1,2,3), \\
\text{and} \quad & x_j \text{ integer} \;\; (j = 1,2,3).
\end{aligned}
$$

Then clearly the optimal integer solution is $x_1 = x_2 = 1$ and $x_3 = 0$, which yields $x_0 = -6$. But the solution $x_1 = x_2 = x_3 = 1/2$ gives x_0 a value of -5. Furthermore, $x_4 = 3 - x_1 - x_3 = 2$ and $x_5 = -3/2 + x_1 + x_2 = 1/2$. so we have $\bar{x}_1 + \bar{x}_2 = 1/2 + 1/2 = 1$; or, the algorithm cannot solve this $0-1$ problem. (Bowman and Nemhauser [90] have shown that a slightly modified version of the Dantzig cut (2) does provide a finite algorithm. A strengthened Bowman/Nemhauser cut is described by Rubin and Graves [493].)

Appendix B
A Variation of the Basic Approach: The Accelerated Euclidean Algorithm

(Martin [415])

An interesting variation of the basic cutting plane approach was proposed by Martin [415]. The technique basically consists of solving a linear program, returning the optimal simplex tableau to one with an all integer (not necessarily feasible) primal solution, resolving the linear program, and so on. The process terminates when an optimal linear programming tableau with an integer primal solution is found or reoptimization is not possible. Transforming the (not integer) optimal tableau to one with an integer primal solution is accomplished by a "composite Gomory cut." Each of these is equivalent to a series of Gomory cuts. The latter inequalities are derived from the same row in successive tableaux. (We are not concerned with feasibility.) The number of Gomory inequalities in the series is the number of cuts required to convert the source row at the optimal linear programming tableau to a row with an integer pivot element. After the form of the final Gomory cut is known we can obtain the composite inequality by expressing the last cut in terms of the nonbasic variables at the current optimal tableau by successive substitutions (see Example 4.4). However, since the same pivot column is used for all Gomory cuts in the series, a simpler procedure which retrieves the composite inequality is available. Also, during these iterations only the source row need be updated as only the final form of the classical cut is required. Finally, to ensure integer constrained Gomory slack variables, the initial tableau must be all integer. We list Martin's algorithm. The computations appear in Example 4.4.

Phase A Apply a simplex algorithm to the current tableau. If optimization is not possible there is no integer solution–terminate. If the primal solution is all integer, the optimal solution has been found–terminate. Otherwise, go to Phase B.

Phase B (Finding an all integer primal solution.)

Step 1 Select a row, indexed by v, with a noninteger constant element. Generate a Gomory cut from v. Determine p, the dual simplex pivot column index.

Step 2 Update row v by pivoting on the element in the current Gomory cut in column p. If the pivot element indexed by vp becomes integer, go to Step 4. Otherwise, go to Step 3.

Step 3 Derive a Gomory cut from the updated row v. Go to Step 2.

Step 4 Find the composite cut indexed by r which after one pivot on the element in its row in column p will convert the original row v to its form found in Step 2.

Step 5 Append r to the optimal simplex tableau (from Phase A) and pivot on the element in its row in column p. (The entire tableau is transformed.) If the new tableau has an integer primal solution, go to Phase A. Otherwise, return to Step 1. (Another composite cut is required.)

Example 4.4 (Martin [415])

$$\text{Maximize } 2x_1 + 3x_2 = x_0$$
$$\text{subject to } 2x_1 + 5x_2 \leq 8$$
$$3x_1 + 2x_2 \leq 9,$$
$$\text{and } x_1, \quad x_2 \geq 0, \text{ integer.}$$

The first optimal simplex tableau, with nonnegative slack variables x_3 and x_4, appears in Tableau #1. The s_5 row, a composite cut, is added later. $\mathbf{B}$ is the current dual basis and circles denote pivot elements.

#1	1	$-x_4$	$-x_3$
x_0	76/11	4/11	5/11
x_1	29/11	5/11	-2/11
x_2	6/11	-2/11	3/11
x_3	0	0	-1
x_4	0	-1	0
s_5	-2/11	-3/11	(-1/11)

$$\mathbf{B} = \begin{pmatrix} 1 & 0 & 0 \\ 9 & 3 & 2 \\ 8 & 2 & 5 \end{pmatrix}$$

$$|\mathbf{B}| = 11$$

We select row 1 to generate the series of cuts ($v = 1$). Then the x_3 column is the pivot column ($p = 2$). Since only row 1 is transformed we use a partial tableau. It takes five Gomory cuts to obtain an integer pivot element.

#1a	1	$-x_4$	$-x_3$
x_1	29/11	5/11	-2/11
s_1	-7/11	-5/11	(-9/11)

$|\mathbf{B}| = 11$

#1b	1	$-x_4$	$-s_1$
x_1	25/9	5/9	-2/9
s_2	-7/9	-5/9	(-7/9)

$|\mathbf{B}| = 11(9/11) = 9$

#1c	1	$-x_4$	$-s_2$	
x_1	3	$5/7$	$-2/7$	$|\mathbf{B}| = 9(7/9) = 7$
s_3	0	$-5/7$	$\boxed{-5/7}$	

#1d	1	$-x_4$	$-s_3$	
x_1	3	1	$-2/5$	$|\mathbf{B}| = 7(5/7) = 5$
s_4	0	0	$\boxed{-3/5}$	

#1e	1	$-x_4$	$-s_4$	
x_1	3	1	$-2/3$	$|\mathbf{B}| = 5(3/5) = 3$
s_5	0	0	$\boxed{-1/3}$	

#1f	1	$-x_4$	$-s_5$	
x_1	3	1	-2	$|\mathbf{B}| = 3(1/3) = 1.$

Now we must obtain the composite constraint which yields in one pivot the x_1 row in Tableau #1f from the x_1 row in Tableau #1. Clearly, since the effect of the composite inequality is equivalent to the successive application of s_1 through s_5, it can be obtained by writing s_5 in terms of x_3 and x_4. This can be done by using the Gomory slack equalities in Tableaux #1e through #1a. That is, since

$$
\begin{aligned}
s_5 &= 1/3 s_4, \\
s_4 &= 3/5 s_3, \\
s_3 &= 5/7 x_4 + 5/7 s_2, \\
s_2 &= -7/9 + 5/9 x_4 + 7/9 s_1, \\
s_1 &= -7/11 + 5/11 x_4 + 9/11 x_3,
\end{aligned}
$$

we may substitute backwards to obtain

$$s_5 = -2/11 - 3/11(-x_4) - 1/11(-x_3).$$

This is the composite cut. However, a computationally more efficient procedure is available. Suppose we denote the matrix associated with

Tableau #k as $\mathbf{A}^k$. Then recall that $\mathbf{A}^k \mathbf{E} = \mathbf{A}^{k+1}$, where $\mathbf{E}$ is a square $(n+1) \times (n+1)$ matrix having identity rows everywhere except in row $p+1$. (The $p+1$st column of the matrix is the pivot column.) Row $p+1$ has the form

$$\left(\frac{a_{r0}}{a_{rp}}, \ldots, \frac{-1}{a_{rp}}, \ldots, \frac{a_{rn}}{a_{rp}} \right),$$

where a_{rp} is the pivot element. We may use this result to obtain the composite cut.

The x_1 row of Tableau #1 is $(29/11, 5/11, -2/11)$ and the x_1 row of Tableau #2 ought to be $(3, 1, -2)$. Denote the row of the composite inequality in Tableau #1 as (a_{r0}, a_{r1}, a_{r2}); then a_{r2} is the pivot. Thus,

$$E = \begin{pmatrix} 1 & 0 & 0 \\ 0 & 1 & 0 \\ a_{r0} & a_{r1} & -1 \\ \hline a_{r2} & a_{r2} & a_{r2} \end{pmatrix}$$

and since $\mathbf{A}^1 \mathbf{E} = \mathbf{A}^2$, we must have

$$(29/11, 5/11, -2/11) \begin{pmatrix} 1 & 0 & 0 \\ 0 & 1 & 0 \\ a_{r0} & a_{r1} & -1 \\ \hline a_{r2} & a_{r2} & a_{r2} \end{pmatrix} = (3, 1, -2).$$

The equality yields $a_{r0} = -2/11, a_{r1} = -3/11$, and $a_{r2} = -1/11$. The new row is appended to Tableau #1. A pivot is made on $-1/11$, the result is the all integer Tableau #2.

#2	1	$-x_4$	$-s_5$	
x_0	6	-1	5	
x_1	3	1	-2	
x_2	0	-1	3	$\|\mathbf{B}\| = 11(3/11) = 1$
x_3	2	(3)	-11	
x_4	0	-1	0	
s_5	0	0	-1	

A primal pivot step returns Tableau #2 to optimality in Tableau #3.

#3	1	$-x_3$	$-s_5$	
x_0	20/3	1/3	4/3	
x_1	7/3	−1/3	5/3	
x_2	2/3	1/3	−2/3	$|\mathbf{B}| = 1(3) = 3$
x_3	0	−1	0	
x_4	2/3	1/3	−11/3	
s_5	0	0	−1	
s_2'	−2/3	(−1/3)	−1/3	

Again x_1 is selected to generate the Gomory cuts. The second column in the tableau ($p = 1$) becomes the pivot column. It requires two cuts to obtain an integer pivot element. The computations appear in partial Tableaux #3a through #3c.

#3a	1	$-x_3$	$-s_5$	
x_1	7/3	−1/3	5/3	$|\mathbf{B}| = 3$
s_1'	−1/3	(−2/3)	−2/3	

#3b	1	$-s_1'$	$-s_5$	
x_1	5/2	−1/2	2	$|\mathbf{B}| = 3(2/3) = 2$
s_2'	−1/2	(−1/2)	0	

#3c	1	$-s_2'$	$-s_5$	
x_1	3	−1	2	$|\mathbf{B}| = 2(1/2) = 1.$

To obtain the composite cut s_2' we use

$$(7/3, -1/3, 5/3) \begin{pmatrix} 1 & 0 & 0 \\ \dfrac{a_{r0}}{a_{r1}} & \dfrac{-1}{a_{r1}} & \dfrac{a_{r2}}{a_{r1}} \\ 0 & 0 & 1 \end{pmatrix} = (3, -1, 2).$$

The solution is $a_{r0} = -2/3$, $a_{r1} = -1/3$, and $a_{r2} = -1/3$. We introduce

$$s_2' = -2/3 - 1/3(-s_3) - 1/3(-s_5)$$

to Tableau #3. Subsequent to a pivot on $-1/3$, the (all-integer) integer optimal Tableau #4 is found.

#3c	1	$-s_2'$	$-s_5$	
x_0	6	1	1	
x_1	3	-1	2	
x_2	0	1	-1	$\lvert \mathbf{B} \rvert = 3(1/3) = 1$.
x_3	2	-3	1	
x_4	0	1	-4	Integer optimal tableau
s_5	0	0	-5	
s_2'	0	-1	0	

Concluding Remarks

1. In the example we actually knew, prior to adjoining each composite cut and pivoting on an element in its row, that the subsequent tableau would be all integer. This follows since each series of Gomory cuts or each composite cut reduced $\lvert \mathbf{B} \rvert$ to 1, and from a previous discussion (Section 4.6) if $\lvert \mathbf{B} \rvert$ is 1 the tableau is necessarily all integer. Using this assertion it is a simple matter to show that we may always obtain a tableau with an all integer primal solution from the previous optimal simplex tableau.

2. We have arbitrarily selected source rows in the example. How-

ever, we could have used a criterion discussed previously (Section 4.7). It is not clear as to which rule, if any, is best.

3. With regard to retrieving the composite constraint we could expedite the computations by deriving formulas from $\mathbf{A}^k \mathbf{E} = \mathbf{A}^{k+1}$. Let $(a_{v0}, \ldots, a_{vn})$ be the source row in $\mathbf{A}^k$ and $(a'_{v0}, \ldots a'_{vn})$ be the transformed row in $\mathbf{A}^{k+1}$. If the coefficients of the composite cut are $(a_{r0}, \ldots, a_{rn})$ and the pivot is a_{rp}, we have

$$a_{rp} = \frac{-a_{vp}}{a'_{vp}}$$

and

$$a_{rj} = (a_{vj} - a'_{vj}) \frac{a_{rp}}{a_{vp}} \quad (j = 0, \ldots, n; j \neq p).$$

Appendix C
Geometrically Derived Cuts

(Balas [26], Young [613])

Inequalities which must be satisfied by every solution to the integer or mixed integer program, different from the Gomory fractional cut (1) (Section 4.3), can be obtained from a geometrical analysis. These inequalities, sometimes referred to as *convexity or intersection cuts*, were originally proposed by Hoang Tui [579] (1964) in context of the minimization of a concave function over a convex polyhedron. The first specializations for the integer and mixed integer program are accredited to Balas [26] (1969), and Young [613] (1968). Since then a substantial amount of work has been done in refining and extending the earlier results. Principal contributors in this area include Egon Balas, Claude-Alain Burdet, Fred Glover, and Richard Young.

We explain the basic concepts for the integer program by utilizing an earlier example. Most of the details and proofs are left to the problems and references. The discussion assumes, unless otherwise specified, that the integer program is of the form: Maximize $\mathbf{cx}$, subject to $\mathbf{Ax} \geq \mathbf{b}, \mathbf{x} \geq \mathbf{0}$ and integer.

Example 4.5 (from Example 4.1)

The integer program and the optimal linear programming tableau are below. The linear programming optimal solution $\mathbf{x}^*$ appears in Fig. 4.10, where the dots indicate integer points $\mathbf{x} = (x_1, x_2)$.

$$
\begin{array}{rrrcrl}
\text{Maximize} & -4x_1 & - & 5x_2 & = & x_0 \\
\text{subject to} & -x_1 & - & 4x_2 & \leq & -5 \quad (x_3) \\
& -3x_1 & - & 2x_2 & \leq & -7 \quad (x_4) \\
\text{and} & x_1, & & x_2 & \geq & 0, \quad \text{integer.}
\end{array}
$$

#3	1	$(-x_4)$	$(-x_3)$
x_0	$-112/10$	$11/10$	$7/10$
x_1	$18/10$	$-4/10$	$2/10$
x_2	$8/10$	$1/10$	$-3/10$
x_3	0	0	-1
x_4	0	-1	0

Consider the smallest square with integer vertices containing the optimal linear programming solution $\mathbf{x}^* = (18/10, 8/10)$. Its coordinates are the points $(1, 0), (2, 0), (2, 1)$, and $(1, 1)$ (Fig. 4.10). Suppose we circumscribe a circle about this square so that each of these coordinates touches the circle in only one point (Fig. 4.10). The circle intersects the boundary of the feasible region in the two points $\mathbf{u}$ and $\mathbf{v}$. It is evident from the diagram that the inequality

$$\beta \mathbf{x} \geq \beta_0, \tag{1}$$

whose hyperplane $\beta bf x = \beta_0$ passes through $\mathbf{u}$ and $\mathbf{v}$, does not eliminate any integer solution and is thus a valid cutting plane.

To find the coefficients $\beta = (\beta_1, \beta_2)$ and β_0, we first obtain the points $\mathbf{u}$ and $\mathbf{v}$. As these vectors lie on the boundary of the feasible region, from Tableau #3 it follows that

Figure 4.10: Graph for Example 4.5

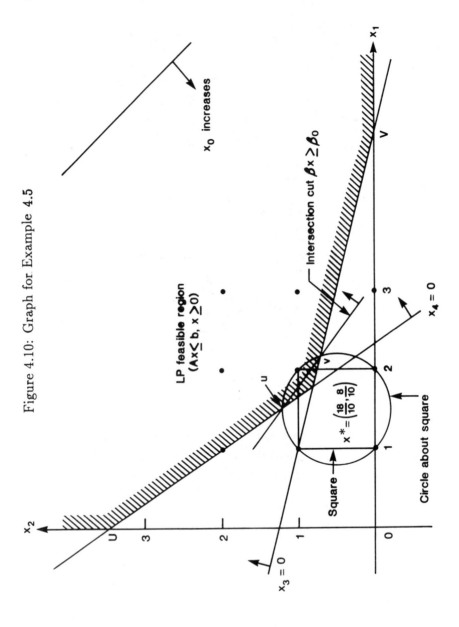

$$\mathbf{u} = (u_1, u_2) = \left(\tfrac{18}{10}, \tfrac{8}{10} - \lambda_1\left(-\tfrac{4}{10}, \tfrac{1}{10}\right)\right) = \left(\tfrac{18}{10} + \tfrac{4}{10}\lambda_1, \tfrac{8}{10} - \tfrac{1}{10}\lambda_1\right) \qquad (2)$$

and

$$\mathbf{v} = (v_1, v_2) = \left(\tfrac{18}{10}, \tfrac{8}{10} - \lambda_2\left(\tfrac{2}{10}, -\tfrac{3}{10}\right)\right) = \left(\tfrac{18}{10} - \tfrac{2}{10}\lambda_2, \tfrac{8}{10} + \tfrac{3}{10}\lambda_2\right) \qquad (3)$$

for some positive value for λ_1 and λ_2. [14] These points also lie on the circle whose center $\mathbf{c}$ is $(3/2, 1/2)$ with radius $r = \sqrt{2}/2$. Thus, points $\mathbf{x}$ satisfying

$$\|\mathbf{x} - \mathbf{c}\| = \tfrac{\sqrt{2}}{2}, \qquad (4)$$

are on the circle. [15] Or, in terms of $\mathbf{u}$ and $\mathbf{v}$ we must have

$$\|\mathbf{u} - \mathbf{c}\| = \left\| \left(u_1 - \frac{3}{2}, u_2 - \frac{1}{2} \right) \right\| = \frac{\sqrt{2}}{2}$$

and

$$\|\mathbf{v} - \mathbf{c}\| = \left\| \left(v_1 - \frac{3}{2}, v_2 - \frac{1}{2} \right) \right\| = \frac{\sqrt{2}}{2}$$

which means

$$(u_1 - \tfrac{3}{2})^2 + (u_2 - \tfrac{1}{2})^2 = \tfrac{1}{2} \qquad (5)$$

---

[14] Equations (2) and (3) represent the rays emanating from $\mathbf{x}^*$ toward the extreme points $\mathbf{u}$ and $\mathbf{v}$, respectively (Fig. 4.10). From Tableau #3, to obtain extreme point $\mathbf{u}$ ($\mathbf{v}$), λ_1 (λ_2) is set to $8/10/1/10 = 8$ ($18/10/2/10 = 9$) which maintains a primal basic feasible solution and makes x_2 (x_1) a nonbasic variable and x_4 (x_3) a basic variable. The points $\mathbf{u}$ and $\mathbf{v}$, being on the line segment joining the vectors $\mathbf{x}^*$ and $\mathbf{u}$, and $\mathbf{x}^*$ and $\mathbf{v}$ respectively, correspond to having $0 < \lambda_1 < 8$ in (2) and $0 < \lambda_2 < 9$ in (3).

[15] The Euclidean distance, written $\|\mathbf{y}\| = \sqrt{\mathbf{y} \cdot \mathbf{y}}$, from the circle's center $\mathbf{c}$ to one of the square's vertices is the radius. In our case,

$$\mathbf{r} = \|(3/2, 1/2) - (1, 0)\| = \sqrt{(3/2 - 1)^2 + (1/2 - 0)^2} = \sqrt{2}/2.$$

Any point $\mathbf{x}$ whose Euclidean distance from $\mathbf{c}$ is $\sqrt{2}/2$ is on the circle; thus we have (4).

and

$$(v_1 - \tfrac{3}{2})^2 + (v_2 - \tfrac{1}{2})^2 = \tfrac{1}{2} \tag{6}$$

Using (2) in (5) and (3) in (6) yields a quadratic equation in λ_1 and λ_2, respectively. Solving and substituting into (2) and (3) produces (approximately) the points $\mathbf{u} = (1.6, 1.2)$ and $\mathbf{v} = (2.1, 0.8)$. As $\mathbf{u}$ and $\mathbf{v}$ satisfy $\beta\mathbf{x} = \beta_1 x_1 + \beta_2 x_2 = \beta_0$ we have

$$\left. \begin{array}{c} 1.6\beta_1 + 1.2\beta_2 = \beta_0 \\ 2.1\beta_1 + 0.8\beta_2 = \beta_0 \end{array} \right\} \tag{7}$$

Solving for β_1 and β_2 in terms of β_0 gives $\beta_1 = .323\beta_0$ and $\beta_2 = .404\beta_0$. Substituting in (1) yields

$$.323x_1 + .404x_2 = 1$$

and thus the intersection cut is

$$.323x_1 + .404x_2 \geq 1, \tag{8}$$

since the point (3.2) satisfied (8) (see Fig. 4.10).

Remarks

1. Inequalities (1) can be used in place of Gomory fractional cuts. However, to prove that the resulting cutting plane algorithm converges, it is necessary to add a Gomory cut every finite number of iterations (Problem 4.31).

2. The linear programming feasible region may appear in Fig. 4.11, where the intersection cut is drawn in each situation. Notice that the boundary of the LP feasible region is extended so that the points $\mathbf{u}$ and $\mathbf{v}$, and thus the intersection cut, can

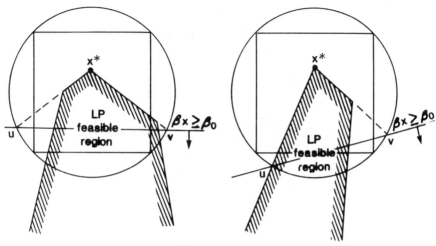

Figure 4.11

be obtained. In any case, it can be shown that the inequality $\beta\mathbf{x} \geq \beta_0$ does not eliminate any integer solution (Problem 4.33).

3. Any convex region which intersects the unit hypercube (the square in higher dimensions) about $\mathbf{x}^*$ in its vertices can be used to generate an intersection cut (Fig. 4.12). It is apparent

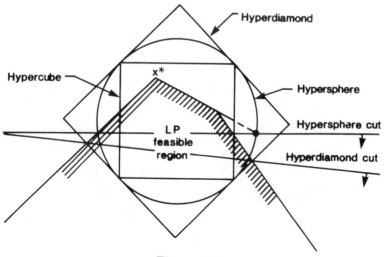

Figure 4.12

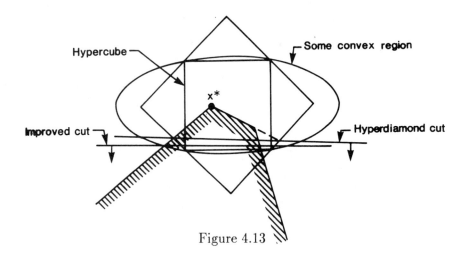

Figure 4.13

that the hyperdiamond (Fig. 4.12) usually produces a stronger
inequality than the hypersphere (circle). Moreover, it requires
the solution of linear (rather than quadratic) equations. The
computations for finding the intersection cut under the hyperdiamond, which was originally described by Balas *et al.* [00],
is in Problem 4.34. Other geometric regions may be used to
obtain stronger inequalities although they usually require substantially more effort to obtain (Fig. 4.13). (Why?)

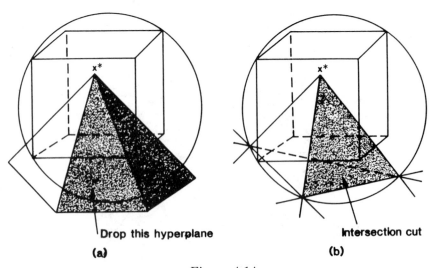

Figure 4.14

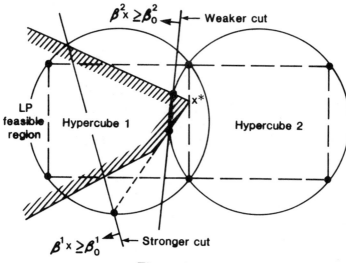

$$\beta^2 x \geq \beta_0^2 \quad \longleftarrow \text{Weaker cut}$$

Weaker cut

LP feasible region

Hypercube 1

Hypercube 2

x*

$$\beta^1 x \geq \beta_0^1 \quad \longleftarrow \text{Stronger cut}$$

Stronger cut

Figure 4.15

4. A nondegenerate extreme point or basic feasible solution $\mathbf{x}^* = (x_1, \ldots, x_n)$ for the convex polyhedron defined by $\mathbf{A}\mathbf{x} \leq \mathbf{b}, \mathbf{x} \geq \mathbf{0}$ ($\mathbf{A}$ is m by n) occurs at the intersection of exactly n hyperplanes of the form $\mathbf{a}^i x = b_i$ or $x_j = 0$, where $\mathbf{a}^i$ is the ith row of $\mathbf{A}$, $\mathbf{b} = (b_i)$ ($i = 1, \ldots, m$), and $\mathbf{x} = (x_j)$ ($j = 1, \ldots, n$). When $\mathbf{x}^*$ is degenerate it is intersected by more than n hyperplanes. (Why?) In this case, the equations corresponding to the basic variables at value 0 should be dropped prior to calculating the intersection cut (Problem 4.36). For example, Fig. 4.14(a) illustrates the case in which $\mathbf{x}^*$ is the intersection of four hyperplanes in three space. This means that at least one of the corresponding slack variables (in $\mathbf{a}^i\mathbf{x} \leq b_i$) is 0 and basic. (Why?) If its associated hyperplane, as indicated in Fig. 4.14(a), is dropped then the intersection cut, as in Fig. 4.14(b), is obtained.

5. If $\mathbf{x}^*$ has k integer components there are 2^k possible unit hypercubes. (Why?) The strongest cut will be obtained from the one with the largest intersection with the feasible region. For example, in Fig. 4.15, $\mathbf{x}^* = (x_1, x_2)$ and x_1 is integer. Then there are two possible hypercubes, but as $\beta^1\mathbf{x} \geq \beta_0^1$ yields a stronger cut than $\beta^2\mathbf{x} \geq \beta_0^2$ hypercube 1 should be used.

Chapter 5

Dual Fractional Mixed Integer Programming

5.1 The Basic Approach

The cutting plane algorithm for the mixed integer program, developed by Ralph Gomory in 1960 and presented in this chapter, is a direct extension of the cutting plane algorithm discussed in Chapter 4. Again the intent is to whittle the feasible region down to one whose optimal vertex has integer values for the integer constrained variables. As before, the cutting plane technique utilizes the dual simplex method and allows fractional numbers in computation and is thus classified as "dual fractional." For ease of reference, we rewrite the basic approach. The first tableau is *not* required to be all integer.

Step 1 Solve the mixed integer program as a linear one, If it is infeasible, so is the mixed integer problem–terminate. If the optimal solution is integer in the integer constrained variables, the mixed integer program is solved–terminate. Otherwise, go to Step 2.

Step 2 From a row corresponding to an integer constrained variable which does not have an integral value, derive a new inequality constraint which "cuts off" the current optimal point but does not eliminate any mixed integer solution. Add the new inequality to the bottom of the simplex tableau which then exhibits primal infeasibility. (The slack variable associated with the new row will be negative.) Go to Step 3.

Step 3 Reoptimize using the dual (lexicographic) simplex method. If the new linear program is infeasible (the dual will be unbounded),

	1	$-x_1$	$\cdots$	$-x_I$	$-x_{I+1}$	$\cdots$	$-x_n$	
$x_0 =$	a_{00}	a_{01}	$\cdots$	a_{0I}	$a_{0,I+1}$	$\cdots$	a_{0n}	
$x_1 =$	0	-1						≥ 0
$\vdots$	$\vdots$	$\vdots$	$\ddots$				$\vdots$	
$x_I =$	0			-1				≥ 0
$x_{I+1} =$	0				-1			≥ 0
$\vdots$	$\vdots$	$\vdots$				$\ddots$	$\vdots$	
$x_n =$	0						-1	≥ 0
$x_{n+1} =$	$a_{n+1,0}$	$a_{n+1,1}$	$\cdots$	$a_{n+1,I}$	$\cdots$		$a_{n+1,n}$	≥ 0
$\vdots$	$\vdots$	$\vdots$		$\vdots$	$\vdots$		$\vdots$	
$x_{n+m} =$	$a_{n+m,0}$	$a_{n+m,1}$	$\cdots$	$a_{n+m,I}$	$\cdots$		$a_{n+m,n}$	≥ 0

Table 5.1: Simplex Tableau

the mixed integer problem has no solution–terminate. If the new optimal solution is integer in the integer constrained variables, the mixed integer program is solved–terminate. Otherwise, go to Step 2.

To simplify correspondence with the preceding chapter we adopt the notation used previously and display the mixed problem in the simplex tableau as in Table 5.1, where $x_1, \ldots, x_I$ $(0 < I < n)$ are the integer constrained variables. Also, to ensure convergence we require the value of x_0 to be integer. (If x_0 is not integer constrained then the process need not always converge; see White [598]. A part of his results appears in Problem 5.11.)

5.2 The Form of the Cut

At the optimal simplex tableau $a_{0j} \geq 0$ $(j = 1, \ldots, n)$, $a_{i0} \geq 0$ $(i = 1, \ldots, n + m)$ and $\boldsymbol{\alpha}_j \overset{\mathcal{L}}{\succ} \mathbf{0}$ $(j = 1, \ldots, n)$. Further, if a_{i0} $(i = 0, 1, \ldots, I)$ is integer, the mixed integer program is solved. Otherwise, there is a source row v $(0 \leq v \leq I)$ with $x_v = a_{v0} + \sum_{j=1}^{n} a_{vj}(-x_{J(j)})$ and a_{v0} fractional. Then the $(k$th$)$ Gomory cut has the form

$$x_{n+m+k} = -f_{v0} + \sum_{j=1}^{n}(-g_{vj})(-x_{J(j)}) \geq 0, \qquad (1)$$

where

$$
g_{vj} =
\begin{cases}
a_{vj} & \text{if } a_{vj} \geq 0 \text{ and } x_{J(j)} \text{ is a continuous variable} \\[2ex]
\dfrac{f_{v0}}{f_{v0}-1} a_{vj} & \text{if } a_{vj} < 0 \text{ and } x_{J(j)} \text{ is a continuous variable,} \\[2ex]
f_{vj} & \text{if } f_{vj} \leq f_{v0} \text{ and } x_{J(j)} \text{ is an integer variable,} \\[2ex]
\dfrac{f_{v0}}{1-f_{v0}}(1-f_{vj}) & \text{if } f_{vj} > f_{v0} \text{ and } x_{J(j)} \text{ is an integer variable,}
\end{cases}
$$

and $f_{vj} = a_{vj} - [a_{vj}]$ $(j = 0, 1, \ldots, n)$. Observe that $0 < f_{v0} < 1$ and thus when the inequality is adjoined to the bottom of the tableau, primal infeasibility is introduced. (The Gomory slack variable, $x_{m+n+k} = -f_{v0} < 0$.) Also, it can be shown that (1) is implied by the source row and integrality requirements, and that it has properties which guarantee a finite algorithm. However, before doing this we present an example.

5.3   An Illustration

We resolve the first problem used to illustrate the fractional integer programming algorithm (Chapter 4, Section 4.4). Of course, some of the variables are now allowed to take on noninteger values. A circled entry in a tableau denotes the pivot element and an arrow ($\rightarrow$) indicates the source row. The latter is the constraint with the largest fractional part in the constant column. In case of ties the largest number is selected.

Example 5.1 (Balinski [50])

$$\text{Maximize} \quad - 4x_1 \quad - \quad 5x_2 \quad = x_0$$

$$\text{subject to} \quad - \quad x_1 \quad - \quad 4x_2 \quad \leq -5 \quad (x_3)$$
$$- \quad 3x_1 \quad - \quad 2x_2 \quad \leq -7 \quad (x_4)$$
$$x_1, \quad x_2 \quad \geq 0$$
$$\text{and} \qquad x_0, \quad x_1 \quad \text{integer.}$$

The linear program can be written in standard form by adding non-negative slack variables x_3 and x_4 to the two inequality constraints. This problem has been solved in Chapter 4 (Section 4.4). The (first) optimal simplex tableau appears in Tableau #3. It takes three cuts to solve the problem. The computations which follows appear in Tableaux #3 through #6.

#3	1	$(-x_4)$	$(-x_3)$	Cut derivation:
x_0	$-112/10$	$11/10$	$7/10$	$f_{10} = 18/10 - [18/10]$
$\rightarrow x_1$	$18/10$	$-4/10$	$2/10$	$= 8/10$
x_2	$8/10$	$1/10$	$-3/10$	$g_{11} = \frac{8/10}{(8/10-1)}(-4/10)$
x_3	0	0	-1	$= 16/10$
x_4	0	-1	0	$g_{12} = 2/10$
x_5	$-8/10$	$\boxed{-16/10}$	$-2/10$	$x_5 = -\frac{8}{10} - \frac{16}{10}(-x_4)$
				$-\frac{2}{10}(-x_3)$

#4	1	$(-x_5)$	$(-x_3)$	Cut derivation:
$\rightarrow x_0$	$-188/16$	$11/16$	$9/16$	$f_{00} = -188/16 - [-188/16]$
x_1	2	$-4/16$	$4/16$	$= 4/16$
x_2	$12/16$	$1/16$	$-5/16$	$g_{01} = 11/16$
x_3	0	0	-1	$g_{02} = 9/16$
x_4	$8/16$	$-10/16$	$2/16$	$x_6 = -\frac{4}{16} - \frac{11}{16}(-x_5)$
x_5	0	-1	0	$-\frac{9}{16}(-x_3)$
x_6	$-4/16$	$\boxed{-11/16}$	$-9/16$	

#5	1	$(-x_6)$	$(-x_3)$
x_0	-12	1	0
$\to x_1$	$23/11$	$-4/11$	$5/11$
x_2	$8/11$	$1/11$	$-4/11$
x_3	0	0	-1
x_4	$8/11$	$-10/11$	$7/11$
x_5	$4/11$	$-16/11$	$9/11$
x_6	0	-1	0
x_7	$-1/11$	$-4/110$	$\boxed{-5/11}$

Cut derivation:

$$f_{10} = 23/11 - [23/11]$$
$$= 1/11$$
$$g_{11} = \frac{1/11}{(1/11-1)}(-4/11)$$
$$= 4/110$$
$$g_{12} = 5/11$$
$$x_7 = -\tfrac{1}{11} - \tfrac{4}{110}(-x_6)$$
$$-\tfrac{5}{11}(-x_3)$$

#6	1	$(-x_6)$	$(-x_7)$
x_0	-12	1	0
x_1	2	$-2/5$	1
x_2	$4/5$	$6/50$	$-4/5$
x_3	$1/5$	$4/50$	$-11/5$
x_4	$3/5$	$-48/50$	$7/5$
x_5	$1/5$	$-76/50$	$9/5$
x_6	0	-1	0
x_7	0	0	-1

Optimal tableau

Tableau #6 displays the optimal mixed integer solution: $x_0 = -12$, $x_1 = 2$, and $x_2 = 4/5$. Expressing the added inequalities in terms of the original nonbasic variables x_1 and x_2 gives

$$
\begin{aligned}
x_5 &= -13 + 5x_1 + 4x_2 \geq 0, \\
x_6 &= -12 + 4x_1 + 5x_2 \geq 0, \\
x_7 &= -14/5 + (3/5)x_1 + 2x_2 \geq 0.
\end{aligned}
$$

Observe that the Gomory slack variable is not necessarily an *integer*

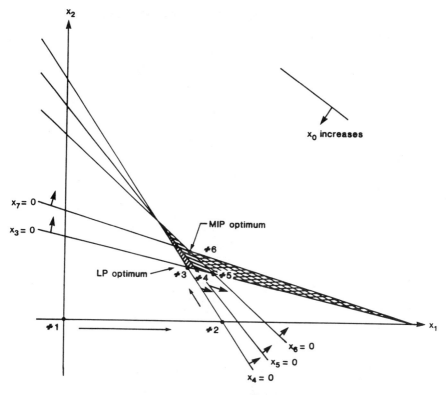

Figure 5.1: Graph for Example 5.1

linear combination of the original nonbasic variables. Hence, it does *not* have to be integer in every mixed integer solution. Therefore, as these constraints (rows) are not eligible to generate inequalities, it is reasonable to adopt the strategy of dropping the Gomory slack variable and its defining row immediately after the slack variable corresponds to the pivot column. Figure 5.1 relates the computations to (x_1, x_2) space.

5.4   The Derivation of the Cut

Let the source row be (row subscripts are omitted)

$$x = a_0 + \sum_{j=1}^{n} a_j(-x_{J(j)}),$$

and let $f_0 = a_0 - [a_0]$. Note that, as $a_0 > 0$ and noninteger, $f_0 > 0$. Also, since x is an integer variable we have $x \equiv 0 \pmod 1$ (see Chapter 4, Section 4.5); hence

$$0 \equiv a_0 + \sum_{j=1}^{n} a_j(-x_{J(j)}) \pmod 1. \tag{2}$$

Now, $a_0 > 0$, and we may add $0 \equiv -[a_0] \pmod 1$ to (2) to reduce a_0 to its fractional part f_0, as indicated by

$$0 \equiv f_0 + \sum_{j=1}^{n} a_j(-x_{J(j)}) \pmod 1. \tag{3}$$

Relation (3) may be written as

$$\sum_{j=1}^{n} a_j x_{J(j)} \equiv f_0 \pmod 1. \tag{4}$$

Now, *suppose* the left-hand side of (4) is positive. Then for it to differ from f_0 by an integer amount it must be equal to f_0, or $1 + f_0$, or $2 + f_0$, etc.; hence, we have

$$\sum_{j=1}^{n} a_j x_{J(j)} \geq f_0.$$

Then for any feasible solution $x_{J(j)} \geq 0$ $(j = 1, \ldots, n)$,

$$\sum_{j \in P} a_j x_{J(j)} \geq \sum_{j=1}^{n} a_j x_{J(j)} \geq f_0, \tag{5}$$

where $P = \{j | a_j \geq 0, \ j = 1, \ldots, n\}$. If, on the other hand, the left-hand side of (4) is negative, it has to equal $-1 + f_0, -2 + f_0$, etc.; or, in this case,

$$\sum_{j=1}^{n} a_j x_{J(j)} \leq -1 + f_0.$$

Then for any feasible solution $x_{J(j)} \geq 0$ $(j = 1, \ldots, n)$,

$$\sum_{j \in N} a_j x_{J(j)} \leq \sum_{j=1}^{n} a_j x_{J(j)} \leq -1 + f_0, \tag{6}$$

where $N = \{j | a_j < 0, j = 1, \ldots, n\}$. Multiplying the outside terms of (6) by the negative number $f_0/(-1 + f_0)$ gives

$$\sum_{j \in N} \left(\frac{f_0}{-1 + f_0} \right) a_j x_{J(j)} \geq f_0. \tag{7}$$

Now, the left hand side of (4) is either positive or negative. It cannot be zero since f_0 is a proper fraction. Hence, at least one of the inequalities (5) or (7) is true. But in any feasible solution each of these has a nonnegative left-hand side. Therefore, we have

$$\sum_{j \in P} a_j x_{J(j)} + \sum_{j \in N} \left(\frac{f_0}{-1 + f_0} \right) a_j x_{J(j)} \geq f_0. \tag{8}$$

Or, after introducing a nonnegative Gomory slack variable x', (8) is

$$x' = -f_0 + \sum_{j \in P} a_j x_{J(j)} + \sum_{j \in N} \left(\frac{f_0}{-1 + f_0} \right) a_j x_{J(j)} \geq 0. \tag{9}$$

The last inequality eliminates the current optimum, for at $x_{J(j)} = 0$ $(j = 1, \ldots, n)$, $x' = -f_0 < 0$. Also, by its derivation, it is satisfied by every mixed integer solution. Hence, (9) is a valid cut. It can, however, be strengthened if some of the nonbasic variables are integer constrained. Consider that situation. Then our objective is to change the coefficients of these variables by integer amounts in (3) so as to produce the smallest possible coefficients in (9). (This will yield a hyperplane $x' = 0$ which intersects each $x_{J(j)}$ axis farther from the origin and hence is a stronger cut.) Consider any nonbasic integer constrained variable x_t with $a_t \neq 0$. Now, adding or subtracting integer multiples of x_t to or from (3) will not invalidate that congruence relation. Thus we have the option of initially placing t in the set P or in the set N and then deriving inequality (9). If t is placed (or is) in P, its smallest coefficient possible in (9) is f_t, which is obtained by changing a_t to f_t in (3). On the other hand, if we put (or leave) t in N, its smallest coefficient possible in (9) is

$$\left(\frac{f_0}{-1+f_0}\right)(f_t-1),$$

and this is obtained by changing a_t to the negative number $f_t - 1$ in (3). (The last assertion follows, since for t to be in the set N, a_t must be equal to $-1 + f_t$, or $-2 + f_t$, or $-3 + f_t$, etc. As $f_0/(-1 + f_0)$ is a fixed negative number the value of $(f_0/(-1+f_0))a_t$ is smallest when a_t is $-1 + f_t$.) Thus, the best strategy is to have t in the set P when

$$f_t \le \frac{f_0}{-1+f_0}(f_t-1), \tag{10}$$

and to have t in the set N otherwise. But for (10) to be true we have $f_t(-1+f_0) \ge f_0(f_t-1)$ or $f_t(1-f_0) \le f_0(1-f_t)$, which means $f_t \le f_0$.

Therefore, for each nonbasic integer variable x_t in (3), we initially change its coefficient a_t to f_t when $f_t \le f_0$ and to $f_t - 1$ when $f_t > f_0$. The result is the strengthened mixed integer cut:

$$x' = -f_0 + \sum_{j=1}^{n} g_j x_{J(j)} \ge 0, \tag{11}$$

where

$$g_j = \begin{cases} a_j & \text{if } a_j \ge 0 \text{ and } x_{J(j)} \text{ is a continuous variable.} \\[2ex] \dfrac{f_0}{f_0-1}a_j & \text{if } a_j < 0 \text{ and } x_{J(j)} \text{ is a continuous variable.} \\[2ex] f_j & \text{if } f_j \le f_0 \text{ and } x_{J(j)} \text{ is an integer variable.} \\[2ex] \dfrac{f_0}{1-f_0}(1-f_j) & \text{if } f_j > f_0 \text{ and } x_{J(j)} \text{ is an integer variable.} \end{cases}$$

To clarify the discussion we derive the form of relation (3) for a given source row. All congruence relations are understood to be modulo 1.

Example 5.2 Suppose the source row is

$$x = 23/10 + (14/10)(-x_1) + (12/10)(-x_2) - (11/10)(-x_3)$$
$$+ (18/10)(-x_4),$$

where x, x_1, and x_2 are integer variables. Then, since $x \equiv 0$, we have

$$0 \equiv 23/10 + (14/10)(-x_1) + (12/10)(-x_2)$$
$$- (11/10)(-x_3) + (18/10)(-x_4). \tag{2}'$$

Adding $0 \equiv -2$ to (2)' reduces $23/10$ to its fractional part $3/10$. Thus,

$$0 \equiv 3/10 + (14/10)(-x_1) + (12/10)(-x_2) - (11/10)(-x_3)$$
$$+ (18/10)(-x_4). \tag{3}'$$

As $f_1 = 14/10 - [14/10] = 4/10 > 3/10$ and x_1 is an integer variable, we add $0 \equiv -2(-x_1)$ to (3)'. Also, since $f_2 = 2/10 < 3/10$ and x_2 is an integer variable, we add $0 \equiv -1(-x_2)$ to (3)'. The two additions give

$$0 \equiv 3/10 - (6/10)(-x_1) + (2/10)(-x_2) - (11/10)(-x_3)$$
$$+ (18/10)(-x_4).$$

Hence, $P = \{2,4\}$ and $N = \{1,3\}$. We then proceed as explained to derive inequality (9), or equivalently (11).

5.5 Applying the Mixed Cut to the Integer Program (Salkin [504])

Suppose all the nonbasic variables in a current optimal simplex tableau are integer constrained. Then we find that the mixed integer Gomory cut for that tableau reduces to

$$x^M = -f_0 + \sum_{j \in J^1} f_j x_{J(j)} + \sum_{j \in J^2} \left(\frac{f_0}{1 - f_0} \right) (1 - f_j) x_{J(j)} \geq 0, \tag{12}$$

where $J^1 = \{j | f_j \leq f_0\}$ and $J^2 = \{j | f_j > f_0\}$. Since the $x_{J(j)}$ ($j = 1, \ldots, n$) are integer constrained we could use the classical integer cut

$$x^I = -f_0 + \sum_{j=1}^{n} f_j x_{J(j)} \geq 0 \tag{13}$$

in place of (12). (The derivation of (13) is in Chapter 4, Section 4.5.) Interestingly, however, inequality (12) is usually stronger than (13). We prove this result.

Theorem 5.1 *Suppose that the $x_{J(j)}$ $(j = 1, \ldots, n)$ are integer variables and J^2 in (12) is not empty. Then $x^M \geq 0$ is a stronger inequality than $x^I \geq 0$.*

Proof: Since the union of the sets J^1 and J^2 contains all the indices $j = 1, \ldots, n$, we have

$$x^I = -f_0 + \sum_{j \in J^1} f_j x_{J(j)} + \sum_{j \in J^2} f_j x_{J(j)} \geq 0.$$

Also

$$x^M = -f_0 + \sum_{j \in J^1} f_j x_{J(j)} + \sum_{j \in J^2} \left(\frac{f_0}{1 - f_0} \right) (1 - f_j) x_{J(j)} \geq 0.$$

Now, if $j \in J^2, f_j > f_0$; or

$$f_0 \left(\frac{1 - f_j}{1 - f_0} \right) < f_0 < f_j \quad \text{since} \quad \frac{1 - f_j}{1 - f_0} < 1.$$

Thus, x^M is a stronger inequality than x^I because it has smaller coefficients of $x_{J(j)}$ when j is in J^2. That is, for $j \in J^2$, the hyperplane $x^M = 0$ intersects the coordinate axis $x_{J(j)}$ at a distance farther from the origin than the corresponding intersection with the hyperplane $x^I = 0$. For j in J^1, the coefficients are the same and hence the hyperplanes intersect each of the $x_{J(j)}$ axis at the same point. ∎

    The assertion seems to suggest that inequality (12) ought to replace (13) in the fractional integer programming algorithm (Chapter 4). Remember, however, that the Gomory slack variable in the mixed integer cut x^M is not necessarily integer in every integer solution. Thus, after a pivot on an element in its row, a continuous variable, namely x^M, appears as a nonbasic variable. Or, the integer problem becomes a mixed integer program and mixed integer cuts must then be used.

These may, in fact, be relatively weak inequalities. But note that the *first* inequality is extracted from nonbasic variables which are all integer constrained. Therefore, if the first mixed integer inequality (12) is substantially stronger than the integer inequality (13), the substitution may yield a computationally more efficient algorithm. This is very apparent in Example 5.3.

Example 5.3 Consider the second integer program solved in Chapter 4, Section 4.4. The problem is

$$
\begin{array}{llrcrcrl}
\text{Maximize} & x_1 & + & x_2 & = & x_0 & & \\
\text{subject to} & -4x_1 & + & x_2 & \leq & -1 & & (x_3) \\
& 4x_1 & + & x_2 & \leq & 3 & & (x_4) \\
\text{and} & x_1, & & x_2 & \geq & 0, & \text{integer.} &
\end{array}
$$

With nonnegative slack variables x_3 and x_4, the (first) optimal simplex tableau appears below.

#4	1	$(-x_3)$	$(-x_4)$	
x_0	12/8	3/8	5/8	
$\to x_1$	4/8	$-1/8$	1/8	$f_{10} = 4/8$
x_2	1	4/8	4/8	$f_{11} = 7/8$
x_3	0	-1	0	$f_{12} = 1/8$
x_4	0	0	-1	
x^M	$-4/8$	$\boxed{-1/8}$	$-1/8$	

Suppose x_1 is the source row. Then, $f_0/(f_0 - 1) = -1$ and the mixed integer cut is

$$
x^M = -4/8 + (1/8)x_3 + (1/8)x_4 \geq 0.
$$

After appending it to Tableau #4 and pivoting on $-1/8$ we obtain Tableau #5. The x_2 row in the latter tableau indicates that the dual problem is unbounded or the primal linear program has no feasible solution. Hence the integer program has no solution. (The linear programming feasible region is in Figure 5.2.)

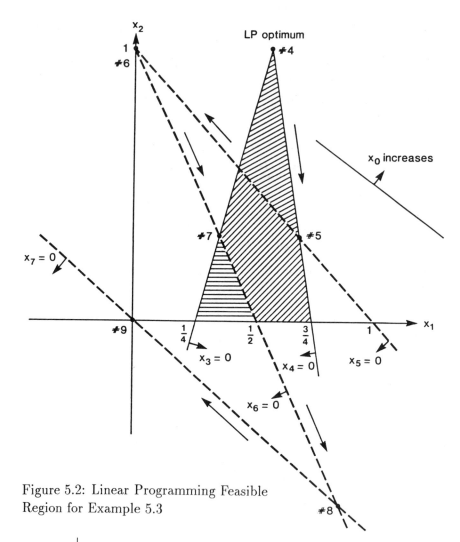

Figure 5.2: Linear Programming Feasible
Region for Example 5.3

#5	1	$(-x^M)$	$(-x_4)$
x_0	0	3	$1/4$
x_1	1	-1	$1/4$
x_2	-1	4	0
x_3	4	-8	1
x_4	0	0	-1
x^M	0	-1	0

Suppose we express x^M in terms of x_1 and x_2. The result is $x^M = -1/4 - (1/4)x_2 \geq 0$, or $x_2 \leq -1$. Clearly, as illustrated in Figure 5.2, the last inequality cuts off the entire linear programming feasible region. (This requires three integer cuts. For example, if x_1 is always the source row, these are $x_1^I = -1 + 3x_1 - x_2 \geq 0$, $x_2^I = -1 + 2x_1 - x_2 \geq 0$, and $x_3^I = -1 + x_1 - x_2 \geq 0$, cf. Section 4.4.)

5.6 Finiteness

To show that the algorithm produces the optimal mixed integer vertex after a finite number of tableaux it is necessary, as in the integer programming case, to assume a lower bound for x_0. This variable must also be integer constrained. It is assumed that the constraints are ordered so that the integer variables $(x_0, x_1, \ldots, x_I)$ appear first in the tableau, and also that the cut is generated from the first eligible source row. In addition, we need the following lemma applicable at optimal tableaux. A prime denotes an element subsequent to a pivot.

Lemma 5.1 *Let row v generate the pivot row (cut) and let p be the pivot column index. If $a_{vp} > 0$, then $a'_{v0} \leq [a_{v0}]$.*

Proof: Since the pivot is g_{vp} (see Section 5.4, Eq. (11) or Section 2.7), we have

$$a'_{v0} = a_{v0} - \frac{f_{v0} a_{vp}}{g_{vp}}. \tag{14}$$

Now, if x_p is a continuous variable, then $g_{vp} = a_{vp}$, as $a_{vp} > 0$. Thus (14) is

$$a'_{v0} = a_{v0} - f_{v0} = [a_{v0}].$$

If x_p is an integer variable and $f_{vp} \leq f_{v0}$, then $g_{vp} = f_{vp}$. In this case since $a_{vp} \geq f_{vp}$, (14) becomes

$$a'_{v0} = a_{v0} - \frac{f_{v0}}{f_{vp}} a_{vp} \leq a_{v0} - f_{v0} = [a_{v0}].$$

Finally, if x_p is an integer variable and $f_{vp} > f_{v0}$, then

$$g_{vp} = \frac{f_{v0}}{1 - f_{v0}}(1 - f_{vp}),$$

and (14) is

$$a'_{v0} = a_{v0} - \frac{f_{v0}a_{vp}(1 - f_{v0})}{f_{v0}(1 - f_{vp})}$$

$$= a_{v0} - a_{vp}\left(\frac{1 - f_{v0}}{1 - f_{vp}}\right).$$

Now, $f_{vp} > f_{v0}$ means $(1 - f_{v0})/(1 - f_{vp}) > 1$. Also, $a_{vp} \geq f_{vp} > f_{v0}$, so that $a'_{v0} < a_{v0} - f_{v0} = [a_{v0}]$, and the assertion is proved. ∎

The finiteness proof is now very similar to the one presented in Chapter 4 (Section 4.9) for the integer program. That is, suppose the algorithm is not finite, then there exists an infinite sequence of tableaux whose 0 columns are lexicographically decreasing. Write this sequence as

$$\alpha_0^k \overset{\mathcal{L}}{\succ} \alpha_0^{k+1} \overset{\mathcal{L}}{\succ} \alpha_0^{k+2} \quad \ldots \text{etc.}$$

Since α_0^k decreases lexicographically, its first component, a_{00}, is nonincreasing, so it must eventually remain fixed at some integer value. For suppose it did not; then row 0 generates the pivot row and by the lemma, $a'_{00} \leq [a_{00}]$. (Since the pivot element cannot be 0, a_{0p} must be positive and so the lemma can be used.) Or, each time row 0 is used as the source row, a_{00} is decreased at least to the next integer. Equivalently, it can be noninteger for only a finite number of tableaux; otherwise, it would eventually fall below its lower bound. Let k' be the iteration at which a_{00} is fixed for all iterations $\geq k'$.

Now that the first component in the 0 column is fixed we know, since this column is lexicographically decreasing, that a_{10} is nonincreasing for all $k \geq k'+1$. Further, a_{10} must remain nonnegative, because otherwise row 1 would become the pivot row; then if column p is the pivot column index we would have

$$a'_{00} = a_{00} - \frac{a_{0p}a_{10}}{a_{1p}}.$$

Now, $a_{10} < 0$ and $a_{1p} < 0$; hence $a_{10}/a_{1p} > 0$. Then a_{0p} must be 0,

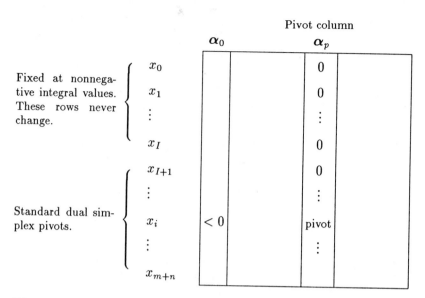

Figure 5.3: The Algorithm Reduces to the Dual Simplex Method

since, by assumption, $a'_{00} = a_{00}$. This contradicts the lexicographic positivity of $\boldsymbol{\alpha}_p$. (If $a_{1j} \geq 0$ for $j = 1, \ldots, n$, the current linear program has no feasible solution.)

To continue the proof we can show that there is an iteration $k'' \geq k' + 1$ such that a_{10} remains fixed at a nonnegative integer value for all iterations greater than or equal to k''. To do this, we repeat the argument used to show that a_{00} remains fixed. The only point worth mentioning is that $a_{1p} > 0$, since $\boldsymbol{\alpha}_p \overset{\mathcal{L}}{\succ} 0, a_{1p} \neq 0$, and $a_{0p} = 0$ for all $k \geq k' + 1$, and thus the lemma can be used.

We then successively show that the second, third, fourth, etc. components of the 0 column become integral. This is continued until the first $I + 1$ rows in the 0 column become integer. At this time either the remaining components in the 0 column are nonnegative—in which case the computations end—or there is at least one row i ($I + 1 \leq i \leq m+n$) with $a_{i0} < 0$. Consider the latter situation. Unless the tableau now exhibits infeasibility,[1] there is a pivot column $\boldsymbol{\alpha}_p$ ($p \geq 1$) with $a_{ip} < 0$. But a pivot operation will lexicographically decrease $\boldsymbol{\alpha}_o$, whose first $I + 1$ elements are fixed at their integral lower bounds. Thus, to maintain the first $I + 1$ values in $\boldsymbol{\alpha}_o$, the first $I + 1$ elements in the pivot column must always be 0 (Fig. 5.3). So the first $I + 1$ rows remain the

---

[1] An example of such a program appears in Problem 5.12

same and are never eligible pivot (or source) rows. Or the algorithm now becomes the standard dual lexicographic simplex method which is known to converge (cf. Chapter 2).

Problems

Problem 5.1 Using Gomory's mixed integer algorithm, solve

$$
\begin{array}{rrrcl}
\text{minimize} & x_1 & + & x_2 & = & x_0 \\
\text{subject to} & 4x_1 & + & 3x_2 & \geq & 6, \\
& & & 2x_2 & \geq & 1, \\
& x_1, & & x_2 & \geq & 0, \\
\text{and} & x_0, & & x_2 & \text{integer.}
\end{array}
$$

Relate the tableaux to the computations in (x_1, x_2) space.

Problem 5.2 Using Gomory's mixed integer algorithm, solve

$$
\begin{array}{rrrrrrrrrcl}
\text{maximize} & 2x_1 & + & 3x_2 & + & 4x_3 & + & 7x_4 & = & x_0 \\
\text{subject to} & 4x_1 & + & 6x_2 & - & 2x_3 & + & 8x_4 & = & 16, \\
& -x_1 & + & 2x_2 & - & 6x_3 & + & 7x_4 & = & 3, \\
& x_1, & & x_2, & & x_3, & & x_4 & \geq & 0, \\
\text{and} & x_0, & & x_1 & \text{integer.}
\end{array}
$$

(Except for all the variables being integer constrained, this is Problem 4.8, Chapter 4.)

Problem 5.3 Consider the "weak form" of the mixed integer cut (derived in Section 5.4)

$$
x' = -f_0 + \sum_{j \in P} a_j x_{J(j)} + \sum_{j \in N} \left(\frac{f_0}{-1 + f_0} \right) a_j x_{J(j)} \geq 0, \tag{1}
$$

where $P = \{j \mid a_j > 0, j = 1, \ldots, n\}$ and $N = \{j \mid a_j < 0, j = 1, \ldots, n\}$. The inequality (1) is derived from a source row

$$
x = a_0 + \sum_{j=1}^{n} a_j (-x_{J(j)}),
$$

where x is an integer constrained variable. Show that inequality (1) with the Basic Approach (Section 5.1) produces a convergent algorithm.

Problem 5.4   After finding a simplex tableau which displays the mixed integer optimal solution, is it possible to continue to add cuts so as to eventually obtain a mixed integer optimal tableau whose rows corresponding to the integer variables are all integer? Explain.

Problem 5.5   Solve the integer problem

$$
\begin{array}{lrcrcl}
\text{minimize} & x_1 & + & x_2 & = & x_0 \\
\text{subject to} & 4x_1 & + & 10x_2 & \geq & 12, \\
 & 10x_1 & + & 4x_2 & \geq & 12, \\
\text{and} & x_1, & & x_2 & \geq & 0, \text{ integer,}
\end{array}
$$

using the mixed integer cut. (This program appears in Problem 4.5, Chapter 4.) Relate the tableaux to the computations in (x_1, x_2) space.

Problem 5.6   Under what circumstances will the Gomory slack variable in a mixed integer cut (9) be an integer constrained variable? (*Hint:* Variables which can be expressed as all-integer linear combinations of integer constrained variables are integer constrained themselves.)

Problem 5.7   Suppose we adopt the strategy of dropping Gomory slack variables and their defining rows subsequent to their use as pivot rows. Can this procedure yield a less efficient algorithm than the one that does not omit added inequalities? Explain.

Problem 5.8   Can a source row be any integer linear combination of rows in the simplex tableau? Explain.

Problem 5.9   Show that it requires at least three integer cuts to eliminate the entire linear programming feasible region in Example 5.3.

Problem 5.10   Suppose row v generates a mixed integer cut which is used as the pivot row. Let p be the pivot column index and let a prime denote an element subsequent to the pivot. If $a_{vp} < 0$, show that $a'_{v0} \geq \lfloor a_{v0} \rfloor + 1$. Can this result be used to simplify the convergence proof?

Problem 5.11 (Failure of algorithm to converge when x_0 is not integer constrained; White [598]) Consider the mixed integer program

$$
\text{maximize} \quad x_1 \ + \ x_2 \ + \ x_3 \ = \ x_0
$$

$$
\begin{array}{llllllll}
\text{subject to} & 3x_1 & + & 2x_2 & + & 2x_3 & \leq & 12, \\
& 4x_1 & + & 6x_2 & + & 3x_3 & \leq & 24, \\
& x_1 & + & x_2 & + & 3x_3 & \leq & 9, \\
& x_1, & & x_2, & & x_3 & \geq & 0, \\
\text{and} & x_1, & & x_2 & & \text{integer}. &
\end{array}
$$

with nonnegative slack variables $x_4, x_5,$ and x_6. The (first) optimal tableau is

	1	$(-x_4)$	$(-x_5)$	$(-x_6)$
x_0	123/23	4/23	2/23	3/23
x_1	30/23	15/23	−4/23	−6/23
x_2	51/23	−9/23	7/23	−1/23
x_3	42/23	−2/23	−1/23	10/23
x_4	0	−1	0	0
x_5	0	0	−1	0
x_6	0	0	0	−1

Show that if the first eligible row is selected as the source row, Gomory's mixed integer algorithm will fail to converge for this program. (*Hint:* Add mixed integer cuts to obtain the tableau

#1	1	$(-s_1)$	$(-s_2)$	$(-x_6)$
x_0	109/21	4/21	4/21	2/21
x_1	9/7	9/7	−5/7	−5/14
x_2	2	−1	1	0
x_3	40/21	−2/21	−2/21	19/42
x_4	7/21	−35/21	7/21	7/42
x_5	8/7	8/7	−20/7	8/14
x_6	0	0	0	−1
s_1	0	−1	0	0
s_2	0	0	−1	0

where s_1 and s_2 are slack variables corresponding to the first and second added inequalities. Then show by induction on t, where s_t is the tth Gomory slack variable, that the above tableau is the first array in an infinite series of tableaux of the form below. Note that the $t = 1$ case is Tableau #1.)

#t	1	$(-s_{2t-1})$	$(-s_{2t})$	$(-x_6)$
x_0	$\frac{60t+49}{12t+9}$	$\frac{4}{12t+9}$	$\frac{4}{12t+9}$	$\frac{2}{12t+9}$
x_1	$\frac{4t+5}{4t+3}$	$\frac{4t+5}{4t+3}$	$-\frac{4t+1}{4t+3}$	$-\frac{4t+1}{8t+6}$
x_2	2	-1	1	0
x_3	$\frac{24t+16}{12t+9}$	$\frac{-2}{12t+9}$	$\frac{-2}{12t+9}$	$\frac{12t+7}{24t+18}$
x_4	$\frac{12t-5}{12t+9}$	$\frac{12t+23}{12t+9}$	$\frac{12t-5}{12t+9}$	$\frac{12t-5}{24t+18}$
x_5	$\frac{8t}{4t+3}$	$\frac{8t}{4t+3}$	$-\frac{8t+12}{4t+3}$	$\frac{8t}{8t+6}$
x_6	0	0	0	-1
s_1	$\frac{2t-2}{4t+3}$	$-\frac{2t+5}{4t+3}$	$\frac{2t-2}{4t+3}$	$\frac{t-1}{4t+3}$
s_2	$\frac{2t-2}{4t+3}$	$\frac{2t-2}{4t+3}$	$-\frac{2t+5}{4t+3}$	$\frac{t-1}{4t+3}$
s_3	$\frac{2t-4}{4t+3}$	$-\frac{2t+7}{4t+3}$	$\frac{2t-4}{4t+3}$	$\frac{1}{4t+3}$
$\vdots$	$\vdots$	$\vdots$	$\vdots$	$\vdots$
s_{2t-2}	$\frac{2}{4t+3}$	$\frac{2}{4t+3}$	$-\frac{4t+1}{4t+3}$	$\frac{1}{4t+3}$

Problem 5.12   By solving the mixed integer program below, show that it is possible for the simplex tableau to exhibit infeasibility *after* all the integer constrained variables have reached their lower bounds.

$$\begin{array}{lrcrcl}
\text{Maximize} & 8x_1 & + & 2x_2 & = & x_0 \\
\text{subject to} & 3x_1 & + & 2x_2 & \leq & 1, \\
& 7x_1 & + & x_2 & \geq & 2, \\
& x_1, & & x_2 & \geq & 0, \\
\text{and} & x_0, & & x_1 & & \text{integer.}
\end{array}$$

Relate the computations to (x_1, x_2) space.

Chapter 6

Dual All-Integer Integer Programming

6.1   The Basic Approach

The cutting plane algorithm for the integer program presented in this chapter was developed by Ralph Gomory in 1960. Its similarity to the fractional method (Chapter 4) is principally due to the utilization of the lexicographic dual simplex method and to the maintenance of lexicographically positive columns in the tableau. However, the basic approach is different from the fractional technique. There is no optimization, generation of a constraint, reoptimization, etc. Rather, inequalities are generated at each iteration starting with the very *first*. Furthermore, each of these constraints is used as the pivot row, and is constructed so that it has integral coefficients and the pivot is -1. The initial tableau is assumed to be all integer and lexicographic dual feasible.[1] Hence, successive tableaux are also all integer and lexicographic dual feasible. The primal integer solution proceeds towards feasibility, and since dual feasibility is maintained, optimality is reached when it is attained.[2]

Note that the method is a direct extension of the classical dual simplex algorithm (cf. Chapter 2). The essential difference is that the pivot row in the all integer algorithm is generated at each iteration and ensures a -1 pivot. Since the technique employs the dual simplex method and maintains all integer tableaux, it is referred to as "dual all-integer." We list the steps.

---

[1] This is assumed for ease of exposition; the tableau in the general case may have rational nonintegral entries (see Problem 6.19).

[2] A feasible integer solution will be distinguished from an integer solution, the former having all *nonnegative* integral values.

Step 1 Start with an all-integer simplex tableau which contains a lexicographic dual feasible solution. Go to Step 2.

Step 2 Select a primal infeasible row v (i.e., $a_{v0} < 0, v \neq 0$). If none exists, the tableau exhibits the optimal integer solution–terminate. Otherwise go to Step 3.

Step 3 Designate the pivot column α_p $(p = 1, \ldots, n)$ to be the lexicographically smallest among those having $a_{vj} < 0$. If none exists (i.e., $a_{vj} \geq 0$ for $j = 1, \ldots, n$), there is no integer feasible solution–terminate. Go to Step 4.

Step 4 Derive an all integer inequality from row v which is not satisfied at the current primal solution. (Its slack will be negative.) It must also have a -1 coefficient in column α_p. Append it to the bottom of the tableau and label it the pivot row. Perform a dual simplex pivot operation and return to Step 2.

Of course, we have to show that a valid cut as described in Step 4 can always be obtained. It must also produce a convergent algorithm. The form and derivation of such an inequality is presented shortly. The latter discussion also shows that the pivot column selection rule is the same as in the lexicographic dual simplex method when the pivot row is the derived inequality (this will justify Step 3). Before going into these details we give a brief geometric interpretation and some notation.

As in all cutting plane algorithms, the intent of the all-integer algorithm is to reduce the (linear programming) feasible region to one whose optimum vertex is all-integer. However, unlike the fractional algorithm, primal solutions are never feasible until the optimal one is found. So successive not-optimal tableaux correspond to points outside the feasible region. Also the kth primal solution $(k \geq 2)$ will be on the hyperplane defined by the $k^{th} - 1$ cut. This follows since at iteration $k, x_{m+n+k-1}$, the Gomory slack variable corresponding to the $k^{th} - 1$ cut is nonbasic and thus at value 0.

To conform with the preceding chapters we use the Beale tableau (cf. Section 2.7) displayed in Table 6.1. In addition, some of the notation introduced in Chapter 4 is employed. In particular, $[y]$ is the largest integer $\leq y, \alpha_j$ $(j = 0, 1, \ldots, n)$ is the first $m + n + 1$ components in

Columns $\rightarrow$	$\boldsymbol{\alpha}_0$	$\boldsymbol{\alpha}_1$	$\cdots$	$\boldsymbol{\alpha}_n$	
Variables	1	$(-x_1)$	$\cdots$	$(-x_n)$	
$\downarrow$					
$x_0 =$	a_{00}	a_{01}	$\cdots$	a_{0n}	
$x_1 =$	0	-1	$\cdots$	0	≥ 0
$\vdots$	$\vdots$	$\vdots$	$\ddots$	$\vdots$	$\vdots$
$x_n =$	0			-1	≥ 0
$x_{n+1} =$	$a_{n+1,0}$	$a_{n+1,1}$	$\cdots$	$a_{n+1,n}$	≥ 0
$\vdots$	$\vdots$			$\vdots$	$\vdots$
$x_{n+m} =$	$a_{n+m,0}$	$a_{n+m,1}$	$\cdots$	$a_{n+m,n}$	≥ 0

Table 6.1: The Beale Tableau

column j of the tableau, and $J(j)$ is the jth index in the current set of indices J corresponding to the nonbasic variables.

6.2　The Form of the Cut

Let the generating row be any row v of the simplex tableau with $a_{v0} < 0$. That is,

$$x_v = a_{v0} + \sum_{j=1}^{n} a_{vj}(-x_{J(j)}) \tag{1}$$

Then the all-integer cut is

$$x' = \left[\frac{a_{v0}}{\lambda}\right] + \sum_{j=1}^{n}\left[\frac{a_{vj}}{\lambda}\right](-x_{J(j)}) \geq 0, \tag{2}$$

where x' is a nonnegative Gomory slack variable and λ is a positive number found by the rules below.

6.2.1   The Rules for Finding λ

Step 1 With v as the generating row, let α_p be the lexicographically smallest column among those having $a_{vj} < 0$ $(j = 1, \ldots, n)$.

Step 2 Let $u_p = 1$, and for every $j \geq 1$ $(j \neq p)$ with $a_{vj} < 0$, let u_j be the largest integer such that

$$\left(\frac{1}{u_j}\right) \alpha_j \overset{\mathcal{L}}{\succ} \alpha_p.$$

Note that $u_j \geq 1$ and $\alpha_p \overset{\mathcal{L}}{\succ} \mathbf{0}$.

Step 3 For each $a_{vj} < 0$ $(j \geq 1)$, set $\lambda_j = -a_{vj}/u_j$. Note that λ_j is not necessarily an integer.

Step 4 Set $\lambda = $ maximum λ_j. Note that $\lambda \geq \lambda_p = -a_{vp}/u_p \geq 1$, since $u_p = 1$ and $-a_{vp}$ is a positive integer.

The rules for finding λ, and hence the cut, may seem somewhat complicated. However, we can show that inequality (2) has the properties described previously. In particular, note that $[a_{v0}/\lambda] < 0$ or the generated row is a legitimate pivot row. Also observe that $-1 \leq a_{vp}/\lambda < 0$, $a_{vp} < 0$, and $\lambda \geq \lambda_p = -a_{vp}$; hence the pivot element $[a_{vp}/\lambda] = -1$. Furthermore, we shall prove that if row v corresponds to the first primal infeasible row every finite number of iterations, the algorithm converges to an optimal integer point. Before deriving the cut (2) from the source row (1) and explaining the rules for finding λ, we present two examples. The finiteness proof appears in a later section.

6.3   Illustrations

The first problem was solved in Chapter 4 to illustrate the fractional integer programming algorithm. The second appears in Gomory's paper. The arrow ($\rightarrow$) next to the tableau corresponds to the source row. The latter is the most infeasible row.[3] A circled element denotes the pivot and p is the pivot column index.

---

[3] We shall see that such a rule does not necessarily yield a convergent algorithm. Nevertheless, it is often used. Other source row selection criteria are discussed in Section 6.6.

Example 6.1 (Balinski [50])

$$
\begin{array}{rrrcll}
\text{Maximize} & -4x_1 & - & 5x_2 & = & x_0 \\
\text{subject to} & -x_1 & - & 4x_2 & \le & -5, \quad (x_3) \\
& -3x_1 & - & 2x_2 & \le & -7, \quad (x_4) \\
\text{and} & x_1, & & x_2 & \ge & 0, \quad \text{integer.}
\end{array}
$$

With nonnegative slack variables x_3 and x_4 we have Tableau #1.

#1	1	$(-x_1)$	$(-x_2)$	
x_0	0	4	5	$(x_1 = x_2 = 0)$
x_1	0	-1	0	$p = 1$
x_2	0	0	-1	$u_1 = 1, u_2 = 1$
x_3	-5	-1	-4	$\lambda_1 = 3, \lambda_2 = 2$
$\to x_4$	-7	-3	-2	$\lambda = \max(\lambda_1, \lambda_2) = 3$
x_5	-3	$\boxed{-1}$	-1	

We use x_4 as the source row, and since $4 < 5$, column 1 is the pivot column. Then $u_1 = 1$ and u_2 is the largest integer such that

$$
\left(\frac{1}{u_2}\right) \alpha_2 \overset{\mathcal{L}}{\succ} \alpha_1.
$$

Using the first terms in α_1 and α_2 we must have $(1/u_2)\, 5 \ge 4$; as $(1/1)5 > 4$, and $(1/2)5 < 4$, it follows that $u_2 = 1$. Therefore, $\lambda_1 = 3/1$ and $\lambda_2 = 2/1$, or $\lambda = \text{maximum}\, (\lambda_1, \lambda_2) = 3$. The derived inequality is

$$
x_5 = \left[\frac{-7}{3}\right] + \left[\frac{-3}{3}\right](-x_1) + \left[\frac{-2}{3}\right](-x_2) \ge 0;
$$

it is appended to Tableau #1. A pivot yields Tableau #2.

#2	1	$(-x_5)$	$(-x_2)$	
x_0	-12	4	1	
x_1	3	-1	1	$(x_1 = 3, x_2 = 0)$
x_2	0	0	-1	$p = 2$
$\rightarrow x_3$	-2	-1	-3	$u_1 = 3, u_2 = 1$
x_4	2	-3	1	$\lambda_1 = 1/3, \lambda_2 = 3$
x_5	0	-1	0	$\lambda_1 = \max(\lambda_1, \lambda_2) = 3$
x_6	-1	-1	$\boxed{-1}$	

As x_3 is the only infeasible row, it is the source row. Since $1 < 4$, column 2 is the pivot column. Then $u_2 = 1$, and u_1 is the largest integer for which $(1/u_1)\alpha_1 \overset{\mathcal{L}}{\succ} \alpha_2$, or $(1/u_1)4 \geq 1$, which means that u_1 can be at most 4. But for $u_1 = 4$, the second term in α_1 and α_2, namely -1 and 1, respectively, gives $(1/4)(-1) < 1$. This violates $(1/u_1)\alpha_1 \overset{\mathcal{L}}{\succ} \alpha_2$, and thus $u_1 = 3$. Hence, the inequality is

$$x_6 = \left[\frac{-2}{3}\right] + \left[\frac{-1}{3}\right](-x_5) + \left[\frac{-3}{3}\right](-x_2) \geq 0.$$

Or

$$x_6 = -1 - 1(-x_5) - 1(-x_2) \geq 0.$$

After adding it to Tableau #2 and pivoting we obtain the all-integer integer optimal Tableau #3.

#3	1	$(-x_5)$	$(-x_6)$	
x_0	-13	3	1	$(x_1 = 2, x_2 = 1)$
x_1	2	-2	1	
x_2	1	1	-1	Integer optimal
x_3	1	2	-3	
x_4	1	-4	1	
x_5	0	-1	0	
x_6	0	0	-1	

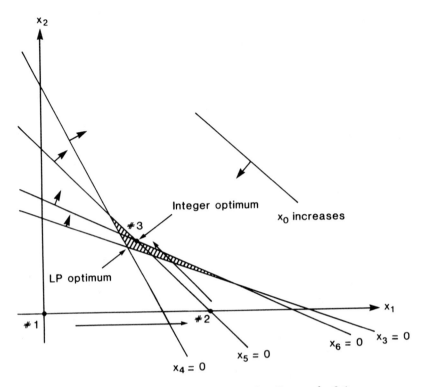

Figure 6.1: Graphical Illustration for Example 6.1

Expressing the added inequalities $x_5 \geq 0$ and $x_6 \geq 0$ in terms of the original nonbasic variables x_1 and x_2 yields

$$x_5 = -3 + x_1 + x_2 \geq 0 \quad \text{and} \quad x_6 = -4 + x_1 + 2x_2 \geq 0.$$

Of course, since each tableau is all-integer and the coefficients of the generated constraints are all-integer, we expect all-integer inequalities. The tableaux are related to (x_1, x_2) space in Fig. 6.1.

Example 6.2 (Gomory [277])

$$
\begin{array}{llrcrcrcll}
\text{Maximize} & -10x_1 & - & 14x_2 & - & 21x_3 & = x_0 \\
\text{subject to} & 8x_1 & + & 11x_2 & + & 9x_3 & \geq & 12, & (x_4) \\
& 2x_1 & + & 2x_2 & + & 7x_3 & \geq & 14, & (x_5) \\
& 9x_1 & + & 6x_2 & + & 3x_3 & \geq & 10, & (x_6) \\
\text{and} & x_1, & & x_2, & & x_3 & \geq & 0, & \text{integer.}
\end{array}
$$

With nonnegative slack variables x_4, x_5, and x_6 we have Tableau #1. The computations proceed as in the previous example. It takes four cuts and five tableaux to solve the program.

#1	1	$(-x_1)$	$(-x_2)$	$(-x_3)$	
x_0	0	10	14	21	$p = 1$
x_1	0	-1	0	0	$u_1 = 1$
x_2	0	0	-1	0	$u_2 = 1, u_3 = 2$
x_3	0	0	0	-1	$\lambda_1 = (2/u_1) = 2$
x_4	-12	-8	-11	-9	$\lambda_2 = (2/u_2) = 2$
$\to x_5$	-14	-2	-2	-7	$\lambda_3 = (7/u_3) = 7/2$
x_6	-10	-9	-6	-3	$\lambda = \dfrac{7}{2}$
x_7	-4	$\boxed{-1}$	-1	-2	

#2	1	$(-x_7)$	$(-x_2)$	$(-x_3)$	
x_0	-40	10	4	1	$p = 3$
x_1	4	-1	1	2	$u_1 = 9$
x_2	0	0	-1	0	$u_3 = 1$
x_3	0	0	0	-1	$\lambda_1 = 2/9$
x_4	20	-8	-3	7	$\lambda_3 = 3/1 = 3$
$\to x_5$	-6	-2	0	-3	$\lambda = 3$
x_6	26	-9	3	15	
x_7	0	-1	0	0	
x_8	-2	-1	0	$\boxed{-1}$	

#3	1	$(-x_7)$	$(-x_2)$	$(-x_8)$	
x_0	-42	9	4	1	
x_1	0	-3	1	2	$p = 1$
x_2	0	0	-1	0	$u_1 = 1$
x_3	2	1	0	-1	$\lambda_1 = 24/1 = 24$
x_4	6	-15	-3	7	$\lambda = 24$
x_5	0	1	0	-3	
$\rightarrow x_6$	-4	-24	3	15	
x_7	0	-1	0	0	
x_8	0	0	0	-1	
x_9	-1	$\boxed{-1}$	0	0	

#4	1	$(-x_9)$	$(-x_2)$	$(-x_8)$	
x_0	-51	9	4	1	
x_1	3	-3	1	2	
x_2	0	0	-1	0	$p = 3$
x_3	1	1	0	-1	$u_3 = 1$
x_4	21	-15	-3	7	$\lambda_3 = 3/1 = 3$
$\rightarrow x_5$	-1	1	0	-3	$\lambda = 3$
x_6	20	-24	3	15	
x_7	1	-1	0	0	
x_8	0	0	0	-1	
x_9	0	-1	0	0	
x_{10}	-1	0	0	$\boxed{-1}$	

#5	1	$(-x_9)$	$(-x_2)$	$(-x_{10})$
x_0	-52	9	4	1
x_1	1	-3	1	2
x_2	0	0	-1	0
x_3	2	1	0	-1
x_4	14	-15	-3	7
x_5	2	1	0	-3
x_6	5	-24	3	15
x_7	1	-1	0	0
x_8	1	0	0	-1
x_9	0	-1	0	0
x_{10}	0	0	0	-1

$(x_0 = -52, x_1 = 1,$
$x_2 = 0, x_3 = 2)$

Integer optimal tableau

6.4   Derivations: The Form of the Cut, the Pivot Column, and the λ Selection Rules

6.4.1   The Form of the Cut

Let us consider the derivation of the first cut. To see that the derivation is valid for successive inequalities we will need only to note that the pivot is always -1. Or equivalently, the simplex tableaux remain all-integer and lexicographic dual feasible. Remember that we assume that the first tableau exhibits these properties.

Designate any row (or integer linear combination of rows)

$$x = a_0 + \sum_{j=1}^{n} a_j(-x_{J(j)}) \tag{3}$$

with $a_0 < 0$ as the source row. (Row indices are being omitted.) If none exists, the tableau is integer optimal. To obtain the form of the cut we first represent every coefficient in (3), including the coefficient 1 of x, as a product term plus a remainder.

Let a be any number and λ a positive number. Set

$$f = \frac{a}{\lambda} - \left[\frac{a}{\lambda}\right];$$

then $0 \leq f < 1$. Defining $r = \lambda f$ yields

$$\frac{r}{\lambda} = \frac{a}{\lambda} - \left[\frac{a}{\lambda}\right], \quad \text{or} \quad a = \left[\frac{a}{\lambda}\right]\lambda + r.$$

As $0 \leq f < 1$ and $\lambda > 0$, we also have $0 \leq r < \lambda$. Thus, with regard to the coefficients in (3), we may write

$$a_j = \left[\frac{a_j}{\lambda}\right]\lambda + r_j \quad (j = 0, 1, \ldots, n) \tag{4}$$

and

$$1 = \left[\frac{1}{\lambda}\right]\lambda + r, \tag{5}$$

where $\lambda > 0, 0 \leq r_j < \lambda$, and $0 \leq r < \lambda$. Substituting (4) and (5) into (3), and placing all the remainder terms except r_0 on the left-hand side, gives

$$\sum_{j=1}^{n} r_j x_{J(j)} + rx = r_0 + \lambda \left\{ \left[\frac{a_0}{\lambda}\right] + \sum_{j=1}^{n} \left[\frac{a_j}{\lambda}\right](-x_{J(j)}) \right.$$
$$\left. + \left[\frac{1}{\lambda}\right](-x) \right\}. \tag{6}$$

In equality (6) let the terms in the curly brackets be denoted by x'. That is,

$$x' = \left[\frac{a_0}{\lambda}\right] + \sum_{j=1}^{n} \left[\frac{a_j}{\lambda}\right](-x_{J(j)}) + \left[\frac{1}{\lambda}\right](-x). \tag{7}$$

Then for any nonnegative integer solution satisfying (3), and hence (6), we need to show that x' is a nonnegative integer. Clearly, for integer values of the variables we will have x' integer since all the coefficients in

(7) are integer. To see that x' must also be nonnegative it is necessary to examine (6). For any feasible solution the left-hand side in that equality is nonnegative. Suppose at the same time that x' is a negative integer, i.e., $-1, -2, -3$, etc. As $r_0 < \lambda$, we would then have the right-hand side of (6), i.e., $r_0 + \lambda x'$, negative. Hence, a contradiction. Therefore we have just shown that x' defined by (7) is a new nonnegative integer variable. Moreover, when $\lambda > 1$, $[1/\lambda] = 0$ and (7) reduces to

$$x' = \left[\frac{a_0}{\lambda}\right] + \sum_{j=1}^{n} \left[\frac{a_j}{\lambda}\right] (-x_{J(j)}) \geq 0. \tag{8}$$

Since $a_0 < 0$ means $[a_0/\lambda] < 0$, inequality (8) may be appended to the bottom of the simplex tableau and used as the pivot row. Further, for λ sufficiently large the pivot element is guaranteed to be -1. For example, setting

$$\lambda \geq \mathop{\text{maximum}}_{a_j < 0} |a_j| \quad \text{yields} \quad \left[\frac{a_j}{\lambda}\right] = -1$$

for *every* $a_j < 0$. We can, however, devise rules to specify a λ that gives a -1 pivot and also increases the convergence rate. Before discussing this we must know how to choose the pivot column.

6.4.2   Choosing the Pivot Column

If the coefficients in the pivot row are denoted by b_j $(j = 0, 1, \ldots, n)$, the pivot column $\boldsymbol{\alpha}_p$ in the lexicographic dual simplex method satisfies

$$\frac{\boldsymbol{\alpha}_p}{-b_p} \mathrel{\overset{\mathcal{L}}{\prec}} \frac{\boldsymbol{\alpha}_j}{-b_j} \tag{9}$$

for every j $(j \neq 0)$ with $b_j < 0$. Now, with reference to the all-integer algorithm, we will have $b_p = -1$ and $b_j = [a_j/\lambda]$. Hence, by (9), the pivot column $\boldsymbol{\alpha}_p$ must satisfy

$$\boldsymbol{\alpha}_p \mathrel{\overset{\mathcal{L}}{\prec}} \frac{\boldsymbol{\alpha}_j}{-\left[\dfrac{a_j}{\lambda}\right]}$$

for every j $(j \neq 0)$ with $a_j/\lambda < 0$. As $-[a_j/\lambda]$ is a positive integer we have

$$\alpha_p \overset{\mathcal{L}}{\prec} \frac{\alpha_j}{-\left[\frac{a_j}{\lambda}\right]} \overset{\mathcal{L}}{\preceq} \alpha_j. \tag{10}$$

Since, for $\lambda > 0$, $a_j/\lambda < 0$ if and only if $a_j < 0$, we have just shown that the pivot column is the lexicographically smallest column which contains a negative element in the pivot row and hence the source row. We can now choose λ more advantageously.

6.4.3 λ Selection Rules

The objective when choosing λ is to obtain a -1 pivot and also to yield a pivot row which will result in the greatest lexicographic decrease in the 0 column. The latter objective is important because it will increase the convergence rate. Now suppose several values of λ give a -1 pivot. Then the smallest of these values for λ should be chosen, since it results in the largest lexicographic decrease in α_0. This follows because the updated 0 column is $\alpha'_0 = \alpha_0 + [a_0/\lambda]\,\alpha'_p$, and $[a_0/\lambda]$ is a negative integer whose absolute value increases as λ decreases. Using this result and the two objectives we can explain the necessary computations to find λ.

If λ_p is the pivot column, we have by (10) that

$$\alpha_p \overset{\mathcal{L}}{\prec} \frac{\alpha_j}{-\left[\frac{a_j}{\lambda}\right]} \tag{11}$$

for every j ($j \neq 0$) with $a_j < 0$. Now, let $u_p = 1$ and let u_j ($j \neq p$) be the largest integer for which

$$\alpha_p \overset{\mathcal{L}}{\prec} \frac{\alpha_j}{u_j}. \tag{12}$$

Since $-\left[\frac{a_j}{\lambda}\right]$ is an integer in (11) and u_j is the largest integer in (12), we have

$$-\left[\frac{a_j}{\lambda}\right] \leq u_j \tag{13}$$

The smallest positive λ satisfying (13) is

$$\lambda_j = \frac{-a_j}{u_j}. \tag{14}$$

Then, to guarantee (11), or equivalently, to maintain lexicographically positive columns α_j $(j = 1, \ldots, n)$, we must have

$$\lambda \geq \text{maximum } \lambda_j. \tag{15}$$

The minimum λ which satisfies (15) is $\lambda = \text{maximum } \lambda_j$. Thus, to find a λ which satisfies the two objectives, we do the following calculations:

1. Let α_p be the lexicographically smallest column with $a_p < 0$ $(p \neq 0)$.

2. For each $a_j < 0$ $(j \neq 0)$ let $u_p = 1$ and let u_j $(j \neq p)$ be the largest integer satisfying (12).

3. For each $a_j < 0$ $(j \neq 0)$ find λ_j as defined by (14).

4. Set $\lambda = \text{maximum } \lambda_j$.

Remarks

1. Since the equations $x_j = -1(-x_j)$ $(j = 1, \ldots, n)$ are used, columns 1 through n in the simplex tableau are linearly independent. Hence, the pivot column α_p is uniquely determined.

2. $\lambda_p \geq 1$ since $\lambda_p = -a_p/u_p = -a_p \geq 1$.

3. If λ turns out to be 1 the source row is the pivot row, and a new inequality is not added. (Of course, the pivot element a_p equals -1.)

4. If the first element in α_p is zero and the first element in α_j is not zero, then u_p $(j \neq p)$ is arbitrarily large. In this case, λ_j $(j \neq p)$ is arbitrarily small and may be taken as 0, since λ_p will always be larger.

5. If $a_p \leq a_j < 0$, then $\lambda_p \geq \lambda_j$, since

$$\lambda_p = -a_p \geq -a_j \geq \frac{-a_j}{u_j} = \lambda_j.$$

Thus, λ_j need not be calculated whenever $a_p \leq a_j < 0$.

6. By its selection, λ need not be an integer.

6.5   Some Properties of the Added Inequalities

6.5.1   The Relative Strength of the Added Inequalities

The selection of λ as previously described does produce a -1 pivot and also results in the largest change in the 0 column. It does not, however, yield an inequality that is as strong as possible. In fact, the new inequality may be weaker than the row (or inequality) from which it is derived. To illustrate, suppose the source row is

$$x = -4 - 3(-x_1) - 5(-x_2) \geq 0; \tag{16}$$

for $\lambda = 2$ the derived inequality is

$$x' = -2 - 2(-x_1) - 3(-x_2) \geq 0; \tag{17}$$

for $\lambda = 3$ the inequality is

$$x' = -2 - 1(-x_1) - 2(-x_2) \geq 0; \tag{18}$$

and for $\lambda = 4$ it is

$$x' = -1 - 1(-x_1) - 2(-x_2) \geq 0. \tag{19}$$

Inequality (16) is stronger than (17), which, in turn, is stronger than (19). Constraint (16), however, is weaker than (18). This is verified by Fig. 6.2.

As indicated by Wilson [602], it is sometimes possible to increase the value of λ, found by the usual computations, so as to obtain a stronger all-integer cut. Consider a derived inequality (row indices are omitted).

$$x' = \left[\frac{a_0}{\lambda}\right] + \sum_{j=1}^{n} \left[\frac{a_j}{\lambda}\right] (-x_{J(j)}) \geq 0. \tag{20}$$

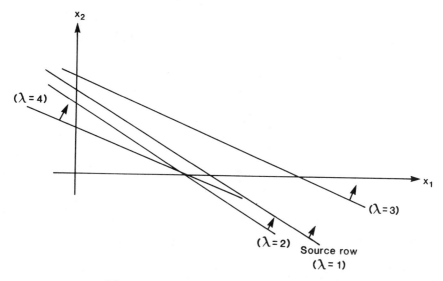

Figure 6.2: Strength of the Inequalities

Then, if λ can be increased so that $[a_0/\lambda]$ and $[a_j/\lambda]$, for $a_j > 0$, do not change, we will have a stronger inequality when some or all of the remaining $[a_j/\lambda]$ $(a_j < 0)$ change. This is true because the larger the λ the smaller in absolute value are the coefficients in (20). Thus, a change in $[a_j/\lambda]$ $(a_j < 0)$ places the hyperplane $x' = 0$ farther away from the origin in $x_{J(j)}$ space, resulting in an improved inequality. To see this, note that the intersection of the hyperplane with the $x_{J(j)}$ axis occurs when $x_{J(j)} = [a_0/\lambda]/[a_j/\lambda]$. As $[a_0/\lambda] < 0$ we have, when $a_j < 0$, $[a_0/\lambda]/[a_j/\lambda]$ increasing as λ increases. Since $[a_0/\lambda]$ remains the same, the stronger inequality produces the same change in the 0 column as the standard inequality.

To illustrate, consider the partial tableau below where x_0 is the objective function and x is the source row.

	1	$(-x_1)$	$(-x_2)$	$(-x_3)$	$(-x_4)$
x_0	20	1	3	4	4
x	-20	-7	-8	-15	18

Then, $\alpha_p = \alpha_1$, $u_1 = 1, u_2 = 2, u_3 = 3$, and $\lambda_1 = 7, \lambda_2 = 4, \lambda_3 = 5$, or $\lambda = 7$. Thus, the Gomory cut is

$$x' = -3 - 1(-x_1) - 2(-x_2) - 3(-x_3) + 2(-x_4) \geq 0 \quad (\lambda = 7).$$

Now, for $7 \leq \lambda < 10$, the value of $[-20/\lambda]$ will be -3, and for $7 \leq \lambda \leq 9$ we have $[18/\lambda] = 2$. Thus, if we take $\lambda = \text{minimum } (9, 10 - \epsilon) = 9$ (ϵ is a small positive number) we have the stronger cut

$$x' = -3 - 1(-x_1) - 1(-x_2) - 2(-x_3) + 2(-x_4) \geq 0 \quad (\lambda = 9).$$

6.5.2 A Second Derivation of the Fractional Cut, and Relation to the All-Integer Cut

During the derivation of the form of the all-integer cut (Section 6.4, equation (7)) we obtained the equation ($\lambda > 0$)

$$x' = \left[\frac{a_0}{\lambda}\right] + \sum_{j=1}^{n} \left[\frac{a_j}{\lambda}\right](-x_{J(j)}) + \left[\frac{1}{\lambda}\right](-x), \qquad (21)$$

where the source row is

$$x = a_0 + \sum_{j=1}^{n} a_j(-x_{J(j)}); \qquad (22)$$

x' is a nonnegative integer variable. Then the all-integer cut was obtained by selecting $\lambda > 1$. Since x' is a nonnegative integer for any positive λ, we may take $\lambda = 1$. Then, $1/\lambda = 1$, and using (22), equality (21) becomes

$$x' = [a_0] + \sum_{j=1}^{n}[a_j](-x_{J(j)}) - \left\{ a_0 + \sum_{j=1}^{n} a_j(-x_{J(j)}) \right\},$$

or

$$x' = -f_0 + \sum_{j=1}^{n}(-f_j)(-x_{J(j)}).$$

The last equality is the Gomory fractional cut, which was derived by a different procedure in Chapter 4. Thus, both the fractional ($\lambda = 1$) and the all-integer ($\lambda > 1$) cut can be obtained using the same derivation.

Composite cutting plane algorithms using both fractional and all-integer cuts are available. For example, we could first use the simplex method to get to, or near, an optimal continuous solution. Then from that point on we add, depending on which is applicable, either fractional or all-integer cuts. (Unless the tableau is all-integer and optimal at least one is available.) Another possibility would be to first solve the linear program. Add fractional cuts to return the tableau to an all-integer, not necessarily primal or dual feasible, tableau.[4] Then if the tableau does not exhibit dual feasibility, a simplex algorithm will return it to a new linear programming optimum. Fractional cuts can again be applied so as to regain an all-integer tableau. When an all-integer tableau displays dual feasibility, all-integer cuts may be added. These composite approaches can easily be shown to converge. The details are left to the problems.

6.6   Algorithm Strategies

6.6.1   Choosing the Source Row

As we shall see in the next section, any source row selection rule which, if the constant term of a row is negative and remains negative will eventually select that row, will support a finite algorithm.[5] Examples of such rules are:

1. Select the first eligible row. In other words, select the first row with a negative 0 column term.

2. Select the first eligible row at the current iteration, the second eligible row at the next iteration, etc. That is, select an eligible row by a cyclic process.

3. Randomly select an eligible row.

Observe that the most negative constant term rule does not necessarily have the required property. In particular, it is possibe for a row to have and maintain a negative constant term that is never the most negative value. (The reader is asked to find such a program in Problem 6.11.)

---

[4]Techniques which employ fractional cuts to return a tableau to an all-integer one are available. For example, the procedure explained in Appendix B of Chapter 4 could be used.

[5]Any rule that selects the first eligible row every finite number of tableaux will automatically have this property (Problem 6.17).

Other more sophisticated rules are can be developed for example, utilizing the magnitude of the coefficients in each column. One such criterion is discussed in Problem 6.13., and a second more complicated one appears in Gomory [277].

6.6.2   Dropping Inequalities

As in the fractional algorithm, it is standard practice to drop the derived inequality when its slack variable becomes a basic variable. Since the generated row is used as the pivot row only once, i.e., at its introduction, no information is lost in this respect when it is omitted. But dropping inequalities may eliminate possible source rows. However, since this is usually unimportant and omitting new inequalities has the obvious advantage of limiting the tableau's row size, it seems reasonable to adopt this strategy.

6.7   Finiteness

To show that the algorithm is finite we assume that x_0, the objective function, is bounded below. Then, as in the proof of the fractional algorithm, we suppose that the process is not finite and appeal to the lexicographic decreasing nature of the 0 column to obtain a contradiction. The details are below.

If the algorithm is not finite there exists an infinite sequence of tableaux, indexed by k, in which the 0 column decreases lexicographically.[6] That is,

$$\alpha_0^k \overset{\mathcal{L}}{\succ} \alpha_0^{k+1} \overset{\mathcal{L}}{\succ} \alpha_0^{k+2}, \; etc.$$

As each component of α_0 is integer, its first component a_{00} must decrease by integer amounts. Thus, to remain above its lower bound it must eventually remain fixed. Then, for the 0 column to continue decreasing lexicographically, a_{10}, its second component, must decrease.

---

[6]This is easy to see. Suppose the pivot column is α_p^k and the pivot element is $[a_p/\lambda] = -1$. Then after pivoting we have $\alpha_0^{k+1} = \alpha_0^k + [a_0/\lambda]\alpha_p^k$. As $[a_0/\lambda]$ is a negative number and $\alpha_p^k \overset{\mathcal{L}}{\succ} 0$, we have $\alpha_0^{k+1} \overset{\mathcal{L}}{\prec} \alpha_0^k$.

We show that it must eventually remain fixed at a nonnegative integer. For suppose a_{10} falls below 0; then row 1 becomes an eligible source row. If it is selected to generate the cut, the pivot row is

$$x' = \left[\frac{a_{10}}{\lambda}\right] + \sum_{j=1}^{n} \left[\frac{a_{1j}}{\lambda}\right] (-x_{J(j)}),$$

where $[a_{10}/\lambda] < 0$. If a_{1j} $(j = 1, \ldots, n)$ is nonnegative, the integer program is infeasible. Otherwise there is an index p such that $a_{1p} < 0$. Since λ is chosen so that the pivot $[a_{1p}/\lambda] = -1$, a_{00} would decrease after pivoting to $a'_{00} = a_{00} + a_{0p}[a_{10}/\lambda]$, where $a_{0p} > 0$, and $[a_{10}/\lambda] < 0$, which is a contradiction. Thus a_{10} must eventually remain fixed at a nonnegative integer. The argument can now be repeated verbatim for a_{20}. That is, since a_{10} is fixed, a_{20} is bounded below by 0, and therefore it also must eventually remain constant. The argument is repeated for all indices i $(i = 0, 1, \ldots, m + n)$.

Observe that the proof required that the first eligible source row will eventually be selected to generate the cut. Hence, any source row selection rule that has this property will support a finite algorithm.

6.7.1   If Only the Cost Row Is All-Integer

A careful examination of the proof reveals that only the initial value of x_0 and the cost row need be all integer. This is because we initially pose the program in the tableau.

	1	$(-x_1)$	$\ldots$	$(-x_n)$	
x_0	a_{00}	a_{01}	$\ldots$	a_{0n}	$\leftarrow$ cost row
x_1	0	-1	$\ldots$	0	
$\vdots$	$\vdots$	$\vdots$	$\ddots$	$\vdots$	
x_n	0	0	$\ldots$	-1	
x_{n+1}	$a_{n+1,0}$	$a_{n+1,1}$	$\ldots$	$a_{n+1,n}$	
$\vdots$	$\vdots$	$\vdots$		$\vdots$	
x_{n+m}	$a_{n+m,0}$	$a_{n+m,1}$	$\ldots$	$a_{n+m,n}$	

Now, if a_{0j} $(j = 0, 1, \ldots, n)$ is all-integer the first $n + 1$ rows of the tableau remain all integer. Hence the finiteness proof goes through unchanged for the first $n+1$ components in the 0 column. As $x_{n+1}, \ldots,$ x_{n+m} are now not necessarily integer variables, [7] two possibilities can occur after the first $n + 1$ components in α_o are fixed. Either $a_{n+i,0} \geq$ 0 $(i = 1, \ldots, m)$, or there is an i such that $a_{n+i,0} < 0$. In the first case the tableau clearly exhibits optimality and the proof ends. We claim that the second possibility cannot occur unless the integer program is infeasible. This is true, since if $a_{n+i,0} < 0$, row $n + i$ can be selected as a dual simplex pivot row. If all $a_{n+i,j}$ $(j = 1, \ldots, n)$ are nonnegative, the integer program is infeasible. Otherwise there is a pivot column α_p $(\alpha_p \overset{\mathcal{L}}{\succ} o)$ with $a_{n+i,p} < 0$. Then for the first $n + 1$ components in α_o to remain the same after pivoting, we must have zeroes in the first $n + 1$ rows of α_p. But for any tableau $\mathbf{A}^k$, we have $\mathbf{A}^k = \mathbf{A}^1 \mathbf{B}^{-1}$, where $\mathbf{B}$ is the dual basis associated with tableau $\mathbf{A}^k$, and $\mathbf{A}^1$ is the initial matrix. Also, the first $n + 1$ rows of $\mathbf{A}^1$ have the form

$$\begin{pmatrix} 0 & \mathbf{c} \\ 0 & -\mathbf{I} \end{pmatrix}$$

where $\mathbf{c} = (a_{01}, \ldots, a_{0n})$ and $\mathbf{I}$ is an identity matrix. Thus, since the determinants of $\mathbf{B}^{-1}$ and $-\mathbf{I}$ are not zero, it is impossible to have a column α_p $(p \geq 1)$ in $\mathbf{A}^k$ with zeroes in its first $n + 1$ rows. Hence, we have a contradiction. We have just shown that after the first $n + 1$ components in the 0 column are fixed, either the tableau is integer optimal or it displays infeasibility. In either case the finiteness proof goes through. (A different analysis can be used to show that even the x_0 row need not have integer coefficients; see Problem 6.19.)

Problems

Problem 6.1   List the major similarities and differences betwen the fractional algorithm and the all-integer algorithm. What are the relative advantages and disadvantages of each approach?

---

[7]The derivation of the all-integer cut did not require the basic variable in the source row to be integer constrained. Thus, rows corresponding to the slack variables $(x_{n+1}, \ldots, x_{n+m})$ can be used as source rows.

Problem 6.2 Solve the following problem by the all-integer algorithm.

$$
\begin{aligned}
\text{Maximize} \quad & -2x_1 \; - \; 5x_2 \; = \; x_0 \\
\text{subject to} \quad & -2x_1 \; - \; 2x_2 \; \leq \; -9, \\
& -2x_1 \; - \; 6x_2 \; \leq \; -22, \\
\text{and} \quad & x_1, \quad\;\; x_2 \; \geq \; 0, \quad \text{integer.}
\end{aligned}
$$

Relate the computations to (x_1, x_2) space.

Problem 6.3 Solve the integer program appearing in Problem 4.6 by the all-integer algorithm.

Problem 6.4 Solve the integer program appearing in Problem 4.7 by the all-integer algorithm.

Problem 6.5 Solve the integer program appearing in Problem 4.8 by the all-integer algorithm.

Problem 6.6 Can an all-integer cut be redundant? That is, at its introduction could it be implied by existing constraints? If so, can we always select a source row so that the cut is not redundant? Give an illustration.

Problem 6.7 With regard to Example 6.1 (Section 6.3):

a) What is the primal and dual basis associated with each tableau?

b) Can you geometrically relate each tableau to a dual space? Explain.

Problem 6.8 The source row

$$
x = a_0 + \sum_{j=1}^{n} a_j(-x_{J(j)})
$$

generates the inequality

$$
x' = \left[\frac{a_0}{\lambda}\right] + \sum_{j=1}^{n} \left[\frac{a_j}{\lambda}\right](-x_{J(j)}) \geq 0. \tag{1}
$$

We know (Section 6.6) that if λ, found by the usual calculations, can be increased so that $[a_0/\lambda]$, and for those $a_j > 0$, $[a_j/\lambda]$ does not change, a stronger cut will be obtained. Find the possibly larger value of λ, say λ^*, in terms of the coefficients in (1), and solve the integer program

$$
\begin{array}{llrrrrr}
\text{Minimize} & - & x_1 & - & 3x_2 & - & 6x_3 & - & 4x_4 \\
\text{subject to} & & 7x_1 & + & 8x_2 & + & 15x_3 & - & 18x_4 & \geq & 20, \\
& & 10x_1 & + & 7x_2 & + & 5x_3 & - & 15x_4 & \geq & 15, \\
& - & 5x_1 & + & 9x_2 & + & 10x_3 & + & 7x_4 & \geq & 11, \\
\text{and} & & \multicolumn{7}{l}{x_1,\ x_2,\ x_3,\ x_4 \geq 0, \quad \text{integer.}}
\end{array}
$$

with the all-integer algorithm using both the value of λ found by the usual calculations and the value of λ^* developed above. Compare the effect of increasing the value of λ on the convergence rate of the all-integer algorithm applied to the above problem.

Problem 6.9   Prove that each of the two composite (fractional all-integer) algorithms described at the end of Section 6.5 converges. Give another composite algorithm. Why does your method converge?

Problem 6.10   Solve the following integer program by the all-integer algorithm. Do *not* initially clear fractions.

$$
\begin{array}{llrll}
\text{Minimize} & x_1 & + & x_2 & = & x_0 \\
\text{subject to} & x_1 & + & (5/2)x_2 & \geq & 3, \\
& x_1 & + & (2/5)x_2 & \geq & 6/5, \\
\text{and} & x_1, & & x_2 & \geq & 0, \quad \text{integer.}
\end{array}
$$

Relate the computations to (x_1, x_2) space.

Problem 6.11   We know (Section 6.7) that any source row selection rule which eventually selects the first infeasible row supports a finite algorithm. Prove, by finding an example, that "the most infeasible row" rule does not have this property.

Problem 6.12   Give two source row selection rules that have not appeared in the chapter which guarantee a convergent algorithm.

Problem 6.13   Consider the following source row selection rule:

Step 1 Rank each column α_j $(j = 1, \ldots, n)$ in order of lexicographic descending size. Denote the jth column rank by $C(j)$.

Step 2 To each of the eligible source rows i $(i = 1, \ldots, m+n)$ assign the rank $R(i) = \underset{j \in J}{\text{maximum}} \; C(j)$, where $J = \{j \mid j \neq 0, a_{ij} < 0\}$.

Step 3 Select the source row v, using $R(v) = \text{minimum} \; R(i)$, where the minimum is over the eligible source rows (i.e., where $R(i)$ is defined). Ties are broken arbitrarily.

   a) What are the advantages of using such a rule?

   b) Does the rule support a finite algorithm?

Problem 6.14   Solve Example 6.2 (Section 6.3) using the source row selection rule described in Problem 6.13.

Problem 6.15   Is there an upper limit on the number of cuts which may be obtained from a particular tableau? Explain.

Problem 6.16   Is it possible to extend the all-integer algorithm to allow it to solve mixed integer programs? What problems arise?

Problem 6.17   Show that any source row rule which selects the first eligible row every finite number of tableaux must imply that for any row i an infinite sequence of tableaux with $a_{i0} < 0$ cannot occur. (*Hint:* Use the finiteness argument in Section 6.7.)

Problem 6.18   Suppose the source row is indexed by v and the pivot column by p. If, in the all integer cut, λ is always set to $-a_{vp}$, does the algorithm necessarily converge?

Problem 6.19   (Salkin, Shroff, and Mehta [518]) Prove that the all-integer algorithm will converge when applied to a tableau with rational, but not necessarily integral, coefficients. The x_0 row is also allowed to have fractional entries. (*Hint:* Set the problem up in our standard simplex tableau. Let h_i $(i = 0, 1, \ldots, m + n)$ be the least common denominator (LCD) for row i. That is, when row i is multiplied by h_i it becomes all integer. Let $\mathbf{A}_I^k$ and $\mathbf{A}_F^k$ denote the kth tableau when the all integer algorithm is applied to an initial tableau with all integral and fractional data, respectively. Show, by induction on k, that tableau $\mathbf{A}_I^k$ can be obtained from tableau $\mathbf{A}_F^k$ by multiplying each of its rows by h_i. As the sequence $\mathbf{A}_I^k$ converges it follows that $\mathbf{A}_F^k$ converges.)

Problem 6.20 [8] Suppose a constraint in a lexicographic dual feasible all-integer tableau has the form (row indices are omitted)

$$x = a_0 + \sum_{j=1}^{n} a_j \left(-x_{J(j)} \right) \geq 0 \tag{2}$$

where $a_j \geq 0$ for all $j \neq 0$ and $j \neq k, a_k < 0 \ (k \neq 0)$, and $a_0 < 0$.

(a) Show that $x_{J(k)} \geq \langle a_0/a_k \rangle$ is a constraint which must be satisfied at the current tableau. That is,

$$s = -\langle \frac{a_0}{a_k} \rangle - 1(-x_{J(k)}) \geq 0 \tag{3}$$

is a valid all integer cut. Here, s is a nonnegative integer constrained slack variable and $\langle y \rangle$ means the smallest integer $\geq y$.

(b) Suppose at every (not optimal) tableau each constraint having $a_0 < 0$ is of form (2). Then at each of these tableaux a cut (3) may be generated. Prove that if this cut replaces the Gomory all-integer one, the resulting dual all-integer algorithm can converge. (*Hint:* Select the first row of type (2) as the generating row. Then, as in the finiteness proof using the Gomory cut, appeal to the lexicographic decreasing nature of the 0 column.)

Problem 6.21 Suppose a lexicographic dual feasible-primal infeasible (all integer) tableau does not exhibit a constraint (2) as described in the previous problem. Show that unless the problem is infeasible, the following instructions will produce a tableau possessing such an inequality. (*Hint:* Prove that the cut $x'' \geq 0$ does not eliminate any integer feasible solutions, and that at least the source row for (5) becomes, subsequent to a pivot, a constraint of type (2).)

INSTRUCTION 1 Select a row with $a_0 < 0$ (row indices are omitted). From this row generate the Gomory all integer cut

---

[8]Problems 6.20 through 6.22 discuss some of the easier parts of Glover's work [238]. In particular, Problem 6.22 is Glover's "Bound Escalation Algorithm" in its simplest form. The reader is referred to the paper for more sophisticated versions.

$$x' = \left[\frac{a_0}{\lambda}\right] + \sum_{j=1}^{n} \left[\frac{a_j}{\lambda}\right] (-x_{J(j)}) \geq 0 \qquad (4)$$

by the standard method (Section 6.2). If (4) cannot be obtained, i.e., $a_j \geq 0$ $(j = 1, \ldots, n)$, the problem has no feasible solution–terminate the calculations.

INSTRUCTION 2 Set $[a_0/\lambda] = 0$ in (4) and adjoin the resulting equation

$$x'' = 0 + \sum_{j=1}^{n} \left[\frac{a_j}{\lambda}\right] (-x_{J(j)}) \qquad (5)$$

to the bottom of the tableau. Note that $x'' = x' - [a_0/\lambda] \geq 0$ since $[a_0/\lambda] < 0$.

INSTRUCTION 3 Perform a dual simplex pivot step (updating the whole tableau) on the coefficient in (5) in the lexicographically smallest column among those having $a_j < 0$ $(j \neq 0)$. Observe that by the derivation of (4) the pivot element must be -1.

Problem 6.22 Prove that the following dual all-integer algorithm converges. (*Hint:* Show that intermingling cuts of the form (5), Problem 6.21, with those of type (3), Problem 6.20 produces an algorithm which maintains dual feasibility and a lexicographic decreasing 0 column. Then make the necessary assumptions and follow the finiteness proof of Gomory's algorithm.)

Step 1 Start with a dual lexicographic feasible all integer tableau. (The identity equations $x_j = -1(-x_j)$ are included in the tableau.) If such a program cannot be found the problem is infeasible–terminate. Go to Step 2.

Step 2 If $a_{i0} \geq 0$ for all $i \geq 1$ the tableau is optimal–terminate. Otherwise, determine if there is an i with $a_{i0} < 0$ whose row is of the form (2) described in Problem 6.20. If such a row exists, go to Step 4. If not, go to Step 3.

Step 3 Obtain a tableau which has a constraint of type (2) by performing Instructions 1 through 3 described in Problem 6.21. If these computations indicate that the problem is infeasible, terminate. Go to Step 4.

Step 4 Generate from the row (2) an all-integer cut of the form (3). Adjoin it to the bottom of the current tableau. Go to Step 5.

Step 5 Perform a dual simplex pivot step on the -1 coefficient of $(-x_{J(k)})$ in the derived row (3). Return to Step 2.

Problem 6.23 Resolve Example 6.2 (Section 6.3) using the algorithm proposed in Problem 6.22.

Chapter 7

Primal All-Integer Integer Programming

7.1 Introduction

As Gomory's dual all-integer integer programming algorithm is an extension of the dual simplex method (Chapter 6), the primal all- integer integer programming algorithm is an extension of the primal simplex method. Specifically, the procedure requires an all integer primal feasible initial tableau.[1] It adds a Gomory cut at each iteration, starting with the very first, so as to maintain an all integer tableau and *primal* feasibility. When dual feasibility is reached the tableau is integer optimal and the integer program is solved. As the algorithm uses Gomory all-integer cuts it cannot solve a mixed integer program.

The "Direct Algorithm", a primal technique, was first described by Ben-Israel and Charnes [81]. That algorithm, however, frequently required the solution of a sometimes difficult integer programming "auxiliary problem." Richard Young in [612] proposed a different primal algorithm, labeled the "Rudimentary Primal Algorithm," which avoided the auxiliary problem. However, the technique was somewhat complicated and difficult to implement. Glover [243] proposed a pseudo primal-dual algorithm which utilizes Gomory's dual-all integer method [277] with a variation of Young's primal all-integer technique [612]. Subsequently, Glover [244] and Young [614], drawing from their and each other's work, developed simplified primal integer programming algorithms. The contents of these papers are essentially built on the same foundations and, in fact, often overlap. There are, however, some differences between the methods. For one thing, they do not employ

---

[1] Actually the tableau may have fractional entries; see Problem 7.7

the same algorithm strategies, tableau format, or notation. A more important difference is in the use of a "reference equation" introduced to find the pivot column and in the rules which select the source row from which the Gomory cut or pivot row is generated. Although both papers present the general necessary properties of the reference row and the source row selection rule, Young, for ease of exposition and implementation, outlines an algorithm with a particular reference row and source row selection criterion. Glover, on the other hand, does not specialize these, thus creating a more general approach. Hence, we shall essentially present the contents of Glover's paper. The simplified algorithm outlined by Young is then treated as a special case.

7.2   The Tableau, the Rudimentary Primal Algorithm

We write the integer program as

$$\text{maximize} \quad a_{00} - \sum_{j=1}^{n} a_{0j} x_j \;=\; x_0$$

$$\text{subject to} \quad \sum_{j=1}^{n} a_{ij} x_j \;\leq\; a_{i0} \qquad (i = n+1, \ldots, n+m)$$

$$\text{and} \quad x_j \;\geq\; 0, \quad \text{integer} \quad (j = 1, \ldots, n).$$

With the nonnegative slack variables $x_{n+1}, \ldots, x_{n+m}$ the program may be exhibited in the following tableau.

	1	$(-x_1)$	$\ldots$	$(-x_n)$	
$x_0 =$	a_{00}	a_{01}	$\ldots$	a_{0n}	
$x_1 =$	0	-1			≥ 0
$\vdots$	$\vdots$		$\ddots$		$\vdots$
$x_n =$	0			-1	≥ 0
$x_{n+1} =$	$a_{n+1,0}$		$\ldots$	$a_{n+1,n}$	≥ 0
$\vdots$	$\vdots$			$\vdots$	$\vdots$
$x_{n+m} =$	$a_{n+m,0}$		$\ldots$	$a_{n+m,n}$	≥ 0

(If the tableau is all-integer, the slack variables are integer constrained.)

As we shall see, the redundant constraints $x_j = -(-x_j)$ $(j = 1, \ldots, n)$ are added to ensure a unique selection of the pivot column. To conform with the preceding chapters we have adopted the same tableau as before. However, Young [614] uses the above tableau without the identity rows. These are adjoined only when needed (see Problem 7.1). Glover [244], on the other hand, uses the identity equalities but places them after the original constraints. As we shall see, this can influence the choice of the pivot column and thus may affect the convergence rate.

7.2.1   The Rudimentary Algorithm

Suppose the tableau is primal feasible and all-integer. That is, each a_{ij} is integer and $a_{n+i,0} \geq 0$ $(i = 1, \ldots, m)$. Then, if the tableau is also dual feasible (equivalently, $a_{0j} \geq 0$ for $j = 1, \ldots, n$), we have the optimal solution to the integer program in its zero column. If the dual problem is not feasible, there are columns indexed by j $(j \geq 1)$ with $a_{0j} < 0$. Suppose, as in the primal simplex method, we let the pivot column correspond to the most negative of these, say a_{0p}. Then, if we continued using the primal simplex algorithm, we could find a pivot row indexed by v so that[2]

$$\theta_p = \frac{a_{v0}}{a_{vp}} = \underset{a_{ip} > 0}{\text{minimum}} \left\{ \frac{a_{i0}}{a_{ip}} \right\}; \tag{1}$$

θ_p is the largest value of the nonbasic variable x_p for which a *feasible* primal solution is maintained. If $a_{ip} \leq 0$ $(i = 1, \ldots, n+m)$, the integer program has an unbounded solution (Problem 7.3) and the computations terminate. Suppose, instead of using (1) to determine a pivot row, we find a row satisfying

$$\left[\frac{a_{i0}}{a_{ip}} \right] \leq \theta_p, \quad a_{ip} > 0, \tag{2}$$

where, as in earlier chapters, $[y]$ means the largest integer $\leq y$. Clearly, one such row exists, namely row v. The equation for row i satisfying (2) may be written as

---

[2]Since linear programming anticycling procedures will not be applicable, break ties in (1) arbitrarily; see Problem 7.7.

$$x = a_{i0} + \sum_{j=1}^{n} a_{ij}(-x_j). \tag{3}$$

Let (3) serve as a generating row from which the Gomory all-integer cut (4) is derived (see Chapter 6, Section 6.2 and Section 6.4):

$$s = \left[\frac{a_{i0}}{\lambda}\right] + \sum_{j=1}^{n} \left[\frac{a_{ij}}{\lambda}\right](-x_j) \geq 0. \tag{4}$$

Then observe what happens when λ is set equal to the positive number a_{ip} and (4) is used as the pivot row. In particular, the pivot is $[a_{ip}/a_{ip}] = 1$ and the tableau retains primal feasibility since, using (2), we have

$$\left[\frac{a_{i0}}{a_{ip}}\right] \bigg/ \left[\frac{a_{ip}}{a_{ip}}\right] = \left[\frac{a_{i0}}{a_{ip}}\right] \leq \theta_p.$$

Hence, a simple extension of the primal simplex algorithm yields a technique which maintains primal feasible all-integer tableaux (since the pivot element is always 1) given that the initial tableau has these properties. Also, since after each pivot $-a_{0p}[a_{i0}/a_{ip}]$, a nonnegative number, is added to a_{00}, the value of x_0 never decreases. Unfortunately, however, this method, termed the Rudimentary Primal Algorithm (RPA), does not necessarily converge (see Appendix A). Nevertheless, it is a foundation from which a convergent algorithm can be developed. Hence we outline the algorithm and present an example.

7.2.2   The Rudimentary Primary Algorithm

Step 1 Start with a primal feasible all-integer tableau. If such a tableau cannot be found the integer program is infeasible–terminate. Otherwise, go to Step 2.

Step 2 If $a_{0j} \geq 0$ $(j = 1, \ldots, n)$, the tableau is integer optimal – terminate. Otherwise, find the pivot column indexed by p using

$$a_{0p} = \min_{a_{0j} < 0} a_{0j} \quad (j \geq 1).$$

Break ties arbitrarily. Go to Step 3.

Step 3 If $a_{ip} \leq 0$ $(i = 1, \ldots, n + m)$, the integer problem has an unbounded solution–terminate. Otherwise, find the row, indexed by v, utilizing

$$\theta_p = \frac{a_{v0}}{a_{vp}} = \underset{a_{ip} > 0}{\text{minimum}} \left\{ \frac{a_{i0}}{a_{ip}} \right\}$$

Arbitrarily break ties. Go to Step 4.

Step 4 From a row i with $a_{ip} > 0$ that satisfies $[a_{i0}/a_{ip}] \leq \theta_p$, generate a Gomory all-integer cut. Set the parameter λ in the cut equal to a_{ip}; let the derived row be the pivot row. Perform a primal simplex pivot step (cf. Section 2.7) and return to Step 2.

To illustrate the RPA we solve the first problem used by Young [612]. For reasons similar to the ones given in the previous chapter (Section 6.6) the pivot rows or Gomory cuts are not kept (see Problem 7.4).[3] Also, the source row is always taken as the natural primal simplex pivot row (i.e., row v).

Example 7.1 (Young [612])

$$
\begin{array}{llrcrcl}
\text{Maximize} & 3x_1 & + & x_2 & = & x_0 \\
\text{subject to} & 2x_1 & + & 3x_2 & \leq & 6, \\
 & 2x_1 & - & 3x_2 & \leq & 3, \\
\text{and} & x_1, & & x_2 & \geq & 0, & \text{integer.}
\end{array}
$$

With nonnegative slack variables x_3 and x_4 we have Tableau #1. As $-3 < -1$, column 1 is the pivot column. The x_4 row generates the cut since $3/2 = \text{minimum}(3/2, 6/2)$. The Gomory slack variable is denoted by s_1. Its row is used as a pivot row, and after a pivot we obtain Tableau #2. The computations are then repeated. It requires three more cuts (their slack variables are s_2, s_3, and s_4) to finish the problem. These appear in Tableaux #2 through #5. In the tableau, the source row is indicated by a $\rightarrow$, the cut is added at the bottom, and the circled entry corresponds to the pivot element.

---

[3]Since this strategy is being adopted it is difficult to relate the Gomory slack variable to a row of the tableau. Therefore, the symbol s_t (rather than x_{m+n+1}) will denote the Gomory slack variable in the tth cut.

pivot column
↓

#1	1	$(-x_1)$	$(-x_2)$	
x_0	0	−3	−1	
x_1	0	−1	0	$(x_1 = x_2 = 0)$
x_2	0	0	−1	$p = 1, \lambda = 2$
x_3	6	2	3	Cut:
→ x_4	3	2	−3	$s_1 = \left[\frac{3}{2}\right] + \left[\frac{2}{2}\right](-x_1) + \left[\frac{-3}{2}\right](-x_2)$
s_1	1	①	−2	$= 1 + (-x_1) - 2(-x_2)$

pivot column
↓

#2	1	$(-s_1)$	$(-x_2)$	
x_0	3	3	−7	
x_1	1	1	−2	$(x_1 = 1, x_2 = 0)$
x_2	0	0	−1	$p = 2, \lambda = 7$
→ x_3	4	−2	7	Cut:
x_4	1	−2	1	$s_2 = \left[\frac{4}{7}\right] + \left[\frac{-2}{7}\right](-s_1) + \left[\frac{7}{7}\right](-x_2)$
s_2	0	−1	①	$= 0 - (-s_1) + (-x_2)$

pivot column
↓

#3	1	$(-s_1)$	$(-s_2)$	
x_0	3	−4	7	
x_1	1	−1	2	$(x_1 = 1, x_2 = 0)$
x_2	0	−1	1	$p = 1, \lambda = 5$
→ x_3	4	5	−7	Cut:
x_4	1	−1	−1	$s_3 = \left[\frac{4}{5}\right] + \left[\frac{5}{5}\right](-s_1) + \left[\frac{-7}{5}\right](-s_2)$
s_3	0	①	−2	$= 0 + (-s_1) - 2(-s_2)$

pivot column
↓

#4	1	$(-s_3)$	$(-s_2)$
x_0	3	4	-1
x_1	1	1	0
x_2	0	1	-1
$\rightarrow x_3$	4	-5	3
x_4	1	1	-3
s_4	1	-2	①

$(x_1 = 1, x_2 = 0)$

$p = 2, \lambda = 3$

Cut:

$s_4 = \left[\frac{4}{3}\right] + \left[\frac{-5}{3}\right](-s_3) + \left[\frac{3}{3}\right](-s_2)$

$= 1 - 2(-s_3) + (-s_2)$

#5	1	$(-s_3)$	$(-s_4)$
x_0	4	2	1
x_1	1	1	0
x_2	1	-1	1
x_3	1	1	-3
x_4	4	-5	3

$(x_1 = 1, x_2 = 1)$

Integer optimal

The optimal integer solution appears in Tableau #5 and is $x_0 = 4$, with $x_1 = x_2 = x_3 = 1$ and $x_4 = 4$. Writing the added inequalities in terms of x_1 and x_2, the original nonbasic variables, yields

$$
\begin{aligned}
s_1 &= 1 - x_1 + 2x_2 \geq 0, \\
s_2 &= 1 - x_1 + x_2 \geq 0, \\
s_3 &= 1 - x_1 \geq 0, \\
s_4 &= 2 - x_1 - x_2 \geq 0.
\end{aligned}
$$

Clearly, these must be all-integer inequalities. (Why?) The tableaux are related to (x_1, x_2) space in Fig. 7.1.

Observe that Tableaux #2, #3, and #4 correspond to the same point. This happened because the constant element in the Gomory cut was 0 in Tableaux #2 and #3. Thus, after pivoting, the zero column did not change. In fact, this may occur for an infinite number of tableaux, in which case the RPA will not converge (see Mathis [418] or Haas,Mathur,

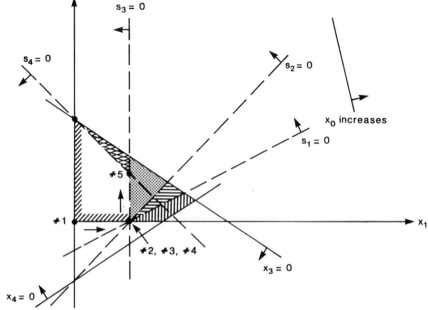

Figure 7.1: Graph for Example 7.1

and Salkin[302]). An example of such a program appears in Appendix A. In the next section we modify the RPA to eliminate the possibility of this phenomenon and hence guarantee a finite algorithm.

7.3 A Convergent Algorithm: The Simplified Primal Algorithm (SPA)

The convergent algorithm developed in this section, entitled the Simplified Primal Algorithm (SPA), is essentially the Rudimentary Algorithm with three modifications. The first is the adjoining of a primal feasible all-integer equation, called the *reference row*, indexed by L, whose coefficients a_{Lj} $(j = 0, 1, \ldots, n)$ satisfy certain properties with relation to the tableau's columns. The second is the selection of the pivot column, which is accomplished by utilizing the coefficients a_{Lj}. The third modification is concerned with the selection of the source row. This is usually done by reference to the size of the coefficients in the pivot column. We present the details. For ease of exposition, the first two modifications are discussed simultaneously.

7.3.1 Modifications 1 and 2: Introducing a Reference Row and Selecting the Pivot Column

Initially we adjoin to the bottom of the tableau an all-integer constraint or reference row which is satisfied by every integer feasible solution. This row, indexed by L, is written as:

$$x_L = a_{L0} + \sum_{j=1}^{n} a_{Lj}(-x_j) \geq 0, \tag{5}$$

where x_L is a nonnegative slack variable. The coefficients in (5) must also satisfy certain other properties. Before presenting these we need to determine the pivot column.

The Pivot Column

For each column $\boldsymbol{\alpha}_j$ $(j \geq 1)$ in the tableau having $a_{Lj} \neq 0$, we will define the vector $\mathbf{r}_j = \boldsymbol{\alpha}_j / a_{Lj}$. The pivot column $\boldsymbol{\alpha}_p$ will then be selected so that

$$\mathbf{r}_p \overset{\mathcal{L}}{\prec} \mathbf{r}_j \text{ for every } j \text{ with } a_{Lj} > 0. \tag{6}$$

Observe that, since the minus identity matrix is always present in the tableau, there cannot exist two columns $\boldsymbol{\alpha}_s, \boldsymbol{\alpha}_k$ with $s \neq k$ $(s, k \geq 1)$ such that $\mathbf{r}_s = \mathbf{r}_k$. Hence, (6) determines the pivot column uniquely. Also, note that if there exists an $a_{Lj} > 0$ $(j \geq 1)$ with $\boldsymbol{\alpha}_j \overset{\mathcal{L}}{\prec} \mathbf{0}$, then $\mathbf{r}_p$ exists and is lexicographically negative.

The Reference Row

To agree with the pivot column selection rule and to guarantee finiteness, we require that the coefficients in (5) satisfy

$$\text{(I)} \quad \boldsymbol{\alpha}_j \overset{\mathcal{L}}{\prec} \mathbf{0} \Longrightarrow a_{Lj} > 0 \quad (j = 1, \ldots, n)$$

$$\text{and} \quad \text{(II)} \quad a_{Lj} < 0 \Longrightarrow \mathbf{r}_j \overset{\mathcal{L}}{\prec} \mathbf{r}_p \quad (j = 1, \ldots, n).$$

Requirement (I) ensures that each column representing a dual variable with a nonpositive value will be considered as a candidate for the pivot column. In particular, if the tableau is not optimal, the pivot column always exists. The first and second requirements with $a_{Lp} > 0$ guarantee that

$$a_{Lj}\boldsymbol{\alpha}_p \overset{\mathcal{L}}{\prec} a_{Lp}\boldsymbol{\alpha}_j$$

for all $j \geq 1$ (cf. Lemma 7.3). As we shall see, these facts are sufficient for convergence.

Note that an equation with all positive coefficients will automatically satisfy (I) and (II). In particular, we can take $a_{Lj} = 1$ $(j = 1, \ldots, n)$; then to ensure that inequality (5) is always satisfied, a_{L0} is set equal to an upper bound for $\sum_{j=1}^{n} x_j$. (This is the reference row used by Young [614].) A less obvious set of a_{Lj} coefficients that have the prescribed properties is described in Section 7.4.

Interestingly, if the coefficients in the reference row satisfy (I) and (II) in the first tableau, they also satisfy these requirements in successive tableaux (cf Theorem 7.6). Using this result it can also be shown that $\bar{\mathbf{r}}_{\bar{p}} \overset{\mathcal{L}}{\succ} \mathbf{r}_p$ (cf. Corollary 7.5.1), where bars correspond to terms in the next tableau ($\bar{p}$ is the pivot column index at the next tableau). Finally, by employing an "acceptable source row selection rule" we can prove that within a finite number of iterations, $\mathbf{r}_p \overset{\mathcal{L}}{\succ} 0$, which will mean that the tableau is optimal. The proofs of these results are left to Section 7.7. We now discuss acceptable source row selection rules.

7.3.2   Modification 3: Acceptable Source Row Selection Rules

Recall from the RPA that once the pivot column $\boldsymbol{\alpha}_p$ has been found we may determine

$$\theta_p = \underset{a_{ip} > 0}{\text{minimum}} \left(\frac{a_{i0}}{a_{ip}}\right).$$

Then, to maintain primal feasibility, only constraints which satisfy $[a_{i0}/a_{ip}] \leq \theta_p$ with $a_{ip} > 0$ $(i \geq 1)$ can be selected as the generating row. Let $V(p)$ be the set of these "legitimate source rows." Specifically,

$$V(p) = \left\{ i \middle| 0 \leq \left[\frac{a_{i0}}{a_{ip}}\right] \leq \theta_p, a_{ip} > 0, i = 1, \ldots, m+n, L \right\}.$$

If there is only one element in $V(p)$ there is no choice as to the generating row. However, when more than one such row is available we must, to ensure convergence, invoke an "acceptable source row selection rule." The following are acceptable rules.

Rule 1 Choose any row in $V(p)$ that ensures $a_{Lp} \leq a_{L0}$ at finite intervals (i.e., after a finite number of tableaux), and which also periodically reduces a_{ip}/a_{Lp} for the smallest $i \geq 1$ such that $(a_{ip}/a_{Lp}) > a_{i0}$.

To implement Rule 1 we utilize the following result.

Theorem 7.1 *Let the source row be indexed by i. Thus the pivot row is*

$$s = \left[\frac{a_{i0}}{a_{ip}}\right] + \sum_{j=1}^{n} \left[\frac{a_{ij}}{a_{ip}}\right] (-x_{J(j)}).$$

(J is the set of indices corresponding to the current nonbasic variables and $J(j)$ is the jth element in J.) Then, subsequent to the pivot,

$$\bar{a}_{ip} = -a_{ip} \quad and$$
$$\bar{a}_{ij} < a_{ip} \quad for \ j \neq p.$$

(Bars denote elements after pivoting.)

Proof: Since the pivot element $[a_{ip}/a_{ip}] = 1$, we have $\bar{a}_{ip} = a_{ip}/(-1) = -a_{ip}$. Also, for $j \neq p$,

$$\bar{a}_{ij} = a_{ij} - a_{ip}\left[\frac{a_{ij}}{a_{ip}}\right] < a_{ij} - a_{ip}\left(\frac{a_{ij}}{a_{ip}} - 1\right) = a_{ip}.$$

The inequality sign appears since $a_{ip} \geq 1$ and $y \geq [y] > y - 1$ for any number y. ∎

Now, with reference to Rule 1, suppose $a_{ip}/a_{Lp} > a_{i0}$. As $a_{Lp} > 0$ and integral, it follows that $a_{Lp} \geq 1$; thus

$$a_{ip} \geq \frac{a_{ip}}{a_{Lp}} > a_{i0}.$$

Using $a_{ip} > a_{i0}$ with $a_{i0} \geq 0$ gives

$$\left\lfloor \frac{a_{i0}}{a_{ip}} \right\rfloor < 1, \quad \text{or} \quad \left\lfloor \frac{a_{i0}}{a_{ip}} \right\rfloor = 0 \leq \theta_p < 1.$$

That is, row i has to be a legitimate source row as long as $a_{ip} > a_{i0}$. Observe that when $a_{ip} \leq a_{i0}$ we have $a_{ip}/a_{Lp} \leq a_{i0}$, since $a_{Lp} > 1$. Thus, if initially $a_{ip}/a_{Lp} > a_{i0}$ (which implies $a_{ip} > a_{i0}$), and subsequently a_{ip} becomes less than or equal to a_{i0}, then a_{ip}/a_{Lp} has been reduced.

Suppose that so long as $a_{ip}/a_{Lp} > a_{i0}$, row i is selected as the source row. Then during these iterations the zero column in the tableaux will remain the same. (This follows since $[a_{i0}/a_{ip}]$, the element in the pivot row and zero column will be 0.) Hence, using the previous theorem, we have

$$a_{ip} - a_{i0} > \bar{a}_{i\bar{p}} - \bar{a}_{i0}, \tag{7}$$

where $\bar{p}$ is the pivot column index at the tableau obtained after pivoting on $[a_{ip}/a_{ip}]$. (Inequality (7) is true because $a_{ip} > \bar{a}_{i\bar{p}}$ and $a_{i0} = \bar{a}_{i0}$. Note that p cannot equal $\bar{p}$ since $\bar{a}_{0p} > 0$.) Therefore, (7) means that after each pivot the difference $a_{ip} - a_{i0}$ decreases, and because all terms are integral, eventually $a_{ip} - a_{i0}$ will become nonpositive–in which case $a_{ip} \leq a_{i0}$. By the previous discussion this implies that a_{ip}/a_{Lp} has been reduced. Of course, the tableau can become dual feasible during the process, in which case the algorithm terminates. Hence, we have shown that Rule 1 can be implemented by repeatedly selecting the first row i which has $a_{ip}/a_{Lp} > a_{i0}$ as the source equation and by continuing to choose this row until a_{ip}/a_{Lp} is reduced or until row i becomes ineligible. (If $i = L$ we do not stop using this row when a_{L0}/a_{Lp} is decreased. It is selected so long as $a_{Lp} > a_{L0}$.)

Suppose now we consider rules which have the following implication. (These are the ones used by Young [614].)

> For any row i, tableaux must occur at finite
> intervals in which $a_{ip} \leq a_{i0}$. $\tag{8}$

First observe that (8) does not require $a_{ip} \leq a_{i0}$ for all i in any tableau. In addition, as shown by Salkin [505], rules which satisfy (8) will also satisfy Rule 1. This follows since we have shown that $a_{ip}/a_{Lp} > a_{i0}$ implies $a_{ip} > a_{i0}$. Thus a rule with this implication must eventually

select the smallest $i \geq 1$ with $a_{ip}/a_{Lp} > a_{i0}$. This rule will then ensure
for this i that a_{ip} becomes less than or equal to a_{i0}. But when $a_{ip} \leq$
a_{i0}, we have $a_{ip}/a_{Lp} \leq a_{i0}$; or a_{ip}/a_{Lp} has been reduced. Hence, the
requirements of Rule 1 are met. Note that when $i = L$, implication (8)
and Rule 1 coincide.

We have just demonstrated that Rule 1 contains weaker requirements than those having the implication (8). Therefore, it suffices to
show that Rule 1 supports a finite algorithm. Then other rules implying
Rule 1 will automatically guarantee convergence. The finiteness proof
for Rule 1 is deferred to Section 7.7.

As the reader probably suspects, rules having the implication can
be devised using Theorem 7.1. In particular, any rule which eventually
selects a row i that has $a_{ip} > a_{i0}$ and continually uses this row as
the source row until $a_{ip} \leq a_{i0}$ will satisfy the implication. (Why?)
For completeness, we give two such rules which are from Young [614].
Recall that $V(p)$ is the set of legitimate source rows; that is, those that
will yield a cut or pivot row which maintains primal feasibility.

Rule 1(a)

(a) Randomly select a source row i from $V(p)$. Go to (b).

(b) Repeatedly select the row i chosen in Step (a) so long as
   $i \in V(p)$. In the first tableau for which $i \notin V(p)$ go to (a).

Rule 1(b)

(a) Select as the source row the row i from $V(p)$ which has been
   a member of $V(p)$ through the longest sequence of preceding
   iterations. Arbitrarily break ties. Go to (b).

(b) Repeatedly select the row chosen in Step (a) so long as $i \in V(p)$.
   In the first tableau for which $i \notin V(p)$ go to (a).[4]

During a previous discussion it was mentioned that the $\mathbf{r}_p$ column

---

[4]Actually this and, in fact, any acceptable rule need only be invoked when there
exist rows with $a_{ip} > a_{i0}$. At other times, any row i from $V(p)$ may generate the cut.
This is in the spirit of distinguishing between iterations during which $[a_{i0}/a_{ip}] = 0$
and those which have $[a_{i0}/a_{ip}] \geq 1$. The former type of iterations are referred to as
"stationary cycles" and the latter as "transition cycles" by Young [614]; see Problem
7.11.

increases lexicographically, and when it becomes lexicographically positive the tableau is optimal. Of course, this is not very obvious. A more evident push toward convergence is to have the pivot column α_p lexicographically increase. This suggests Glover's [244] Rule 2, which also produces a finite algorithm.

Rule 2 At finite intervals designate the row from $V(p)$ that will result in $\bar{a}_{0\bar{p}} > a_{0p}$ as the source row. (Bars denote terms in the next tableau.) Continue to use this rule until there are none that satisfy it. When that occurs select any row from $V(p)$ as the generating row.

Implementing Rule 2 may require computing for a given legitimate source row, $\bar{a}_{Lj}$ for $j = 1, \ldots, n$; and then for positive $\bar{a}_{Lj}$ finding $\bar{a}_{0j}$ and then $\bar{a}_{0\bar{p}}$. If $\bar{a}_{0\bar{p}}$ is not greater than a_{0p}, another legitimate source row must be selected and the computations repeated. Specifically, we might do the following:

(a) Select the first index v in $V(p)$ as the source row. Go to (b).

(b) Generate the pivot row from row v. With α_p as the pivot column, find the last n components in row L at the next tableau. For each $\bar{a}_{Lj} > 0$ find $\bar{a}_{ij}$ for $i = 0, 1, \ldots$, so that the lexicographically smallest column $\bar{\mathbf{r}}_j = \bar{\alpha}_j / \bar{a}_{Lj}$ can be determined. Label this the pivot column $\alpha_{\bar{p}}$. Go to (c).

(c) If the first element in $\bar{\alpha}_{\bar{p}}$, that is, $\bar{a}_{0\bar{p}}$, yields $\bar{a}_{0\bar{p}} > a_{0p}$, row v satisfies Rule 2, and therefore update the entire tableau. Find the new set of legitimate source rows and go to (a). If $\bar{a}_{0\bar{p}} \le a_{0p}$, restore the original elements a_{ij} in the tableau and select the next row from $V(p)$. Go to (b). If all rows from $V(p)$ have been considered, go to (d).

(d) For a finite number of tableaus, designate any row from the set of legitimate source rows as the generating row. Then, return to (a).[5]

Observe that if we can show Rule 2 results in a finite algorithm, it is

---

[5]This may seem computationally inefficient but keep in mind that integer arithmetic is very fast in a computer. In fact, Arnold and Bellmore [11] indicate that this rule seems to perform better than most.

impossible to remain at step (d) for an infinite number of tableaux, since if this did occur, the tableau could never exhibit dual feasibility. The finiteness proof with Rule 2 is left to Section 7.7.

When the L row is a legitimate source row it may, if selected, automatically imply Rule 2. To justify this we need the next result.

Theorem 7.2 *Suppose* $a_{0p} < 0$ *and* $\bar{\mathbf{r}}_{\bar{p}} \overset{\mathcal{L}}{\succ} \mathbf{r}_p$, *then* $\bar{a}_{L\bar{p}} < a_{Lp}$ *implies* $\bar{a}_{0\bar{p}} > a_{0p}$.

Proof: By its definition, $\bar{\mathbf{r}}_{\bar{p}} = \bar{\boldsymbol{\alpha}}_{\bar{p}}/\bar{a}_{L\bar{p}}$ where $\bar{a}_{L\bar{p}} > 0$, and $\mathbf{r}_p = \boldsymbol{\alpha}_p/a_{Lp}$ where $a_{Lp} > 0$. Thus, $\bar{\mathbf{r}}_{\bar{p}} \overset{\mathcal{L}}{\succ} \mathbf{r}_p$ means $a_{Lp}\bar{\boldsymbol{\alpha}}_{\bar{p}} \overset{\mathcal{L}}{\succ} \bar{a}_{L\bar{p}}\boldsymbol{\alpha}_p$, which implies $a_{Lp}\bar{a}_{0\bar{p}} \geq \bar{a}_{L\bar{p}}a_{0p}$. But $a_{Lp} > 0, a_{0p} < 0$, and by hypothesis, $\bar{a}_{L\bar{p}} < a_{Lp}$; hence $\bar{a}_{0\bar{p}}/a_{0p} \leq \bar{a}_{L\bar{p}}/a_{Lp} < 1$. As $a_{0p} < 0$, we then have $\bar{a}_{0\bar{p}} > a_{0p}$. $\blacksquare$

Recall that if the requirements (I) $\boldsymbol{\alpha}_j \overset{\mathcal{L}}{\prec} \mathbf{0}$ implies $a_{Lj} > 0$ and (II) $a_{Lj} < 0$ implies $\mathbf{r}_j \overset{\mathcal{L}}{\prec} \mathbf{r}_p$ are initially satisfied by the L row, then we can show that it always contains them. Thus, unless the tableau is optimal, $a_{0p} < 0$. Also, using (I) and (II), we can show that $\bar{\mathbf{r}}_{\bar{p}} \overset{\mathcal{L}}{\succ} \mathbf{r}_p$. That is, until the optimal tableau is reached, the supposition of the theorem is true. The proofs of these facts are in Section 7.7.

Now, with regard to implementing Rule 2, suppose $a_{Lp} > a_{L0}$. Then, since $[a_{L0}/a_{Lp}] = 0$, row L must be a legitimate source row. If it is used as the generating row, by Theorem 7.1 we have $a_{Lp} > \bar{a}_{L\bar{p}}$, and thus, by Theorem 7.2, $\bar{a}_{0\bar{p}} > a_{0p}$. That is, when $a_{Lp} > a_{L0}$, row L, if selected as the source row, will automatically satisfy Rule 2.

We have presented some acceptable source row selection rules. Others, such as variations of those given here, are possible. These become more apparent after examining the finiteness proofs. Several are suggested in the problems and another is in Section 7.6. We now combine the modifications to the Rudimentary Primal Algorithm and formally describe the Simplified Primal Algorithm.[6]

7.3.3 The Simplified Primal Algorithm (SPA)

Step 1 Start with a primal feasible all-integer tableau.[7] If such a tableau cannot be found, the integer program is infeasible–

---

[6]Young's [614] version of this algorithm appears in Problem 7.8.

[7]Salkin, Shroff, and Mehta [518] have shown that rational coefficients will suffice; see Problem 7.7

terminate. Define a reference row, indexed by L, which does not eliminate any integer feasible solutions and also has the properties (I) $\boldsymbol{\alpha}_j \overset{\mathcal{L}}{\prec} 0$ implies $a_{Lj} > 0$ and (II) $a_{Lj} < 0$ implies $\mathbf{r}_j \overset{\mathcal{L}}{\prec} \mathbf{r}_p$ (p is the pivot column index). Adjoin this row to the bottom of the tableau. Go to Step 2.

Step 2 For each j with $a_{Lj} \neq 0$ define $\mathbf{r}_j = \boldsymbol{\alpha}_j / a_{Lj}$. Select the pivot column indexed by p so that $\mathbf{r}_p \overset{\mathcal{L}}{\prec} \mathbf{r}_j$ for every $j \geq 1$ with $a_{Lj} > 0$. If $\mathbf{r}_p \overset{\mathcal{L}}{\succ} 0$ or it cannot be defined, the tableau is optimal–terminate. Go to Step 3.

Step 3 Define the set of legitimate source rows

$$V(p) = \left\{ i | 0 \leq \left[\frac{a_{i0}}{a_{ip}} \right] \leq \theta_p, a_{ip} > 0, i = 1, \ldots, m + n, L \right\},$$

where $\theta_p = \text{minimum} \{ (a_{i0}/a_{ip}) | a_{ip} > 0 \}$. Select a source row v from $V(p)$ according to an acceptable source row selection rule. Generate from this row an all-integer Gomory cut where $\lambda = a_{vp}$. Use this constraint as the pivot row. Go to Step 4.

Step 4 Perform a primal simplex pivot operation and return to Step 2.

We have mentioned on several occasions that the $\mathbf{r}_p$ column increases lexicographically from tableau to tableau and that when it becomes lexicographically positive the optimal solution has been reached. As the selection of the pivot column $\boldsymbol{\alpha}_p$ and the magnitude of the elements in $\mathbf{r}_p$ depend on the coefficients in the reference row, the initial determination of the coefficients a_{Lj} ($j = 0, 1, \ldots, n$) may have a marked influence on the algorithm's convergence rate. Up to this time we noted that one permissible reference equation is

$$x_L = a_{L0} + \sum_{j=1}^{n} (-x_j),$$

where a_{L0} is an upper bound on the $\sum_{j=1}^{n} x_j$. The next section discusses an alternate permissible equation which can sometimes reduce the number of iterations required to obtain an optimal solution.

7.4   A Second Reference Equation: Using the Dual Variables

The reference row, as already mentioned, should be a constraint which does not eliminate any integer feasible solutions. Its coefficients a_{Lj} ($j = 1, \ldots, n$) must also satisfy (I) $\boldsymbol{\alpha}_j \overset{\mathcal{L}}{\prec} \mathbf{0}$ implies $a_{Lj} > 0$ and (II) $a_{Lj} < 0$ implies $\mathbf{r}_j \overset{\mathcal{L}}{\prec} \mathbf{r}_p$, where $\mathbf{r}_j = \boldsymbol{\alpha}_j / a_{Lj}$. We now proceed to develop such a row.

To simplify the notation, we put the identity equalities in the last n rows of the tableau. That is, the second through $m+1$ rows of the tableau contain the initial constraints. Then if we temporarily drop the identity equations and the a_{00} term, the original integer program may be written as

$$\begin{aligned}
\text{maximize} \quad & \mathbf{cx} = x_0 \\
\text{subject to} \quad & \mathbf{Ax} \leq \mathbf{b}, \quad \mathbf{x} \geq \mathbf{0}, \\
\text{and} \quad & \mathbf{x} \quad \text{integer};
\end{aligned}$$

here $\mathbf{c} = (-a_{0j})$, $\mathbf{x} = (x_j)$, $\mathbf{A} = (a_{ij})$, and $\mathbf{b} = (a_{i0})$ ($i = 1, \ldots, m$, and $j = 1, \ldots, n$). Omitting the $\mathbf{x}$ integer requirements and taking the dual of the resulting linear program (cf. Chapter 2) yields

$$\begin{aligned}
\text{maximize} \quad & \mathbf{wb} \\
\text{subject to} \quad & \mathbf{wA} \geq \mathbf{c} \\
\text{and} \quad & \mathbf{w} \geq \mathbf{0},
\end{aligned}$$

where $\mathbf{w} = (w_1, \ldots, w_m)$. If we label the columns of $\mathbf{A}$ as $\boldsymbol{\alpha}_j$ ($j = 1, \ldots, n$), then column j of the tableau or $\boldsymbol{\alpha}_j$ has the form

$$\begin{pmatrix} a_{0j} \\ \mathbf{a}_j \\ -\mathbf{e}_j \end{pmatrix} \quad \text{for } j = 1, \ldots, n,$$

where the unit column $\mathbf{e}_j$, is the jth column of an n by n identity matrix. Also, if $a_{0j} \geq 0$ for some $j \geq 1$, we shall assume that the coefficients

initially satisfy $a_{1j} > 0$ if $a_{0j} \geq 0$ and $a_{1j} = 0$ if $a_{0j} < 0$ $(j = 1, \ldots, n)$.[8]
We can now present the following result.

Theorem 7.3 *Let $\bar{w}$ be a feasible solution to the dual linear program.
Then the constraint*

$$x_L = a_{L0} + \sum_{j=1}^{n} a_{Lj}(-x_j) \geq 0,$$

*where $a_{L0} = \bar{w}b$ and $a_{Lj} = \bar{w}\alpha_j$, is a permissible reference row. That
is, it has all the required properties.*

Proof: (Property I: $\alpha_j \overset{\mathcal{L}}{\prec} 0 \implies a_{Lj} > 0$). Since $\bar{w}$ is a feasible solution
to the dual problem we have $\bar{w}A \geq c$, or equivalently,

$$\bar{w}\alpha_j = a_{Lj} \geq -a_{0j} \quad (j = 1, \ldots, n). \tag{9}$$

Now, by the assumption that $a_{0j} = 0$ implies $a_{1j} > 0$, we have that
$\alpha_j \overset{\mathcal{L}}{\prec} 0$ implies $a_{0j} < 0$; then, using (9), we have that $\alpha_j \overset{\mathcal{L}}{\prec} 0$ implies
$a_{Lj} > 0$, which is property (I).

(Property II: $a_{Lj} < 0 \implies r_j \overset{\mathcal{L}}{\prec} r_p$). By inequality (9), if $a_{Lj} < 0$,
then $a_{0j}/a_{Lj} < -1$, and if $a_{Lj} > 0$, then $a_{0j}/a_{Lj} \geq -1$. In particular,
since $a_{Lp} > 0$, we have $a_{0p}/a_{Lp} \geq -1$.[9] Thus, $a_{Lj} < 0$ implies $a_{0j}/a_{Lj} \leq$
a_{0p}/a_{Lp}. As these fractions are the first terms in r_j and r_p, respectively,
either $r_j \overset{\mathcal{L}}{\prec} r_p$ or $a_{0j}/a_{Lj} = a_{0p}/a_{Lp}$. The first case is property II; that

---

[8]The assumption is necessary to support the proof of Theorem 7.3. If the coefficients do not initially satisfy the assumption, then a constraint of the form

$$S = b_0 + \sum_{j=1}^{n} b_j(-x_j) \geq 0,$$

where $b_j = 1$ if $a_{0j} \geq 0$ and $b_j = 0$ if $a_{0j} < 0$, can be inserted in the first row of
the tableau. Naturally, b_0 is an upper bound on $\sum x_j$, where the sum is taken over
those j such that $b_j = 1$.

[9]We are assuming that an $a_{Lj} > 0$ $(j \neq 0)$ exists. If $a_{Lj} \leq 0$ $(j = 1, \ldots, n)$, the
tableau must be optimal; see Corollary 7.6.1.

is, $a_{Lj} < 0$ implies $\mathbf{r}_j \overset{\mathcal{L}}{\prec} \mathbf{r}_p$. Suppose $a_{0j}/a_{Lj} = a_{0p}/a_{Lp}$. Using $a_{Lj} < 0$ with (9) gives $a_{0j} > 0$. Then, by the assumption $a_{1j} > 0$ if $a_{0j} \geq 0$, the second term in $\mathbf{r}_j$ or $a_{1j}/a_{Lj} < 0$. But the second term in $\mathbf{r}_p$ or a_{1p}/a_{Lp} is 0 because $a_{0p} < 0$, which means (by the assumption $a_{1j} = 0$ whenever $a_{0j} < 0$) that $a_{1p} = 0$. Hence, again $a_{Lj} < 0$ implies $\mathbf{r}_j \overset{\mathcal{L}}{\prec} \mathbf{r}_p$. This verifies property (II).

To show that the reference row does not eliminate any integer feasible solutions, consider the linear programming constraints $\mathbf{Ax} \leq \mathbf{b}$. As $\bar{\mathbf{w}} \geq \mathbf{0}$ we have $\bar{\mathbf{w}}\mathbf{Ax} \leq \bar{\mathbf{w}}\mathbf{b}$. But $\bar{\mathbf{w}}\mathbf{A} = (a_{L1}, \ldots, a_{Ln})$ and $\bar{\mathbf{w}}\mathbf{b} = a_{L0}$. Thus, the last inequality means that $\sum_{j=1}^{n} a_{Lj}x_j \leq a_{L0}$, or after adding a nonnegative slack variable x_L, the constraint becomes

$$x_L = a_{L0} + \sum_{j=1}^{n} a_{Lj}(-x_j) \geq 0,$$

which is the reference row. We have just shown that the constraint $x_L \geq 0$ is implied by the inequalities $\mathbf{Ax} \leq \mathbf{b}$; and therefore every feasible solution to $\mathbf{Ax} \leq \mathbf{b}$ satisfies $\sum_{j=1}^{n} a_{Lj}x_j \leq a_{L0}$. In particular, every integer feasible solution to $\mathbf{Ax} \leq \mathbf{b}$ gives x_L a nonnegative value.

Finally, integer values for the coefficients a_{Lj} $(j = 0, \ldots, n)$ have always been used up to this time. Clearly, as they are currently defined, the coefficients will not necessarily be integers. They will, however, be rational. Actually, it is not really necessary for x_L to be an integer variable as the convergence arguments using integer a_{Lj} can be extended to rational a_{Lj}.[10] Nevertheless, if an all-integer tableau is desired, we can clear fractions in the reference row and use the resulting all integer constraint (see Example 7.2). It is a simple matter to show that if $x_L \geq 0$ satisfies the properties necessary of a reference row, then $kx_L \geq 0$ has these properties, where k is a positive number; see Problem 7.17. ∎

To make this and the previous discussions concrete, two examples are presented. The first appears in Glover [244] and the second in Young [614]. We follow their computations and strategies.

---

[10]The derivation of the all-integer cut only required the source row's slack variable to be nonnegative. That is, it did not have to be integer constrained (see Section 6.4, Chapter 6). Thus the reference row is an eligible source row.

7.5   Illustrations (SPA)

We shall solve two examples. In each case the pivot or cut rows will not be kept. A circle will denote the pivot element, s_j the jth Gomory slack variable, p the pivot column index, and $V(p)$ the set of legitimate source rows.

Example 7.2 (Glover [244])   The problem is

$$
\begin{array}{lrcrcrcl}
\text{maximize} & 4x_1 & + & 6x_2 & + & 3x_3 & = & x_0 \\
\text{subject to} & x_1 & + & 2x_2 & & & \leq & 5, \quad (x_4) \\
& 9x_1 & + & 2x_2 & - & 4x_3 & \leq & 8, \quad (x_5) \\
& -3x_1 & - & 2x_2 & + & 2x_3 & \leq & 1, \quad (x_6) \\
& -5x_1 & + & 4x_2 & + & 6x_3 & \leq & 16, \quad (x_7) \\
\text{and} & x_1, & & x_2, & & x_3 & \geq & 0, \quad \text{integer.}
\end{array}
$$

Before giving the first tableau we note that the reference row defined by the dual variables, as described in the previous section, will be employed. However, to maintain all-integer tableaux the constraint will initially be cleared of fractions. So to obtain the L row we omit the x_j integer ($j = 1, 2, 3$) requirements from the program and construct its dual as below.

$$
\begin{array}{lrcrcrcrl}
\text{minimize} & 5w_1 & + & 8w_2 & + & w_3 & + & 16w_4 \\
\text{subject to} & w_1 & + & 9w_2 & - & 3w_3 & - & 5w_4 & \geq 4, \\
& 2w_1 & + & 2w_2 & - & 2w_3 & + & 4w_4 & \geq 6, \\
& & - & 4w_2 & + & 2w_3 & + & 6w_4 & \geq 3, \\
\text{and} & w_1, & & w_2, & & w_3, & & w_4 & \geq 0.
\end{array}
$$

We choose the optimal dual feasible solution to determine the coefficients of the reference row.[11] After solving the dual problem we get $w_1 = 0, w_2 = 39/34, w_3 = 0$, and $w_4 = 43/34$. Then,

$$
a_{L0} = (0, 39/34, 0, 43/34) \begin{pmatrix} 5 \\ 8 \\ 1 \\ 16 \end{pmatrix} = 1000/34
$$

---

[11]This does not mean that this feasible solution produces a better reference row than another. Interestingly, however, when the optimal **w** is used the equation may become what is termed a strongest "surrogate constraint"; see Chapter 9, Section 9.3.7.

and

$$(a_{L1}, a_{L2}, a_{L3}) = (0, 39/34, 0, 43/34) \begin{pmatrix} 1 & 2 & 0 \\ 9 & 2 & -4 \\ -3 & -2 & 2 \\ -5 & 4 & 6 \end{pmatrix}$$

$$= 1/34(136, 250, 102).$$

Hence, the L constraint is

$$(136/34)x_1 + (250/34)x_2 + (102/34)x_3 \le 1000/34;$$

or, after multiplying by 17 and adding the nonnegative slack variable x_L, the reference row is

$$x_L = 500 + 68(-x_1) + 125(-x_2) + 51(-x_3) \ge 0.$$

Adjoining this row to the constraint set and adding the nonnegative slack variables $x_4, x_5, x_6,$ and x_7 to the four original inequalities yields Tableau #1.

#1	1	$(-x_1)$	$(-x_2)$	$(-x_3)$	
x_0	0	−4	−6	−3	
x_4	5	1	2	0	$p = 3,$
x_5	8	9	2	−4	$\theta_p = 1/2$
x_6	1	−3	−2	2	$V(p) = \{3\}$
x_7	16	−5	4	6	$\lambda = 2$
x_L	500	68	125	51	
x_1	0	−1	0	0	
x_2	0	0	−1	0	
x_3	0	0	0	−1	
s_1	0	−2	−1	①	

source row → x_6

cut → s_1

Note that a_{01}, a_{02}, and a_{03} are negative. Hence, it is unnecessary to consider the coefficients in the x_4 row, or to insert a new constraint between x_0 and x_4 to support the lexicographic ordering explained in the previous section. Finally, we shall select the source row according to a slight variation of Glover's Rule 2. Specifically, whenever possible, the legitimate source row which maximizes $\bar{a}_{0\bar{p}}$ is picked. Ties are broken by selecting the row which maximizes the sum of the negative $\bar{a}_{0j}$ $(j \geq 1)$ and then by selecting the smallest index. The computations appear in the seven tableaux.

#2	1	$(-x_1)$	$(-x_2)$	$(-s_1)$	
x_0	0	-10	-9	3	
x_4	5	1	2	0	$p = 1,$
x_5	8	1	-2	4	$\theta_p = 1$
x_6	1	①	0	-2	$V(p) = \{3\}$
x_7	16	7	10	-6	$\lambda = 1$
x_L	500	170	176	-51	
x_1	0	-1	0	0	
x_2	0	0	-1	0	
x_3	0	-2	-1	1	

source row → (points to x_6 row)

Since $\lambda = 1$, a cut is not added to Tableau #2. In the next tableau there are three legitimate source rows; namely rows 2, 4, and 5. If row 2 is used, then $\lambda = 6$, $\bar{p} = 1$ and $\bar{a}_{o\bar{p}} = -7$; when row 4 is picked, $\lambda = 8$, $\bar{p} = 1$ and $\bar{a}_{o\bar{p}} = -7$; finally row 5 yields $\lambda = 289$, $\bar{p} = 2$ and $\bar{a}_{o\bar{p}} = -9$. A tie exists between rows 2 and 4. Row 2 would result in $\bar{a}_{01} = -7, \bar{a}_{02} = -26$ and $\bar{a}_{03} = 17$; and if row 4 is chosen, we get $\bar{a}_{01} = -7, \bar{a}_{02} = 8$ and $\bar{a}_{03} = 17$. As $-7 > -7 - 26$, row 4 is selected. Similar computations are performed at Tableaux #4, #5, and #6.

#3	1	$(-x_6)$	$(-x_2)$	$(-s_1)$	
x_0	10	10	−9	−17	
x_4	4	−1	2	2	
x_5	7	−1	−2	6	$p = 3,$
x_6	0	−1	0	0	$\theta_p = 9/8$
source row → x_7	9	−7	10	8	$V(p) = \{2,4,5\}$
x_L	330	−170	176	289	$\lambda = 8$
x_1	1	1	0	−2	
x_2	0	0	−1	0	
x_3	2	2	−1	−3	
cut → s_2	1	−1	1	①	

#4	1	$(-x_6)$	$(-x_2)$	$(-s_2)$	
x_0	27	−7	8	17	
x_4	2	1	0	−2	
source row → x_5	1	5	−8	−6	$p = 1,$
x_6	0	−1	0	0	$\theta_p = 1/5$
x_7	1	1	2	−8	$V(p) = \{2,5\}$
x_L	41	119	−113	−289	$\lambda = 5$
x_1	3	−1	2	2	
x_2	0	0	−1	0	
x_3	5	−1	2	3	
cut → s_3	0	①	−2	−2	

The primal feasible solution exhibited in Tableau #4 is optimal. Nevertheless, three more iterations are needed to achieve dual feasibility.

#5	1	$(-s_3)$	$(-x_2)$	$(-s_2)$	
x_0	27	7	-6	3	
x_4	2	-1	2	0	
x_5	1	-5	2	4	$p = 2,$
x_6	0	1	-2	-2	$\theta_p = 41/125$
x_7	1	-1	4	-6	$V(p) = \{2,4,5\}$
source row → $\quad x_L$	41	-119	125	-51	$\lambda = 125$
x_1	3	1	0	0	
x_2	0	0	-1	0	
x_3	5	1	0	1	
cut → $\quad s_4$	0	-1	$\boxed{1}$	-1	

#6	1	$(-s_3)$	$(-s_4)$	$(-s_2)$	
x_0	27	1	6	-3	
x_4	2	1	-2	2	
x_5	1	-3	-2	6	$p = 3,$
x_6	0	-1	2	-4	$\theta_p = 1/6$
x_7	1	3	-4	-2	$V(p) = \{2,5\}$
source row → $\quad x_L$	41	6	-125	74	$\lambda = 74$
x_1	3	1	0	0	
x_2	0	-1	1	-1	
x_3	5	1	0	1	
cut → $\quad s_5$	0	0	-2	$\boxed{1}$	

#7	1	$(-s_3)$	$(-s_4)$	$(-s_5)$	
x_0	27	1	0	3	
x_4	2	1	2	-2	
x_5	1	-3	10	-6	Integer Optimal
x_6	0	-1	-6	4	$x_0 = 27, x_1 = 3,$
x_7	1	3	-8	2	$x_2 = 0, x_3 = 5$
x_L	41	6	23	-74	
x_1	3	1	0	0	
x_2	0	-1	-1	1	
x_3	5	1	2	-1	

Tableau #7 exhibits the optimal integer solution. It is worth mentioning that, for this problem, the reference row reduced the number of iterations. In particular, Glover [244] points out that it would have taken 15 iterations to obtain the optimal tableau if we have used $x_L = 8 + (-x_1) + (-x_2) + (-x_3)$ with the same source row selection rules.

Example 7.3 (Young [614]) [12] The problem is

$$
\begin{array}{llllllll}
\text{maximize} & x_1 & + & x_2 & + & x_3 & = & x_0 \\
\text{subject to} & -4x_1 & + & 5x_2 & + & 2x_3 & \leq & 4, & (x_4) \\
& -2x_1 & + & 5x_2 & & & \leq & 5, & (x_5) \\
& 3x_1 & - & 2x_2 & + & 2x_3 & \leq & 6, & (x_6) \\
& 2x_1 & - & 5x_2 & & & \leq & 1, & (x_7) \\
\text{and} & x_1, & & x_2, & & x_3 & \geq & 0, & \text{integer.}
\end{array}
$$

In this example we shall use $x_1 + x_2 + x_3 \leq 10$ as the L constraint. The number 10 was selected by inspection of the constraints. Also, the source row will be selected using Rule 1(b). Recall that this se-

---

[12]We are following the SPA as listed in Section 7.3. Had Young's variation of this algorithm (see Problem 7.8) been used it would require a total of 19 tableaux to solve this example. These computations appear in his paper [614].

lects the row which has been in $V(p)$ the longest sequence of preceding tableaux. Ties are broken arbitrarily. With nonnegative slack variables $x_4, x_5, x_6,$ and x_7 we have the first tableau. It requires a total of seven iterations to reach optimality.

#1	1	$(-x_1)$	$(-x_2)$	$(-x_3)$	
x_0	0	−1	−1	−1	
x_4	4	−4	5	2	
x_5	5	−2	5	0	$p = 1,$
x_6	6	3	−2	2	$\theta_p = 1/2$
source row → x_7	1	2	−5	0	$V(p) = \{4\}$
x_L	10	1	1	1	$\lambda = 2$
x_1	0	−1	0	0	
x_2	0	0	−1	0	
x_3	0	0	0	−1	
cut → s_1	0	①	−3	0	

#2	1	$(-s_1)$	$(-x_2)$	$(-x_3)$	
x_0	0	1	−4	−1	
x_4	4	4	−7	2	
x_5	5	2	−1	0	$p = 2,$
source row → x_6	6	−3	7	2	$\theta_p = 6/7$
x_7	1	−2	1	0	$V(p) = \{3\}$
x_L	10	−1	4	1	$\lambda = 7$
x_1	0	1	−3	0	
x_2	0	0	−1	0	
x_3	0	0	0	−1	
cut → s_2	0	−1	①	0	

#3	1	$(-s_1)$	$(-s_2)$	$(-x_3)$	
x_0	0	-3	4	-1	
x_4	4	-3	7	2	
x_5	5	1	1	0	$p = 1,$
source row → x_6	6	4	-7	2	$\theta_p = 3/2$
x_7	1	-1	-1	0	$V(p) = \{3\}$
x_L	10	3	-4	1	$\lambda = 4$
x_1	0	-2	3	0	
x_2	0	-1	1	0	
x_3	0	0	0	-1	
cut → s_3	1	①	-2	0	

#4	1	$(-s_3)$	$(-s_2)$	$(-x_3)$	
x_0	3	3	-2	-1	
x_4	7	3	1	2	
source row → x_5	4	-1	3	0	$p = 2,$
x_6	2	-4	1	2	$\theta_p = 4/3$
x_7	2	1	-3	0	$V(p) = \{2\}$
x_L	7	-3	2	1	$\lambda = 3$
x_1	2	2	-1	0	
x_2	1	1	-1	0	
x_3	0	0	0	-1	
cut → s_4	1	-1	①	0	

#5	1	$(-s_3)$	$(-s_4)$	$(-x_3)$	
x_0	5	1	2	−1	
x_4	6	4	−1	2	
x_5	1	2	−3	0	$p = 3,$
x_6	1	−3	−1	2	$\theta_p = 1/2$
x_7	5	−2	3	0	$V(p) = \{3\}$
x_L	5	−1	−2	1	$\lambda = 2$
x_1	3	1	1	0	
x_2	2	0	1	0	
x_3	0	0	0	−1	
s_5	0	−2	−1	(1)	

source row → x_6

cut → s_5

#6	1	$(-s_3)$	$(-s_4)$	$(-s_5)$	
x_0	5	−1	1	1	
x_4	6	8	1	−2	
x_5	1	2	−3	0	$p = 1,$
x_6	1	1	1	−2	$\theta_p = 1/2$
x_7	5	−2	3	0	$V(p) = \{1, 2\}$
x_L	5	1	−1	−1	$\lambda = 2$
x_1	3	1	1	0	
x_2	2	0	1	0	
x_3	0	−2	−1	1	
s_6	0	(1)	−2	0	

source row → x_5

cut → s_6

Observe that at Tableau #6 we arbitrarily selected the second row (i.e., the x_5 equation) as the generating row.[13] Then at Tableau #7, row 1

---

[13]It turns out that the x_4 equation would have produced a pivot row that yields dual feasibility at the next tableau.

reappears in $V(p)$ (and row 2 does not) so it is selected as the source row. Tableau #8 is optimal with $x_0 = 5, x_1 = 3, x_2 = 2$, and $x_3 = 0$.

#7	1	$(-s_6)$	$(-s_4)$	$(-s_5)$	
x_0	5	1	−1	1	
x_4	6	−8	17	−2	
x_5	1	−2	1	0	$p = 4,$
x_6	1	−1	3	−2	$\theta_p = 1/3$
x_7	5	2	−1	0	$V(p) = \{1, 3\}$
x_L	5	−1	1	−1	$\lambda = 17$
x_1	3	−1	3	0	
x_2	2	0	1	0	
x_3	0	2	−5	1	
s_7	0	−1	①	−1	

source row → (row x_4)

cut → (row s_7)

#8	1	$(-s_6)$	$(-s_7)$	$(-s_5)$	
x_0	5	0	1	0	
x_4	6	9	−17	15	
x_5	1	−1	−1	1	Integer Optimal
x_6	1	2	−3	1	$x_0 = 5, x_1 = 3,$
x_7	5	1	1	−1	$x_2 = 2, x_3 = 0$
x_L	5	0	−1	0	
x_1	3	2	−3	3	
x_2	2	1	−1	1	
x_3	0	−3	5	−4	

Before turning to the convergence proofs we discuss an interesting phenomenon which may sometimes allow the algorithm to terminate before reaching dual feasibility.

7.6   Optimality without Dual Feasibility

Clearly, the tableau is optimal when dual feasibility is reached. However, as Glover [244] has shown (cf. Example 7.3), it is sometimes possible to terminate the computations prior to attaining dual feasibility and yet have an optimal primal solution. This is suggested by the following result.

Theorem 7.4 *Suppose the current tableau does not exhibit a dual feasible solution and the reference row has properties (I) and (II). Then*

$$a_{00} + \left[-a_{L0} \frac{a_{0p}}{a_{Lp}} \right]$$

is an upper bound[14] for the optimal value of x_0.

Proof: As usual, select the pivot column so that $\mathbf{r}_p \overset{\mathcal{L}}{\prec} \mathbf{r}_j$ for every $j \geq 1$ with $a_{Lj} > 0$. Then suppose to each equation i $(i = 0, 1, \ldots, m + n, L)$ in the current tableau an $(-a_{ip}/a_{0p})(-t_{n+1})$ term is added, where t_{n+1} is a nonnegative variable. In addition, to ensure that the problem with the extra column has the same set of feasible solutions as the real problem, we introduce the inequalities $x_{m+n+2} = -(-t_{n+1}) \geq 0$ and $x_{m+n+3} = -t_{n+1} \geq 0$ as the last two rows in the tableau. This ensures that $t_{n+1} = 0$ in every primal solution.

Now observe that the created $(n+1)^{st}$ column can serve as the pivot column, since the ith element in $\mathbf{r}_{n+1}$ is

$$\frac{-a_{ip}/a_{0p}}{-a_{Lp}/a_{0p}} = \frac{a_{ip}}{a_{Lp}};$$

that is, $\mathbf{r}_{n+1} = \mathbf{r}_p$. Suppose this column is chosen instead of $\boldsymbol{\alpha}_p$. Further, assume that the reference equation L is selected as the source row even though primal feasibility may be lost. Subsequent to the pivot (see Figure 7.2) we have $\bar{a}_{00}$ equal to

$$a_{00} + \left[-a_{L0} \frac{a_{0p}}{a_{Lp}} \right].$$

But by Theorem 7.1, $a_{L,n+1} > \bar{a}_{Lj}$ for every j different from $n + 1$. In particular, $a_{L,n+1} > \bar{a}_{L\bar{p}}$, where $\bar{p}$ is the pivot column index at the next tableau. The last inequality with Theorem 7.2 means that $\bar{a}_{0\bar{p}} >$

---

[14]It may be possible to obtain a smaller upper bound; see Problem 7.19.

Pivot column
↓

		1	$-x_{J(1)}$	$\cdots$	$-x_{J(p)}$	$\cdots$	$-x_{J(n)}$	$-t_{n+1}$
	x_0	a_{00}	$\cdots$		a_{0p}	$\cdots$	$\cdots$	$\dfrac{-a_{0p}}{a_{0p}} = -1$
	x_1	a_{10}	$\cdots$	$\cdots$	a_{1p}	$\cdots$	$\cdots$	$\dfrac{-a_{1p}}{a_{0p}}$
	$\vdots$	$\vdots$			$\vdots$			$\vdots$
	x_n	a_{n0}	$\cdots$	$\cdots$	a_{np}	$\cdots$	$\cdots$	$\dfrac{-a_{np}}{a_{0p}}$
	$\vdots$	$\vdots$			$\vdots$			$\vdots$
	x_{m+n}							
Source row $\rightarrow$	x_L	a_{L0}	$\cdots$	$\cdots$	a_{Lp}	$\cdots$	$\cdots$	$\dfrac{-a_{Lp}}{a_{0p}}$
	x_{m+n+2}	0	$\cdots$	$\cdots$	0	$\cdots$	$\cdots$	-1
	x_{m+n+3}	0	$\cdots$	$\cdots$	0	$\cdots$	$\cdots$	$+1$
Cut and pivot row $\lambda = \dfrac{-a_{Lp}}{a_{0p}} \rightarrow$	s	$\left[-a_{L0}\dfrac{a_{0p}}{a_{Lp}}\right]$	$\cdots$	$\cdots$	$-a_{0p}$	$\cdots$	$\cdots$	1

↑
Pivot
element

Figure 7.2: Developing Acceptable Source Row Selection Rule 3

$a_{0,n+1} = -1$; or since the tableau is all integer $\bar{a}_{0\bar{p}} \geq 0$. Thus we have just shown that the next tableau will be dual feasible, which means, by the dual theorem of linear programming, that

$$x_0 \leq \bar{a}_{00} = a_{00} + \left[-a_{L0}\frac{a_{0p}}{a_{Lp}}\right].$$

When the value of x_0 is a_{00} and $-a_{L0}(a_{0p}/a_{Lp}) < 1$ or $(a_{0p}/a_{Lp}) > (-1/a_{L0})$, it is clear from the theorem that $\boldsymbol{\alpha}_o$ contains the optimal integer solution. (If $a_{L0} = 0$, $-a_{L0}(a_{0p}/a_{Lp})$ is automatically less than 1.) This not only allows early termination of the algorithm but, as Glover [244] indicates, it permits another acceptable source row selection rule. In particular, suppose in addition to the reference row, the tableau is initially augmented by

$$x_q = a_{q0} + \sum_{j=1}^{n}(-a_{0j})(-x_j) \geq 0,$$

where the a_{0j}'s are the last n coefficients in the cost row, and $-a_{q0}$ is a nonpositive lower bound for $\sum_{j=1}^{n} a_{0j}x_j$. Then the following is an acceptable source row selection rule.

Rule 3 Replace $a_{Lp} \leq a_{L0}$ by $-a_{0p} \leq a_{q0}$ in Rule 1 (Section 7.3.2) and terminate when

$$a_{L0} = 0 \quad \text{or} \quad \frac{a_{0p}}{a_{Lp}} > \frac{-1}{a_{L0}}.$$

We now show that this rule and the ones given in Section 7.3 produce a finite algorithm.

7.7   Convergence

Before proving convergence for each acceptable source row selection rule we need the following results. As usual, a bar will denote an element subsequent to a pivot operation, α_p will be the pivot column, and L the reference row index.

Lemma 7.1 *For each pair of indices i, L,*

$$\begin{aligned}
a_{Lj}a_{ip} &< a_{Lp}a_{ij} &\Leftrightarrow& \quad \bar{a}_{Lj}a_{ip} &<& a_{Lp}\bar{a}_{ij}, \\
a_{Lj}a_{ip} &= a_{Lp}a_{ij} &\Leftrightarrow& \quad \bar{a}_{Lj}a_{ip} &=& a_{Lp}\bar{a}_{ij}, \\
a_{Lj}a_{ip} &> a_{Lp}a_{ij} &\Leftrightarrow& \quad \bar{a}_{Lj}a_{ip} &>& a_{Lp}\bar{a}_{ij},
\end{aligned}$$

Proof: By definition, for any j we have $\bar{\alpha}_j = \alpha_j - K\alpha_p$, where $K = [a_{vj}/a_{vp}]$ and v is the source row index. Thus,

$$\bar{a}_{Lj}a_{ip} = (a_{Lj} - Ka_{Lp})a_{ip} = a_{Lj}a_{ip} - Ka_{Lp}a_{ip}$$

and

$$a_{Lp}\bar{a}_{ij} = a_{Lp}(a_{ij} - Ka_{ip}) = a_{Lp}a_{ij} - Ka_{Lp}a_{ip},$$

Therefore,

$$\bar{a}_{Lj}a_{ip} - a_{Lp}\bar{a}_{ij} = a_{Lj}a_{ip} - a_{Lp}a_{ij},$$

The assertions are evident by the last equality. ∎

Lemma 7.2 *For any index j,*

$$a_{Lj}\boldsymbol{\alpha}_p \stackrel{\mathcal{L}}{\prec} a_{Lp}\boldsymbol{\alpha}_j \Leftrightarrow \bar{a}_{Lj}\boldsymbol{\alpha}_p \stackrel{\mathcal{L}}{\prec} a_{Lp}\bar{\boldsymbol{\alpha}}_j,$$
$$a_{Lj}\boldsymbol{\alpha}_p = a_{Lp}\boldsymbol{\alpha}_j \Leftrightarrow \bar{a}_{Lj}\boldsymbol{\alpha}_p = a_{Lp}\bar{\boldsymbol{\alpha}}_j,$$
$$a_{Lj}\boldsymbol{\alpha}_p \stackrel{\mathcal{L}}{\succ} a_{Lp}\boldsymbol{\alpha}_j \Leftrightarrow \bar{a}_{Lj}\boldsymbol{\alpha}_p \stackrel{\mathcal{L}}{\succ} a_{Lp}\bar{\boldsymbol{\alpha}}_j,$$

Proof: The lemma is the vector extension of the previous one and is an immediate consequence of it. ∎

Lemma 7.3 *If the current tableau is not optimal and the coefficients in its reference row satisfy (I) $\boldsymbol{\alpha}_j \stackrel{\mathcal{L}}{\prec} \mathbf{0} \Rightarrow a_{Lj} > 0$ and (II) $a_{Lj} < 0 \Rightarrow \mathbf{r}_j \stackrel{\mathcal{L}}{\prec} \mathbf{r}_{p,}$, then for any $j = 1, \ldots, n \ (j \neq p), a_{Lp}\boldsymbol{\alpha}_j \stackrel{\mathcal{L}}{\succ} a_{Lj}\boldsymbol{\alpha}_p$.*

Proof: To prove the assertion we first note that $a_{Lp} > 0$ and then consider the sign of a_{Lj}. When $a_{Lj} < 0$, property (II) means that $\mathbf{r}_j \stackrel{\mathcal{L}}{\prec} \mathbf{r}_p$, or equivalently, $a_{Lp}\boldsymbol{\alpha}_j \stackrel{\mathcal{L}}{\succ} a_{Lj}\boldsymbol{\alpha}_p$. If $a_{Lj} = 0$, we have by (I) that $\boldsymbol{\alpha}_j \stackrel{\mathcal{L}}{\succeq} \mathbf{0}$. But $\boldsymbol{\alpha}_j$ cannot be the zero vector since the columns of any tableau are linearly independent. Thus $a_{Lp}\boldsymbol{\alpha}_j \stackrel{\mathcal{L}}{\succ} \mathbf{0} = a_{Lj}\boldsymbol{\alpha}_p$. Finally, the definition of $\mathbf{r}_p$ implies that $\mathbf{r}_j \stackrel{\mathcal{L}}{\succ} \mathbf{r}_p$ for all $j \neq 0, p$ with $a_{Lj} > 0$. This proves the assertion. ∎

Theorem 7.5 *If the current tableau is not optimal and the coefficients in its reference row satisfy properties (I) and (II) (as in Lemma 7.3), then $\bar{\mathbf{r}}_j \stackrel{\mathcal{L}}{\succ} \mathbf{r}_p$ for all $j \neq 0, p$ such that $\bar{a}_{Lj} > 0$.*

Proof: By Lemma 7.3, we have $a_{Lp}\alpha_j \overset{\mathcal{L}}{\succ} a_{Lj}\alpha_p$ for all $j \neq 0, p$. This means, by Lemma 7.2, that $a_{Lp}\bar{\alpha}_j \overset{\mathcal{L}}{\succ} \bar{a}_{Lj}\alpha_p$ Thus, when $\bar{a}_{Lj} > 0$, it follows that $\bar{r}_j \overset{\mathcal{L}}{\succ} r_p$, since $a_{Lp} > 0$. ∎

Corollary 7.5.1 *If the hypothesis of Theorem 7.5 is met, then* $\bar{r}_{\bar{p}} \overset{\mathcal{L}}{\succ} r_p$.

Proof: This follows by the theorem, since $\bar{p} \neq 0, p$. ∎

Theorem 7.6 *If the current tableau is not optimal and the coefficients in its source row satisfy (I)* $\alpha_j \overset{\mathcal{L}}{\prec} 0 \Rightarrow a_{Lj} > 0$ *and (II)* $a_{Lj} < 0 \Rightarrow r_j \overset{\mathcal{L}}{\prec} r_p$, *then* $(\bar{I})$ $\bar{\alpha}_j \overset{\mathcal{L}}{\prec} 0 \Rightarrow \bar{a}_{Lj} > 0$ *and* $(\overline{II})$ $\bar{a}_{Lj} < 0 \Rightarrow \bar{r}_j \overset{\mathcal{L}}{\prec} \bar{r}_{\bar{p}}$ *for* $j = 1, \ldots, n$ $(j \neq p)$.

Proof: Since the hypothesis of Lemma 7.3 is satisfied we have $a_{Lp}\alpha_j \overset{\mathcal{L}}{\succ} a_{Lj}\alpha_p$ for all $j \neq 0, p$; and again, by Lemma 7.2, $a_{Lp}\bar{\alpha}_j \overset{\mathcal{L}}{\succ} \bar{a}_{Lj}\alpha_p$. But $a_{Lp} > 0$, and as the tableau is not optimal, $\alpha_p \overset{\mathcal{L}}{\prec} 0$. Thus $\bar{a}_{Lj} \leq 0$ means that $\bar{a}_{Lj}\alpha_p \overset{\mathcal{L}}{\succeq} 0$, which implies $a_{Lp}\bar{\alpha}_j \overset{\mathcal{L}}{\succeq} 0$ or $\bar{\alpha}_j \overset{\mathcal{L}}{\succeq} 0$. Hence, $(\bar{I})$ is true.

To obtain $(\overline{II})$, divide $a_{Lp}\bar{\alpha}_j \overset{\mathcal{L}}{\succ} \bar{a}_{Lj}\alpha_p$ by $a_{Lp}\bar{a}_{Lj}$ $(j \neq p)$. If $\bar{a}_{Lj} < 0$, this yields $\bar{r}_j \overset{\mathcal{L}}{\prec} r_p$, which implies, by the previous corollary, that $\bar{r}_j \overset{\mathcal{L}}{\prec} \bar{r}_{\bar{p}}$ for all $j \neq p$ with $\bar{a}_{Lj} < 0$. When $j = p$, $\bar{a}_{Lp} = -a_{Lp} < 0$ and $\bar{r}_p = r_p$; again by the corollary, this implies that $\bar{r}_p \overset{\mathcal{L}}{\prec} \bar{r}_{\bar{p}}$. This proves $(\overline{II})$. ∎

Corollary 7.6.1 *The first tableau in which* $r_p \overset{\mathcal{L}}{\succ} 0$ *or* r_p *cannot be defined (i.e.,* $a_{Lj} \leq 0$ *for* $j = 1, \ldots, n$) *is optimal.*

Proof: By Theorem 7.6, the reference row maintains property (I) and (II). Thus by (I), $a_{Lj} \leq 0$ implies $\alpha_j \overset{\mathcal{L}}{\succ} 0$. Consequently, if $a_{Lj} \leq 0$ for all $j \geq 1$ the tableau is dual feasible and hence optimal. Also, for all $a_{Lj} > 0, r_j \overset{\mathcal{L}}{\succ} r_p \overset{\mathcal{L}}{\succ} 0$. As $r_j = \alpha_j/a_{Lj}$, this implies that $\alpha_j \overset{\mathcal{L}}{\succ} 0$, and again the tableau is optimal. ∎

We have just shown that if the reference row initially possesses properties (I) and (II) then it will always maintain them. Also, that the $\mathbf{r}_p$ column lexicographically increases and when it cannot be defined or becomes lexicographically positive the tableau exhibits optimality. We now demonstrate that for each acceptable source row selection rule this optimal condition will occur after a finite number of tableaux. For ease of reference, each rule is rewritten. Remember that $V(p)$ is the set of legitimate source rows, that is, those that will maintain primal feasibility.

7.7.1 Finiteness under Rule 1

Rule 1 Choose any row in $V(p)$ that ensures $a_{Lp} \leq a_{L0}$ at finite intervals and also periodically reduces a_{ip}/a_{Lp} for the smallest $i \geq 1$ such that $a_{ip}/a_{Lp} > a_{i0}$.

A pivot operation results in a nonpositive integer multiple of the pivot column being added to every other column. As the pivot column is lexicographically negative, this quantity is lexicographically nonnegative. Thus the constant column, in particular, either undergoes a lexicographic increase or it remains the same. Therefore, in a bounded problem,[15] $\boldsymbol{\alpha}_0$ can change only a finite number of times whether the algorithm is finite or not. This means that each component of $\boldsymbol{\alpha}_0$ can change only a finite number of times. Consequently, each of its elements a_{i0} $(i = 0, 1, \ldots, m + n, L)$ is bounded above by some number U_i. In particular, $a_{L0} \leq U_L$ for all tableaux.

Now suppose the algorithm is not finite. Then there exists an infinite sequence of tableaux during which $\mathbf{r}_p$ lexicographically increases while remaining lexicographically negative. This means that a_{0p}/a_{Lp}, the first element in $\mathbf{r}_p$, is negative and nondecreasing. That is, $a_{0p}/a_{Lp} \geq t_0$ (some negative number) during the infinite sequence. We prove that a_{0p}/a_{Lp} must eventually remain fixed at some negative number. Consider the tableaux during which $a_{Lp} \leq a_{L0}$, or equivalently, $a_{Lp} \leq a_{L0} \leq U_L$. By Rule 1 such tableaux must occur at finite intervals. As there is an infinite sequence of iterations this means that $a_{Lp} \leq a_{L0}$ for an

---

[15]We are, in reality, assuring this by using either of the two reference rows discussed in this chapter. In particular, the first one, by its definition, bounds x_0, and the other requires that the dual, hence the primal, linear program be feasible. This also bounds the optimal value of the integer feasible solution x_0, since it can never exceed the optimal value of the linear programming objective function.

infinite subsequence of tableaux. Now if a_{0p}/a_{Lp} does not eventually remain constant it will increase at various tableaux in the subsequence. But during these iterations $a_{0p} \leq U_L$ and since a_{0p}/a_{Lp} is always $\geq t_0$, we have that the fraction a_{0p}/a_{Lp} can only take on a finite number of values.[16] Thus eventually all its possibilities will be exhausted and it must remain fixed at some negative number. (If it becomes positive, $\mathbf{r}_p \overset{\mathcal{L}}{\succ} \mathbf{0}$, and the proof ends.)

Now that the first element in $\mathbf{r}_p$ is fixed a similar argument can be applied to show that its second element a_{1p}/a_{Lp} must eventually remain constant. Then a repetition implies that the third component remains fixed; etc. However, since $\mathbf{r}_p$ increases lexicographically for an infinite sequence of tableaux, at least one of its components must increase indefinitely. Suppose for $i = 0, 1, \ldots, q-1$ that a_{ip}/a_{Lp} remains fixed and a_{qp}/a_{Lp} increase indefinitely. This means that after some point, a_{qp}/a_{Lp} will become and remain larger than $U_q \geq a_{q0}$. Then, by Rule 1, eventually a_{qp}/a_{Lp} will be reduced. This contradicts the lexicographic increasing nature of $\mathbf{r}_p$. Hence we have a contradiction, and the algorithm converges.

7.7.2 Finiteness under Rule 2

Rule 2 At finite intervals designate the row from $V(p)$ that will result in $\bar{a}_{0\bar{p}} > a_{0p}$ as the source row. Continue to use this rule until there are no rows that satisfy it. When this occurs, select any i from $V(p)$ as the generating row.

Again suppose the algorithm is not finite. Then there is an infinite sequence of tableaux during which $\mathbf{r}_p$ lexicographically increases. The pivot column also remains lexicographically negative. Rule 2 then implies that eventually there will exist an infinite subsequence of tableaux during which all legitimate source rows result in $\bar{a}_{0\bar{p}} = a_{0p}$. By an earlier discussion (Section 7.3) this means that $a_{Lp} \leq a_{L0}$ at tableaux in this infinite subsequence. (If $a_{Lp} > a_{L0}$, row L may be selected as a generating row. Then, by Theorem 7.1, $a_{Lp} > \bar{a}_{L\bar{p}}$, which implies by Theorem 7.2 that $\bar{a}_{0\bar{p}} > a_{0p}$.) But, as mentioned in the previous finiteness proof, each component of $\boldsymbol{\alpha}_o$ is bounded above. That is, $a_{i0} \leq U_i$ for $i = 0, \ldots, m + n, L$. Thus, for an infinite number of

---

[16]This is evident since a_{Lp} can equal $1, 2, \ldots, U_L$ and $a_{0p} \geq t_0 a_{Lp}$ means that a_{0p} can equal $[t_0 a_{Lp}], [t_0 a_{Lp}]+1, [t_0 a_{Lp}]+2, \ldots, -1$. Thus there are at most $U_L |[t_0 a_{Lp}]|$ possibilities for the ratio a_{0p}/a_{Lp}.

tableaux, $a_{Lp} \leq a_{L0} \leq U_L$. Hence, we can repeat word-for-word the argument used in the previous proof to show that there is a row index q such that the first $q - 1$ elements in $\mathbf{r}_p$ eventually remain fixed and the qth element in $\mathbf{r}_p$, a_{qp}/a_{Lp}, increases indefinitely. This means that at some point a_{qp}/a_{Lp} is greater than $U_q \geq a_{q0}$. As $a_{Lp} > 0$ and a_{qp} is now positive we have

$$a_{qp} \geq \frac{a_{qp}}{a_{Lp}} > U_q \geq a_{q0} \quad \text{or} \quad a_{qp} > a_{q0}.$$

The last inequality indicates that row q is (and will remain) an eligible source row. When it is selected, $a_{qp} > \bar{a}_{q\bar{p}}$ (by Theorem 7.1), which in conjunction with $\bar{a}_{q\bar{p}}/\bar{a}_{L\bar{p}} > a_{qp}/a_{Lp}$ (equivalently $a_{Lp} > (a_{qp}/\bar{a}_{q\bar{p}})\bar{a}_{L\bar{p}}$) implies that $a_{Lp} > \bar{a}_{L\bar{p}}$. But by Theorem 7.2 this means that $\bar{a}_{0\bar{p}} > a_{0p}$. This contradicts that $\bar{a}_{0\bar{p}} = a_{0p}$ during the infinite subsequence, which completes the proof.

7.7.3   Finiteness under Rule 3

Rule 3 In addition to the reference row the equation

$$x_q = a_{q0} + \sum_{j=1}^{n} (-a_{0j})(-x_j),$$

where $-a_{q0}$ is a nonpositive lower bound for $\sum_{j=1}^{n} a_{0j}x_j$, is added to the initial tableau. Then Rule 3 is the same as Rule 1 except that $a_{Lp} \leq a_{L0}$ is replaced by $-a_{0p} \leq a_{q0}$, and the algorithm terminates when $a_{L0} = 0$ or $a_{0p}/a_{Lp} > -1/a_{L0}$ (which indicates that the tableau is optimal; see Theorem 7.4).

If, at finite intervals, $a_{Lp} \leq C$, some finite constant, then the convergence proof is the same as for Rule 1. Otherwise, if the algorithm does not converge, eventually a_{Lp} becomes and remains greater than $C = U_q U_L$, where U_q is an upper bound for a_{q0} and U_L is an upper bound for a_{L0}. That is, after some initial period there is an infinite sequence of tableaux with $a_{Lp} > U_q U_L \geq a_{q0} a_{L0}$. By Rule 3 there exists an infinite subsequence of these tableaux at which $-a_{0p} \leq a_{q0}$. Using $a_{Lp} > a_{q0} a_{L0}$ with the last inequality gives $a_{Lp} > a_{L0}(-a_{0p})$, or $a_{0p}/a_{Lp} > -1/a_{L0}$, since a_{Lp} and a_{L0} are positive. This contradicts the nonconvergence assumption. Hence, the algorithm with Rule 3 is finite.

Problems

Unless otherwise stated, all notation, tableaux format, and symbols are the same as in the chapter. In particular, RPA means Rudimentary Primal Algorithm and SPA means Simplified Primal Algorithm.

Problem 7.1   For each original nonbasic variable x_j $(j = 1, \ldots, n)$, we introduce the identity equation $x_j = -1(-x_j)$ to the initial simplex tableau. Where in the development of the SPA is this fact used? Suppose these equalities are not initially added, but rather, as Young [614] suggests, that we introduce an equation $x_j = -1(-x_j)$ whenever x_j is an original nonbasic variable whose column has been selected as the current pivot column. Explain why this strategy is allowed.

Problem 7.2   The RPA may not converge because the constant element in the cut or pivot row may be 0 for an infinite number of tableaux (see Appendix A). Explain why this difficulty cannot be resolved by applying anticycling procedures of the primal simplex method.

Problem 7.3   Show that if an integer program has one feasible integer solution and its associated linear program has an unbounded solution, then the integer problem must also have an unbounded solution. Illustrate geometrically.

Problem 7.4   In the RPA and SPA we have not kept the pivot rows or cuts. What information might be lost because of this?

Problem 7.5   Given the bounded integer program of the form

$$
\begin{array}{lrcll}
\text{maximize} & a_{01}x_1 & + & a_{02}x_2 & = & x_0 \\
\text{subject to} & a_{11}x_1 & + & a_{12}x_2 & \leq & a_{10}, \\
& a_{21}x_1 & + & a_{22}x_2 & \leq & a_{20}, \\
\text{and} & x_1, & & x_2 & \geq & 0, \text{ integer},
\end{array}
$$

where a_{10} and a_{20} are nonnegative.

a) Suppose the RPA is applied to this problem. If the source row is always the same, show that the algorithm converges.

b) Does the RPA converge when the source rows are not necessarily the same?

c) Are your arguments in (a) and (b) valid for any rational values of a_{10} and a_{20}?

Problem 7.6   Solve the integer program

$$
\begin{array}{llll}
\text{maximize} & x_1 & + & x_2 & = & x_0 \\
\text{subject to} & 2x_1 & & & \leq & 7, \\
& -x_1 & + & x_2 & \leq & 0, \\
\text{and} & x_1, & & x_2 & \geq & 0, \text{ integer,}
\end{array}
$$

using the RPA. Relate the computations to (x_1, x_2) space.

Problem 7.7 (Salkin, Shroff, and Mehta [518]) Show that an all integer initial simplex tableau is not necessary for the convergence of the SPA. (*Hint*: Proceed as suggested in Problem 6.19, Chapter 6.)

Problem 7.8 (Young's SPA [614]) Consider the following primal all integer algorithm for a bounded integer program. The initial tableau is assumed primal feasible and all integer. For simplicity it also contains the identity equations $x_j = -1(-x_j)$ $(j = 1, \ldots, n)$.

Step 1 If $a_{0j} \geq 0$ for $j = 1, \ldots, n$, the tableau is optimal— terminate. Otherwise, for *each* $j \geq 1$ with $a_{0j} < 0$, find

$$
\theta_j = \operatorname*{minimum}_{\substack{a_{ij} > 0 \\ i \neq 0}} \frac{a_{i0}}{a_{ij}}.
$$

If $\theta_j \geq 1$ for at least one j, delete the reference row, when present, from the current tableau and go to Step 2.

If $\theta_j < 1$ for all j, introduce the reference equation, $x_L = a_{L0} + \sum_{j=1}^{n}(-x_{J(j)})$ to the bottom of the current tableau provided that it does not contain one. (a_{L0} is an upper bound for $\sum_{j=1}^{n} x_{J(j)}$.) In any case, go to Step 3.

Step 2 ("Transition cycle") Select one j which has $\theta_j > 0$ and designate it as the pivot column. Choose a legitimate source row i and derive the all-integer cut with $\lambda = a_{ij}$. Perform a primal simplex pivot operation and return to Step 1.

Step 3 ("Stationary cycle") Define the columns $\mathbf{r}_j$ and the pivot column as in the SPA. Choose a legitimate source row i by an acceptable source row selection rule. Derive the all-integer cut with $\lambda = a_{ij}$. Perform a primal simplex pivot operation and return to Step 1.

a) Suppose that Step 2 is entered at finite intervals. Prove that the algorithm converges. (*Hint*: During a transition cycle, a_{00} must increase by at least 1.)

b) Prove that Young's algorithm converges when source row selection Rule 1 (Section 7.3) is used. (*Hint*: Distinguish between intervals of stationary and transition cycles. Show that it is impossible to have an infinite sequence of stationary cycles. This can be done by a convergence proof similar to the one for Rule 1 which appears in Section 7.7.)

c) Resolve Example 7.1 (Section 7.2), using Young's algorithm with any acceptable source row selection rule.

d) What are possible advantages and disadvantages of the algorithm outlined here when compared with the one given in Section 7.3? In particular, can the redefining of the reference equation at the beginning of each sequence of stationary cycles increase the convergence rate?

Problem 7.9   Solve the integer program

$$
\begin{array}{rlrlll}
\text{maximize} & 3x_1 & + & x_2 & = & x_0 \\
\text{subject to} & 2x_1 & + & 3x_2 & \leq & 6 \\
& 2x_1 & - & 3x_2 & \leq & 3 \\
\text{and} & x_1, & & x_2 & \geq & 0 \text{ and integer,}
\end{array}
$$

using the SPA. Choose any source row selection rule. Relate the computations to (x_1, x_2) space.

Problem 7.10   Solve the integer program

$$
\begin{array}{rlrlrlll}
\text{maximize} & x_1 & - & x_2 & + & 2x_3 & = & x_0 \\
\text{subject to} & 2x_1 & + & 4x_2 & - & x_3 & \leq & 20, \\
& -8x_1 & + & x_2 & + & 3x_3 & \leq & 10, \\
& 2x_1 & - & 9x_2 & + & 8x_3 & \leq & 6, \\
\text{and} & x_1, & & x_2, & & x_3 & \geq & 0 \text{ and integer,}
\end{array}
$$

using the SPA. Choose any acceptable source row selection rule.

Problem 7.11   Suppose we modify the SPA to distinguish between "stationary cycles," or iterations at which $[a_{v0}/a_{vp}] = 0$, and "transition cycles," or iterations at which $[a_{v0}/a_{vp}] > 1$. (p is the pivot column index selected by

$$a_{0p} = \begin{array}{c} \text{maximum} \\ a_{0j} < 0 \\ j \neq 0 \end{array} |a_{0j}|$$

and row v satisfies

$$\frac{a_{v0}}{a_{vp}} = \begin{array}{c} \text{minimum} \\ a_{ip} > 0 \\ i \neq 0 \end{array} \frac{a_{i0}}{a_{ip}}.)$$

Explain why the SPA will converge if an acceptable source row selection rule is applied only during stationary cycles. During transition cycles any legitimate source row may be used to derive the cut. Compare the modified SPA suggested here to Young's [614] version which appears in Problem 7.8.

Problem 7.12   Suppose that at finite intervals the SPA is employed for a finite number of iterations. When the SPA is not used, the RPA is followed. Does this composite primal algorithm necessarily converge? If not, what modification to the above procedure will guarantee finiteness? (*Hint*: See Problem 7.13.)

Problem 7.13   Which of the following strategies supports a finite SPA? Explain.

Rule A Apply any acceptable source row selection rule at finite intervals for a finite number of iterations.

Rule B Define stationary and transition cycles as in Problem 7.11. Apply any acceptable source row selection rule at finite intervals until at least one transition cycle occurs.

Problem 7.14   With regard to source row selection Rule 1 (Section 7.3):

    a) Show that it is unnecessary to require a_{ip}/a_{Lp} to be decreased unless it has been nondecreasing for some finite duration.

b) Prove that when U_L and U_i replace a_{L0} and a_{i0}, respectively, the resulting rule yields a convergent algorithm; U_L is an upper bound for a_{L0} and U_i an upper bound for a_{i0}.

Problem 7.15   Suppose in source row selection Rule 2 (Section 7.3), $\bar{a}_{L\bar{p}} < a_{Lp}$ replaces $\bar{a}_{0\bar{p}} > a_{0p}$. Prove that the resulting rule produces a finite algorithm. How can this rule be implemented? (*Hint:* Use a proof similar to the one for Theorem 7.2 to show that if equation i is the source row, then $\bar{a}_{L\bar{p}} < a_{Lp}$ whenever $\bar{a}_{i\bar{p}}/\bar{a}_{L\bar{p}} > a_{ip}/a_{Lp}$.)

Problem 7.16   Show that the following source row selection rule ensures convergence:

At finite intervals choose any legitimate row that ensures $a_{Lp} \leq a_{L0}$, and periodically select any source row i such that

$$\frac{a_{ip}}{a_{Lp}} > a_{i0} \quad \text{and} \quad \frac{\bar{a}_{i\bar{p}}}{\bar{a}_{L\bar{p}}} \geq \frac{a_{ip}}{a_{Lp}}.$$

Continue this selection as long as possible, then select any legitimate source row. (*Hint:* Follow a proof similar to the one in Problem 7.15.)

Problem 7.17   Suppose the reference equation

$$x_L = a_{L0} + \sum_{j=1}^{n} a_{Lj}(-x_{J(j)}) \geq 0$$

has the properties (I) $\alpha_j \stackrel{\mathcal{L}}{\prec} 0 \Rightarrow a_{Lj} > 0$ and (II) $a_{Lj} < 0 \Rightarrow r_j \stackrel{\mathcal{L}}{\prec} r_p$ (p is the pivot column index). Then for any positive integer k show that kx_L maintains properties (I) and (II).

Problem 7.18   Resolve Example 7.2, using the equation

$$x_L = \frac{50}{17} + \frac{68}{17}(-x_1) + \frac{125}{17}(-x_2) + \frac{51}{17}(-x_3) \geq 0$$

as the reference row. Do not initially clear fractions.

Problem 7.19   Suppose the current tableau is not optimal and the reference row has properties (I) and (II) (as in Problem 7.17). Then for $K > 0$ prove that $a_{00} + K[\gamma_0/K]$, where K satisfies $K[\gamma_j/K] \geq a_{0j}$ $(j = 0, 1, \ldots, n)$ and $\gamma_j = -a_{Lj}(a_{0p}/a_{Lp})$, is an upper bound for the optimal value of x_0. (*Hint:* Reprove Theorem 7.4 where the coefficients in the $(n + 1)$st or created column have the form $-Ka_{ip}/a_{0p}$.) What is the smallest upper bound for x_0 implied by the above assertion?

Problem 7.20   With regard to Rule 3 (Section 7.6), is it possible for the element in the reference row and zero column to have value 0? Explain.

Problem 7.21   Prove that if an acceptable source row selection rule is used, then at finite intervals there must exist a pair of adjacent tableaux for which $(a_{0p}/a_{Lp}) < (\bar{a}_{0\bar{p}}/\bar{a}_{L\bar{p}})$. (*Hint:* Appeal to the results given in Section 7.7)[17]

Problem 7.22   Using the assertion given in Problem 7.21 show that a_{0p}/a_{Lp} becomes nonnegative after a finite number of iterations. (*Hint:* Show that any acceptable source row selection rule implies that all negative values of a_{0p}/a_{Lp} will eventually be used up.)

Problem 7.23   Explain the differences and similarities between the primal all-integer algorithm (SPA) presented in this chapter and the dual all-integer algorithm presented in the last one.

Problem 7.24   Show that if the RPA is applied to the integer program,

$$
\begin{array}{rrrrrl}
\text{maximize} & x_1 & - & x_2 & + 2x_3 & = x_0 \\
\text{subject to} & 2x_1 & + & 4x_2 & - 5x_3 & \leq 20, \\
 & -6x_1 & + & x_2 & + 3x_3 & \leq 10, \\
 & 4x_1 & - & 9x_2 & + 8x_3 & \leq 6, \\
\text{and} & x_1, & & x_2, & x_3 & \geq 0 \text{ and integer,}
\end{array}
$$

it may not converge. (*Hint:* Follow the arguments used in Appendix A.)

Problem 7.25   Consider the integer program with constraints as

---

[17]The assertions given in Problems 7.21 and 7.22 are proved in Young [614]. Essentially, he proves convergence by first showing that r_j increases lexicographically and then by using these facts.

in the problem of Appendix A, but with the objective function $x_0 = x_1 - x_2 + 2x_3$.

a) Show using the method in Appendix A, that the RPA fails to converge on this problem.

b) Show that x_0 may be made arbitrarily large.

c) Show that the following modified RPA will converge in three tableaux to an unbounded solution (cf. (b) above) on this program: Select column 1 rather than column 3 as the pivot column in Tableau #0, then apply the RPA as usual. Mathis[418] indicates that if this rule which selects a pivot column that results, whenever possible, in a "transition cycle", this problem and the one appearing in Problem 7.24 will converge.

Problem 7.26   Solve the problem in Appendix A graphically. (*Hint:* Apply the transformations $y_1 = 2x_1 - x_2$, and $y_2 = -x_2 + x_3$ to show that $x_0/b < 1$.)

Appendix A
Cycling in Rudimentary Primal Algorithm

(Mathis [418], Haas,Mathur, and Salkin[302])

Suppose we attempt to solve the integer program

$$
\begin{array}{lrcrcrcll}
\text{maximize} & & - & bx_2 & + & bx_3 & = & x_0 & \\
\text{subject to} & -8x_1 & + & x_2 & + & 3x_3 & \leq & 10 & (x_4), \\
& 2x_1 & - & 9x_2 & + & 8x_3 & \leq & 6 & (x_5), \\
\text{and} & x_1, & & x_2, & & x_3 & \geq & 0 & \text{and integer,}
\end{array}
$$

where $b > 0$ is an arbitrary constant, using the RPA (see Section 7.2). With nonnegative slack variables x_4 and x_5 we have Tableau #0. Column 3 is the unique possible pivot column (indexed by p), and since row 5 gives the minimum value for the ratio a_{i0}/a_{i3} ($i \neq 0$) when a_{i3} is positive, it serves as the generating row. The all-integer cut or pivot row (s_0) is derived, and after a pivot operation Tableau #1 is obtained. The calculations are repeated and the sequence of Tableaux #0 through #7 are created. At each iteration there is a unique eligible source row. The pivot element is circled.

#0	1	$(-x_1)$	$(-x_2)$	$(-x_3)$
x_0	0	0	b	$-b$
x_1	0	-1	0	0
x_2	0	0	-1	0
x_3	0	0	0	-1
x_4	10	-8	1	3
source row → x_5	6	2	-9	8
cut $(\lambda = 8)$ → s_0	0	0	-2	①

#1	1	$(-x_1)$	$(-x_2)$	$(-s_0)$
x_0	0	0	$-b$	b
x_1	0	-1	0	0
x_2	0	0	-1	0
x_3	0	0	-2	1
x_4	10	-8	7	-3
source row → x_5	6	2	7	-8
cut $(\lambda = 7)$ → s_1	0	0	①	-2

#2	1	$(-x_1)$	$(-s_1)$	$(-s_0)$
x_0	0	0	b	$-b$
x_1	0	-1	0	0
x_2	0	0	1	-2
x_3	0	0	2	-3
source row → x_4	10	-8	-7	11
x_5	6	2	-7	6
cut $(\lambda = 11)$ → s_2	0	-1	-1	①

#3	1	$(-x_1)$	$(-s_1)$	$(-s_2)$
x_0	0	$-b$	0	b
x_1	0	-1	0	0
x_2	0	-2	-1	2
x_3	0	-3	-1	3
x_4	10	3	4	-11
source row → x_5	6	8	-1	-6
cut $(\lambda = 8)$ → s_3	0	①	-1	-1

#4	1	$(-s_3)$	$(-s_1)$	$(-s_2)$
x_0	0	b	$-b$	0
x_1	0	1	-1	-1
x_2	0	2	-3	0
x_3	0	3	-4	0
x_4	10	-3	7	-8
source row → x_5	6	-8	7	2
cut $(\lambda = 7)$ → s_4	0	-2	①	0

#5	1	$(-s_3)$	$(-s_4)$	$(-s_2)$
x_0	0	$-b$	b	0
x_1	0	-1	1	-1
x_2	0	-4	3	0
x_3	0	-5	4	0
source row → x_4	10	11	-7	-8
x_5	6	6	-7	2
cut $(\lambda = 11)$ → s_5	0	①	-1	-1

#6	1	$(-s_5)$	$(-s_4)$	$(-s_2)$
x_0	0	b	0	$-b$
x_1	0	1	0	-2
x_2	0	4	-1	-4
x_3	0	5	-1	-5
x_4	10	-11	4	3
source row → $\quad x_5$	6	-6	-1	8
cut $(\lambda = 8)$ → $\quad s_6$	0	1	-1	(1)

#7	1	$(-s_5)$	$(-s_4)$	$(-s_6)$
x_0	0	0	$-b$	b
x_1	0	-1	-2	2
x_2	0	0	-5	4
x_3	0	0	-6	5
x_4	10	-8	7	-3
x_5	6	2	7	-8
s_1	0	0	1	-2

Observe that the x_4 and x_5 rows are the same in Tableaux #1 and #7. In addition, only these equations were eligible as the source row. Also, note that the pivot column decreases lexicographically. Suppose we define $f_i = 2a_{i1} + 4a_{i2} + 4a_{i3}$ for $i = 0, 1, \ldots, 5$, where a_{ij} is the (i, j) element in Tableau #1. Then $f_0 = 0$, $f_1 = -2$, $f_2 = -4$, $f_3 = -4$, and $f_4 = f_5 = 0$, and it easily verified that the coefficients in Tableau #7 have the form $\boldsymbol{\alpha}_0, \boldsymbol{\alpha}_1, \boldsymbol{\alpha}_2 + \mathbf{F}, \boldsymbol{\alpha}_3 - \mathbf{F}$ where $\boldsymbol{\alpha}_j$ $(j = 0, 1, 2, 3)$ is column j of Tableau #1 and

$$\mathbf{F} = \begin{pmatrix} f_0 \\ \vdots \\ f_5 \end{pmatrix}.$$

Also, if we continue generating tableaux it is a simple matter to show that

   i) the columns of Tableau #13 are α_0, α_1, $\alpha_2 + 2\mathbf{F}$, $\alpha_3 - 2\mathbf{F}$,

   ii) the sequence of cuts, as well as the source row and pivot column, are the same in every sixth tableau, and

   iii) only the x_4 and x_5 equations are eligible as source rows.

Because of this we can prove (by induction on $N = 1 + 6n$) that the columns of Tableau $\#1 + 6n$, where $n = 0, 1, 2, \ldots$ are $\alpha_0, \alpha_1, \alpha_2 + n\mathbf{F}, \alpha_3 - n\mathbf{F}$ as illustrated below.

$\#1 + 6n$	1	$(-s_5)$	$(-s_4)$	$(-s_6)$
x_0	0	0	$-b$	b
x_1	0	-1	$-2n$	$2n$
x_2	0	0	$-1 - 4n$	$4n$
x_3	0	0	$-2 - 4n$	$1 + 4n$
x_4	10	-8	7	-3
x_5	6	2	7	-8

Thus, as n increases the first element in the pivot column is negative (i.e., $-b$), so dual feasibility can never be attained, and an unbounded sequence of tableaux is created.

Chapter 8

Branch and Bound Enumeration

8.1 Introduction

In an integer programming formulation of any real world problem, it can be reasonably assumed that each integer variable x_j is bounded above[1] by some finite integer u_j. So for n integer variables there are exactly $N = \prod_{j=1}^n (u_j + 1)$ different nonnegative integer vectors $\mathbf{x} = (x_1, \ldots, x_n)$. Thus a natural procedure for solving an integer program is to enumerate each of the possibilities, and, for the mixed integer problem, successively solve the resulting linear programs. Of course, the trouble with this approach is that the number N, although finite, is usually quite large, thus making it computationally impossible to explicitly enumerate all the integer vectors. However, criteria derived from the integrality, non-negativity, and other constraints may often indicate that certain sets of related integer vectors cannot produce improved solutions. Therefore these vectors do not have to be examined; we say that these sets have been implicitly enumerated. The algorithms presented in this chapter and the next one solve the mixed integer or integer program by enumerating, either explicitly or implicitly, all possible integer vectors. We shall present the various enumeration schemes and derive criteria for implicit enumeration. The efficiency of the algorithms will naturally depend principally on the effectiveness of these criteria.

---

[1]If necessary, a bounding constraint such as $\sum_j x_j \leq M$, where M is a large positive number, can be added to the constraint set.

Background

This chapter contains an exposition of the classical Land and Doig [383] enumeration algorithm (1960) and of the variations which have since appeared in the literature. The method described in [383] was specialized for the traveling salesman problem by Little, Murty, Sweeney, and Karel (1963) [400] (Problem 8.39).[2] They termed the specialized procedure "branch and bound", which, as mentioned in Balinski [50], is also an apt designation for the Land and Doig algorithm. In 1964 Thompson [563] presented an algorithm for the integer program, "The Stopped Simplex Method," which is a consequence of Land and Doig's work (Problem 8.37). A year later, Dakin [153] proposed a simple, yet interesting, variation of the Land and Doig algorithm (Section 8.5). Beale and Small [62] (as described more fully in Beale [61]) extended the Dakin [153] method to include the linear programming post-optimization procedures suggested by Driebeek [177] (Section 8.7). Later, Tomlin [566, 567], described a refined version of the Beale and Small [62] algorithm (Section 8.7). The papers by Balinski [50], Balinski and Spielberg [51], Geoffrion and Marsten [229], and Lawler and Wood [390] are survey articles which also contain expositions of the Land and Doig [383] method. A general description of branch and bound may be found in Agin [2], Balas [21] and Mitten [431]. A graph theoretic interpretation appears in Bertier and Roy [82]. The underlying properties which characterize an algorithm to be a branch and bound procedure appeared in Spielberg [543] (1968), and subsequently in Colmin and Spielberg [139], and Balinski and Spielberg [51] (Section 8.7). A procedure which specializes the Land and Doig approach to the zero-one case (Section 8.7) and incorporates the work of Driebeek [177] and also the algorithm branching strategies proposed by Spielberg [543] is described in Davis, Kendrick, and Weitzman [164] (Problem 8.34). The general branch and bound algorithm has been customized by many authors to efficiently solve mixed integer problems with special structure (see, e.g., Mathur, Salkin, et al. [419, 420]). Johnson et al. [353] have recently designed and implemented a branch and bound based algorithm to efficiently solve large 0-1 mixed integer programming problems with a sparse coefficient matrix.

---

[2]One formulation of the traveling salesman problem is in Section 1.2.2 (Chapter 1); other formulations, theoretical results, and algorithms are described in Chapter 15.

8.2   The Problem, Notation, and the Basic Result

8.2.1   The Problem, Notation

To simplify the discussion, we shall, unless otherwise specified, consider the mixed integer program (MIP).[3] Specifically, the problem under consideration is

MIP   maximize z

subject to

$$\mathbf{cx} + \mathbf{dy} - z = 0 \tag{1}$$
$$\mathbf{Ax} + \mathbf{Dy} \leq \mathbf{b} \tag{2}$$
$$\mathbf{x} \geq 0, \ \mathbf{y} \geq 0 \tag{3}$$

and

$$\mathbf{x} = (x_j) \text{ integer}; \tag{4}$$

where $\mathbf{A}$ is m by n, $\mathbf{D}$ is m by n', and all other terms are of appropriate size.

The vectors $(\mathbf{x}, \mathbf{y}, z)$ satisfying (1), (2), and (3) define the set S. The linear program: Maximize z subject to $(\mathbf{x}, \mathbf{y}, z)$ in S, will be denoted by LP^0. A point or vector $\mathbf{x}$ with ℓ $(0 \leq \ell \leq n)$ of its components fixed at nonnegative integer values is labeled $\mathbf{x}^\ell$. The point $\mathbf{x}^\ell$ has $n - \ell$ not-fixed or free components. The linear program LP^0 with the additional constraint that the ℓ fixed components in $\mathbf{x}^\ell$ are fixed at their integer values, is also a linear program in the free $\mathbf{x}$ and the $\mathbf{y}$ variables, and will be referred to as LP^ℓ, or more simply, as the subproblem at $\mathbf{x}^\ell$. The feasible region corresponding to the linear program LP^ℓ is denoted by S^ℓ; equivalently, S^ℓ is the set of points $(\mathbf{x}^\ell, \mathbf{y}, z)$ satisfying (1), (2), and (3). Notice that S^ℓ is an $(n - \ell) + n' + 1$ dimensional set in $n + n' + 1$ space, and also that $S^0 = S$. The optimal solution to the (linear programming) subproblem at $\mathbf{x}^{\ell+1}$, where $\mathbf{x}^{\ell+1}$ is defined from $\mathbf{x}^\ell$ by setting one of its free variables x_k to the nonnegative integer t, is denoted by $z(\ell, k, t)$. Its value includes the costs of the $\ell + 1$ fixed variables in $\mathbf{x}^{\ell+1}$. Observe that $z(\ell, k, t)$ is the optimal solution to the linear program LP^ℓ with the additional constraint $x_k = t$. Finally, the number z^ℓ $(\ell = 0, 1, \ldots, n - 1)$ will be an upper bound for any mixed integer programming solution found from the point $\mathbf{x}^\ell$.[4]

---

[3]For the integer program, the symbol $\mathbf{y}$, which represents the vector of continuous variables, and the word mixed should be omitted from expressions in which they appear.

[4]Note that it is possible for two different points to have the same value for (ℓ, k, t). For example, suppose $\mathbf{x} = (x_1, x_2, x_3, x_4)$ and one point is defined by setting x_1 to 1, x_2 to 1, and then x_4 to 1, and a second is obtained by fixing x_1 to 2, x_3 to 2, and then x_4 to 1. In both cases, $(\ell, k, t) = (2, 4, 1)$, but the point $(1, 1, x_3, 1)$ is not the same as $(2, x_2, 2, 1)$. However, no ambiguity will be encountered.

8.2.2   The Basic Result and Geometric Interpretations

Earlier we mentioned that the efficiency of the enumeration algorithm depends principally on its ability to implicitly enumerate a large fraction of the possible integer vectors. The Land and Doig [383] and related methods derive their implicit enumeration criteria mainly from Theorem 8.1, which describes the structure of S^ℓ. Remember that during an enumeration, integer variables will successively be set equal to nonnegative integers. This will eventually produce the point $\mathbf{x}^\ell$ or the linear program LP^ℓ with the feasible region S^ℓ. For convenience, a mixed integer programming solution $\mathbf{x}$, which can be produced from $\mathbf{x}^\ell$ by fixing its $n - \ell$ free components to nonnegative integer values, will be called a solution "derived from (the point) $\mathbf{x}^\ell$ ". As each free variable in $\mathbf{x}^\ell$ is fixed, the resulting linear program becomes more restrictive as its feasible region is a subset of S^ℓ. Thus, every mixed integer programming solution derived from $\mathbf{x}^\ell$ must be in S^ℓ; this basic fact will suggest implicit enumeration tests.

As is illustrated in Figure 8.1, we define the upper (lower) boundary of S^ℓ $(0 \le \ell \le n)$ to be the set of points on the boundary of the polygon S^ℓ above (below) the line segment joining the points (x_k^M, z^M) and (x_k^L, z^L), where x_k^M (x_k^L) is a maximum (minimum) value for x_k when $(\mathbf{x}^\ell, \mathbf{y}, z)$ is in S^ℓ and z^M (z^L) is the corresponding value of z. Then, since the projection of a convex set is convex (Problem 8.1), we have the following result.

Theorem 8.1 *(Land and Doig [383]) If S^ℓ $(0 \le \ell \le n)$ is not empty it defines a convex polyhedral set in $n + n' + 1$ space. Moreover, the projection of this set onto any of the $(n - \ell)$ (x_k, z) coordinate planes, where x_k is a free variable, is a convex polygon whose upper (lower) boundary is a concave (convex) function of the x_k axis.*

To illustrate, consider the projection of S^ℓ onto an (x_k, z) coordinate plane as in Figure 8.1. Since, as we have said, every solution to the MIP derived from $\mathbf{x}^\ell$ must satisfy constraints (1), (2), and (3), or equivalently, must be in S^ℓ, the following remarks are evident.

Remarks

1. If the set S^ℓ is empty, constraints (1), (2), and (3) cannot be satisfied simultaneously with the ℓ components of $\mathbf{x}$ fixed at their current value,

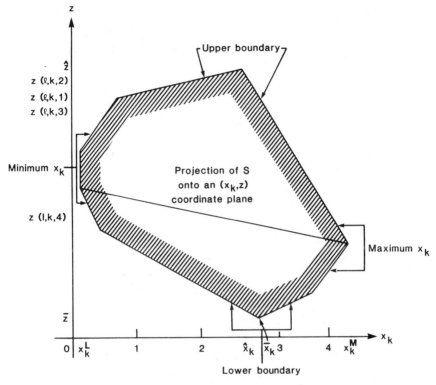

Figure 8.1: Projection of S^ℓ

that is, a mixed integer programming solution cannot be derived from the point $\mathbf{x}^\ell$. Note that S^ℓ is empty if and only if LP^ℓ has no feasible solution. Hence, if the first linear program LP^0 has no feasible solution, the MIP has no integer feasible solution.

2. The largest value z can take with $(\mathbf{x}^\ell, \mathbf{y}, z)$ in S^ℓ corresponds to the highest point on the polygon and is denoted by $\hat{z}$. This number, found by solving LP^ℓ, is an upper bound for any mixed integer programming solution derived from $\mathbf{x}^\ell$. In particular, the optimal solution to the MIP cannot exceed the optimal solution to LP^0. The x_k value in the optimal linear programming solution is $\hat{x}_k$.

Similarly, the lowest point on the polygon $\bar{z}$ can be found by solving: Minimize z, subject to $(x^\ell, \mathbf{y}, z)$ in S^ℓ. The corresponding value of x_k is $\bar{x}_k$.

3. The boundary of the polygon to the right of the line segment de-

termined by the two points $(\hat{x}_k, \hat{z})$ and $(\bar{x}_k, \bar{z})$ may be traced out by solving the linear program: Maximize x_k, subject to $(\mathbf{x}^\ell, \mathbf{y}, z)$ in S^ℓ, for all fixed values of z between $\bar{z}$ and $\hat{z}$. Analogously, the boundary to the left of this line may be found by solving the linear program: Minimize x_k, subject to $(\mathbf{x}^\ell, \mathbf{y}, z)$ in S^ℓ, for all fixed values of z between $\bar{z}$ and $\hat{z}$.[5] Each of these problems can be solved using straightforward parametric linear programming (see Example 8.1).

4. If the linear program: Maximize z, subject to $(\mathbf{x}^\ell, \mathbf{y}, z)$ in S^ℓ, has an unbounded solution, then the point (x_k, z) is "at infinity," and the polygon is unbounded from above. This occurs in Problem 8.4.

5. Figure 8.1 indicates that the largest possible value of z with x_k integer occurs when $x_k = 2$ and $z^\ell = z(\ell, k, 2)$. Consequently, an upper bound on the mixed integer programming solution found from the point $\mathbf{x}^\ell$ is $z(\ell, k, 2)$. In general, since the upper boundary of S^ℓ is concave, any solution to the MIP which is produced from $\mathbf{x}^\ell$ cannot exceed z^ℓ, equal to the maximum of $z(\ell, k, [\hat{x}_k])$ and $z(\ell, k, [\hat{x}_k] + 1)$, where $[h]$ is the largest integer $\leq h$.

6. Figure 8.1 indicates that there are no points (x_k, z) with $x_k = 0$ or $x_k = 5, 6, 7, \ldots$ that intersect the projection of S^ℓ. It follows that there are no points $(\mathbf{x}^\ell, \mathbf{y}, z)$ in S^ℓ which have the kth component of $\mathbf{x}$ at any of these integer values. Hence, a mixed integer programming solution found from $\mathbf{x}^\ell$ must have $x_k = 1, 2, 3$, or 4. This means that the linear program LP^ℓ will have no feasible solution when $x_k \leq 0$ or $x_k \geq 5$. In general, if LP^ℓ has no feasible solution when x_k, a free variable in $\mathbf{x}^\ell$, equals $[\hat{x}_k] + I_1$ or $[\hat{x}_k] - I_2$, (I_1 and I_2 are nonnegative integers) then LP^ℓ will have no feasible solution when $x_k \geq [\hat{x}_k] + I_1$ or when $x_k \leq [\hat{x}_k] - I_2$, respectively (see Figure 8.2). Thus, $[x_k] + I_1$ and $[x_k] - I_2$ are the upper and lower bounds, respectively, on the variable x_k for deriving a mixed integer programming solution from x^ℓ.

To make the remarks concrete an example is presented.

---

[5] The boundary can also be determined by solving the pair of linear programs: Maximize z and Minimize z over the common feasible region S^ℓ for fixed values of x_k between x_k^L and x_k^M; see Problem 8.2).

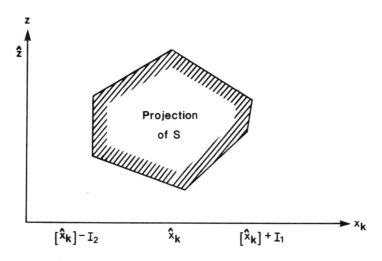

Figure 8.2: To be in $S^\ell : [\hat{x}_k] - I_2 < x_k < [\hat{x}_k] + I_1$

Example 8.1

Maximize z

subject to

$$
\begin{array}{llllllll}
x_1 & + & x_2 & + & y_1 & - & z & = & 0 & \quad (a) \\
2x_1 & + & x_2 & + & y_1 & & & \leq & 6 & \quad (b) \\
x_1 & + & 2x_2 & + & y_1 & & & \leq & 6 & \quad (c) \\
x_1 & + & x_2 & + & 2y_1 & & & \leq & 6 & \quad (d) \\
x_1 & + & x_2 & + & y_1 & & & \leq & 4 & \quad (e) \\
x_1, & & x_2, & & y_1 & & & \geq & 0 & \quad (f)
\end{array}
$$

and $x_1, \quad x_2$ integer.

First let us find the projection of the convex set $S^0 = S$ which is defined by the intersection of the constraints (a) through (f), onto the (x_1, z) plane. This is equivalent to solving the pairs of linear programs: Maximize x_1, and Minimize x_1, subject to the common feasible region (x_1, x_2, y_1, z) in S for fixed values of z. We note that the smallest possible value for z satisfying (a) through (f) is 0 (when $x_1 = x_2 = y_1 = 0$) and inequality (e) limits z to at most 4. Moreover, the point $(x_1, x_2, y_1, z) = (2, 2, 0, 4)$ is in S. So for z between 0 and 4 inclusively, there exist values for x_1 in S. The first linear program with the nonnegative slack variables $y_2, y_3,$ and y_4, appears below. Inequality (e) has been dropped because we shall only consider $z \leq 4$.

Maximize x_1
subject to $x_1 + x_2 + y_1 \qquad\qquad = z$ (a),
$2x_1 + x_2 + y_1 + y_2 \qquad = 6$ (b),
$x_1 + 2x_2 + y_1 \qquad + y_3 = 6$ (c),
$x_1 + x_2 + 2y_1 \qquad\quad + y_4 = 6$ (d),
and $x_1, x_2 \geq 0, y_j \geq 0 \quad (j = 1, 2, 3, 4)$,

where $0 \leq z \leq 4$. Since we wish to maximize x_1, an obvious set of basic variables is x_1, y_2, y_3, and y_4. Hence, x_1 is eliminated from equalities (b), (c), and (d) by making the substitution $x_1 = z - x_2 - y_1$. This produces the equivalent problem

maximize x_1
subject to $x_1 + x_2 + y_1 \qquad\qquad = z$ (a),
$-x_2 - y_1 + y_2 \qquad\quad = 6 - 2z$ $(b) - 2(a)$,
$x_2 \qquad + y_3 \quad = 6 - z$ $(c) - \ \ (a)$,
$y_1 \qquad + y_4 = 6 - z$ $(d) - \ \ (a)$,
and $x_1, x_2 \geq 0, y_j \geq 0 \quad (j = 1, 2, 3, 4)$,

where $0 \leq z \leq 4$. Posing this program in our usual Beale's simplex tableau yields tableau #1, which exhibits dual feasibility.

Pivot
Column

#1	1	$(-x_2)$	$(-y_1)$
x_1	z	1	1
x_2	0	-1	0
y_1	0	0	-1
y_2	$6 - 2z$	$\boxed{-1}$	-1
y_3	$6 - z$	1	0
y_4	$6 - z$	0	1

Pivot row $(3 < z \leq 4) \rightarrow$ is indicated at the y_2 row.

Now, for $0 \leq z \leq 3$, the tableau is also primal feasible and hence optimal with $x_1 = z$. This is line segment L_1 in Figure 8.3. When $3 < z \leq 4$, $6 - 2z < 0$, and the y_2 equality may serve as a pivot row in

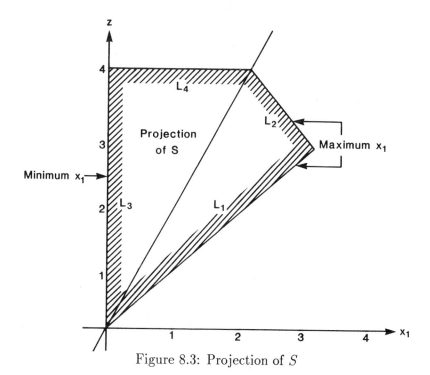

Figure 8.3: Projection of S

the dual simplex algorithm. Computing the minimum of $1/|-1|$ and $1/|-1|$ yields a tie; but $0/|-1| > -1/|-1|$, so the x_2 column is the pivot column. After a pivot on the circled entry we obtain Tableau #2.

#2	1	$(-y_2)$	$(-y_1)$
x_1	$6 - z$	1	0
x_2	$-6 + 2z$	-1	1
y_1	0	0	-1
y_2	0	-1	0
y_3	$12 - 3z$	1	-1
y_4	$6 - z$	0	1

The second tableau is optimal when $3 \leq z \leq 4$ with $x_1 = 6 - z$. This equation is designated as L_2 in Figure 8.3.

To find the rest of the projection we consider the linear program

$$\begin{array}{llll}
\text{minimize} & x_1 \\
\text{subject to} & x_1 + x_2 + y_1 & = z & (a), \\
& 2x_1 + x_2 + y_1 + y_2 & = 6 & (b), \\
& x_1 + 2x_2 + y_1 \quad\ + y_3 & = 6 & (c), \\
& x_1 + x_2 + 2y_1 \quad\ + y_4 = 6 & & (d), \\
\text{and} & x_1, x_2 \ge 0, y_j \ge 0 \quad (j = 1, 2, 3, 4),
\end{array}$$

where $0 \le z \le 4$. As we wish to minimize x_1, which must always remain nonnegative, it may be worthwhile to start with x_1 as a nonbasic variable. Then an obvious set of basic variables is y_1, y_2, y_3, and y_4. Eliminating y_1 from equalities (b), (c), and (d) yields the equivalent problem

$$\begin{array}{llll}
\text{maximize} & -x_1 \\
\text{subject to} & x_1 + x_2 + y_1 & = z & (a), \\
& x_1 \quad\quad\ + y_2 & = 6 - z & (b) - (a), \\
& x_2 \quad\quad\ + y_3 & = 6 - z & (c) - (a), \\
& -x_1 - x_2 \quad\quad\ + y_4 = 6 - 2z & & (d) - 2(a), \\
\text{and} & x_1, x_2 \ge 0, y_j \ge 0 \quad (j = 1, 2, 3, 4),
\end{array}$$

where $0 \le z \le 4$. Since x_1 is to be a nonbasic variable, it cannot serve as the variable that defines the objective function. However, from equality (b)--(a), $y_2 - (6 - z) = -x_1$, so maximize $-x_1$ can be replaced by maximize y_2. This allows us to form the first simplex tableau.

#1	1	$(-x_1)$	$(-x_2)$
y_2	$6 - z$	1	0
x_1	0	-1	0
x_2	0	0	-1
y_1	z	1	1
y_3	$6 - z$	0	1
Pivot row $(3 < z \le 4) \rightarrow$ y_4	$6 - 2z$	-1	$\boxed{-1}$

Now for $0 \le z \le 3$, Tableau #1 is optimal with $x_1 = 0$. For $3 < z \le 4$, a pivot on the circled entry is performed.

#2	1	$(-x_1)$	$(-y_4)$
y_2	$6 - z$	1	0
x_1	0	-1	0
x_2	$-6 + 2z$	1	-1
y_1	$6 - z$	0	1
y_3	$12 - 3z$	-1	1
y_4	0	0	-1

The resulting tableau is optimal for $3 \leq z \leq 4$. Thus we have the vertical line segment L_3 in the figure. Finally, since the maximum value of z is 4 and the polygon is convex, the horizontal line segment L_4 must close the region (see Problem 8.3).

The diagram indicates that the optimal solution to the MIP is bounded above by 4 (in fact, we have shown that it is 4), and x_1 cannot have any integer value other than $0, 1, 2$, or 3. Thus, had LP^0 been successively solved with x_1 fixed at integral values, we would have obtained $z^0 = z(0, 1, 0) = z(0, 1, 1) = z(0, 1, 2) = 4 > z(0, 1, 3)$, and no feasible solution when $x_1 = 4, 5, \dots$. Equivalently, the subproblem at the point $\mathbf{x}^1$, defined from $\mathbf{x}^0 = (x_1, x_2)$, by fixing x_1 at integer values, would have the same optimal solutions for $x_1 = 0, 1$, and 2, a smaller solution when $x_1 = 3$, and no feasible solution otherwise. An upper bound for any mixed integer programming solution found from $\mathbf{x}^0$ is 4.

8.3   The Enumeration Tree, Algorithm Formulation, An Example

8.3.1   The Tree, Algorithm Formulation

Enumerative algorithms are usually easier to understand if they are pictorially related to a *tree* composed of *nodes* and *branches*. A node corresponds to a point $\mathbf{x}^\ell$ and a branch links the node $\mathbf{x}^\ell$ with $\mathbf{x}^{\ell+1}$, where the latter point is defined from the previous one by fixing one of its $n - \ell$ free variables to a nonnegative integer. As a free variable x_k can usually take on several values, it is possible to have many branches emanating from the node $\mathbf{x}^\ell$. With regard to Fig. 8.4, the

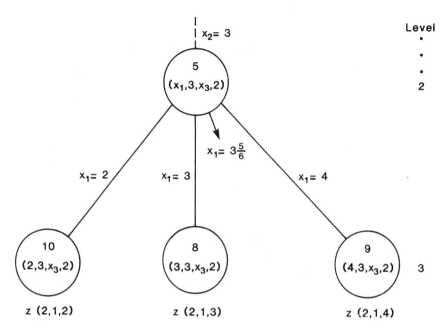

Figure 8.4: Branch and Bound Tree

nodes numbered $8, 9$, and 10 are created by setting x_1, a free variable in node 5, to $3, 4$ and 2, respectively. Nodes such as these three which have not as yet been used to create other nodes, or equivalently, which lack emanating branches, are called *dangling*. For example, node 5 is not dangling. Also, the *level number* ℓ refers to the number of fixed variables (see Fig. 8.4).

Suppose that $x_1 = 3\frac{5}{6}$ in the optimal solution to the subproblem at node 5. Then by fixing x_1 at the next lowest and highest integer, nodes 8 and 9 are created (Fig. 8.4). Solving the subproblem or linear program at each node yields $z(2,1,3)$ and $z(2,1,4)$. The larger of these two numbers, say $z(2,1,3)$ is an upper bound z^2 for any mixed integer programming solution derived from node 5. By fixing x_1 at 2, node 10 (which is to the immediate left of node 8) is created and $z(2,1,2)$ can be determined. Note that if node 8 is ever eliminated from consideration, z^2 may be replaced by the maximum of $z(2,1,2)$ and $z(2,1,4)$. (Why?) If the current best mixed integer solution z^* is at least as large as $z(2,1,3)$, then none of the nodes which can be traced back to node 5 (or are descendants of node 5) can be a candidate for an improved solutions to the MIP. Similarly, if $z^* \geq z(2,1,2)$ ($z^* \geq z(2,1,4)$), then all nodes to the left (right) of node 8 may be implicitly enumerated. So if the current best solution is at least as large as both $z(2, 1, 2)$ and $z(2, 1, 4)$, and $z^* < z(2, 1, 3)$, node 8 alone needs to be examined at level 3. Thus we may say that only those nodes at which the optimal linear programming solution exceeds z^*, the current best solution for the MIP, should be eligible to create new nodes. In other words, these are the only ones that should be explicitly examined or remain on the list of dangling nodes. When a linear program at a node is infeasible, then all subproblems at nodes either to the left or right of it are also infeasible. Hence, in this case, by taking $z(\ell, k, t)$ arbitrarily very small, the discussion goes through unchanged.

The Land and Doig algorithm repeatedly makes use of the previous remarks which, as mentioned earlier, are a consequence of Theorem 8.1 (see Remarks 5 and 6). In particular, unless LP^0 is infeasible or it solves the MIP, the initial node $\mathbf{x}^0$, the point $\mathbf{x}$ with all variables free, is labeled dangling. At any step, from the list of dangling nodes the one $\mathbf{x}^\ell$ with the largest linear programming solution is selected. An integer variable which is not integral in this optimal solution is fixed at the next lowest and highest integers to create two new nodes. The subproblem is resolved at each one of these two nodes. The larger of the two optimal solutions is z^ℓ, an upper bound for any mixed integer programming solution derived from $\mathbf{x}^\ell$. A newly created node is then labeled dangling if its subproblem solution exceeds the current best one for the MIP. Whenever an improved solution to the MIP is found it is recorded and each dangling node whose optimal linear programming solution does not exceed it is dropped from the list. Prior to creating new nodes from $\mathbf{x}^\ell$, a node either to its left or right is created so that if $\mathbf{x}^\ell$ is eliminated from consideration a new value for z^ℓ can be immediately determined. When the list of dangling nodes becomes empty the algorithm either terminates with the optimal solution to the MIP or it indicates that none exists.

8.3.2   An Example

Before outlining the basic approach, the example given in the Land and Doig [383] paper is presented. The purpose of this problem is to illustrate the underlying principles of the algorithm numerically. The linear programming calculations are omitted.

Example 8.2 (Land and Doig [383])   Consider the mixed integer program[6]

---

[6]Numbers are being rounded to the nearest integer.

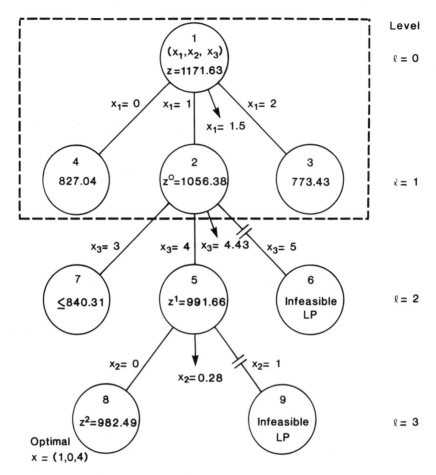

Figure 8.5: Branch and Bound Tree

$$\text{maximize} \quad 78x_1 + 77x_2 + 90x_3 + 97y_1 + 31y_2$$
$$\text{subject to} \quad 11x_1 + 4x_2 - 41x_3 + 44y_1 + 7y_2 \leq 82,$$
$$-87x_1 + 33x_2 + 24x_3 + 14y_1 - 13y_2 \leq 77,$$
$$61x_1 + 69x_2 + 69x_3 - 57y_1 + 23y_2 \leq 87,$$
$$x_1, x_2, x_3 \geq 0, \quad y_1, y_2 \geq 0,$$
$$\text{and} \quad x_1, x_2, x_3 \text{ integer.}$$

The steps outlined below relate to the tree in Fig. 8.5.

NODE 1 (x_1, x_2, x_3) The optimal solution to the linear problem with all variables free is

$$\hat{z} = 1171.63 \quad \begin{aligned} x_1 &= 1.5027 \\ x_2 &= 0 \\ x_3 &= 5.0452 \end{aligned} \quad \begin{aligned} y_1 &= 6.1892 \\ y_2 &= 0. \end{aligned}$$

Now we select any free variable to be fixed to a nonnegative value to create nodes at the next level. In this example, we pick x_1, as it is the furthest from an integer value.[7] It is set to 1 and 2, creating the two nodes $(1, x_2, x_3)$, node 2, and $(2, x_2, x_3)$, node 3.

NODE 2 $(1, x_2, x_3)$ The optimal linear programming solution is

$$z(0, 1, 1) = 1056.358 \quad \begin{aligned} x_1 &= 1 \\ x_2 &= 0 \\ x_3 &= 4.4278 \end{aligned} \quad \begin{aligned} y_1 &= 5.5032 \\ y_2 &= 1.4855 \end{aligned}$$

NODE 3 $(2, x_2, x_3)$ The optimal linear programming solution is $z(0, 1, 2) = 773.432$. (As is true for several other nodes, it will be unnecessary to know the optimal values of the variables.)

NODE 4 $(0, x_2, x_3)$ The upper bound for any solution to the MIP found from $\mathbf{x}^0$ is thus $1056.358 = z(0, 1, 1) = z^0$. Now the node to the left of node 2 is created by setting x_1 to 0 and $z(0, 1, 0)$ is found. Its value is 827.04. The tree currently consists of nodes $1, 2, 3$, and 4 (see Fig. 8.5). Since the linear programming solution at node 2 is the largest and its optimal solution is $x_1 = 1, x_2 = 0, x_3 = 4.4278$, x_3 is set to 4 and 5, creating the two nodes $(1, x_2, 4)$, node 5, and $(1, x_2, 5)$, node 6.

NODE 5 $(1, x_2, 4)$ The optimal linear programming solution is

$$z(1, 3, 4) = 991.6577 \quad \begin{aligned} x_1 &= 1 \\ x_2 &= 0.2849 \\ x_3 &= 4 \end{aligned} \quad \begin{aligned} y_1 &= 5.1498 \\ y_2 &= 1.0384 \end{aligned}$$

NODE 6 $(1, x_2, 5)$ The linear program is infeasible. Thus an upper bound for any solution to the MIP found from node 2 is $z(1, 3, 4) = 991.6577 = z^1$.

---

[7]Branching rules are discussed in Section 8.7.

The current list of dangling nodes is

node	$\ell+1,k,t$	$z(\ell,k,t)$
3	$1,1,2$	773.4320
4	$1,1,0$	827.0400
5	$2,3,4$	991.6577

NODE 7 $(1, x_2, 3)$ As 991.6577 is the largest value, node 5, $(1, x_2, 4)$, is selected. Node 7 is created to the left of 5 by fixing x_3 at 3 (Fig. 8.5). The optimal linear programming solution is $z(1,3,3) = 840.3124$. As $x_2 = 0.2849$ in the optimal linear programming solution at node 5, it is set to 0 and 1, creating node 8, or $(1,0,4)$, and node 9, or $(1,1,4)$.

NODE 8 $(1,0,4)$ The linear programming solution is feasible and, of course, integer in the integer constrained variables. That is,

$$z(2,2,0) \;=\; 982.4933 \quad \begin{aligned} x_1 &= 1 & y_1 &= 5.0709 \\ x_2 &= 0 & y_2 &= 1.6974 \\ x_3 &= 4 \end{aligned}$$

is our first solution to the MIP with $z^* = 982.4933$. (the values of the variables should also be kept.)

NODE 9 $(1,1,4)$ The linear program is infeasible. Thus, $z^2 = z(2,2,0) = 982.4933$. The current list of dangling nodes is

node	$\ell+1,k,t$	$z(\ell,k,t)$
3	$1,1,2$	773.4320
4	$1,1,0$	827.0400
7	$2,3,3$	840.3124

Because $z^* = 982.4933$ and it exceeds all $z(\ell,k,t)$ values it is optimal.[8] Note that $\hat{z} \geq z^0 \geq z^1 \geq z^2$, or $1171.63 \geq 1056.358 \geq 991.6577 \geq 982.4933$.

---

[8]Unfortunately, in this example the first mixed integer solution found is optimal. This, of course, is not guaranteed and usually will not be true. When a mixed integer programming solution is found, only some of the dangling nodes are normally removed from the list and those which can possibly produce an improved solution to the MIP still have to be considered; see Example 8.3.

8.4   The Basic Approach, A Second Example

By this time the reader should be relatively aware of how and why the algorithm works. For completeness, a general outline of the Land and Doig procedure is given. A formal proof of convergence is left to Problem 8.8, which essentially asks for a rewording of earlier comments. The terminology and notation of the previous section is adopted. It is assumed that each integer variable is bounded above.

8.4.1   The Basic Approach (Land and Doig [383])

Step 1 (Initialization) Set z^*, the current best solution to the MIP, to some arbitrarily small or predetermined value.[9] The initial node, with all variables free, is $\mathbf{x}^0 = (x_1, \ldots, x_n)$. Solve LP^0. If it is infeasible, so is the MIP–terminate. If its optimal solution is integer in the integer constrained variables, the MIP is solved–terminate. Otherwise, set $\mathbf{x}^\ell = \mathbf{x}^0$ and go to Step 2.

Step 2 (Branching) From the optimal linear programming solution $(\hat{\mathbf{x}}, \hat{\mathbf{y}}, \hat{z})$ at $\mathbf{x}^\ell$ pick a variable x_k which does not satisfy the integer requirement. By fixing x_k at $[\hat{x}_k]$ and $[\hat{x}_k] + 1$, define two nodes from $\mathbf{x}^\ell$. Solve the subproblem at each of these nodes. Label dangling those created nodes whose optimal linear programming solution exceeds z^*. Check each node for an improved solution to the MIP. If one is found, it is recorded, and all dangling nodes with a subproblem solution not exceeding it are dropped from the list. Go to Step 3.

Step 3 (Termination test) If the current list of dangling nodes is empty, either the optimal solution to the MIP, z^*, has been found, or no solution exists–terminate. Otherwise, go to Step 4.

Step 4 (Bounding) Determine the dangling node $\mathbf{x}^\ell$ which has the largest optimal solution. Break ties arbitrarily.[10] Suppose the point

---

[9]It may be known apriori that z^* cannot be smaller than some value. The closer it is to the optimal solution the shorter the enumeration. However, if this value exceeds the optimal solution the algorithm will incorrectly indicate that no mixed integer solution exists.

[10]Since the larger the level number, the closer a point may be to a mixed integer solution, it may be better to choose the node at the highest level.

$\mathbf{x}^{\ell-1}$ defines $\mathbf{x}^{\ell}$, the selected node, by setting $x_k = t$. (This means the optimal solution to LP^{ℓ} is $z(\ell - 1, k, t)$.) Set $z^{\ell-1}$, an upper bound for any mixed integer programming solution found from $\mathbf{x}^{\ell-1}$, equal to $z(\ell - 1, k, t)$ at $\mathbf{x}^{\ell}$. Create a node to the immediate left or right of $\mathbf{x}^{\ell}$ so that if another dangling node created from $\mathbf{x}^{\ell-1}$ is eventually selected, a new (not higher) value for $z^{\ell-1}$ may be found. Remove $\mathbf{x}^{\ell}$ from the list of dangling nodes and go to Step 2.

Remarks

1. If the linear program at the node created by fixing a not integral $\hat{x}_k$ at its next lowest integer value $[\hat{x}_k]$, and that at the node created by fixing x_k at its next highest integer value $[\hat{x}_k] + 1$ are both infeasible, there cannot exist a solution to the MIP with x_k integer emanating from the node $\mathbf{x}^{\ell}$. In particular, if this occurs at level 0 ($\ell = 0$) the MIP has no feasible solution. Geometrically the upper boundary of S^{ℓ} is as in Fig. 8.6.

2. The upper bound $z^{\ell-1}$, equal to some $z(\ell - 1, k, t)$ value, for any mixed integer programming solution found from $\mathbf{x}^{\ell-1}$, is replaced (usually by a lower value) when it is ascertained that there cannot exist an improved mixed integer solution emanating from $\mathbf{x}^{\ell}$. Whenever $z^{\ell-1} \leq z^{*}$, node $\mathbf{x}^{\ell-1}$ and all points defined from it cannot be candidates for an improved solution to the MIP. Hence, prior to branching

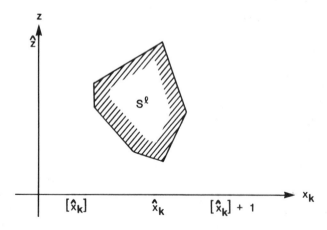

Figure 8.6: Infeasible MIP

from $\mathbf{x}^\ell$, the purpose of creating a level ℓ node to the left or right of it is to have the immediate ability to revise $z^{\ell-1}$, usually downward. This is done by replacing $z^{\ell-1}$ by the maximum of the subproblem solution at the nodes on either side of $\mathbf{x}^\ell$. For example, consider the part of the tree in Fig. 8.5 that is enclosed by dashed lines. Here $\ell = 1$ and $\mathbf{x}^{\ell-1}$ is node 1. Nodes 2 and 3 are created from node 1 and $z^{\ell-1} = z^0 = \text{maximum}(1056.358, 773.432) = 1056.358$. Prior to branching from x^1 or node 2, node 4 is created. If it is ever determined that node 2 cannot produce an improved mixed integer programming solution, z^0 is lowered to the maximum $(773.432, 827.04) = 827.04$.

3. When a subproblem at level n is feasible, it produces a solution to the MIP. Hence, if each integer variable has a finite upper bound, the algorithm converges, since, in the worst case, all nodes of the tree are enumerated explicitly, and the nodes in the last row represent all possible combinations of integers.

4. To reduce computational effort it is probably worthwhile to explicitly introduce the inequality $\mathbf{cx}^\ell + \mathbf{dy} \geq z^* + \epsilon$, where ϵ is a small positive number, to each linear program LP^ℓ. By doing this, a feasible linear program always indicates a dangling node, and only improved mixed integer solutions can be found.

5. If node $\mathbf{x}^\ell$ is defined from $\mathbf{x}^{\ell-1}$ by setting x_k, a free variable in $\mathbf{x}^{\ell-1}$, to t, it is a simple matter to modify the optimal simplex tableau at node $\mathbf{x}^{\ell-1}$ to produce the first tableau at $\mathbf{x}^\ell$.[11]

6. The continual branching and bounding process suggests the branch and bound designation for the algorithm.

8.4.2 A Second Example

As a second illustration, the integer program which appeared in Chapters 4 and 6 is resolved. The computational details, including a post-

---

[11]The usually large number of dangling nodes prohibits the storage of all optimal tableaux. Thus, using the optimal simplex tableau at $\mathbf{x}^{\ell-1}$ to produce the first simplex tableau at $\mathbf{x}^\ell$ is usually not possible.

optimization technique, are left to Appendix A. The superscript ˆ is being omitted.

Example 8.3 (Balinski [50]) Consider the integer program

$$
\begin{array}{rrrcll}
\text{maximize} & -4x_1 & - & 5x_2 & = & z \\
\text{subject to} & -x_1 & - & 4x_2 & \leq & -5, \qquad (x_3) \\
& -3x_1 & - & 2x_2 & \leq & -7, \qquad (x_4) \\
& x_1, & & x_2 & \geq & 0 \qquad \text{and integer.}
\end{array}
$$

Step 1 The initial node is $\mathbf{x}^0 = (x_1, x_2)$. The subproblem solution is $z = -112/10, x_1 = 18/10, x_2 = 8/10$, and $x_3 = x_4 = 0$.

Step 2 x_1 is set to $[18/10] = 1$ to define node 2 or $(1, x_2)$. The subproblem solution is $z = -14, x_1 = 1, x_2 = 2, x_3 = 4$, and $x_4 = 0$; x_1 is set to $[18/10] + 1 = 2$ to define node 3 or $(2, x_2)$. Its subproblem solution is $z = -47/4, x_1 = 2, x_2 = 3/4, x_3 = 0$, and $x_4 = 2/4$. Node 2 yields an integer programming solution with $z = -14$; hence $z^* = -14$; node 3 is labeled dangling because $-47/4 > -14$.

Steps 3,4 $z^0 = -47/4$; and node 4 or $(3, x_2)$ to the right of node 3 (Fig. 8.7) is created. Its subproblem solution has $z = -58/4$.

Step 2 In the optimal linear programming solution at node 3, $x_2 = 3/4$. So by fixing x_2 to 0 and 1, nodes 5 and 6 are created. The subproblem at node 5 is infeasible. The subproblem at node 6 produces the improved integer solution $z = -13$ with $x_1 = 2$ and $x_2 = 1$; hence $z^* = -13$.

Step 3 As node 4 has $z = -58/4 < -13$, there are no dangling nodes and the process terminates. Fig. 8.7 relates the computations to the tree.

8.5   A Variation of the Basic Approach (Dakin [153])

Each time in the Land and Doig algorithm a node is created and labeled dangling, its number, optimal linear programming solution, and other parameters have to be kept (Problem 8.17). As many dangling

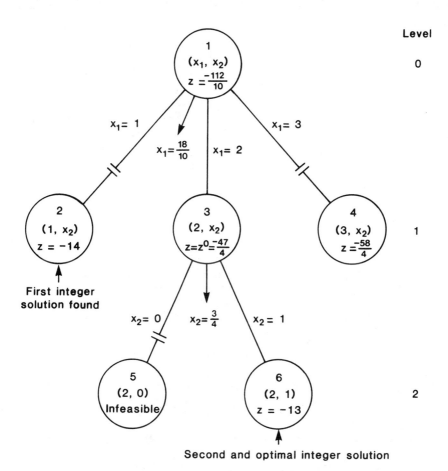

Figure 8.7: Land and Doig Algorithm

nodes are possible, this may mean that the technique, when computer implemented, would involve excessive storage requirements. To alleviate this difficulty, Dakin [153] suggested that only two nodes be created from each dangling one. If an optimal linear programming solution has $x_k = t$, where t is not integral, then the first node is created by introducing the inequality $x_k \leq [t]$ and the second is defined by constraining $x_k \geq [t] + 1$. This is in contrast to creating nodes by setting $x_k = [t]$ and $x_k = [t] + 1$ and then defining other nodes to the left or right of these. When the Dakin procedure is adopted, the tree appears as in Fig. 8.8. Notice that it is possible for more than one constraint to operate on a variable at a time. For example, at

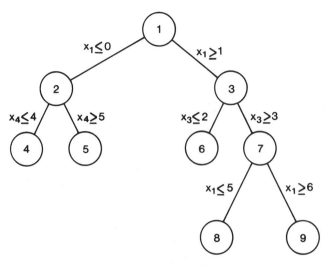

Figure 8.8: The Dakin Variation

node 8 we have $x_1 \geq 1, x_3 \geq 3$, and $x_1 \leq 5$, and at node 9 we have $x_1 \geq 1, x_3 \geq 3$, and $x_1 \geq 6$. The $x_1 \geq 1$ inequality may be omitted from computations at node 9 and at other points which can be traced back to it. Also, observe that constraining a variable to be less than or equal to 0, or greater than or equal to its upper bound, is equivalent to fixing it at 0, or at its upper bound, respectively. In Fig. 8.8, $x_1 = 0$ is in force at node 2.

As in the Land and Doig algorithm, the initial node is the point $\mathbf{x}^0$ with all variables free. The node is labeled dangling if its linear program is feasible and does not solve the MIP. At any point in the algorithm the dangling node $\mathbf{x}^\ell$ with the largest optimal linear programming solution is selected to define two nodes at the next level. The subproblem at each of these nodes is solved. The largest of these two solutions is z^ℓ, an upper bound for any solution to the MIP found from $\mathbf{x}^\ell$. A newly created node is labeled dangling if its optimal linear programming solution exceeds the current best solution z^* for the MIP. If an improved mixed integer solution is found, z^* is updated, and those dangling nodes with an optimal subproblem solution not exceeding it are dropped from the list of dangling nodes. The procedure terminates when the list of dangling nodes is empty. Therefore, except for notation modifications,[12] the creation of only two nodes from each selected one,

---

[12]For example, $\mathbf{x}^\ell$ means the point $\mathbf{x}$ where ℓ variables have inequalities operating on them.

and the omission of creating a node to the left or right of the selected one, the Dakin variation is essentially the same as the Land and Doig algorithm outlined in the previous section. A formal listing is left to Problem 8.18.

The Dakin algorithm converges, since in the worst case nodes could be created until the permitted range of all integer variables is reduced to zero, in which case they must take on integer values. In particular, if an optimal mixed integer solution has $x_1 = a_1, x_2 = a_2$, etc. the procedure can, if necessary, create the node which has $x_1 \geq a_1, x_1 \leq a_1, x_2 \geq a_2, x_2 \leq a_2$, etc. For example, with regard to Fig. 8.8, suppose the optimal linear programming solution at node 9 had $x_1 = 6.3$. Then two nodes would be created, one with $x_1 \leq 6$ and the other with $x_1 \geq 7$. At the node with $x_1 \geq 6$ and $x_1 \leq 6$ active, we have $x_1 = 6$. To clarify the discussion, we resolve the previous integer programming example. Computational details are left to Appendix A.

Example 8.4 (from Example 8.3)

The subproblem at node 1 or $\mathbf{x}^0 = (x_1, x_2)$ produces the solution $z = -112/10, x_1 = 18/10, x_2 = 8/10$, and $x_3 = x_4 = 0$.

Node 2 or $(x_1 \leq 1, x_2)$ is created by constraining $x_1 \leq 1$, and node 3 or $(x_1 \geq 2, x_2)$ by introducing the inequality $x_1 \geq 2$.

Node 2 produces the integer solution $z = -14, x_1 = 1, x_2 = 2, x_3 = 4$, and $x_4 = 0$

The subproblem solution at node 3 has $z = -47/4, x_1 = 2$, and $x_2 = 3/4$. As $-47/4 > -14, z^0$ is set to $-47/4$.

Node 4 or $(x_1 \geq 2, x_2 \leq 0)$ is created by constraining $x_2 \leq 0$ (equivalently $x_2 = 0$) in the linear program at node 3, and node 5 or $(x_1 \geq 2, x_2 \geq 1)$ is defined by adjoining the inequality $x_2 \geq 1$ to the constraint set at node 3. The subproblem at the former node has $z \leq -20 < z^* = -14$, and the one at the latter yields the improved integer solution $z = -13, x_1 = 2, x_2 = 1$, and $x_3 = x_4 = 1$. As there are no dangling nodes, this solution is optimal. The tree appears in Figure 8.9.

8.6   Specialization for the Zero-One Problem

Suppose now we consider the problem in which each integer variable is either 0 or 1; that is, the upper bound $u_j = 1$ for every variable x_j.

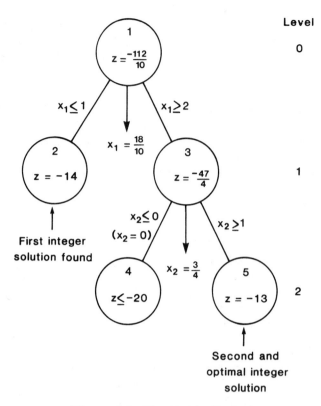

Figure 8.9: The Dakin Variation

Since $x_j \leq 0$ and $x_j \geq 1$ are then equivalent to $x_j = 0$ and $x_j = 1$, respectively, the Dakin variation is precisely the same as the Land and Doig algorithm. By fixing an integer variable to 0 and 1, we define two nodes from the selected one. An upper bound for any mixed integer programming solution found from the selected node is just the larger of the two optimal linear programming solutions. (As x_j cannot exceed 1 in any solution to the MIP, the constraints $x_j \leq 1$ are added to every linear program.) Specialization of the Land and Doig algorithm is straightforward, and the details are left to Problem 8.28. As an illustration, a possible solution tree appears in Fig. 8.10. It may be explained as follows: Nodes 2 and 3 are defined from node 1 and the subproblem at each one of these is solved. Both nodes are labeled dangling, and the optimal solution at node 2 is larger than the one at node 3. So node 2 is no longer dangling, and nodes 4 and 5 are created.

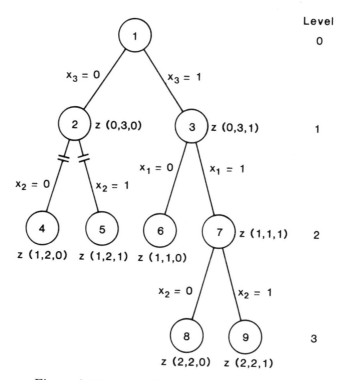

Figure 8.10: A tree for the zero-one problem

The subproblem at each node is infeasible, thus only node 3 remains dangling. It is removed from the list and nodes 6 and 7 are created–and so on.

8.7 Node Selection, Branching Rules, and Penalties

8.7.1 Node Selection (Spielberg [543])

In the Land and Doig algorithm the dangling node with the largest optimal linear programming solution is selected to create new nodes. The intent of this strategy is to quickly find a "good" mixed integer solution. However, as already mentioned, this selection procedure may require excessive computer storage. A second rule is to select the most recently created dangling node with the largest subproblem solution.

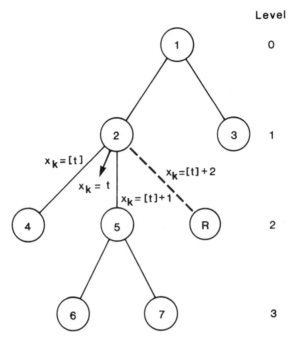

Figure 8.11: A Second Node Selection Rule

For example, suppose in Fig. 8.11 that the optimal solution at node 2 is larger than the one at node 3. Then nodes 4 and 5 are created. The optimal solution at node 5 is larger than the one at node 4 (although it may be less than the optimal solution at node 3), so nodes 6 and 7 are defined–and so on. When this forward process stops, say at level 3, the enumeration reverts back to some level ℓ, here level 2, and the node R is created (Fig. 8.11). Its linear program is solved. The optimal solution at node R is compared with the other dangling nodes at the same level (here node 4). The point with the larger solution is selected and the forward process is restarted from this node. Since "good" solutions for the MIP may not be found early, this process may tend to examine more nodes than in the previous case. However, during the computations it is easier to retrieve and label a dangling node. Therefore, a simpler bookkeeping scheme to keep track of the enumeration (and usually less computer storage) is required. (Problem 8.32 asks for a listing of the Land and Doig algorithm under this node selection rule and for a scheme to keep track of the enumeration.)

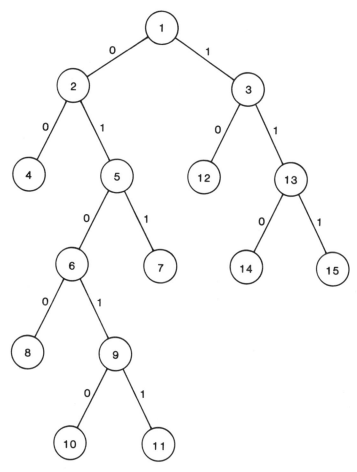

Figure 8.12: Tree

For the zero-one problem the rule just described produces a tree as
in Fig. 8.12. Since the nodes can be "flipped over", the tree may also be
pictured as in Fig. 8.13. The latter diagram suggests the name "single-
branch/branch and bound" for the modified Land and Doig algorithm.
Without the change the Land and Doig procedure may then be called
"multibranch/branch and bound." As discussed in Spielberg [543] and
Colmin and Spielberg [139], an algorithm for a zero-one problem whose
solution tree is of one of these two forms is of the branch and bound
type. Both node selection rules are also described in Davis, Kendrick,
and Weitzman [164].

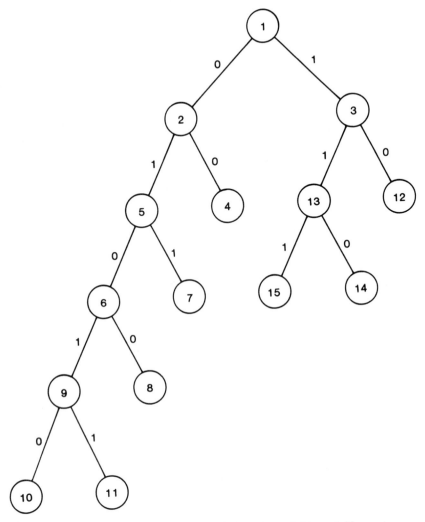

Figure 8.13: Single-Branch/Branch and Bound Tree

8.7.2   Branching Rules and Penalties (Beale and Small [62], Tomlin [566, 567]

At the selected dangling node, the optimal linear programming solution usually has several integer variables which are not integral. In the Land and Doig algorithm one of these variables is to be fixed at the next lowest and highest integer to define two nodes. The question is which of these variables should be picked. Up to this time we have arbitrarily picked the one with the largest fractional component. Another possibility would be to round each integer variable that is not integral to the next lowest and highest integers, solve the subproblem at the pair of nodes, and select the variable that produces the node with the largest optimal linear programming solution. Although this procedure probably tends to reduce the number of nodes which are explicitly examined, the additional computations will almost surely make it unworthwhile.[13]

The optimal linear programming tableau can also be used to devise branch selection rules. A not-satisfied constraint can be derived for each integer variable that is not integral. Then, by performing one dual simplex pivot, we may select a variable according to the size of the decrease in the objective function. More precisely, suppose the optimal solution has the basic integer variable $x_k = n_k + f_k$, where n_k is a nonnegative integer and f_k is a proper fraction. As either $x_k \leq n_k$ or $x_k \geq n_k + 1$, we may introduce the inequality

$$x_k^D = -x_k + n_k \geq 0 \quad \text{or} \quad x_k^U = x_k - n_k - 1 \geq 0$$

to the optimal simplex tableau. If $x_k = (n_k + f_k) + \sum_j a_{kj}(-x_{J(j)})$, where $x_{J(j)}$'s are the current nonbasic variables, then

$$x_k^D = -f_k + \sum_j (-a_{kj})(-x_{J(j)}) \quad \text{and} \quad x_k^U = (f_k - 1) + \sum_j a_{kj}(-x_{J(j)}).$$

Since $-f_k$ and $f_k - 1$ are negative, the x_k^D or x_k^U equality may serve as a pivot row in the dual simplex method. After one pivot on an

---

[13]An analogous strategy exists in the simplex algorithm. The vector which leaves the basis is determined from the $\theta_j \bar{c}_j$ rather than from the reduced cost $\bar{c}_j$, where

$$\theta_j = \underset{i}{\text{minimum}} \left\{ \frac{x_{Bi}}{y_{ij}}, y_{ij} > 0 \right\},$$

where x_{Bi} is the current value of the ith basic variable and y_{ij} is the ith scalar when the original column j is written as a linear combination of the current basic vectors. The latter rule, at the expense of more computations, will usually reduce the number of visited extreme points.

	1	$-x_{J(1)}$ ...	$-x_{J(G)}$ ...	$-x_{J(q)}$ ...	$-x_{J(r)}$ ...	$-x_{J(n)}$	
$z =$	a_{00}	a_{01} ...	a_{0G} ...	a_{0q} ...	a_{0r} ...	a_{0n}	
$x_k =$	$n_k + f_k$	a_{k1} ...	a_{kG} ...	a_{kq} ...	a_{kr} ...	a_{kn}	≥ 0
pivot on circled elements yields							
$P_k^D \to x_k^D =$	$-f_k$	$-a_{k1}$	...	$\boxed{-a_{kq}}$	...	$-a_{kn}$	≥ 0
$P_k^U \to x_k^U =$	$f_k - 1$	a_{k1}	...		$\boxed{a_{kr}}$ ..	a_{kn}	≥ 0
$P_k^G \to x_k^G =$	$-f_k$	$-g_{k1}$ ...	$\boxed{-g_{kG}}$	...		$-g_{kn}$	≥ 0

Figure 8.14: Penalty Calculations

element in the x_k^D row (Fig. 8.14), the objective function a_{00} decreases
by $-f_k(a_{0q}/a_{kq})$, where

$$\frac{a_{0q}}{a_{kq}} = \operatorname*{minimum}_{j\,:\,-a_{kj} < 0} \frac{a_{0j}}{a_{kj}}.$$

Or, subsequent to a pivot on a coefficient in the x_k^U row, a_{00} decreases
by $(f_k - 1)(a_{0r}/a_{kr})$, where

$$\frac{a_{0r}}{-a_{kr}} = \operatorname*{minimum}_{j\,:\,-a_{kj} < 0} \frac{a_{0j}}{-a_{kj}}.$$

Let us define

$$P_k^D = -f_k(a_{0q}/a_{kq}) \quad \text{and} \quad P_k^U = (f_k - 1)(a_{0r}/a_{kr});$$

then P_k^D (P_k^U) can be thought of as the smallest penalty for rounding
x_k down (up). Notice that P_k^D and P_k^U are nonnegative. As variable
x_k must be integer, a_{00} minus the minimum of P_k^D and P_k^U is an upper
bound for any mixed integer programming solution found from the
selected node. Thus, to produce an improved mixed integer solution
we must have

$$a_{00} - \max_{j \,:\, f_j > 0} \text{imum} \,(\text{ minimum}(P_j^D, P_j^U)) > z^*, \tag{5}$$

where z^* is the current best solution and the maximum is taken over those basic integer variables which are not integral. If (5) does not hold, the node should not remain dangling.

If it is possible to produce a better solution, the variable with the smallest associated penalty could be chosen. This branching rule will hopefully create a path to a "good" mixed integer solution. However, when several $a_{0j} = 0 \,(j \neq 0), a_{0q}$ or a_{0r} is 0 and there are many ties for the smallest penalty. In this case the criterion breaks down and the choice is somewhat arbitrary. At the other extreme, we may select the variable which gives the largest penalty. The intent here is to eliminate the current node from consideration as quickly as possible. (Why?)

The method described in Beale and Small [62] is essentially the Dakin [153] variation of the Land and Doig algorithm where the node and branch selection rules are based on penalty calculations. In particular, the variable with the largest penalty is found. Suppose it is P_k^U. Then two nodes, one with $x_k \leq n_k$ and the second with $x_k \geq n_k + 1$, are created; and the node opposite to the highest penalty (in this case the one with $x_k \leq n_k$) is selected. The intent of this procedure is to have dangling nodes which can hopefully produce only poor mixed integer solutions, and thus when a "good" solution is found these nodes can be dropped from the list either immediately or after only a few iterations. This node and branching criterion was used earlier by Little, Murty, Sweeney, and Karel [400] in their branch and bound algorithm for the traveling salesman problem.

It is clear that the efficiency of the enumeration may rely heavily on the effect of inequality (5), and hence on the magnitude of the penalties. Recall from Chapter 5 (Section 5.4) that if

$$x_k = (n_k + f_k) + \sum_{j=1}^{n} a_{kj}(-x_{J(j)})$$

is an integer variable which is not integral (or $f_k > 0$), we may derive the Gomory mixed integer cut

$$x^G = -f_k + \sum_{j=1}^{n}(-g_{kj})(-x_{J(j)}) \geq 0, \tag{6}$$

where

$$
g_{kj} = \begin{cases}
a_{kj} & \text{if } a_{kj} \geq 0 \text{ and } x_{J(j)} \text{ is a continuous variable,} \\[2ex]
\dfrac{f_k}{f_k - 1} a_{kj} & \text{if } a_{kj} < 0 \text{ and } x_{J(j)} \text{ is a continuous variable,} \\[2ex]
f_{kj} & \text{if } f_{kj} \leq f_k \text{ and } x_{J(j)} \text{ is an integer variable,} \\[2ex]
\dfrac{f_k}{1 - f_k}(1 - f_{kj}) & \text{if } f_{kj} > f_k \text{ and } x_{J(j)} \text{ is an integer variable,}
\end{cases}
$$

and $f_{kj} = a_{kj} - [a_{kj}]$ for all $j \geq 1$. As $x^G = -f_k < 0$ when the nonbasic variables are set to 0, equality (6) may serve as a pivot row in the dual simplex algorithm. Subsequent to a single pivot, a_{00} decreases by the nonnegative number

$$
P_k^G = \frac{f_k a_{0G}}{g_{kG}}, \quad \text{where} \quad \frac{a_{0G}}{g_{kG}} = \underset{j\,:\,-g_{kj}<0}{\text{minimum}} \frac{a_{0j}}{g_{kj}}.
$$

Using the definition of g_{kj} and $f_k < 1$, it is simple to show that $P_k^G \geq \text{minimum}\,(P_k^D, P_k^U)$ (Problem 8.33); or we may compute P_j^G for each integer constrained basic variable which is not at an integer value and then replace inequality (5) by an even stronger condition; namely, if

$$
a_{00} - \underset{j\,:\,f_j>0}{\text{maximum}}\, P_j^G > z^* \tag{7}
$$

does not hold, the current node should not remain dangling. Expression (7) was derived by Tomlin [566, 567].

Up to this time we have not made use of the fact that some of the *nonbasic* variables may be required to be integer. At present each has value 0. But if a nonbasic integer constrained variable, say $x_{J(t)}$, is not to remain at 0, a_{00} must be decreased by at least a_{0t}. Thus, if for any nonbasic integer variable $a_{00} - a_{0j} \leq z^*$, then $x_{J(j)}$ must be fixed at 0 in any solution to the MIP found from the current node. More generally, $x_{J(j)}$ is bounded above by $\bar{u}_j$, where $\bar{u}_j$ is the largest nonnegative integer for which $a_{00} - \bar{u}_j a_{0j} > z^*$. Mathur and Salkin [419, 420, 422] used this criterion in developing efficient penalties for knapsack-like problems. (In Problem 8.35 the reader is asked to explain how these bounds can be used to shorten the Land and Doig or Dakin enumeration.)

Besides finding upper bounds it is sometimes possible to calculate stronger penalties than P_j^D and P_j^U by realizing that the nonbasic variable associated with the pivot column in the dual simplex method may be required to be integer. Thus, if after a pivot it is not integer, we have to pay an additional penalty. The details are left to Problem 8.37. Post-optimization and making use of nonbasic variables for the zero-one integer program is also discussed in Salkin and Spielberg [519] and in Davis, Kendrick, and Weitzman [164] (Problem 8.34). Other branch and node selection stragegies, as well as the use of penalties, appear in Benichou et al. [80].

Finally, according to Tomlin [568], the Dakin branch and bound technique with penalty calculations appears to be the most effective integer programming algorithm in practice. Problems with up to 4000 rows and 130 zero-one variables have been "solved" in times of the order of multiples of two or three of the first linear program solution time [204]. However, in many instances computer storage limitations required nodes to be dropped, so the best solution found was not necessarily optimal. Finally, Johnson, et al [353] discusses the implementation aspects of a large $0 - 1$ mixed integer programming algorithm which uses a branch and bound procedure.

Problems

Problem 8.1

a) Show that for a two dimensional convex set it is always possible to find two points on its boundary such that the boundary on one side of these points is a convex function of one of the axes and the boundary on the other side is a concave function of the same axis. How can the two points which divide the boundary be found? Can these results be generalized to an n dimensional convex set?

b) Prove that the projection of a convex set is convex.

Problem 8.2 Explain why the polygon appearing in Fig. 8.1 can be found by solving the pair of linear programs:

Minimize and maximize z,

subject to the common feasible region $\quad \mathbf{cx} + \mathbf{dy} - z = 0,$

$$\mathbf{Ax} + \mathbf{Dy} \qquad \leq \mathbf{b},$$
$$\mathbf{x} \geq 0, \quad \mathbf{y} \geq 0;$$

for all values of x_k. In particular, show why the approach used in Example 8.1 is equivalent to the one suggested here.

Problem 8.3 Given the mixed integer program appearing in Example 8.1:

a) Show that the maximum value of z is 4 for all x_1 in the closed interval $[0,2]$.

b) Find the projection of S onto the (x_2, z) plane.

c) Knowing the (x_1, z) and (x_2, z) projections, can you explain what the polyhedral set S looks like?

d) Is there a point (x_1, x_2, y_1, z) such that either x_1 or x_2, but not both, is integer, and the coordinate (x_1, z) or (x_2, z), with x_1 or x_2 integer, is in the projection of S onto the corresponding plane? Use a geometric argument to explain the question.

In Problems 8.4 through 8.7, S is the convex set of points $(\mathbf{x}, \mathbf{y})$ which satisfy constraints (1), (2), and (3).

Problem 8.4 Given the mixed integer program

maximize $\quad z$

subject to
$$
\begin{array}{llllll}
2x_1 + & x_2 + 3y_1 + 4y_2 - z = 0 & \quad (1) \\
x_1 + 3x_2 - & y_1 - 2y_2 & \leq 16 \\
-x_1 + 2x_2 + & y_1 - & y_2 & \leq 4
\end{array} \Bigg\} \quad (2)
$$
$$x_1, \quad x_2, \quad y_1, \quad y_2 \geq 0 \qquad (3)$$

and $\quad x_1, x_2$ integer.

Find the projection of S onto the (x_1, z) plane. What does the projection indicate?

Problem 8.5 Given the mixed integer program

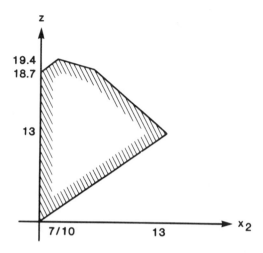

Figure 8.15: Projection of S onto the (x_2, z) plane for Problem 8.5

$$
\begin{aligned}
\text{maximize} \quad & z \\
\text{subject to} \quad & 5x_1 + x_2 + 4y_1 - z = 0 \quad &(1) \\
& 2x_1 + 3y_1 \leq 10 \\
& 4x_1 + y_1 \leq 11 \\
& 3x_1 + x_2 + 3y_1 \leq 13 \\
& x_1, \quad x_2, \quad y_1 \geq 0 \quad &(3)
\end{aligned}
$$

with (2) denoting the braced group of three inequalities.

and $\quad x_1, x_2$ integer.

Show that the projection of S onto the (x_2, z) plane is the set graphed in Fig. 8.15.

Problem 8.6 Given the mixed integer program

$$
\begin{aligned}
\text{maximize} \quad & z \\
\text{subject to} \quad & x_1 - 2x_2 + y_1 - z = 0 \quad &(1) \\
& x_1 + 2x_2 + y_1 = 11 \\
& 2x_1 \leq 9 \\
& x_1, \quad x_2, \quad y_1 \geq 0 \quad &(3)
\end{aligned}
$$

and $\quad x_1, x_2$ integer.

Find the projection of S onto the (x_1, z) plane using the procedure described in Problem 8.2.

Problem 8.7   Given the mixed integer program

$$
\begin{aligned}
\text{maximize} \quad & z \\
\text{subject to} \quad & x_1 + x_2 + y_1 - z = 0 & (1) \\
& \left.\begin{aligned} x_1 + x_2 \quad\quad\ & \leq 3 \\ x_2 + y_1 & \leq 5 \end{aligned}\right\} & (2) \\
& x_1, \quad x_2, \quad y_1 \quad \geq 0 & (3) \\
\text{and} \quad & x_1, \ x_2 \ \text{integer.}
\end{aligned}
$$

Find the projection of S onto the (x_1, z) plane.

Problem 8.8   Prove that the Land and Doig algorithm outlined in Section 8.4 converges to an optimal solution in a finite number of steps. Assume that each integer variable is bounded above.

Problem 8.9   Explain how the Land and Doig algorithm (Section 8.4) can be modified so that all optimal solutions are found. Would the resulting technique necessarily require more computation time?

Solve the following mixed integer programs by the Land and Doig algorithm. In each case relate the computations to the tree.

Problem 8.10   The MIP appearing in Example 8.1.

Problem 8.11   The MIP given in Problem 8.4.

Problem 8.12   The MIP given in Problem 8.5.

Problem 8.13   The MIP given in Problem 8.6.

Problem 8.14   The MIP given in Problem 8.7.

Problem 8.15   Suppose for a particular mixed integer program that the projection of S onto any (x_j, z) plane always produces a convex set whose upper boundary is horizontal. How does this affect the Land and Doig algorithm? Devise such an example.

Problem 8.16   Suppose that we have an optimal simplex tableau and that it is necessary to incorporate the additional constraint $x_j = t_j, x_j \geq t_j$, or $x_j \leq t_j$, where t_j is any number. Explain how this can be done (a) when x_j is a nonbasic variable, and (b) when x_j is a basic variable. In case (b) do *not* increase the row size of the tableau. Does the simplex method necessarily converge when using your procedure? Generalize the technique to allow for introducing several new constraints simultaneously. (*Hint:* For case (b) proceed as in Example 8.2, but do not keep linearly dependent rows. When primal feasibility is reached, check to see if the dependent rows are satisfied. If not, generate the violated ones and add them to the tableau, dropping their counterparts. Reoptimize to obtain a new primal feasible solution. Check for dependent rows, and so on.)

Problem 8.17 Suppose you are about to write a computer program for the Land and Doig algorithm.

a) Draw a flow chart of the technique.

b) Explain what parameters would be needed to keep track of the enumeration. In particular, include what would be the initial values and how the parameters would be updated.

c) Using Example 8.3, numerically detail the bookkeeping scheme you developed in (b).

Problem 8.18 Formally outline the Dakin variation (Section 8.5) of the Land and Doig algorithm. Prove that the algorithm converges. Where in the listing of the algorithm is Theorem 8.1 used? Suppose all optimal solutions are desired. How is the algorithm modified?

Problem 8.19 Suppose you are about to write a computer program for the Dakin algorithm outlined in Problem 8.18. Answer questions (a), (b), and (c) appearing in Problem 8.17. Will the Dakin variatior usually require less computer storage than the Land and Doig algorithm?

Problem 8.20 For each of the following algorithms, what is the maximum number of nodes that can appear in the tree? Assume each integer variable x_j $(j = 1, \ldots, n)$ is bounded above by the positive integer u_j.

a) The Land and Doig algorithm.

b) The Land and Doig algorithm for the zero-one program.

c) The Dakin algorithm.

d) The Dakin algorithm for the zero-one program.

Problem 8.21 Explain the similarities and differences between the linear programming post optimization procedures that occur in the Land and Doig algorithm and the one that is needed in the Dakin variation. (*Hint*: See Appendix A.)

Solve the following mixed integer programs by the Dakin algorithm. In each case relate the computations to the tree.

Problem 8.22 As in Example 8.1.

Problem 8.23 As in Problem 8.4.

Problem 8.24 As in Problem 8.5.

Problem 8.25 As in Problem 8.6.

Problem 8.26 As in Problem 8.7.

Problem 8.27 Explain why branch and bound is an appropriate designation for the Dakin algorithm.

Problem 8.28 Specialize the outline of the Land and Doig algorithm (Section 8.4) for the *zero-one* mixed integer or integer program. Is Theorem 8.1 necessary to prove convergence in the zero-one case?

Problem 8.29 Suppose you are about to write a computer program for the Land and Doig algorithm for the zero-one mixed integer program. Answer questions (a), (b), and (c) appearing in Problem 8.17. Also explain how the bookkeeping simplifies when compared with the scheme for the general mixed integer program.

Solve the following zero-one mixed integer programs by the Land and Doig algorithm. In each case relate the computations to the tree.

Problem 8.30

$$\text{Minimize} \quad 2x_1 \; + \; 4x_2 \; + \; y_1 \; + \; y_2 \; = \; z$$

$$\text{subject to } 2x_1 \quad - \quad 3x_2 \quad + \quad y_1 \quad - \quad y_2 \quad \leq \quad -5/2,$$
$$x_1 \quad + \quad 5x_2 \quad - \quad y_1 \quad + \quad 3y_2 \quad \leq \quad 7,$$
$$y_1, \qquad y_2 \quad \geq \quad 0,$$

and $\quad x_1,\ x_2 = 0 \ \text{ or } \ 1.$

Problem 8.31

$$\text{Minimize} \quad 10x_1 \quad - \quad x_2 \quad - \quad 3y_1 \quad - \quad 3y_2 \quad = \quad z$$
$$\text{subject to} \quad -10x_1 \quad + \quad 4x_2 \quad + \quad 10y_1 \quad - \quad 10y_2 \quad = \quad -7,$$
$$y_1, \qquad y_2 \quad \geq \quad 0,$$

and $\quad x_1,\ x_2 = 0 \ \text{ or } \ 1.$

Problem 8.32 Suppose we adopt the "most recently created dangling node" node selection rule as explained in Section 8.7.

a) Explain how the outline of the Land and Doig algorithm (Section 8.4) is modified under this rule.

b) How is the Dakin variation (Section 8.5) changed under this rule? Draw the associated tree.

c) In case (a) and then (b) discuss a bookkeeping scheme for computer implementation which will keep track of the enumeration. Would substantial computer storage usually be required?

d) Resolve Example 8.1, using the algorithm outlined in either (a) or (b).

Problem 8.33 Show that $P_j^G \geq$ minimum (P_j^D, P_j^U), where the penalties are defined in Section 8.7. How can the penalty P_j^G be used to devise branch selection rules in the Land and Doig algorithm? In the Dakin variation? Explain the motivation behind each rule you give.

Problem 8.34 Given a zero-one mixed integer program.

a) Find the down penalty P_j^D and the up penalty P_j^U as defined in Section 8.7.

b) Explain how these and condition (5) (Section 8.7) may be used

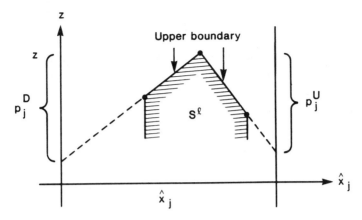

Figure 8.16: Penalty Calculation

in the Land and Doig algorithm to increase computational efficiency.

c) Show how the penalties P_j^D and P_j^U can be found via a geometric argument. (*Hint*: By Theorem 8.1 the maximum z is a piecewise linear function of the abscissa or x_j axis. Compute the z intercept for each of the two half lines originating at the maximal value of x_j and lying partially on the upper boundary. The intercept to the left with $x_j = 0$ gives P_j^D; the intercept to the right with $x_j = 1$ yields P_j^U. See Fig. 8.16 and perhaps Davis, Kendrick, and Weitzman [164].)

Problem 8.35 How can information be extracted from each nonbasic integer variable in an optimal linear programming solution so as to reduce computational effort in the Land and Doig algorithm? In the Dakin variation? (*Hint*: For each of the nonbasic integer variables derive the upper bound u_j as discussed in Section 8.7. Then incorporate this bound in the enumeration.) Illustrate your answer by resolving Example 8.3; draw the solution tree. Has the computational effort been reduced?

Problem 8.36 (Tomlin [566, 567]) Show that it is sometimes possible to replace the down and up penalties P_j^D and P_j^U as defined in Section 8.7 by the stronger forms

$$\hat{P}_j^D = \text{maximum } (a_{0q}, P_j^D) \text{ and}$$

$$\hat{P}_j^U = \text{maximum } (a_{0r}, P_j^U),$$

where q or r is the pivot column index in the dual simplex method (Fig. 8.14). (*Hint*: If $x_{J(q)}$ or $x_{J(r)}$ is a nonbasic integer variable which is not to remain at 0, it must be increased to at least one. After a pivot it has value $\hat{P}_j^D/a_{0q}$ or $\hat{P}_j^U/a_{0r}$.)

Problem 8.37 (Thompson [563]) Write the integer program as

maximize	z				
subject to	$\mathbf{cx}$	$-$	z	$=$ 0	(4)
	$\mathbf{Ax}$			$\leq$ **b**,	(5)
	$\mathbf{x}$			$\geq$ **0**,	(6)
and	$\mathbf{x}$	integer,			(7)

where **c** is integer and z is an unrestricted integer variable. Suppose in the Land and Doig algorithm that z is always selected to create the level 1 nodes. (When it is integer, only one node is created.) Explain how the basic approach (Section 8.4) is modified. If $\hat{z}$ is the maximal value of z without (7), and it is ascertained that there is no integer solution with $z = [\hat{z}]$ or $z = [\hat{z}] + 1$, show that it is always possible to determine a positive integer L, not necessarily 1, such that there is no integer solution z in the closed interval

$$[[\hat{z}], [\hat{z}] - (L+1)] \text{ or } [[\hat{z}] + 1, [\hat{z}] + L].$$

How can this be used to increase computational efficiency?

Problem 8.38 Solve the integer program

maximize	z						
subject to	$4x_1$	$+$	$5x_2$	$-$	z	$=$	0
	$3x_1$	$+$	$2x_2$			$\leq$	10,
	x_1	$+$	$4x_2$			$\leq$	11,
	$3x_1$	$+$	$3x_2$			$\leq$	13,
	$x_1,$		x_2			$\geq$	0,
and	$x_1,$		x_2		integer,		

by the algorithm outlined in Problem 8.37. Relate the computations to the tree.

Problem 8.39 In Chapter 1 we derived the integer program below which represents the traveling salesman problem.

$$\text{Minimize} \quad \sum_{\substack{i=0}}^{n} \sum_{\substack{j=1 \\ j \neq i}}^{n+1} d_{ij} x_{ij} \tag{8}$$

$$\text{subject to} \quad \sum_{i=0}^{n} x_{ij} = 1 \qquad (j = 1, \ldots, n+1), \tag{9}$$

$$\sum_{j=1}^{n+1} x_{ij} = 1 \qquad (i = 0, 1, \ldots, n), \tag{10}$$

$$\alpha_i - \alpha_j + (n+1)x_{ij} \leq n \quad \begin{array}{l} (i = 0, 1, \ldots, n \\ j = 1, \ldots, n+1, \ i \neq j), \end{array} \tag{11}$$

$$\text{and} \quad x_{ij} = 0 \text{ or } 1 \qquad \begin{array}{l} (i = 0, 1, \ldots, n \\ j = 1, \ldots, n+1, \ i \neq j), \end{array} \tag{12}$$

where $x_{0,n+1} = 0$, and $d_{ij} \geq 0$ for all i, j $(i \neq j)$.

a) Give what appears to be an efficient branch and bound algorithm for this problem which solves the integer program defined by the expressions (8), (9), (10), and (12) at each node in place of the linear programming subproblem. (*Hint*: If the optimal solution to the *integer* programming subproblem violates the inequalities (11), subtours exist and the node remains dangling. The branching variable may be selected so as to break a pair of subtours. Note that without the inequalities (11) the integer program is an assignment problem which can be solved by the Hungarian method or other linear programming techniques. When a solution to the assignment problem exists, the subtours can be joined together manually so that a traveling salesman solution appears, and thus an upper bound on the value of the objective function is available. When solved, the assignment problem produces a lower bound.)

b) Using the technique outlined in (a), solve the following $n = 5$ city problem.

i \ j	1	2	3	4	5	
1	–	3	4	2	1	
2	6	–	2	5	3	
3	4	1	–	4	2	$= (d_{ij})$
4	3	4	6	–	1	
5	2	3	3	1	–	

Appendix A
Computational Details of Examples 8.3 and 8.4

The problem is

$$
\begin{array}{rlll}
\text{maximize} & -4x_1 - 5x_2 &= z & \\
\text{subject to} & -x_1 - 4x_2 &\leq -5, & (x_3) \\
& -3x_1 - 2x_2 &\leq -7, & (x_4) \\
\text{and} & x_1, \quad x_2 &\geq 0 & \text{and integer.}
\end{array}
$$

Example 8.3: Land and Doig Algorithm (cf Section 8.4)

Step 1 The initial node is $\mathbf{x}^0 = (x_1, x_2)$. The optimal linear programming solution with nonnegative slack variables x_3 and x_4 is in Tableau #1 (see Chapter 4).

#1	1	$(-x_4)$	$(-x_3)$	
z	$-112/10$	$11/10$	$7/10$	$z = -112/10$
x_1	$18/10$	$-4/10$	$2/10$	$x_1 = 18/10, x_2 = 8/10$
x_2	$8/10$	$1/10$	$-3/10$	
x_3	0	0	-1	
x_4	0	-1	0	

Step 2 As x_1 must be integer, it is fixed at $[18/10]$ and $[18/10] + 1$ to define node 2 or $(1, x_2)$, and node 3 or $(2, x_2)$ (Figure 8.7, Section 8.4).

To solve the node 2 linear program we need to incorporate the constraint $x_1 = 1$, or equivalently $x_1 \leq 1$ and $x_1 \geq 1$, into Tableau #1. As

$$x_1 = 18/10 + (4/10)x_4 - (2/10)x_3,$$

we have

$$1 - x_1 = -8/10 - (4/10)x_4 + (2/10)x_3.$$

Thus, defining $\bar{x}_1^1 = 1 - x_1$, and requiring $\bar{x}_1^1 \geq 0$ yields $x_1 \leq 1$. Similarly,

$$\bar{\bar{x}}_1^1 = -\bar{x}_1^1 = -1 + x_1 = 8/10 + (4/10)x_4 - (2/10)x_3 \geq 0$$

enforces $x_1 \geq 1$. Since $x_1 \geq 1$ implies $x_1 \geq 0$, the x_1 row is omitted and Tableau #1 becomes

#2a	1	$(-x_4)$	$(-x_3)$
z	$-112/10$	$11/10$	$7/10$
$\bar{x}_1^1$	$-8/10$	$4/10$	$\boxed{-2/10}$
$\bar{\bar{x}}_1^1$	$8/10$	$-4/10$	$2/10$
x_2	$8/10$	$1/10$	$-3/10$
x_3	0	0	-1
x_4	0	-1	0

The tableau is dual feasible and primal infeasible. Hence, using the Dual Simplex Algorithm (cf. Chapter 2), we pivot on the circled element and the tableau subsequently regains optimality. Notice that it is never really necessary to explicitly carry both the $\bar{x}_1^1$ and $\bar{\bar{x}}_1^1$ row. This is true because if one row is present, the other one can be found whenever needed. As discussed in Problem 8.16, it is generally not necessary to explicitly carry linearly dependent rows. Also note that in any feasible solution to this problem $x_1 = 1$, meaning that at the optimal solution both variables $\bar{x}_1^1$ and $\bar{\bar{x}}_1^1$ must have value 0.

#2b	1	$(-x_4)$	$(-\bar{x}_1^1)$	
z	-14	$5/2$	$7/2$	$z(\ell = 0, k = 1, t = 1) = -14$
$\bar{x}_1^1$	0	0	-1	$\bar{x}_1 = 1 - x_1 = 0 \Rightarrow x_1 = 1$
$\bar{\bar{x}}_1^1$	0	0	1	$x_2 = 2$
x_2	2	$-1/2$	$-3/2$	
x_3	4	-2	$-10/2$	
x_4	0	-1	0	

Similarly, to solve the linear program at node 3, the constraint $x_1 = 2$, or $x_1 \leq 2$ and $x_1 \geq 2$, is added to Tableau #1. This is done by replacing the x_1 constraint with the two constraints

$$\bar{x}_1^2 = 2 - x_1 = 2/10 - (4/10)x_4 + (2/10)x_3$$

and

$$\bar{\bar{x}}_1^2 = -\bar{x}_1^2 = -2 + x_1 = -2/10 + (4/10)x_4 - (2/10)x_3.$$

Equivalently, we may add 1 to the $\bar{x}_1^1$ row and -1 to the $\bar{\bar{x}}_1^1$ in Tableau #2a. In any case, we have Tableau #3a. Note that $\bar{x}_1^2 \geq 0$ and $\bar{\bar{x}}_1^2 \geq 0$ guarantees that $x_1 = 2$. After one dual pivot step the tableau exhibits optimality.

#3a	1	$(-x_4)$	$(-x_3)$
z	$-112/10$	$11/10$	$7/10$
$\bar{x}_1^2$	$2/10$	$4/10$	$-2/10$
$\bar{\bar{x}}_1^2$	$-2/10$	$\left(-4/10\right)$	$2/10$
x_2	$8/10$	$1/10$	$-3/10$
x_3	0	0	-1
x_4	0	-1	0

#3b	1	$(-\bar{\bar{x}}_1^2)$	$(-x_3)$
z	$-47/4$	$11/4$	$5/4$
$\bar{x}_1^2$	0	1	0
$\bar{\bar{x}}_1^2$	0	-1	0
x_2	$3/4$	$1/4$	$-1/4$
x_3	0	0	-1
x_4	$2/4$	$-10/4$	$-2/4$

$z(0,1,2) = -47/4$

$x_1 = 2, x_2 = 3/4$

Node 2 corresponds to the mixed integer solution $z^* = -14, x_1 = 1$, and $x_2 = 2$. As $-47/4 > -14$, only node 3 is labeled dangling. (Remember that we could add the constraint $z = -4x_1 - 5x_2 > -14$ or $-4x_1 - 5x_2 \geq -13$ to tableau #1. However, to avoid further computations, this is not done.)

Step 3,4 Node 3 or $(2, x_2)$, is selected. $z^0 = z(0,1,2) = -47/4$. Node 4 or $(3, x_2)$ is created to the right of node 3 (Fig. 8.7). By introducing the constraint equations $\bar{x}_1^3 = 3 - x_1$ and $\bar{\bar{x}}_1^3 = -\bar{x}_1^3 = -3 + x_1$ in place of the x_1 row in Tableau #1, or equivalently by adding 1 to the $\bar{x}_1^2$ row and -1 to the $\bar{\bar{x}}_1^2$ row in Tableau #3a, we have Tableau #4a. A dual pivot step produces an optimal tableau.

#4a	1	$(-x_4)$	$(-x_3)$
z	$-112/10$	$11/10$	$7/10$
$\bar{x}_1^3$	$12/10$	$4/10$	$-2/10$
$\bar{\bar{x}}_1^3$	$-12/10$	$(-4/10)$	$2/10$
x_2	$8/10$	$1/10$	$-3/10$
x_3	0	0	-1
x_4	0	-1	0

#4b	1	$(-\bar{\bar{x}}_1^3)$	$(-x_3)$
z	$-58/4$	$11/4$	$5/4$
$\bar{x}_1^3$	0	1	0
$\bar{\bar{x}}_1^3$	0	-1	0
x_2	$2/4$	$1/4$	$-1/4$
x_3	0	0	-1
x_4	3	$-10/4$	$-2/4$

$z(1,1,3) = -58/4$

$x_1 = 3, x_2 = 2/4$

Step 2 The optimal linear programming solution at node 3 had $x_1 = 2$ and $x_2 = 3/4$ (Tableau #3b). Thus, by setting $x_2 = 0$ and $x_2 = 1$, node 5 or $(2,0)$, and node 6 or $(2,1)$, respectively, are created (Figure 8.7). As all the variables are fixed, the linear program at each of these nodes could be solved by direct substitution. However, to continue illustrating the post-optimization procedure, we introduce the inequality $\bar{x}_2^1 = -x_2 = -3/4 + (1/4)\bar{\bar{x}}_1^2 - (1/4)x_3 \geq 0$ to Tableau #3b, which corresponds to node $(2, x_2)$. In this case, two new constraints are not necessary since the inequality $x_2 \geq 0$ is already in the tableau. The result is Tableau #5a, and after a pivot, Tableau #5b is obtained.

#5a	1	$(-\bar{\bar{x}}_1^2)$	$(-x_3)$
z	$-47/4$	$11/4$	$5/4$
$\bar{x}_1^2$	0	1	0
$\bar{\bar{x}}_1^2$	0	-1	0
x_2	$3/4$	$1/4$	$-1/4$
$\bar{x}_2^1$	$-3/4$	$(-1/4)$	$1/4$
x_3	0	0	-1
x_4	$2/4$	$-10/4$	$-2/4$

#5b	1	$(-\bar{\bar{x}}_2^1)$	$(-x_3)$
z	-20	11	4
$\bar{x}_1^2$	-3	4	1
$\bar{\bar{x}}_1^2$	3	-4	-1
x_2	0	1	0
$\bar{x}_2^1$	0	-1	0
x_3	0	0	-1
x_4	8	-10	-3

$z(1, 2, 0) \leq -20$

In Tableau #5b, a further pivot must decrease the value of z, and since $z^* = -14$, this node will not be labeled dangling. So the simplex computations are terminated. To find the initial tableau at node 6 or $(2, 1)$, we could introduce inequalities $\bar{x}_2^2 = 1 - x_2 \geq 0$ and $\bar{\bar{x}}_2^2 = -1 + x_2 \geq 0$ in place of the x_2 inequality in Tableau #3b. Alternatively, in Tableau #5b we may simply add 1 times the $(-\bar{x}_2^1)$ column to the constant column and omit the $(-\bar{x}_2^1)$ column and x_2 and $\bar{x}_2^1$ rows. This is allowed because each row of this tableau is an equality of the form $a + b(-\bar{x}_2^1) + c(-x_3)$. Therefore, fixing x_2 at 1 is equivalent to setting $\bar{x}_2^1 = -1$. This amounts to adding b to a. After this addition, we have Tableau #6a. Two dual pivot steps yield primal feasibility and hence optimality in Tableau #6c.

#6a	1	$(-x_3)$
z	-9	4
$\bar{x}_1^2$	1	1
$\bar{\bar{x}}_1^2$	-1	-1
x_3	0	-1
x_4	-2	$\boxed{-3}$

#6b	1	$(-x_4)$
z	$-35/3$	$4/3$
$\bar{x}_1^2$	$1/3$	$1/3$
$\bar{\bar{x}}_1^2$	$-1/3$	$\boxed{-1/3}$
x_3	$2/3$	$-1/3$
x_4	0	-1

#6c	1	$(-\bar{\bar{x}}_1^2)$	
z	-13	4	$z(1,2,1) = -13$
$\bar{x}_1^2$	0	1	$x_1 = 2, x_2 = 1$
$\bar{\bar{x}}_1^2$	0	-1	
x_3	1	-1	
x_4	1	-3	

Node 6 gives the improved mixed integer solution $z^* = -13, x_1 = 2$ and $x_2 = 1$. As there are no other dangling nodes, the solution $z = -13, x_1 = 2$ and $x_2 = 1$ is optimal. Figure 8.7 relates the computations to the tree.

Example 8.4: The Dakin Variation (cf. Section 8.5)

At the first node (x_1, x_2) the optimal linear programming solution with nonnegative slacks x_3 and x_4 is given in Tableau #1.

NODE 1 (x_1, x_2)

#1	1	$(-x_4)$	$(-x_3)$
z	$-112/10$	$11/10$	$7/10$
x_1	$18/10$	$-4/10$	$2/10$
x_2	$8/10$	$1/10$	$-3/10$
x_3	0	0	-1
x_4	0	-1	0

Node 2 is created by introducing the inequality $x_1 \leq [18/10] = 1$ and node 3 by constraining $x_1 \geq [18/10] + 1 = 2$. As $x_1 \leq 1$ is equivalent to $\bar{x}_1 = 1 - x_1 \geq 0$ we commence the node 2 linear program with Tableau #2a.[14] Similarly, since $x_1 \geq 2$ is the same as $\bar{\bar{x}}_1 = -2 + x_1 \geq 0$, the node 3 linear program is solved staring with Tableau #3a. We then have the usual pivoting on the circled entries. In both cases, one dual simplex iteration yields the optimal solution in Tableau #2b and #3b, respectively.

NODE 2 $(x_1 \leq 1, x_2)$

#2a	1	$(-x_4)$	$(-x_3)$
z	$-112/10$	$11/10$	$7/10$
x_1	$18/10$	$-4/10$	$2/10$
x_2	$8/10$	$1/10$	$-3/10$
x_3	0	0	-1
x_4	0	-1	0
$1 - x_1 = \bar{x}_1$	$-8/10$	$4/10$	$-2/10$

#2b	1	$(-x_4)$	$(-\bar{x}_1)$
z	-14	$5/2$	$7/2$
x_1	1	0	1
x_2	2	$-1/2$	$-3/2$
x_3	4	-2	$-10/2$
x_4	0	-1	0
$\bar{x}_1$	0	0	-1

---

[14]For clarity, we are explicitly adding constraints to the tableau. As discussed in Problem 8.16, it is not necessary to increase the tableau row size. Also, for

NODE 3 $(x_1 \geq 2, x_2)$

#3a	1	$(-x_4)$	$(-x_3)$
z	$-112/10$	$11/10$	$7/10$
x_1	$18/10$	$-4/10$	$2/10$
x_2	$8/10$	$1/10$	$-3/10$
x_3	0	0	-1
x_4	0	-1	0
$-2 + x_1 = \bar{\bar{x}}_1$	$-2/10$	$\boxed{-4/10}$	$2/10$

#3b	1	$(-\bar{\bar{x}}_1)$	$(-x_3)$
z	$-47/4$	$11/4$	$5/4$
x_1	2	-1	0
x_2	$3/4$	$1/4$	$-1/4$
x_3	0	0	-1
x_4	$2/4$	$-10/4$	$-2/4$
$\bar{\bar{x}}_1$	0	-1	0

Node 2 produces the integer solution $z = -14$, with $x_1 = 1$ and $x_2 = 2$. Thus it is recorded and only node 3 is dangling. Node 4 is created by constraining $x_2 \leq [3/4] = 0$; node 5 is defined by introducing the inequality $x_2 \geq [3/4]+1 = 1$. Thus the node 4 (node 5) linear program is initiated from Tableau #3b with the additional inequality $\bar{x}_2 = -x_2 \geq 0$ ($\bar{\bar{x}}_2 = -1 + x_2 \geq 0$).

---

simplicity we are ignoring the superscript notation for the complementing variables $\bar{x}_i$ and $\bar{\bar{x}}_i$.

NODE 4 $(x_1 \geq 2, 0)$

#4a	1	$(-\bar{\bar{x}}_1)$	$(-x_3)$
z	$-47/4$	$11/4$	$5/4$
x_1	2	-1	0
x_2	$3/4$	$1/4$	$-1/4$
x_3	0	0	-1
x_4	$2/4$	$-10/4$	$-2/4$
$\bar{\bar{x}}_1$	0	-1	0
$-x_2 = \bar{x}_2$	$-3/4$	$\boxed{-1/4}$	$1/4$

#4b	1	$(-\bar{x}_2)$	$(-x_3)$
z	-20	11	4
x_1	5	-4	-1
x_2	0	1	0
x_3	0	0	-1
x_4	8	-10	-3
$\bar{\bar{x}}_1$	3	-4	-1
$\bar{x}_2$	0	-1	0

NODE 5 $(x_1 \geq 2, x_2 \geq 1)$

#5a	1	$(-\bar{\bar{x}}_1)$	$(-x_3)$
z	$-47/4$	$11/4$	$5/4$
x_1	2	-1	0
x_2	$3/4$	$1/4$	$-1/4$
x_3	0	0	-1
x_4	$2/4$	$-10/4$	$-2/4$
$\bar{\bar{x}}_1$	0	-1	0
$-1 + x_2 = \bar{\bar{x}}_2$	$-1/4$	$1/4$	$\boxed{-1/4}$

#5b	1	$(-\bar{\bar{x}}_1)$	$(-\bar{\bar{x}}_2)$
z	-13	4	5
x_1	2	-1	0
x_2	1	0	-1
x_3	1	-1	-4
x_4	1	-3	-2
$\bar{\bar{x}}_1$	0	-1	0
$\bar{\bar{x}}_2$	0	0	-1

Since $z = -20 < -14$, node 4 is not labeled dangling. Node 5 gives the improved mixed integer solution $z = -13$, with $x_1 = 2$ and $x_2 = 1$. As no nodes are dangling it is optimal. The tree representing the computations appears in Fig. 8.9. Comparing the tree in Fig. 8.7 with this one indicates that for this problem the Dakin variation is more efficient.

Chapter 9

Search Enumeration

9.1 Introduction

As in branch and bound, search algorithms either explicitly or implicitly enumerate all possible integer vectors. The literature devoted to search techniques is confined principally to the zero-one integer program and, in some instances, to the zero-one mixed integer program. Extensions for the general integer and mixed integer program are left to the problems. For ease of exposition, the integer program is considered first.

Background

An enumerative procedure unlike the branch and bound algorithm introduced by Land and Doig [383] was first proposed by Egon Balas [41] in 1963. The technique, referred to as the "additive algorithm," is described more fully by Balas in [19]. This work was elaborated on by Glover [239], and later by Lemke and Spielberg [391]. Linear programming, absent in earlier search algorithms, was computer implemented by Geoffrion [223], and by Salkin and Spielberg [519]. Papers which describe implicit enumeration tests include: the linear programming post optimization work of Driebeek [177], which was implemented by Shareshian and Spielberg [536] and extended by Shareshian [534]; the implementation and uses of "surrogate constraints" by Geoffrion [223] and Balas [42], which was introducd by Glover [239] and surveyed by him in the use of Lagrangian relaxation suggested by Geoffrion [223] and Fisher [199]; the generation and use of other inequalities implied by the constraint set by Bradley, Hammer, and Wolsey [100], Johnson and Spielberg [354], and by Spielberg [547], [548]; and a refinement of the original Balas tests by Glover and Zionts [270] and Lemke and

Spielberg [391]. Relaxing the originally rigid branching rules, suggested by Glover and Zionts [270], was formally proposed by Spielberg [543], and later by Guignard and Spielberg [297], [298], [299], and Tuan [576] (Sections 9.3, 9.4). Allowing the enumeration to start at a point different from zero appeared in Salkin [502], [503], Salkin and Spielberg [519], and Spielberg [544] (Section 9.5). A general discussion of the additive algorithm is in Geoffrion [222], and somewhat similar schemes are discussed in Lawler and Bell [388] and in Golomb and Baumert [274]. Computational results (Appendix B) for general zero-one problems appear in Fleischman [200], Freeman [207], Geoffrion [223], Jambekar and Steinberg [343], Lemke and Spielberg [391], Petersen [458], Salkin [519] and Johnson et al. [353]. Survey articles which discuss the additive algorithm and extensions include Balinski and Spielberg [51], Geoffrion and Marsten [229], and Salkin [507].

9.2   The Basic Approach, the Tree

Consider the zero-one integer program[1]

$$
\begin{array}{lll}
\text{minimize} & \mathbf{cx} \;=\; z \\
\text{subject to} & \mathbf{Ax} \;\le\; \mathbf{b}, \\
& 0 \le \mathbf{x} \le \mathbf{e}, \\
\text{and} & x_j \text{ integer} \quad (j = 1, \ldots, n);
\end{array}
$$

where $\mathbf{A}$ is an m by n matrix, $\mathbf{c}$ the cost row, $\mathbf{b}$ a vector of constants, and $\mathbf{e}$ a column of ones. To solve this problem, search algorithms enumerate either explicitly or implicitly all 2^n possible zero-one vectors $\mathbf{x}$. As in branch and bound (Chapter 8), the enumerative procedure may be related to a *tree* composed of *nodes* and *branches*. A node corresponds to a zero-one value of $\mathbf{x}$. A branch joins two nodes, reachable from one another in one step via the algorithm's branching rule. The two nodes differ in the *state* of one variable, where a variable can be in one of three states: fixed at 1, fixed at 0, or free (i.e., not fixed). A new node is defined by fixing a variable to 1 (a *forward step*), and a node is revisited after fixing a variable to 0 (a *backward step*). The branching

---

[1]For ease of exposition, in this chapter and the next, a minimization rather than a maximization problem is considered. This format agrees with virtually all the related literature in the respective fields.

rules are relatively rigid, since a single path in the tree is pursued until there are no free variables in its last node. Then one or more backward steps are taken until a first node is revisited so that forward steps from it will create new nodes. As we shall see, if we define the *level* of a node as the number of variables that are set to 1, the enumerative process can be implemented by keeping track of which variables are fixed at 0 and at what *level* the fixing occurred. We list the basic approach as suggested by Lemke and Spielberg [391], and borrow their explanation. The point x^ℓ is the node x with ℓ variables fixed at 1. The enumeration usually starts from the point x^0 with all variables free.

9.2.1 The Basic Approach

    i) Fix a free variables x_k from x^ℓ (initially $x^\ell = x^0$) at value 1.

    ii) Resolve the subproblem in the remaining free variables; then

    iii) fix x_k at value 0 (or *cancel* x_k at level ℓ); and

    iv) repeat this process for the problem with x_k fixed at 0.

The basic approach implies a general state of the search, in which some ℓ variables are currently fixed at 1, some additional C of the variables are canceled back to 0, and we are concerned with resolving the current integer programming *subproblem* in the remaining $f = n - (\ell + C)$ free variables. Resolving the current subproblem means enumerating, either explicitly or implicitly, each of the 2^f points x whose $(\ell + C)$ fixed or currently canceled variables have their values set at 1 or 0, respectively.

9.2.2 The Tree

In terms of a tree, Figure 9.1 depicts the enumerations from the current point x^ℓ with ℓ components fixed at 1. The diagram may be explained as follows. Point x^ℓ was reached for the first time on a forward step from $x^{\ell-1}$. Corresponding to the f currently free variables, there are currently f *permissible branches* from x^ℓ. On a *forward step* from x^ℓ, the algorithm selects some free variable x_k to be fixed at 1; this defines $x^{\ell+1}$. After the subproblem, in $f - 1$ free variables, associated with $x^{\ell+1}$ has been resolved, x_k is canceled at level ℓ to the value 0, and

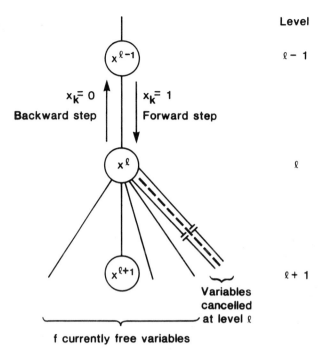

Figure 9.1: Search Tree

another permissible branch from the current point $\mathbf{x}^\ell$ is considered. The subproblem associated with $\mathbf{x}^\ell$ is resolved when all branches from $\mathbf{x}^\ell$ have been canceled, in which case (1) x_k is canceled at level $\ell - 1$; (2) all variables canceled at level ℓ revert to the status free; and (3) the process reverts to $\mathbf{x}^{\ell-1}$ at level $\ell - 1$. The search is complete when all variables are canceled at level 0. Reverting to $\mathbf{x}^{\ell-1}$ defines the *backward step* from $\mathbf{x}^\ell$.

To illustrate a complete enumeration which shows that the process can enumerate all 2^n possibilities we consider any three-variable integer program. The initial point is $\mathbf{x}^0 = (x_1, x_2, x_3)$ and we arbitrarily assume that the algorithm selects the free variable with the smallest index to set to one. The procedure is tablulated in Table 9.1, where the $2^3 = 8$ possible zero-one vectors are indicated by arrows in the rightmost column. The corresponding tree appears in Fig. 9.2. In that diagram the circled numbers denote the order of occurrences, and the integer solutions are numbered in the nodes as they appear.

Occurrence		Step	Level	State of the node[†]	
	–	–	0	(x_1, x_2, x_3)	
1	$x_1 = 1$	forward	1	$(1, x_2, x_3)$	
2	$x_2 = 1$	forward	2	$(1, 1, x_3)$	
3	$x_3 = 1$	forward	3	$(1, 1, 1)$	$\leftarrow 1$
4	$x_3 = 0$	backward	2	$(1, 1, 0_2)$	$\leftarrow 2$
5	$x_2 = 0, x_3$ free	backward	1	$(1, 0_1, x_3)$	
6	$x_3 = 1$	forward	2	$(1, 0_1, 1_1)$	$\leftarrow 3$
7	$x_3 = 0$	backward	1	$(1, 0_1, 0_1)$	$\leftarrow 4$
8	$x_1 = 0, x_2$ free, x_3 free	backward	0	$(0_0, x_2, x_3)$	
9	$x_2 = 1$	forward	1	$(0_0, 1, x_3)$	
10	$x_3 = 1$	forward	2	$(0_0, 1, 1)$	$\leftarrow 5$
11	$x_3 = 0$	backward	1	$(0_0, 1, 0_1)$	$\leftarrow 6$
12	$x_2 = 0, x_3$ free	backward	0	$(0_0, 0_0, x_3)$	
13	$x_3 = 1$	forward	1	$(0_0, 0_0, 1)$	$\leftarrow 7$
14	$x_3 = 0$	backward	0	$(0_0, 0_0, 0_0)$	$\leftarrow 8$
15	termination				

[†]0_ℓ means cancelled at level ℓ.

Table 9.1: Complete Search Enumeration

Up to this time we have explained how the search algorithm can explicitly enumerate all the 2^n possible zero-one vectors $\mathbf{x}$. But, as we have stressed, the efficiency of an enumerative technique relies crucially on its ability to enumerate large numbers of points or portions of the tree *implicitly*. Thus, a natural question is: where in a search algorithm, do we incorporate the criteria designed for implicit enumeration, and what are these tests? The answer is that the criteria are applied to the current subproblem (in itself a zero-one integer program); or, since the subproblem may be associated with a node, the tests are applied at the current node. The tests are inequalities or other criteria derived from the fact that we are seeking *improved zero-one solutions* from the current node. They may immediately resolve the entire subproblem (for example, by showing that the current constraints are inconsistent), or indicate that certain free variables must have value 0 or 1. These criteria, together with the algorithm's branching rules and the order in which they appear, will be called the *point or node algorithm*. Hence, as indicated in Figure 9.3, a search algorithm can be thought of as a

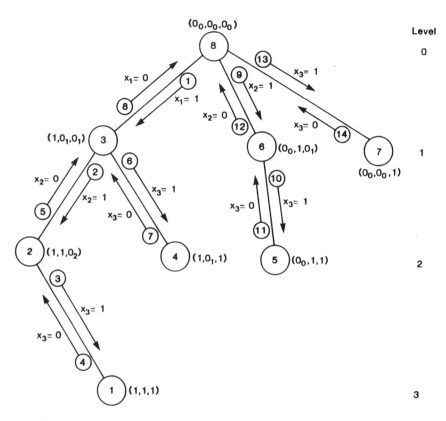

Figure 9.2: Complete Search Enumeration

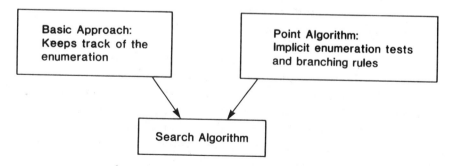

Figure 9.3: Composition of a Search Algorithm

composition of the basic approach with a point algorithm. Implicit enumeration criteria and branching rules which can be included in a point algorithm are discussed in the next two sections. Keep in mind that the efficiency of the search technique depends principally on their effectiveness. A sample enumeration algorithm and the details necessary for its computer implementation appears in Appendix A.

9.3   The Point Algorithm: Implicit Enumeration Criteria

Before discussing implicit enumeration criteria, some notation must be introduced. The subproblem at the point $\mathbf{x}^\ell$ is:

$$
\begin{aligned}
\text{minimize} \quad & \mathbf{c}\mathbf{x}^\ell \;=\; z \\
\text{subject to} \quad & \mathbf{A}\mathbf{x}^\ell \;\le\; \mathbf{b}, \\
& 0 \le \mathbf{x}^\ell \le \mathbf{e}, \\
\text{and} \quad & x_j = 0 \text{ or } 1 \ \ (\text{for } j = 1, \ldots, n).
\end{aligned}
$$

As before, $\mathbf{A}$ is an m by n matrix, $\mathbf{c}$ the cost row, $\mathbf{b}$ a vector of constants, and $\mathbf{e}$ a column of ones. The point $\mathbf{x}^\ell$ contains ℓ variables fixed at 1, C variables fixed at 0, and $f = n - (\ell + C)$ free variables. If the C variables at 0 are dropped from the subproblem, and the ℓ columns whose variables are fixed at 1 are subtracted from the right-hand side, we have the equivalent subproblem:

$$
\begin{aligned}
\text{minimize} \quad & \mathbf{c}^f \mathbf{x}^f \;=\; z - z^\ell \\
\text{subject to} \quad & \mathbf{A}^f \mathbf{x}^f \;\le\; \mathbf{b}^\ell, \\
& 0 \le \mathbf{x}^f \le \mathbf{e}, \\
\text{and} \quad & x_j = 0 \text{ or } 1 \ \ (\text{for } j \in F) \,;
\end{aligned}
$$

where $\mathbf{x}^f = (x_j)$ is the vector corresponding to the free variables, F is the set of indices of the free variables, $\mathbf{c}^f$ and $\mathbf{A}^f = (\mathbf{a_j})$ are the associated costs and columns of $\mathbf{A}$, a_{ij} is the ith component of the column $\mathbf{a_j}$, $\mathbf{b}^\ell = (b_i^\ell)$ is the updated $\mathbf{b}$ column, and z^ℓ is the sum of the costs of the ℓ variables fixed at 1. This subproblem is denoted by P^ℓ. The value of *any* zero-one solution to P^ℓ will be understood to include the cost z^ℓ of the ℓ fixed variables. The associated linear program, obtained by dropping the constraints $x_j = 0$ or 1 $(j \in F)$ is labeled LP^ℓ and its *optimal* solution also including the constant z^ℓ, is labeled by $\hat{z}$. The current minimal integer solution (which initially may be set to the sum of the positive costs) is denoted by z^*.

9.3.1 Ceiling Tests

The constant value in the objective function at node $\mathbf{x}^\ell$ is z^ℓ. This value can be decreased by, at most, the sum over the free variables of the negative costs. Thus it is possible to find an improved integer solution from the node $\mathbf{x}^\ell$ only when

$$z^\ell + \sum_{j \in \bar{F}} c_j \le z^*, \tag{1}$$

where $\bar{F} = \{j \in F \mid c_j < 0\}$. This means that a free variable having a positive cost should not be set to 1 if it increases z^ℓ so that (1) cannot be satisfied. In terms of the basic approach, cancel at level ℓ any $j \in F$ for which

$$c_j + z^\ell + \sum_{j \in \bar{F}} c_j \ge z^*. \tag{2}$$

Example 9.1   Suppose $z^* = 8, z^\ell = 10$, and the objective function corresponding to the free variables is $30x_3 + 2x_{17} - 5x_{20}$. Then $\bar{F} = \{20\}$ and it is possible to obtain an integer solution below 8 since $10 + (-5) <$ 8. However, this would be impossible if x_3 were set to 1 because $c_3 = 30$ and $30 + 10 + (-5) > 8$; so x_3 is canceled at level ℓ.

Nonnegative Costs

In Section 9.5 it is shown that a zero-one integer program can always be converted to one with nonnegative costs. In this case, the set $\bar{F}$ is always null and inequality (1) is

$$z^\ell < z^*. \tag{1$'$}$$

Now, suppose that after reaching node $\mathbf{x}^\ell$ on a forward step an improved integer solution is found. Then z^* is reduced to z^ℓ and further nodes defined from $\mathbf{x}^\ell$ can never produce an integer solution below this new value. Thus, when all the costs are nonnegative, feasibility at a node (i.e., $\mathbf{b}^\ell \geq \mathbf{0}$) always signals for an immediate backward step. This suggests that it is advantageous to have problems with nonnegative costs. However, as explained in Section 9.5, this is not necessarily so.

9.3.2 Infeasibility Test

The current subproblem's constraints, with the nonnegative slack variables $\mathbf{s}^\ell = (s_i)$, may be written as

$$\mathbf{s}^\ell = \mathbf{b}^\ell - \mathbf{A}^f \mathbf{x}^f \geq 0. \tag{3}$$

Since x_j must not exceed 1, the largest possible value for each slack variable s_i is

$$P_i = b_i^\ell - \sum_{j \in F^-} a_{ij}, \tag{4}$$

where $F^- = \{j \in F | a_{ij} < 0\}$. Thus, for a zero-one solution to be possible, we must have

$$P_i \geq 0 \quad \text{for} \quad i = 1, \dots, m. \tag{5}$$

Observe that $P_i \geq b_i^\ell$, and thus for $b_i^\ell \geq 0$, inequality (5) is automatically satisfied. Therefore, using (4) we may find the number P_i for each constraint which has $b_i^\ell < 0$. If one or more of these values is negative a backward step is justified.

9.3.3 Cancellation Zero Test

Suppose a free variable has a (large) positive coefficient in some row. Then setting this variable to 1 reduces b_i^ℓ and may result in a $P_i < 0$. Therefore, for any $j \in F$ with $a_{ij} > 0$, cancel x_j at level ℓ if

$$P_i - a_{ij} < 0 \quad \text{for some } i \ (1 \leq i \leq m). \tag{6}$$

If (6) indicates a cancellation, a column $\mathbf{a_j}$ is omitted from $\mathbf{A}^f$, and consequently the values of P_i (as defined by (4)) may change. So after each cancellation, P_i may be updated, and (5) and (6) retested. The following example illustrates the continual updating and interaction between the infeasibility and cancellation tests.

Example 9.2 Consider constraints i and $\bar{i}$ appearing in a subproblem.

$$
\begin{array}{llllllll}
i: & -4x_3 & + & 50x_6 & + & x_7 & \le & -3 \\
\bar{i}: & 20x_3 & - & 20x_6 & - & x_7 & \le & 5.
\end{array}
$$

In our notation, $F = \{3,6,7\}, P_i = -3 - (-4) = 1$, and $P_{\bar{i}} = 5 - (-20 - 1) = 26$; so it is possible (at first glance) to satisfy constraints i and $\bar{i}$. But $P_i = 1$, and

$$
1 - a_{i6} = 1 - 50 < 0;
$$

so cancel x_6 at level 1. (Clearly, setting x_6 to 1 makes it impossible to satisfy inequality i while having x_3 at most 1.) Now, $P_i = 1$, but $P_{\bar{i}}$ is reduced to $6 = 5 - (-1)$. Test (6) applied to the updated row $\bar{i} : 20x_3 - x_7 \le 5$ allows us to cancel x_3. Presently, $P_i = -3 < 0$ and a backward step is justified. After canceling x_3 and x_6, constraint $i : x_7 \le -3$ is obviously inconsistent with $x_7 = 0$ or 1.

9.3.4   Cancellation One Test

The number P_i is the sum of b_i^ℓ and minus the sum of the negative coefficients of the free variables in row i. Suppose one of these coefficients was omitted when computing P_i (i.e., a free variable was temporarily fixed at 0). If this resulted in $P_i < 0$, then for it to be possible to produce a zero-one solution the omitted variable must take the value 1.[2] Hence, for any $j \in F$ with $a_{ij} < 0$, if

$$
P_i + a_{ij} < 0 \quad \text{for some } i \ (1 \le i \le m), \tag{7}
$$

x_j must have value 1 in any zero-one solution produced from node $\mathbf{x}^\ell$. If $b_i^\ell \ge 0$, observe that $P_i + a_{ij} \ge b_i^\ell \ge 0$ for any $a_{ij} < 0$, or (7) would never indicate that a variable must have value 1.

---

[2] A different derivation of (7) is suggested in Problem 9.7. Other feasibility and cancellation tests are suggested by Hammer and Nguyen [309].

Example 9.3 Suppose row i is

$$i : 5x_2 - 2x_3 - 4x_4 - x_6 \le -5.$$

Then $P_i = -5 - (-2 - 4 - 1) = 2 > 0$ and $2 + (-4) < 0$; so x_4 must be 1 in any zero-one solution found from $\mathbf{x}^\ell$. If another test results in canceling x_4, a backward step is justified.

9.3.5 Linear Programming

The subproblem is denoted by P^ℓ and its associated linear program by LP^ℓ. Since the set of zero-one points satisfying the constraints of P^ℓ is contained in the feasible region defined by the constraints of LP^ℓ, the following observations are evident.

i) If the optimal solution to LP^ℓ is integer (in the free $\mathbf{x}$ variables), this solution is optimal for P^ℓ.

ii) If LP^ℓ has no feasible solution, then there is no zero-one solution to P^ℓ.

iii) The optimal value of the objective function $\hat{z}$ for the linear program is a lower bound on the value z for *any* zero-one solution to the subproblem; that is, $\hat{z} \le z$.

Suppose LP^ℓ is solved by the dual simplex algorithm.[3] Then the value of the objective function is nondecreasing, and we may use (iii) to derive a ceiling test which may signal for a backward step prior to terminating the simplex computations. That is:

iv) If at any dual simplex iteration the value of the linear program's objective function exceeds the current best integer solution z^*, a backward step is allowed.

---

[3]The dual linear program has the form: Maximize $\mathbf{w}^1 \mathbf{b}^\ell + \mathbf{w}^2 \mathbf{e}$, subject to $\mathbf{w}^1 \mathbf{A}^J + \mathbf{w}^2 \mathbf{I} \ge -\mathbf{c}$ and $\mathbf{w}^1, \mathbf{w}^2 \ge 0$. So when $\mathbf{c} \ge 0$, a basic feasible solution to the dual problem is initially $\mathbf{w}^2 = -\mathbf{c}, \mathbf{w}^1 = 0$. In Section 9.5 we show that it is relatively easy to control the sign of the costs and hence obtain this initial dual feasible solution.

Current value of variable x_j	Additional Constraint $x_j = 0$	$x_j = 1$	Action when $\hat{\hat{z}} \geq z*$
0	-	√	x_j is canceled
1	√	-	x_j must have value 1
fractional	-	√	x_j is canceled
fractional	√	-	x_j must have value 1
fractional	√	√	backward step

√ means constraint imposed.

Table 9.2: LP Post-Optimization Procedures

9.3.6 Post-Optimization, Penalties

After solving the linear program LP^ℓ, we may consider the effect on the minimal value $\hat{z}$ of the objective function, of forcing each integer variable to assume a value of 0 (or 1). Evidently such a requirement imposes an additional constraint on the problem, and, upon reoptimization, a new value for the objective function $\hat{\hat{z}} \geq \hat{z}$ is obtained. The difference $\hat{\hat{z}} - \hat{z}$ is the penalty for forcing the variable to have value 0 (or 1). If a variable is constrained to a value that it currently equals, $\hat{z}$ does not change and the penalty is zero. When the added requirement results in an infeasible linear program, the penalty can be thought of as infinite.

The new value $\hat{\hat{z}}$ may be compared with z^*. If $\hat{\hat{z}} \geq z^*$, the variable under consideration must either have value 1 or it is canceled, depending on whether it had been forced to be 0 or 1, respectively. Furthermore, if a variable must have value 1 and is also canceled, a backward step is justified. These comments are summarized in Table 9.2.

---

[4]For computer implementation, the optimal tableau must be stored so that the computations may be performed for different variables. Instead of completely reoptimizing, we can substantially reduce storage requirements by performing a *single* pivot step. Of course, the resulting penalty may be relatively small. A more complete discussion of this strategy appears in Salkin and Spielberg [519].

To actually compute the penalties we first consider (as in Chapter 8) whether x_j is a basic or nonbasic variable. If it is basic and fractional in the optimal linear programming tableau, an additional constraint is adjoined, and dual simplex iterations are necessary to regain optimality.[4] Now, suppose that $(a_{01}, \ldots, a_{0n})$ denotes the cost row in the first optimal linear programming tableau. Then, if x_j is a nonbasic variable, the penalty for forcing it to 1 is a_{0j}, and to 0 is zero. Thus, if $\bar{\mathbf{x}}^f$ is a particular zero-one vector value or "completion" for the free variables, a "total penalty" for this point is given by

$$p(\bar{\mathbf{x}}^f) = \text{maximum} \left\{ \begin{array}{c} \text{maximum } p_j, \sum_{j \in NB} a_{0j} \\ j \in B \end{array} \right\}, \qquad (8)$$

where $B = \{j \in F \mid x_j \text{ is a basic variable}\}$, $p_j = \hat{\hat{z}} - \hat{z}$ is the penalty for forcing a basic variable to take its value in $\bar{\mathbf{x}}^f$ and $NB = \{j \in F \mid x_j$ is a nonbasic variable which is at value 1 in $\bar{\mathbf{x}}^f\}$. Expression (8) may be used in a variety of ways. First, if $\hat{z} + p(\bar{\mathbf{x}}^f) \geq z^*$, the point reached by setting $\mathbf{x}^f$ to $\bar{\mathbf{x}}^f$ cannot produce an improved zero-one solution and should be ignored. More importantly, if relatively few completions of $\mathbf{x}^f$ are possible[5] we can, in a reasonable amount of time, compute the total penalty for each one. Then if a zero-one value for $\mathbf{x}^f$ yields a node which can be traced back to one which is implicitly enumerated, all other nodes produced by completions with the same or higher penalty may be implicitly enumerated. As the level of the enumeration increases, this results in cancellations, variables which must have value 1, and backward steps. To reach a good initial solution, this suggests branching toward nodes with lower penalties first. (Using penalties for branching purposes is discussed in the next section.) A more detailed discussion of penalties and their uses in search algorithms may be found in Driebeek [177], Shareshian and Spielberg [536], Shareshian [534], and in Problem 9.8.

9.3.7   Surrogate Constraints

As defined by Glover [246], a surrogate constraint is an inequality implied by the constraints of an integer program and designed to capture useful information that cannot be extracted from the parent constraints

---

[5] For example, when there are relatively few (free) integer variables.

individually, but which is nevertheless a consequence of their conjunction. The development and use of such constraints, first proposed by Glover in [239], has been extended by Balas [42], Geoffrion [223] and by Karwan and Rardin[359].

The integer program at a node $\mathbf{x}^\ell$ (ignoring superscripts) is P^ℓ :

$$
\begin{aligned}
&\text{minimize} && \mathbf{cx} \\
&\text{subject to} && \mathbf{Ax} \ \leq \ \mathbf{b}, \\
& && 0 \leq \mathbf{x} \leq \mathbf{e}, \\
&\text{and} && \mathbf{x} \ \text{integer.}
\end{aligned}
$$

Since $\mathbf{Ax} \leq \mathbf{b}$ implies $\mathbf{b} - \mathbf{Ax} \geq \mathbf{0}$, we have for a nonnegative weighting vector $\mathbf{u}$ and nonnegative number T that

$$
\mathbf{u(b - Ax)} + T \geq 0 \tag{9}
$$

is a surrogate constraint. A value of $\mathbf{u}$ is selected which satisfies a most useful or a "strongest" surrogate constraint definition as given in Glover [239], [246], Balas [42], or Geoffrion [223]. It has been shown by Glover [246] that $\mathbf{u}$ comprises the optimal values of the variables of the (sometimes slightly modified) dual linear program to LP^ℓ. Both Glover [270] and Balas [42] take $T = 0$; Geoffrion [223], however, defines it as $z^* - \mathbf{cx}$ (in which case a strict inequality in (9) is allowed).

Constraints of this type can be used in several ways. At each node in the enumeration a new surrogate constraint can be generated and explicitly attached to the current constraint set. (After backward steps, those not relevant to the current subproblem are dropped.) The contention is that the implicit enumeration tests applied to the surrogate constraints will be *most* effective. (The procedure has been implemented by Geoffrion [223] with apparent success; see Appendix B). Another way of using surrogate constraints follows from the fact that the *knapsack problem*

$$
\begin{aligned}
&\text{minimize} && \mathbf{cx} \\
&\text{subject to} && \mathbf{u(b - Ax)} \geq 0, \\
& && 0 \leq \mathbf{x} \leq \mathbf{e}, \\
&\text{and} && \mathbf{x} \ \text{integer,}
\end{aligned}
$$

produces a lower bound on any zero-one solution to the integer programming subproblem. The subproblem can thus be solved by finding the minimal solution to the knapsack problem which also satisfies $\mathbf{Ax} \leq \mathbf{b}$. (This is the basis of the Balas "Filter Algorithm" [42]; see Problem 9.15, and Djerdjour, Mathur, Salkin's [173] algorithm for Quadratic Multi-dimensional Knapsack Problem) Other uses of surrogate constraints are left to Problem 9.9. For completeness, Geoffrion's definition of a strongest surrogate constraint is now given and it is shown that the weighting vector in a strongest constraint consists of the optimal dual variables of the associated linear program. Several other definitions and derivations of strongest surrogate constraints appear in Glover [246]. The Balas [42] and Glover [239] definition may also be found in Problems 9.10 and 9.12, respectively.

Example 9.4 (Geoffrion [223]) Given the following definition of a strongest surrogate constraint, find the weighting vector $\mathbf{u}$ which yields a strongest surrogate constraint.

Definition 9.1 The surrogate constraint $\mathbf{u}^1(\mathbf{b} - \mathbf{Ax}) + (z^* - \mathbf{cx}) > 0$ is said to be stronger (at the current node) than the surrogate constraint $\mathbf{u}^2(\mathbf{b} - \mathbf{Ax}) + (z^* - \mathbf{cx}) > 0$ if the maximum of the left-hand side of the first constraint is less than the maximum of the left-hand side of the second constraint, the maximum being taken over all zero-one values of the free variables.[6]

Finding a strongest surrogate constraint is then the problem of minimizing, over all $\mathbf{u} \geq \mathbf{0}$, the expression

$$\left.\begin{array}{cl} \text{maximize} & \mathbf{u}(\mathbf{b} - \mathbf{Ax}) + (z^* - \mathbf{cx}) \\ \mathbf{x} & \\ \text{subject to} & x_j = 0 \text{ or } 1 \quad \text{for all (free) variables.} \end{array}\right\} \quad (10)$$

---

[6]The set of zero-one solutions $\mathbf{x}$ to $\mathbf{Ax} \leq \mathbf{b}$ or $\mathbf{b} - \mathbf{Ax} \geq \mathbf{0}$ with $\mathbf{cx} < z^*$ or $z^* - \mathbf{cx} > 0$, is contained in the set of zero-one solutions to the surrogate constraint $\mathbf{u}(\mathbf{b} - \mathbf{Ax}) + (z^* - \mathbf{cx}) > 0$, where $\mathbf{u}$ is a fixed nonnegative vector. As the maximum value (over binary vectors $\mathbf{x}$) of the left-hand side of the surrogate constraint decreases, it becomes less likely for a $0-1$ vector $\mathbf{x}$ to satisfy this inequality, and thus the size of the set of binary solutions to the surrogate constraint is usually reduced. In this case, it is more likely for zero-one solutions $\mathbf{x}$ to $\mathbf{u}^1(\mathbf{b} - \mathbf{Ax}) + (z^* - \mathbf{cx}) > 0$ to also satisfy $\mathbf{Ax} \leq \mathbf{b}$ and $\mathbf{cx} < z^*$, in which case $\mathbf{x}$ is an improved integer feasible solution.

Since, for any $\mathbf{u} \geq \mathbf{0}$,

$$\mathbf{u}(\mathbf{b} - \mathbf{A}\mathbf{x}) + (z^* - \mathbf{c}\mathbf{x}) = (\mathbf{u}\mathbf{b} + z^*) - (\mathbf{u}\mathbf{A} + \mathbf{c})\mathbf{x},$$

problem (10) is equivalent to

$$(\mathbf{u}\mathbf{b} + z^*) \quad - \quad \left. \begin{array}{l} \underset{\mathbf{x}}{\text{minimize}} \quad \bar{\mathbf{c}}\mathbf{x} \\[1em] \text{subject to} \quad x_j = 0 \text{ or } 1, \end{array} \right\} \tag{11}$$

where $\bar{\mathbf{c}} = (\bar{c}_j) = \mathbf{u}\mathbf{A} + \mathbf{c}$. As the minimal value for $\bar{\mathbf{c}}\mathbf{x}$ may be found by taking $x_j = 1$ if $\bar{c}_j < 0$ and $x_j = 0$ otherwise, and this result is not altered when $x_j = 0$ or 1 is replaced by $0 \leq x_j \leq 1$ (or $\mathbf{0} \leq \mathbf{x} \leq \mathbf{e}$), problem (11) is equivalent to

$$(\mathbf{u}\mathbf{b} + z^*) \quad - \quad \left. \begin{array}{lrcl} \text{minimize} & (\mathbf{u}\mathbf{A} + \mathbf{c})\mathbf{x} \\ \text{subject to} & \mathbf{I}\mathbf{x} & \leq & \mathbf{e}, \\ \text{and} & \mathbf{x} & \geq & \mathbf{0}, \end{array} \right\} \tag{12}$$

where $\mathbf{I}$ is an identity matrix and $\mathbf{e}$ is a vectors of 1's. Taking the dual of problem (12) gives

$$(\mathbf{u}\mathbf{b} + z^*) \quad + \quad \left. \begin{array}{lrcl} \text{minimize} & \mathbf{w}\mathbf{e} \\ \text{subject to} & \mathbf{w}\mathbf{I} & \geq & -(\mathbf{u}\mathbf{A} + \mathbf{c}), \\ \text{and} & \mathbf{w} & \geq & \mathbf{0}. \end{array} \right\} \tag{13}$$

Using (13) in place of (10), the problem of finding the strongest surrogate constraint is thus

$$z^* \quad + \quad \left. \begin{array}{lrcrcl} \text{minimize} & \mathbf{u}\mathbf{b} & + & \mathbf{w}\mathbf{e} \\ \text{subject to} & \mathbf{u}\mathbf{A} & + & \mathbf{w}\mathbf{I} & \geq & -\mathbf{c}, \\ \text{and} & \mathbf{u} \geq \mathbf{0}, & \mathbf{w} & \geq & \mathbf{0}. \end{array} \right\} \tag{14}$$

Remember that the associated linear program LP^ℓ has the form

$$\begin{array}{lrcl} \text{minimize} & \mathbf{c}\mathbf{x} \\ \text{subject to} & \mathbf{A}\mathbf{x} & \leq & \mathbf{b}, \\ & \mathbf{x} & \leq & \mathbf{e}, \\ \text{and} & \mathbf{x} & \geq & \mathbf{0}. \end{array}$$

Assigning dual variables $\mathbf{u}$ to the first set of constraints and $\mathbf{w}$ to the bounding inequalities, and passing to the dual, yields problem (14). Thus, the optimal dual variables $\mathbf{u}$ of the associated linear program LP^ℓ comprise the weighting vector which yields a strongest surrogate constraint under the given definition.

9.4   The Point Algorithm: Branching Strategies

The selection of the free variable to fix at 1 may greatly influence the algorithm's efficiency. Choosing one free variable may produce a direct path to an optimal or near optimal solution and a rapid return to the current node, while selecting another may lead the enumeration far from any optimal point. Making a poor branch choice, especially early in the search, may thus result in the fruitless enumeration of huge numbers of points and in the algorithm's failure to solve the problem in a reasonable length of time. Before describing actual branch selection rules we show that it is sometimes possible to eliminate some of the candidates.

When a forward step from the node $\mathbf{x}^\ell$ is to be made, one of the f free variables will be selected and set to 1. This suggests that each free variable must eventually be chosen, so that f nodes will be created from $\mathbf{x}^\ell$ (cf. Figure 9.1). However, we are often able to find a subset of the free variables in which at least one variable must have value 1 for an improved integer solution to be possible from $\mathbf{x}^\ell$. This means that only these variables need be considered as branch candidates, and once they all have been canceled a backward step to $\mathbf{x}^{\ell-1}$ is allowed. We now derive such subsets and make these ideas more precise.

9.4.1   Preferred Sets (Lemke and Spielberg [391])

Suppose a current constraint i, written as

$$\sum_{j \in F} a_{ij}x_j \le b_i^\ell, \tag{15}$$

has $b_i^\ell < 0$ and integral.[7] Inequality (15) may be written as

$$\sum_{j \in F^+} a_{ij}x_j + \sum_{j \in F^-} a_{ij}x_j \le b_i^\ell, \tag{16}$$

---

[7]Without loss of generality, we may assume that $\mathbf{c}$, $\mathbf{b}$, and $\mathbf{A}$ are initially all-integer.

where $F^+ = \{j \in F \mid a_{ij} > 0\}$ and $F^- = \{j \in F \mid a_{ij} < 0\}$. Multiplying this last expression by -1 and rearranging terms gives

$$\sum_{j \in F^-} (-a_{ij})x_j \geq -b_i^\ell + \sum_{j \in F^+} a_{ij}x_j \geq -b_i^\ell; \tag{17}$$

the second inequality sign follows from the requirement that each $x_j \geq 0$, hence $\sum_{j \in F^+} a_{ij}x_j \geq 0$. As $-b_i^\ell$ is positive and integral it must be at least 1. So, from (17), we have

$$\sum_{j \in F^-} (-a_{ij})x_j \geq 1. \tag{18}$$

For expression (18) to be true with each variable either zero or one, we must have at least one x_j equal to one. Thus, the constraint (15) implies the inequality

$$\sum_{j \in F^-} x_j \geq 1, \tag{19}$$

where $F^- \subset F$. As an example, suppose row i is

$$5x_2 - 2x_3 - 4x_4 - x_6 \leq -5. \tag{15$'$}$$

This implies

$$2x_3 + 4x_4 + x_6 \geq 5 + 5x_2 \geq 5. \tag{17$'$}$$

or

$$x_3 + x_4 + x_6 \geq 1. \tag{19$'$}$$

Notice that once inequality (17) has been found, it is sometimes possible to reduce the set F^- even further. In particular, in (17)$'$, since x_6 may not exceed 1, we have

$$2x_3 + 4x_4 \geq 5 - x_6 \geq 4;$$

also, $x_3 \leq 1$ yields

$$4x_4 \geq 4 - 2x_3 \geq 2,$$

which means

$$x_4 \geq 1/2,$$

and thus

$$x_4 = 1.$$

This process, termed *complete reduction* by Lemke and Spielberg [391], subtracts from (17) the terms $a_{ij}x_j$ ($a_{ij} < 0$) in order of decreasing a_{ij} until b_i^ℓ will become negative. The resulting inequality has the form

$$\sum_{j \in P} x_j \geq 1, \tag{20}$$

where $P \subset F^- \subset F$ is called the *preferred set* and an x_j with j in P, a *preferred variable*.

As already mentioned, only the preferred variables need be considered as branch candidates, and once they have all been canceled (at level ℓ) a backward step may be taken. Thus it seems advantageous to find the smallest possible preferred set. A strategy in this context is to perform complete reduction on each constraint i which has $b_i^\ell < 0$, and choose a preferred set with a minimal number of entries. If, as in the example, a preferred set has only one element, then only the corresponding single preferred variable is selected for branching. Once a backward step to $\mathbf{x}^\ell$ is taken, an immediate return to node $\mathbf{x}^{\ell-1}$ is justified. Observe that when all b_i^ℓ are nonnegative (i.e., an integer solution has been found), the complete reduction process breaks down and the set of preferred variables is the set of free variables.

We have shown previously that it is sometimes possible to establish that certain free variables must have value 1 for an improved integer feasible solution to be possible from the current node and in that case each preferred set has one element. These variables may serve as the next series of branches. If a (smallest) preferred set contains more than one element, a choice must be made of which variable to use for branching. A simple rule is to select the preferred variable with the least cost coefficient. This criterion, however, totally ignores the current constraints and may lead the enumeration away from an improved solution. A rule designed to direct the search toward, or not far away from a solution is given by Balas [19] and is discussed in the following section.

9.4.2   The Balas Test

If a free variable x_j is set to 1, the constant term b_i^ℓ in each constraint i becomes $b_i^\ell - a_{ij}$. When $b_i^\ell - a_{ij} \geq 0$, constraint i is satisfied, and thus a measure of the total "constraint infeasibility" (with $x_j = 1$) is given by

$$v_j = \sum_{i \in M_j} (b_i^\ell - a_{ij}), \tag{21}$$

where $M_j = \{i \mid b_i^\ell - a_{ij} < 0, \quad i = 1, \ldots, m\}$, and v_j is defined as 0 if M_j is null. Therefore, to rapidly reach or return to a zero-one solution it is reasonable to branch on the preferred variable which maximizes v_j.[8] Observe that $v_j \leq 0$, and when $v_j = 0$ a solution is found by fixing x_j at 1, and the other free variables at 0.

Example 9.5   Consider the current constraints

$$
\begin{array}{rcrcrcl}
-3x_{10} & - & x_{12} & - & 5x_{14} & \leq & -1, \\
8x_{10} & - & 3x_{12} & + & x_{14} & \leq & 5, \\
-3x_{10} & - & 4x_{12} & - & x_{14} & \leq & -4,
\end{array}
$$

where $F = \{10, 12, 14\}$. Complete reduction on row 1 produces the preferred set $\{10, 12, 14\}$, and on row 3 the set $\{10, 12\}$. Computing v_j for the smallest preferred set yields

$$
\begin{array}{rcl}
v_{10} & = & -3 - 1 = -4 \\
v_{12} & = & 0;
\end{array}
$$

or x_{12} is set to 1 and as a consequence the constraints become satisfied with $x_{10} = x_{14} = 0$.

---

[8]As some of the costs may be negative, it is possible to find a better binary solution by branching from a feasible node (i.e., $\mathbf{b}^\ell \geq \mathbf{0}$). In this instance, a natural objective is to introduce minimal infeasibility, and thus quickly return to another feasible node. Thus, the maximum v_j rule may still be used.

9.4.3   Other Branching Rules

When the linear program LP^ℓ is solved, penalties, as described in Section 9.3.6, may be computed. An obvious branching rule is to pick the preferred variable with the least penalty; this hopefully will lead the enumeration to a "good" or optimal solution. Another possibility is to select the preferred variable with the largest penalty. This criterion might result in a rapid cancellation of the selected variable (which may cause other cancellations) and a quick return to the previous node. As the smallest penalty rule seeks an optimal solution and the largest penalty rule strives towards cancellations and backward steps, a natural dynamic strategy is the following. Use the first rule until it is felt that a near optimal or optimal solution has been found and then switch to the second rule. In general, a dynamic branching strategy which initially uses a criterion designed to seek a optimal solution (such as the Balas test or least penalty) and later changes to one that yields cancellations and backward steps seems very attractive.[9] Other branching rules and their objectives are left to Problems 9.18 and 9.19.

9.5   The Generalized Origin, Restarts (Spielberg [544], Salkin and Spielberg [519])

The enumeration is initiated from the node $\mathbf{x}^0$ with all variables free. On a forward step, a zero-one variable is changed from the status free to fixed at 1, and the free variables have value 0 when the enumeration reveals a zero-one solution. In this context, the state "free" may be thought of as "free at value 0" and the origin node $\mathbf{x}^0$ as the zero vector. [10]

Often to gain nonnegativity in the cost row and increase the efficiency of the ceiling test (Section 9.3.1), the substitution $x_j = 1 - \bar{x}_j$, where $\bar{x}_j = 1 - x_j$ is the complement of x_j, is made for each variable x_j with $c_j < 0$. Doing this causes the term $c_j x_j$ to become $-c_j \bar{x}_j + c_j$ and $-c_j > 0$. The complemented variable, when free, has value 0. However, if $\bar{x}_j = 0$, we have $1 - x_j = 0$ or $x_j = 1$. Thus, in terms of the original variables, the state free for a complemented variable may be thought of

---

[9]Computational experience suggests that a near optimal or optimal solution is quite often found early in the enumeration; see [223], [391], and [502].

[10]The search algorithm of Balas [19] is introduced in a linear programming format and the free variables are treated as the nonbasic ones. This presentation perhaps more clearly illustrates the free-at-value-0 concept.

as free at value 1. If a complemented variable is selected for branching, it is set to 1 or the original variable is set to 0.

In general, the origin node $\mathbf{x}^0$ may contain complemented variables (e.g., so that the costs are nonnegative). In this case, the search starts from the point $\mathbf{x}^0 = (x_j^0)$ where $x_j^0 = 0$ if the variable x_j^0 is in its original form, and $x_j^0 = 1$ if x_j has been complemented. The zero origin or the one determined by the negative costs may initialize the enumeration far from the minimal point in terms of distance in the tree and the value of the objective function. Since the optimal linear programming solution is often close to an optimal integer point, it seems advantageous to complement those basic variables which are greater than some selected value a $(0 \le a < 1)$ in the minimal LP^0 solution and use the resulting point as the origin. Nevertheless, as this origin may also be far from an optimal integer node, it may be worthwhile to restart the enumeration when a point near the minimal one is thought to have been found. Nodes corresponding to improved zero-one solutions seem to be the natural candidates for new origins.

The potential efficacy of restarting the entire enumeration is based on the feeling that if we initiate a search using as origin the minimal $\mathbf{x}$, say $\mathbf{x}^*$, the tests for implicit enumeration would be most effective and the algorithm would terminate relatively quickly. Continuing this conjecture, it might further be supposed that commencing the enumeration at some (integer solution) $\mathbf{x}$ close to $\mathbf{x}^*$ would also result in an efficient algorithm. In this regard, it has been observed empirically by Lemke and Spielberg [391] that when the search is started at the point determined by the negative costs, and never restarted, the effort required to ascertain optimality is discouragingly substantial, whereas Salkin and Spielberg [519] furnish empirical evidence that an enumeration started at a point determined by the linear program and restarted at an improved zero-one solution is much more implicit.[11] Observe that on restarting, some search redundancy may be introduced (that is, the same point could be enumerated more than once). However, by limiting the number of restarts the process remains finite. (Salkin and Spielberg [519] report that this is of minor importance, and the overall work is substantially reduced when the linear programming origin and single restart strategy is adopted.) To make the discussion more concrete an example is presented.

---

[11]A description of their algorithm appears in Appendix A, and of their computational experience in Appendix B.

Example 9.6 (Salkin [502]) Suppose we wish to solve the following integer program by the basic search approach, with a node algorithm that includes the ceiling test and the branch selection rule which selects the free variable with the least cost coefficient:

$$\begin{aligned}
\text{minimize} \quad & -5x_1 \; - \; 4x_2 \; - \; 3x_3 \; = \quad z \\
\text{subject to} \quad & x_1 \; + \; 2x_2 \; - \; 3x_3 \; \leq \; -1 \\
& 2x_1 \; - \; 3x_2 \; + \; 5x_3 \; \leq \quad 3 \\
\text{and} \quad & x_j \; = \; 0 \text{ or } 1 \; (j = 1, 2, 3).
\end{aligned}$$

Figure 9.4 portrays the tree search when the origin is $(1, 1, 1)$, which enforces nonnegativity in the cost row. (As before, the encircled numbers denote the order of occurrences.) To incorporate this origin, and produce the equivalent problem actually solved, we substitute $1 - \bar{x}_1, 1 - \bar{x}_2$, and $1 - \bar{x}_3$, for x_1, x_2, and x_3, respectively. Then $\bar{\mathbf{x}} = (\bar{x}_1, \bar{x}_2, \bar{x}_3) =$

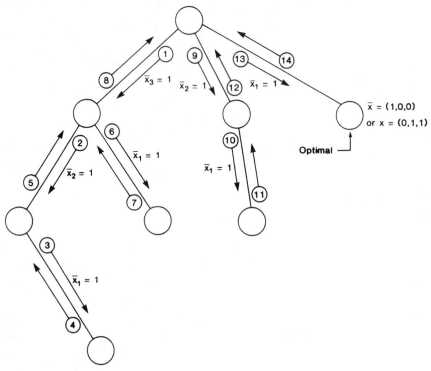

Figure 9.4: Tree Search with Origin $(1, 1, 1)$

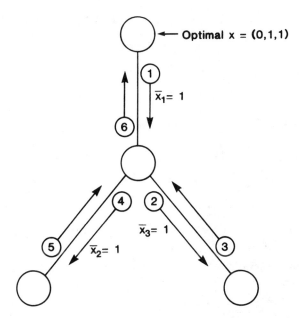

Figure 9.5: Tree Search with Origin $(0, 1, 1)$

0 means $\mathbf{x} = (1, 1, 1)$ and vice versa. The example with the new origin is

$$
\begin{array}{llrrrrrl}
\text{minimize} & 5\bar{x}_1 & + & 4\bar{x}_2 & + & 3\bar{x}_3 & - & 12 = z \\
\text{subject to} & -\bar{x}_1 & - & 2\bar{x}_2 & + & 3\bar{x}_3 & \leq & -1 \\
& -2\bar{x}_1 & + & 3\bar{x}_2 & - & 5\bar{x}_3 & \leq & -1 \\
\text{and} & \bar{x}_j & = & 0 \text{ or } 1 & (j = 1, 2, 3).
\end{array}
$$

Figure 9.5 depicts the tree search when the origin is $(0, 1, 1)$, or the optimal value for $\mathbf{x}$. Replacing x_2 and x_3 with $1 - \bar{x}_2$ and $1 - \bar{x}_3$, respectively, transforms the example to the equivalent one below. The right-hand side is, of course, nonnegative. (Why?)

$$
\begin{array}{llrrrrrl}
\text{Minimize} & -5x_1 & + & 4\bar{x}_2 & + & 3\bar{x}_3 & - & 7 = z \\
\text{subject to} & x_1 & - & 2\bar{x}_2 & + & 3\bar{x}_3 & \leq & 0 \\
& 2x_1 & + & 3\bar{x}_2 & - & 5\bar{x}_3 & \leq & 1 \\
\text{and} & x_1 & = & 0 \text{ or } 1, & \bar{x}_j = 0 \text{ or } 1 & (j = 2, 3).
\end{array}
$$

Comparing the two diagrams indicates that the search is indeed more implicit when started at the optimal integer solution.

9.6   Search Enumeration in $0-1$ Mixed Integer Programming (Lemke and Spielberg [391])

Relatively little work has been done on applying the search procedure to the zero-one mixed integer program. The basic approach, introduced by Lemke and Spielberg in [391], is very simple. A search enumeration over the integer variables is performed. Each time a node is explicitly examined, a linear program in *only the continuous variables* is solved; that is, prior to the simplex computations, the free variables are fixed at 0. Thus, a feasible linear program always produces a mixed integer feasible solution.

Unless each continuous variable is bounded above by a relatively small number, the natural extensions of the integer programming implicit enumeration criteria (Section 9.3) become quite inefficient and, in most cases, are probably not worth implementing.[12] This suggests that 2^n linear programs (each in only the continuous variables) must always be solved and that the algorithm could not be efficient except when applied to problems with relatively few integer variables. However, this is not necessarily so. It is possible, for example, to derive constraints valid at any node (e.g., from the linear program) in only the zero-one variables. Then at each node, an integer programming point algorithm could be applied to these constraints, which may result in many cancellations or free variables which must have value 1, and thus a largely implicit enumeration. The general procedure appears in Figure 9.6. We now derive constraints in only the integer variables.

Suppose we write the zero-one mixed integer program as

$$
\begin{aligned}
\text{minimize} \quad & \mathbf{cx} \; + \; \mathbf{dy} \; = \; z \\
\text{subject to} \quad & \mathbf{Ax} \; + \; \mathbf{Dy} \; \leq \; \mathbf{b}, \\
& \mathbf{0} \leq \mathbf{x} \leq \mathbf{e}, \; \mathbf{y} \geq \mathbf{0}, \\
\text{and} \quad & \mathbf{x} \text{ integer},
\end{aligned}
$$

where $\mathbf{c}, \mathbf{A}, \mathbf{b}$, and $\mathbf{e}$ have already been defined, $\mathbf{D} = (d_{ij})$ is an m by n' matrix, and $\mathbf{y} = (y_j)$ is the vector of continuous variables. Now, for

---

[12]For example, the number P_i defined by equation (4) and necessary to the infeasibility and cancellation tests is increased (and hence weakened) by the term $\sum_{j \in N_i} d_{ij} u_j$, where $N_i = \{ j \mid d_{ij} < 0, \; j = 1, \ldots, n' \}$, d_{ij} is the row i coefficient of the continuous variable y_j, u_j is its upper bound, and there are n' continuous variables.

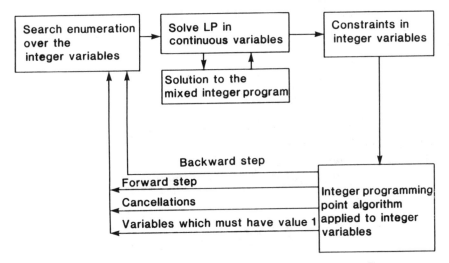

Figure 9.6: Search Algorithm for the 0-1 Mixed Integer Program

any fixed zero-one vector $\mathbf{x}$, the mixed integer program becomes the ordinary linear program

$$
\begin{array}{lrcl}
[P] & \text{minimize} & \mathbf{dy} = & z - \mathbf{cx} \\
& \text{subject to} & \mathbf{Dy} \leq & \mathbf{b} - \mathbf{Ax} \\
& \text{and} & \mathbf{y} \geq & \mathbf{0}.
\end{array}
$$

Its dual is

$$
\begin{array}{lr}
[D] & \text{minimize} \quad \mathbf{u}(\mathbf{b} - \mathbf{Ax}) \\
& \text{subject to} \quad \mathbf{u} \in U = \{\mathbf{u} \mid \mathbf{uD} + \mathbf{d} \geq \mathbf{0}, \mathbf{u} \geq \mathbf{0}\}.
\end{array}
$$

If $(\mathbf{x}, \mathbf{y})$ is to be an improved mixed integer solution, we must have

$$
\mathbf{cx} + \mathbf{dy} < z^* \quad \text{or} \quad z^* - \mathbf{cx} - \mathbf{dy} > 0
$$

and

$$
\mathbf{Ax} + \mathbf{Dy} \leq \mathbf{b} \quad \text{or} \quad \mathbf{b} - \mathbf{Ax} - \mathbf{Dy} \geq \mathbf{0},
$$

where z^* is the value of the current minimal solution. Multiplying the second equation by a nonnegative vector $\mathbf{u}$ and adding it to the first yields

$$z^* - \mathbf{cx} - \mathbf{dy} + \mathbf{ub} - \mathbf{uAx} - \mathbf{uDy} > 0$$

or

$$(z^* + \mathbf{ub}) - (\mathbf{uD} + \mathbf{d})\mathbf{y} - (\mathbf{uA} + \mathbf{c})\mathbf{x} > 0. \tag{22}$$

At each node which is explicitly enumerated, a linear program of the form [P] is solved. The simplex computations generate extreme points of the polyhedron U. (If it is null, the mixed integer program has no solution.) As this means $\mathbf{uD} + \mathbf{d} \geq \mathbf{0}$, for nonnegative $\mathbf{y}$ we have

$$(z^* + \mathbf{ub}) - (\mathbf{uA} + \mathbf{c})\mathbf{x} \;\geq\; (z^* + \mathbf{ub}) - (\mathbf{uA} + \mathbf{c})\mathbf{x}$$
$$-(\mathbf{uD} + \mathbf{d})\mathbf{y} > 0 \tag{23}$$

or

$$(\mathbf{c} + \mathbf{uA})\mathbf{x} < z^* + \mathbf{ub}. \tag{24}$$

Therefore, whenever a linear program [P] is solved it yields dual extreme points $\mathbf{u}$ and constraints of type (24) in only the $\mathbf{x}$ variables. As the set U is independent of $\mathbf{x}$, these inequalities are valid at any node.[13]

It is conceivable that a huge number of dual extreme points, and thus constraints of type (24), can be generated at each node. Although storing some constraints may prove useful later in the search, it is computationally impossible to maintain and test more than a relatively small number of these inequalities. Nevertheless, it seems natural to generate at least one constraint, using the *optimal* dual extreme point, for each linear program that is solved. In this case, inequality (24), which may be written as $(z^* - \mathbf{cx}) + \mathbf{u}(\mathbf{b} - \mathbf{Ax}) > 0$, is a surrogate constraint (Section 9.3.7).

Finally, it is possible for the set U to be unbounded. If it is, the simplex computations may terminate with a dual extreme point $\mathbf{u}^0$ and direction $\mathbf{v} \geq \mathbf{0}$ such that the points $\mathbf{u} = \mathbf{u}^0 + \theta\mathbf{v}$, for $\theta \geq 0$, constitute a ray or half-line in U with $\mathbf{u}(\mathbf{b} - \mathbf{Ax}) \to -\infty$ as $\theta \to +\infty$. As

---

[13]A more direct derivation of (24) utilizes linear programming duality theory. In particular, from the dual linear programs [P] and [D] we have $-\mathbf{dy} \leq \mathbf{u}(\mathbf{b} - \mathbf{Ax})$; but $-\mathbf{dy} = \mathbf{cx} - z$, so $(\mathbf{c} + \mathbf{uA})\mathbf{x} \leq z + \mathbf{ub}$. Using $z = \mathbf{cx} + \mathbf{dy} < z^*$ in the last inequality gives (24).

$$u(b - Ax) = (u^0 + \theta v)(b - Ax) = (u^0 + \theta v)b - (u^0 + \theta v)Ax,$$

it follows that

$$u(b - Ax) = u^0(b - Ax) + \theta v(b - Ax). \tag{25}$$

Now, $u^0(b - Ax)$ is a fixed vector and thus, for $\theta \geq 0$,

$$v(b - Ax) < 0. \tag{26}$$

Therefore, if the minimum of $u(b - Ax)$ for $u \in U$ is unbounded along an extreme ray of U with direction v, then $v(b - Ax) < 0$. Conversely, if there exists an extreme ray in U with direction v such that $v(b - Ax) < 0$, then the minimum of $u(b - Ax)$ over $u \in U$ is unbounded along this ray. (This follows because for some extreme point u in $U, u = u^0 + \theta v \in U$ for any $\theta \geq 0$, and (25) is true.)

Now, if problem [D] is unbounded, then by the dual theorem of linear programming problem [P] has no feasible solution and thus the mixed integer program has no feasible solution. Thus, a necessary and sufficient condition on x to admit feasible solutions y to the mixed integer program is that

$$v(b - Ax) \geq 0. \tag{27}$$

A constraint of type (27) can be formed for each ray direction v. These inequalities are valid at every node and can be collected and tested like the previous ones.

Problems

Unless otherwise specified, the notation used throughout the chapter is continued.

Problem 9.1   Using the notation adopted in Section 9.2, perform a complete enumeration (as in Table 9.1) for a four variable zero-one integer program. List the sixteen possible zero-one vectors and relate the computations to a tree. Do the same for a four integer variable zero-one mixed integer program.

Problem 9.2   Explain how the basic search approach (Section 9.2) would be extended to solve general, rather than zero-one, integer programs. Describe the corresponding tree. What are some efficient implicit enumeration criteria and branch selection rules? Is it possible to use Theorem 8.1 (Chapter 8) to derive these criteria?

Problem 9.3   Answer the same series of questions as in Problem 9.2 when the basic approach for the zero-one mixed integer program (Section 9.6) is extended to solve the general mixed integer program.

Problem 9.4

a. List and explain the differences and similarities between branch and bound enumeration and search enumeration for the following problems:

1) the zero-one integer program,

2) the zero-one mixed integer program.

b. Under what circumstances are the two approaches almost the same (except for terminology) and perhaps computationally equivalent? (*Hint:*   Devise a point algorithm for the search enumeration which includes the criteria which are explicitly part of the branch and bound algorithm.)

c. For programs (1) and (2) in (a), which enumerative approach seems more efficient: branch and bound or search? Justify your answer intuitively.

Problem 9.5   In general, how can a point algorithm be modified so that all alternate optima may be found? Illustrate by modifying at least two specific implicit enumeration tests. Why is it often useful to find alternate optimal solutions? What price is paid for adopting this strategy?

Problem 9.6 (Double Ceiling Test)   Suppose, at a node x^ℓ, c^1 and c^2 are the first and second smallest costs corresponding to the free variables. Under what circumstances is it valid to say that to produce an improved solution (from $\mathbf{x}^\ell$) we must have

$$z^\ell + c^1 + c^2 + \sum_{j \in \bar{F}} c_j < z^*, \tag{1}$$

where $\bar{F} = \{j \in F | c_j < 0\}$. What implicit enumerations tests can be deduced from inequality (1)? What calculations are necessary for their implementation? Do you think it would be worthwhile to incorporate these criteria in a point algorithm?

Problem 9.7 Show that when the cancellation one test (Section 9.3.4) indicates that a variable must have value 1, it is equivalent to performing complete reduction (on the constraint under consideration) and obtaining a preferred set (Section 9.5.1) with one entry. Prove that the converse is also true. Does this preclude the use of both the cancellation one test and complete reduction in the same point algorithm?

Problem 9.8 (Multiple Penalties, Shareshian [534]) Suppose the penalty (as defined in Section 9.3.6) for forcing x_j to have value K is denoted by $p_j(K)$. Now, define $p_{ij}(K, L)$ to be the penalty for *simultaneously* forcing x_i and x_j to take on the integral values K and L, respectively.

a) Show that $p_{ij}(K, L) \geq \text{maximum } (p_i(K), p_j(L))$.

b) Explain how $p_{ij}(K, L)$ is computed.

c) How can these multiple penalties be used for implicit enumeration?

d) What is an upper bound on the number of penalties $p_{ij}(K, L)$? Does this suggest that these penalties are not worth computing?

Problem 9.9 List three uses of a surrogate constraint (Section 9.3.7) not discussed in the chapter. In each case, explain how the surrogate constraint can curtail the length of the search. When compared with the computations necessary for their implementation, would your criteria be worthwhile?

Problem 9.10 (Strongest Surrogate Constraint, Balas [42]) Given the following definition for relative strength of a surrogate constraint, find the weighting vector **u** which yields a strongest surrogate constraint.

Definition: The surrogate constraint $\mathbf{u}^1(\mathbf{b} - \mathbf{Ax}) \geq 0$ is stronger than the surrogate constraint $\mathbf{u}^2(\mathbf{b} - \mathbf{Ax}) \geq 0$ if the minimum **cx**, subject to $\mathbf{u}^1(\mathbf{b} - \mathbf{Ax}) \geq 0$, and $\mathbf{0} \leq \mathbf{x} \leq \mathbf{e}$, is greater than the minimum **cx**, subject to $\mathbf{u}^2(\mathbf{b} - \mathbf{Ax}) \geq 0$, and $\mathbf{0} \leq \mathbf{x} \leq \mathbf{e}$.

Problem 9.11 Show that if $\mathbf{u}(\mathbf{b} - \mathbf{Ax}) \geq 0$ is a strongest surrogate constraint under the definition given in Problem 9.10, then the surrogate constraint $\mathbf{u}(\mathbf{b} - \mathbf{Ax}) + (z^* - \mathbf{cx}) \geq 0$ is a strongest surrogate constraint under the definition given in Section 9.3.7, and conversely.

Problem 9.12 (Strongest Surrogate Constraint, Glover [239]) Suppose the strongest surrogate constraint definition appearing in Problem 9.10 is changed so that the "...minimum **cx**, subject ..." parts include the "**x** integer constraints". In context of using a surrogate constraint to capture useful information about the parent constraints, why is this definition better? Explain why it is difficult to determine a strongest surrogate constraint according to this definition. Does this suggest disregarding it and using a definition that is easier to implement?

Problem 9.13 (Composite Infeasibility-Ceiling Test, Glover and Zionts [270]) An integer solution may be found from the current node $\mathbf{x}^\ell$ only if $P_i = b_i^\ell - \sum_{j \in F^-} a_{ij} \geq 0$ for each constraint i, where $F^- = \{j \in F | a_{ij} < 0\}$ (see Section 9.3.2). But it is possible for P_i to be nonnegative only at the expense of forcing the objective function too high. Prove that the following condition, which detects whether such a situation exists, is true. If $c_j \geq 0$ $(j = 1, \ldots, n)$ and there exists an i such that $b_i^\ell < 0$ and $b_i^\ell(c_j/a_{ij}) \geq z^* - z^\ell$ for all $j \in F$, then there cannot exist an improved integer solution found from $\mathbf{x}^\ell$. If the infeasibility test (Section 9.3.2) is used in a point algorithm, does it necessarily follow that it is advantageous to incorporate this criterion also?

Problem 9.14 (Feasibility Test, Balas [19]) Suppose at a node $\mathbf{x}^\ell$ there exist free variables x_j which have $a_{ij} \geq 0$ whenever $b_i^\ell < 0$. Under what circumstances can these variables be canceled? Does this suggest meaningful implicit enumeration tests?

Problem 9.15 (Filter Algorithm, Balas [42]) Suppose the integer program is

$$P \quad \text{minimize} \quad \mathbf{cx}$$
$$\text{subject to} \quad \mathbf{Ax} \leq \mathbf{b},$$
$$\mathbf{0} \leq \mathbf{x} \leq \mathbf{e},$$
$$\text{and} \quad \mathbf{x} \ \text{integer}.$$

Consider the one constraint (knapsack) problem

$$F \quad \text{minimize} \quad \mathbf{cx}$$
$$\text{subject to} \quad \mathbf{uAx} \leq \mathbf{ub},$$
$$\mathbf{0} \leq \mathbf{x} \leq \mathbf{e},$$
$$\text{and} \quad \mathbf{x} \ \text{integer},$$

where $\mathbf{u}$ is a nonnegative weighting vector.

a) Prove that one way to solve problem P is to find the minimal integer solution to the knapsack problem F which also satisfies $\mathbf{Ax} \leq \mathbf{b}$.

b) If the solution procedure suggested in (a) is adopted, why is it natural to choose $\mathbf{u}$ as an optimal solution of the dual linear program associated with problem P?

c) The scheme suggested in (a) can be implemented by performing a search enumeration on problem F and checking to see whether the integer solutions found satisfy $\mathbf{Ax} \leq \mathbf{b}$. Explain what branching strategies can be adopted so that the *first* integer solution to problem F that satisfies $\mathbf{Ax} \leq \mathbf{b}$ is optimal.

Problem 9.16   At any node $\mathbf{x}^\ell$ which has $b_i^\ell < 0$ for at least one constraint i, we may derive an inequality of the form

$$\sum_{j \in P_i} x_j \geq 1, \tag{2}$$

where $P_i \subset F$ is the preferred set for row i (see Section 9.4.1).

a) Suppose at each node an inequality (2) is generated for each constraint i which has $b_i^\ell < 0$. Explain how these inequalities can be used for implicit enumeration.

b) Suppose that at a node $\mathbf{x}^\ell$ which has at least one $b_i^\ell < 0$ the integer program (or set covering problem)

$$[S^\ell] \quad \text{minimize} \quad \mathbf{cx}^\ell$$

$$\text{subject to} \quad \sum_{j \in P_i} x_j \geq 1 \quad \text{for each } i \text{ with } b_i^\ell < 0,$$

$$0 \leq \mathbf{x}^\ell \leq \mathbf{e},$$

$$\text{and} \quad \mathbf{x}^\ell \text{ integer},$$

is defined. Show that one way of resolving the current subproblem P^ℓ is to find a minimal solution to S^ℓ which satisfies $A\mathbf{x}^\ell \geq \mathbf{b}$. (The original problem P appears in Problem 9.15.) How can this procedure be implemented and does it seem efficient? In particular, what if $\ell = 0$ and all variables are free?

Problem 9.17 (A Reduction Process, Guignard and Spielberg [298], Spielberg [547]) Consider a current subproblem constraint $\sum_{j \in F} a_{ij} x_j \geq b_i^\ell$, which has $P_i = b_i^\ell - \sum_{j \in F^-} a_{ij} x_j \geq 0$ and $F^- = \{j \in F | a_{ij} < 0\}$. Then, by substituting $1 - \bar{x}_j$ for x_j, introduce the complementing variable $\bar{x}_j = 1 - x_j$ for each x_j with $j \in F$. The resulting inequality is

$$\sum_{j \in F} a_{ij}^+ x_j^+ \leq P_i, \qquad (3)$$

where $a_{ij}^+ = -a_{ij}$ and $x_j^+ = \bar{x}_j$ if $j \in F^-$, and $a_{ij}^+ = a_{ij}$ and $x_j^+ = x_j$ if $j \in F - F^-$.

a) How can inequality (3) be used to cancel x_j^+ variables? What does this mean in terms of the x_j variables?

b) Does utilizing constraints of type (3) preclude the use of the complete reduction process (Section 9.4.1)? Illustrate by first performing complete reduction on the constraint $-5x_1 + 3x_2 - 4x_3 + 2x_4 \leq -7$ and then by generating an inequality of type (3).

Problem 9.18 (State Enumeration, Spielberg [543], [548], Guignard and Spielberg [298]) At a node $\mathbf{x}^\ell$ the index set F corresponding to the free variables may be divided into three mutually exclusive subsets $F0, F1$, and FF. On forward steps the variables in $F0$ are fixed at zero and those in $F1$ at one. The variables in FF are chosen as branch variables only if no other choice is possible.

a) How can the basic search approach be simply modified so that a forward step can be equivalent to fixing a free variable at value 0. (*Hint:* Use complementing variables.)

b) What is the motivation for adopting the above branching procedure? In particular, is there a tendency toward greater implicit enumeration and thus a more efficient search algorithm?

c) Is the generalized origin concept (Section 9.5) related to this splitting of F strategy?

Problem 9.19 List three branching rules or strategies not discussed in the chapter. What is the motivation for using each rule? Can they be implemented simply?

Problem 9.20 *A priori* information about which variables "most probably" are active (or at 1) in a particular integer programming model is often available in a real world environment. Explain how such information can be used to reduce the length of the enumeration.

Problem 9.21 It is possible to derive many implicit enumeration tests and branching rules. Does a point algorithm which includes numerous tests and branching strategies necessarily work better than one with fewer criteria?

The integer programs in Problems 9.22 through 9.26 have the form

$$
\begin{array}{lll}
\text{minimize} & \mathbf{cx} & \\
\text{subject to} & \mathbf{Ax} \leq \mathbf{b}, & \\
\text{and} & x_j = 0 \text{ or } 1 & (j = 1, \dots, n).
\end{array}
$$

Solve each problem by the basic search approach and any point algorithm. Describe the composition of the point algorithm and draw the corresponding tree.

Problem 9.22 (Balas [19]) $n = 5$

$$c = \begin{pmatrix} -5 & 7 & 10 & -3 & 5 \end{pmatrix}$$

$$A = \begin{pmatrix} 1 & 3 & -5 & 1 & 4 \\ -2 & -6 & 3 & -2 & -2 \\ 0 & 1 & -2 & -1 & 1 \end{pmatrix}$$

$$b = \begin{pmatrix} 0 \\ -4 \\ -2 \end{pmatrix}$$

Problem 9.23 (Balas [19]) $n = 10$

$$c = \begin{pmatrix} -10 & 7 & -1 & 12 & -2 & -8 & 3 & 1 & -5 & -3 \end{pmatrix}$$

$$A = \begin{pmatrix} -3 & -12 & 8 & 1 & 0 & 0 & 0 & 0 & 7 & -2 \\ 0 & 1 & 10 & 0 & 5 & -1 & 7 & 1 & 0 & 0 \\ -5 & -3 & 1 & 0 & 0 & 0 & 0 & -2 & 0 & -1 \\ 5 & 3 & -1 & 0 & 0 & 0 & 0 & 2 & 0 & 1 \\ 0 & 0 & 4 & -2 & 0 & 5 & 1 & -9 & 2 & 0 \\ 0 & 9 & 0 & -12 & 7 & -6 & 0 & 2 & 15 & 3 \\ 8 & 5 & -2 & -7 & 1 & 0 & -5 & 0 & 10 & 0 \end{pmatrix}$$

$$b = \begin{pmatrix} 8 \\ 13 \\ -6 \\ 6 \\ 8 \\ 12 \\ 16 \end{pmatrix}$$

(*Hint:* The minimal value of the objective function is 23 and x_2, x_3, and x_4 must be 0.)

Problem 9.24 (Balas [19]) $n = 9$

$$c = \begin{pmatrix} 4 & 2 & 1 & 5 & 3 & 6 & 1 & 2 & 3 \end{pmatrix}$$

$$A = \begin{pmatrix} 3 & 5 & -2 & -1 & 0 & -1 & 0 & -4 & 2 \\ 6 & -2 & 0 & -2 & 2 & -4 & 3 & 0 & 0 \\ 0 & 5 & 0 & 3 & -3 & 6 & -1 & 0 & -2 \\ -5 & 0 & 1 & 0 & 5 & 0 & -2 & 1 & -1 \end{pmatrix}$$

$$b = \begin{pmatrix} -1 \\ -3 \\ 2 \\ -8 \end{pmatrix}$$

(*Hint:* This problem has no feasible integer solution.)

Problem 9.25 (Balas [19]) $n = 12$

$$c = \begin{pmatrix} 5 & 1 & 3 & 2 & 6 & 4 & 7 & 2 & 4 & 1 & 1 & 5 \end{pmatrix}$$

$$A = \begin{pmatrix}
-1 & 3 & -12 & 0 & -1 & 7 & -1 & 0 & 0 & 3 & -5 & -1 \\
3 & -7 & 0 & 1 & 6 & 0 & 0 & 0 & 0 & 0 & 0 & 0 \\
11 & 0 & 1 & -7 & 0 & -1 & 2 & 1 & -5 & 0 & 9 & 0 \\
0 & 5 & 6 & 0 & -12 & 7 & 0 & 3 & 1 & -8 & 0 & 5 \\
-7 & -1 & -5 & 3 & 1 & -8 & 0 & -2 & 7 & 1 & 0 & -7 \\
-2 & 0 & 0 & -4 & 0 & 0 & -3 & -5 & -1 & 0 & 1 & 1
\end{pmatrix}$$

$$b = \begin{pmatrix} -6 \\ 1 \\ -4 \\ 8 \\ -7 \\ -9 \end{pmatrix}$$

(*Hint:* An integer feasible solution is $x_j = 1$ for $j = 3, 4, 8, 10, 12$ and $x_j = 0$ otherwise.)

Problem 9.26 (Balas [42]) $n = 20$

$$c^t = \begin{pmatrix} 3 \\ 2 \\ 5 \\ 8 \\ 6 \\ 9 \\ 11 \\ 4 \\ 5 \\ 6 \\ 11 \\ 2 \\ 8 \\ 5 \\ 8 \\ 7 \\ 3 \\ 9 \\ 2 \\ 4 \end{pmatrix} \quad A^t = \begin{pmatrix}
-6 & -1 & 3 \\
5 & 3 & 6 \\
-8 & 3 & 1 \\
-3 & 4 & -3 \\
0 & 1 & -5 \\
-1 & 0 & 6 \\
-3 & 4 & -9 \\
-8 & -1 & 6 \\
-9 & -6 & 3 \\
3 & 0 & -9 \\
-8 & 8 & -6 \\
6 & 0 & -3 \\
-3 & 1 & -6 \\
-8 & -5 & -6 \\
-6 & -4 & 6 \\
7 & -1 & 2 \\
6 & -9 & -7 \\
-2 & -7 & -6 \\
3 & 2 & 0 \\
7 & 2 & 7
\end{pmatrix} \quad b = \begin{pmatrix} -21 \\ -10 \\ -14 \end{pmatrix}$$

(*Hint:* The minimal solution must have $x_1 = x_2 = x_3 = x_4 = 1$ and the value of the objective function does not exceed 23.)

Problem 9.27 (Driebeek[177]) Explain the concept and uses of total penalties (Section 9.3.6) in context of a search enumeration applied to a zero-one mixed integer program. Illustrate by solving

$$
\begin{array}{rrrrrrr}
\text{minimize} & -10x_1 & + & x_2 & + & 3y_1 & + & 3y_2 \\
\text{subject to} & 10x_1 & - & 4x_2 & - & 10y_1 & + & 10y_2 & = & 7, \\
& & & & & y_1, & & y_2 & \geq & 0, \\
\text{and} & x_1, & & x_2 & = & \text{0 or 1.}
\end{array}
$$

Problem 9.28 Consider the zero-one mixed integer program.

$$
\begin{array}{ll}
\text{minimize} & \mathbf{cx} + \mathbf{dy} \\
\text{subject to} & \mathbf{Ax} + \mathbf{Dy} \leq \mathbf{b}, \\
& 0 \leq \mathbf{x} \leq \mathbf{e}, \ \mathbf{y} \geq \mathbf{0}, \\
\text{and} & \mathbf{x} \ \text{integer.}
\end{array}
$$

a) Given the following definition of a strongest surrogate constraint, find the weighting vector $\mathbf{u}$ which yields a strongest surrogate constraint.

Definition (Geoffrion [223]) The surrogate constraint

$$
\mathbf{u}^1(\mathbf{b} - \mathbf{Ax} - \mathbf{Dy}) + (z^* - \mathbf{cx} - \mathbf{dy}) > 0
$$

is said to be stronger (at the current node) than the surrogate constraint

$$
\mathbf{u}^2(\mathbf{b} - \mathbf{Ax} - \mathbf{Dy}) + (z^* - \mathbf{cx} - \mathbf{dy}) > 0
$$

if the maximum of the left-hand side of the first constraint is less than the maximum of the left-hand side of the second constraint; the maximum being taken over the continuous variables and the zero-one values of the free variables. (To avoid additional notation, the superscript l has been dropped.)

b) In context of a search enumeration (Section 9.6), how can surrogate constraints be used?

Problem 9.29 Consider the composite branch and bound/search algorithm concept. That is, the enumeration is allowed at any time to switch from the branch and bound type to the search form, or vice versa.

a) What is the purpose of using a composite algorithm?

b) How is such an algorithm implemented? Would it be difficult to keep track of the enumeration?

c) Answer (a) and (b) for a composite (branch and bound and/or search) enumeration-cutting plane algorithm. That is, any sub-problem that appears during the enumeration may be solved by a cutting plane algorithm.

Problem 9.30 (This problem appears in Golomb and Baumert [274], and is also discussed in Netto [445].) By complete enumeration find the number of ways of placing four chess queens on a 4×4 chessboard so that no two attack each other; that is, so that no two of them are on a common row, column, or diagonal of the board. Do the same for eight queens on an 8×8 chessboard. (*Hint:* Formulate constraints in zero-one variables.)

Problem 9.31   (Lexicographic Ordering, Lawler and Bell [388]) For each of the 2^n possible zero-one vectors $\mathbf{x} = (x_1, \ldots, x_n)$, let

$$n(\mathbf{x}) = x_1 2^{n-1} + x_2 2^{n-2} + \ldots + x_n 2^0.$$

Also, define $\mathbf{x}^1 \leq \mathbf{x}^2$ if and only if $x_j^1 \leq x_j^2$ for $j = 1, \ldots, n$.

a) Suppose all zero-one n vectors are listed in numerical order; that is,

$$
\begin{array}{ll}
(0, \ldots, 0, 0, 0) & n(\mathbf{x}) = 0 \\
(0, \ldots, 0, 0, 1) & n(\mathbf{x}) = 1 \\
(0, \ldots, 0, 1, 0) & n(\mathbf{x}) = 2 \\
(0, \ldots, 0, 1, 1) & n(\mathbf{x}) = 3 \\
(0, \ldots, 1, 0, 0) & n(\mathbf{x}) = 4, etc.
\end{array}
$$

Immediately following an arbitrary vector $\mathbf{x}$ there may or may not exist a number of vectors $\mathbf{x}'$ with the property that $\mathbf{x} \leq$

$\mathbf{x}'$. (For example, immediately following $\mathbf{x} = (0,1,0,0)$ are $(0,1,0,1)$, $(0,1,1,0)$, and $(0,1,1,1)$, each of which is greater than $\mathbf{x}$). Let x^* denote the first vector following $\mathbf{x}$ in the numerical ordering that has the property $\mathbf{x} \not\leq \mathbf{x}^*$. Show that, for any given nonzero $\mathbf{x}$, $\mathbf{x}^*$ is calculated as follows:

Treat $\mathbf{x}$ as a binary number

1) subtract 1 from $\mathbf{x}$

2) using Boolean arithmetic (in particular, $1 + 1 = 1$), add the components of $\mathbf{x}$ and $\mathbf{x} - 1$ to obtain $\mathbf{x}^* - 1$

3) add 1 to obtain $\mathbf{x}^*$.

For example, let $\mathbf{x} = 0101100$, then $n(\mathbf{x}) = 44$ and $n(\mathbf{x} - 1) = 43$; so $\mathbf{x} - 1 = 0101011$ and $\mathbf{x}^* - 1 = 0101111$. As $n(\mathbf{x}^* - 1) = 47$, $n(\mathbf{x}^*) = 48$ or $\mathbf{x}^* = 0110000$. (*Hint:* $\mathbf{x}^* - 1$ is greater than each of $\mathbf{x}, \mathbf{x} + 1, \ldots, \mathbf{x}^* - 2$.)

b) Consider the *not* necessarily linear integer program in two constraints

$$\begin{array}{ll}
\text{minimize} & g_0(\mathbf{x}) \\
\text{subject to} & -g_1(\mathbf{x}) \geq 0, \\
& g_2(\mathbf{x}) \geq 0, \\
\text{and} & x_j = 0 \ \text{ or } \ 1 \quad \text{for all } j,
\end{array}$$

where each of the functions g_0, g_1, and g_2 is assumed to be monotone nondecreasing in each of the variables $x_1, x_2, \ldots, x_n$. Suppose the enumerations procedure of examining the 2^n possible solution vectors is performed in numerical order, beginning with $\mathbf{x} = (0, 0, \ldots, 0)$ and ending with $\mathbf{x} = (1, 1, \ldots, 1)$. If $\hat{\mathbf{x}}$ is the current best integer solution, $g_0(\hat{\mathbf{x}})$ its value, and $\mathbf{x}$ is the current vector being examined, justify the following implicit enumeration tests

1) If $g_0(\mathbf{x}) \geq g_0(\hat{\mathbf{x}})$, skip to $\mathbf{x}^*$.

2) If $\mathbf{x}$ is a feasible solution, i.e., $-g_1(\mathbf{x}) \geq 0$ and $g_2(\mathbf{x}) \geq 0$, skip to $\mathbf{x}^*$.

3) If either $-g_1(\mathbf{x}) \not\geq 0$ or $g_2(\mathbf{x}^* - 1) \not\geq 0$, skip to $\mathbf{x}^*$.

c) Using the procedure suggested in (b), first perform a complete enumeration of all (16) possible vectors and thus solve

$$\begin{array}{ll}
\text{minimize} & g_0(\mathbf{x}) = 3x_1 + 2x_2 + x_3 + x_4 \\
\text{subject to } -g_1(\mathbf{x}) = & -x_2x_3 - x_4 + 1 \geq 0, \\
& g_2(\mathbf{x}) = 2x_1 + x_2x_3 + x_4 - 3 \geq 0, \\
\text{and} & x_1, \ldots, x_4 = 0 \text{ or } 1.
\end{array}$$

Then resolve the problem using the three tests appearing in (b).

d) Can a point algorithm of a search procedure be devised so that the branching rule will enumerate as discussed in (b)? If so, can the tests described there be incorporated in the point algorithm? Is it possible to devise other implicit enumeration tests and thus can the resulting search algorithm be made more efficient? Is it applicable to any zero-one linear integer program?

(A more complete discussion, which includes the generalization of the procedure outlined in (b) to the general integer program with many constraints, appears in Lawler and Bell [388]. A refinement of their algorithm may be found in Dragan [174].)

Problem 9.32

a) Explain why the basic search approach can be used to solve the most general zero-one program

$$\begin{array}{ll}
\text{minimize} & f_0(\mathbf{x}, \mathbf{y}) \\
\text{subject to} & f_i(\mathbf{x}, \mathbf{y}) \geq 0 \qquad (i = 1, \ldots, m), \\
& \mathbf{y} \geq \mathbf{0}, \\
\text{and} & \mathbf{x} = (x_1, \ldots, x_n) \quad \text{a zero-one vector,}
\end{array}$$

where the f_i $(i = 0, \ldots, m)$ are any real valued functions.

b) Given each of the following particular functions, is it, in general, possible to formulate an efficient search algorithm for the resulting program?

i) f_0 and f_i $(i = 1, \ldots, m)$ are "ill-behaved" functions.

ii) f_0 is concave and the f_i are convex.

iii) f_0 is concave and the f_i are linear.

c) Repeat (a) and (b) with x_j $(j = 1, \ldots, n)$ allowed to be any nonnegative integer.

Using the basic search approach and any point algorithm you wish, solve the nonlinear integer programs in Problems 9.33 and 9.34. (These appear in Lawler and Bell [388].)

Problem 9.33

$$
\begin{array}{llllllll}
\text{Minimize} & x_1^2 & + & x_2^2 & + & x_3^2 & + & x_4^2 & + & x_5^2 \\
\text{subject to} & x_1 & + & x_2 & + & & & x_4 & & & \geq 4, \\
& & & x_2 & + & 2x_3 & & & & & \geq 3, \\
& x_1 & & & & & & & + & 2x_5 & \geq 5, \\
& x_1 & + & x_2 & + & 2x_3 & & & & & \leq 6, \\
& 2x_1 & & & + & x_3 & & & & & \leq 4, \\
& x_1 & & & & & & & + & 4x_5 & \leq 13,
\end{array}
$$

$$0 \leq x_j \leq 3 \quad (j = 1, \ldots, 5),$$

and $\qquad x_j$ integer $(j = 1, \ldots, 5)$.

(*Hint:* The minimal value of the objective function is atleast 8.)

Problem 9.34

$$
\begin{array}{lllll}
\text{Minimize} & x_1x_7 & + 3x_2x_6 & + x_3x_5 & + 7x_4 \\
\text{subject to} & x_1 & + x_2 & + x_3 & \geq 6, \\
& x_4 & + x_5 & + 6x_6 & \geq 8, \\
& x_1x_6 & + x_2 & + 3x_5 & \geq 7, \\
& 4x_2x_7 & + 3x_4x_5 & & \geq 25, \\
& 3x_1 & + 2x_3 & + x_5 & \geq 7, \\
& 3x_1x_3 & + 6x_4 & + 4x_5 & \leq 20, \\
& 4x_1 & + 2x_3 & + x_6x_7 & \leq 15, \\
& & & x_4 & \leq 15, \\
& & & x_5 & \leq 15, \\
& & & x_j & \leq 7 \quad (j = 1, 2, 3, 6),
\end{array}
$$

and $\qquad x_j \geq 0$ and integer $(j = 1, \ldots, 7)$.

(*Hint:* A minimal solution has $x_2 = 4$, $x_4 = 0$, and the other x_j's equal to 1 or 2.)

For each of the integer programs in Problems 9.35 through 9.37, follow the directions preceding Problem 9.22. Also relate the computation to the bookkeeping vectors σ and η, and the slack vector s, described in Appendix A.

Problem 9.35   As in Problem 9.22.

Problem 9.36   As in Problem 9.24.

Problem 9.37   As in Problem 9.25.

Problem 9.38   Discuss what is involved in computer programming: a search algorithm? A branch and bound algorithm? Is it simpler to computer implement a branch and bound procedure or a search algorithm? In terms of storage requirements and updating, which type of enumeration is more efficient?

Problem 9.39   Solve the integer program appearing in Problem 9.27 by the basic search approach and the point algorithm described in Appendix A and Fig. 9.6.

Problem 9.40   Given the computational experience outlined in Appendix B, can conclusions be drawn as to which algorithm is best (second best, etc.)? What implicit enumeration tests in each algorithm seem most effective.

Appendix A
A Sample Search Algorithm and Its Implementation

(Salkin and Spielberg [519])

In this section, the search algorithm formulated and computer programmed by Salkin and Spielberg [519] is described. The procedure is essentially an extension of the one developed earlier by Lemke and Spielberg [391], which in turn is an elaboration of the original Balas additive algorithm [19]. Computational experience with the Salkin and

Spielberg code, compared with other published results, is presented in the next appendix.

A.1 The Basic Approach; the Point Algorithm

The basic approach is a search enumeration with a generalized origin and restart capability. If a point thought to be near or at the minimal one is known *a priori*, it serves as the initial origin. (For example, we could use a solution obtained from a previous computer run or published material.) Otherwise, the integer program is solved as a linear one, and values greater than the input parameter a ($0 \leq a < 1$) in its solution are set to 1 in the origin and the others are at 0. The enumeration is then restarted at the first integer solution guaranteed to be near the optimal one by a "low" *a priori* upper bound on any integer solution which is also an input parameter.[14]

On reaching a node for the first time (i.e., on a forward step) the point algorithm applies the cancellation zero test and the infeasibility criterion (Sections 9.3.2 and 9.3.3) to the free variables and the current constraints, respectively. Then, assuming these two tests do not result in a backward step, the associated linear program is solved.[15] If it is infeasible or exceeds the current ceiling, the search reverts to the previous node. If it produces a zero-one solution, it is recorded, z^* is updated, and a backward step is taken. Otherwise, the value of the minimal linear programming solution is used as a lower bound on the best integer solution possible from the current node; the constraints $x_j = 1$ and $x_j = 0$ are then successively imposed on the free variable not at 1 or 0 in the optimal simplex tableau. (In actual computation, to reduce storage requirements and execution time, only one pivot step is performed. The authors, however, do not discount the possible advantage in pivoting to regain primal feasibility.) Then, if the tableau exhibits infeasibility or the value of the objective function exceeds the ceiling z^*, the variable under consideration is canceled or designated as one that must have value 1, depending on whether the additional constraint is $x_j = 1$, or $x_j = 0$, respectively. When a variable is canceled and must have value 1, a backward step is taken.

---

[14]The algorithm has the capability of restarting several times. However, because of the considerable amount of time involved in restarting, the authors felt that one restart, near the minimal solution, is best.

[15]The algorithm can, if desired, omit the linear program at nodes corresponding to certain levels (controlled by an input vector). It also allows for linear programming before, rather than after, the cancellation and infeasibility criteria.

Using the ceiling test (Section 9.3.1) the point algorithm then ascertains that z can be made less than z^*. Assured of this, it proceeds in selecting the next branch.

If the linear program reveals variables which must have value 1, these variables determine the next series of branches. (Observe that if one of these is canceled, a return to the node prior to the one in which they are produced is justified.) Otherwise, at "infeasible nodes" (i.e., a $b_i^{\ell} < 0$), the algorithm generates via complete reduction (Section 9.4.1) the preferred set for each infeasible row, and applies the Balas test (Section 9.4.2) to the smallest one. (In ties for v_j, the smallest index is selected.) If all b_i^{ℓ} are nonnegative, the Balas test is applied to all the free variables.

On a backward step (i.e., a return to a previously visited node) the linear program is not solved again, and the preferred set is already known.

The general structure of the algorithm appears in Fig. 9.7.

A.2 The Bookkeeping Scheme [16]

A sequence vector $\boldsymbol{\sigma}$ records which steps have been taken so far; σ_1 is taken as 0. For example, $\boldsymbol{\sigma} = (0, 3, 1)$ means that $x_3 = 1, x_1 = 1$, and all other x_j are zero (either canceled or free).

A state vector $\boldsymbol{\eta}$ of n components (initially all ones) records which branches are permitted (in which case $\eta_j = 1$) or not (then $\eta_j \leq 0$). Whenever it is ascertained that a branch is not permissible, the corresponding η_j element is set to zero. On forward steps each nonpositive component of $\boldsymbol{\eta}$ is reduced by one; on backward steps these components are increased by one. This procedure ensures that the state vector is always up to date. The $\boldsymbol{\eta}$ and $\boldsymbol{\sigma}$ vectors keep track of the basic search approach in the computer program.

During the algorithm, the slack vector $\mathbf{s}$ (initially set to $\mathbf{b}$) is updated: e.g., $x_j \rightarrow 1$ leads to $s_i \rightarrow s_i - a_{ij}$ for all i. Feasibility is attained whenever $s_i \geq 0$ for all i.

To simplify test procedures, the variables are always ordered so that $c_{j+1} \geq c_j$ for all j. A detailed outline of the algorithm and a numerical example appear in Salkin and Spielberg [519] and in Salkin [502].

---

[16]This procedure first appeared in Lemke and Spielberg [391].

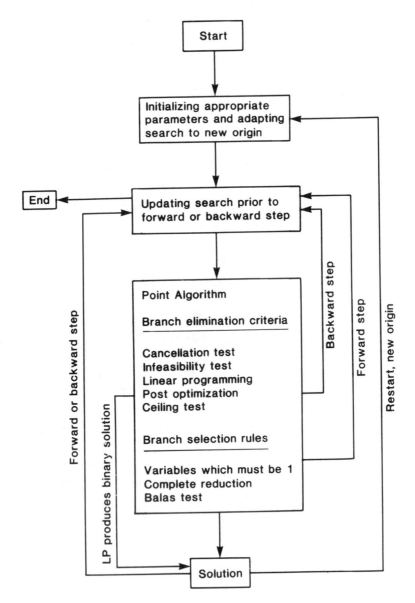

Figure 9.7: General Structure of the Algorithm

Appendix B
Computational Experience

A series of 11 integer programming problems, ranging in size from 6 rows and 12 columns to 28 rows and 89 columns, have been solved by several different search algorithms. The results of these efforts are summarized in Table 9.3. The different machines used make comparisons somewhat inaccurate. All times are in minutes. Each computer program and sample problem was stored entirely in core.

Problem A is a test problem of Balas [19]; B and E are two of the nine IBM test problems located in Haldi [306]. Problems F, J, and K were sent to Lemke and Spielberg by M. S. Sidrow at Texaco, and problem H by W. Acuri at IBM. The remaining ones (C, D, G, I) originated with Petersen [458] and have nonnegative constraint matrices.

Recently Crowder, Johnson, and Padberg [151] has developed an enumerative algorithm for the large scale zero-one integer programming problems. Their computational results are summerized in Table 9.4.

B.1 Program Descriptions

Petersen [458] computer programmed the original Balas additive algorithm [19]. Essentially it is the basic search approach with a point algorithm that includes the infeasibility test, the ceiling test, the feasibility test described in Problem 9.14, and the Balas test (Sections 9.3.3 and 9.3.4). As these tests and the bookkeeping scheme only require additions and subtractions in computations, we have an "additive" algorithm. Also tried were variations of the additive algorithm, including a more efficient bookkeeping scheme which keeps track of the enumeration (suggested by Glover [239] and used by Geoffrion [222]), and some of the implicit enumeration criteria appearing in Glover [239] and in Glover and Zionts [270]. In all versions, variables with negative cost coefficients are complemented to gain nonnegativity in the cost row. Earlier computational experience with the Balas additive algorithm also appears in Fleischman [200] and in Freeman [207].

The Lemke and Spielberg technique [391] has a slightly more sophisticated point algorithm than the one used by Petersen. It contains the infeasibility and cancellation zero test, the ceiling test, complete reduc-

	Problem		Petersen SDS/930		Lemke-Spielberg IBM 360/40		Salkin-Spielberg IBM 360/50		Geoffrion IBM 7044	
No.	Rows	Columns	Time	Iterations	Time	Iterations	Time	Iterations	Time	Iterations
A	6	12	–	–	1.00	17	0.015	8	–	–
B	15	15	–	–	–	–	0.056	4	–	–
C	10	20	4.0	1805	–	–	0.209	44	0.04	27
D	10	28	3.0	1041	–	–	0.663	93	0.24	181
E	31	31	–	–	> 50	> 70000	4.082	38	2.00	209
F	28	35	–	–	5.50	1279	0.842^b	65	0.19	35
G	5	39	5.0	2628	–	–	0.359	85	0.43	143
H	12	44	–	–	5.00	963	0.663	91	1.38	447
I	5	50	36.0	16778	–	–	1.354^b	213	0.46	115
J	37	74	–	–	> 30^a	> 71000	75.549^c	1039	> 10	> 517
K	28	89	–	–	> 30^a	> 65100	39.750^d	2251	–	–

a) Computed on IBM 7094
b) An upper bound on the value of the solution was assumed a priori.
c) Solution ascertained to be within 15 of optimal.
d) The origin was obtained from Lemke and Spielberg [391] and the solution was ascertained to be within 136 of the optimal one.

Table 9.3: Comparison of Results

Problem Number	Number of Variables	Number of Constraints	Density	Time*
1	33	16	17.4	0.69
2	40	24	11.3	0.18
3	195	134	5.0	10.07
4	282	222	1.5	12.72
5	290	206	1.3	0.85
6	527	157	1.9	0.91
7	1550	94	7.4	1.50
8	1939	109	4.8	14.99
9	2655	147	3.4	6.75
10	2734	739	0.4	54.42

*CPU time in minutes on IBM 370/168 - MVS

Table 9.4: Computational Results for Crowder et. al. Algorithm

tion, and the Balas test. Again, to achieve a nonnegative cost vector, complementing variables are used. The bookkeeping scheme appears in Appendix A.

The Salkin and Spielberg procedure [519] is the same as the previous one except that it has a generalized origin and restart capability (Section 9.5) and uses linear programming with post-optimization ability. Because the dynamic origin strategy is adopted, the cost vector may have negative entries. A more detailed description of the algorithm is in Appendix A. Test results of the Salkin and Spielberg code applied to specialized integer programming models may be found in Ashour and Char [16] and in Hwang, et al. [337].

The search procedure formulated by Geoffrion [222] contains a point algorithm which includes the ceiling test, linear programming and surrogate constraints, and the Balas test. A new surrogate constraint is generated by the optimal dual variables whenever possible, and is explicitly added to the constraint set. However, to reduce storage requirements, only the last four constraints are kept and used. A nonnegative cost row is also initially enforced. A somewhat similar algorithm is described by Jambekar and Steinberg [343], [344].

Chapter 10

Partitioning in Mixed Integer Programming

10.1   Introduction

Using linear programming duality theory, it is possible to show (cf. Section 10.2) that any mixed integer program can be written as an integer program. This means it may be worth solving a mixed integer problem (MIP) by solving its equivalent integer program, particularly if the number of integer variables in the MIP is very small. However, it is often computationally impossible to obtain all the constraints of the integer program, since to construct each constraint we need the numerical value of an extreme point or an extreme ray in a convex polyhedron, and there are as many constraints as there are extreme points and extreme rays. As the convex polyhedron may have a vast number of extreme points, all the constraints cannot usually be enumerated. To overcome this difficulty, J. F. Benders [79] proposed a technique in which the equivalent integer program is solved after generating only a subset of its constraints. That is, the remaining "implicitly enumerated" constraints do not further constrain the integer program. The Benders procedure partitions (hence the name "partitioning algorithm") the mixed integer program into an integer and a linear program, consisting respectively of the integer and the continuous variables of the original problem. The "partitioning algorithm" works by successively solving a linear program and an integer program. The linear program produces an extreme point or an extreme ray and a single constraint for the integer program. Also, the value of the linear programming optimal solution gives an upper bound for the optimal solution to the mixed integer program. When solved, the integer program, which is the mixed integer program's equivalent when it has all its constraints, yields a nondecreasing lower bound. When the two bounds coincide, the optimal mixed integer solution has been found and the process terminates.

Before developing the procedure, we note that Benders' [79] algorithm is applicable to a more general problem. In particular, it solves any mathematical program in which the variables can be partitioned into two subsets so that when the variables in one are assigned numerical values the problem reduces to a linear program. A mixed integer program, of course, fits this description and is the only type of problem that we will consider. Nevertheless, an outline of the general development (which is clearly detailed in Lasdon [385]) appears in Problems 10.15 and 10.16.[1] We also note that one of the first to specialize Benders' algorithm for the mixed integer program was Balinski [50], and an application of the specialized version to the uncapacitated plant location problem may be found in Balinski and Wolfe [52] (see Appendix A). Geoffrion [228] recently used a specialized partitioning algorithm to solve a multiproduct distribution model (see Problem 10.23). A special purpose enumerative technique for the integer programs that appear during the partitioning algorithm is explained in Balas [42]. Related work is in Rardin and Unger [473] and in Rutten [498].

10.2   Posing the Mixed Integer Program as an Integer Program

Consider the mixed integer program (written as a *minimization problem* with *greater than or equal to* constraints)

$$[\text{MIP}] \quad \begin{array}{ll} \text{minimize} & \mathbf{cx} + \mathbf{dy} \\ \text{subject to} & \mathbf{Ax} + \mathbf{Dy} \geq \mathbf{b}, \\ & \mathbf{x} \geq 0, \mathbf{y} \geq 0, \\ \text{and} & \mathbf{x} \quad \text{integer}, \end{array}$$

where $\mathbf{A}$ is an m by n matrix, $\mathbf{D}$ is an m by n' matrix, and the other terms are appropriate m, n, or n' vectors. If we let X denote the set of all feasible nonnegative integer vectors $\mathbf{x}$, then [MIP] may be written as

---

[1] Geoffrion [225] has extended Benders' results to the case where, for a subset of the variables fixed, the mathematical program simplifies to one for which an efficient solution technique is available (see Problem 10.17).

$$\text{minimize } (\mathbf{cx} + \text{minimize } (\mathbf{dy} \mid \mathbf{Dy} \geq \mathbf{b} - \mathbf{Ax}, \mathbf{y} \geq \mathbf{0})) \quad (1)$$
$$\mathbf{x} \in X$$

For a fixed $\mathbf{x}$, the minimization problem in the inner parentheses is the linear program

[L]   minimize $\mathbf{dy}$
           subject to $\mathbf{Dy} \geq \mathbf{b} - \mathbf{Ax}$
           and $\mathbf{y} \geq \mathbf{0}$;

its dual is

[DL]   maximize $\mathbf{u}(\mathbf{b} - \mathbf{Ax})$
           subject to $\mathbf{uD} \leq \mathbf{d}$
           and $\mathbf{u} \geq \mathbf{0}$.

If the set $U = \{\mathbf{u} \mid \mathbf{uD} \leq \mathbf{d}, \mathbf{u} \geq \mathbf{0}\}$ is empty, the dual problem [DL] is infeasible, and from duality theory, the primal problem [L] has no feasible solution or is unbounded. In the latter case, the original mixed integer problem is unbounded. However, in any real world situation, the mixed integer problem is bounded. Therefore, we shall assume that the set U is nonempty. Also, observe that the convex polyhedron U is independent of $\mathbf{x}$ and thus, no matter what the value of $\mathbf{x}$, the maximum over U of $\mathbf{u}(\mathbf{b} - \mathbf{Ax})$ occurs at an extreme point of U or it grows without bound along one of its extreme rays. But U has a finite number of extreme points and extreme rays. Denote the extreme points by $\mathbf{u}^p$ $(p = 1, \ldots, P)$ and the directions of the extreme rays (which can be enumerated by finding all extreme rays of $\mathbf{uD} \leq \mathbf{0}, \mathbf{u} \geq \mathbf{0}$; see Problem 10.1) by $\mathbf{v}^s$ $(s = 1, \ldots, S)$. Now, if for some $\mathbf{x}$ there exists a $\mathbf{v}^s$ such that $\mathbf{v}^s(\mathbf{b} - \mathbf{Ax}) > 0$, then the maximum of $\mathbf{u}(\mathbf{b} - \mathbf{Ax})$ for $\mathbf{u} \in U$ is unbounded. On the other hand, if the maximum of $\mathbf{u}(\mathbf{b} - \mathbf{Ax})$ for $\mathbf{u} \in U$ is unbounded, then for this $\mathbf{x}$ there is a direction $\mathbf{v}^s$ such that $\mathbf{v}^s(\mathbf{b} - \mathbf{Ax}) > 0$. When problem [DL] is unbounded, it means, by duality theory in linear programming (cf. Chapter 2), that problem [L] has no feasible solution and thus the mixed integer program has no solution for that $\mathbf{x}$. Therefore,

$$\mathbf{v}^s(\mathbf{b} - \mathbf{Ax}) \leq \mathbf{0} \quad (s = 1, \ldots, S) \quad (2)$$

provides necessary and sufficient conditions on $\mathbf{x}$ for feasible solutions

$\mathbf{y}$ to exist for the mixed integer program.[2] The set X of feasible $\mathbf{x}$ vectors can thus be defined as

$$\mathbf{v}^s(\mathbf{b} - \mathbf{Ax}) \leq \mathbf{0} \quad (s = 1, \ldots, S), \quad \text{and} \quad \mathbf{x} \geq \mathbf{0} \text{ and integer}$$

Also, for any $\mathbf{x} \in X$, the inner minimization in (1) is equal to the maximum of the dual problem [DL]. Thus the mixed integer program is

$$\underset{\mathbf{x}}{\text{minimize}} \quad \left(\mathbf{cx} + \underset{p = 1, \ldots, P}{\text{maximum}} \quad \mathbf{u}^p(\mathbf{b} - \mathbf{Ax}) \right)$$

$$\text{subject to} \quad \mathbf{v}^s(\mathbf{b} - \mathbf{Ax}) \leq \mathbf{0} \quad (s = 1, \ldots, S),$$

$$\text{and} \quad \mathbf{x} \geq \mathbf{0} \text{ and integer.}$$

Letting

$$z = \mathbf{cx} + \underset{p = 1, \ldots, P}{\text{maximum}} \quad \mathbf{u}^p(\mathbf{b} - \mathbf{Ax}),$$

we have $z \geq \mathbf{cx} + \mathbf{u}^p(\mathbf{b} - \mathbf{Ax})$ for $p = 1, \ldots, P$, and the mixed integer program is equivalent to the integer program

$$[I] \quad \underset{\mathbf{x}, z}{\text{minimize}} \quad z$$

$$\text{subject to} \quad z \geq \mathbf{cx} + \mathbf{u}^p(\mathbf{b} - \mathbf{Ax}) \quad (p = 1, \ldots, P)$$

$$0 \geq \mathbf{v}^s(\mathbf{b} - \mathbf{Ax}) \quad (s = 1, \ldots, S)$$

$$\text{and} \quad \mathbf{x} \geq \mathbf{0} \text{ and integer.}$$

To clarify the discussion we transform a small mixed integer program to an integer program.

---

[2] A similar result is shown with more detail in the latter part of Section 9.6, Chapter 9. (In that section the MIP has a slightly different format so that the inequality in (2) becomes reversed.)

Example 10.1   The mixed integer program is

$$[\text{MIP}] \quad \begin{aligned} &\text{minimize} & x &+ y_1 & &+ & y_3 \\ &\text{subject to} & 2x &+ y_1 &- 6y_2 &- & 5y_3 &\geq 1, \\ & & 3x &- y_1 &+ y_2 &+ & 2y_3 &\geq 2, \\ & & x &\geq 0, & y_1, y_2, y_3 &\geq 0, \\ &\text{and} & x & \text{ integer.} \end{aligned}$$

Problem [L] is

$$\begin{aligned} &\text{minimize} & y_1 & & &+ & y_3 \\ &\text{subject to} & y_1 &- 6y_2 &- & 5y_3 &\geq 1 - 2x, \\ & & -y_1 &+ y_2 &+ & 2y_3 &\geq 2 - 3x, \\ &\text{and} & & & y_1, y_2, y_3 &\geq & 0. \end{aligned}$$

Problem [DL] is

$$\begin{aligned} &\text{maximize} & u_1(1 - 2x) + u_2(2 - 3x) = z_D \\ &\text{subject to} & u_1 &- & u_2 &\leq 1 \\ & & -6u_1 &+ & u_2 &\leq 0 \\ & & -5u_1 &+ & 2u_2 &\leq 1 \\ &\text{and} & & u_1, & u_2 &\geq 0. \end{aligned} \Bigg\} U$$

The set U appears in Fig. 10.1. It has extreme points $\mathbf{u}^1 = (1,0), \mathbf{u}^2 = (0,0)$, and $\mathbf{u}^3 = (1/7, 6/7)$, and two extreme rays with directions $\mathbf{v}^1 = (1,1)$ and $\mathbf{v}^2 = (2/5, 1)$.

Thus, the mixed integer program is equivalent to the integer program

$$\begin{aligned} [\text{I}] \quad &\text{minimize} & z \\ &x, z \\ &\text{subject to} & z \geq x + (1 - 2x) = 1 - x, & \quad (a) \\ \\ & & z \geq x, & \quad (b) \\ \\ & & z \geq x + (1/7)(1 - 2x) + (6/7)(2 - 3x) = 13/7 - (20/7)x, & \quad (c) \\ \\ & & 0 \geq (1 - 2x) + (2 - 3x) = 3 - 5x, & \quad (d) \\ \\ & & 0 \geq (2/5)(1 - 2x) + 1(2 - 3x) = 12/5 - (19/5)x, & \quad (e) \\ \\ &\text{and} & x \geq 0 \text{ and integer.} \end{aligned}$$

Plotting the five inequalities in (x, z) space (Fig. 10.2) reveals the

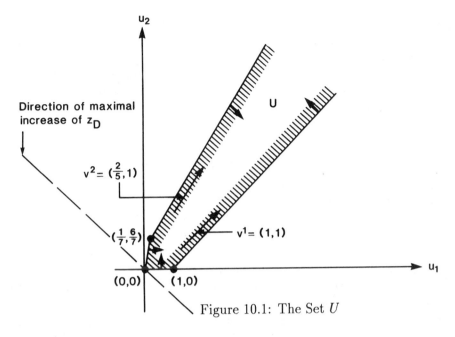

Figure 10.1: The Set U

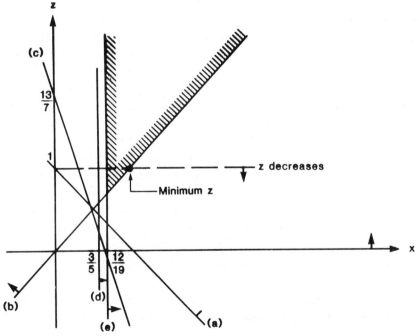

Figure 10.2: Problem [I] in (x, z) Space

optimal solution $x = 1$ and $z = 1$. To find the optimal vector $\mathbf{y} = (y_1, y_2, y_3)$, problem [L] is solved with $x = 1$. By inspection, since the right-hand side in problem [L] becomes $1 - 2(1) = -1$ and $2 - 3(1) = -1$, any vector $\mathbf{y} = (0, y_2, 0)$ with $0 \leq y_2 \leq 1/6$ is optimal. Note that for $x = 1$, $u_1(1 - 2x) + u_2(2 - 3x) = -u_1 - u_2$, and thus the optimal solution to problem [DL] is $u_1 = u_2 = 0$ with the objective function $-u_1 - u_2 = 0$.

10.3   The Partitioning Algorithm

As we have already mentioned, the polyhedron U usually has a vast number of extreme points, and therefore it is virtually impossible to generate all of the constraints of the integer program [I]. However, its derivation suggests a procedure for solving the mixed integer program. Note that for a fixed nonnegative integer vector $\mathbf{x}$ we can solve the linear program [DL]. As this problem is more restricted than the mixed integer program, the value of its maximal solution when added to $\mathbf{c}\mathbf{x}$ is an upper bound on the optimal solution to the mixed integer program. (If problem [DL] has an unbounded solution, its value can be thought of as infinite.) Furthermore, problem [DL], when solved, reveals an extreme point of U or the direction of one of its extreme rays, and thus an inequality for the integer program [I]. Suppose the integer program is solved with this one constraint. Its optimal solution is a lower bound on the best solution to the mixed integer program. (Why?) Also, it yields a nonnegative integer vector $\mathbf{x}$ for problem [DL], which, when solved, produces a new extreme point or ray direction and thus a second inequality for the integer problem. The integer program, now with two constraints, is solved and an at-least-as-good lower bound and nonnegative integer vector $\mathbf{x}$ is found. With the new value for $\mathbf{x}$, the linear program [DL] can again be solved, and so forth. The process terminates when the lower bound equals the upper bound. To find the optimal $\mathbf{y}$ vector, problem [L] is solved with $\mathbf{x}$ equal to its optimal value. The procedure is outlined below and in Fig. 10.3; z^u (z^l) denotes the best upper (lower) bound for the optimal mixed integer programming solution.

Step 1 (Initialization) Select a nonnegative integer value for the vector $\mathbf{x}$, say $\bar{\mathbf{x}}$, and set $z^u(z^l)$ arbitrarily large (small). Go to Step 2.

Step 2 (Linear Programming Phase) Solve the linear program

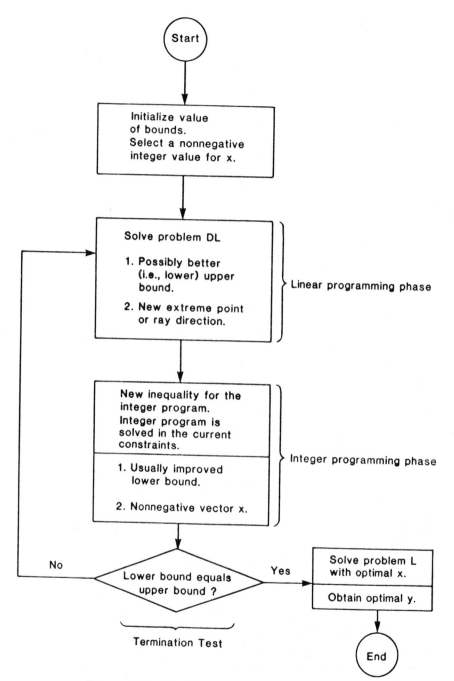

Figure 10.3: Benders' Partitioning Algorithm

[DL]   maximize $\mathbf{u}(\mathbf{b} - \mathbf{Ax})$
       subject to $\mathbf{u} \in U = \{\mathbf{u} | \mathbf{u}\mathbf{D} \leq \mathbf{d}, \mathbf{u} \geq \mathbf{0}\}$,

with $\mathbf{x}$ fixed at $\bar{\mathbf{x}}$. This yields either an optimal extreme point $\mathbf{u}^P$ or identifies an extreme ray direction $\mathbf{v}^s$ along which [DL] is unbounded. In the extreme point case, replace the value of z^u by $\mathbf{c}\bar{\mathbf{x}} + \mathbf{u}^P(\mathbf{b} - \mathbf{A}\bar{\mathbf{x}})$ whenever $z^u > \mathbf{c}\bar{\mathbf{x}} + \mathbf{u}^P(\mathbf{b} - \mathbf{A}\bar{\mathbf{x}})$. Go to Step 3.

Step 3 (Integer Programming Phase) Solve the integer program

$[IR]$ minimize z
         $\mathbf{x},z$

   subject to $z \geq \mathbf{cx} + \mathbf{u}^p(\mathbf{b} - \mathbf{Ax})$ $(p = 1, \ldots, P^l)$
                $0 \geq \mathbf{v}^s(\mathbf{b} - \mathbf{Ax})$ $(s = 1, \ldots, S^l)$
   and $\mathbf{x} \geq \mathbf{0}$ and integer,

where P^l (S^l) is the number of extreme points (extreme rays) found so far from Step 2. Observe that problem [IR] is problem [I] with only some of its contraints. Hence, problem [IR] is a relaxation of problem [I] and the value of any solution to problem [IR] is a lower bound on the optimal solution to the MIP. Set z^l equal to the minimal value of z and $\bar{\mathbf{x}}$ equal to the optimal $\mathbf{x}$ vector.[3] Go to Step 4.

Step 4 (Termination test) If $z^l < z^u$, go to Step 2. Otherwise $(z^l = z^u)$, $\bar{\mathbf{x}}$ is optimal. In this case, solve problem [L]

   minimize $\mathbf{dy}$
   subject to $\mathbf{Dy} \geq \mathbf{b} - \mathbf{A}\bar{\mathbf{x}}$
   and $\mathbf{y} \geq \mathbf{0}$

to obtain the optimal $\mathbf{y}$ vector $\bar{\mathbf{y}}$. Then $\mathbf{x} = \bar{\mathbf{x}}$, $\mathbf{y} = \bar{\mathbf{y}}$, and $z^l = z^u = \mathbf{c}\bar{\mathbf{x}} + \mathbf{d}\bar{\mathbf{y}}$ solves the mixed integer program–terminate.

Before discussing some of the algorithmic properties and proving convergence, we resolve the previous example.

Example (from Example 10.1) The mixed integer program and problems [L] and [DL] appear in Example 10.1.

---

[3]It is possible for the minimum value of z to be unbounded below with components of $\mathbf{x}$ taking on any integer values which satisfy the constraints. In this case, any integer feasible solution may be selected; see the illustrative example.

Step 1 $x = 0, z^u \to +\infty, z^l \to -\infty.$

Step 2 Problem [DL] with $x = 0$ is

$$
\begin{array}{rrcrcl}
\text{maximize} & u_1 & + & 2u_2 & & \\
\text{subject to} & u_1 & - & u_2 & \leq & 1 \\
& -6u_1 & + & u_2 & \leq & 0 \\
& -5u_1 & + & 2u_2 & \leq & 1 \\
\text{and} & u_1, & & u_2 & \geq & 0.
\end{array}
$$

The solution is unbounded (see Fig. 10.1), and suppose the simplex computations yield the extreme ray whose direction is $\mathbf{v} = (1, 1)$.

Step 3 As $1(1 - 2x) + 1(2 - 3x) = 3 - 5x$, problem [IR] is

$$
\begin{array}{ll}
\underset{x}{\text{minimize}} & z \\
\\
\text{subject to} & 0 \geq 3 - 5x \\
\text{and} & x \geq 0 \text{ and integer.}
\end{array}
$$

The minimal value of z can be made arbitrarily small independent of x. As $0 \geq 3 - 5x$ for $x = 1, 2, 3, \ldots$, suppose x is set equal to M, any positive integer; z^l remains arbitrarily small.

Step 4 $\to$ Step 2 Problem [DL] with $x = M$ is

$$
\begin{array}{ll}
\text{maximize} & u_1(1 - 2M) + u_2(2 - 3M) \\
\text{subject to} & (u_1, u_2) \in U.
\end{array}
$$

The optimal solution is $u_1 = u_2 = 0$; z^u is set to M.

Step 3 As $x + 0(1 - 2x) + 0(2 - 3x) = x$, problem [IR] is

$$
\begin{array}{llcl}
\text{minimize} & z & & \\
\text{subject to} & 0 & \geq & 3 - 5x, \\
& z & \geq & x, \\
\text{and} & x & \geq & 0 \text{ and integer.}
\end{array}
$$

The optimal solution is $x = 1$ and $z = 1$; hence $z^l = 1$.

Step 4 → Step 2[4] Problem [DL] with $x = 1$ is

$$\begin{aligned} \text{maximize} \quad & -u_1 - u_2 \\ \text{subject to} \quad & (u_1, u_2) \in U. \end{aligned}$$

The optimal solution is $u_1 = u_2 = 0$, and thus z^u is set to $1 - 0 - 0 = 1$.

Step 3 Problem [IR] does not change; $x = 1$ and $z = 1$ are still optimal.

Step 4 As $z^l = z^u, x = 1$ with $z = 1$ solves the mixed integer program. To find $\mathbf{y} = (y_1, y_2, y_3)$, problem [L] is solved with $x = 1$. This gives $y_1 = y_3 = 0$ and any y_2 such that $0 \leq y_2 \leq 1/6$.

10.4 Properties of the Partitioning Algorithm

1. In the next section we show that unless the optimal mixed integer solution is found,

> i) an optimal solution to the integer program in Step 3 never repeats itself.

> ii) each time the linear program [DL] in Step 2 is solved, a *new* extreme point (or extreme ray) of U is generated.

2. Unless the bounding inequality $z \geq z^l$ is explicitly introduced to each integer program, all generated nonredundant "z inequalities" have to be kept. In particular, an inequality $z \geq \mathbf{cx} + \mathbf{u}^p(\mathbf{b} - \mathbf{Ax})$ *cannot* be dropped from the integer program in Step 3 if its slack variable is positive in an optimal integer programming solution. This follows, since an inequality may have eliminated integer points which correspond to lower values of z. Thus, if such an inequality were dropped, optimal solutions to subsequent integer programs could yield worse lower bounds and consequently the algorithm might not converge.

For example, suppose that in Fig. 10.4 $\mathbf{x}^0$ is in the optimal solution to the mixed integer program, z^0 is the optimal value of the objective function, and the current z inequalities define the region **ABCD**. The point $(\mathbf{x}^1, z^1)$ corresponds to the integer vector $\mathbf{x}$ which minimizes z.

---

[4]If $M = 1$ is selected, $z^u = M = 1 = z^l$, and problem [L] would now be solved. We are supposing that this is not the case.

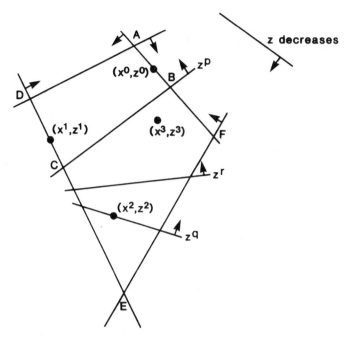

Figure 10.4: Dropping Inequalities Without $z \geq z^l$ Constraint

Since this coordinate is not intersected by the hyperplane $s = z^p - \mathbf{cx} - \mathbf{u}^p(\mathbf{b} - \mathbf{Ax}) = 0$, the slack variable s in the optimal solution to the current integer program is positive. If the $z^p \geq \mathbf{cx} + \mathbf{u}^p(\mathbf{b} - \mathbf{Ax})$ inequality is dropped, suppose the region is expanded to **AFED** (Fig. 10.4). As the minimal value of z can now be reduced to z^2 at the integer point $\mathbf{x}^2$, it is conceivable that subsequently generated z inequalities (denoted by z^q and z^r) may reduce the region so that $(\mathbf{x}^1, z^1)$ solves the integer program in Step 3 for a second time. By remark 1(i), this would mean that $\mathbf{x}^1$ is the optimal $\mathbf{x}$ for the mixed integer program.

To avoid this situation, we may introduce the inequality $z \geq z^l$ in each integer program that appears in Step 3. By doing this, we may drop the nonbinding z inequalities and often substantially reduce computer storage problems. Note, however, that if this strategy is adopted the lower bounds may increase at a slower rate, and thus the algorithm may take longer to converge.

To illustrate, suppose that in the $z^p \geq 0$ inequality in Fig. 10.4 is dropped and the $z \geq z^l$ constraint is introduced in the integer program (Fig. 10.5). At the next visit to the integer program the $z^t \geq 0$

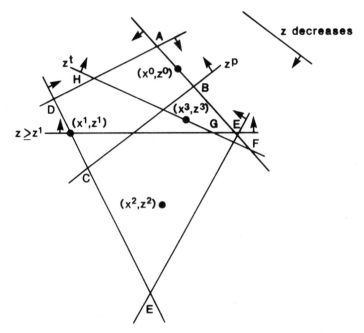

Figure 10.5: Dropping Inequalities With $z \geq z^l$ Constraint

inequality has been generated, so that the feasible region is **AEGH**. The optimal solution to this integer program is $(\mathbf{x}^3, z^3)$. However, had the z^p inequality been kept, the feasible region would be **ABJH**, and the mixed integer program's optimal solution $(\mathbf{x}^0, z^0)$ would have been discovered (Fig. 10.5). (A numerical example illustrating these phenomena is in Appendix A, Section A.3; these results are also discussed in [506].)

3. Every time an integer vector $\mathbf{x}$ yields an optimal vector $\mathbf{u}$ for problem [DL], a mixed integer solution, and thus an upper bound for the minimal mixed integer solution, is found. Therefore, if computer time or storage becomes excessive, computations can be stopped and the current best solution z^u can be used. Notice that it is at most $z^u - z^l$ away from the true optimum.

4. The partitioning aspect of the algorithm preserves the structure of the matrix $\mathbf{D}$ in the linear program [L], and hence in its dual prob-

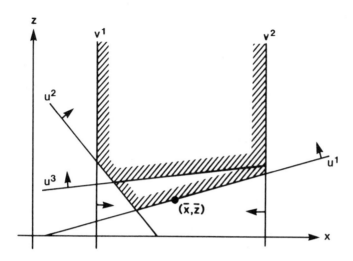

Figure 10.6: Cutting Plane Flavor of the Partitioning Algorithm

lem [DL]. Thus, for example, if **D** is a transportation type constraint matrix, then every occurrence of problem [DL] involves the dual of a transportation problem.

5. The computational efficiency of the algorithm depends principally on the number of extreme points and extreme rays that have to be found. If this number is relatively large, the approach will not be efficient.

6. The partitioning algorithm has a cutting plane flavor. In particular, each time an increased lower bound is obtained from the integer program, the most recently added z inequality cuts off the previous optimal integer solution. For example, in Fig. 10.6, u^1 and u^2 (v^1 and v^2) are extreme point (ray) inequalities. The optimal solution to this problem is denoted by $\bar{z}$ and $\bar{x}$. Problem [DL] then yields a new extreme point u^3 and thus an inequality for the integer program which cuts off $\bar{z}$ and $\bar{x}$).

7. If the integer variables **x** in MIP must also satisfy some class of side constraints (such as an upper bound on each variable) then Benders' algorithm remains unchanged except that the integer program [I] must include these constraints.

10.5   Finiteness

To simplify the proof of convergence we shall assume that the convex polyhedron U is bounded.[5] Since it has a finite number of extreme points it suffices to show that each pass through Step 2 produces a new vertex $\mathbf{u}^p$ of U until the optimal solution to the mixed integer program is found.

Suppose the first t ($t \geq 1$) extreme points of the polyhedron U, say $\mathbf{u}^1, \mathbf{u}^2 \ldots, \mathbf{u}^t$ that are generated from problem [DL] in Step 2 are unequal. Then the integer program in Step 3 is

$$
\begin{array}{llll}
[\text{IR}] & \text{minimize} & z & \\
& \text{subject to} & z \geq & \mathbf{cx} + \mathbf{u}^i(\mathbf{b} - \mathbf{Ax}) \quad (i = 1, \ldots, t) \\
& \text{and} & \mathbf{x} \geq & \mathbf{0} \quad \text{and integer.}
\end{array}
$$

Let the optimal solution to this problem be z^l, $\bar{\mathbf{x}}$. That is, for some k ($1 \leq k \leq t$)

$$
z^l = \mathbf{c}\bar{\mathbf{x}} + \mathbf{u}^k(\mathbf{b} - \mathbf{A}\bar{\mathbf{x}}) \geq \mathbf{c}\bar{\mathbf{x}} + \mathbf{u}^i(\mathbf{b} - \mathbf{A}\bar{\mathbf{x}}) \quad (i = 1, \ldots, t). \quad (3)
$$

As z^l is a lower bound for the optimal solution z^0 to the mixed integer program, we have $z^0 \geq z^l$; or, from (3)

$$
\mathbf{u}^k(\mathbf{b} - \mathbf{A}\bar{\mathbf{x}}) \leq z^0 - \mathbf{c}\bar{\mathbf{x}}. \quad (4)
$$

Now, upon revisiting Step 3, problem [DL]: Maximize $\mathbf{u}(\mathbf{b} - \mathbf{A}\bar{\mathbf{x}})$, for $\mathbf{u} \in U$ is solved. If $\mathbf{u}^{t+1}$ is a vertex of U which maximizes $(\mathbf{b} - \mathbf{A}\bar{\mathbf{x}})$, we have from linear programming duality theory that

$$
\mathbf{u}^{t+1}(\mathbf{b} - \mathbf{A}\bar{\mathbf{x}}) = \mathbf{d}\bar{\mathbf{y}}, \quad (5)
$$

where $\bar{\mathbf{y}}$ is the optimal solution to problem [L]:

---

[5] If U is not bounded it may be possible to initially enumerate its extreme rays and add the resulting ray inequalities $\mathbf{v}^s(\mathbf{b} - \mathbf{Ax}) \leq 0$, which are sufficient to close U, to the $\mathbf{Ax} + \mathbf{Dy} \leq \mathbf{b}$ constraints. Another possibility is to introduce the inequality $\sum_i u_i \leq M$, where M is a very large positive number, to the $\mathbf{uD} \leq \mathbf{d}$ constraints. In this case, U is bounded with an increased number of extreme points, some of which may have components taking on very large numerical values.

minimize $\mathbf{dy}$,
subject to $\mathbf{Dy} \geq \mathbf{b} - \mathbf{A\bar{x}}$,
and $\mathbf{y} \geq \mathbf{0}$.

As $(\mathbf{\bar{x}}, \mathbf{\bar{y}})$ is a solution for the mixed integer program, $z^u = \mathbf{c\bar{x}} + \mathbf{d\bar{y}} \geq z^0$; equivalently $z^0 - \mathbf{c\bar{x}} \leq \mathbf{d\bar{y}}$. Combining the last inequality with (4) and (5) gives

$$\mathbf{u}^k(\mathbf{b} - \mathbf{A\bar{x}}) \leq z^0 - \mathbf{c\bar{x}} \leq \mathbf{d\bar{y}} = \mathbf{u}^{t+1}(\mathbf{b} - \mathbf{A\bar{x}}). \tag{6}$$

Now, if equality holds, $z^0 - \mathbf{c\bar{x}} = \mathbf{d\bar{y}}$, or $(\mathbf{\bar{x}}, \mathbf{\bar{y}})$ solves the mixed integer program.[6] Otherwise,

$$\mathbf{u}^k(\mathbf{b} - \mathbf{A\bar{x}}) < \mathbf{u}^{t+1}(\mathbf{b} - \mathbf{A\bar{x}}), \tag{7}$$

in which case $\mathbf{u}^k \neq \mathbf{u}^{t+1}$. But from the inequality in (3),

$$\mathbf{u}^k(\mathbf{b} - \mathbf{A\bar{x}}) \geq \mathbf{u}^i(\mathbf{b} - \mathbf{A\bar{x}}) \quad (i = 1, \ldots, t).$$

By (7) this means

$$\mathbf{u}^i(\mathbf{b} - \mathbf{A\bar{x}}) < \mathbf{u}^{t+1}(\mathbf{b} - \mathbf{A\bar{x}}) \quad (i = 1, \ldots, t);$$

or $\mathbf{u}^{t+1} \neq \mathbf{u}^i$ $(i = 1, \ldots, t)$. And thus, unless the optimal mixed integer solution is found, a new extreme point is generated each time problem [DL] is solved. Therefore, in the worst case, all the extreme points are enumerated and the mixed integer program's equivalent integer program is solved, which proves that the algorithm converges.

The proof of convergence also shows that until the mixed integer program is solved, an optimal solution to the integer program in Step 3 cannot repeat itself.[7] For suppose to the contrary that $\mathbf{\bar{x}}$ solves the integer program twice. This means that $\mathbf{u}(\mathbf{b} - \mathbf{A\bar{x}})$, the objective function for problem [DL], appears twice, and as its feasible region U is independent of $\mathbf{x}$, optimal extreme points $\mathbf{u}^l$ $(1 \leq l \leq t)$ and $\mathbf{u}^{t+1}$ are found

---

[6]Note that $z^l = z^u$, and the algorithm terminates computations, since $z^l - \mathbf{c\bar{x}} = \mathbf{u}^k(\mathbf{b} - \mathbf{A\bar{x}}) = z^0 - \mathbf{c\bar{x}} = z^u - \mathbf{c\bar{x}}$, or $z^l = z^u$.

[7]This does not mean that the lower bound z^l always increases. It is possible to have the same lower bound for different values of $\mathbf{x}$ (see the Appendix, Section A.3).

so that

$$\mathbf{u}^l(\mathbf{b} - \mathbf{A}\bar{\mathbf{x}}) = \mathbf{u}^{t+1}(\mathbf{b} - \mathbf{A}\bar{\mathbf{x}}). \tag{8}$$

Now, using the same argument as before, we can show that there exists an index k $(1 \leq k \leq t)$ such that

$$\mathbf{u}^k(\mathbf{b} - \mathbf{A}\bar{\mathbf{x}}) \geq \mathbf{u}^i(\mathbf{b} - \mathbf{A}\bar{\mathbf{x}}) \quad (i = 1, \dots, t), \tag{9}$$

and also that

$$\mathbf{u}^{t+1}(\mathbf{b} - \mathbf{A}\bar{\mathbf{x}}) \geq \mathbf{u}^k(\mathbf{b} - \mathbf{A}\bar{\mathbf{x}}). \tag{10}$$

Combining inequalities (9) and (10) and using (8) gives

$$\mathbf{u}^{t+1}(\mathbf{b} - \mathbf{A}\bar{\mathbf{x}}) = \mathbf{u}^k(\mathbf{b} - \mathbf{A}\bar{\mathbf{x}}), \tag{11}$$

which by expression (6) means that $\bar{\mathbf{x}}$ and the resulting $\bar{\mathbf{y}}$ solve the mixed integer program.

We have thus shown that until the optimal mixed integer solution is discovered a new extreme point $\mathbf{u}^i$ of U is generated after each visit to Step 2. This means that for each nonoptimal vector $\mathbf{x}$ which solves the integer program appearing in Step 3, the maximum of $\mathbf{u}(\mathbf{b} - \mathbf{A}\mathbf{x})$ for $\mathbf{u} \in U$ occurs at a different vertex $\mathbf{u}^i$. Thus it is impossible to have the situation appearing in Fig. 10.7.

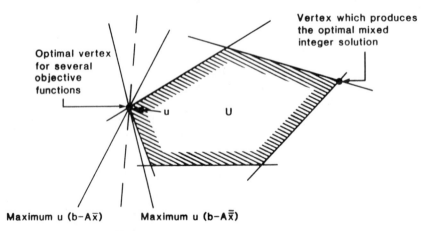

Figure 10.7: Impossible Situation

Finally, note that when the upper bound z^u decreases, a nonredundant constraint $z \geq \mathbf{cx} + \mathbf{u}^i(\mathbf{b} - \mathbf{Ax})$, is introduced to the integer program in Step 2. (Why?)

Problems

Problem 10.1

a) Show that finding the extreme rays of $U = \{\mathbf{u} \geq \mathbf{0} | \mathbf{uD} \leq \mathbf{d}\}$ is equivalent to finding all extreme rays of $U^0 = \{\mathbf{u} \geq \mathbf{0} | \mathbf{uD} \leq \mathbf{0}\}$.

b) Find the extreme rays of U when

$$\mathbf{D} = \begin{pmatrix} -1 & 1 \\ 1 & -2 \end{pmatrix} \quad \text{and} \quad \mathbf{d} = \begin{pmatrix} -1 \\ 2 \end{pmatrix}$$

by finding the extreme rays of U^0.

Solve the following mixed integer programs by first finding, and then solving, the equivalent integer programs.

Problem 10.2

$$
\begin{array}{llrcrcl}
\text{Maximize} & x & + & y & & \\
\text{subject to} & -x & + & 2y & \leq & 8, \\
& 2x & + & y & \leq & 16, \\
& x & & & \leq & 6, \\
\text{and} & x, & & y & \geq & 0 \text{ and } x \text{ integer.}
\end{array}
$$

Problem 10.3

$$
\begin{array}{llrcrcl}
\text{Maximize} & 5y & + & 4x_1 & = & x_2 \\
\text{subject to} & 2y & + & 3x_1 & \geq & 7, \\
& 4y & + & x_1 & \geq & 5, \\
\text{and} & y, & & x_1 & \geq & 0 \text{ and } x_1, x_2 \text{ integer.}
\end{array}
$$

Problem 10.4

$$\begin{array}{llrcll}
\text{Maximize} & 2x & + & y & & \\
\text{subject to} & 2x & + & 2y & \leq & 3, \\
& x & + & 3y & \geq & 2, \\
\text{and} & x, & & y & \geq & 0 \text{ and } x \text{ integer.}
\end{array}$$

Problem 10.5   Express the following mixed integer program as a zero-one integer program. Solve the zero-one problem by a search enumeration and then obtain the optimal solution to the original problem.

$$\begin{array}{llrcll}
\text{Maximize} & x & + & \frac{3}{2}y & & \\
\text{subject to} & 2x & + & 3y & \leq & 6, \\
& x & + & 4y & \geq & 4, \\
\text{and} & x, & & y & \geq & 0 \text{ and } x \text{ integer.}
\end{array}$$

Problem 10.6   Suppose a mixed integer program is

$$\begin{array}{llrcll}
\text{Maximize} & \mathbf{cx} & + & \mathbf{dy} & & \\
\text{subject to} & \mathbf{Ax} & + & \mathbf{Dy} & = & \mathbf{b}, \\
& \mathbf{x}, & & \mathbf{y} & \geq & \mathbf{0}, \\
\text{and} & \mathbf{x} & \text{integer.} & & &
\end{array}$$

Explain how Benders' partitioning algorithm is modified because of the equality condition. Would it be computationally more efficient to use your modified algorithm rather than convert each equality constraint to two inequalities and use the partitioning algorithm given in Section 10.3?

Problem 10.7   In Benders' partitioning algorithm we have the option of initially ensuring bounding of the convex polyhedron $U = \{\mathbf{u} \geq \mathbf{0} | \mathbf{uD} \leq \mathbf{d}\}$ by adjoining the inequality $\sum_i u_i \leq M$, where M is a very large positive number, to the $\mathbf{uD} \leq \mathbf{d}$ constraints. If this bounding inequality is not employed, extreme ray inequalities (1) (Section 10.2) may have to be generated. Under what circumstances might it be computationally more efficient to initially bound U?

Problem 10.8   Rewrite Bender's partitioning algorithm (Section 10.3) so that Step 3 is interchanged with Step 2. Why does the resulting algorithm necessarily converge? Could interchanging the two steps substantially affect the convergence rate?

Problem 10.9 What are characteristics of the integer programs that appear in Step 3 of the partitioning algorithm which might suggest a solution procedure? In particular, does an enumerative algorithm seem more adaptable than a cutting plane approach? If an enumerative algorithm is to be developed, why does it seem natural to incorporate a generalized origin capability (Chapter 9, Section 9.5)?

Solve each of the mixed integer programs in Problems 10.10 through 10.14 using Benders' partitioning algorithm.

Problem 10.10 The mixed integer program appearing in Example 10.1 (Section 10.2), except that the constraint $u_1 + u_2 \leq M$, where M is a large positive number, is introduced to bound the polyhedron U.

Problem 10.11 The mixed integer program appearing in problem 10.3. Verify the optimal solution graphically.

Problem 10.12

$$
\begin{array}{lrcrcrcl}
\text{Minimize} & x_1 & + & x_2 & + & \frac{5}{2}y & & \\
\text{subject to} & 3x_1 & + & 2x_2 & + & y & \geq & 5, \\
& -x_1 & + & x_2 & + & 4y & \geq & 7, \\
& x_1 & - & x_2 & + & 2y & \geq & 4, \\
\text{and} & x_1, & & x_2, & & y & \geq & 0 \text{ and } x_1, x_2 \text{ integer.}
\end{array}
$$

Problem 10.13

$$
\begin{array}{lrcrcl}
\text{Maximize} & 5x & + & y & & \\
\text{subject to} & 4x & + & 3y & \leq & 6, \\
& 3x & + & y & \geq & 2, \\
\text{and} & x, & & y & \geq & 0 \text{ and } x \text{ integer.}
\end{array}
$$

Verify the optimal solution graphically.

Problem 10.14

$$
\begin{array}{lrcrcrcrcl}
\text{Maximize} & 2x & + & 3y_1 & + & 4y_2 & + & 7y_3 & & \\
\text{subject to} & 4x & + & 6y_1 & - & 2y_2 & + & 8y_3 & = & 16, \\
& -x & + & 2y_1 & - & 6y_2 & + & 7y_3 & = & 3, \\
\text{and} & x, & & y_1, & & y_2, & & y_3 & \geq & 0 \text{ and } x \text{ integer.}
\end{array}
$$

Problem 10.15 Consider the mathematical programming problem

[P1]   minimize $f(\mathbf{x}) + \mathbf{dy}$
       subject to $\mathbf{F(x)} + \mathbf{Dy} \geq \mathbf{b}$
       and $\mathbf{y} \geq 0, \mathbf{x} \in X,$

where $\mathbf{x}$ is an n vector, f a scalar valued function of $\mathbf{x}$, $\mathbf{d}$ and $\mathbf{y}$ n' vectors, $\mathbf{F}$ an m vector whose components are functions of $\mathbf{x}$, $\mathbf{D}$ an m by n' matrix, $\mathbf{b}$ an m vector, and X an arbitrary subset of E^n (the set of all n vectors).

a) Using Farkas' lemma ([577]) which says: There exists a vector $\mathbf{t} \geq 0$ satisfying $\mathbf{Ht} = \mathbf{a}$ if and only if $\mathbf{wa} \geq 0$ for all $\mathbf{w}$ satisfying $\mathbf{wH} \geq 0$; show that a necessary and sufficient condition on $\mathbf{x} \in X$ to admit nonnegative $\mathbf{y}$ satisfying $\mathbf{Dy} \geq \mathbf{b} - \mathbf{F(x)}$ is that

$$\mathbf{v}^s(\mathbf{b} - \mathbf{F(x)}) \leq 0 \quad (s = 1, \ldots, S),$$

where the $\mathbf{v}^s$ are the extreme rays of the polyhedral cone $U^0 = \{\mathbf{u} \geq 0 | \mathbf{uD} \leq 0\}$.

b) Using (a), show that problem [P1] may be written as

[I]   minimize z
      subject to $z \geq f(\mathbf{x}) + \mathbf{u}^p(\mathbf{b} - \mathbf{F(x)}) \, (p = 1, \ldots, P),$
                   $0 \geq \mathbf{v}^s(\mathbf{b} - \mathbf{F(x)}) \, (s = 1, \ldots, S),$
      and $\mathbf{x} \in X,$

where $\mathbf{u}^p$ ($\mathbf{v}^s$) is an extreme point (extreme ray direction) of the convex polyhedron $U = \{\mathbf{u} \geq 0 | \mathbf{uD} \leq \mathbf{d}\}$ and U has P (S) extreme points (extreme rays). (*Hint*: By problem 10.1, $\mathbf{v}^s$ are the exrtreme ray directions of U. Then proceed as in Section 10.2, noting that by linear programming duality theory,

$$\underset{\mathbf{y}}{\text{minimum}} \ \{\mathbf{dy} | \mathbf{Dy} \geq \mathbf{b} - \mathbf{F(x)}, \mathbf{y} \geq 0)\} =$$

$$\underset{\mathbf{u}}{\text{maximum}} \ \{\mathbf{u}(\mathbf{b} - \mathbf{F(x)}) | \mathbf{u} \in U\}.)$$

Problem 10.16   Given that the mathematical program [P1] is equivalent to the program [I] (both are given in Problem 10.15) explain how Benders' partitioning algorithm can solve problem [P1]. Rewrite the partitioning algorithm (Section 10.3) so that it solves problem [P1].

Explain why the rewritten algorithm converges. Are any assumptions on X, $f(\mathbf{x})$, and the components of $\mathbf{F}(\mathbf{x})$ necessary ?

Problem 10.17 (Geoffrion[225]) Consider the mathematical programming problem

$$
\begin{array}{ll}
[\text{P1}] & \underset{\mathbf{x},\mathbf{y}}{\text{maximize}} \quad g(\mathbf{x},\mathbf{y}) \\
& \text{subject to} \quad \mathbf{G}(\mathbf{x},\mathbf{y}) \geq \mathbf{0} \\
& \text{and} \qquad\quad \mathbf{x} \in X, \; \mathbf{y} \in Y,
\end{array}
$$

where $\mathbf{x}$ is an n vector, $\mathbf{y}$ an n' vector, $g(\mathbf{x},\mathbf{y})$ a scaler valued function of $\mathbf{x}$ and $\mathbf{y}$, $\mathbf{G}$ an m vector whose components are functions of $\mathbf{x}$ and $\mathbf{y}$, X a subset of E^n and Y a subset of $E^{n'}$ (E^k is the set of all k vectors).

a) For each of the following cases explain why a partitioning algorithm seems to be a natural solution procedure for problem [P1].

   i) For fixed $\mathbf{x}$, [P1] separates into a number of independent optimization problems, each involving a different subvector of $\mathbf{y}$.

   ii) For fixed $\mathbf{x}$, [P1] assumes a well known structure (such as the classical transportation form) for which efficient solution procedures are available.

   iii) [P1] is not a concave program in $\mathbf{x}$ and $\mathbf{y}$ jointly, but fixing $\mathbf{x}$ renders it so in $\mathbf{y}$.

b) For cases (i), (ii), and (iii) appearing in (a), explain how you would derive a partitioning algorithm for problem [P1]. Would nonlinear programming duality theory (e.g., as in [224]) be used?

(A detail solution of (a) and (b) is in Geoffrion[225].)

Problem 10.18 In Benders' partitioning algorithm is it possible to determine when the integer program in Step 3 will produce a strictly improved lower bound? In particular, if the integer program always has a unique optimal solution, does the lower bound always increase? If the integer program has several alternate optimal solutions, does the lower bound necessarily remain the same for the several passes through Step 3?

Problems 10.19 through 10.22 relates to Appendix A.

Problem 10.19   For an m plant n customer plant location problem, we showed in Appendix A (Section A.2) that the optimal solution $(\mathbf{u}, \mathbf{v})$ to the linear program [DL] which appears in Step 2 of Benders' partitioning algorithm is given by the expression (5). That is,

$$
\left.
\begin{array}{ll}
v_j = c_{i(j),j} & (j = 1, \ldots, n) \\
u_{i(j)} = 0 & (j = 1, \ldots, n) \\
\text{and} \quad u_i = \text{maximum } (0, m_i) & (i = 1, \ldots, m, \text{ and} \\
& \quad i \neq i(j) \text{ for } j = 1, \ldots, n).
\end{array}
\right\} (5)
$$

where

$$
m_i = \begin{array}{c} \text{maximum} \\ j = 1, \ldots, n \end{array} \left(c_{i(j),j} - c_{ij} \right)
$$

and $i(j)$ corresponds to the open plant i which serves customer j with the smallest shipping cost. If $m = n$, under what circumstances do the optimal $u_i = v_j$? (This always occurs in the example given in Section A.3.)

Problem 10.20   Consider a plant location problem in which some plants cannot service all of the customers. That is, in model [P1] which appears in Appendix A, it is possible to have $n_i < n$, where n_i is the number of customers that plant i can service and there are n customers. Specialize Benders' partitioning algorithm for this variation of the plant location problem.

Problem 10.21   Suppose that in the plant location problem [P1] given in Appendix A, the

$$
\sum_{j=1}^{n} y_{ij} \leq n_i x_i \quad (i = 1, \ldots, m)
$$

inequalities are replaced by the constraints $y_{ij} \leq x_i$ for all i, j.

    a) With regards to the model, why are the two constraints equivalent?

b) What is the reformulated problem's equivalent integer program? What is the form of the linear program [DL] which appears in Step 2 of Benders' partitioning algorithm? (*Hint:* The n_i term is dropped and the u_i variables are replaced by u_{ij} variables.)

c) What is the optimal solution to the linear program [DL] found in part (b)? How is it computed? Is there a relationship between the optimal values of the variables given here and those given in expression (5) (Problem 10.19) for the linear program [DL] given in Section A.2 of the Appendix?

d) Summerize how Benders' partitioning algorithm is specialized for this version of the plant location problem.

e) When using Benders' algorithm for the plant location problem, does it seem computationally more efficient to use the $y_{ij} \leq x_i$ inequalities rather than the $\sum_j y_{ij} \leq n_i x_i$ constraints, or vice versa? In particular, are the lower bounds (generated in Step 3) substantially weaker when one form of the model is used as compared with the other?

f) Resolve the example appearing in Appendix A, Section A.3, using the algorithm outlined in part (c).

Problem 10.22

a) Using the partitioning algorithm, solve the following four plant, four customer plant location problem which has $f_i = 7$ for $i = 1, 2, 3, 4$ and

$$
(c_{ij}) = \begin{pmatrix} 0 & 2 & 20 & 18 \\ 12 & 0 & 8 & 6 \\ 20 & 8 & 0 & 6 \\ 18 & 6 & 6 & 0 \end{pmatrix}.
$$

b) Resolve the example under the assumption that plants 1 and 3 can each service at most one customer.

Problem 10.23   Consider the multicommodity distribution model

(6) - (13) in Chapter 1. Specialize and subsequently list the steps of
the Benders' partitioning algorithm applied to this problem. Discuss
the advantages of using Benders' algorithm over other approaches (e.g.,
enumeration or cutting plane).

Appendix A
Application of the Partitioning Algorithm to the Uncapacitated Plant Location Problem

(Balinski and Wolfe [52])

The special structure of the uncapacitated plant location problem permits us to obtain, very simply, the optimal solutions to the linear programs [L] and [DL] that appear during Benders' partitioning algorithm.
Recall from Chapter 1 that the uncapacitated plant location problem[8]
is

$$[P1] \quad \text{minimize} \quad \sum_{i=1}^{m} f_i x_i \; + \; \sum_{i=1}^{m}\sum_{j=1}^{n} c_{ij} y_{ij}$$

$$\text{subject to} \quad n_i x_i \; - \; \sum_{j=1}^{n} y_{ij} \; \geq \; 0 \quad (i = 1,\dots,m),$$

$$\sum_{i=1}^{m} y_{ij} \; = \; 1 \quad (j = 1,\dots,n),$$

$$y_{ij} \; \geq \; 0 \quad (\text{all } i,j),$$

$$\text{and} \quad x_i = 0 \text{ or } 1 \quad (i = 1,\dots,m),$$

where f_i is a fixed cost associated with plant i, x_i is 1(0) if plant i
is open (closed), c_{ij} is a nonnegative shipping cost, y_{ij} is the fraction
of the jth customer's demand satisfied by plant i, n_i is the number of
customers that can be satisfied by plant i (which for ease of exposition
will be taken as n[9]), and there are m plants and n customers.

---

[8]A slightly different version of the uncapacitated plant location problem is discussed in Balinski and Wolfe [52]. Their model and results appear in Problem 10.21.
Program [P1] was selected because it was used before. The plant location problem
is discussed more extensively in Chapter 14.

[9]The case where $n_i < n$ is left to Problem 10.20.

A.1 Problem [P1]'s Equivalent Integer Program [I]

Notice that for a fixed zero-one vector $\mathbf{x}$, problem [P1] reduces to the linear problem [L]. Its dual is problem [DL].

$$[L] \quad \underset{y_{ij}}{\text{minimize}} \quad \sum_{i=1}^{m} \sum_{j=1}^{n} c_{ij} y_{ij}$$

$$\text{subject to} \quad -\sum_{j=1}^{n} y_{ij} \geq -n_i x_i \quad (i = 1, \ldots, m),$$

$$\sum_{i=1}^{m} y_{ij} = 1 \quad (j = 1, \ldots, n),$$

$$\text{and} \quad y_{ij} \geq 0 \quad (\text{all } i, j).$$

$$[DL] \quad \underset{u_i, v_j}{\text{maximize}} \quad \sum_{j=1}^{n} v_j - \sum_{i=1}^{m} n_i x_i u_i$$

$$\text{subject to} \quad v_j - u_i \leq c_{ij} \quad \begin{array}{l}(i = 1, \ldots, m \\ \text{and } j = 1, \ldots, n),\end{array}$$

$$u_i \geq 0 \quad (i = 1, \ldots, m),$$

$$\text{and} \quad v_j \text{ unrestricted} \quad (j = 1, \ldots, n).$$

Now consider the convex polyhedron

$$U = \{(u_1, \ldots, u_m, v_1, \ldots, v_n) | v_j - u_i \leq c_{ij}$$
$$\text{for all } i, j, \text{ and } u_i \geq 0 \text{ for } i = 1 \ldots, m\}.$$

It is independent of $\mathbf{x}$ and has P extreme points (S extreme ray directions), each having components (u_i^P, v_j^P) $((u_i^S, v_j^S))$. Then, by the analysis used in Section 10.2 we have that the mixed integer problem [P1] is equivalent to the integer program

$$[I] \quad \text{minimize} \quad z$$

$$\text{subject to} \quad z \geq \sum_{i=1}^{m} f_i x_i + \sum_{j=1}^{n} v_j^p - \sum_{i=1}^{m} n_i x_i u_i^p \quad (p = 1, \ldots, P),$$

$$0 \geq \sum_{j=1}^{n} v_j^s - \sum_{i=1}^{m} n_i x_i u_i^s \quad (s = 1, \ldots, S),$$

and $x_i = 0$ or 1 $(i = 1, \ldots, m)$.

Remember that the second set or extreme ray inequalities are necessary
and sufficient conditions on $\mathbf{x}$ to admit feasible solutions $\mathbf{y} = (y_{ij})$ to
problem [L] and thus problem [P1]. However, a careful examination
of the linear program [L] shows that for any zero-one vector $\mathbf{x}$, except
$\mathbf{x} = \mathbf{0}$, there is a feasible solution $\mathbf{y}$. That is, if at least one plant is open,
all the customers can be serviced. (Why?) Thus, the ray inequalities
can be dropped so long as $\mathbf{x} \neq \mathbf{0}$ is enforced.

A.2 Solving the Linear Program [DL]

To solve problem [DL] we first solve its dual problem [L]. Then, using
complementary slackness properties, optimal values for the u_i and v_j
will be found.

For a fixed zero-one vector $\mathbf{x}$ $(\mathbf{x} \neq \mathbf{0})$ the linear program [L] is a
transportation problem which may be solved by inspection. In particu-
lar, when $x_i = 1$ plant i is open and can service each of the n customers.
As $c_{ij} \geq 0$ and shipments cannot be made from closed plants $(x_i = 0)$,
an optimal solution is to service each customer j from the open plant
$i(j)$ with the cheapest shipping cost. That is, if we let

$$I_1 = \{i | x_i = 1\}, \tag{1}$$

an optimal solution to problem [L] is

$$y_{i(j),j} = 1 \quad (j = 1, \ldots, n)$$

and

$$y_{ij} = 0 \quad \text{otherwise},$$

where

$$c_{i(j),j} = \min_{i \in I_1} c_{ij} \quad (j = 1, \ldots, n). \tag{2}$$

To obtain the corresponding dual solution, from linear programming,
complementary slackness implies that $v_j - u_{i(j)} = c_{i(j),j}$, or

$$v_j = c_{i(j),j} + u_{i(j)} \quad (j = 1, \ldots, n). \tag{3}$$

Substituting for v_j in the dual problem [DL], and noting that $x_i = 1$ if $i \in I_1$ and $x_i = 0$ otherwise, gives the equivalent linear program

$$[DL'] \quad \text{maximize} \quad \sum_{j=1}^{n} u_{i(j)} - \sum_{i \in I_1} n_i u_i$$

$$\text{subject to} \quad u_{i(j)} - u_i \leq c_{ij} - c_{i(j),j} \quad \text{(all } i,j)$$

$$\text{and} \qquad u_i \geq 0 \qquad\qquad (i = 1, \ldots, m),$$

where the constant $\sum_{j=1}^{n} c_{i(j),j}$ has been omitted from the objective function. Now for each $j, i(j)$ corresponds to some plant i, where i must be in the set I_1. Suppose we define N_i as the number of times the index i is an $i(j)$ and as 0 if the index i is never an $i(j)$; then

$$\sum_{j=1}^{n} u_{i(j)} = \sum_{i \in I_1} N_i u_i.$$

Or, problem [DL'] may be rewritten as

$$[DL''] \quad \text{maximize} \quad \sum_{i \in I_1} (N_i - n_i) u_i$$

$$\text{subject to} \quad u_i \geq c_{i(j),j} - c_{ij} + u_{i(j)} \quad (i = 1, \ldots, m \quad \text{and } j = 1, \ldots, n)$$

$$\text{and} \qquad u_i \geq 0 \qquad\qquad (i = 1, \ldots, m).$$

Since N_i can be at most n_i, $N_i - n_i$ is nonpositive, and thus maximizing the objective function will make the u_i as small as possible. As $u_{i(j)}$ must be nonnegative and it can only increase the value of u_i, an optimal solution to problem [DL''], and thus problem [DL], has $u_{i(j)} = 0$ for every $i(j)$. Then, for each index i, u_i must be nonnegative, and it must also satisfy $u_i \geq c_{i(j),j} - c_{ij}$ for each $j = 1, \ldots, n$. Therefore, if we let

$$\left. \begin{array}{l} m_i = \quad \text{maximum} \quad \left(c_{i(j),j} - c_{ij} \right) \\ \qquad\quad j = 1, \ldots, n \\ \text{for each } i = 1, \ldots, m, \ i \neq i(j) \ \text{for all } j = 1, \ldots, n, \end{array} \right\} \quad (4)$$

an optimal solution is $u_i = \text{maximum } (0, m_i)$. Summarizing and using (3), we have that an optimal solution to problem [DL] is

$$
\left.
\begin{array}{lll}
v_j = c_{i(j),j} & (j = 1, \ldots, n) & \\
u_{i(j)} = 0 & (j = 1, \ldots, n) & \\
\text{and} \quad u_i = \text{maximum } (0, m_i) & (j = 1, \ldots, m, i \neq i(j)) & \\
& \text{for all } (j = 1, \ldots, n), &
\end{array}
\right\} (5)
$$

where $i(j)$ is defined by (2) and m_i by (4).

An upper bound for the optimal solution for the plant location problem is from problem [DL']

$$
z^u = \sum_{i \in I_1} f_i + \sum_{j=1}^{n} c_{i(j),j} - \sum_{i \in I_1} n_i u_i, \tag{6}
$$

which by linear programming duality theory is also equal to

$$
z^u = \sum_{i \in I_1} f_i + \sum_{j=1}^{n} c_{i(j),j}. \tag{7}
$$

A.3 Summary of the Partitioning Algorithm

The linear programs that appear in Step 2 are solved by inspection, using (5). The optimal extreme point solution, whose components are (u_i, v_j), yields a z inequality for the integer program in Step 3 and also an upper bound z^u defined by (6) or (7). After solving the integer program a new zero-one vector $\mathbf{x}$ and a lower bound z^l are obtained. If $z^l = z^u$, the optimal solution has been found and the computations end. Otherwise $(z^l < z^u)$, the linear program [DL] is again solved by inspection, and a new extreme point of U is found, and so forth. Extreme ray inequalities are never generated, since $\mathbf{x} \neq \mathbf{0}$ is enforced in Step 3 and is never selected as the initial value for $\mathbf{x}$ in Step 1.

A.4 An Illustrative Example [10]

Consider the three plant, three customer plant location problem which has $f_i = 7$, $n_i = n = 3$ for $i = 1, 2, 3$, and

---

[10]This problem represents part of the one appearing in Balinski [50].

$$c_{ij} = \begin{pmatrix} 0 & 12 & 20 \\ 12 & 0 & 8 \\ 20 & 8 & 0 \end{pmatrix}.$$

Step 1 We choose $\mathbf{x} = (1,1,0)$, z^u is set to $+\infty$ and z^l to $-\infty$.

Step 2 (#1) $I_1 = \{1,2\}$, so $i(1) = 1, i(2) = 2$, and $i(3) = 2$. Thus, by (5).

$$v_1 = c_{11} = 0 \quad v_2 = c_{22} = 0 \quad v_3 = c_{23} = 8$$

$$u_1 = 0 \qquad u_2 = 0$$

$$m_3 = \text{maximum}(-20, -8, 8) = 8 \text{ so } u_3 = 8$$

and $(7+7) + (0+0+8) - 0 = 22 \rightarrow z^u$.

Step 3 (#1) The integer program is

$$\begin{array}{ll} \text{minimize} & z \\ \text{subject to} & z \geq 8 + 7x_1 + 7x_2 - 17x_3 \\ \text{and} & x_i = 0 \text{ or } 1 \quad (i = 1,2,3). \end{array}$$

By inspection, the optimal solution is $\mathbf{x} = (0,0,1)$ and $z = -9 < 22 = z^u$.

Step 2 (#2) $I_1 = \{3\}$, so $i(1) = i(2) = i(3) = 3$. Or, by (5).

$$v_1 = 20 \quad v_2 = 8 \quad v_3 = 0 \quad u_3 = 0$$

$$m_1 = 20 \quad \text{so } u_1 = 20 \quad \text{and } m_2 = 8 \quad \text{so } u_2 = 8$$

As $7 + (20 + 8 + 0) - 0 = 35 > 22$, z^u remains unchanged.

Step 3 (#2) The integer program is

$$\begin{array}{ll} \text{minimize} & z \\ \text{subject to} & z \geq 8 + 7x_1 + 7x_2 - 17x_3 \\ & z \geq 28 - 53x_1 - 17x_2 + 7x_3 \\ \text{and} & x_i = 0 \text{ or } 1 \quad (i = 1,2,3). \end{array}$$

The optimal solution, found by complete enumeration (see Table 10.1), is $\mathbf{x} = (1,0,1)$ and $z^l = -2 < 22 = z^u$. Notice that the second inequality is not binding when $\mathbf{x} = (1,0,1)$.[11]

Step 2 (#3) $I_1 = \{1,3\}$, so $i(1) = 1, i(2) = 3$, and $i(3) = 3$. Or, using (5), $v_1 = 0$, $v_2 = 8, v_3 = 0, u_1 = 0, u_2 = 8$, and $u_3 = 0$. As $(7+7) + (0+8+0) - 0 = 22$, z^u remains at 22.

Step 3 (#3) The integer program is

$$
\begin{array}{llrrrrr}
\text{minimize} & z \\
\text{subject to} & z \geq & 8 + & 7x_1 + & 7x_2 & - & 17x_3 \\
& z \geq & 28 - & 53x_1 & - 17x_2 & + & 7x_3 \\
& z \geq & 8 + & 7x_1 & - 17x_2 & + & 7x_3 \\
\text{and} & x_i = 0 \text{ or } 1 & (i = 1,2,3).
\end{array}
$$

The optimal solution is $\mathbf{x} = (1,1,1)$ and $z^l = 5 < 22$ (see Table 10.1).

Step 2 (#4) $I_1 = \{1,2,3\}$, so $i(1) = 1, i(2) = 2$, and $i(3) = 3$. Or, by (5), $v_1 = 0, v_2 = 0, v_3 = 0, u_1 = 0, u_2 = 0, u_3 = 0$. As $(7+7+7) + (0+0+0) - 0 = 21 < 22$. z^u is reduced to 21.

Step 3 (#4) The integer program is

$$
\begin{array}{llrrrrr}
\text{minimize} & z \\
\text{subject to} & z \geq & 8 + & 7x_1 + & 7x_2 & - & 17x_3 \\
& z \geq & 28 - & 53x_1 & - 17x_2 & + & 7x_3 \\
& z \geq & 8 + & 7x_1 & - 17x_2 & + & 7x_3 \\
& z \geq & & 7x_1 & + 7x_2 & + & 7x_3 \\
\text{and} & x_i = 0 \text{ or } 1 & (i = 1,2,3).
\end{array}
$$

As indicated in Table 10.1, there are two optimal solutions: $\mathbf{x} = (1,0,0)$ and $\mathbf{x} = (0,1,0)$ with $z^l = 15 < 21 = z^u$; $\mathbf{x} = (1,0,0)$ is selected.

Step 2 (#5) $I_1 = \{1\}$, so $i(1) = i(2) = i(3) = 1$. Or, by (5), $v_1 = 0, v_2 = 12, v_3 = 20, u_1 = 0, u_2 = 12$, and $u_3 = 20$. As $7 + (0+12+20) - 0 = 39 > 21$, z^u remains at 21.

---

[11]If not binding inequalities are dropped and the constraints $z \geq z^l$ are not introduced, the algorithm may not converge (see Salkin [506]).

x	Integer Program at Iteration						
	#1	#2	#3	#4	#5	#6	#7
$1,0,0$	15	15	15	(15)	39	39	39
$0,1,0$	15	15	15	15	(15)	27	27
$0,0,1$	(-9)	35	35	35	35	35	35
$1,1,0$	22	22	22	22	22	22	22
$1,0,1$	-2	(-2)	22	22	22	22	22
$0,1,1$	-2	18	18	18	18	(18)	26
$1,1,1$	5	5	(5)	21	21	21	(21)

Table 10.1: Minimum z values

Step 3 (#5) The integer program is

$$
\begin{aligned}
\text{minimize} \quad & z \\
\text{subject to} \quad & z \geq 8 + 7x_1 + 7x_2 - 17x_3 \\
& z \geq 28 - 53x_1 - 17x_2 + 7x_3 \\
& z \geq 8 + 7x_1 - 17x_2 + 7x_3 \\
& z \geq 7x_1 + 7x_2 + 7x_3 \\
& z \geq 32 + 7x_1 - 29x_2 - 53x_3 \\
\text{and} \quad & x_i = 0 \text{ or } 1 \quad (i = 1,2,3).
\end{aligned}
$$

The optimal solution is $\mathbf{x} = (0,1,0)$ and $z^l = 15 < 21 = z^u$.

Step 2 (#6) $I_1 = \{2\}$, so $i(1) = i(2) = i(3) = 2$. Or, by (5), $v_1 = 12, v_2 = 0, v_3 = 8, u_1 = 12, u_2 = 0$, and $u_3 = 8$. Since $7 + (12 + 8) = 27 > 21$, z^u is unchanged.

Step 3 (#6) The integer program is the same as the one in Step 3 (#5), except that the inequality

Iteration #	Step 2: Linear Program						Obj func.	Step 3: Linear Program								
	u_1	u_2	u_3	v_1	v_2	v_3		z^u	$z \geq ()$	$+()x_1$	$+()x_2$	$+()x_3$	x_1	x_2	x_3	z^l
Initialization								$+\infty$								$-\infty$
#1	0	0	8	0	0	8	22	22	8	7	7	-17	1	1	0	-9
#2	20	8	0	20	8	0	35	22	28	-53	-17	7	0	0	1	-2
#3	0	8	0	0	8	0	22	22	8	7	-17	7	1	0	1	5
#4	0	0	0	0	0	0	21	21	0	7	7	7	1	1	0	15
#5	0	12	20	0	12	20	39	21	32	7	-29	-53	0	0	0	15
#6	12	0	8	12	0	8	27	21	20	-29	7	-17	0	1	1	18
#7	12	0	0	12	0	0	26	21	12	-29	7	7	1	1	1	21

Table 10.2: Summary of Computations

$$z \geq 20 - 29x_1 + 7x_2 - 17x_3 \qquad (a)$$

is added to the constraint set. The optimal solution is $\mathbf{x} = (0, 1, 1)$ and $z^l = 18 < 21 = z^u$.

Step 2 (#7) $I_1 = \{2, 3\}$, so $i(1) = 2, i(2) = 2$, and $i(3) = 3$. Or, by (5), $v_1 = 12, v_2 = 0, v_3 = 0, u_1 = 12, u_2 = 0$, and $u_3 = 0$. Since $(7 + 7) + 12 = 26 > 21$, z^u remains at 21.

Step 3 (#7) The integer program is as in Step 3 (#5) with inequality (a) and also the new one

$$z \geq 12 - 29x_1 + 7x_2 + 7x_3 \qquad (b)$$

The optimal solution is $\mathbf{x} = (1, 1, 1)$ and $z^l = 21$. As $z^l = z^u = 21, \mathbf{x} = (1, 1, 1)$ is optimal.

The computations are summarized in Tables 10.1 and 10.2. In Table 10.2 a row gives (from left to right) the iteration number (where an iteration is an execution of Steps 2 and 3), the optimal solution $(\mathbf{u}, \mathbf{v})$ to the current linear program [DL] (determined using (5)), the optimal value of its objective function, the best upper bound, the coefficients of the inequality constraint for the integer program generated by the optimal $(\mathbf{u}, \mathbf{v})$, the optimal solution $\mathbf{x}$ to the current integer program (found from Table 10.1), and the optimal value of its objective function. In the first table an entry z_{ij} is the smallest value z can have in the integer program at iteration j when $\mathbf{x}$ is fixed at the value indicated in row i. A circled entry corresponds to the optimal solution to the integer program at iteration j. Observe that if for a given $i, z_{ij} = z_{i,j+1}$, we have that the most recently generated z inequality had no effect on the minimal value of z for the zero-one $\mathbf{x}$ corresponding to row i. Thus, column $j + 1$ may be found using column j and the new z inequality. In particular, if the inequality indicates that the minimum value of z has to be larger than $z_{i,j}$, we set $z_{i,j+1}$ equal to the smallest number so that equality holds in this constraint. Otherwise $z_{i,j+1}$ is set to z_{ij}. Of course, in computations a table such as this can rarely be stored.

Finally, notice that for this problem, Benders' algorithm worked very poorly. As indicated in Table 10.2, all $2^n - 1$ possible zero-one vectors $\mathbf{x}$ $(\mathbf{x} \neq 0)$ were enumerated. However, Geoffrion and Graves [228] presents some encouraging results of Benders' algorithm applied to a more general variation of plant location problem.

Chapter 11

Group Theory in Integer Programming

11.1 Introduction

In 1960 Ralph Gomory [276] showed that any integer program may be solved by first solving its linear program, adjoining a new constraint which cuts off the current linear programming solution, reoptimizing, and so on. This process, referred to as a fractional cutting plane algorithm, was extensively investigated in Chapter 4. In the same article Gomory indicated that the coefficient vectors of the derived inequalities form a finite set which is closed under the operation of addition when the arithmetic operations are taken modulo 1 (i.e., integer parts are dropped). Such a set forms what is called a *group*. Furthermore, as the group described in the Gomory article is commutative,[1] it is termed an *abelian group*. Gomory [278] also explained that this group can have at most D elements, where D is the absolute value of the determinant of the current linear programming basis, and that it can often be generated by one element, in which case the group is called a *cyclic group*.[2] Using these results, Gomory [278] in 1965 showed that by relaxing nonnegativity, but not integrality constraints on certain variables, an integer program may be transformed to one whose columns of constraint coefficients and the right-hand side are elements of an abelian group. If this group problem is solved and its solution yields nonnegative values for the variables of the original problem, then we can show that the integer program has been solved. A sufficient condition which guarantees that this will always occur, as well as an algorithm for solving the group problem, is introduced in [278]. A later paper of

---

[1] That is, for all elements g_1 and g_2 in the group, $g_1 + g_2 = g_2 + g_1$.

[2] If g generates the group, then repeatedly adding g to itself causes all elements of the group to appear.

Gomory [279] discusses the geometry of the group problem's feasible region and indicates that its bounding (or defining) hyperplanes will yield cutting planes which are often very strong. Both articles ([278], [279]) are greatly expanded upon in [280]. The Gomory papers [278], [279], [280] indicated the need to devise algorithms which solve the group problem, and to further explore the bounding hyperplanes or *faces* of the polyhedron corresponding to the group problem's feasible region.

We will show that the group problem can be treated as an integer program with one constraint[3] and also as a network problem, where a shortest route is desired. Algorithms designed to solve the integer programming formulations are usually of the dynamic programming type.[4] There are several techniques for solving a standard shortest route problem (e.g., [171],[195],[202],[336],[601]). However, as we shall see, the network which is derived from the group problem has special structure, and algorithms which exploit it are described by Hu [335], Shapiro [531], Chen and Zionts [129], and Morito [434]. The use of enumerative algorithms is discussed by Glover [241], Glover and Litzler [265], Gorry and Shapiro [287], and by Shapiro[532].

As the number of elements in the group increases, the algorithms naturally become substantially less efficient. Papers which describe techniques that may allow solving smaller group problems include those by Gorry, Shapiro, and Wolsey [288], Wolsey [603], [604], Shapiro [532], and Hefley and Thomas [317].

The polyhedron which corresponds to the group problem's feasible region has drawn considerable interest. Principally, the investigations are concerned with characterizing and generating the faces of the polyhedron, since, as we will show, they will yield strong and thus useful inequalities. Research in this direction includes the work of Gomory [279], [280], [281], White [599], Glover [241], [249], [250], Devine and Glover [170], and Rubin and Links [494]. Gomory and Johnson [284], [285] have shown that the faces previously generated can be used to give valid inequalities for any integer program. The use of these inequalities in a branch and bound and in a search enumeration algorithm has been investigated by Johnson and Spielberg [354] and by Guignard and Spielberg [300].

Special attention has been given to the zero-one integer program. In particular, Balas [43] has shown that the sufficient condition proposed

---

[3]Referred to as a knapsack problem.

[4]Network problems and dynamic programming are discussed in later sections.

by Gomory [278], which guarantees that the optimal solution to the group problem produces an optimal solution to the integer program, is never satisfied when the variables are zero-one. A dynamic programming algorithm for the group problem representing the zero-one integer program may be found in White [599], and a branch and bound algorithm for the zero-one problem is described by Shapiro [530] (Problem 11.73).

A problem similar to the group problem may also be constructed from a mixed integer program. A discussion of the problem and its polyhedron appears in Wolsey [603], [605], in Gomory and Johnson [284],[285], and in Johnson [352].

Several classical integer programming models have been studied in a context of specially structured algorithms for their equivalent group problem. The plant location problem has been treated by Wolsey [603], and another fixed charge problem by Kennington and Unger [364] and by Tompkins and Unger [570]. White [599] and Mathur and Salkin [419] discuss the case where multiple choice constraints are present, and the traveling salesman problem is considered by Baxter [58].[5]

Computational experience with algorithms for the group problem is reported by Chen and Zionts [129], Hefley and Thomas [317], Shapiro [532], White [599], and more recently by Morito [434]. A summary of these results appears in Appendix C.

Other related articles include the surveys by Kortanek and Jeroslow [375], Chen and Zionts [129], and the discussion by Burdet [115] which shows that cuts obtained from a purely geometrical analysis (Chapter 4, Appendix C) and those obtained utilizing group theory are essentially drawn from the same theoretical foundation. In the next section we introduce the group problem. The following part presents algorithms for this problem and several examples are solved. A network formulation is also given. Certain properties of the group are then discussed which allows us to derive a second equivalent group problem. The subsequent section discusses the polyhedron which represents its feasible region. The last part shows how the faces of related polyhedra can yield valid inequalities for any integer program.

---

[5]The fixed charge and plant location problems are considered in Chapter 14. Some of the specializations are also discussed in the problems.

11.2   The Group Minimization Problem

Consider the integer program

P1   maximize $\mathbf{c'x'}$

subject to $\mathbf{A'x'} \leq \mathbf{b}$

$\mathbf{x'} \geq \mathbf{0}$ and integer,

where $\mathbf{A'}$ is an m by n integer matrix, $\mathbf{c'}$ is an n integer vector, and $\mathbf{b}$ is an m integer vector. After adding m nonnegative slack variables $\mathbf{s}$, one to each inequality, we have the equivalent problem

P2   maximize $\mathbf{cx}$

subject to $\mathbf{Ax} = \mathbf{b}$

$\mathbf{x} \geq \mathbf{0}$ and integer,

where $\mathbf{A} = (\mathbf{A'}, \mathbf{I})$ is an m by $m + n$ matrix whose rank is m, $\mathbf{I}$ is an m by m identity matrix, $\mathbf{c} = (\mathbf{c'}, \mathbf{0})$, and $\mathbf{x} = (\mathbf{x'}, \mathbf{s})$. Now suppose $\mathbf{B}$ is a basis whose columns are from $\mathbf{A}$, and that we rearrange the terms in the previous problem so that it is the same as

P3   maximize $\mathbf{c_B x_B} \quad + \quad \mathbf{c_N x_N}$

subject to $\mathbf{Bx_B} \quad + \quad \mathbf{Nx_N} = \mathbf{b}$ (1)

$\mathbf{x_B} \geq 0, \quad \mathbf{x_N} \geq 0$ and integer,

where $\mathbf{x_B}$ ($\mathbf{x_N}$) are the basic (nonbasic) variables associated with $\mathbf{B}$ ($\mathbf{N}$), and the costs corresponding to the basic (nonbasic) variables are $\mathbf{c_B}$ ($\mathbf{c_N}$). As $\mathbf{B}$ is a basis, it has an inverse, and thus in $\mathbf{Bx_B} + \mathbf{Nx_N} = \mathbf{b}$ we may solve for $\mathbf{x_B}$ to obtain

$$\mathbf{x_B} = \mathbf{B^{-1}b} - \mathbf{B^{-1}Nx_N} \qquad (2)$$

Therefore, problem P3 (and hence the original problem P1) is equivalent to

P4   maximize $\mathbf{c_B B^{-1} b} - (\mathbf{c_B B^{-1} N} - \mathbf{c_N}) \mathbf{x_N}$

$$\text{subject to} \quad \mathbf{x_B} \;=\; \mathbf{B^{-1}b} - \mathbf{B^{-1}Nx_N}$$

$$\mathbf{x_B} \;\geq\; \mathbf{0} \text{ and integer} \tag{3}$$

$$\mathbf{x_N} \;\geq\; \mathbf{0} \text{ and integer.}$$

Now the constant term $\mathbf{c_B B^{-1} b}$ does not influence the maximization, so it can be dropped. Furthermore, $\mathbf{x_B}$ integer is equivalent to saying that $\mathbf{x_B} \equiv \mathbf{0}$ modulo 1.[6] Or, by (2)

$$\mathbf{B^{-1}b} - \mathbf{B^{-1}Nx_N} \equiv \mathbf{0} \pmod 1$$

which is the same as

$$\mathbf{B^{-1}Nx_N} \equiv \mathbf{B^{-1}b} \pmod 1. \tag{4}$$

Substituting (4) for (3) in problem P4 and changing the maximization to a minimization yields

$$\text{P5} \quad \text{minimize} \quad (\mathbf{c_B B^{-1} N} - \mathbf{c_N})\mathbf{x_N}$$
$$\text{subject to} \quad \mathbf{B^{-1}Nx_N} \equiv \mathbf{B^{-1}b} \qquad \pmod 1$$
$$\mathbf{x_N} \geq \mathbf{0}, \text{ integer}$$
$$\text{and} \quad \mathbf{x_B} = \mathbf{B^{-1}b} - \mathbf{B^{-1}Nx_N} \geq \mathbf{0}.$$

Notice that if $\mathbf{x_N}$ satisfies (4), it yields an integral $\mathbf{x_B}$ which satisfies (1) and vice versa. Thus we may solve the original problem P1 by solving P5. Also, note that the costs in problem P5 are of the form $\bar{c}_j = \mathbf{c_B B^{-1} a_j} - c_j$, where $\mathbf{a_j}$ is some column from $\mathbf{A}$ in $\mathbf{N}$ and c_j is the corresponding cost. So the $\bar{c}_j$ are just the reduced costs or the values of the dual surplus variables.[7] Suppose the basis $\mathbf{B}$ yields a dual feasible

---

[6]Given two numbers a and b, a is said to be congruent to b modulo K, written $a \equiv b \pmod K$, if there exists an integer i such that $a - b = iK$. Two vectors of equal size are congruent $(\bmod K)$ if and only if their corresponding elements are congruent $(\bmod K)$. Thus $\mathbf{x_B} \equiv \mathbf{0}$ modulo 1, if and only if $\mathbf{x_B}$ is an integer vector.

[7]Taking the dual of the linear program of problem P1 gives: Minimize $\mathbf{wb}$, subject to $\mathbf{wA} \geq \mathbf{c}$ and $\mathbf{w} \geq \mathbf{0}$. For the primal solution $\mathbf{x_B} = \mathbf{B^{-1}b}$, the corresponding dual solution is $\mathbf{w} = \mathbf{c_B B^{-1}}$, the value of the jth surplus variable w_{sj} in $\mathbf{wA} \geq \mathbf{c}$ is $w_{sj} = \mathbf{wa_j} - c_j = \mathbf{c_B B^{-1} a_j} - c_j$ when $\mathbf{w} = \mathbf{c_B B^{-1}}$.

solution, that is, $\bar{\mathbf{c}} = (\bar{c}_j) = \mathbf{c_B} \mathbf{B}^{-1}\mathbf{N} - \mathbf{c_N} \geq \mathbf{0}$. Also, suppose we *drop the nonnegativity requirements* on $\mathbf{x_B}$. Then problem P5 becomes

> P6   minimize $\mathbf{cx_N}$
>
> subject to $\mathbf{B}^{-1}\mathbf{Nx_N} \equiv \mathbf{B}^{-1}\mathbf{b} \pmod{1}$
>
> and $\mathbf{x_N} \geq \mathbf{0}$, integer.

Notice that $\mathbf{B}^{-1}\mathbf{Nx_N}$ is equivalent to $\sum_{j=1}^n \boldsymbol{\alpha}_j x_{J(j)}$, where $\boldsymbol{\alpha}_j$ is the jth column of $\mathbf{B}^{-1}\mathbf{N}$ and $x_{J(j)}$ is the jth $(1 \leq j \leq n)$ nonbasic variable. Let us denote $\mathbf{B}^{-1}\mathbf{b}$ by the column $\boldsymbol{\alpha}_o$, then the congruence relationship becomes

$$\sum_{j=1}^n \boldsymbol{\alpha}_j x_{J(j)} \equiv \boldsymbol{\alpha}_o \pmod{1}. \tag{5}$$

As two vectors are congruent (modulo 1) if and only if the corresponding elements are congruent (modulo 1) we actually have m congruence equations in (5), thus we may add or subtract multiples of $x_{J(j)} \equiv 0 \pmod{1}$ to each equation without destroying the congruence relationship so that every column $\boldsymbol{\alpha}_j$ has nonnegative entries less than one. Similarly, the elements in $\boldsymbol{\alpha}_o$ may be reduced to nonnegative fractions by adding or subtracting multiples of $0 \equiv 1 \pmod{1}$ to each component. We denote the columns that have the fractional parts as $\bar{\boldsymbol{\alpha}}_j$ $(j = 0, 1, \ldots, n)$. As a result, problem P6 is equivalent to

P7 minimize $\displaystyle\sum_{j=1}^n \bar{c}_j x_{J(j)}$

subject to $\displaystyle\sum_{j=1}^n \bar{\boldsymbol{\alpha}}_j x_{J(j)} \equiv \bar{\boldsymbol{\alpha}}_0 \pmod{1}$

$x_{J(j)} \geq 0$ and integer $(j = 1, \ldots, n),$

$\left.\rule{0pt}{5em}\right\}$ Group minimization problem (GMP)

where $\bar{c}_j \geq 0$ and each column $\bar{\boldsymbol{\alpha}}_j$ $(j = 0, 1, \ldots, n)$ satisfies $\mathbf{0} \leq \bar{\boldsymbol{\alpha}}_j < \mathbf{e}$ (a column of ones). Problem P7 is sometimes referred to as the *group minimization problem* (GMP). To solidify the discussion we derive a GMP.

Example 11.1 (Gomory [280]) [8] Consider the integer program

$$P1 \quad \text{maximize} \quad 2x_1 + x_2 + 4x_3 + 3x_4 + x_5$$

$$\text{subject to} \qquad 2x_2 + x_3 + 4x_4 + 2x_5 \leq 41 \ (x_6)$$

$$3x_1 - 4x_2 + 5x_3 + x_4 - x_5 \leq 47 \ (x_7)$$

$$x_j \geq 0 \text{ and integer } (j = 1, \ldots, 5),$$

where x_6 and x_7 are the corresponding slack variables. When P1 is solved as a linear program, the basic variables are x_1 and x_2. So suppose we choose

$$\mathbf{B} = \begin{pmatrix} 0 & 2 \\ 3 & -4 \end{pmatrix}$$

which, of course, yields nonnegative reduced costs.[9] After adding slack variables x_6 and x_7 we have

$$\mathbf{B}^{-1} = \begin{pmatrix} \frac{4}{6} & \frac{2}{6} \\ \frac{3}{6} & 0 \end{pmatrix}, \quad \mathbf{B}^{-1}\mathbf{b} = \begin{pmatrix} \frac{4}{6} & \frac{2}{6} \\ \frac{3}{6} & 0 \end{pmatrix} \begin{pmatrix} 41 \\ 47 \end{pmatrix} = \begin{pmatrix} 43 \\ 20\frac{3}{6} \end{pmatrix} = \alpha_o,$$

$$\mathbf{B}^{-1}\mathbf{N} = \begin{pmatrix} \frac{4}{6} & \frac{2}{6} \\ \frac{3}{6} & 0 \end{pmatrix} \begin{pmatrix} 1 & 4 & 2 & 1 & 0 \\ 5 & 1 & -1 & 0 & 1 \end{pmatrix}$$

$$= \begin{pmatrix} 2\frac{2}{6} & 3 & 1 & \frac{4}{6} & \frac{2}{6} \\ \frac{3}{6} & 2 & 1 & \frac{3}{6} & 0 \end{pmatrix} = (\alpha_j),$$

$$\mathbf{X_N} = (x_3, x_4, x_5, x_6, x_7) = (x_{J(j)}),$$

$$\mathbf{c_B}\mathbf{B}^{-1}\mathbf{b} = (2, 1) \begin{pmatrix} 43 \\ 20\frac{3}{6} \end{pmatrix} = 106\frac{3}{6},$$

---

[8]This is not quite the example appearing in the Gomory article; $a_{23} = 5$ is replacing $a_{23} = 4$, and $c_3 = 4$ is replacing $c_3 = 1$.

[9]We shall see that it seems best, in the sense that the basic variables are more likely to be nonnegative, to choose a basis which yields an *optimal* linear programming solution.

and

$$c_B B^{-1} N - c_N = (2,1) B^{-1} N - c_N = (\tfrac{7}{6}, 5, 2, \tfrac{11}{6}, \tfrac{4}{6});$$

so problem P6 is

minimize $\quad \tfrac{7}{6}\ x_3 + \ 5\ x_4 + \ 2\ x_5 + \tfrac{11}{6}\ x_6 + \tfrac{4}{6}\ x_7$

subject to $\begin{pmatrix} 2\tfrac{2}{6} \\ \\ \tfrac{3}{6} \end{pmatrix} x_3 + \begin{pmatrix} 3 \\ \\ 2 \end{pmatrix} x_4 + \begin{pmatrix} 1 \\ \\ 1 \end{pmatrix} x_5 + \begin{pmatrix} \tfrac{4}{6} \\ \\ \tfrac{3}{6} \end{pmatrix} x_6 + \begin{pmatrix} \tfrac{2}{6} \\ \\ 0 \end{pmatrix} x_7 \equiv \begin{pmatrix} 43 \\ \\ 20\tfrac{3}{6} \end{pmatrix}$ (mod 1)

$x_3, x_4, x_5, x_6, x_7 \geq 0$ and integer.

Hence, the group minimization problem P7 is

minimize $\quad \tfrac{7}{6}\ x_{J(1)} + \ 5\ x_{J(2)} + \ 2\ x_{J(3)} + \tfrac{11}{6}\ x_{J(4)} + \tfrac{4}{6}\ x_{J(5)}$

subject to $\begin{pmatrix} \tfrac{2}{6} \\ \\ \tfrac{3}{6} \end{pmatrix} x_{J(1)} + \begin{pmatrix} 0 \\ \\ 0 \end{pmatrix} x_{J(2)} + \begin{pmatrix} 0 \\ \\ 0 \end{pmatrix} x_{J(3)} + \begin{pmatrix} \tfrac{4}{6} \\ \\ \tfrac{3}{6} \end{pmatrix} x_{J(4)} + \begin{pmatrix} \tfrac{2}{6} \\ \\ 0 \end{pmatrix} x_{J(5)} \equiv \begin{pmatrix} 0 \\ \\ \tfrac{3}{6} \end{pmatrix}$ (mod 1)

$x_{J(1)}, x_{J(2)}, x_{J(3)}, x_{J(4)}, x_{J(5)} \geq 0$ and integer.

11.2.1   The Group $G(\bar{\alpha})$

The example suggests several important results. Note that $D = |\text{determinant } \mathbf{B}| = 6$ and that every column $\bar{\alpha}_j$ has elements of the form I_j/D, where I_j is a nonnegative integer less than D. In order to prove that this is always the case, we proceed as in Section 4.6 (Chapter 4), which explains why every element in problem P6 may be written as an integer divided by D. In fact, a careful examination of the derivation of problem P7 reveals that the coefficients in each congruence equation are precisely those that appear in the Gomory fractional cut (which are either zero or proper fractions, see Chapter 4), where the source row corresponds to the row in which the basic variable appeared (Problem 11.15).

To see this in our example, first note that $x_B = B^{-1}b - B^{-1}Nx_N$; thus, we have

	1	$-x_{J(1)}$	$-x_{J(2)}$	$-x_{J(3)}$	$-x_{J(4)}$	$-x_{J(5)}$	
source row: $x_{B_1} = x_1 =$	43	$2\frac{2}{6}$	3	1	$\frac{4}{6}$	$\frac{2}{6}$	
Gomory cut: $s_1 =$	0	$\frac{2}{6}$	0	0	$\frac{4}{6}$	$\frac{2}{6}$	≥ 0
source row: $x_{B_2} = x_2 =$	$20\frac{3}{6}$	$\frac{3}{6}$	0	1	$\frac{3}{6}$	0	
Gomory cut: $s_2 =$	$\frac{3}{6}$	$\frac{3}{6}$	0	0	$\frac{3}{6}$	0	≥ 0,

where s_1 and s_2 are nonnegative (Gomory) slack variables.

Now suppose we represent the components of $\bar{\alpha}_j$ $(j = 1, \ldots, n)$ as I_{ij}/D, where I_{ij} is an integer $(i = 1, \ldots, m)$ and $0 \leq I_{ij} < D$. Let us add column $\bar{\alpha}_r$ to column $\bar{\alpha}_s$, where the addition is taken modulo 1. That is, the components of the resulting vector are of the form $((I_{ir} + I_{is})/D) - [(I_{ir} + I_{is})/D]$, where $[y]$ means the largest integer $\leq y$. As the terms are either proper fractions or zero, the result is another vector whose components are (I_{ij}/D), where $0 \leq I_{ij} < D$. Also, observe that if $\bar{\alpha}_j$ is added (modulo 1) to itself enough times we will obtain the 0 vector, since after D additions its components are $D(I_{ij}/D) = I_{ij} \equiv 0 \pmod 1$ for $i = 1, \ldots, m$.[10] Thus we have shown that the set of columns of the form (I_j/D), where I_j is a nonnegative integer $< D$, is closed under addition (modulo 1). We shall denote the set of vectors generated by additions (modulo 1) of the n $\bar{\alpha}_j$'s in problem P7 as $G(\bar{\alpha})$. It is a simple matter to show that (additions are modulo 1)

i) if $\bar{\alpha}_r, \bar{\alpha}_s, \bar{\alpha}_t$, in $G(\bar{\alpha})$ then $(\bar{\alpha}_r + \bar{\alpha}_s) + \bar{\alpha}_t = \bar{\alpha}_r + (\bar{\alpha}_s + \bar{\alpha}_t)$.

ii) for any $\bar{\alpha}_r$ in $G(\bar{\alpha})$ there exists an element 0 in $G(\bar{\alpha})$ such that

---

[10]For convenience, x added to itself D times, each addition taken modulo K, will be written as $Dx \pmod K$ or just Dx. This is somewhat imprecise but should not create any ambiguity.

$$\bar\alpha_r + 0 = 0 + \bar\alpha_r = \bar\alpha_r.$$

iii) for every element $\bar\alpha_r$ in $G(\bar\alpha)$ there is an element $\bar\alpha_s$ in $G(\bar\alpha)$ such that $\bar\alpha_s + \bar\alpha_r = \bar\alpha_r + \bar\alpha_s = 0$.

A set such as $G(\bar\alpha)$, which is closed under addition (in our case, modulo 1) and satisfies (i), (ii), and (iii), is called a *group*.[11] Furthermore, as $\bar\alpha_r + \bar\alpha_s = \bar\alpha_s + \bar\alpha_r$ for any $\bar\alpha_r, \bar\alpha_s$ in the group, the addition is commutative and $G(\bar\alpha)$ is termed an *abelian group*. This explains why problem P7 is referred to as the group minimization problem.

To find the abelian group generated by the 5 $\bar\alpha_j$'s in our example, first consider multiples of $\bar\alpha_1$.

$$\lambda = 1 \quad \lambda = 2 \quad \lambda = 3 \quad \lambda = 4 \quad \lambda = 5 \quad \lambda = 6$$

$$\lambda\bar\alpha_1 = \begin{pmatrix} \frac{2}{6} \\ \frac{3}{6} \end{pmatrix} \begin{pmatrix} \frac{4}{6} \\ 0 \end{pmatrix} \begin{pmatrix} 0 \\ \frac{3}{6} \end{pmatrix} \begin{pmatrix} \frac{2}{6} \\ 0 \end{pmatrix} \begin{pmatrix} \frac{4}{6} \\ \frac{3}{6} \end{pmatrix} \begin{pmatrix} 0 \\ 0 \end{pmatrix}$$

$$\bar\alpha_1 \qquad\qquad \bar\alpha_0 \qquad \bar\alpha_5 \qquad \bar\alpha_4 \quad \bar\alpha_2 = \bar\alpha_3$$

Notice that $7\bar\alpha_1 = \bar\alpha_1$, $8\bar\alpha_1 = 2\bar\alpha_1, \ldots, (D + \lambda)\bar\alpha_1 = \lambda\bar\alpha_1$, so that the distinct elements in $G(\bar\alpha)$, generated by multiples of $\bar\alpha_1$ are the six listed above. Furthermore, for every $j, \bar\alpha_j = \lambda\bar\alpha_1$ for some λ $(1 \le \lambda \le 6)$, so that $G(\bar\alpha)$ is precisely these six elements. This follows because for nonnegative integers β and $\gamma, \beta\bar\alpha_j + \gamma\bar\alpha_k$ for any $j \ne k$ $(j, k = 2, 3, 4, 5)$ is the same as $\lambda_1\bar\alpha_1 + \lambda_2\bar\alpha_1 = \bar\lambda\bar\alpha_1$, where $1 \le \bar\lambda \le 6$, so that the other $\bar\alpha_j$'s $(j \ne 1)$ cannot yield additional elements. As $\bar\alpha_1$ generates the entire group, $G(\bar\alpha)$ is a *cyclic group*, where a cyclic group is a group containing at least one element which generates the entire group.

Suppose we also consider multiples of the other $\bar\alpha_j$'s. As $(D + \lambda)\bar\alpha_j = \lambda\bar\alpha_j$, we only need to list values for $\lambda = 1, 2, 3, 4, 5$, and 6.

	$\lambda = 1$	$\lambda = 2$	$\lambda = 3$	$\lambda = 4$	$\lambda = 5$	$\lambda = 6$
$\lambda\bar\alpha_2$ $\lambda\bar\alpha_3$	$\begin{pmatrix} 0 \\ 0 \end{pmatrix}$	$\begin{pmatrix} 0 \\ 0 \end{pmatrix}$	$\begin{pmatrix} 0 \\ 0 \end{pmatrix}$	$\begin{pmatrix} 0 \\ 0 \end{pmatrix}$	$\begin{pmatrix} 0 \\ 0 \end{pmatrix}$	$\begin{pmatrix} 0 \\ 0 \end{pmatrix}$

---

[11]A general definition of a group appears in Problem 11.1

$$\lambda\bar{\alpha}_4 \qquad \begin{pmatrix}\frac{4}{6}\\[2pt]\frac{3}{6}\end{pmatrix} \quad \begin{pmatrix}\frac{2}{6}\\[2pt]0\end{pmatrix} \quad \begin{pmatrix}0\\[2pt]\frac{3}{6}\end{pmatrix} \quad \begin{pmatrix}\frac{4}{6}\\[2pt]0\end{pmatrix} \quad \begin{pmatrix}\frac{2}{6}\\[2pt]\frac{3}{6}\end{pmatrix} \quad \begin{pmatrix}0\\[2pt]0\end{pmatrix}$$

$$\lambda\bar{\alpha}_4 \;=\; \bar{\alpha}_4 \qquad \bar{\alpha}_5 \qquad \bar{\alpha}_0 \qquad\qquad \bar{\alpha}_1 \qquad \bar{\alpha}_2 = \bar{\alpha}_3$$

$$\lambda\bar{\alpha}_5 \qquad \begin{pmatrix}\frac{2}{6}\\[2pt]0\end{pmatrix} \quad \begin{pmatrix}\frac{4}{6}\\[2pt]0\end{pmatrix} \quad \begin{pmatrix}0\\[2pt]0\end{pmatrix} \quad \begin{pmatrix}\frac{2}{6}\\[2pt]0\end{pmatrix} \quad \begin{pmatrix}\frac{4}{6}\\[2pt]0\end{pmatrix} \quad \begin{pmatrix}0\\[2pt]0\end{pmatrix}$$

Note that $\bar{\alpha}_4$ also generates $G(\bar{\alpha})$, and that $\bar{\alpha}_5$ produces only three distinct elements of $G(\bar{\alpha})$. The zero vector $\bar{\alpha}_2 = \bar{\alpha}_3$ yields only itself. Suppose we let $G(\bar{\alpha}_j)$ be the set of elements (vectors) generated by each $\bar{\alpha}_j$ $(j = 1,\ldots,n)$. Then it is a simple matter to show that $G(\bar{\alpha}_j)$ is a *cyclic subgroup* of $G(\bar{\alpha})$, meaning that $G(\bar{\alpha}_j)$ is a cyclic group whose elements are also in $G(\bar{\alpha})$. In our case.

$$G(\bar{\alpha}_2) = G(\bar{\alpha}_3) = \{0\} \qquad G(\bar{\alpha}_1) = G(\bar{\alpha}_4) = G(\bar{\alpha})$$

and

$$G(\bar{\alpha}_5) = \left\{ \begin{pmatrix} 2/6 \\ 0 \end{pmatrix} \begin{pmatrix} 4/6 \\ 0 \end{pmatrix} \begin{pmatrix} 0 \\ 0 \end{pmatrix} \right\}.$$

To determine whether an element $\bar{\alpha}_j$ generates the group $G(\bar{\alpha})$ we examine its components. As

$$\bar{\alpha}_j = \frac{1}{D} \begin{pmatrix} I_{1j} \\ I_{2j} \\ \vdots \\ I_{mj} \end{pmatrix},$$

where $0 \le I_{ij} < D$, it is a simple matter to show that:

i) If there is no common factor other than 1 between D and the nonzero I_{ij} $(i = 1, \ldots, m)$, then $\bar{\alpha}_j$ generates $G(\bar{\alpha})$, or $G(\bar{\alpha}_j) = G(\bar{\alpha})$.

ii) If there is a common factor, then the order of $G(\bar{\alpha}_j) = D/\lambda = d_j$, where λ is the greatest common divisor of $(D, I_{1j}, \ldots, I_{mj})$.

iii) The zero element can only generate the group consisting of the zero, and thus the order of $G(0)$ is always 1.

The proofs of (i), (ii), and (iii) follows directly from the modulo 1 addition and are left to Problem 11.9. Relating to our example:

j	$D\bar{\alpha}_j$	common factor λ	order of the subgroup d_j
1	$\begin{pmatrix} 2 \\ 3 \end{pmatrix}$	1	6
2,3	$\begin{pmatrix} 0 \\ 0 \end{pmatrix}$	−	1
4	$\begin{pmatrix} 4 \\ 3 \end{pmatrix}$	1	6
5	$\begin{pmatrix} 2 \\ 0 \end{pmatrix}$	2	$6/2 = 3$

We have demonstrated that either an element $\bar{\alpha}_j$ generates the whole group–in which case $G(\bar{\alpha})$ is cyclic–or that each element generates a subgroup $G(\bar{\alpha}_j)$ of order d_j. If there is no element $\bar{\alpha}_j$ which can generate all of $G(\bar{\alpha})$, it means that some elements of $G(\bar{\alpha})$ can only be obtained by adding nonzero terms contained in certain cyclic subgroups $G(\bar{\alpha}_j)$, where these subgroups have no elements in common except the zero. (Otherwise there is a cycle subgroup $G(\bar{\alpha}_j) = G(\bar{\alpha})$.) Therefore, in this situation, every nonzero element in $G(\bar{\alpha})$ can be expressed as a sum of one or more nonzero elements, each in cyclic subgroups $G(\bar{\alpha}_j)$ which are disjoint except for the zero element. When every nonzero element in a group can be written this way, we say the group can be written as a *direct sum of cyclic subgroups*: where "direct" infers that the subgroups have only the zero element in common. Thus we may say that *either $G(\bar{\alpha})$ is a cyclic group or it can be expressed as a direct sum of cyclic subgroups*. As there are d_j terms in each subgroup, there are at most $\Pi_j d_j$ different possible group elements, where the product is over those j such that $G(\bar{\alpha}_j)$ sum to $G(\bar{\alpha})$. In a later section (Theorem 11.8), we will show that either

$$\prod_j d_j = D$$

or

$$\prod_j d_j \text{ divides}^{12} D.$$

That is, the order of $G(\bar{\alpha})$ equals D or it divides D. Also, as we shall see, it equals D whenever **A** contains an identity matrix. For ease of exposition we are assuming that, unless otherwise specified, **A** contains an identity matrix.

The previous example also suggests another interesting and useful result. In particular, note that $\bar{\alpha}_0$ is in $G(\bar{\alpha})$, since $\bar{\alpha}_0 = 3\bar{\alpha}_1 = 3\bar{\alpha}_4$. In general, $\bar{\alpha}_0$ in $G(\bar{\alpha})$ means that there exists a set of n integers β_j $(0 \leq \beta_j < D)$ such that

$$\bar{\alpha}_0 = \sum_{j=1}^n \beta_j \bar{\alpha}_j.$$

As $x_{J(j)} = \beta_j$ $(j = 1,\ldots,n)$ is a feasible solution to the GMP, we may say that the GMP has at least one solution if and only if $\bar{\alpha}_0$ is in $G(\bar{\alpha})$. To see this using our example, observe that $\bar{\alpha}_0$ is not in $G(\bar{\alpha}_5)$. So suppose $G(\bar{\alpha}) = G(\bar{\alpha}_5)$; that is, the original integer program only consisted of variables x_1, x_2, x_4, x_5, and x_7. Then the congruence relationship in the GMP is:

$$\begin{pmatrix} 0 \\ 0 \end{pmatrix} x_4 + \begin{pmatrix} 0 \\ 0 \end{pmatrix} x_5 + \begin{pmatrix} \frac{2}{6} \\ \frac{6}{0} \end{pmatrix} x_7 \equiv \begin{pmatrix} 0 \\ \frac{3}{6} \end{pmatrix} \pmod{1},$$

---

^{12}By a divides b, written $a|b$, we mean that b/a is an integer.

which can never be satisifed, and the integer program has no integer solution. Note that without the variable x_3 and the slack variable x_6, the first constraint in the original problem becomes $2x_2 + 4x_4 + 2x_5 = 41$, which can never be satisifed with $x_2, x_4,$ and x_5 integer. (Why?)

11.2.2   Solving the GMP

The objective function of the integer program is equivalent to $c_B B^{-1} b +$ minimum $\bar{c} x_N$, where $\bar{c} \geq 0$. Since $x_B = B^{-1} b - B^{-1} N x_N$, we may view $B^{-1} N x_N$ as a correction factor in the sense that we wish to increase the components of x_N from their current value of zero by integer amounts so that x_B, currently equal to $B^{-1} b$, becomes integer. As $\bar{c} \geq 0$, we would like to make this increase as small as possible: equivalently we would like to minimize $\bar{c} x_N$ while maintaining a feasible solution to the GMP. Thus *the group minimization, when solved, yields the least expensive correction that makes* x_B *integer.* So if $x_B = B^{-1} b - B^{-1} N x_N$ turns out to be nonnegative, we have an optimal solution to the original integer program. Note that if $x_B = B^{-1} b$ is already integer, $\bar{\alpha}_0 = 0$, and an optimal solution to the GMP is $x_N = 0$. This means that the linear program has solved the integer program provided that $x_B \geq 0$. Because of this, and since, in general, we wish to have $B^{-1} b - B^{-1} N x_N \geq 0$, it seems natural to select B to be an optimal linear programming basis.

   The previous discussion suggests that it is worth devising algorithms to solve the GMP and hope x_B turns out to be nonnegative. In the next section we present techniques for solving problem P7. In addition, it is later shown that if x_B is nondegenerate (i.e., $x_B = B^{-1} b > 0$) and the components of b are sufficiently large, then $x_B = B^{-1} b - B^{-1} N x_N \geq 0$ for all feasible solutions x_N to the GMP. Moreover, if x_B has negative components we shall devise a procedure which finds a least expensive correction which gives a feasible solution x_N to the GMP so that $x_B \geq 0$. Therefore, we may always solve the integer program by finding a certain feasible solution to its equivalent group minimization algorithm. Keep in mind that the GMP is identical to the integer program except for the relaxation of the nonnegativity constraints on the basic variables.

11.3   Solving the Group Minimization Problem

As we have just pointed out, it is desirable to solve the GMP and from its minimal solution find the basic variables. If they turn out to be nonnegative we have solved the integer program. Recall that the GMP is

$$\text{minimize} \quad \sum_{j=1}^{n} \bar{c}_j x_{J(j)}$$

$$\text{subject to} \quad \sum_{j=1}^{n} \bar{\alpha}_j x_{J(j)} = \bar{\alpha}_0$$

$$x_{J(j)} \geq 0 \text{ and integer} \quad (j = 1, \dots, n),$$

where $\bar{c}_j \geq 0$, $x_{J(j)}$ is the jth nonbasic variable, $\bar{\alpha}_j$ $(j = 0, 1, \dots, n)$ is a column whose entries are 0 or proper fractions, and the summation is taken (modulo 1)[13]. The coefficient columns $\bar{\alpha}_j$ $(j \geq 1)$ are just the columns of $\mathbf{B}^{-1}\mathbf{N}$ reduced modulo 1. Thus, if the difference of two columns $\bar{\alpha}_r$ and $\bar{\alpha}_s$ in $\mathbf{B}^{-1}\mathbf{N}$ is an integer column, then $\bar{\alpha}_r$ and $\bar{\alpha}_s$ are the same. As $\bar{c}_r$ and $\bar{c}_s$ are nonnegative, this means that either $x_{J(r)}$ or $x_{J(s)}$, depending on which has the larger cost, will be 0 in any minimal solution to the GMP. Thus the column and variable with the greater cost may be dropped. If $\bar{c}_r = \bar{c}_s$, we may arbitrarily select the column and variable to be omitted.[14] Also suppose that several columns of $\mathbf{B}^{-1}\mathbf{N}$ are integer; then each of these becomes a column of zeroes in the group problem. As a zero column does not contribute to the right-hand side $\bar{\alpha}_0$, and the costs are nonnegative, we may delete all zero columns. Note that this implies that all nonbasic variables associated with integer columns of $\mathbf{B}^{-1}\mathbf{N}$ are 0 in any minimal solution to the GMP. Also, remember that $\bar{\alpha}_0$ is not $\mathbf{0}$ since $\mathbf{x_B} = \mathbf{B}^{-1}\mathbf{b}$ is not all integer. To summarize the discussion we have the following result.

Theorem 11.1 *If $\bar{c} \geq 0$, then, without loss of generality, we may assume when solving the GMP that the columns $\bar{\alpha}_j$ $(j = 1, \dots, n)$ are distinct and none are equal to zero.*

---

[13]This is identical to saying $\sum_j \bar{\alpha}_j x_{J(j)} \equiv \bar{\alpha}_0 \pmod{1}$.

[14]This is not quite true. It may be computationally preferable to keep a particular one; see Problem 11.14.

For ease of exposition, we shall, prior to solving the GMP, always drop the higher cost duplicate columns and those of all zeroes. However, for simplicity, we will always index j in the GMP from 1 to n.

11.3.1   Dynamic Programming Algorithms

Let us think of the equalities $\sum_{j=1}^{n} \bar{\alpha}_j x_{J(j)} = \bar{\alpha}_0$ as one constraint. Then we may develop the following iterative scheme referred to as a dynamic programming procedure. Suppose we define for an index k ($1 \leq k \leq n$) and fixed (m vector) $\mathbf{g}$ in $G(\bar{\alpha})$, the following reduced problem

$$f(k, \mathbf{g}) = \text{minimum} \quad \sum_{j=1}^{k} \bar{c}_j x_{J(j)}$$

$$\text{subject to} \quad \sum_{j=1}^{k} \bar{\alpha}_j x_{J(j)} = \mathbf{g}$$

$$x_{J(j)} \geq 0 \text{ and integer} \quad (j = 1, \ldots, k),$$

where the sum is modulo 1. Note that $f(k, \mathbf{g})$ is the minimal value of the objective function to the GMP using the first k variables and with the right-hand side $\bar{\alpha}_0$ replaced by $\mathbf{g}$. Therefore, $f(n, \bar{\alpha}_0)$ is the value of the minimal solution to the GMP. Also, as $\bar{c}_j \geq 0$, $f(k, \mathbf{0}) = 0$ for any $k = 1, \ldots, n$. We shall also let $f(0, \mathbf{g})$ be arbitrarily large for any $\mathbf{g}$. Defining $f(k, \mathbf{g})$ this way allows us to develop the following relationship: Suppose that for some index k and all right-hand sides $\mathbf{g}$ in $G(\bar{\alpha})$, we know $f(k - 1, \mathbf{g})$. Using this information we would like to determine $f(k, \mathbf{g})$. As the expanded problem differs in only the variable $x_{J(k)}$, we must decide whether it should be fixed at 0, or should be at least 1. Naturally, we would use $x_{J(k)}$ only if it can yield an improved value for the objective function. Thus,

i) $x_{J(k)} = 0$ if $f(k, \mathbf{g}) = f(k - 1, \mathbf{g}) \leq \bar{c}_k + f(k, \mathbf{g} - \bar{\alpha}_k)$,

ii) $x_{J(k)} \geq 1$ if $f(k, \mathbf{g}) = \bar{c}_k + f(k, \mathbf{g} - \bar{\alpha}_k) < f(k - 1, \mathbf{g})$,

where (ii) says that $x_{J(k)}$ should be at least 1 if its cost added to the minimal cost of the resulting problem (with the right-hand side reduced by $\bar{\alpha}_k$) is smaller than the minimal cost using only the first $k - 1$ variables. Observe that in this expression we use $f(k, \mathbf{g} - \bar{\alpha}_k)$ instead of $f(k - 1, \mathbf{g} - \bar{\alpha}_k)$ since we want to allow $x_{J(k)}$ to be used more than once. As we are seeking minimal solutions, (i) and (ii) give

$$f(k, \mathbf{g}) = \text{ minimum } \{f(k-1, \mathbf{g}), \bar{c}_k + f(k, \mathbf{g} - \bar{\alpha}_k)\}. \tag{6}$$

Expression (6), referred to as a recursive relationship,[15] allows us to compute $f(k, \mathbf{g})$ for all $k = 1, \ldots, n$ and $\mathbf{g}$ in $G(\bar{\alpha})$, provided that every k generates the group $G(\bar{\alpha})$; equivalently, $G(\bar{\alpha}_k) = G(\bar{\alpha})$ for every k. As will be shown, if $\bar{\alpha}_k$ does not generate $G(\bar{\alpha})$, we may not be able to find $\mathbf{g} - \bar{\alpha}_k$ in terms of $\bar{\alpha}_k$. This is necessary since if $\mathbf{g} = \lambda \bar{\alpha}_k$, then

$$f(k, \mathbf{g} - \bar{\alpha}_k) = f(k, (\lambda - 1)\bar{\alpha}_k),$$

which will be known from a previous calculation (Example 11.2).

The procedure starts with $f(1, \mathbf{0}) = 0$ and finds $f(1, \mathbf{g}) = \bar{c}_1 + f(1, \mathbf{g} - \bar{\alpha}_1)$ for all $\mathbf{g}$ in $G(\bar{\alpha})$. This is used to find $f(2, \mathbf{g})$ for all $\mathbf{g}$, and so on. The computations are carried in an array containing $f(k, \mathbf{g})$ values, where the rows correspond to increasing indices k and the columns are associated with multiples of a single generating element (e.g.) $\bar{\alpha}_1$. The optimal value of the objective function may be found in row n and column λ, where $\bar{\alpha}_0 = \lambda \bar{\alpha}_1$. To retrieve the values of the variables we use a bookkeeping scheme borrowed from Hu [336], which keeps track of the index of the last variable used. Specifically, define $j(k, \mathbf{g})$ to be the index of the last variable used in making up $f(k, \mathbf{g})$. Thus, $j(1, \mathbf{g}) = 1$ for all $\mathbf{g}$, and for $k \geq 2$ define

$$j(k, \mathbf{g}) = j(k-1, \mathbf{g}) \quad \text{if} \quad x_{J(k)} = 0 \quad \text{or} \quad f(k, \mathbf{g}) = f(k-1, \mathbf{g}),$$

$$j(k, \mathbf{g}) = k \qquad \text{if} \quad x_{J(k)} \geq 1 \quad \text{or} \quad f(k, \mathbf{g}) = \bar{c}_k + f(k, \mathbf{g} - \bar{\alpha}_k).$$

These indices are entered in the array under the $f(k, \mathbf{g})$ values. To retreive the solution making up $f(n, \bar{\alpha}_0 = \lambda \bar{\alpha}_1)$, initialize all nonbasic variables $x_{J(k)}$, $k = 1, \ldots, n$ at value 0. Now we pick out the $j(n, \lambda \bar{\alpha}_1)$ entry. This yields an index t, which means that variable $x_{J(t)}$ is used and we increment $x_{J(t)}$ by 1. Now we find the index of the last variable making up $f(t, \lambda \bar{\alpha}_1 - \bar{\alpha}_t = \beta \bar{\alpha}_1)$, which is $j(t, \beta \bar{\alpha}_1) = r$. Or $x_{J(r)}$ is incremented by 1 and we decrease $\beta \bar{\alpha}_1$ by $\bar{\alpha}_r$ and repeat the process until the right-hand side is reduced to zero. To illustrate the technique we solve the example considered in several earlier chapters.

---

[15]This type of solution approach, referred to as a dynamic programming, originally appeared in Bellman [67]. Recursive relationships are the basis of most dynamic programming algorithms. A detailed discussion of dynamic programming may be found in Bellman and Dreyfus [71].

Example 11.2 (Balinski [50]) ($\mathbf{x_B} \geq \mathbf{0}$ and each $G(\bar{\alpha}_j) = G(\bar{\alpha})$)

$$
\begin{array}{rrrrrrl}
\text{P2} \quad \text{maximize} & - & 4x_1 & - & 5x_2 \\
\text{subject to} & - & x_1 & - & 4x_2 & + & x_3 & & = -5 \\
& - & 3x_1 & - & 2x_2 & & & + x_4 & = -7 \\
\end{array}
$$
$$x_1, x_2, x_3, x_4 \geq 0 \text{ and integer.}$$

From Example 4.1 (Chapter 4) the optimal linear programming solution contains basic variables x_1 and x_2; so

$$
\mathbf{B} = \begin{pmatrix} -1 & -4 \\ -3 & -2 \end{pmatrix}, \qquad D = |\det \mathbf{B}| = 10,
$$

$$
\mathbf{B}^{-1} = \begin{pmatrix} 2/10 & -4/10 \\ -3/10 & 1/10 \end{pmatrix}, \quad \mathbf{B}^{-1}\mathbf{b} = \begin{pmatrix} 18/10 \\ 1/10 \end{pmatrix},
$$

$$
\mathbf{N} = \begin{pmatrix} 1 & 0 \\ 0 & 1 \end{pmatrix} = \mathbf{I}, \qquad \mathbf{B}^{-1}\mathbf{N} = \mathbf{B}^{-1}\mathbf{I} = \mathbf{B}^{-1},
$$

$$
\mathbf{c_B}\mathbf{B}^{-1}\mathbf{N} - \mathbf{c_N} = \mathbf{c_B}\mathbf{B}^{-1} = (7/10, 11/10).
$$

Or the group problem is

$$
\text{GMP} \quad \text{minimize} \quad \tfrac{7}{10} \, x_{J(1)} + \tfrac{11}{10} \, x_{J(2)}
$$

$$
\text{subject to} \quad \begin{pmatrix} \frac{2}{10} \\ \frac{7}{10} \end{pmatrix} x_{J(1)} + \begin{pmatrix} \frac{6}{10} \\ \frac{1}{10} \end{pmatrix} x_{J(2)} \equiv \begin{pmatrix} \frac{8}{10} \\ \frac{8}{10} \end{pmatrix} \quad (\text{mod } 1)
$$

$$
x_{J(1)}, x_{J(2)} \geq 0, \text{ and integer,}
$$

where $x_{J(1)} = x_3$ and $x_{J(2)} = x_4$. As $D\bar{\alpha}_1 = \begin{pmatrix} 2 \\ 7 \end{pmatrix}$ and $D\bar{\alpha}_2 = \begin{pmatrix} 6 \\ 1 \end{pmatrix}$, there is no common factor other than one between the components in each $\bar{\alpha}_j$. Thus, by an earlier discussion, $G(\bar{\alpha}_1) = G(\bar{\alpha}_2) = G(\bar{\alpha})$ which has $D = 10$ elements. These are tabulated in Fig. 11.1. To solve the problem using the recursive relationship (6) we have $f(k, \mathbf{0}) = 0$.

k	$\lambda\bar{\alpha}_1$ $\gamma\bar{\alpha}_2$	$\lambda=1$ $\gamma=7$	$\lambda=2$ $\gamma=4$	$\lambda=3$ $\gamma=1$	$\lambda=4$ $\gamma=8$	$\lambda=5$ $\gamma=5$	$\lambda=6$ $\gamma=2$	$\lambda=7$ $\gamma=9$	$\lambda=8$ $\gamma=6$	$\lambda=9$ $\gamma=3$	$\lambda=10$ $\gamma=10$
	g	$\begin{pmatrix}\frac{2}{10}\\\frac{7}{10}\end{pmatrix}$	$\begin{pmatrix}\frac{4}{10}\\\frac{4}{10}\end{pmatrix}$	$\begin{pmatrix}\frac{6}{10}\\\frac{1}{10}\end{pmatrix}$	$\begin{pmatrix}\frac{8}{10}\\\frac{8}{10}\end{pmatrix}$ $\bar{\alpha}_0$	$\begin{pmatrix}0\\\frac{5}{10}\end{pmatrix}$	$\begin{pmatrix}\frac{2}{10}\\\frac{2}{10}\end{pmatrix}$	$\begin{pmatrix}\frac{4}{10}\\\frac{9}{10}\end{pmatrix}$	$\begin{pmatrix}\frac{6}{10}\\\frac{6}{10}\end{pmatrix}$	$\begin{pmatrix}\frac{8}{10}\\\frac{3}{10}\end{pmatrix}$	$\begin{pmatrix}0\\0\end{pmatrix}$
1	$f(1,\mathbf{g})$	$\frac{7}{10}$	$\frac{14}{10}$	$\frac{21}{10}$	$\frac{28}{10}$	$\frac{35}{10}$	$\frac{42}{10}$	$\frac{49}{10}$	$\frac{56}{10}$	$\frac{63}{10}$	0
	$j(1,\mathbf{g})$	1	1	1	1	1	1	1	1	1	–
2	$f(2,\mathbf{g})$	$\frac{7}{10}$	$\frac{14}{10}$	$\frac{11}{10}$	$\frac{18}{10}$	$\frac{25}{10}$	$\frac{22}{10}$	$\frac{29}{10}$	$\frac{36}{10}$	$\frac{33}{10}$	0
	$j(2,\mathbf{g})$	1	1	2	2	2	2	2	2	2	–

Figure 11.1: Computations for Example 11.2

So for $k = 1$,

$$f(1,1\bar{\alpha}_1) = \bar{c}_1 + f(1,0) \qquad = \bar{c}_1 \qquad j(1,\bar{\alpha}_1) = 1,$$
$$f(1,2\bar{\alpha}_1) = \bar{c}_1 + f(1,2\bar{\alpha}_1 - \bar{\alpha}_1) = 2\bar{c}_1 \qquad j(1,2\bar{\alpha}_1) = 1,$$
$$\vdots \qquad\qquad\qquad\qquad \vdots \qquad\qquad \vdots$$
$$f(1,9\bar{\alpha}_1) = \bar{c}_1 + f(1,9\bar{\alpha}_1 - \bar{\alpha}_1) = 9\bar{c}_1 \qquad j(1,9\bar{\alpha}_1) = 1.$$

These numbers are entered consecutively in row 1 of the array in Fig. 11.1. Now, for $k = 2$ by (6), we have that

$$f(2,\mathbf{g}) = \text{minimum } \{f(1,\mathbf{g}), \bar{c}_2 + f(2,\mathbf{g} - \bar{\alpha}_2)\}.$$

The only value of $f(2,\mathbf{g})$ that is currently known is $f(2,0)$. So start with $\mathbf{g} = \bar{\alpha}_2 = 3\bar{\alpha}_1$, since $\mathbf{g} - \bar{\alpha}_2 = 0$, and

$$f(2,\bar{\alpha}_2) = \text{minimum } \{f(1,3\bar{\alpha}_1), \bar{c}_2 + f(2,0)\}.$$
$$= \text{minimum } \left\{\tfrac{21}{10}, \tfrac{11}{10} + 0\right\} = \tfrac{11}{10} \quad j(2,\bar{\alpha}_2) = 2.$$

The value $11/10$ is therefore written in the second row and third column of the array. Once $f(2,\bar{\alpha}_2)$ is known, we can find $f(2,2\bar{\alpha}_2)$ and so on.

$$f(2, 2\bar{\alpha}_2) = \text{minimum } \{f(1, 6\bar{\alpha}_1), \bar{c}_2 + f(2, \bar{\alpha}_2)\}.$$
$$= \text{minimum } \left\{\frac{42}{10}, \frac{11}{10} + \frac{11}{10}\right\} = \frac{22}{10} \quad j(2, 2\bar{\alpha}_2) = 2.$$

$$\vdots$$

$$f(2, 9\bar{\alpha}_2) = \text{minimum } \{f(1, 7\bar{\alpha}_1), \bar{c}_2 + f(2, 8\bar{\alpha}_2)\}.$$
$$= \text{minimum } \left\{\frac{49}{10}, \frac{11}{10} + \frac{18}{10}\right\} = \frac{29}{10} \quad j(2, 9\bar{\alpha}_2) = 2.$$

As $\bar{\alpha}_0 = 4\bar{\alpha}_1 = 8\bar{\alpha}_2$, the minimal value of the objective function is $f(2, 4\bar{\alpha}_1) = 18/10$. To obtain the values of the variables we have

$$j(2, 4\bar{\alpha}_1) = 2 \quad \text{or} \quad x_{J(2)} = 1 \quad \text{and} \quad 4\bar{\alpha}_1 - \bar{\alpha}_2 = \bar{\alpha}_1,$$

so

$$j(2, \bar{\alpha}_1) = 1 \quad \text{or} \quad x_{J(2)} = 1 \quad \text{and} \quad \bar{\alpha}_1 - \bar{\alpha}_1 = 0.$$

Then the solution is $x_3 = 1$ and $x_4 = 1$, which gives

$$\begin{pmatrix} x_1 \\ x_2 \end{pmatrix} = \begin{pmatrix} \frac{18}{10} \\ \frac{8}{10} \end{pmatrix} \begin{pmatrix} \frac{2}{10} & -\frac{4}{10} \\ -\frac{3}{10} & \frac{1}{10} \end{pmatrix} \begin{pmatrix} 1 \\ 1 \end{pmatrix} = \begin{pmatrix} 2 \\ 1 \end{pmatrix} \geq 0.$$

Thus the optimal solution to the integer program is $\mathbf{x} = (2, 1, 1, 1)$ with the value of the objective function equal to

$$\mathbf{c_B B^{-1} b} - \frac{18}{10} = -\frac{112}{10} - \frac{18}{10} = -13.$$

Note that once we have an array which has the values of $f(n = 2, \mathbf{g})$, we may solve the group minimization for any right-hand side provided that the basic variables turn out to be nonnegative. For example, if the right-hand side had been $\begin{pmatrix} -6 \\ -9 \end{pmatrix}$ then

$$\mathbf{B^{-1} b} = \begin{pmatrix} \frac{24}{10} \\ \frac{9}{10} \end{pmatrix}, \quad \text{so} \quad \bar{\alpha}_0 = \begin{pmatrix} \frac{4}{10} \\ \frac{9}{10} \end{pmatrix} = 7\bar{\alpha}_1 \quad \text{and} \quad f(2, 7\bar{\alpha}_1) = \frac{29}{10}.$$

Retrieving the solution, we have $j(2, 7\bar{\alpha}_1) = 2$ or $x_{J(2)} = 1; j(2, 7\bar{\alpha}_1 - \bar{\alpha}_2) = j(2, 4\bar{\alpha}_1)$ or $x_{J(2)} = 2$ since it is used again; $j(2, 4\bar{\alpha}_1 - \bar{\alpha}_2) = j(2, \bar{\alpha}_1) = 1$ or $x_{J(1)} = 1$, and $\bar{\alpha}_1 - \bar{\alpha}_1 = 0$. Now the nonbasic variables $x_3 = 1$ and $x_4 = 2$ yield $x_1 = 3$ and $x_2 = 1$. So the optimal solution to this integer program is $\mathbf{x} = (3, 1, 1, 2)$ and the objective function is

−17. This is a distinct advantage of using the dynamic programming algorithm.

When $G(\bar{\alpha}_j) \neq G(\bar{\alpha})$ for some $\bar{\alpha}_j$ (Hu [336])

In the previous example we were able to use expression (6) to find $f(k, \mathbf{g})$ for all indices k and group elements $\mathbf{g}$. This was possible because each $\bar{\alpha}_j$ generated the group $G(\bar{\alpha})$. Had this not been the case, then we would reach a stage in the computations where the recursive relationship could not be used directly. In particular, suppose $\bar{\alpha}_k$ ($k \geq 2$) generates a subgroup of order $d < D$.[16] Then, using (6), we can find $f(k, \beta\bar{\alpha}_k)$ for $\beta = 1, \ldots, d$, since $f(k - 1, \bar{\alpha}_k)$ are known and $f(k, \beta\bar{\alpha}_k - \bar{\alpha}_k)$ may be found. However, suppose we wish to find $f(k, \mathbf{g})$ for an element $\mathbf{g}$ which cannot be written as a multiple of $\bar{\alpha}_k$; that is, $\mathbf{g}$ is not in $G(\bar{\alpha}_k)$. To do this, consider the following scheme. For ease of reference we restate expression (6):

$$f(k, \mathbf{g}) = \text{minimum} \ \{f(k - 1, \mathbf{g}), \bar{c}_k + f(k, \mathbf{g} - \bar{\alpha}_k)\}.$$

Now, as $\mathbf{g}$ is not in $G(\bar{\alpha}_k)$, it follows that $\mathbf{g} - \bar{\alpha}_k$ is not a multiple of $\bar{\alpha}_k$ and $f(k, \mathbf{g} - \bar{\alpha}_k)$ is unknown. However, we have the value of $f(k - 1, \mathbf{g})$, so why not temporarily assume that $f(k, \mathbf{g}) = f(k - 1, \mathbf{g})$. At worst, this is an overestimate. Then the natural question is: How does this help us computationally? As we know $f(k - 1, \mathbf{t})$ for every group element $\mathbf{t}$, and we have assigned a value to $f(k, \mathbf{g})$, consider

$$f(k, \mathbf{g} + \beta\bar{\alpha}_k) = \min \ \underbrace{\{f(k - 1, \mathbf{g} + \beta\bar{\alpha}_k),}_{\text{known}} \ \underbrace{\bar{c}_k + f(k, \mathbf{g} + \beta\bar{\alpha}_k - \bar{\alpha}_k)\}}_{\text{tentatively estimated}} .(7)$$

Now, assuming that $f(k, \mathbf{g}) = f(k - 1, \mathbf{g})$, we may start with $\beta = 1$ and using (7) find $f(k, \mathbf{g} + \beta\bar{\alpha}_k)$ for each $\beta = 1, \ldots, d$. When $\beta = d$ we have a new estimate for $f(k, \mathbf{g})$. If this value agrees with the original one we will show that it is the correct value. If it differs, the whole process is repeated. By doing this for every group element not reached using (6) directly, we may find $f(k, \mathbf{g})$ for all $\mathbf{g}$ in $G(\bar{\alpha})$. Before showing that this process converges to the correct values we solve Example 11.1.

Example (from Example 11.1) ($\mathbf{x_B} \geq 0$ and $G(\bar{\alpha}_j) \neq G(\bar{\alpha})$ for some column $\bar{\alpha}_j$)

---

[16]If $k = 1$, take $f(1, \mathbf{g}) = +\infty$ for $\mathbf{g} \notin G(\bar{\alpha}_1)$, since there is no solution to $\bar{\alpha}_1 x_{J(1)} = \mathbf{g}$, where $x_{J(1)}$ is a nonnegative integer.

From the first example we have the group minimization problem below (the two columns of zeroes have been dropped).

minimize $\quad \frac{7}{6} \ x_3 + \ \frac{11}{6} \ x_6 + \ \frac{4}{6} \ x_7$

subject to $\quad \begin{pmatrix} \frac{2}{6} \\ \frac{3}{6} \end{pmatrix} x_3 + \begin{pmatrix} \frac{4}{6} \\ \frac{3}{6} \end{pmatrix} x_6 + \begin{pmatrix} \frac{2}{6} \\ 0 \end{pmatrix} x_7 \equiv \begin{pmatrix} 0 \\ \frac{3}{6} \end{pmatrix} \quad$ (mod 1)

$$x_3, x_6, x_7 \geq 0 \text{ and integer.}$$

As the group is generated by

$$\bar{\alpha}_1 = \begin{pmatrix} \frac{2}{6} \\ \frac{3}{6} \end{pmatrix} \quad \text{and} \quad \bar{\alpha}_2 = \begin{pmatrix} \frac{4}{6} \\ \frac{3}{6} \end{pmatrix},$$

we proceed as in Example 11.2 to find $f(1,\mathbf{g})$ and $f(2,\mathbf{g})$. This yields the $k = 1$ and $k = 2$ rows of the tableau.

	$\lambda\bar{\alpha}_1$	$\lambda = 1$	$\lambda = 2$	$\lambda = 3$	$\lambda = 4$	$\lambda = 5$	$\lambda = 6$
k	$\gamma\bar{\alpha}_2$	$\gamma = 5$	$\gamma = 4$	$\gamma = 3$	$\gamma = 2$	$\gamma = 1$	$\gamma = 6$
	$\beta\bar{\alpha}_3$	—	$\beta = 2,5$	—	$\beta = 1,4$	—	$\beta = 3,6$
	$\mathbf{g}$	$\begin{pmatrix} \frac{2}{6} \\ \frac{3}{6} \end{pmatrix}$	$\begin{pmatrix} \frac{4}{6} \\ 0 \end{pmatrix}$	$\begin{pmatrix} 0 \\ \frac{3}{6} \end{pmatrix}$	$\begin{pmatrix} \frac{2}{6} \\ 0 \end{pmatrix}$	$\begin{pmatrix} \frac{4}{6} \\ \frac{3}{6} \end{pmatrix}$	$\begin{pmatrix} 0 \\ 0 \end{pmatrix}$
1	$f(1,\mathbf{g})$	$\frac{7}{6}$	$\frac{14}{6}$	$\frac{21}{6}$	$\frac{28}{6}$	$\frac{35}{6}$	0
	$j(1,\mathbf{g})$	1	1	1	1	1	—
2	$f(2,\mathbf{g})$	$\frac{7}{6}$	$\frac{14}{6}$	$\frac{21}{6}$	$\frac{22}{6}$	$\frac{11}{6}$	0
	$j(2,\mathbf{g})$	1	1	1	2	2	—
3	$f(3,\mathbf{g})$	$\frac{7}{6}$	$\frac{8}{6}$	$\frac{15}{6}$	$\frac{4}{6}$	$\frac{11}{6}$	0
	$j(3,\mathbf{g})$	1 From II	3 From I	3 As in II	3 From I	2 or 3 As in II	—

I. Since $\bar\alpha_3$ generates a group of order 3 we may find

$$f(3, \bar\alpha_3) \;=\; \text{minimum}\;\{f(2, \bar\alpha_3), \bar c_3 + f(3, 0)\}$$

$$=\; \text{minimum}\;\{\tfrac{22}{6}, \tfrac{4}{6} + 0\} = \tfrac{4}{6} \quad j(3, \bar\alpha_3) = 3,$$

$$f(3, 2\bar\alpha_3) \;=\; \text{minimum}\;\{f(2, 2\bar\alpha_3), \bar c_3 + f(3, \bar\alpha_3)\}$$

$$=\; \text{minimum}\;\{\tfrac{14}{6}, \tfrac{4}{6} + \tfrac{4}{6}\} = \tfrac{8}{6} \quad j(3, 2\bar\alpha_3) = 3.$$

Naturally $f(3, 3\bar\alpha_3) = f(3, 0)$, $f(3, 4\bar\alpha_3) = f(3, \bar\alpha_3)$, and $f(3, 5\bar\alpha_3) = f(3, 2\bar\alpha_3)$. Thus the group elements corresponding to $\bar\alpha_1, 3\bar\alpha_1$, and $5\bar\alpha_1$ are not reached.

II. To find $f(3, \bar\alpha_1)$, assume it is equal to $f(2, \bar\alpha_1) = \tfrac{7}{6}$. Or, from (7),

$$f(3\bar\alpha_1 + \beta\bar\alpha_3) = \text{minimum}\{f(2, \bar\alpha_1 + \beta\bar\alpha_3), \tfrac{4}{6} + f(3, \bar\alpha_1 + (\beta - 1)\bar\alpha_3)\}.$$

which gives

$$\beta = 1 \quad f(3, \bar\alpha_1 + \bar\alpha_3) \;=\; \text{minimum}\;\{f(2, \bar\alpha_1 + \bar\alpha_3), \tfrac{4}{6} + f(3, \bar\alpha_1)\}$$

$$=\; \text{minimum}\;\{f(2, 5\bar\alpha_1), \tfrac{4}{6} + f(2, \bar\alpha_1)\}$$

$$=\; \text{minimum}\;\{\tfrac{11}{6}, \tfrac{4}{6} + \tfrac{7}{6}\} = \tfrac{11}{6},$$

$$\beta = 2 \quad f(3, \bar\alpha_1 + 2\bar\alpha_3) \;=\; \text{minimum}\;\{f(2, 3\bar\alpha_1), \tfrac{4}{6} + f(3, \bar\alpha_1 + \bar\alpha_3)\}$$

$$=\; \text{minimum}\;\{\tfrac{21}{6}, \tfrac{4}{6} + \tfrac{11}{6}\} = \tfrac{15}{6},$$

$$\beta = 3 \quad f(3, \bar\alpha_1 + 3\bar\alpha_3) \;=\; \text{minimum}\;\{f(2, \bar\alpha_1), \tfrac{4}{6} + f(3, \bar\alpha_1 + 2\bar\alpha_3)\}$$

$$=\; \text{minimum}\;\{\tfrac{7}{6}, \tfrac{4}{6} + \tfrac{15}{6}\} = \tfrac{7}{6}.$$

As this agrees with the estimate, the correct value of $f(3, \bar\alpha_1) = \tfrac{7}{6}$. Note that the third variable is not used in making up $f(3, \bar\alpha_1)$, so $j(3, \bar\alpha_1) = j(2, \bar\alpha_1) = 1$.

To find the values for $3\bar\alpha_1$ and $5\bar\alpha_1$ we repeat the above process for each element. If this is done we obtain

$$f(3, 3\bar\alpha_1) = \tfrac{15}{6}, \quad j(3, 3\bar\alpha_1) = 3,$$

and $f(3, 5\bar{\alpha}_1) = \frac{11}{6}$, $j(3, 5\bar{\alpha}_1) = 2$ or 3.

As $\bar{\alpha}_0 = 3\bar{\alpha}_1$, we have that $f(3, 3\bar{\alpha}_1) = \frac{15}{6}$ and $x_6 = x_7 = 1$. Substituting those values back in Example 11.1, we obtain $\mathbf{x_N} = (x_3, x_4, x_5, x_6, x_7) = (0, 0, 0, 1, 1)$ and $\mathbf{x_B} = (x_1, x_2) = (42, 20)$. Thus, the integer program has been solved.[17]

To prove that the estimating procedure gives the correct value for $f(k, \mathbf{g})$ after a finite number of steps, we first note that by setting $f(k, \mathbf{g}) = f(k - 1, \mathbf{g})$ we can only overestimate $f(k, \mathbf{g})$. Thus, if in (7) an $f(k, \mathbf{g} + \beta\bar{\alpha}_k)$ is in reality equal to an $f(k - 1, \mathbf{g} + \beta\bar{\alpha}_k)$ for some β $(1 \leq \beta \leq d)$, then it will be the minimum in (7). This means that from that point on the correct values for $f(k, \mathbf{g} + \beta\bar{\alpha}_k)$ will be obtained. We claim that $f(k, \mathbf{g} + \beta\bar{\alpha}_k)$ will equal $f(k - 1, \mathbf{g} + \beta\bar{\alpha}_k)$ for some β during the first d steps. Equivalently, the process must converge to the correct value sometime between step d and step $2d$ inclusively. [18]

To prove the claim, suppose the opposite. That is, the real value of

$$f(k, \mathbf{g} + \beta\bar{\alpha}_k) = \bar{c}_k + f(k, \mathbf{g} + \beta\bar{\alpha}_k - \bar{\alpha}_k) \tag{8}$$

for $\beta = 1, 2, \ldots, d$. Then by (8)

$$
\begin{aligned}
f(k, \mathbf{g}) &= f(k, \mathbf{g} + d\bar{\alpha}_k) \\
&= f(k, \mathbf{g} + d\bar{\alpha}_k - \bar{\alpha}_k) + \bar{c}_k \\
&= f(k, \mathbf{g} + d\bar{\alpha}_k - 2\bar{\alpha}_k) + 2\bar{c}_k \\
&\vdots \\
&= f(k, \mathbf{g}) + d\bar{c}_k.
\end{aligned}
$$

---

[17]Notice that in this problem, the computations in II produce the values of $f(3, 3\bar{\alpha}_1)$ and $f(3, 5\bar{\alpha}_1)$, since $\bar{\alpha}_1 + 2\bar{\alpha}_3 = 3\bar{\alpha}_1$ (the $\beta = 2$ calculation) and $\bar{\alpha}_1 + \bar{\alpha}_3 = 5\bar{\alpha}_1$ (the $\beta = 1$ calculation). This is not always the case, and several computations, as in II, may be necessary (see Problem 11.17).

Also, note that $j(3, 5\bar{\alpha}_1) = 2$ or 3 indicates that for $\bar{\alpha}_0 = 3\bar{\alpha}_1$ there are alternate optimal solutions to the group problem. In this case, $j(3, 5\bar{\alpha}_1) = 2$ gives $x_3 = 0$, and $x_6 = x_7 = 1$, or an optimal solution to the GMP is $\mathbf{x_N} = (x_3, x_4, x_5, x_6, x_7) = (0, 0, 0, 1, 1)$, which also solves the integer program, since $\mathbf{x_B} = (x_1, x_2) = (42, 20) \geq 0$; $j(3, 5\bar{\alpha}_1) = 3$ produces $x_3 = 1, x_6 = 0$ and $x_7 = 2$, or a second optimal solution to the GMP is $\mathbf{x_N} = (1, 0, 0, 0, 2)$; this solution is also optimal for the integer program since $\mathbf{x_B} = (40, 20) \geq 0$.

[18]The former case happens when $\beta = 0$ (i.e., the initital estimate was correct) and the latter case occurs when $\beta = d$.

This is a contradiction unless $\bar{c}_k = 0$. In this case, consider

$$\mathbf{g} = \sum_{j=1}^{k-1} \bar{\alpha}_j x_{J(j)} + \beta \bar{\alpha}_k$$

for any β $(1 \le \beta \le d)$. By assumption, $\mathbf{g}$ cannot be written as a multiple of $\bar{\alpha}_k$ alone. So for each β.

$$\mathbf{t} = \sum_{j=1}^{k-1} \bar{\alpha}_j \bar{x}_{J(j)} \ne \mathbf{0},$$

where the $\bar{x}_{J(j)}$'s are fixed. Now

$$f(k-1, \mathbf{t}) \quad = \quad \text{minimum} \quad \sum_{j=1}^{k-1} \bar{c}_j x_{J(j)},$$

$$\text{subject to} \quad \sum_{j=1}^{k-1} \bar{\alpha}_j x_{J(j)} \quad = \quad \mathbf{t},$$

$$x_{J(j)} \ge 0 \text{ and integer } (j = 1, \ldots, k-1)$$

and

$$f(k, \mathbf{t} + \beta \bar{\alpha}_k) \quad = \quad \text{minimum} \quad \sum_{j=1}^{k-1} \bar{c}_j x_{J(j)},$$

$$\text{subject to} \quad \sum_{j=1}^{k-1} \bar{\alpha}_j x_{J(j)} + \bar{\alpha}_k x_{J(k)} \quad = \quad \mathbf{t} + \beta \bar{\alpha}_k,$$

$$x_{J(j)} \ge 0 \text{ and integer } (j = 1, \ldots, k).$$

So if $\bar{x}_{J(1)}, \ldots, \bar{x}_{J(k-1)}$ makes up $f(k-1, \mathbf{t})$, then $\bar{x}_{J(1)}, \ldots, \bar{x}_{J(k-1)}, \bar{x}_{J(k)} = \beta$ make up $f(k, \mathbf{t} + \beta \bar{\alpha}_k)$ and vice versa. Since $\bar{c}_k = 0$, this means that $f(k-1, \mathbf{t}) = f(k, \mathbf{t} + \beta \bar{\alpha}_k)$ for each β $(1 \le \beta \le d)$. For $\beta = d$ we have

$$f(k-1, \mathbf{t}) = f(k, \mathbf{t}) \quad \text{or} \quad f(k-1, \mathbf{g} - \beta \bar{\alpha}_k) = f(k, \mathbf{g} - \beta \bar{\alpha}_k).$$

Equivalently, $f(k-1, \mathbf{g} + \bar{\beta}\bar{\boldsymbol{\alpha}}_k) = f(k, \mathbf{g} + \bar{\beta}\bar{\boldsymbol{\alpha}}_k)$, where $\bar{\beta} = d - \beta$ ($1 \leq \bar{\beta} \leq d$), since $d\bar{\boldsymbol{\alpha}}_k = \mathbf{0}$ and $1 \leq \beta \leq d$. This is a contradiction because we have assumed that (8) is always true. Therefore, we have that the estimating procedure works.

11.3.2   A Sufficient Condition for $\mathbf{x}_{\mathbf{B}} \geq 0$

Recall that the group minimization problem is equivalent to the integer program from which it is derived except for the relaxation of the nonnegativity constraints on the basic variables. The optimal solution to the GMP gives a set of integer values for the nonbasic variables $\mathbf{x}_{\mathbf{N}}$. If $\mathbf{x}_{\mathbf{B}} = \mathbf{B}^{-1}\mathbf{b} - \mathbf{B}^{-1}\mathbf{N}\mathbf{x}_{\mathbf{N}} \geq 0$, then $\mathbf{x}_{\mathbf{B}}, \mathbf{x}_{\mathbf{N}}$ is an optimal integer solution. Although this need not always occur, we may nevertheless derive sufficient conditions for which $\mathbf{x}_{\mathbf{B}} \geq 0$ for any solution $\mathbf{x}_{\mathbf{N}}$. One such condition is described below and another is left to Problem 11.18. Before proceeding we should mention that the discussion assumes that $\mathbf{N}$ includes *all* nonbasic columns, including those which may have been dropped (because of duplications or zero columns) in the group minimization problem.

Consider the cone $K_{\mathbf{B}}$ generated by the columns of $\mathbf{B}$. That is, $K_{\mathbf{B}}$ is the set of points which can be written as a nonnegative linear combination of the columns of $\mathbf{B}$; Equivalentlly, a nonnegative solution $\mathbf{x}_{\mathbf{B}}$ to $\mathbf{B}\mathbf{x}_{\mathbf{B}} = \bar{\mathbf{b}}$ (any m column) exists if and only if $\bar{\mathbf{b}}$ is in $K_{\mathbf{B}}$. As the linear programming solution ($\mathbf{x}_{\mathbf{N}} = \mathbf{0}$) is feasible, we have that $\mathbf{b}$ is in the cone. So for $\mathbf{b} - \mathbf{N}\mathbf{x}_{\mathbf{N}}$ to be in $K_{\mathbf{B}}$ we must have the length of $\mathbf{N}\mathbf{x}_{\mathbf{N}}$, written[19] $\|\mathbf{N}\mathbf{x}_{\mathbf{N}}\|$, small enough so that when the vector $\mathbf{N}\mathbf{x}_{\mathbf{N}}$ is subtracted from the vector $\mathbf{b}$, the resulting point is inside $K_{\mathbf{B}}$ (Figure 11.2).

Suppose we define $K_{\mathbf{B}}(d)$ to be the subset of points in $K_{\mathbf{B}}$ at a Euclidean distance of d or more away from the boundary points of $K_{\mathbf{B}} = K_{\mathbf{B}}(0)$ (Figure 11.2). Then if $d \geq \|\mathbf{N}\mathbf{x}_{\mathbf{N}}\|$ and $\mathbf{b}$ is in $K_{\mathbf{B}}(d)$ we have that $\mathbf{b} - \mathbf{N}\mathbf{x}_{\mathbf{N}}$ is in $K_{\mathbf{B}}$, so that $\mathbf{x}_{\mathbf{B}} = \mathbf{B}^{-1}(\mathbf{b} - \mathbf{N}\mathbf{x}_{\mathbf{N}}) \geq 0$ (Fig. 11.2). To get an upper bound on $\|\mathbf{N}\mathbf{x}_{\mathbf{N}}\|$ we use a result whose proof is postponed until Section 11.4.2. In particular, we can show that every optimal solution $x_{J(1)}, \ldots, x_{J(n)}$ to the group problem

---

[19] $\|\mathbf{a}\| = \sqrt{\sum_i a_i^2}$, where $\mathbf{a} = (a_i)$; that is, $\|\mathbf{a}\|$ is the Euclidean distance from the point $\mathbf{a}$ to the origin.

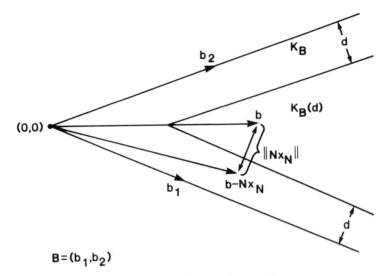

$$B=(b_1,b_2)$$

Figure 11.2: The Cone K_B

$$\text{minimize} \quad \sum_{j=1}^{n} \bar{c}_j x_{J(j)},$$

$$\text{subject to} \quad \sum_{j=1}^{n} \bar{\alpha}_j x_{J(j)} \equiv \mathbf{g_0} \qquad (\text{mod } 1),$$

$$x_{J(j)} \geq 0 \text{ and integer} \quad (j = 1, \ldots, n),$$

where $\mathbf{g_0}$ is *any* element in the group, must satisfy

$$\sum_{j=1}^{n} x_{J(j)} \leq D - 1, \tag{9}$$

where $D = |\det \mathbf{B}|$. Also, for $\mathbf{N} = (\mathbf{a}_{J(j)})$, we have

$$\| \mathbf{N}\mathbf{x_N} \| = \| \sum_{j=1}^{n} \mathbf{a}_{J(j)} x_{J(j)} \| \leq \sum_{j=1}^{n} \| \mathbf{a}_{J(j)} \| \, |x_{J(j)}|.$$

As any solution to the GMP is nonnegative, $x_{J(j)} \geq 0$ for every j, or

$$\| \mathbf{N}\mathbf{x_N} \| \leq \sum_{j=1}^{n} \| \mathbf{a}_{J(j)} \| x_{J(j)} \leq \ell_m \sum_{j=1}^{n} x_{J(j)} \leq \ell_m(D-1);$$

where ℓ_m is the length of the longest nonbasic column (i.e., maximum$_j$ $\|\mathbf{a}_{J(j)}\|$), and the last inequality is from (9). As $\|\mathbf{Nx_N}\| \le \ell_m(D-1)$ for every minimal solution $\mathbf{x_N}$ to the GMP, we have that $\mathbf{x_B} \ge \mathbf{0}$, provided $\mathbf{b}$ is in $K_{\mathbf{B}}(d)$, where $d \ge \ell_m(D-1)$. We may summarize the discussion in the following result.

Theorem 11.2 *If* $\mathbf{b}$ *is in* $K_{\mathbf{B}}(\ell_m(D-1))$, *where* $D = |\det \mathbf{B}|$ *and* ℓ_m *is the length of the longest nonbasic column, then* $\mathbf{x_B} = \mathbf{B}^{-1}\mathbf{b} - \mathbf{B}^{-1}\mathbf{Nx_N} \ge \mathbf{0}$ *for every optimal solution* $\mathbf{x_N}$ *to the group minimization problem.*

The above theorem suggests several remarks.

Remarks

1. If $D = 1, \ell_m(D-1) = 0$ and the theorem reduces to having $\mathbf{b}$ in the cone spanned by the columns of $\mathbf{B}$. That is, for any right-hand side $\mathbf{b}$ in $K_{\mathbf{B}}$, the linear programming solution $\mathbf{x_B} = \mathbf{B}^{-1}\mathbf{b}$ is integer. This is easy to see since $\mathbf{x_B} = \mathbf{B}^{-1} = \pm\mathbf{B}^{+}\mathbf{b}$ and $\mathbf{B}^{+}$, the adjoint of $\mathbf{B}$, is integer. As D increases from 1, the region $K_{\mathbf{B}}(\ell_m(D-1))$ moves farther away from the vertex and boundaries of the cone $K_{\mathbf{B}}$ and it is less likely that $\mathbf{b}$ will satisfy the sufficiency condition. In this sense the integer program becomes more difficult as D increases.

2. If the current linear programming solution $\mathbf{x_B} = \mathbf{B}^{-1}\mathbf{b}$ is degenerate, then it lies on certain boundaries of the cone, and unless $D = 1$ the sufficiency condition will never be satisfied. This can be used to show that when the integer variables are restricted to be 0 or 1 the sufficiency condition cannot hold (see Balas [43]).

3. Suppose $\mathbf{x_B} = \mathbf{B}^{-1}\mathbf{b}$ is nondegenerate and $\mathbf{b}$ does not lie in $K_{\mathbf{B}}(\ell_m(D-1))$. Then for a sufficiently large scalar λ, $\lambda\mathbf{b}$ will be inside $K_{\mathbf{B}}(\ell_m(D-1))$ (Fig. 11.3). Hence, we may say that the integer program

$$\begin{array}{lll} A & \text{maximize} & \mathbf{cx} \\ & \text{subject to} & \mathbf{Ax} = \lambda\mathbf{b} \\ & & \mathbf{x} \ge \mathbf{0} \text{ and integer,} \end{array}$$

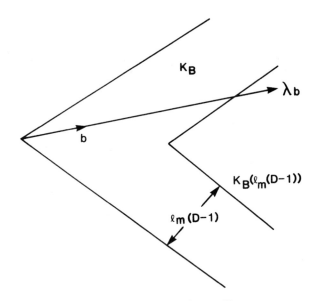

Figure 11.3: The Cone $K_\mathbf{B}$

may be solved as a group minimization problem whenever λ is sufficiently large, provided that $\mathbf{x_B}$ is nondegenerate. Problem A is often referred to as the asymptotic problem.

4. The term $\ell_m(D-1)$ is independent of **b**. Thus, once the basic columns have been selected, we may determine $K_\mathbf{B}(\ell_m(D-1))$. The GMP will solve the integer program whenever **b** is in this cone.

To make the discussion concrete, we reconsider the Gomory example.

Example   (from Example 11.1)

$$\mathbf{N} = \begin{pmatrix} 1 & 4 & 2 & 1 & 0 \\ 5 & 1 & -1 & 0 & 1 \end{pmatrix},$$

or $\ell_1 = \sqrt{26}$, $\ell_2 = \sqrt{17}$, $\ell_3 = \sqrt{5}$, $\ell_4 = 1$, $\ell_5 = 1$, and thus $\ell_m = \sqrt{26}$.

$$\mathbf{B} = \begin{pmatrix} 0 & 2 \\ 3 & -4 \end{pmatrix} \quad D = 6 \quad \text{and} \quad \ell_m(D-1) = 5\sqrt{26} \approx 25.5.$$

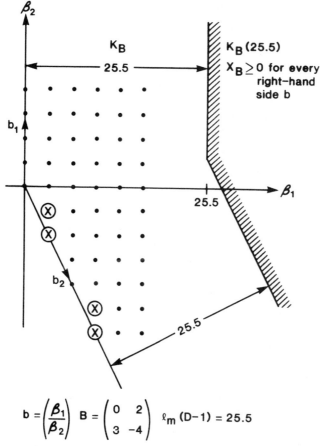

Figure 11.4: The Cone $K_{\mathbf{B}}(25.5)$

The cone $K_{\mathbf{B}}$ and the region $K_{\mathbf{B}}(25.5)$ are depicted in Fig. 11.4. As $\mathbf{b} = \begin{pmatrix} 41 \\ 47 \end{pmatrix}$ lies in $K_{\mathbf{B}}(25.5), \mathbf{x_B} \geq \mathbf{0}$. From Example 11.1 the GMP is

$$\text{minimize} \quad \tfrac{7}{6} \; x_3 + \; \tfrac{11}{6} \; x_6 + \; \tfrac{4}{6} \; x_7$$

$$\text{subject to} \quad \begin{pmatrix} \tfrac{2}{6} \\ \tfrac{3}{6} \end{pmatrix} x_3 + \begin{pmatrix} \tfrac{4}{6} \\ \tfrac{3}{6} \end{pmatrix} x_6 + \begin{pmatrix} \tfrac{2}{6} \\ 0 \end{pmatrix} x_7 \equiv \mathbf{g_0} \pmod 1$$

$$x_3, x_6, \text{ and } x_7 \geq 0 \text{ and integer,}$$

where $\mathbf{g_0} = \begin{pmatrix} 0 \\ \tfrac{3}{6} \end{pmatrix}$ and is obtained from $\mathbf{b} = \begin{pmatrix} 41 \\ 47 \end{pmatrix}$.

Note that for *any* b in $K_B(25.5)$, $x_B = B^{-1}(b - Nx_N) \geq 0$, where x_N is the minimal solution to the above group problem with g_0 obtained from b (or more precisely, $B^{-1}b$).

As Gomory [280] indicates, this problem is highly feasible in the sense that x_B turns out to be nonnegative for most right-hand sides b not satisfying the sufficiency condition; that is, for most b in the band separating K_B with $K_B(25.5)$. Each heavy dot in Fig. 11.4 is a right-hand side b which yields a group element g_0 for the group problem (representing the *Gomory example*) whose minimal solution x_N gives $x_B = B^{-1}(b - Nx_N) \geq 0$. The circles with crosses in them indicate right-hand sides for which some of the basic variables turn out to be negative. In the next section we discuss what to do when this occurs.

11.3.3   An Enumeration Algorithm (Shapiro [532])

The value of the objective function for the integer program is $c_B B^{-1}b - \bar{c}x_N$. When x_N, the optimal solution to the group problem which minimizes $\bar{c}x_N$ yields $x_B \geq 0$, we have solved the integer program. If one or more of the basic variables turns out to be negative, we have to seek the smallest value of $\bar{c}x_N$ for which $x_B \geq 0$ and x_N satisfies the constraints of the GMP. Thus, we could develop a set of recursive relationships to find the rth best solution to the group problem and hence be able to solve the integer program. However, the expressions are quite complicated and difficult to implement (see White [599]). Instead of this approach we introduce a branch and bound enumeration which is similar to the Dakin [153] variation discussed in Chapter 8.

Recall from Chapter 8 that the intent of an enumeration algorithm is to explicitly or implicitly examine all possible integer solutions. The scheme presented here inspects the integer solutions to the group problem by successively adding constraints of the form $x_{J(j)} \geq K$ $(j = 1, \ldots, n)$, where K starts at 0 and is incremented by 1. In particular, relating the procedure to a tree composed of nodes and branches, a node corresponds to a vector x_N with a greater than or equal to inequality acting on each variable, and two nodes are joined by a branch whenever the inequality acting on one particular variable has been incremented by 1 and the others are the same (Fig. 11.5). A branch then naturally corresponds to introducing that inequality.

The basic approach works as follows. The origin node corresponds to the vector x_N with the inequalities $x_{J(j)} \geq 0$ $(j = 1, \ldots, n)$; n nodes

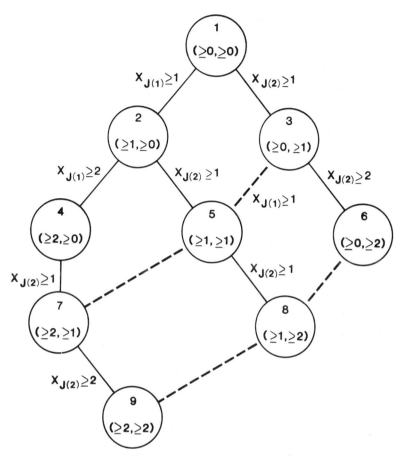

Figure 11.5: The Branch and Bound Tree

are created by adjoining the constraint $x_{J(j)} \geq 1$ one at a time to each variable, starting (e.g.) with $j = 1$. From each of these n nodes, as many more distinct nodes as possible are created by incrementing by 1, each inequality acting on the variables at each of the nodes. This process is then repeated. A complete enumeration tree for a two (nonbasic) variable problem appears in Fig. 11.5. For convenience, we have assumed that each variable will not exceed two and that the algorithm selects the higher nodes (starting from the left) to create nodes from. Dashed lines represent nodes which have not been created because duplication would result. Notice that, as each $x_{J(j)}$ can have three values, there are $3 \times 3 = 9$ possible combinations for $(x_{J(1)}, x_{J(2)})$. These are numbered in the figure in order of their creation.

It is easy to see that if each $x_{J(j)} \geq 0$ have an upper bound λ_j, then the basic approach will yield the $T = \Pi_{j=1}^n(\lambda_j + 1)$ nodes. So at each node we may replace the inequalities by equalities to obtain a value for $\mathbf{x_N}$. Then $\mathbf{x_B}$ may be found by using $\mathbf{B^{-1}b - B^{-1}Nx_N}$. By recording the values of the objective function for those yielding $\mathbf{x_B} \geq \mathbf{0}$, we can find the optimal solution to the integer program. Naturally, for T large this approach cannot be used directly. That is, we would have to be able to implicitly enumerate large numbers of points. To do this, observe that the constraints on the variables become more restrictive as we proceed down the tree. Suppose we solve the group problem at every node with the inequality constraints active on each variable. This produces a lower bound for the value of $\mathbf{\bar{c}x_N}$ for any solution $\mathbf{x_N}$ to the group problem found at lower nodes. Thus, if in the current best integer solution $(\mathbf{\hat{x}_B}, \mathbf{\hat{x}_N})$ we have $\hat{z} = \mathbf{\bar{c}\hat{x}_N} \leq \mathbf{\bar{c}x_N}$, then the node under consideration (and all nodes which can be traced back to it) may be dropped. By doing this we consider only those nodes which can possibly produce an improved integer solution. Naturally, every time a group problem is solved and $\mathbf{\bar{c}x_N} < \hat{z}$, we compute $\mathbf{x_B}$. If it is nonnegative, $\hat{z}$ is updated to this new value and the current node is dropped. The process terminates when no nodes are left which can possibly produce an improved integer solution. Morito [434] used the basic bookkeeping scheme with a particular branching rule and information from the optimal solution to the group problem, to create a more efficient branch and bound algorithm. The details of this method are in Appendix D. Before presenting an example several remarks are in order.

Remarks

1. We have purposely avoided discussing node selection and branch (or variable) selection rules. Criteria motivated by a discussion similar to the one appearing in Section 8.7 (Chapter 8) may be contrived. (For example, to reduce computer storage we may create nodes so as to proceed down the tree.) Some possibilities are left to Problem 11.25.

2. Other implicit enumeration criteria are possible. For example, we may use the inequality $\sum_{j=1}^n x_{J(j)} \leq D-1$ to curtail the number of nodes which are created. Incidentally, this inequality bounds each nonbasic variable from above and thus the algorithm converges. A discussion of other criteria is left to Problem 11.25.

3. The enumeration is over *all* possible integer combinations for x_N. Thus, even though some columns and variables may not appear in the group problem they cannot be ignored in the enumeration. In particular, the optimal solution to the integer program may have some of these variables positive. Note that this does not contradict an earlier statement which said that in every minimal solution to the *group problem* these may be set to zero. When a branch corresponds to a variable not in the group problem, the inequality, of course, has no affect on the group problem's minimal solution at that node.

4. To define one node—and hence one group problem—from another we add constraints of the form $x_{J(j)} \geq K$. As in earlier chapters, we can do this quite simply by defining the complementing variable $\bar{x}_{J(j)} = x_{J(j)} - K \geq 0$. That is, enforcing the inequality is equivalent to substituting $\bar{x}_{J(j)} + K$ for $x_{J(j)}$. Computationally this reduces to subtracting $K\bar{\alpha}_j$ from the right–hand side of the congruence relationship and reducing the result modulo 1. Note that this is just another element in the group $G(\bar{\alpha})$. Thus, if we have the minimal solution to the GMP for all right–hand sides, we need only pick out the appropriate value. To retrieve the value of $x_{J(j)}$ we add K to $\bar{x}_{J(j)}$. In addition, to obtain the lower bound, $K\bar{c}_j$ for each complemented variable j, must be added to the minimal value of the objective function.

5. It is desirable to obtain all optimal solutions to each group problem because only some of them may yield nonnegative values for the basic variables. This is equivalent to recording more than one index when ties occur in computing the minimum in the recursive relationship (6).

To illustrate these comments and the enumerative procedure we solve an example. The initial group problem is solved using the recursive relationship (6) and all optimal solutions for each right–hand side are found. The node with the largest value of the minimal solution to the group problem is selected to create nodes from. In case of ties the lowest node in the tree is chosen. Branches are created according to increasing indices.

Example 11.3 (White [599]) ($x_B \not\geq 0$ and $G(\bar{\alpha}_j) \neq G(\bar{\alpha})$ for some column $\bar{\alpha}_j$)

Maximize $\qquad\qquad x_2$

subject to $-4x_1 + 4x_2 + x_3 \qquad\qquad = 3,$
$\qquad\qquad 2x_1 - x_2 \qquad + x_4 \qquad = 2,$
$\qquad\qquad x_1 + x_2 \qquad\qquad + x_5 = 6,$
$\qquad\qquad x_1, \ldots, x_5 \geq 0 \text{ and integer.}$

The optimal solution to the linear program gives

$$
\mathbf{B} = \begin{pmatrix} 4 & 0 & -4 \\ -1 & 1 & 2 \\ 1 & 0 & 1 \end{pmatrix}, \quad D = 8, \quad \mathbf{x_B} = \begin{pmatrix} x_2 \\ x_4 \\ x_1 \end{pmatrix} = \begin{pmatrix} \frac{27}{8} \\ \frac{1}{8} \\ \frac{21}{8} \end{pmatrix},
$$

$$
\mathbf{B^{-1}N} = \mathbf{B^{-1}(a_3, a_5)} = \begin{pmatrix} \frac{1}{8} & \frac{4}{8} \\ \frac{3}{8} & -\frac{4}{8} \\ -\frac{1}{8} & \frac{4}{8} \end{pmatrix} \quad \mathbf{\bar{c}} = (\bar{c}_3, \bar{c}_5) = (\tfrac{1}{8}, \tfrac{4}{8});
$$

so the group problem is

minimize $\quad \frac{1}{8} x_{J(1)} + \frac{4}{8} x_{J(2)}$

subject to $\begin{pmatrix} \frac{1}{8} \\ \frac{3}{8} \\ \frac{7}{8} \end{pmatrix} x_{J(1)} + \begin{pmatrix} \frac{4}{8} \\ \frac{4}{8} \\ \frac{4}{8} \end{pmatrix} x_{J(2)} \equiv \begin{pmatrix} \frac{3}{8} \\ \frac{1}{8} \\ \frac{5}{8} \end{pmatrix} \pmod 1$

$x_{J(1)}, x_{J(2)} \geq 0 \text{ and integer.}$

We may solve the above problem by inspection, or use expressions (6) and (7), proceeding as in the Gomory example. (Note that $G(\bar{\alpha}_2) \neq G(\bar{\alpha})$). Using the notation appearing in Section 11.3.1 we obtain the tableau below.

$\lambda\bar{\alpha}_1$	1	2	3	4	5	6	7	8
$\beta\bar{\alpha}_2$	$--$	$--$	$--$	$1,3,5,7$	$--$	$--$	$--$	$2,4,6,8$
g	$\begin{pmatrix}\frac{1}{8}\\\frac{3}{8}\\\frac{7}{8}\end{pmatrix}$	$\begin{pmatrix}\frac{2}{8}\\\frac{6}{8}\\\frac{6}{8}\end{pmatrix}$	$\begin{pmatrix}\frac{3}{8}\\\frac{1}{8}\\\frac{5}{8}\end{pmatrix}$	$\begin{pmatrix}\frac{4}{8}\\\frac{4}{8}\\\frac{4}{8}\end{pmatrix}$	$\begin{pmatrix}\frac{5}{8}\\\frac{7}{8}\\\frac{3}{8}\end{pmatrix}$	$\begin{pmatrix}\frac{6}{8}\\\frac{2}{8}\\\frac{2}{8}\end{pmatrix}$	$\begin{pmatrix}\frac{7}{8}\\\frac{5}{8}\\\frac{1}{8}\end{pmatrix}$	$\begin{pmatrix}0\\0\\0\end{pmatrix}$

k

2	$f(2,\mathbf{g})$	$\frac{1}{8}$	$\frac{2}{8}$	$\frac{3}{8}$	$\frac{4}{8}$	$\frac{5}{8}$	$\frac{6}{8}$	$\frac{7}{8}$	0
	$j(2,\mathbf{g})$	1	1	1	1 or 2	1	1	1	$--$

The enumeration is summarized in Table 11.1, in which $\hat{z}$ is the current minimal value of $\bar{\mathbf{c}}\mathbf{x}_N$ for which $\mathbf{x}_B \geq \mathbf{0}$ (initially set very large), and the dangling nodes are those which remain to be examined. Note that the minimal solution to the (first) group problem gives $x_4 = -1 < 0$ and the enumeration is necessary. Also, note that as the solution to GMP is $x_{J(1)} = 3$ and $x_{J(2)} = 0$, nodes 2,4, and 6 are not really needed and could be omitted during the computations (the procedure for doing this is explained in Appendix D). Figure 11.6 relates the computations to a tree. The nodes are numbered in order of their creation, and the number next to each node is the lower bound on $\bar{\mathbf{c}}\mathbf{x}_N$.

11.4 The Group Minimization Problem Viewed as a Network Problem

11.4.1 Formulation

A *network* is a collection of nodes and arcs that join them. We can represent any group minimization problem in terms of a network as fol-

No.	Node $(x_{j(1)}, x_{j(2)})$ inequalities	The Group Problem Right-hand side g	$f(2.g)$	$K\bar{c}_j$	Lower bound	All optimal x_N x_3	x_5	Integer Solution x_B x_1	x_2	x_4		Dangling nodes
1	$(\geq 0, \geq 0)$	$\bar{\alpha}_0 = 3\bar{\alpha}_1$	3/8	0	3/8	3	0	3	3	-1	—	1
2	$(\geq 1, \geq 0)$	$3\bar{\alpha}_1 - \bar{\alpha}_1 = 2\bar{\alpha}_1$	2/8	1/8	3/8	3	0	3	3	-1	8	1.2
3	$(\geq 0, \geq 1)$	$3\bar{\alpha}_1 - \bar{\alpha}_2 = 7\bar{\alpha}_1$	7/8	4/8	11/8	7	-1	3	2	-2	8	2
						3	-2	2	2	-2	11/8	2.4
4	$(\geq 2, \geq 0)$	$3\bar{\alpha}_1 - 2\bar{\alpha}_1 = \bar{\alpha}_1$	1/8	2/8	3/8	3	0	3	3	-1	11/8	
						3	-1	2	2	-2	11/8	
5	$(\geq 1, \geq 1)$	$3\bar{\alpha}_1 - \bar{\alpha}_1 - \bar{\alpha}_2 = 6\bar{\alpha}_1$	6/8	1/8 + 4/8	11/8	3	0	3	2	0	11/8	4
6	$(\geq 3, \geq 0)$	$3\bar{\alpha}_1 - 3\bar{\alpha}_1 = 0$	0	3/8	3/8	3	-1	3	2	-1	11/8	4.6
						7	-2	2	2	-2	11/8	
7	$(\geq 2, \geq 1)$	$3\bar{\alpha}_1 - 2\bar{\alpha}_1 - \bar{\alpha}_2 = 5\bar{\alpha}_1$	5/8	2/8 + 4/8	11/8	3	0	3	2	0	11/8	6
8	$(\geq 4, \geq 0)$	$3\bar{\alpha}_1 - 4\bar{\alpha}_1 = 7\bar{\alpha}_1$	7/8	4/8	11/8	11	0	4	2	-4	11/8	6
						7	-1	3	2	-2	11/8	
9	$(\geq 3, \geq 1)$	$3\bar{\alpha}_1 - 3\bar{\alpha}_1 - \bar{\alpha}_2 = 4\bar{\alpha}_1$	4/8	3/8 + 4/8	11/8	3	-2	2	2	0	11/8	—

Table 11.1: Computations for Example 11.3

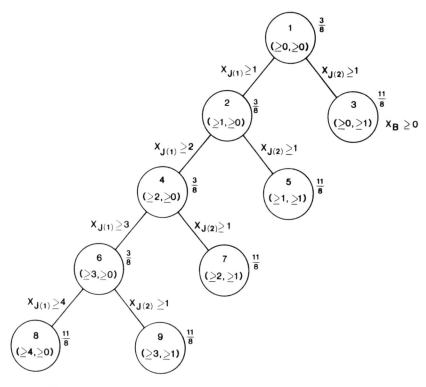

Figure 11.6: Branch and Bound Tree for Example 11.3

lows: Each element in the group $G(\bar{\alpha})$ corresponds to a distinct node. A directed arc[20] (i,k) joins two nodes, representing group elements $\mathbf{g}_i$ and $\mathbf{g}_k$, whenever $\mathbf{g}_k - \mathbf{g}_i$ (mod 1) is equal to some $\bar{\alpha}_j$ $(j \geq 1)$. Traversing an arc from node i to node k therefore corresponds to incrementing $x_{J(j)}$ by one, and so the cost $\bar{c}_j$ is assigned to arc (i,k). Then the group minimization problem reduces to finding a cheapest (or shortest) *route* from the node representing the zero element of the group (denoted by $\mathbf{g}_0$) to the one representing the right–hand side $\bar{\alpha}_0$. Here, a route is a sequence of directed arcs originating from $\mathbf{g}_0$, and if an arc (i,k) is used in the route it means that $x_{J(j)}$ is increased by 1, where $\bar{\alpha}_j \equiv \mathbf{g}_k - \mathbf{g}_i$ (mod 1). Initially (at node $\mathbf{g}_0$), all $x_{J(j)}$ are 0. To illustrate the representation, we reconsider the previous example.

---

[20]An arc which can only be traversed in the direction it indicates.

Example   (from Example 11.3)

The group minimization problem is

$$\text{minimize} \quad \tfrac{1}{8} \ x_{J(1)} + \tfrac{4}{8} \ x_{J(2)}$$

$$\text{subject to} \quad \begin{pmatrix} \tfrac{1}{8} \\ \tfrac{3}{8} \\ \tfrac{7}{8} \end{pmatrix} x_{J(1)} + \begin{pmatrix} \tfrac{4}{8} \\ \tfrac{4}{8} \\ \tfrac{4}{8} \end{pmatrix} x_{J(2)} \equiv \begin{pmatrix} \tfrac{3}{8} \\ \tfrac{1}{8} \\ \tfrac{5}{8} \end{pmatrix} \quad (\text{mod } 1)$$

$$x_{J(1)}, x_{J(2)} \geq 0 \text{ and integer.}$$

For ease of reference, we restate the $D = 8$ elements of $G(\bar{\alpha})$.

$\lambda\bar{\alpha}_1$	1	2	3	4	5	6	7	8
$\beta\bar{\alpha}_2$	–	–	–	$1,3,5,7$	–	–	–	$2,4,6,8$
	g_1	g_2	g_3	g_4	g_5	g_6	g_7	$g_D = g_8 = g_0$
g	$\begin{pmatrix} \tfrac{1}{8} \\ \tfrac{3}{8} \\ \tfrac{7}{8} \end{pmatrix}$	$\begin{pmatrix} \tfrac{2}{8} \\ \tfrac{6}{8} \\ \tfrac{6}{8} \end{pmatrix}$	$\begin{pmatrix} \tfrac{3}{8} \\ \tfrac{1}{8} \\ \tfrac{5}{8} \end{pmatrix}$	$\begin{pmatrix} \tfrac{4}{8} \\ \tfrac{4}{8} \\ \tfrac{4}{8} \end{pmatrix}$	$\begin{pmatrix} \tfrac{5}{8} \\ \tfrac{7}{8} \\ \tfrac{3}{8} \end{pmatrix}$	$\begin{pmatrix} \tfrac{6}{8} \\ \tfrac{2}{8} \\ \tfrac{2}{8} \end{pmatrix}$	$\begin{pmatrix} \tfrac{7}{8} \\ \tfrac{5}{8} \\ \tfrac{1}{8} \end{pmatrix}$	$\begin{pmatrix} 0 \\ 0 \\ 0 \end{pmatrix}$

The network therefore has eight nodes, one for each group element. To find the arcs consider each $\bar{\alpha}_j$ separately. In particular, for $\bar{\alpha}_1$ we have $g_1 - g_0 = \bar{\alpha}_1$, $g_2 - g_1 = \bar{\alpha}_1$, $g_3 - g_2 = \bar{\alpha}_1$, $\ldots$, $g_0 - g_7 = \bar{\alpha}_1$; or eight arcs are created and each is assigned a cost of $\bar{c}_1 = 1/8$ (Fig. 11.7(a)). Also for $\bar{\alpha}_2$ we have the eight arcs associated with

$$\bar{\alpha}_2 = g_4 - g_0 = g_0 - g_4 = g_5 - g_1 = g_1 - g_5 = g_6 - g_2 =$$
$$g_2 - g_6 = g_7 - g_3 = g_3 - g_7$$

each of these arcs is assigned a cost of $\bar{c}_2 = 4/8$ (Fig. 11.7(b)). Thus, the network representing this group problem is a composition of these

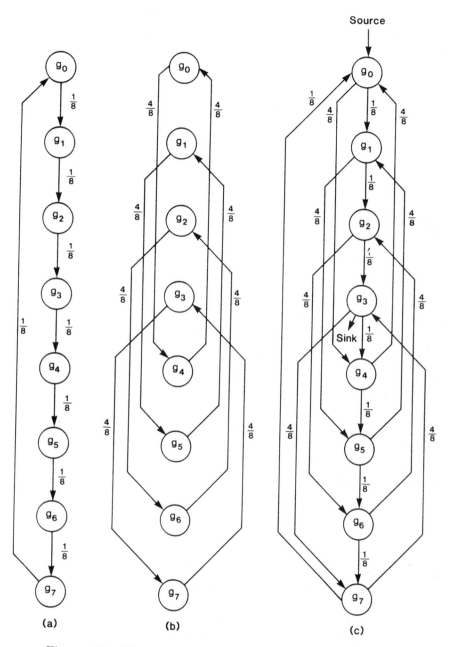

Figure 11.7: Networks Representing the Group Problem

two networks and appears in Fig. 11.7(c). The problem is then to find a minimal cost path from g_0, termed the *source*, to $g_3 = \bar{\alpha}_0$, termed the *sink*.

11.4.2 Solving the Integer Program by Solving the Network Problem

Once a network such as the one in Fig. 11.7(c) is obtained we have to find the shortest route from the source node g_0 to the sink node $\bar{\alpha}_0$. Several well known (enumerative) algorithms are available to do this and are surveyed by Dreyfus in [175]. One technique is described in Appendix A. Suppose a shortest route has been found. Then we have solved the group problem. If the basic variables turn out to be nonnegative, the integer program is solved and we are done. However, when one or more of the basic variables are negative, more work is required. In particular, we could use the enumeration algorithm discussed in Section 11.3.3. The group problem at each node reduces to finding a shortest route from the node g_0 to some node g_k $(0 \leq k \leq D - 1)$, where g_k is the current value of the right-hand side. As standard shortest route algorithms usually produce the shortest route between the source and all nodes in the network (see Appendix A), the first computation yields the shortest routes for all subsequent problems. (Remember that this also occurred when dynamic programming was used.)

An alternate to the enumeration is an algorithm which seeks the smallest shortest route for which $x_B \geq 0$. However, this involves considerable computation and reduces to utilizing the recursive relationships described in White [599]. Since our example may be solved by inspection, we have the (first) shortest route

$$\overset{1/8}{g_0} \rightarrow \overset{1/8}{g_1} \rightarrow \overset{1/8}{g_2} \rightarrow g_3$$

which means $(x_{J(1)}, x_{J(2)}) = (x_3, x_5) = (3, 0)$ and $x_B = (x_1, x_2, x_4) = (3, 3, -1)$. As $x_4 = -1 < 0$, we seek a second best shortest route. It is

$$\overset{4/8}{g_0} \rightarrow \overset{1/8}{g_4} \rightarrow \overset{4/8}{g_5} \rightarrow \overset{1/8}{g_1} \rightarrow \overset{1/8}{g_2} \rightarrow g_3$$

which gives $(x_{J(1)}, x_{J(2)}) = (3, 2)$ and $x_B = (2, 2, 4)$. Therefore, $x = (2, 2, 3, 4, 2)$ is an optimal integer solution.

The structure of the network also suggests some interesting results. For one thing, each node will have as many arcs of the same cost directed into it as it has leaving. This follows because if $\bar{\alpha}_j$ generates a subgroup of order d, then

$$g_k - g_{k-D/d} = g_{k+D/d} - g_k = \bar{\alpha}_j,$$

where $k = 0, 1, \ldots, D$ and $k + D/d$ is taken modulo D. (In the previous example, $D = 8$ and $\bar{\alpha}_2$ generates a subgroup of order 2. So $\bar{\alpha}_2 = g_k - g_{k-4} = g_{k+4} - g_k$ for $k = 0, 1, \ldots, D$. This gives the arcs appearing in Fig. 11.7(b).) Therefore, if arc $(k, k - D/d)$ is created, then the arc $(k + D/d, k)$ is created, and vice versa. Also, several second best (third best, etc.) shortest routes may correspond to the same values of the nonbasic variables, and some may, in fact, contain cycles (or loops). For example, from Fig. 11.7(c), the following are second best shortest routes with value $11/8$ and $\mathbf{x_N} = (3, 2)$.

$$g_0 \xrightarrow{4/8} g_4 \xrightarrow{1/8} g_5 \xrightarrow{4/8} g_1 \xrightarrow{1/8} g_2 \xrightarrow{1/8} g_3$$

$$g_0 \xrightarrow{1/8} g_1 \xrightarrow{4/8} g_5 \xrightarrow{1/8} g_6 \xrightarrow{4/8} g_2 \xrightarrow{1/8} g_3$$

Second best shortest routes without cycles

$$g_0 \xrightarrow{1/8} g_1 \xrightarrow{1/8} g_2 \xrightarrow{4/8} g_6 \xrightarrow{1/8} g_7 \xrightarrow{4/8} g_3$$

$$g_0 \xrightarrow{1/8} g_1 \xrightarrow{1/8} g_2 \xrightarrow{1/8} g_3 \xrightarrow{4/8} g_7 \xrightarrow{4/8} g_3$$

$$g_0 \xrightarrow{4/8} g_4 \xrightarrow{4/8} g_0 \xrightarrow{1/8} g_1 \xrightarrow{1/8} g_2 \xrightarrow{1/8} g_3$$

Second best shortest routes with cycles

$$g_0 \xrightarrow{1/8} g_1 \xrightarrow{4/8} g_5 \xrightarrow{4/8} g_1 \xrightarrow{1/8} g_2 \xrightarrow{1/8} g_3$$

$$g_0 \xrightarrow{1/8} g_1 \xrightarrow{1/8} g_2 \xrightarrow{4/8} g_6 \xrightarrow{4/8} g_2 \xrightarrow{1/8} g_3$$

This suggests that a main concern of an algorithm would be to avoid the duplications. A specially structured procedure which does this and takes the arc structure into account appears in Chen and Zionts [129], and in Morito [434] (which is discussed in Appendix D).

Note that the (first) shortest route need never contain more than $D - 1$ arcs and cannot have more unless some of the costs are zero. This means that every minimal solution $x_{J(1)}, \ldots, x_{J(n)}$ to the group problem can be made to satisfy

$$\sum_{j=1}^{n} x_{J(j)} \leq D - 1, \tag{10}$$

which proves an assertion used in Section 11.3.2. We may also conclude that if $x_B \geq 0$ for every solution x_N which satisfies (10), then the shortest route solution produces an optimal solution for the integer program (Problem 11.34).

11.5   An Equivalent Group Minimization Problem

In this section we develop a second group minimization problem. This group problem may be easier to solve than the one considered thus far. In the next section we show the relationships between the two groups. For ease of exposition, we rewrite the integer program as

$$c_B B^{-1} b \ - \ \text{minimize} \qquad \bar{c} x_N$$
$$\text{subject to} \quad B x_B + N x_N \ = \ b$$
$$x_B \geq 0, \ x_N \geq 0 \ \text{and integer,}$$

where B is an m by m basis selected so that the reduced costs $\bar{c} = c_B B^{-1} N - c_N$ are nonnegative and the other terms are as defined previously. Before proceeding to the derivation we present some useful results.

Lemma 11.1 Let C be any m by m integer matrix with $|\det C| = 1$, and let $x = Cy$, where x and y are m columns. Then there is a one-to-one correspondence between the integer values of x and the integer values of y.

Proof: As the determinant of $\mathbf{C}$ is not zero, $\mathbf{C}^{-1}$ exists, and $\mathbf{C}$ or $\mathbf{C}^{-1}$ is a basis for m space. Since the representation of a vector in terms of a basis is unique, given a vector $\mathbf{y}$ there is a unique $\mathbf{x}$ (using $\mathbf{x} = \mathbf{Cy}$) and vice versa (using $\mathbf{y} = \mathbf{C}^{-1}\mathbf{x}$). As $|\det \mathbf{C}| = 1$, we have $|\det \mathbf{C}^{-1}| = 1$, and therefore $\mathbf{x}$ is integer if and only if $\mathbf{y}$ is integer. ∎

Definition 11.1 An *elementary row operation* is the permutation of two rows, or the addition or subtraction of integer multiples of one row from another. Similarly, an *elementary column operation* is the permutation of two columns, or the addition or subtraction of integer multiples of one column from another.

It is easy to demonstrate that an elementary row operation is equivalent to premultiplying the given matrix by an appropriate integral matrix $\mathbf{R}$ which has $|\det \mathbf{R}| = 1$, and that an elementary column operation is equivalent to postmultiplying the given matrix by an appropriate integral matrix $\mathbf{C}$ with $|\det \mathbf{C}| = 1$. To see this, let the given matrix be denoted by $\mathbf{B}$, and, for simplicitly, suppose the first two rows (columns) are to be altered. Then define $\mathbf{R}$ ($\mathbf{C}$) to have the form

$$\mathbf{R} = \begin{pmatrix} \mathbf{R}^1 & 0 \\ 0 & \mathbf{I}_{m-2} \end{pmatrix} \quad \mathbf{C} = \begin{pmatrix} \mathbf{C}^1 & 0 \\ 0 & \mathbf{I}_{m-2} \end{pmatrix},$$

where $\mathbf{R}^1$ ($\mathbf{C}^1$) is a 2 by 2 submatrix, and $\mathbf{I}_{m-2}$ is an $m - 2$ by $m - 2$ identity matrix. If $\mathbf{R}^1$ ($\mathbf{C}^1$) has the form $\begin{pmatrix} 0 & 1 \\ 1 & 0 \end{pmatrix}$, the two rows (columns) are permuted, and if it is $\begin{pmatrix} 1 & K \\ 0 & 1 \end{pmatrix}$, or $\begin{pmatrix} 1 & 0 \\ K & 1 \end{pmatrix}$, K times the second row (column) is added to the first. In any case, $|\det \mathbf{R}^1| = |\det \mathbf{C}^1| = 1$, and thus $|\det \mathbf{R}| = |\det \mathbf{C}| = 1$. It turns out that any basis $\mathbf{B}$ can be transformed to a diagonal matrix using only elementary row and column operations, and furthermore, the elements on the diagonal divide each other. We state this as the next result.

Theorem 11.3 *From any m by m nonsingular integral matrix $\mathbf{B}$, a diagonal matrix $\hat{\mathbf{B}} = \mathbf{RBC}$ may be constructed, using a finite number of elementary row and column operations. Furthermore,*

$$\hat{\mathbf{B}} = \begin{pmatrix} \epsilon_1 & 0 & \cdots & 0 \\ 0 & \epsilon_2 & \cdots & 0 \\ \vdots & \vdots & \ddots & \vdots \\ 0 & 0 & \cdots & \epsilon_m \end{pmatrix},$$

where ϵ_i are positive integers $(i = 1, \ldots, m)$, ϵ_i divides ϵ_{i+1}
$(i = 1, \ldots, m - 1)$, and $|det\ \mathbf{R}| = |det\ \mathbf{C}| = 1$.

Proof: To show that the diagonalization is possible we give the computational procedure. However, as it is somewhat detailed, we leave the construction to Appendix B. (It is also in MacDuffee [402], and Mostow [435].) Note that if the diagonalization is possible, it follows from the previous discussion that $\hat{\mathbf{B}} = \mathbf{RBC}$ and $|det\mathbf{R}| = |det\ \mathbf{C}| = 1$. In this case, $\mathbf{R}$ ($\mathbf{C}$) is the product of several row (column) transforming matrices. Also, observe that

$$\prod_{i=1}^{m} \epsilon_i = |det\ \hat{\mathbf{B}}| = |det\ \mathbf{RBC}| = |(det\ \mathbf{R})(det\ \mathbf{B})(det\ \mathbf{C})|$$
$$= |det\ \mathbf{B}| = D.$$

Or, $\quad D = \displaystyle\prod_{i=1}^{m} \epsilon_i.$ ∎

The diagonalization process and the lemma allow us to derive a second group minimization problem. Let us drop the nonnegativity requirements on the basic variables and consider the set of solutions to

$$\mathbf{Bx_B} + \mathbf{Nx_N} = \mathbf{b}$$
$$\mathbf{x_B}\ \text{integer},\ \mathbf{x_N} \geq \mathbf{0}\ \text{and integer}.$$

Premultiplying the equality constraints by $\mathbf{R}$ and using $\mathbf{RB} = \hat{\mathbf{B}}\mathbf{C}^{-1}$ gives

$$\hat{\mathbf{B}}\mathbf{C}^{-1}\mathbf{x_B} + \mathbf{RNx_N} = \mathbf{Rb}$$
$$\mathbf{x_B}\ \text{integer},\ \mathbf{x_N} \geq \mathbf{0}\ \text{and integer}.$$

Now, using the lemma with $\mathbf{y_B} = \mathbf{C}^{-1}\mathbf{x_B}$ gives

$$\left. \begin{array}{l} \hat{\mathbf{B}}\mathbf{y_B} + \mathbf{RNx_N} = \mathbf{Rb} \\ \mathbf{y_B}\ \text{integer},\ \mathbf{x_N} \geq \mathbf{0}\ \text{and integer}. \end{array} \right\} \tag{11}$$

The equality is the same as $\hat{\mathbf{B}}\mathbf{y_B} = \mathbf{Rb} - \mathbf{RNx_N}$. But each component

$i = 1, \ldots, m$ of $\hat{\mathbf{B}}\mathbf{y_B}$ is $\epsilon_i y_{\mathbf{B}_i}$, where $\mathbf{y_B} = (y_{\mathbf{B}_i})$. Thus, $\mathbf{y_B}$ will be integer and may take on any integer value if and only if each row i of $\mathbf{Rb} - \mathbf{RNx_N}$ is a multiple of ϵ_i. This is the same as saying

$$\mathbf{Rb} - \mathbf{RNx_N} \equiv 0 \quad \text{mod} \begin{pmatrix} \epsilon_1 \\ \vdots \\ \epsilon_m \end{pmatrix}.$$

Or the constraints (11) are equivalent to

$$\mathbf{RNx_N} \equiv \mathbf{Rb} \quad \text{mod} \begin{pmatrix} \epsilon_1 \\ \vdots \\ \epsilon_m \end{pmatrix}$$

$$\mathbf{x_N} \geq \mathbf{0} \text{ and integer.}$$

Reducing each row of $\mathbf{RN}$ and $\mathbf{Rb}$ modulo ϵ_i gives the equivalent congruence relationship

$$\left. \begin{array}{c} \displaystyle\sum_{j=1}^{n} \boldsymbol{\lambda}_j x_{J(j)} \equiv \boldsymbol{\lambda}_0 \quad \text{mod} \begin{pmatrix} \epsilon_1 \\ \vdots \\ \epsilon_m \end{pmatrix} \\[2em] \mathbf{x_N} = (x_{J(1)}, x_{J(2)}, \ldots, x_{J(n)}) \geq \mathbf{0} \text{ and integer,} \end{array} \right\} \quad (12)$$

where $\boldsymbol{\lambda}_j$ is the jth column of $\mathbf{RN}$ with each component i reduced modulo ϵ_i, and $\boldsymbol{\lambda}_0$ is the reduced version of $\mathbf{Rb}$. Note that each component i of $\boldsymbol{\lambda}_j$ will range between 0 and $\epsilon_i - 1$.

The constraints (12) are equivalent to those of the integer program except that the nonnegativity constraints on the basic variables are not being enforced. Therefore, the integer program, without $\mathbf{x_B} \geq \mathbf{0}$ and the constant term $\mathbf{c_B}\mathbf{B}^{-1}\mathbf{b}$ may be written as

$$\begin{aligned} \text{FGMP} \quad \text{minimize} \quad & \sum_{j=1}^{n} \bar{c}_j x_{J(j)} \\ \text{subject to} \quad & \sum_{j=1}^{n} \boldsymbol{\lambda}_j x_{J(j)} = \boldsymbol{\lambda}_0 \\ & x_{J(j)} \geq 0 \text{ and integer} \quad (j = 1, \ldots, n), \end{aligned}$$

where the addition in row i is taken modulo ϵ_i. Before presenting an example, several remarks are in order.

Remarks

1. It is easy to show that the set $G(\boldsymbol{\lambda})$ generated by the $\boldsymbol{\lambda}_j$'s ($j = 1, \ldots, n$) is an abelian group (Problem 11.37), and so the previous problem is a group minimization problem. Consequently, it may be solved by any of the techniques for the group problem over $G(\bar{\boldsymbol{\alpha}})$.

2. In the next section we shall show that $G(\boldsymbol{\lambda})$ is *isomorphic to* $G(\bar{\boldsymbol{\alpha}})$. That is, for each distinct element in one group there is a corresponding distinct element in the other group. Also, if $\mathbf{g}_i$ and $\mathbf{g}_j$ are in $G(\boldsymbol{\lambda})$ and the corresponding elements in $G(\bar{\boldsymbol{\alpha}})$ are $\mathbf{g}'_i$ and $\mathbf{g}'_j$, respectively, then $\mathbf{g}_k = \mathbf{g}_i \oplus \mathbf{g}_j$ in $G(\boldsymbol{\lambda})$ corresponds to $\mathbf{g}'_k = \mathbf{g}'_i \oplus \mathbf{g}'_i$, in $G(\bar{\boldsymbol{\alpha}})$, where $\oplus$ means the addition is as defined in the appropriate group. Therefore, for each $\boldsymbol{\lambda}_j$ in $G(\boldsymbol{\lambda})$, there is a distinct $\bar{\boldsymbol{\alpha}}_j$, and vice versa. So once the mapping (or *isomorphism*) which sends elements from one group to the other is known, one group problem may be obtained from the other. The isomorphic property also means that $x_{J(1)}, \ldots, x_{J(n)}$ is a solution to the group problem over $G(\bar{\boldsymbol{\alpha}})$ if and only if it is a solution to the group problem over $G(\boldsymbol{\lambda})$ (Problem 11.38).

3. We will show that $G(\boldsymbol{\lambda})$ is isomorphic to a "factor group", and thus the new problem is referred to as the factor group minimization problem (FGMP). To obtain problem FGMP we have to construct $\hat{\mathbf{B}}$, which, as indicated in Appendix B, may involve considerable effort. However, in the next section we shall show that $G(\boldsymbol{\lambda})$ is either a cyclic group or that it can be represented by a sum of m cyclic subgroups each of order ϵ_i. Since ϵ_i divides ϵ_{i+1} and $\Pi_{i=1}^{m}\epsilon_i = D$, this means that if $G(\boldsymbol{\lambda})$ is cyclic, $\epsilon_1 = \epsilon_2 = \cdots = \epsilon_{m-1} = 1$ and $\epsilon_m = D$. Note that the only group of order 1 is the one with the zero element, and so the groups which correspond to $\epsilon_i = 1$ may be dropped in the representation. Also, as the components of $\boldsymbol{\lambda}_j$ are nonnegative and at most $\epsilon_i - 1$, we have that the modulo 1 equations are of the form $\sum_{j=1}^{n} 0 x_{J(j)} \equiv 0 \pmod{1}$, which may be omitted. Therefore, if $G(\boldsymbol{\lambda})$ is a cyclic group, there is only one meaningful equation in problem FGMP and its sum is taken modulo ϵ_m. The computations may then be somewhat simpler.

In general, $G(\lambda)$ is the sum of K $(1 \le K \le m)$ (nonzero) cyclic subgroups. So $\hat{B}$ is of the form (all elements not on the diagonal are zero)

$$\left.\begin{pmatrix} 1 & & & & & \\ & \ddots & & & 0 & \\ & & 1 & & & \\ & & & \epsilon_{m-K+1} & & \\ 0 & & & & \ddots & \\ & & & & & \epsilon_m \end{pmatrix}\begin{array}{c}\left.\vphantom{\begin{pmatrix}1\\1\\1\end{pmatrix}}\right\} m-K \\ \left.\vphantom{\begin{pmatrix}1\\1\end{pmatrix}}\right\} K\end{array}\right.$$

where $\epsilon_i > 1$ for $i = m - K + 1, \ldots, m$, and thus there are $m - K$ zero equations which may be dropped. Note that each active equation is modulo $\epsilon_i \le D$.

We now derive the factor group problem for the Gomory example and remark about its form in context of the comments.

Example (from Example 11.1)

$$\text{Maximize } 2x_1 + \ x_2 + 4x_3 + 3x_4 + \ x_5$$
$$\text{subject to} \qquad 2x_2 + \ x_3 + 4x_4 + 2x_5 + x_6 \qquad = 41$$
$$3x_1 - 4x_2 + 5x_3 + \ x_4 - \ x_5 \qquad + x_7 = 47$$
$$x_j \ge 0 \text{ and integer } (j = 1, \ldots, 7).$$

As in Example 11.1, select

$$\mathbf{B} = \begin{pmatrix} 0 & 2 \\ 3 & -4 \end{pmatrix}.$$

Then, from Appendix B, we will find that

$$\mathbf{R} = \begin{pmatrix} 0 & 1 \\ 1 & 2 \end{pmatrix}, \quad \mathbf{C} = \begin{pmatrix} -1 & 4 \\ -1 & 3 \end{pmatrix}, \quad \text{and}$$

$$\hat{\mathbf{B}} = \begin{pmatrix} 1 & 0 \\ 0 & 6 \end{pmatrix} = \begin{pmatrix} 0 & 1 \\ 1 & 2 \end{pmatrix}\begin{pmatrix} 0 & 2 \\ 3 & -4 \end{pmatrix}\begin{pmatrix} -1 & 4 \\ -1 & 3 \end{pmatrix}$$

$$\mathbf{RN} = \begin{pmatrix} 0 & 1 \\ 1 & 2 \end{pmatrix}\begin{pmatrix} 1 & 4 & 2 & 1 & 0 \\ 5 & 1 & -1 & 0 & 1 \end{pmatrix}$$

$$= \begin{pmatrix} 5 & 1 & -1 & 0 & 1 \\ 11 & 6 & 0 & 1 & 2 \end{pmatrix}$$

$$\mathbf{Rb} = \begin{pmatrix} 0 & 1 \\ 1 & 2 \end{pmatrix}\begin{pmatrix} 41 \\ 47 \end{pmatrix} = \begin{pmatrix} 47 \\ 135 \end{pmatrix}.$$

Taking row 1 of $\mathbf{RN}$ and $\mathbf{Rb}$ modulo 1, and row 2 modulo 6 yields

$$(\boldsymbol{\lambda}_j) = \begin{pmatrix} 0 & 0 & 0 & 0 & 0 \\ 5 & 0 & 0 & 1 & 2 \end{pmatrix} \quad \text{and} \quad \boldsymbol{\lambda}_0 = \begin{pmatrix} 0 \\ 3 \end{pmatrix}.$$

So problem FGMP is (the costs $\bar{c}_j$ are from the first example)

$$\text{min.} \quad \tfrac{7}{6}x_{J(1)} + 5x_{J(2)} + 2x_{J(3)} + \tfrac{11}{6}x_{J(4)} + \tfrac{4}{6}x_{J(5)}$$

$$\text{s.t.} \quad \begin{pmatrix} 0 \\ 5 \end{pmatrix}x_{J(1)} + \begin{pmatrix} 0 \\ 0 \end{pmatrix}x_{J(2)} + \begin{pmatrix} 0 \\ 0 \end{pmatrix}x_{J(3)} + \begin{pmatrix} 0 \\ 1 \end{pmatrix}x_{J(4)} + \begin{pmatrix} 0 \\ 2 \end{pmatrix}x_{J(5)} \equiv \begin{pmatrix} 0 \\ 3 \end{pmatrix}\text{mod}\begin{pmatrix} 1 \\ 6 \end{pmatrix}$$

$x_{J(j)} \geq 0$ and integer $(j = 1,\dots,5)$.

Remarks

1. Notice that $D = |\det \mathbf{B}| = 6 = 1(6)$. Since $\epsilon_1 = 1$ and $\epsilon_2 = D$, $G(\boldsymbol{\lambda})$, is cyclic and may be generated by any $\boldsymbol{\lambda}_j$ whose nonzero component has no common factor with D other than 1. (Why?) In our case, either $\boldsymbol{\lambda}_1$ or $\boldsymbol{\lambda}_4$ may be used and

$$G(\boldsymbol{\lambda}) = \left\{ \begin{pmatrix} 0 \\ 1 \end{pmatrix} \begin{pmatrix} 0 \\ 2 \end{pmatrix} \begin{pmatrix} 0 \\ 3 \end{pmatrix} \begin{pmatrix} 0 \\ 4 \end{pmatrix} \begin{pmatrix} 0 \\ 5 \end{pmatrix} \begin{pmatrix} 0 \\ 0 \end{pmatrix} \right\}$$

$\boldsymbol{\lambda}_4$	$2\boldsymbol{\lambda}_4$	$3\boldsymbol{\lambda}_4$	$4\boldsymbol{\lambda}_4$	$5\boldsymbol{\lambda}_4$	$6\boldsymbol{\lambda}_4$
$5\boldsymbol{\lambda}_1$	$4\boldsymbol{\lambda}_1$	$3\boldsymbol{\lambda}_1$	$2\boldsymbol{\lambda}_1$	$\boldsymbol{\lambda}_1$	$6\boldsymbol{\lambda}_1$
$\boldsymbol{\lambda}_5$			$2\boldsymbol{\lambda}_5$		$3\boldsymbol{\lambda}_5$

2. As the group is cyclic and there are two equations, the first one has all zero coefficients and may be dropped. Columns with all zeroes may also be omitted and the problem to be solved is

FGMP minimize $\frac{7}{6}x_{J(1)} + \frac{11}{6}x_{J(4)} + \frac{4}{6}x_{J(5)}$

subject to $5x_{J(1)} + x_{J(4)} + 2x_{J(5)} \equiv 3 \pmod 6$

$x_{J(1)},\ x_{J(4)},\ x_{J(5)} \geq 0$ and integer.

3. There are two ways of thinking about this group problem in relation to the former one. The first is to note that the modulo 6 equation is the same as

$$\frac{5}{6}x_{J(1)} + \frac{1}{6}x_{J(4)} + \frac{2}{6}x_{J(5)} \equiv \frac{3}{6} \pmod 1,$$

which is the sum (modulo 1) of the two equations in the group problem over $G(\bar{\alpha})$ (see Example 11.1). We may also relate the $\bar{\alpha}_j$ to the λ_j. In particular,

$$
\begin{array}{cccccc}
\bar{\alpha}_1 & \lambda_1 & \bar{\alpha}_2 & \lambda_2 & \bar{\alpha}_3 & \lambda_3 \\
\begin{pmatrix} \frac{2}{6} \\ \frac{3}{6} \end{pmatrix} \leftrightarrow \begin{pmatrix} 0 \\ 5 \end{pmatrix}; &
\begin{pmatrix} 0 \\ 0 \end{pmatrix} \leftrightarrow \begin{pmatrix} 0 \\ 0 \end{pmatrix}; &
\begin{pmatrix} 0 \\ 0 \end{pmatrix} \leftrightarrow \begin{pmatrix} 0 \\ 0 \end{pmatrix};
\end{array}
$$

$$
\begin{array}{cccc}
\bar{\alpha}_4 & \lambda_4 & \bar{\alpha}_5 & \lambda_5 \\
\begin{pmatrix} \frac{4}{6} \\ \frac{3}{6} \end{pmatrix} \leftrightarrow \begin{pmatrix} 0 \\ 1 \end{pmatrix}; &
\begin{pmatrix} \frac{2}{6} \\ 0 \end{pmatrix} \leftrightarrow \begin{pmatrix} 0 \\ 2 \end{pmatrix}.
\end{array}
$$

Or, an element

$$\begin{pmatrix} I_{1j}/D \\ I_{2j}/D \end{pmatrix} \text{ in } G(\bar{\alpha}) \text{ becomes } \begin{pmatrix} 0 \\ I_{1j} + I_{2j} \pmod D \end{pmatrix} \text{ in } G(\lambda).$$

So, in this example, to obtain problem FGMP, the first equation in the group problem (over $G(\bar{\alpha})$) is added (modulo 1) to the second, and its first equation is dropped. Observe that the zero columns remain as zero columns.

11.6   The Isomorphic Factor Group $G(\mathbf{A})/G(\mathbf{B})$

In this section we first present some definitions and useful well-known results. We then define a group which is isomorphic to $G(\boldsymbol{\lambda})$ and show it is isomorphic to $G(\bar{\boldsymbol{\alpha}})$. This will mean $G(\boldsymbol{\lambda})$ is isomorphic to $G(\bar{\boldsymbol{\alpha}})$. The subgroup decomposition of $G(\boldsymbol{\lambda})$ is derived and is used to show that the order of either group equals $D = |\det \mathbf{B}|$ or it divides D. Unless otherwise specified, G and G' represent general abelian groups.

11.6.1   Definitions, Results

Mappings

Definition 11.2   Let f be a mapping from G to G', namely f maps every element of G to an element of G', written $f : G \to G'$. Then the set of all elements in G which are mapped to the zero of G' is called the *kernel* of f. Formally, the kernel of f is $\{g \in G | f(g) = 0 \text{ in } G'\}$.

Definition 11.3   The *image* of the mapping $f : G \to G'$ is the set of all elements g' in G' for which $f(g) = g'$ for some g in G. The map f is *surjective* (onto) if the image of f is equal to all of G'. This means that given any element g' in G', there exists an element g in G such that $f(g) = g'$.

Definition 11.4   A map $f : G \to G'$ is said to be *injective* (one-to-one) if whenever g_1, g_2 are in G and $g_1 \neq g_2$, then $f(g_1) \neq f(g_2)$.

Definition 11.5   A *homomorphism* $f : G \to G'$ is a map having the property that for all g_1 and g_2 in G, $f(g_1 \oplus g_2) = f(g_1) \oplus f(g_2)$, where $\oplus$ indicates addition as defined in the appropriate group.

Definition 11.6   An *isomorphism* (f) between groups $(G$ and $G')$ is a homomorphism $f : G \to G'$ such that f is both one-to-one and onto. If an isomorphism exists between two groups they are said to be *isomorphic*.

Cosets

Definition 11.7   Let H be a subgroup of the group G and let g be an element of G. The set of all elements $g + h$ with h in H is called a coset of H in G. It is denoted by $g + H$ (or $H + g$). [21]

Lemma 11.2 *Let $g_1 + H$ and $g_2 + H$ be cosets of H in the group G. Either these cosets are equal, or they have no element in common.*

Proof: Suppose the cosets have one element in common. In particular, let h_1 and h_2 be elements of H for which $g_1 + h_1 = g_2 + h_2$. Then $g_1 = g_2 + (h_2 - h_1)$, where $h_2 - h_1$ is in the group H (by property (iii) of a group). Now, for any element $g_1 + h$ in $g + H$ we have

$$g_1 + h = g_2 + (h_2 - h_1) + h = g_2 + (h_2 - h_1 + h).$$

But $h_2 - h_1 + h$ is an element of H. Thus $g_1 + H$ is a subset of $g_2 + H$. Similarly, we can show that $g_2 + H$ is a subset of $g_1 + H$ and hence the cosets are equal. ∎

Lemma 11.3 *Let H be a finite subgroup of G. Then the number of elements in the coset $g + H$ is equal to the number of elements in H.*

Proof: Suppose h_1 and h_2 are distinct elements of H Then $g + h_1$ and $g + h_2$ are distinct elements of $g + H$, because $g + h_1 = g + h_2$ means that $h_1 = h_2$. So each element h_i in H defines a distinct element $g + h_i$ in the coset $g + H$, and conversely. ∎

Theorem 11.4 *Let H be a subgroup of the finite group G and let $(G : H)$ be the number of distinct cosets of H in G. Then the order of G equals $(G : H)$ (order of H).*

Proof: Consider any element g in G. As $g = g + 0$ and 0 is in H, we have that every element in G lies in some coset, namely $g + H$. (Note that this does not mean that there are as many distinct cosets as there are elements in G. Several of the cosets $g + H$ may be equal.)

Let the set of distinct cosets of H in G be denoted by $\{g_i + H\}$, where g_i is in G. By the first lemma, each coset $g_i + H$ has no element in common, and by the second lemma, the number of elements in $g_i + H$

---

[21] If the group is not commutative, we have to distinguish between left and right cosets. A more general definition and discussion of this section may be found in Fuchs [18] or Lang [384]. Also, in all expressions, addition is as defined in the group containing the elements being summed.

is precisely the order of H elements. Therefore, each distinct coset contains the order of H different elements of G each of the form $g_i + h$ for h in H. As there are $(G : H)$ cosets, we have $(G : H)$ (order of H) different elements of G. By definition, this quantity cannot exceed the order of G, and since every element of G is in some coset it cannot be less than the order of G. Thus, the equality in the assertion. ∎

Corollary 11.4.1 The order of H divides the order of G.

Factor Groups

Theorem 11.5 *Let $f : G \to G'$ be a homomorphism and suppose its kernel is the subgroup H of G. Let g' be an element in G' which is in the image of f; that is, $g' = f(g)$ for some g in G. Then the set of elements in G which is mapped to the element g' is precisely the coset $g + H$.*

Proof: We first note that the kernel of $f = \{g \in G | f(g) = 0\}$ is a subgroup of G. This follows since f is a homomorphism, so for g_1 and g_2 in its kernel

$$f(g_1 + g_2) = f(g_1) + f(g_2) = 0 + 0 = 0,$$

or $g_1 + g_2$ is in the kernel. The arithmetic properties (i), (ii), and (iii) (see Problem 11.1) are also evident.

To prove the remaining part of the assertion we have to show that an element in the coset is mapped to g' and that an element mapped to g' is in the coset. In the first case, let g_1 be in the coset $g + H$. Then $g_1 = g + h$ for some h in H. As f is a homomorphism and its kernel is H, we have

$$f(g_1) = f(g + h) = f(g) + f(h) = g' + 0 = g.$$

On the other hand, suppose that g_1 is any element in G mapped to g' in G'. Then

$$f(g_1 - g) = f(g_1) - f(g) = g' - g' = 0.$$

Therefore, $g_1 - g$ lies in the kernel of f and must be equal to some h in H. As $g_1 - g = h$ implies $g_1 = g + h$, we have that g_1 is in the coset $g + H$. ∎

Theorem 11.6 *Let H be a subgroup of (an abelian group) G. Suppose g_1 and g_2 are elements of G. Define addition between the cosets $g_1 + H$ and $g_2 + H$ as*

$$(g_1 + H) + (g_2 + H) = (g_1 + g_2) + (H + H) = (g_1 + g_2) + H$$

(where addition is as defined in G). Then the collection of all cosets of H in G, denoted by G/H, is an abelian group. G/H is referred to as the factor group *of G by H.*

Proof: Let $g_1 + H$ and $g_2 + H$ be two distinct elements of G/H. Then for any h_1, h_2 in H,

$$(g_1 + h_1) + (g_2 + h_2) = (g_1 + g_2) + (h_1 + h_2) = g + h$$

and thus the group is closed under addition. Also, for three elements $g_1 + H, g_2 + H$, and $g_3 + H$ in the factor group, we have

$$((g_1 + H) + (g_2 + H)) + (g_3 + H) = (g_1 + H) + ((g_2 + H) + (g_3 + H)),$$

which is property (i) (Problem 11.1). To show (ii) we note that $(g_1 + H) + H = g_1 + H$, and therefore H is the zero element of the group. Since $(g_1 + H) + (-g_1 + H) = H$, we have (iii). Finally, G/H is an abelian group, since

$$(g_1 + H) + (g_2 + H) = (g_1 + g_2) + H = g_2 + g_1 + H = (g_2 + H) + (g_1 + H),$$

where $g_1 + g_2 = g_2 + g_1$ follows from the fact that g_1 and g_2 are from the abelian group G. ∎

Corollary 11.6.1   Let H be a subgroup of G and consider the map $f : g \rightarrow G/H$ which to each element g in G associates the coset $f(g) = g + H$. Then f is a homomorphism, and its kernel is precisely H.

Proof: To show that f is a homomorphism we note that

$$f(g_1 + g_2) = (g_1 + g_2) + H = (g_1 + H) + (g_2 + H) = f(g_1) + f(g_2)$$

for any two elements g_1 and g_2. For elements h in H, $f(h) = H$, and thus every element in H (the zero element of G/H) is in the kernel of f. Similarly, if g_1 in G is in the kernel of f, then $f(g_1) = H$, which means $g_1 + H = H$, or g_1 is in H. ∎

Corollary 11.6.2   Let $f : G \to G'$ be a homomorphism whose kernel is H. Then G/H is isomorphic to the image of f, under the map f^* which associates with each coset $g + H$ in G/H the element $f(g)$ in G'.

Proof: We want to show that $f^* : G/H \to$ (Image of f) is an isomorphism, where $f^*(g + H) = f(g)$. To do this we prove that f^* is a homomorphism, and that it is also both surjective and injective.

(Homomorphism)   Let $g_1 + H$ and $g_2 + H$ be two cosets of G/H. Then

$$f^*(g_1 + H + g_2 + H) = f^*(g_1 + g_2 + H) = f(g_1 + g_2).$$

But f is a homomorphism, so

$$f(g_1 + g_2) = f(g_1) + f(g_2) = f^*(g_1 + H) + f^*(g_2 + H)$$

or f^* is a homomorphism.

(Surjective)   Let g' be an element in the image of f. Then there is an element g in G for which $f(g) = g'$. As $f(g) = f^*(g + H)$, $f^*(g + H) = g'$ and f^* maps onto.

(Injective)   Let $g_1 + H$ and $g_2 + H$ be two distinct cosets of G/H and suppose $f^*(g_1 + H) = f^*(g_2 + H)$. Then

$$f(g_1) = f(g_2) \text{ or } f(g_1) - f(g_2) = f(g_1 - g_2) = 0,$$

and thus $g_1 - g_2$ is in the kernel of f or H. As $g_1 - g_2 = h$ for some element h in H, we have $g_1 = g_2 + h$ or the two cosets have an element in common. By Lemma 11.2 this means that they must be equal. Hence, the contradiction and f^* maps one-to-one. ∎

11.6.2   The Isomorphic Groups $G(\bar{\alpha})$ and $G(\lambda)$

We shall now use the previous assertions to show that the group $G(\boldsymbol{\lambda})$ and $G(\bar{\alpha})$ are isomorphic. To do this we need to define the sets

1. $G(\mathbf{I})$ as the collection of all integer m vectors.

2. $G(\mathbf{A})$ as the collection of all integer linear combinations of the

columns $(\mathbf{a}_j)$ of the m by n constraint matrix $\mathbf{A}$. That is, $\mathbf{a}$ is in $G(\mathbf{A})$ whenever there exist integers n_j such that $\mathbf{a} = \sum_{j=1}^{n} n_j \mathbf{a}_j$.

3. $G(\mathbf{B})$ as the collection of all integer linear combinations of the columns of the m by m basis matrix $\mathbf{B}$ whose columns are from $\mathbf{A}$. Note that for any integer m vector $\mathbf{y}$ we have $\mathbf{y} = \mathbf{B}\boldsymbol{\lambda}$ for some scalars $\boldsymbol{\lambda}$. However, the components of $\boldsymbol{\lambda}$ may not be integral.

It is a simple matter to show that each of the sets forms an abelian group. Also, by their construction, $G(\mathbf{B}) \subset G(\mathbf{A}) \subset G(\mathbf{I})$ and $G(\mathbf{A}) = G(\mathbf{I})$ whenever $\mathbf{A}$ contains an identity matrix.

Let us consider the map f which sends elements from $G(\mathbf{A})$ to elements in the group $G(\bar{\boldsymbol{\alpha}})$. That is,

$$f : G(\mathbf{A}) \to G(\bar{\boldsymbol{\alpha}}).$$

Recall that to do this an element in $G(\mathbf{A})$ is premultiplied by $\mathbf{B}^{-1}$ and the resulting column is reduced modulo 1. As these operations are distributive, we have for two elements $\mathbf{a}^1$ and $\mathbf{a}^2$ in $G(\mathbf{A})$ that

$$
\begin{aligned}
f(\mathbf{a}^1 + \mathbf{a}^2) &= [\text{mod } 1](\mathbf{B}^{-1}(\mathbf{a}^1 + \mathbf{a}^2)) = [\text{mod } 1](\mathbf{B}^{-1}\mathbf{a}^1 + \mathbf{B}^{-1}\mathbf{a}^2) \\
&= [\text{mod } 1]\mathbf{B}^{-1}\mathbf{a}^1 \oplus [\text{mod } 1]\mathbf{B}^{-1}\mathbf{a}^2 \\
&= f(\mathbf{a}^1) \oplus f(\mathbf{a}^2),
\end{aligned}
$$

where $[\text{mod } 1]$ means that the column is reduced modulo 1, and $\oplus$ indicates that the addition is in $G(\bar{\boldsymbol{\alpha}})$. Thus, f is a homomorphism. Note that if $\mathbf{A} = (\mathbf{B}, \mathbf{N})$, f sends columns from $\mathbf{B}$ into the zero of $G(\bar{\boldsymbol{\alpha}})$. We now prove that the kernel of f is $G(\mathbf{B})$. This will allow us to apply the last corollary and obtain a principal result.

Lemma 11.4 *The kernel of f is $G(\mathbf{B})$.*

Proof: Let $\mathbf{a}$, an element of $G(\mathbf{A})$, be in the kernel of f. This is true if and only if ($\Leftrightarrow$)

$$[mod1](\mathbf{B}^{-1}\mathbf{a}) = 0$$
$$\Leftrightarrow \quad \mathbf{B}^{-1}\mathbf{a} = \boldsymbol{\gamma} \quad (\boldsymbol{\gamma} = (\gamma_i) \text{ is some integer } m \text{ vector})$$
$$\Leftrightarrow \quad \mathbf{a} = \mathbf{B}\boldsymbol{\gamma} = \sum_{i=1}^{m} \mathbf{b}_i \gamma_i \quad (\mathbf{B} = (\mathbf{b}_i))$$
$$\Leftrightarrow \quad \mathbf{a} \text{ is in } G(\mathbf{B}), \text{ which proves the assertion.} \quad \blacksquare$$

Now we are in a position to use Corollary 11.6.2. In particular, let $G = G(\mathbf{A})$, $G' = G(\bar{\boldsymbol{\alpha}})$, and $H = G(\mathbf{B})$. Since the image of f, by definition, is $G(\bar{\boldsymbol{\alpha}})$ and its kernel is $G(\mathbf{B})$, we have the next result.

Theorem 11.7 $G(\mathbf{A})/G(\mathbf{B})$ *is isomorphic to* $G(\bar{\boldsymbol{\alpha}})$ *under the map* f^* *which associates with each coset* $\mathbf{a} + G(\mathbf{B})$ *in* $G(\mathbf{A})/G(\mathbf{B})$ *the element* $f(\mathbf{a})$ *in* $G(\bar{\boldsymbol{\alpha}})$.

Elements of the factor group $G(\mathbf{A})/G(\mathbf{B})$ are the cosets $\mathbf{a} + G(\mathbf{B})$, where $G(\mathbf{B})$ has absorbed integer multiples of the columns of $\mathbf{B}$ from the vector $\mathbf{a}$. In general, it is difficult to find the cosets in computations, since these integer multiples of columns from $\mathbf{B}$ are hard to identify in a given vector of $G(\mathbf{A})$. However, suppose we first diagonalize $\mathbf{B}$ as discussed in the preceding section. This causes the remaining columns of $\mathbf{A}$ to be premultiplied by the matrix, which defines the elementary row operations.[22] The cosets can then easily be identified. To illustrate, reconsider Example 11.1.

Example (from Example 11.1)

$$\mathbf{B} = (\mathbf{a}_1, \mathbf{a}_2) = \begin{pmatrix} 0 & 2 \\ 3 & -4 \end{pmatrix}$$

was diagonalized to

$$\hat{\mathbf{B}} = \begin{pmatrix} 1 & 0 \\ 0 & 6 \end{pmatrix}.$$

$\mathbf{N}$ became

$$\hat{\mathbf{N}} = \begin{pmatrix} 5 & 1 & -1 & 0 & 1 \\ 11 & 6 & 0 & 1 & 2 \end{pmatrix},$$

and thus $\mathbf{A}$ is transformed to

---

[22]From Section 11.5 or Appendix B we see that only the row operations affect the nonbasic columns. The column operations do not alter the matrix $\mathbf{N}$

$$\hat{\mathbf{A}} = \begin{pmatrix} 1 & 0 & | & 5 & 1 & -1 & 0 & 1 \\ 0 & 6 & | & 11 & 6 & 0 & 1 & 2 \end{pmatrix}.$$

$$\underbrace{}_{\hat{\mathbf{B}}} \qquad \underbrace{}_{\hat{\mathbf{N}}}$$

Now, a column of $\hat{\mathbf{A}}$ belongs to a coset $\hat{\mathbf{a}} + G(\hat{\mathbf{B}})$. The value of $\hat{\mathbf{a}}$ is generally not unique. For example, the column $\binom{5}{11}$ is in the coset

$$\begin{pmatrix} \hat{a}_1 \\ \hat{a}_2 \end{pmatrix} + \left(r \begin{pmatrix} 1 \\ 0 \end{pmatrix} + s \begin{pmatrix} 0 \\ 6 \end{pmatrix} \right),$$

where $\hat{a}_1, \hat{a}_2, r,$ and s are integers, and $\hat{a}_1 + r = 5$ and $\hat{a}_2 + 6s = 11$. However, we are selecting $\hat{\mathbf{a}}$ so that its components are nonnegative and as small as possible. If this is done the cosets obtained from the columns of $\hat{\mathbf{A}}$ are written as

Columns of $\hat{\mathbf{A}}$	Coset						
	$\hat{\mathbf{a}}$	$+$			$G(\hat{\mathbf{B}})$		
$\begin{pmatrix} 1 \\ 0 \end{pmatrix}$	$\begin{pmatrix} 0 \\ 0 \end{pmatrix}$		$1\begin{pmatrix} 1 \\ 0 \end{pmatrix}$	$+$		$0\begin{pmatrix} 0 \\ 6 \end{pmatrix}$	
$\begin{pmatrix} 0 \\ 6 \end{pmatrix}$	$\begin{pmatrix} 0 \\ 0 \end{pmatrix}$		$0\begin{pmatrix} 1 \\ 0 \end{pmatrix}$	$+$		$1\begin{pmatrix} 0 \\ 6 \end{pmatrix}$	
$\begin{pmatrix} 5 \\ 11 \end{pmatrix}$	$\begin{pmatrix} 0 \\ 5 \end{pmatrix}$		$5\begin{pmatrix} 1 \\ 0 \end{pmatrix}$	$+$		$1\begin{pmatrix} 0 \\ 6 \end{pmatrix}$	
$\begin{pmatrix} 1 \\ 6 \end{pmatrix}$	$\begin{pmatrix} 0 \\ 0 \end{pmatrix}$		$1\begin{pmatrix} 1 \\ 0 \end{pmatrix}$	$+$		$1\begin{pmatrix} 0 \\ 6 \end{pmatrix}$	
$\begin{pmatrix} -1 \\ 0 \end{pmatrix}$	$\begin{pmatrix} 0 \\ 0 \end{pmatrix}$		$-1\begin{pmatrix} 1 \\ 0 \end{pmatrix}$	$+$		$0\begin{pmatrix} 0 \\ 6 \end{pmatrix}$	
$\begin{pmatrix} 0 \\ 1 \end{pmatrix}$	$\begin{pmatrix} 0 \\ 1 \end{pmatrix}$		$0\begin{pmatrix} 1 \\ 0 \end{pmatrix}$	$+$		$0\begin{pmatrix} 0 \\ 6 \end{pmatrix}$	
$\begin{pmatrix} 1 \\ 2 \end{pmatrix}$	$\begin{pmatrix} 0 \\ 2 \end{pmatrix}$		$1\begin{pmatrix} 1 \\ 0 \end{pmatrix}$	$+$		$0\begin{pmatrix} 0 \\ 6 \end{pmatrix}$	

As the cosets are members of a group, we may take integer multiples of them to obtain the remaining elements. It is easy to see that $\begin{pmatrix} 0 \\ 5 \end{pmatrix}$ or $\begin{pmatrix} 0 \\ 1 \end{pmatrix}$ may be used to generate all the cosets $\hat{\mathbf{a}} + G(\hat{\mathbf{B}})$ of the group which are written as (without the $G(\hat{\mathbf{B}})$ term)

$$\begin{pmatrix} 0 \\ 1 \end{pmatrix} \quad \begin{pmatrix} 0 \\ 2 \end{pmatrix} \quad \begin{pmatrix} 0 \\ 3 \end{pmatrix} \quad \begin{pmatrix} 0 \\ 4 \end{pmatrix} \quad \begin{pmatrix} 0 \\ 5 \end{pmatrix} \quad \begin{pmatrix} 0 \\ 0 \end{pmatrix}.$$

Note that we have found factor group $G(\hat{\mathbf{A}})/G(\hat{\mathbf{B}})$ rather than $G(\mathbf{A})/G(\mathbf{B})$. Because $\hat{\mathbf{A}} = \mathbf{RA}$ and $\hat{\mathbf{B}} = \mathbf{RBC}$, where $\mathbf{R}$ and $\mathbf{C}$ define the row and column transformations, respectively, we have that the two factor groups are isomorphic. Since the factor group $G(\hat{\mathbf{A}})/G(\hat{\mathbf{B}})$ is isomorphic to the factor group $G(\mathbf{A})/G(\mathbf{B})$, which in turn is isomorphic to $G(\bar{\alpha})$, we have that $G(\hat{\mathbf{A}})/G(\hat{\mathbf{B}})$ is *isomorphic to* $G(\bar{\alpha})$. (The reader is asked to define the composite map in Problem 11.42.)

It is clear from the way $G(\hat{\mathbf{A}})/G(\hat{\mathbf{B}})$ is constructed that the $\hat{\mathbf{a}}$ generate the group $G(\boldsymbol{\lambda})$, over which our second group minimization problem has been defined. Note that the six cosets $\hat{\mathbf{a}}+G(\hat{\mathbf{B}})$, which generate the group $G(\hat{\mathbf{A}})/G(\hat{\mathbf{B}})$ correspond to the six generators $\hat{\mathbf{a}}$ found for $G(\boldsymbol{\lambda})$ in Example 11.1. Therefore, $G(\hat{\mathbf{A}})/G(\hat{\mathbf{B}})$ is isomorphic to $G(\boldsymbol{\lambda})$. We summarize the discussion thus far.

Corollary 11.7.1 Let $\mathbf{B}$ be diagonalized to $\hat{\mathbf{B}}$ and the constraint matrix $\mathbf{A}$ be transformed to $\hat{\mathbf{A}}$. Then the factor group $G(\hat{\mathbf{A}})/G(\hat{\mathbf{B}})$ is isomorphic to the group $G(\boldsymbol{\lambda})$ and thus $G(\boldsymbol{\lambda})$ is isomorphic to $G(\bar{\alpha})$.

Let us return to the group $G(\mathbf{A})/G(\mathbf{B})$. Suppose this group only contained one coset, namely $\mathbf{0} + G(\mathbf{B})$. The only way this can happen is if $G(\mathbf{B}) = G(\mathbf{A})$. If $\mathbf{A}$ contains an identity matrix, it means that $\mathbf{B}$ is an identity matrix. In this case the basic variables $\mathbf{x_B} = \mathbf{B}^{-1}\mathbf{b} = \mathbf{b}$ are integer and there is no group minimization problem over $G(\boldsymbol{\lambda})$ or $G(\bar{\alpha})$ to be solved. To see this, note that, by the corollary, both $G(\boldsymbol{\lambda})$ and $G(\bar{\alpha})$ have only one element, namely the zero vector. Thus the constraints in either problem are of the form

$$\sum_{j=1}^{n} \mathbf{0} x_{J(j)} = \mathbf{0}$$

which means that the minimal solution to either GMP is $\mathbf{x_N} = (x_{J(j)}) = \mathbf{0}$, and thus $\mathbf{x_B} = \mathbf{B}^{-1}\mathbf{b}$. As the order of $G(\boldsymbol{\lambda})$ or $G(\bar{\alpha})$ increases from one, the number of elements in $G(\mathbf{A})$ which cannot be written in terms of elements in $G(\mathbf{B})$ increases. Therefore, the size of $G(\boldsymbol{\lambda})$ or $G(\bar{\boldsymbol{\alpha}})$ in this sense is a measure of the degree to which $G(\mathbf{B})$ fails to span $G(\mathbf{A})$.

11.6.3   The Subgroup Decomposition of $G(\mathbf{A})/G(\mathbf{B})$

We now show that the factor group $G(\mathbf{A})/G(\mathbf{B})$ may be expressed as a direct sum of cyclic subgroups. (That is, every element in $G(\mathbf{A})/G(\mathbf{B})$ may be written as a sum of terms that are in subgroups which are disjoint except for the zero.) To do this, let us first consider the factor group $G(\mathbf{I})/G(\mathbf{B})$, where $G(\mathbf{I})$ is the collection of all integer m vectors. The discussion becomes much simpler if we first diagonalize the basis $\mathbf{B}$ to $\hat{\mathbf{B}}$. Then the group $G(\mathbf{B})$ becomes $G(\hat{\mathbf{B}})$. We will explore the factor group $G(\mathbf{I})/G(\hat{\mathbf{B}})$, which is isomorphic to $G(\mathbf{I})/G(\mathbf{B})$. A definition of a basis for a group is needed.

Definition 11.8 Let G be an abelian group. When we speak of a *basis* for G we mean a set of elements $v_1, \ldots, v_n$ $(n \geq 1)$ in the group for which every element in G may be written uniquely as $c_1 v_1 + c_2 v_2 + \ldots + c_n v_n$, with integer coefficients c_i $(i = 1, \ldots, n)$; where $c_i v_i$ $(i = 1, \ldots, n)$ means that v_i is added to itself c_i times and each addition is as defined in the group.

The columns of $\mathbf{B}$, which are $\mathbf{b}_1, \ldots, \mathbf{b}_m$, form a basis for $G(\mathbf{B})$. When $\mathbf{B}$ is diagonalized to $\hat{\mathbf{B}}$, the columns $\hat{\mathbf{b}}_1, \ldots, \hat{\mathbf{b}}_m$ are a basis for $G(\hat{\mathbf{B}})$. Notice that $\hat{\mathbf{b}}_i$ is a column of zeroes except for row i, which contains the positive integer ϵ_i. The unit vectors $\mathbf{e}_1, \ldots, \mathbf{e}_m$ where $\mathbf{e}_i$ is an m column with zeroes everywhere except for a 1 in row i, serve as a basis for $G(\mathbf{I})$. As $G(\hat{\mathbf{B}})$ is a subgroup of $G(\mathbf{I})$, every element in $G(\hat{\mathbf{B}})$ may be expressed in terms of the basis $\mathbf{e}_1, \ldots, \mathbf{e}_m$. In particular, $\hat{\mathbf{b}}_i = \epsilon_i \mathbf{e}_i$ for $i = 1, \ldots, m$.

Now consider any element in $G(\mathbf{I})$. It can be expressed as

$$c_{1j} \mathbf{e}_1 + c_{2j} \mathbf{e}_2 + \ldots + c_{mj} \mathbf{e}_m,$$

where c_{ij} are integers $(i = 1, \ldots, m)$. Also, any element in $G(\hat{\mathbf{B}})$ may be written as

$$d_{1k} \hat{\mathbf{b}}_1 + d_{2k} \hat{\mathbf{b}}_2 + \ldots + d_{mk} \hat{\mathbf{b}}_m,$$

where the d_{ik} are integers $(i = 1, \ldots, m)$. But $\hat{\mathbf{b}}_i = \epsilon_i \mathbf{e}_i$ and therefore any element in $G(\hat{\mathbf{B}})$ can be expressed as

$$\epsilon_1 d_{1k}\mathbf{e}_1 + \epsilon_2 d_{2k}\mathbf{e}_2 + \ldots + \epsilon_m d_{mk}\mathbf{e}_m.$$

So any coset in $G(\mathbf{I})/G(\hat{\mathbf{B}})$ has the form

$$c_{1j}\mathbf{e}_1 + c_{2j}\mathbf{e}_2 + \ldots + c_{mj}\mathbf{e_m} + G(\hat{\mathbf{B}}), \tag{13}$$

and any term in this coset may be expressed as

$$c_{1j}\mathbf{e}_1 + c_{2j}\mathbf{e}_2 + \ldots + c_{mj}\mathbf{e}_m + (\epsilon_1 d_{1k}\mathbf{e}_1 + \epsilon_2 d_{2k}\mathbf{e}_2 + \ldots + \epsilon_m d_{mk}\mathbf{e}_m).\tag{14}$$

As the addition is commutative, (14) is the same as

$$(c_{1j} + (\epsilon_1 d_{1k}\mathbf{e}_1)) + (c_{2j} + (\epsilon_2 d_{2k}\mathbf{e}_2)) + \ldots + (c_{mj} + (\epsilon_m d_{mk}\mathbf{e}_m)). \tag{15}$$

Now, as different integers d_{ik} are selected, the $\epsilon_i d_{ik}\mathbf{e}_i = d_{ik}\hat{\mathbf{b}}_i$ term yields the group generated by column $\hat{\mathbf{b}}_i$ of $\hat{\mathbf{B}}$. Suppose we denote this by $G(\hat{\mathbf{b}}_i) = G(\epsilon_i\mathbf{e}_i)$. Then any coset (13) in $G(\mathbf{I})/G(\hat{\mathbf{B}})$ may be expressed as

$$(c_{1j} + G(\epsilon_1\mathbf{e}_1)) + (c_{2j} + G(\epsilon_2\mathbf{e}_2)) + \ldots + (c_{mj} + G(\epsilon_m\mathbf{e}_m)). \tag{16}$$

Now suppose we consider all cosets of $G(\mathbf{I})/G(\hat{\mathbf{B}})$. That is, we allow j to vary so as to obtain all possible c_{ij} ($j = 1, \ldots, m$). Then each term $c_{ij}\mathbf{e}_i + G(\epsilon_i\mathbf{e}_i)$, may be thought of as a coset and an element in the factor group $G(\mathbf{e}_i)/G(\epsilon_i\mathbf{e}_i)$. Thus we have the following result.

Lemma 11.5 *The factor group $G(\mathbf{I})/G(\hat{\mathbf{B}})$ is isomorphic to the direct sum group*

$$G(\mathbf{e}_1)/G(\epsilon_1\mathbf{e}_1) + G(\mathbf{e}_2)/G(\epsilon_2\mathbf{e}_2) + \ldots G(\mathbf{e}_m)/G(\epsilon_m\mathbf{e}_m) \tag{17}$$

under the map which sends a coset (13) in $G(\mathbf{I})/G(\hat{\mathbf{B}})$ to the cosets (16) of the direct sum group.

Proof: The previous discussion demonstrates that the map is onto and one-to-one. Also, since the addition is commutative, it is a simple matter to show that it is a homomorphism. The direct sum is allowed, since by construction each subgroup is disjoint except for the zero element.

∎

Let us examine the factor group $G(\mathbf{e}_i)/G(\epsilon_i \mathbf{e}_i)$. The cosets are of the form $c\mathbf{e}_i + G(\epsilon_i \mathbf{e}_i)$ (where c is a nonnegative integer), and thus the only distinct cosets are (dropping the $G(\epsilon_i \mathbf{e}_i)$ term)

$$\mathbf{e}_i, \; 2\mathbf{e}_i, \; \ldots, \; (\epsilon_i - 1)\mathbf{e}_i, \; \epsilon_i \mathbf{e}_i = \mathbf{0}.$$

This means that $G(\mathbf{e}_i)/G(\epsilon_i \mathbf{e}_i)$ is a cyclic subgroup of order ϵ_i. As the subgroups are disjoint, the direct sum group (17) has precisely $\Pi_{i=1}^m \epsilon_i = |\det \mathbf{B}| = D$ elements, and by the lemma the order of $G(\mathbf{I})/G(\hat{\mathbf{B}})$ is D. Furthermore, since $G(\mathbf{I})/G(\mathbf{B})$ is isomorphic to $G(\mathbf{I})/G(\hat{\mathbf{B}})$, the order of $G(\mathbf{I})/G(\mathbf{B})$ is D.

Before summarizing the results let us return to the factor group $G(\hat{\mathbf{A}})/G(\hat{\mathbf{B}})$. If $\mathbf{A}$ contains an identity matrix, $G(\mathbf{A}) = G(\mathbf{I})$, and thus $G(\hat{\mathbf{A}}) = G(\mathbf{I})$ because $\hat{\mathbf{A}} = \mathbf{RA}$ and $|\det \mathbf{R}| = 1$. In this case, the lemma and subsequent comments are valid for $G(\hat{\mathbf{A}})/G(\hat{\mathbf{B}})$ and therefore for the isomorphic groups $G(\boldsymbol{\lambda})$ and $G(\bar{\boldsymbol{\alpha}})$. If $G(\hat{\mathbf{A}})$ is a subgroup of $G(\mathbf{I})$, then it may be possible to span $G(\hat{\mathbf{A}})$ with only a subset of the unit vectors $\mathbf{e}_i$, and thus only a subset of the elements $\epsilon_i \mathbf{e}_i$ may be needed for $G(\hat{\mathbf{B}})$. Consequently, (13) through (16) may contain less than m terms, and the direct sum group (17) may have less than m cyclic subgroups. However, the lemma and subsequent comments otherwise remain valid. The results are summarized in a principle result stated in terms of $G(\bar{\boldsymbol{\alpha}})$.

Theorem 11.8 *If the constraint matrix $\mathbf{A}$ contains an identity matrix, the group $G(\bar{\boldsymbol{\alpha}})$ is isomorphic to the direct sum group (17), where each subgroup is cyclic and of order ϵ_i. The order of $G(\bar{\boldsymbol{\alpha}})$ is precisely D. When $\mathbf{A}$ does not contain an identity matrix, the group $G(\bar{\boldsymbol{\alpha}})$ is isomorphic to the direct sum group (17), which may have less than m terms.*

To illustrate the discussion, let us take another look at the Gomory problem (Example 11.1). In that problem, $\mathbf{A}$ contained an identity matrix and $\mathbf{B}$ became

$$\hat{\mathbf{B}} = \begin{pmatrix} 1 & 0 \\ 0 & 6 \end{pmatrix}.$$

Thus $G(\bar{\boldsymbol{\alpha}})$ is isomorphic to $G(\mathbf{e}_1)/G(1\mathbf{e}_1) + G(\mathbf{e}_2)/G(6\mathbf{e}_2)$. To see this, note that the order of the first subgroup is 1 and it only contains the zero element. Therefore it may be dropped from the direct sum. The second subgroup has six ($\epsilon_2 = 6$) elements. They are (without the $G(6\mathbf{e}_2)$ term)

$$\begin{pmatrix} 0 \\ 1 \end{pmatrix} \begin{pmatrix} 0 \\ 2 \end{pmatrix} \begin{pmatrix} 0 \\ 3 \end{pmatrix} \begin{pmatrix} 0 \\ 4 \end{pmatrix} \begin{pmatrix} 0 \\ 5 \end{pmatrix} \begin{pmatrix} 0 \\ 0 \end{pmatrix}.$$

In Section 11.6.2 we showed this is the group $G(\boldsymbol{\lambda})$, isomorphic to $G(\bar{\boldsymbol{\alpha}})$ (Corollary 11.7.1). Thus $G(\bar{\boldsymbol{\alpha}})$ is isomorphic to $G(\mathbf{e}_1)/G(1\mathbf{e}_1) + G(\mathbf{e}_2)/G(6\mathbf{e}_2)$.

The illustration indicates that when $\epsilon_i = 1$, the subgroup $G(\mathbf{e}_i)/G(\epsilon_i\mathbf{e}_i)$ may be dropped in the direct sum representation. In particular, if $G(\bar{\boldsymbol{\alpha}})$ is cyclic, $\epsilon_1 = \ldots = \epsilon_{m-1} = 1, \epsilon_m = D$, and thus it is isomorphic to the cyclic subgroup $G(\mathbf{e}_m)/G(D\mathbf{e}_m)$. We now investigate the order of the group when $\mathbf{A}$ does not necessarily contain an identity matrix.

11.6.4   The Order of $G(\bar{\alpha})$ and $G(\lambda)$

From the previous theorem we know that the order of $G(\bar{\boldsymbol{\alpha}})$ is D whenever $\mathbf{A}$ contains an identity matrix. We intend to show that it will divide D otherwise. To do this we use Theorem 11.4, which says that the order of

$$G = \text{(the number of cosets of H in G)(the order of H)}, \quad (18)$$

where H is a subgroup of the group G. We have shown that the order of $G(\mathbf{I})/G(\mathbf{B})$ is D, and we are interested in the order of $G(\bar{\boldsymbol{\alpha}})$ or the isomorphic group $G(\mathbf{A})/G(\mathbf{B})$. Since $G(\mathbf{A})/G(\mathbf{B})$ is a subgroup of $G(\mathbf{I})/G(\mathbf{B})$, we let $G = G(\mathbf{I})/G(\mathbf{B})$ and $H = G(\mathbf{A})/G(\mathbf{B})$ in (18) to obtain

$$\begin{aligned} D &= \text{order } G(\mathbf{I})/G(\mathbf{B}) \\ &= \text{(the number of cosets of } G(\mathbf{A})/G(\mathbf{B}) \text{ in } G(\mathbf{I})/G(\mathbf{B})) \\ &\quad \times \text{(the order of } G(\mathbf{A})/G(\mathbf{B})). \end{aligned}$$

As the order of $G(\mathbf{A})/G(\mathbf{B})$ is equal to the order of $G(\bar{\boldsymbol{\alpha}})$, this equality demonstrates that the order of $G(\bar{\boldsymbol{\alpha}})$ divides D. Furthermore, if $\mathbf{A}$ contains an identity matrix, $G(\mathbf{A})/G(\mathbf{B}) = G(\mathbf{I})/G(\mathbf{B})$, and so the number of cosets of $G(\mathbf{A})/G(\mathbf{B})$ in $G(\mathbf{I})/G(\mathbf{B})$ is one (namely the zero). This re-proves that the order of $G(\bar{\boldsymbol{\alpha}}) = D$ whenever $\mathbf{A}$ contains an identity matrix.

11.7   The Geometry

We will now discuss the solution region to the integer program and the group problem in various spaces. For ease of reference we restate the constraints of the integer program. The objective function has little bearing on this discussion and therefore will not appear. Originally we had the inequalities

$$\mathbf{A}'\mathbf{x}' \leq \mathbf{b}, \quad \mathbf{x}' \geq \mathbf{0} \text{ and integer,} \tag{19}$$

where $\mathbf{A}'$ is m by n. After adding the nonnegative slack variables $\mathbf{s}$ we obtained

$$\mathbf{A}\mathbf{x} = \mathbf{b}, \quad \mathbf{x} \geq \mathbf{0} \text{ and integer,} \tag{20}$$

where $\mathbf{A} = (\mathbf{A}', \mathbf{I})$ is m by $(m + n)$ and $\mathbf{x} = (\mathbf{x}', \mathbf{s})$. To construct the group problem a basis $\mathbf{B}$ was selected from $\mathbf{A}$, and (20) was rewritten as

$$\mathbf{B}\mathbf{x}_{\mathbf{B}} + \mathbf{N}\mathbf{x}_{\mathbf{N}} = \mathbf{b}, \quad \mathbf{x}_{\mathbf{B}} \geq \mathbf{0}, \mathbf{x}_{\mathbf{N}} \geq \mathbf{0} \text{ and integer.} \tag{21}$$

Then $\mathbf{x}_{\mathbf{B}} \geq \mathbf{0}$ was dropped and (21) became

$$\left. \begin{array}{l} \displaystyle\sum_{j=1}^{n} \bar{\alpha}_j x_{J(j)} \equiv \bar{\alpha}_0 \pmod 1, \\[2em] \mathbf{x}_{\mathbf{N}} = \left(x_{J(1)}, \ldots, x_{J(n)} \right) \geq \mathbf{0} \text{ and integer,} \end{array} \right\} \tag{22}$$

where $\bar{\alpha}_j$ is the jth column of $\mathbf{B}^{-1}\mathbf{N}$ reduced modulo 1, and $\bar{\alpha}_0$ is $\mathbf{B}^{-1}\mathbf{b}$ reduced modulo 1. The constraints (22) are those of the group problem.

11.7.1   The Corner Polyhedron ($\mathbf{x}'$ Space)

It is a simple matter to picture the constraints (22) in terms of the region defined by (19) in $\mathbf{x}'$ space. Suppose we temporarily drop the $\mathbf{x}'$ integer constraints in (19). Then the region defined by the inequalities remaining is an n dimensional convex polyhedron P. Adding the $\mathbf{x}'$ in-

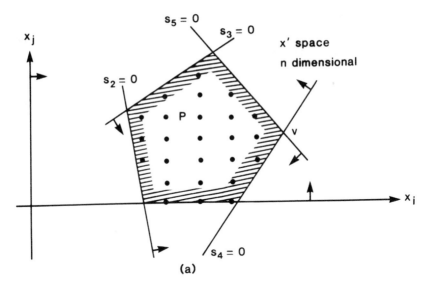

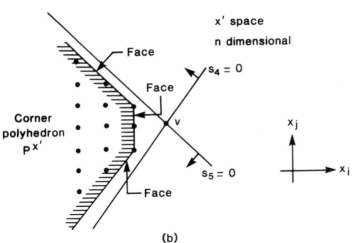

Figure 11.8: Polyhedra in $\mathbf{x}'$ Space

teger requirements yields the lattice points inside P (Fig. 11.8(a)). A basis $\mathbf{B}$ selected from $\mathbf{A}$ produces the basic feasible solution $\mathbf{x_B} = \mathbf{B}^{-1}\mathbf{b}$ to $\mathbf{Ax} = \mathbf{b}$; $\mathbf{x} = (\mathbf{x_B}, \mathbf{0})$ is an extreme point to the convex region defined by $\mathbf{Ax} = \mathbf{b}, \mathbf{x} \geq \mathbf{0}$. A corresponding extreme point in $\mathbf{x}'$ space is obtained by omitting the slack variables $\mathbf{s}$ from $\mathbf{x} = (\mathbf{x_B}, \mathbf{0})$. This is the vertex $\mathbf{v}$ of P in Fig. 11.8(a).

The constraints (22) for the group problem are obtained from those of the integer program by dropping the nonnegativity requirements on the basic variables $\mathbf{x_B}$. In $\mathbf{x}'$ space this is equivalent to dropping those inequalities from $\mathbf{A}'\mathbf{x}' \leq \mathbf{b}$ and $\mathbf{x}' \geq \mathbf{0}$ which are not binding at the vertex $\mathbf{v}$. To see this, denote the constraints of $\mathbf{A}'\mathbf{x}' \leq \mathbf{b}$ as $\mathbf{a}^i\mathbf{x}' \leq b_i$, where $\mathbf{a}^i$ is the ith row of $\mathbf{A}'$ and b_i is the ith component of $\mathbf{b}$, and write $\mathbf{x}' \geq \mathbf{0}$ as $x_j \geq 0$ for $j = 1, \ldots, n$. Then if $\mathbf{a}^i\mathbf{x}' < b_i$, it means that the constraint is not binding and the associated slack variable $s_i > 0$ in $\mathbf{x} = (\mathbf{x_B}, \mathbf{0})$, or s_i is a basic variable. Similarly, if $x_j > 0$, the hyperplane $x_j = 0$ does not intersect $\mathbf{v}$ and x_j is a basic variable. Therefore, disregarding the nonnegativity requirements on the basic variables is equivalent to allowing $\mathbf{a}^i\mathbf{x}'$ or x_j to take on any value when their hyperplane, $s_i = b_i - \mathbf{a}^i\mathbf{x}' = 0$, or $x_j = 0$ does not pass through $\mathbf{v}$. After dropping these hyperplanes we obtain the convex region with vertex $\mathbf{v}$ appearing in Fig. 11.8(b). The integer solutions to (22) are the lattice points inside this region. The convex hull of these integer points may be thought of as the feasible region to the group problem. This hull, denoted by $P^{\mathbf{x}'}$ in Fig. 11.8(b), is termed a *corner polyhedron*. The hyperplanes which bound $P^{\mathbf{x}'}$ are its *faces*. In n space, a face is an $n - 1$ dimensional hyperplane.

Example (from Example 11.2)

As an illustration, let us reconsider Example 11.2. In that problem,

$$\mathbf{A}' = \begin{pmatrix} -1 & -4 \\ -3 & -2 \end{pmatrix} \quad \text{and} \quad \mathbf{b} = \begin{pmatrix} -5 \\ -7 \end{pmatrix}.$$

Therefore, in $\mathbf{x}' = (x_1, x_2)$ space the region P is defined by $-x_1 - 4x_2 \leq -5$, $-3x_1 - 2x_2 \leq -7$, $x_1 \geq 0, x_2 \geq 0$, and appears in Fig. 11.9. The integer solutions are dots inside P. The basis $\mathbf{B}$ selected was $\mathbf{A}'$, and

$$\mathbf{x_B} = (x_1, x_2) = \mathbf{B}^{-1}\mathbf{b} = \begin{pmatrix} 18/10 \\ 8/10 \end{pmatrix}$$

is the vertex $\mathbf{v}$ in the figure. Therefore, the corner polyhedron is obtained by dropping the hyperplanes $x_1 = 0$ and $x_2 = 0$, and taking the convex hull of the integer solutions. $P^{\mathbf{x}'}$ is the shaded area in Fig. 11.9.

Observe that any integer point $\mathbf{x}'$ in $P^{\mathbf{x}'}$ will yield nonnegative integer values for the nonbasic variables which satisfy the congruence

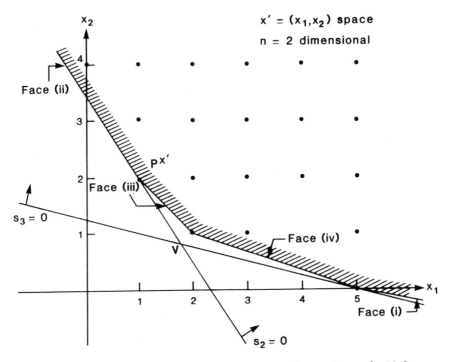

Figure 11.9: Corner Polyhedron (x' Space) for Example 11.2

relationship in (22). Furthermore, a vertex of the corner polyhedron is an optimal solution for the integer program. The vertices are at the intersection of faces. These inequalities are the strongest that can be obtained from the integer program without using the $x_B \geq 0$ requirements (Problem 11.49). Thus, it may be worth finding the face inequalities and using them (e.g.) in a cuttng plane algorithm or as implicit enumeration criteria in a branch and bound algorithm. To facilitate generating these inequalities we examine the corner polyhedron in x_N space.

11.7.2 The Corner Polyhedron (x_N Space)

The region defined by the constraints in (22) can also be plotted directly in x_N space. To do this, recall that the congruence relationship

$$\sum_{j=1}^{n} \bar{\alpha}_j x_{J(j)} \equiv \bar{\alpha}_0 \pmod{1}$$

is equivalent to

$$\sum_{j=1}^{n} f_{ij} x_{J(j)} \equiv f_{i0} \quad (\text{mod } 1) \quad (i = 1, \ldots, m), \tag{23}$$

where f_{ij} is the ith component $(1 \leq i \leq m)$ of $\bar{\alpha}_j$ $(j = 0, 1, \ldots, n)$. As the left-hand side in (23) must differ from the right-hand side by an integer amount, for $x_{J(j)} \geq 0$ the congruence can hold only when the left-hand side is $f_{i0}, 1 + f_{i0}, 2 + f_{i0}$ etc. Or (23) implies

$$\sum_{j=1}^{n} f_{ij} x_{J(j)} \geq f_{i0} \quad (i = 1, \ldots, m). \tag{24}$$

Note that the inequalities (24) are precisely the Gomory fractional cuts (Chapter 4) obtained from the m equalities $\mathbf{x_B} = \mathbf{B}^{-1}\mathbf{b} - \mathbf{B}^{-1}\mathbf{Nx_N}$.

Every $x_{J(1)}, \ldots, x_{J(n)}$ which satisfies (23) will satisfy (24). However, the converse is not true, since $x_{J(1)}, \ldots, x_{J(n)}$ satisfying (24) may yield a value for $\sum_{j=1}^{n} \bar{\alpha}_j x_{J(j)}$ which is not an integer plus $\bar{\alpha}_0$. Therefore, the constraints (22) for the group problem may be viewed in $\mathbf{x_N}$ space by plotting the inequalities (24) along with $\mathbf{x_N} \geq 0$ and integer, and taking the convex hull of the points satisfying (23). The region is termed the corner polyhedron $P^{\times}N$. In Fig. 11.10 the hyperplanes $s^1 = 0$ and $s^2 = 0$ are from (24). The circled lattice points are those which satisfy the congruence relationship, and $P^{\times}N$ is the shaded area. Notice that the polyhedron is unbounded above. This is always true, since if $\mathbf{x_N}$ satisfies (23), then for any integer K, $K|\det \mathbf{B}|$ added to each of its components will satisfy (23). The faces of $P^{\times}N$ are also labeled.

Example (from Example 11.2)

The constraints of the group problem are

$$\begin{pmatrix} \frac{2}{10} \\ \frac{7}{10} \end{pmatrix} x_{J(1)} + \begin{pmatrix} \frac{6}{10} \\ \frac{1}{10} \end{pmatrix} x_{J(2)} \equiv \begin{pmatrix} \frac{8}{10} \\ \frac{8}{10} \end{pmatrix} \quad (\text{mod } 1) \tag{23'}$$

$x_{J(1)}, x_{J(2)} \geq 0$ and integer.

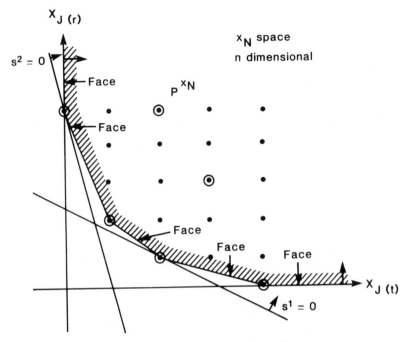

Figure 11.10: Corner Polyhedron

Thus plotting

$$2x_{J(1)} + 6x_{J(2)} \geq 8 \quad (s^1)$$
$$7x_{J(1)} + 1x_{J(2)} \geq 0 \quad (s^2) \tag{24$'$}$$
$$x_{J(1)}, x_{J(2)} \geq 0 \quad \text{and integer}$$

yields Fig. 11.11. The integer points satisfying $(23)'$ are circled and the corner polyhedron P^{x_N} is the convex hull of these points. Notice that in this problem the convex hull is the region defined by the intersection of the constraints $(24)'$ with $x_{J(1)}, x_{J(2)} \geq 0$. This usually will not be true.

11.7.3   Relating the Corner Polyhedra

Points

A point in $\mathbf{x}'$ space can be obtained from a point in $\mathbf{x_N}$ space using

$$\mathbf{x_B} = \mathbf{B}^{-1}\mathbf{b} - \mathbf{B}^{-1}\mathbf{N}\mathbf{x_N}$$

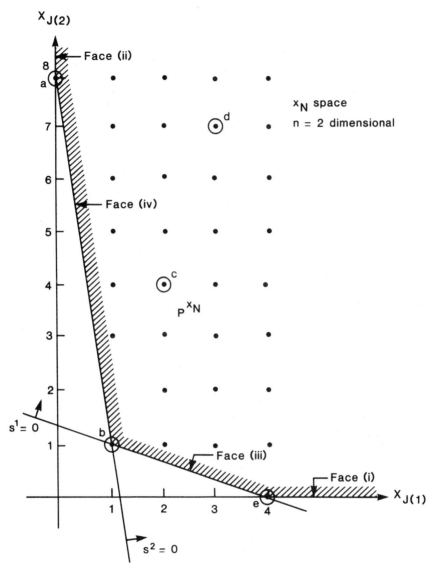

Figure 11.11: Corner Polyhedron (x_N space) for Example 11.2

and deleting the m slack variables from $x = (x_B, x_N)$. Every point in $P^x N$ which satisfies the congruence relationship (23) is a solution to the group problem (circled in Fig. 11.10) and therefore corresponds to a lattice point in $P^x N$ (Fig. 11.8(b)).

Example (from Example 11.2)

Some circled points in Fig. 11.11 are

No.	a	b	c	d	e
$x_{J(1)}$	0	1	2	3	4
$x_{J(2)}$	8	1	4	7	0

Using

$$\mathbf{x_B} = \mathbf{B^{-1}b} - \mathbf{B^{-1}Nx_N} = \begin{pmatrix} 18/10 \\ 8/10 \end{pmatrix} - \begin{pmatrix} 2/10 & -4/10 \\ -3/10 & 1/10 \end{pmatrix} \mathbf{x_N},$$

we obtain

	No.			a	b	c	d	e
x′	x_{B1}	=	x_1	5	2	3	4	1
	x_{B2}	=	x_2	0	1	1	1	2
	$x_{J(1)}$	=	s_1	0	1	2	3	4
	$x_{J(2)}$	=	s_2	8	1	4	7	0
				vertex	vertex	–	–	vertex

Note that each extreme point of $P^{\mathbf{x}}\mathbf{N}$ corresponds to an extreme point of $P^{\mathbf{x'}}$. Before proving that this is always the case, we examine the faces of the polyhedra.

Faces

A face of $P^{\mathbf{x}}\mathbf{N}$ has the form

$$\sum_{j=1}^{n} \bar{\gamma}_j x_{J(j)} = \bar{\gamma}_0, \tag{25}$$

where points $\mathbf{x_N} = (x_{J(1)}, \ldots, x_{J(n)})$ satisfying the face inequality

$$\sum_{j=1}^{n} \bar{\gamma}_j x_{J(j)} \geq \bar{\gamma}_0$$

are in $P^{\mathbf{x}}\mathbf{N}$. We can show that the coefficients $\bar{\gamma}_j$ $(j = 1, \ldots, n)$ are nonnegative. To do this, fix $\bar{\gamma}_0$ and select a point $\mathbf{x_N}$ which satisfies (25). Suppose $\bar{\gamma}_s < 0$, then in (25), for K integer

$$\sum_{\substack{j=1 \\ j \neq s}}^{n} \bar{\gamma}_j x_{J(j)} + \bar{\gamma}_s (x_{J(s)} + K|\det \mathbf{B}|) \geq \bar{\gamma}_0. \tag{26}$$

The inequality (26) is true because $x_{J(s)} \bar{\alpha}_s \equiv (x_{J(s)} + K|\det \mathbf{B}|) \bar{\alpha}_s$ (mod 1), so (23) is satisfied by the new vector $\mathbf{x_N}$ (element $x_{J(s)}$ is replaced by $x_{J(s)} + K|\det \mathbf{B}|$), and therefore it is in $P^{\mathbf{x}}N$, or the face inequality (26) is satisfied. As K increases, $\bar{\gamma}_s < 0$ invalidates the inequality (26), which is a contradiction. Note that $\bar{\gamma}_j \geq 0$ ($j = 1, \ldots, n$) means that $\bar{\gamma}_0 \geq 0$, since

$$\sum_{j=1}^{n} \bar{\gamma}_j x_{J(j)} \geq 0$$

and (25) must be binding for points on the face.

To obtain a face of $P^{\mathbf{x}'}$ from a face (25) of $P^{\mathbf{x}}N$, we examine each nonbasic variable. There are two possibilities. Either $x_{J(j)}$ is some original variable x_k ($1 \leq k \leq n$) or it is a slack variable s_i, in which case it can be expressed as a linear combination of the original variables $\mathbf{x}'$, namely $s_i = b_i - \mathbf{a}^i \mathbf{x}'$. Substituting for each nonbasic variable in (25) produces

$$\sum_{j=1}^{n} \gamma_j x_j = \gamma_0, \tag{27}$$

which is a face for $P^{\mathbf{x}'}$. Equality (25) is a face for $P^{\mathbf{x}}N$ if and only if (27) is a face for $P^{\mathbf{x}'}$, because when $\mathbf{x_N}$ yields a strict inequality (equality) in (25), then the corresponding vector $\mathbf{x}'$ produces a strict inequality (equality) in (26) and vice versa.

Example (from Example 11.2)

The original equality constraints are

$$
\begin{array}{rrrrl}
-x_1 & -4x_2 & +s_1 & & = -5, \\
-3x_1 & -2x_2 & & +s_2 & = -7,
\end{array}
$$

From Fig. 11.11 the faces of $P^{\mathbf{x}}N$ are

    i) $x_{J(1)} = s_1 \geq 0$

    ii) $x_{J(2)} = s_2 \geq 0$

    iii) $2x_{J(1)} + 6x_{J(2)} \geq 8$

    iv) $7x_{J(1)} + 1x_{J(2)} \geq 8.$

Substituting $-5 + x_1 + 4x_2$ for s_1 and $-7 + 3x_1 + 2x_2$ for s_2 produces the corresponding four faces of $P^{\mathbf{x}'}$ (Fig. 11.9)

    i) $x_1 \;+\; 4x_2 \;\geq\; 5$

    ii) $3x_1 \;+\; 2x_2 \;\geq\; 7$

    iii) $x_1 \;+\; x_2 \;\geq\; 3$

    iv) $x_1 \;+\; 3x_2 \;\geq\; 5.$

Extreme Points

We can show that $\mathbf{x}'$ is an extreme point of $P^{\mathbf{x}'}$ if and only if the corresponding $\mathbf{x}_N$ is an extreme point of $P^{\mathbf{x}}N$ using the relationship between the faces. A vertex $\mathbf{x}'$ of $P^{\mathbf{x}'}$ is a basic feasible solution to a system of n face equalities, each of the form

$$\sum_{j=1}^{n} \gamma_j x_j = \gamma_0$$

It follows that $\mathbf{x}_N$ is a feasible solution to the corresponding n face equalities

$$\sum_{j=1}^{n} \bar{\gamma}_j x_j = \bar{\gamma}_0 \tag{28}$$

in $P^{\mathbf{x}}N$. Furthermore, $\mathbf{x}_N$ is a basic solution, since if not, some of the hyperplanes (28) would not intersect $\mathbf{x}_N$, which contradicts that $\mathbf{x}_N$ is a solution to all the face equalities. Thus, $\mathbf{x}_N$ is an extreme point of $P^{\mathbf{x}}N$. The proof in the other direction is identical.

Example (from Example 11.2) $n = 2, \mathbf{x}' = (x_1, x_2)$

$P^X N$ (Figure11.11)		$P^{X'}$(Figure11.9)	
Basic feasible solution to equations		Basic feasible solution to equations	
or vertex	or defining faces	or vertex	or defining faces
$\begin{pmatrix} 0 \\ 8 \\ 1 \\ 1 \\ 4 \\ 0 \end{pmatrix}$	$x_{J(1)} = 0$ $7x_{J(1)} + x_{J(2)} = 8$ $2x_{J(1)} + 6x_{J(2)} = 8$ $7x_{J(1)} + x_{J(2)} = 8$ $x_{J(2)} = 0$ $2x_{J(1)} + 6x_{J(2)} = 8$	$\begin{pmatrix} 5 \\ 0 \\ 2 \\ 1 \\ 1 \\ 2 \end{pmatrix}$	$x_1 + 4x_2 = 5$ $x_1 + 3x_2 = 5$ $x_1 + x_2 = 3$ $x_1 + 3x_2 = 5$ $3x_1 + 2x_2 = 7$ $x_1 + x_2 = 3$

We know show how to obtain faces. To do this we introduce the master (corner) polyhedron.

11.7.4   The Master (Corner) Polyhedron

The corner polyhedron $P^X N$ is the convex hull of solutions to

$$\left. \begin{array}{l} \sum_{j=1}^{n} \bar{\alpha}_j x_{J(j)} \equiv \bar{\alpha}_0 \quad (\text{mod } 1) \\[2mm] x_{J(j)} \geq 0 \text{ and integer, } j = 1, \ldots, n. \end{array} \right\} \qquad (29)$$

We have shown that the $\bar{\alpha}_j (j = 1, \ldots, n)$ generate a finite abelian group $G(\bar{\alpha})$ of order D (Section 11.6.4). (For simplicity, we are assuming that A contains an identity matrix.) Let $g_1, \ldots, g_D$ be the elements of $G(\bar{\alpha})$, where g_D is the zero element. That is, each g_k $(k = 1, \ldots, D)$ is some integer linear combination of the vectors $\bar{\alpha}_j$ $(j = 1, \ldots, n)$. Suppose now we consider the convex hull of solutions to

$$\left. \begin{array}{l} \sum_{k=1}^{D-1} g_k t_k \equiv g_t \quad (\text{mod } 1) \\[4mm] t_k \geq 0 \text{ and integer} \quad (k = 1, \ldots, D-1), \end{array} \right\} \qquad (30)$$

where $\bar{\alpha}_0$ is the group element g_t.[23] This region is termed the *master polyhedron* P^G and is $D-1$ dimensional. Let us investigate the

---

[23]We are assuming that the group problem (29) has a solution and thus $\bar{\alpha}_0$ is an element of $G(\bar{\alpha})$

relationships between the polyhedron P^G and $P^X\mathbf{N}$. To simplify the notation, assume that $\bar{\alpha}_j = \mathbf{g}_j$ for $j = 1,\ldots,n$ and therefore the remaining group elements $\mathbf{g}_j$ $(j = n+1,\ldots,D)$ are either multiples of an $\bar{\alpha}_j$ $(1 \le j \le n)$ or some integer linear combination of at least two of them. ($D \ge n$ because n is the number of distinct nonzero columns $\bar{\alpha}_j$ in the GMP.) Examining the constraint sets (29) and (30) makes it evident that $x_{J(1)}, \ldots, x_{J(n)}$ is in $P^X\mathbf{N}$ if and only if

$$t_1 = x_{J(1)},\ t_2 = x_{J(2)},\ \ldots,\ t_n = x_{J(n)},\ t_{n+1} = 0,\ \ldots,\ t_{D-1} = 0$$

is in P^G. Thus the corner polyhedron $P^X\mathbf{N}$ is the intersection of the master polyhedron P^G with the hyperplanes $t_{n+1} = 0, t_{n+2} = 0,\ldots,t_{D-1} = 0$. Algebraically, this means that by omitting the $t_{n+1} = 0,\ldots,t_{D-1}$ components of a point in P^G we obtain a point in $P^X\mathbf{N}$. It follows that if

$$\sum_{k=1}^{D-1} \bar{\gamma}_k t_k = \bar{\gamma}_0 \tag{31}$$

is a bounding hyperplane or face of P^G, then

$$\sum_{j=1}^{n} \bar{\gamma}_j x_{J(j)} = \bar{\gamma}_0 \tag{32}$$

is a face for $P^X\mathbf{N}$. Consequently, the faces (32) for $P^X\mathbf{N}$ may be obtained from the faces (31) for P^G by dropping the t_k components whose group elements is not an $\bar{\alpha}_j$. [24] The next theorem tells us how to calculate the faces of P^G. It is essentially based on the fact that a face is a $D-2$ dimensional hyperplane spanned by $D-1$ extreme points, each of which is a basic feasible solution to a system of $D-1$ face equalities (31). The proof is quite lengthy and some of the details are discussed in Problems 11.56 to 11.61.

Theorem 11.9 *Suppose $\bar{\gamma}_0 \ne 0$ and define $\gamma'_k = \bar{\gamma}_k/\bar{\gamma}_0$ for $k = 1,\ldots,$ $D-1$. Then $\sum_{k=1}^{D-1} \gamma'_k t_k = 1$ (or (31)) is a face of P^G if and only if $\gamma'_1,\ldots,\gamma'_{D-1}$ is a basic feasible solution to the following system ($\oplus$ means addition as defined in the group):*

i) $\gamma'_t = 1$ if $\mathbf{g}_t \ne \mathbf{g}_D$ (zero element)

---

[24]This is not very precise. As we shall see, redundant inequalities of the form (31) may appear, in which case they are not faces and should also be dropped.

ii) $\gamma_r' + \gamma_s' = 1$ *for* $r = 1, \ldots, D-1$ $r \neq t$ *whenever* $\mathbf{g}_r \oplus \mathbf{g}_s = \mathbf{g}_t$

iii) $\gamma_r' + \gamma_s' - \gamma_k' \geq 0$ *for* $r, s = 1, \ldots, D-1$ *whenever* $\mathbf{g}_r \oplus \mathbf{g}_s = \mathbf{g}_k$, $\mathbf{g}_k \neq \mathbf{g}_D$

iv) $\gamma_k' \geq 0$ *for* $k = 1, \ldots, D-1$.

Moreover, for $\bar{\gamma}_0 = 0$ *the only possible faces are* $t_k = 0$ $(k = 1, \ldots, D-1)$, *and these are faces of* P^G *if and only if the group is not of order two.*

To generate the system of equations (i) through (iv) we have to add group elements and see if their sum is another nonzero group element. We have used the group $G(\bar{\alpha})$ to define the corner polyhedron $P^X\mathrm{N}$ and the master polyhedron P^G. Its elements, $\mathbf{g}_k$, are of the form

$$\frac{1}{D}\begin{pmatrix} I_{1k} \\ \vdots \\ I_{mk} \end{pmatrix},$$

where I_{ik} is an integer satisfying $0 \leq I_{ik} < D$ $(i = 1, \ldots, m)$. In computations, it may be simpler to use the isomorphic group $G(\boldsymbol{\lambda})$ (Section 11.6) whose elements, $\boldsymbol{\lambda}_k$ have integer components usually with many zeroes. Since the groups are isomorphic it follows that the solutions to

$$\left. \begin{aligned} \sum_{k=1}^{n} \boldsymbol{\lambda}_k x_{J(k)} &\equiv \boldsymbol{\lambda}_0 \mod \begin{pmatrix} \epsilon_1 \\ \vdots \\ \epsilon_m \end{pmatrix} \\ x_{J(k)} &\geq 0 \text{ and integer } (k = 1, \ldots, n) \end{aligned} \right\} \tag{33}$$

is the corner polyhedron $P^X\mathrm{N}$, and that the solutions to

$$\left. \begin{aligned} \sum_{k=1}^{D-1} \mathbf{g}_k x_{J(k)} &\equiv \mathbf{g}_t \mod \begin{pmatrix} \epsilon_1 \\ \vdots \\ \epsilon_m \end{pmatrix} \\ x_{J(k)} &\geq 0 \text{ and integer } (k = 1, \ldots, D-1) \end{aligned} \right\} \tag{34}$$

is the master polyhedron P^G. Naturally, in (33) the isomorphic image of $\bar{\alpha}_k$ is $\boldsymbol{\lambda}_k$ $(k = 0, 1, \ldots, n)$, and the basis $\mathbf{B}$ may be diagonalized to

$$\hat{B} = \begin{pmatrix} \epsilon_1 & & 0 \\ & \ddots & \\ 0 & & \epsilon_m \end{pmatrix} \quad \text{with} \quad D = \prod_{i=1}^{m} \epsilon_i.$$

In (34), the g_k $(k = 1, \ldots, D-1)$ are the group elements of $G(\lambda)$, g_t is the group element λ_0, and g_D is the zero element in $G(\lambda)$. The advantage of using $G(\lambda)$ in place of $G(\bar{\alpha})$ becomes especially clear when the group is cyclic, in which case its elements are of the form

$$\begin{pmatrix} 0 \\ \vdots \\ 0 \\ \ell \end{pmatrix},$$

where ℓ is an integer satisfying $0 \leq \ell < D$. To illustrate the theorem and the use of the group $G(\lambda)$ we continue with Example 11.2.

Example (from Example 11.2)

$$\mathbf{B} = \begin{pmatrix} -1 & -4 \\ -3 & -2 \end{pmatrix} \quad \mathbf{N} = \begin{pmatrix} 1 & 0 \\ 0 & 1 \end{pmatrix} = \mathbf{I}, \quad \mathbf{b} = \begin{pmatrix} -5 \\ -7 \end{pmatrix}.$$

$$\quad\;\; \mathbf{a}_1 \;\;\; \mathbf{a}_2 \qquad\qquad \mathbf{a}_3 \;\; \mathbf{a}_4$$

From Appendix B we obtain

$$\hat{B} = \begin{pmatrix} 1 & 0 \\ 0 & 10 \end{pmatrix} = \begin{pmatrix} -1 & 0 \\ -3 & 1 \end{pmatrix} \begin{pmatrix} -1 & -4 \\ -3 & -2 \end{pmatrix} \begin{pmatrix} 1 & -4 \\ 0 & -1 \end{pmatrix} = \mathbf{RBC}.$$

Thus,

$$\mathbf{RN} = \mathbf{RI} = \mathbf{R} = \begin{pmatrix} -1 & 0 \\ -3 & 1 \end{pmatrix} \quad \text{and} \quad \mathbf{Rb} = \begin{pmatrix} 5 \\ 8 \end{pmatrix}.$$

$G(\lambda)$ is a cyclic group of order 10 and

$$\lambda_1 = \begin{pmatrix} -1 \\ -3 \end{pmatrix} \text{(mod)} \begin{pmatrix} 1 \\ 10 \end{pmatrix} = \begin{pmatrix} 0 \\ 7 \end{pmatrix},$$

$$\lambda_2 = \begin{pmatrix} 0 \\ 1 \end{pmatrix} \text{(mod)} \begin{pmatrix} 1 \\ 10 \end{pmatrix} = \begin{pmatrix} 0 \\ 1 \end{pmatrix},$$

$$\lambda_0 = \begin{pmatrix} 5 \\ 8 \end{pmatrix} \text{(mod)} \begin{pmatrix} 1 \\ 10 \end{pmatrix} = \begin{pmatrix} 0 \\ 8 \end{pmatrix}.$$

It is convenient to choose $\mathbf{g}_k = \begin{pmatrix} 0 \\ k \end{pmatrix}$ $(k = 1, \ldots, 10)$, then

$$\underset{\substack{\mathbf{g}_1 \\ \lambda_2}}{\begin{pmatrix} 0 \\ 1 \end{pmatrix}} \underset{\mathbf{g}_2}{\begin{pmatrix} 0 \\ 2 \end{pmatrix}} \underset{\mathbf{g}_3}{\begin{pmatrix} 0 \\ 3 \end{pmatrix}} \underset{\mathbf{g}_4}{\begin{pmatrix} 0 \\ 4 \end{pmatrix}} \underset{\mathbf{g}_5}{\begin{pmatrix} 0 \\ 5 \end{pmatrix}} \underset{\mathbf{g}_6}{\begin{pmatrix} 0 \\ 6 \end{pmatrix}} \underset{\substack{\mathbf{g}_7 \\ \lambda_1}}{\begin{pmatrix} 0 \\ 7 \end{pmatrix}} \underset{\substack{\mathbf{g}_8 \\ \lambda_0}}{\begin{pmatrix} 0 \\ 8 \end{pmatrix}} \underset{\mathbf{g}_9}{\begin{pmatrix} 0 \\ 9 \end{pmatrix}} \underset{\mathbf{g}_{10}}{\begin{pmatrix} 0 \\ 10 \end{pmatrix}} = \begin{pmatrix} 0 \\ 0 \end{pmatrix}$$

From Theorem 11.8 we have the system

i) $\gamma_8' = 1$

ii) $\gamma_1' + \gamma_7' = 1$; $\gamma_2' + \gamma_6' = 1$; $\gamma_3' + \gamma_5' = 1$;
$\gamma_4' + \gamma_4' = 1$; $\gamma_9' + \gamma_9' = 1$

iii) $2\gamma_1' - \gamma_2' \geq 0$, $\gamma_1' + \gamma_2' - \gamma_3' \geq 0$, $\gamma_1' + \gamma_3' - \gamma_4' \geq 0$, $\gamma_1' + \gamma_4' - \gamma_5' \geq 0$, etc. (about 30 inequalities of the form $\gamma_r' + \gamma_s' - \gamma_{r \oplus s}' \geq 0$).

iv) $\gamma_1', \ldots, \gamma_9' \geq 0$.

There are seven basic feasible solutions to this system. They have been calculated by Gomory [280] (using one of the techniques suggested in that article, also see Problem 11.62) and are listed below as coefficients of $\bar{\gamma}_k$. Besides these seven faces of the master polyhedron, we have the nine additional face inequalities $t_k \geq 0$ $(k = 1, \ldots, 9)$.

Face of P^G	$x_{J(2)}$ $\bar{\gamma}_1$	$\bar{\gamma}_2$	$\bar{\gamma}_3$	$\bar{\gamma}_4$	$\bar{\gamma}_5$	$\bar{\gamma}_6$	$x_{J(1)}$ $\bar{\gamma}_7$	$\bar{\gamma}_8$	$\bar{\gamma}_9$	$\geq \bar{\gamma}_0$
1	3	1	4	2	0	3	1	4	2	4
2	2	4	6	3	0	2	4	6	3	6
3	2	4	1	3	5	2	4	6	3	6
4	6	2	3	4	5	6	2	8	4	8
5	1	2	3	4	5	6	7	8	4	8
6	9	8	7	6	5	4	3	12	6	12
7	9	8	2	6	10	4	3	12	6	12

As $\lambda_1 = \mathbf{g}_7$ and $\lambda_2 = \mathbf{g}_1$, the faces for the corner polyhedron $P^X N$ in our example may be obtained from

Inequality number	Coefficients for		$\geq \bar{\gamma}_0$
	$x_{J(1)}$	$x_{J(2)}$	
1	1	3	4
2	4	2	6
3	4	2	6
4	2	6	8
5	7	1	8
6	3	9	12
7	3	9	12
8	1	0	0
9	0	1	0

Plotting these inequalities in $x_{J(1)}, x_{J(2)}$ space (Fig.11.12) indicates that the four faces of $P^\mathbf{x}N$ are inequalities 1 (or 4,6,7), 5, 8, and 9, which agrees with an earlier discussion (Fig.11.11).

Remarks

1. The example indicates that some of the inequalities obtained from the faces of the master polyhedron may be redundant and therefore are not faces of $P^\mathbf{x}N$. (In Fig. 11.12, inequality 2 or 3 is not a face.) In computations, once a list of faces for P^G has been computed we may have to apply a linear programming phase I routine on the relevant columns (i.e., inequalities) to determine which are the redundant inequalities and which are the faces. The objective function in the linear program would assign costs of 0 to the $x_{J(j)}$ variables and -1 to the artificial variable added to each face equation.

2. The inequalities (i) and (ii) may be used to eliminate variables in (iii). If this is done and duplicate inequalities are dropped there will be approximately $D^2/6$ inequalities left in (iii) (Problem 11.64). Even though the coefficient matrix is highly structured, finding all basic feasible solutions still involves a considerable amount of effort. (Possible

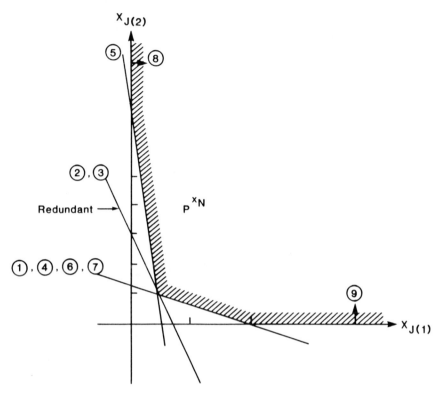

Figure 11.12: Faces of $P^X N$

techniques are discussed in Problem 11.62.) However, Gomory [280] has computed all basic feasible solutions to the system (i) through (iv) for several groups and right-hand side element g_t. The tables appearing in the Gomory article assume that $G(\lambda)$ is used to define the master polyhedron, its order is at most 11, and when the group is cyclic its elements g are of the form

$$\begin{pmatrix} 0 \\ k \end{pmatrix} \quad k = 1, \ldots, D-1.$$

3. Up to this time we have ignored the possibility of having duplicate nonzero columns in (29) or (33). However, a careful examination of the previous discussion and theorem will indicate that duplicate columns may be dropped prior to the face calculation. Once a face for the corner polyhedron has been found, the omitted nonbasic variables are added

to the inequality (32) with a coefficient equal to the one of nonbasic variables with the same group element.

Example (from Example 11.1)

Consider Example 11.1, where the first nonbasic column is changed from $\binom{1}{5}$ to $\binom{2}{3}$, then the optimal basis does not change and

$$\mathbf{B} = \begin{pmatrix} 0 & 2 \\ 3 & -4 \end{pmatrix} \quad \mathbf{N} = \begin{pmatrix} 2 & 4 & 2 & 1 & 0 \\ 3 & 1 & -1 & 0 & 1 \end{pmatrix} \quad \mathbf{b} = \begin{pmatrix} 41 \\ 47 \end{pmatrix}$$

$\mathbf{B}$ was diagonalized to

$$\hat{\mathbf{B}} = \begin{pmatrix} 1 & 0 \\ 0 & 6 \end{pmatrix} = \begin{pmatrix} 0 & 1 \\ 1 & 2 \end{pmatrix} \begin{pmatrix} 0 & 2 \\ 3 & -4 \end{pmatrix} \begin{pmatrix} -1 & 4 \\ -1 & 3 \end{pmatrix} = \mathbf{RBC}.$$

Thus,

$$\mathbf{RN} = \begin{pmatrix} 3 & 1 & -1 & 0 & 1 \\ 8 & 6 & 0 & 1 & 2 \end{pmatrix} \quad \text{and} \quad \mathbf{Rb} = \begin{pmatrix} 47 \\ 135 \end{pmatrix}.$$

$$(\boldsymbol{\lambda}_j) = \begin{pmatrix} 0 & 0 & 0 & 0 & 0 \\ 2 & 0 & 0 & 1 & 2 \end{pmatrix} \quad \boldsymbol{\lambda}_0 = \begin{pmatrix} 0 \\ 3 \end{pmatrix}$$
$$\phantom{(\boldsymbol{\lambda}_j) = }\; x_{J(1)} \quad x_{J(2)} \quad x_{J(3)} \quad x_{J(4)} \quad x_{J(5)}$$

$$G(\boldsymbol{\lambda}) = \left\{ \begin{pmatrix} 0 \\ 1 \end{pmatrix} \begin{pmatrix} 0 \\ 2 \end{pmatrix} \begin{pmatrix} 0 \\ 3 \end{pmatrix} \begin{pmatrix} 0 \\ 4 \end{pmatrix} \begin{pmatrix} 0 \\ 5 \end{pmatrix} \begin{pmatrix} 0 \\ 0 \end{pmatrix} \right\}$$
$$\phantom{G(\boldsymbol{\lambda}) = \{}\; g_1 \quad\;\; g_2 \quad\;\; g_3 \quad\;\; g_4 \quad\;\; g_5 \quad\;\; g_6$$
$$\phantom{G(\boldsymbol{\lambda}) = \{}\; \lambda_4 \quad \lambda_1 = \lambda_5 \quad \lambda_0$$

The faces of the master polyhedron from the Gomory article are

| | $x_{J(4)}$ | $x_{J(5)}$ | | | | |
Face	$\bar{\gamma}_1$	$\bar{\gamma}_2$	$\bar{\gamma}_3$	$\bar{\gamma}_4$	$\bar{\gamma}_5$	$\geq \bar{\gamma}_0$
1	1	0	1	0	1	1
2	2	1	3	2	1	3
3	1	2	3	2	1	3
4	1	2	3	1	2	3

and also $t_1 \geq 0$, $t_2 \geq 0$, $t_3 \geq 0$, $t_4 \geq 0$, and $t_5 \geq 0$. So the faces of the corner polyhedron defined by

$$\lambda_4 x_{J(4)} + \lambda_5 x_{J(5)} \equiv \lambda_0 \quad \mod \begin{pmatrix} 1 \\ 6 \end{pmatrix}$$

$$x_{J(4)} \geq 0, x_{J(5)} \geq 0, \text{ and integer,}$$

are obtained from the inequalities

1.	$x_{J(4)}$		$\geq$	1
2.	$2x_{J(4)}$	$+ \quad x_{J(5)}$	$\geq$	3
3.	$x_{J(4)}$	$+ \quad 2x_{J(5)}$	$\geq$	3
4.	$x_{J(4)}$	$+ \quad 2x_{J(5)}$	$\geq$	3
5.	$x_{J(4)}$		$\geq$	0
6.		$x_{J(5)}$	$\geq$	0.

Dropping the duplicate and redundant inequalities yields

1.	$x_{J(4)}$		$\geq$	1
2.	$2x_{J(4)}$	$+ \quad x_{J(5)}$	$\geq$	3
3.	$x_{J(4)}$	$+ \quad 2x_{J(5)}$	$\geq$	3
6.		$x_{J(5)}$	$\geq$	0.

As $\lambda_1 = \lambda_5$, the faces of the corner polyhedron $P^X N$ (in 5 space) are

1'.		$x_{J(4)}$		$\geq$	1
2'.	$x_{J(1)}$	$+ \quad 2x_{J(4)}$	$+ \quad x_{J(5)}$	$\geq$	3
3'.	$2x_{J(1)}$	$+ \quad x_{J(5)}$	$+ \quad 2x_{J(5)}$	$\geq$	3
6'.	$x_{J(1)}$		$+ \quad x_{J(5)}$	$\geq$	0.

11.7.5 Generating Valid Inequalities from the Faces of Master Polyhedra (Gomory and Johnson [284])

Ralph Gomory and Ellis Johnson [284] have shown that it is possible to derive valid inequalities for *any* integer program using the faces of master polyhedra. We outline some of their simpler results. The details and proofs are left to the references.

Consider a row of an optimal linear programming tableau in which the basic variable is integer constrained but at a nonintegral value. This equation implies

$$\sum_{j=1}^{n} f_j x_{J(j)} \equiv f_0 \pmod{1}, \tag{35}$$

where f_j $(j = 0, 1, \ldots, n)$ is the fractional part of the coefficients in the row under consideration. Each f_j $(j = 0, 1, \ldots, n)$ may be written as I_j/D, where I_j is a nonnegative integer less than $D = |\det \mathbf{B}|$, and $\mathbf{B}$ is an optimal linear programming basis. So (35) may be written as

$$\sum_{j=1}^{n} \frac{I_j}{D} x_{J(j)} \equiv \frac{I_0}{D} \pmod{1}. \tag{36}$$

It turns out that we can use certain master polyhedra to derive inequalities which are implied by (36) and thus valid for the integer program from which the congruence (35) was obtained.

Define G_D to be the cyclic group $G(\boldsymbol{\lambda})$ of order D. Its components are of the form

$$\mathbf{g}_k = \begin{pmatrix} 0 \\ k \end{pmatrix} \quad \text{for } k = 1, \ldots, D,$$

where $\mathbf{g}_D$ is the zero element. For each coefficient I_j/D $(j = 1, \ldots, n)$ in (36), form the subgroup G_{D-1} and find one face (with a positive right-hand side) of the master polyhedron formed by the convex hull of nonnegative integer solutions to

$$\sum_{k=1}^{D-I_j-1} \mathbf{g}_k t_k \equiv \mathbf{g} \pmod{1}, \tag{37}$$

where the $\mathbf{g}_k$ $(k = 1, \ldots, D - I_j)$ are from the group G_{D-I_j}, $\mathbf{g}$ is any group element $\mathbf{g}_k$ $(1 \le k \le D - I_j)$, and addition is with respect to the group $G(\boldsymbol{\lambda})$. This yields the n face inequalities

$$\sum_{k=1}^{D-I_j-1} \bar{\gamma}_{ik} t_k \ge \bar{\gamma}_{i0}, \tag{38}$$

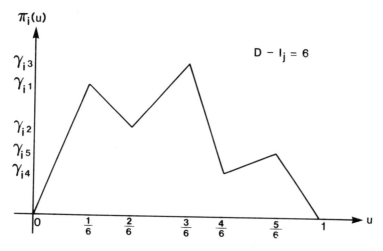

Figure 11.13: Piecewise Linear Function $\pi_i(u)$

where $\bar{\gamma}_{i0} > 0$ for $i = 1, \ldots, n$. For each inequality (38), define the piecewise linear function $\pi_i(u)$ on the unit interval by plotting (Fig. 11.13)

$$\left.\begin{array}{l} \pi_i(0) = 0 \\ \pi_i\left(\frac{k}{D-I_j}\right) = \bar{\gamma}_{ik} \quad \text{for} \quad k = 1, \ldots, D - I_j - 1 \\ \pi_i(1 - \epsilon) = 0, \end{array}\right\} \tag{39}$$

where ϵ is a very small positive number.

Then, using the properties of the function, it can be shown that the congruence relationship (35) implies the valid inequalities

$$\sum_{j=1}^{n} \pi_i\left(\frac{I_j}{D}\right) x_{J(j)} \geq \pi_i\left(\frac{I_0}{D}\right) \quad \text{for} \quad i = 1, \ldots, n, \tag{40}$$

provided that $\pi_i(I_0/D) > 0$.

Example 11.4 (Gomory and Johnson [284])

$$\text{Maximize} \quad -x_1 \quad - \quad x_2 \quad - \quad x_3 \quad - \quad 2x_4 \quad - \quad 4x_5 \quad = \quad x_0$$

$$\text{subject to} \quad x_1 + 2x_2 + x_3 + x_4 + 5x_5 = 10$$

$$3x_1 - 3x_2 + 2x_3 - 3x_4 + 3x_5 = 5$$

$$x_j \geq 0 \text{ and integer } (j = 1, 2, 3, 4, 5).$$

The optimal linear programming basis is $\mathbf{B} = \begin{pmatrix} 1 & 2 \\ 3 & -3 \end{pmatrix}$ and $D = 9$. Writing the nonbasic variables in terms of the basic variables yields

$$x_0 = -\frac{65}{9} - \frac{11}{9}x_3 - \frac{15}{9}x_4 - \frac{3}{9}x_5,$$

$$x_1 = \frac{40}{9} - \frac{7}{9}x_3 + \frac{3}{9}x_4 - \frac{21}{9}x_5,$$

$$x_2 = \frac{25}{9} - \frac{1}{9}x_3 - \frac{6}{9}x_4 - \frac{12}{9}x_5.$$

Suppose the x_1 row is used to derive valid inequalities. Then $x_1 \equiv 0$ (mod 1) means that

$$\frac{7}{9}x_3 - \frac{3}{9}x_4 + \frac{21}{9}x_5 \equiv \frac{40}{9} \quad (\text{mod } 1)$$

or

$$\tfrac{7}{9}x_3 + \tfrac{6}{9}x_4 + \tfrac{3}{9}x_5 \equiv \tfrac{4}{9} \quad (\text{mod } 1) \tag{36'}$$

As $D = 9$, $9 - 7 = 2$, $9 - 6 = 3$, and $9 - 3 = 6$, and thus we need a face of the master polyhedra formed by the cyclic groups G_2, G_3 and G_6 when the right-hand side $\mathbf{g}$ in (37) is some element in each group. In the first case, the group is of order 2 and $\mathbf{g}$ is selected as $\mathbf{g}_1$. The only face of this convex hull is (Problem 11.65)

$$1t_1 \geq 1. \tag{i}$$

When the group is of order 3, $\mathbf{g}$ is selected as $\mathbf{g}_1$ and the only face of this polyhedron that we can use is (Problem 11.66)

$$2t_1 + t_2 \geq 2. \tag{ii}$$

For the third group, G_6, $\mathbf{g}$ is taken as $\mathbf{g}_3$, and from Appendix 5 in the Gomory article [280] there are four faces with a positive right-hand side in the master polyhedron. One of them is

$$1t_1 + 2t_2 + 3t_3 + 1t_4 + 2t_5 \geq 3 \tag{iii}$$

So from (i)

$$\pi_1(u) = \begin{cases} 2u & 0 \le u \le \frac{1}{2} \\ 2 - 2u & \frac{1}{2} \le u < 1. \end{cases}$$

From (ii)

$$\pi_2(u) = \begin{cases} 3u & 0 \le u \le \frac{1}{3} \\ \frac{3}{2} - \frac{3}{2}u & \frac{1}{3} \le u < 1. \end{cases}$$

Finally, from (iii)

$$\pi_3(u) = \begin{cases} 2u & 0 \le u \le \frac{1}{2} \\ 3 - 4u & \frac{1}{2} \le u \le \frac{2}{3} \\ -1 - 2u & \frac{2}{3} \le u \le \frac{5}{6} \\ 4 - 4u & \frac{5}{6} \le u < 1. \end{cases}$$

These functions are plotted in Fig. 11.14. Since the coefficients in (36)′ are 7/9, 6/9, and 3/9, with a right–hand side of 4/9 we have the valid inequalities

$$\pi_i\left(\frac{7}{9}\right) x_3 + \pi_i\left(\frac{6}{9}\right) x_4 + \pi_i\left(\frac{3}{9}\right) x_5 \ge \pi_i\left(\frac{4}{9}\right) \tag{40'}$$

for $i = 1, 2, 3$, which, after simplifying, are

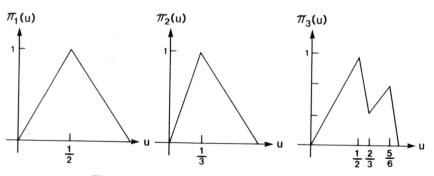

Figure 11.14: Piecewise Linear Function

$$\left.\begin{array}{llll}
(i = 1) & 2x_3 + 3x_4 + 3x_5 \geq 4 \\
(i = 2) & 2x_3 + 3x_4 + 6x_5 \geq 5 \\
(i = 3) & 5x_3 + 3x_4 + 6x_5 \geq 8
\end{array}\right\} \qquad (40)''$$

Remarks

1. The intersection of the hyperplane (40) with each $x_{J(j)}$ axis occurs at the point $\pi_i(I_0/D)/\pi_i(I_j/D)$. The farther this point is from the origin the stronger the inequality. Thus it is desirable to have $\pi_i(I_0/D)$ large relative to $\pi_i(I_j/D)$. In this regard, it is desirable to choose the right–hand side $\mathbf{g}$ in (37) close to I_0/D, since this will tend to keep I_0/D away from the extremities of the unit interval which have $\pi_i(\ )$ near 0 (see Fig. 11.13).

2. The inequalities (40)$''$ are valid for any integer program which has a row that can be used to derive the congruence relationship (36)$'$. Moreover, the functions $\pi_i(u)$ plotted in Fig. 11.14 can be used to produce valid inequalities (40) for any integer program which has constraints that can yield a congruence equation of the form

$$\frac{I_1}{D}x_{J(1)} + \frac{I_2}{D}x_{J(2)} + \frac{I_3}{D}x_{J(3)} \equiv \frac{I_0}{D} \qquad (\text{mod } 1),$$

where $D - I_j$ is 2, 3, and 6 for some ordering of the I_j $(j = 1, 2, 3)$. For example, had (36) been

$$\frac{4}{10}x_{J(1)} + \frac{7}{10}x_{J(2)} + \frac{8}{10}x_{J(3)} \equiv \frac{3}{10} \qquad (\text{mod } 1),$$

then we could obtain the face inequality (iii) for G_6, (ii) for G_3, and (i) for G_2. This would produce the functions $\pi_1(u), \pi_2(u),$ and $\pi_3(u)$ respectively, as in Fig. 11.14. The valid inequalities are

$$\pi_i\left(\frac{4}{10}\right)x_{J(1)} + \pi_i\left(\frac{7}{10}\right)x_{J(2)} + \pi_i\left(\frac{8}{10}\right)x_{J(3)} \geq \pi_i\left(\frac{3}{10}\right) \quad \text{for } i = 1, 2, 3.$$

3. Once the faces of the master polyhedra are known, it is relatively easy to produce the valid inequalities. However, these inequalities are usually not faces of a corner polyhedron and may, in fact, be relatively weak. Nevertheless, they can easily be incorporated in an enumerative or cutting plane algorithm. Computational experience with an enumerative technique by Guignard and Spielberg [300] seems to indicate

that the inequalities are worth using. Their results appear in Chapter 9 (Appendix A) and in Chapter 12 (Appendix A). A discussion of the uses of these inequalities in a branch and bound algorithm appears in Johnson and Spielberg [354].

4. The results appearing in this section have also been extended by Gomory and Johnson [284] to the mixed integer program. A minimization problem for a mixted integer program similar to the GMP for the integer program can be constructed (Problem 11.70). The mixed integer solutions to this problem can be described by a corner polyhedron. A master polyhedron can then be defined naturally. Results which indicate how to find the faces of this polyhedron appear in Gomory and Johnson [284], as well as lists of faces for smaller polyhedra. Other papers by Gomory and Johnson [285] and by Johnson [352] generalize the results to equations whose coefficients yield fractional parts which cannot be put in the form of I_j/D, where I_j is an integer and $0 \le I_j < D$.

Problems

Unless otherwise specified, the notation appearing in the problems conforms to the chapter. In particular, $G(\bar{\alpha})$ denotes the abelian group which is generated by the columns which are the fractional parts of $\mathbf{B}^{-1}\mathbf{N}$ (the basis inverse, the nonbasic columns), $G(\boldsymbol{\lambda})$ is the isomorphic group whose elements are obtained by reducing the m basis $\mathbf{B}$ to diagonal form, and $D = |\det \mathbf{B}|$.

Problem 11.1 Consider the general definition of a group:

A group G is a set, together with a rule which to each pair of elements a, b in G associates an element denoted by ab in G, having the following properties

i) For all a, b, c in G we have associativity, namely $(ab)c = a(bc)$.

ii) There exists an element e of G such that $ea = ae = a$ for all a in G.

iii) If a is an element of G, then there exists an element b of G such that $ab = ba = e$.

a) Using this definition, prove that $G(\bar{\alpha})$ is a group under addition. Do the same for $G(\lambda)$.

b) Is the set of all m by n matrices a group under addition? Under multiplication? What if $m = n$?

For each of the integer programs in Problems 11.2 through 11.8:

a) Find an optimal linear programming basis.

b) Find the group $G(\bar{\alpha})$. State its order and generators.

c) Find the subgroups $G(\bar{\alpha}_j)$ for each nonbasic column j.

d) State the group minimization problem.

e) Draw or describe the representative network.

All problems are of the form:

$$\begin{array}{rl} \text{Maximize} & \mathbf{c'x'} \\ \text{subject to} & \mathbf{A'x'} \leq \mathbf{b} \\ & \mathbf{x'} \geq \mathbf{0} \quad \text{and integer.} \end{array}$$

Problem 11.2 (White [599])

$$\mathbf{c'} = \begin{pmatrix} 6 & 5 \end{pmatrix}$$

$$\mathbf{A'} = \begin{pmatrix} -3 & 2 \\ 7 & 5 \end{pmatrix}, \quad \mathbf{b} = \begin{pmatrix} 3 \\ 15 \end{pmatrix}$$

Problem 11.3 (White [599])

$$\mathbf{c'} = \begin{pmatrix} 1 & 1 & 1 \end{pmatrix}$$

$$\mathbf{A'} = \begin{pmatrix} 2 & 0 & 0 \\ 0 & 2 & 0 \\ 0 & 0 & 2 \\ 1 & 1 & -4 \end{pmatrix}, \quad \mathbf{b} = \begin{pmatrix} 3 \\ 3 \\ 3 \\ -3 \end{pmatrix}$$

Problem 11.4 (White [599])

$$\mathbf{c}' = \begin{pmatrix} 5 & 4 \end{pmatrix}$$

$$\mathbf{A}' = \begin{pmatrix} 2 & -2 \\ 1 & 2 \\ 4 & 3 \end{pmatrix}, \quad \mathbf{b} = \begin{pmatrix} 7 \\ 9 \\ 20 \end{pmatrix}$$

Problem 11.5 (White [599])

$$\mathbf{c}' = \begin{pmatrix} 0 & 1 \end{pmatrix}$$

$$\mathbf{A}' = \begin{pmatrix} -4 & 4 \\ 2 & -1 \\ 1 & 1 \end{pmatrix}, \quad \mathbf{b} = \begin{pmatrix} 3 \\ 2 \\ 6 \end{pmatrix}$$

Problem 11.6 (White [599])

$$\mathbf{c}' = \begin{pmatrix} 0 & 1 \end{pmatrix}$$

$$\mathbf{A}' = \begin{pmatrix} 2 & 0 \\ -2 & 2 \end{pmatrix}, \quad \mathbf{b} = \begin{pmatrix} 3 \\ -1 \end{pmatrix}$$

Problem 11.7 (White [599])

$$\mathbf{c}' = \begin{pmatrix} 8 & 6 & 1 & 1 \end{pmatrix}$$

$$\mathbf{A}' = \begin{pmatrix} 1 & 1 & 0 & 0 \\ 0 & 0 & 1 & 1 \\ 7 & 0 & 8 & 0 \\ -4 & -4 & 0 & 6 \\ 0 & 5 & -4 & -4 \end{pmatrix}, \quad \mathbf{b} = \begin{pmatrix} 1 \\ 1 \\ 6 \\ 6 \\ 5 \end{pmatrix}$$

Problem 11.8 (Gomory [280])

$$\mathbf{c}' = \begin{pmatrix} 1 & 2 & 3 & 1 & 1 \end{pmatrix}$$

$$\mathbf{A}' = \begin{pmatrix} 1 & 0 & 4 & 2 & 1 \\ 4 & 3 & 1 & -4 & -1 \end{pmatrix}, \quad \mathbf{b} = \begin{pmatrix} 41 \\ 47 \end{pmatrix}$$

Problem 11.9   Under what conditions does an element of $G(\bar{\alpha})$: (i) generate $G(\bar{\alpha})$, (ii) generate a subgroup of $G(\bar{\alpha})$, (iii) generate a subgroup with only the zero element. In each case, prove and illustrate your answer.

Problem 11.10   Is it possible for the group $G(\bar{\alpha})$ or one of its iso-morphic groups to have order D when the original constraint matrix $\mathbf{A}$ does *not* contain an identity matrix? If not, prove it; if so, give an example.

Problem 11.11   Is the order of $G(\bar{\alpha})$ or one of its isomorphic groups equal to D when the original constraint matrix $\mathbf{A}$ contains a diagonal, but not identity, matrix? If so, prove it; if not, given an example. Does your argument change when the diagonal elements are equal (but not 1)?

Problem 11.12   When constructing a group minimization prob-lem, why is it essential to choose a basis $\mathbf{B}$ so that the reduced costs $c_{\mathbf{B}}\mathbf{B}^{-1}\mathbf{N} - c_{\mathbf{N}}$ are nonnegative? What difficulties arise when this is not done?

Problem 11.13   How do redundant constraints affect the structure of the group minimization problem? Do they affect the algorithms which solve the group problem? Must they be omitted prior to constructing or solving the group problem? Illustrate your answer.

Problem 11.14   Suppose two nonbasic columns with the same reduced cost are equal to the same group element in either $G(\bar{\alpha})$ or $G(\lambda)$, and one is to be dropped prior to solving the group problem. Are there any reasons for keeping one over the other? Explain. Can both columns be kept? Is there any advantage in doing this?

Problem 11.15   Consider an optimal linear programming tableau representing an integer program. Suppose a Gomory fractional cut

(Chapter 4) is derived from each row of the tableau. Let the coefficients in a cut be

$$\mathbf{F} = (f_0, f_1, \ldots, f_n) \quad \text{where} \quad 0 \leq f_j < 1 \ (j = 0, 1, \ldots, n).$$

Denote the set of all integer linear combinations of the rows $\mathbf{F}$ as $G(\mathbf{F})$.

a) Show that $G(\mathbf{F})$ is an abelian group when addition is modulo 1.

b) Show that $G(\mathbf{F})$ is isomorphic to $G(\bar{\boldsymbol{\alpha}})$. Define the isomorphism which sends elements of $G(\mathbf{F})$ to $G(\bar{\boldsymbol{\alpha}})$.

c) Is $G(\mathbf{F})$ cyclic whenever $G(\bar{\boldsymbol{\alpha}})$ is cyclic?

d) Can $G(\mathbf{F})$ be written as a direct sum of cyclic subgroups?

Problem 11.16 Develop a flow chart for the dynamic programming algorithm discussed in Section 11.3.1. Assume that not every element generates the group and the basic variables turn out to be nonnegative. What parameters have to be stored in a computer program? In particular, must the entire $f(k, \mathbf{g})$ $(k = 1, \ldots, n)$ array be kept?

Problem 11.17 Consider the estimating procedure used in the dynamic programming algorithm when each $\bar{\alpha}_j$ does not generate $G(\bar{\boldsymbol{\alpha}})$ (Section 11.3.1). In terms of the order of each subgroup $G(\bar{\alpha}_j)$, what is an upper bound on the number of times the estimating procedure may have to be used? Under what circumstances can the procedure be used less than this upper bound?

Problem 11.18 Let l_i $(i = 1, \ldots, m)$ be the maximum number in row i of $\mathbf{B}^{-1}\mathbf{N}$. Construct the vector $\mathbf{L} = (l_1, \ldots, l_m)$. Show that if $\mathbf{B}^{-1}\mathbf{b} - (D - 1)\mathbf{L} \geq \mathbf{0}$, then every optimal solution $\mathbf{x_N}$ to the GMP yields $\mathbf{x_B} = \mathbf{B}^{-1}\mathbf{b} - \mathbf{B}^{-1}\mathbf{N}\mathbf{x_N} \geq \mathbf{0}$.

In this chapter we have proved that the asymptotic integer program,

$$
\begin{aligned}
\text{maximum} \quad & \mathbf{cx} \\
\text{subject to} \quad & \mathbf{Ax} = \lambda \mathbf{b} \\
& \mathbf{x} \geq \mathbf{0} \quad \text{and integer,}
\end{aligned}
$$

may be solved by finding an optimal solution to the group minimization problem provided that the scalar λ is sufficiently large and the selected basis $\mathbf{B}$ yields a nondegenerate basic solution $\mathbf{x_B} = \mathbf{B^{-1}b}$. Find the smallest possible λ for the following problems. In each case, select a basis so that the reduced costs are nonnegative and $\mathbf{x_B} > \mathbf{0}$. (*Hint:* See Theorem 11.2)

Problem 11.19 Example 11.3.

Problem 11.20 Example 11.6.

Problem 11.21 Problem 11.2.

Problem 11.22 Problem 11.6.

Problem 11.23 Why can we say that the difficulty in solving an integer program substantially increases as the size of $D = |\det \mathbf{B}|$ increases? Is your answer algorithm dependent? Can the data initially be altered to reduce D? Discuss what is involved in group theoretic algorithms.

Problem 11.24 Formally list the branch and bound algorithm described in Section 11.3.3. Be brief and concise, and construct the algorithm about the following steps: (1) initialization, (2) branching, (3) termination tests, and (4) bounding (see Chapter 8, Section 8.4). Draw a simple flow chart describing the procedure.

Problem 11.25 Discuss node and branch selection criteria for the enumeration algorithm described in Section 11.3.3. In particular, what criteria will reduce computer storage and which rules will tend to get an optimal solution faster at the expense of requiring more computer storage and bookkeeping? What implicit enumeration tests are possible? Can penalties in the sense of Chapter 8 (Section 8.7) be used?

Problem 11.26 Explain why the enumerative approach (Section 11.3.3) without any implicit enumeration (i.e., a GMP is not solved at any node and thus dynamic programming is never used) can be used to solve the integer program when posed as a group minimization problem. Illustrate your answer by solving Problem 11.2. Relate the computations to a tree.

For each of the integer programs in Problems 11.27 through 11.33:

a) Find a group minimization problem.

b) Solve the group problem using dynamic programming.

c) If $\mathbf{x_B} \not\geq 0$, use the enumeration method to solve the integer program. Relate the computations to a tree.

Problem 11.27 Problem 11.2, for all right-hand sides **b**.

Problem 11.28 Problem 11.3.

Problem 11.29 Problem 11.4.

Problem 11.30 Problem 11.5. Check your answer by drawing the representative network and solve the shortest route problem by inspection.

Problem 11.31 Problem 11.6, for all right-hand sides **b**.

Problem 11.32 Problem 11.7.

Problem 11.33 Problem 11.8. Check your answer by drawing the representative network and solve the shortest route problem by inspection.

Problem 11.34 Prove that if all solutions $x_{J(1)}, \ldots, x_{J(n)}$ to the GMP which satisfy $\sum_{j=1}^{n} x_{J(j)} \leq D-1$ also produce $\mathbf{x_B} \geq 0$, then the minimal solution to the group problem solves the integer program. (*Hint*: Think of the shortest route formulation of the GMP.)

Problem 11.35 Suppose we have developed and implemented a shortest route or dynamic programming algorithm which finds the rth best solution to the group minimization problem. Is it ever possible for this algorithm to fail to solve the integer program? In particular, what if duplicate and/or zero columns have been dropped.

Problem 11.36 (Chen and Zionts [129]) In the network formulation of the GMP, show that an upper bound in the number of equivalent integer solutions $x_{J(1)}, \ldots, x_{J(n)}$ obtained from alternative routs of the same length is

$$\left(\sum_{j=1}^{n} x_{J(j)} \right)! \Big/ \sum_{j=1}^{n} (x_{J(j)}!),$$

where $K! = (K)(K-1)(K-2)\ldots(2)(1)$ and K is a positive integer.

Problem 11.37   Show that the set $G(\lambda)$ (Section 11.5) is an abelian group. Is it ever possible for $G(\lambda)$ to be precisely $G(\bar{\alpha})$? Explain.

Problem 11.38   Consider the group minimization problem:

$$\text{Minimize } \sum_{j=1}^{n} \bar{c}_j x_{J(j)},$$

$$\text{subject to } \sum_{j=1}^{n} \mathbf{g}_j x_{J(j)} = \mathbf{g}_t,$$

$$x_{J(j)} \geq 0 \text{ and integer} \quad (j = 1, \ldots, n),$$

where addition is with respect to the abelian group G whose elements are $\mathbf{g}_1, \ldots, \mathbf{g}_n$ and $\mathbf{g}_t$ is in G. Show that the set of nonnegative integer solutions to this problem does not change if the $\mathbf{g}_j$ ($j = 1, \ldots, n$) are replaced by their images in any group isomorphic to G.

Problem 11.39   Let G be a finite abelian group. If G is cyclic, does it necessarily follow that all groups which are isomorphic to it are cyclic? Can G always be written as a direct sum of cyclic subgroups?

Problem 11.40   Suppose the congruence relationship in the group problem over $G(\bar{\alpha})$ is written in terms of modulo D rather than modulo 1 addition. Then, is it true that the m congruence constraints in the group problem over $G(\lambda)$ are always linear integer combinations of some of the constraints over $G(\bar{\alpha})$, where the sum in row i of $G(\lambda)$ is taken modulo ϵ_i? Is a similar decomposition, which yields constraints for $G(\bar{\alpha})$ from $G(\lambda)$, possible? Illustrate.

Problem 11.41   Which of the lemmas and theorems appearing in Section 11.6.1 are no longer valid if the groups are not abelian? Why?

Problem 11.42 Consider the factor groups $G(\hat{\mathbf{A}})/G(\hat{\mathbf{B}})$ and $G(\mathbf{A})/G(\mathbf{B})$ as defined in Section 11.6.2. Prove that $G(\lambda)$ is isomorphic to $G(\mathbf{A})/G(\mathbf{B})$. In particular, define the isomorphism ϕ which sends elements from $G(\lambda)$ to $G(\mathbf{A})/G(\mathbf{B})$. Illustrate the mapping using Example 11.1.

Problem 11.43   Define the composite mapping $\psi = f^* \circ \phi$ which sends elements first from $G(\lambda)$ to $G(\mathbf{A})/G(\mathbf{B})$, then to $G(\bar{\alpha})$. That is,

$$G(\lambda) \xrightarrow{\phi} G(\mathbf{A})/G(\mathbf{B}) \xrightarrow{f^*} G(\bar{\alpha}),$$

and for $\mathbf{g}$ in $G(\boldsymbol{\lambda})$, $\psi(\mathbf{g}) = f^*$ is as defined in Theorem 11.6. Show that ψ is an isomorphism and thus $G(\boldsymbol{\lambda})$ is isomorphic to $G(\bar{\alpha})$.

Problem 11.44 In what sense can we say that the GMP becomes easier to solve as the number of cosets in $G(\mathbf{I})/G(\mathbf{A})$ increases, where $G(\mathbf{I})$ is the set of all integer m vectors? Illustrate your answer.

Problem 11.45 Prove that the corner polyhedron in $\mathbf{x}'$ space (Section 11.7.1) is always a convex region as in Figure 11.8(b). Is it possible to relate the dimension of the space that the region lies in to the maximum number of faces it can have?

For Problems 11.46 through 11.48:

    a) Graph the corner polyhedron in $\mathbf{x}'$ space (Section 11.7.1).

    b) Graph the corner polyhedron in $\mathbf{x_N}$ space (Section 11.7.2).

    c) Find the faces of each polyhedron.

Problem 11.46 Problem 11.2.

Problem 11.47 Problem 11.4.

Problem 11.48 Problem 11.6.

Problem 11.49 The corner polyhedron in $\mathbf{x_N}$ space (Section 11.7.2) was introduced as the convex hull containing the nonnegative integer solutions to

$$\sum_{j=1}^{n} \bar{\alpha}_j x_{J(j)} \equiv \bar{\alpha}_0 \pmod{1}, \tag{1}$$

where the $\bar{\alpha}_j$ are in the group $G(\bar{\alpha})$. Suppose that each $\bar{\alpha}_j$ in (1) is replaced by its image in an isomorphic group. What is the relationship between the resulting convex hull and the corner polyhedron obtained when $G(\bar{\alpha})$ is used? If the isomorphic group is the row group $G(\mathbf{F})$, described in Problem 11.15, can anything be said about the faces of the resulting convex hull in context of the strength of Gomory fractional cuts? Use your answer to show that the faces of the corner polyhedron are the strongest than can be obtained from the integer program without using the nonnegativity requirements on the basic variables.

In several subsequent problems we use the following definition. Let $t_1, \ldots, t_{D-1}$ be an integer point in the master polyhedron P^G (Section 11.7.4) which satisfies

$$\sum_{k=1}^{D-1} \mathbf{g}_k t_k \equiv \mathbf{g}_t \pmod{1}, \tag{2}$$

where $\mathbf{g}_1, \ldots, \mathbf{g}_{D-1}$ are elements of the abelian group G, and $\mathbf{g}_t$ is used to define P^G. We say that $t_1, \ldots, t_{D-1}$ is an *irreducible point* if there does not exist a different nonnegative integer vector $(\bar{t}_1, \ldots, \bar{t}_{D-1})$ whose components satisfy $\bar{t}_k \leq t_k$ $(k = 1, \ldots, D-1)$.

Problem 11.50

    a) Prove that a component t_k $(1 \leq k \leq D-1)$ of an irreducible point cannot have a value exceeding the order of the group.

    b) Show that $t_1, \ldots, t_{D-1}$ is irreducible if and only if any set of integers $\bar{t}_k$ and $\bar{\bar{t}}_k$ $(k = 1, \ldots, D-1)$ which satisfy

$$0 \leq \bar{t}_k \leq t_k, \;\; 0 \leq \bar{\bar{t}}_k \leq t_k, \;\; \text{and} \;\; \sum_{k=1}^{D-1} \mathbf{g}_k \bar{t}_k = \sum_{k=1}^{D-1} \mathbf{g}_k \bar{\bar{t}}_k,$$

    implies that $\bar{t}_k = \bar{\bar{t}}_k$ for $k = 1, \ldots, D-1$.

Problem 11.51   Show that every extreme point of the master polyhedron P^G is irreducible. (*Hint*: Take a vertex t and assume it is reducible. Using Problem 11.50(b), create the two points $t_k - \bar{t}_k + \bar{\bar{t}}_k$ and $t_k + \bar{t}_k - \bar{\bar{t}}_k$, and show that they satisfy (2) – which is a contradiction, since $t = \frac{1}{2}(\bar{t} + \bar{\bar{t}})$.

Problem 11.52   Show that the value of the components t_k $(k = 1, \ldots, D-1)$ of an irreducible point satisfy

$$\prod_{k=1}^{D-1} (1 + t_k) \leq D. \tag{3}$$

where $D = |\det \mathbf{B}|$ is the order of G. (*Hint*: Use Problem 11.50(b), which indicates that $\sum_{k=1}^{D-1} \mathbf{g}_k \bar{t}_k$ is a different group element for each set of nonnegative integers $\bar{t}_k$ satisfying $\bar{t}_k \leq t_k, k = 1, \ldots, D-1$.)

Problem 11.53   Show that the components of an irreducible point $t_1, \ldots, t_{D-1}$ satisfy

$$\sum_{k=1}^{D-1} t_k \leq D - 1, \tag{4}$$

(*Hint:* Use (3) and the fact that if the product of several positive integers is fixed, then the maximum value of these integer variables is attained by setting all but one of them to 1.)

Problem 11.54   Let $P^{\mathbf{x}}\mathbf{N}$, the corner polyhedron (in $\mathbf{x_N}$ space), be the intersection of master polyhedron P^G with the hyperplanes $t_{n+i} = 0$ for $i = 1, \ldots, D-n-1$. Show that $(x_{J(l)}, \ldots, x_{J(n)})$ is an extreme point of $P^{\mathbf{x}}\mathbf{N}$ if and only if

$$(t_1 = x_{J(1)}, \ldots, t_n = x_{J(n)}, t_{n+1} = 0, \ldots, t_{D-1} = 0)$$

is an extreme point of P^G.

Problem 11.55   In equalities (3) and (4), replace t_k by $x_{J(j)}$ and let the product or sum range from $j = 1$ to n in place of $k = 1$ to $D - 1$. Prove that the resulting expressions are valid for an irreducible point $x_{J(1)}, \ldots, x_{J(n)}$ of the corner polyhedron $P^{\mathbf{x}}\mathbf{N}$. (*Hint:* Show that a point is irreducible in the master polyhedron if and only if its image in the corner polyhedron is irreducible. Then appeal to the results of Problems 11.52 and 11.53.)

Problems 11.56 through 11.64 are related to the faces of P^G. For ease of reference, it is redefined here. The master polyhedron P^G is the convex hull of nonnegative integer solutions to

$$\sum_{k=1}^{D-1} \mathbf{g}_k t_k \equiv \mathbf{g}_t, \tag{5}$$

where the elements $\mathbf{g}_k$ $(k = 1, \ldots, D-1)$ are from the group G (which may be taken as $G(\boldsymbol{\lambda})$). Its zero element is $\mathbf{g}_D$, and $\mathbf{g}_t$ is some group element in G. Let T be the set of all nonnegative integer solutions to (5). A face inequality of P^G is written as

$$\sum_{k=1}^{D-1} \bar{\gamma}_k t_k \geq \bar{\gamma}_0,$$

or, more simply, as

$$\bar{\gamma} t \geq \bar{\gamma}_0, \tag{6}$$

where $\bar{\gamma} = (\bar{\gamma}_1, \ldots, \bar{\gamma}_{D-1}) \geq 0, \bar{\gamma}_0 \geq 0$, and $t = (t_1, \ldots, t_{D-1})$.

Problem 11.56   Show that (6) is a face (inequality) of P^G if and only if (i) for every t in $T, \bar{\gamma} t \geq \bar{\gamma}_0$, and (ii) there are n t^i in T which generate the hyperplane $\bar{\gamma} t = \bar{\gamma}_0$. Use Problem 11.6 to illustrate your answer.

Problem 11.57   Prove that for $\bar{\gamma}_0 > 0$, (6) provides a face (inequality) of P^G if and only if $\bar{\gamma}$ is a basic feasible solution to the system of inequalities

$$\bar{\gamma} t \geq \bar{\gamma}_0, \quad \text{for all } t \text{ in } T. \tag{7}$$

(*Hint*: A basic feasible solution to the inequalities (7) is an extreme point which in n space is defined by the intersection of n linearly independent equalities (7). Use this and appeal to (i) and (ii) in the previous problem.)

Problem 11.58   Prove that the only possible face inequalities (6) for P^G with $\bar{\gamma}_0 = 0$ are of the form $t_k \geq 0$. (*Hint*: By Problem 11.56(ii) there are a set of $n-1$ linearly independent vectors t^i on the face. This gives a system of equalities (6) of the form $\bar{\gamma} t^i = 0$, which will mean that only one component of $\bar{\gamma}$ can be positive.)

Problem 11.59   Prove that the condition $t_h \geq 0$ provides a face of P^G if and only if the element g_t in (5) lies in the subgroup G_H of G, generated by the elements g_k ($k = 1, \ldots, D-1, k \neq h$). (*Hint*: If g_t is not in the subgroup, there is no solution to (5) with $t_h = 0$. If g_t lies in the subgroup, a basic feasible solution $\bar{\gamma} = (0, 0, \ldots, \bar{\gamma}_h = 1, 0, \ldots, 0)$ to (7) can be constructed.)

Problem 11.60 Explain how the results appearing in Problems 11.56 through 11.59 can be used to prove Theorem 11.9. What else is needed to complete the proof of that theorem? Outline the remaining steps. (If necessary, see Gomory [279]).

Problem 11.61 Show that $t_k \geq 0$ for $k = 1, \ldots, D - 1$ are face inequalities of the master polyhedron P^G if and only if the group G is not of order 2. (*Hint*: Show that $\mathbf{g}_t$ in (5) lies in the subgroup G_h, described in Problem 11.59 if and only if G is not of order 2.)

Problem 11.62 Discuss procedures for generating all basic feasible solutions to a system of equations. How might linear programming be used? Should there be any preference in selecting the objective function? Need all the equations in the system explicitly appear? Relate your discussion to implementing either Theorem 11.9 or the assertion appearing in Problem 11.57.

Problem 11.63 Show that the master polyhedron does not change when the group elements $\mathbf{g}_1, \ldots, \mathbf{g}_{D-1}$ and $\mathbf{g}_t$ in (5) are replaced by their images in an isomorphic group.

Problem 11.64 In Theorem 11.9 show that once (i) and (ii) are used to eliminate variables in (iii) and duplicate inequalities are dropped, there are approximately $D^2/6$ inequalities left in (iii). How many nonzero entries are in each row of the resulting system? In what sense do the coefficients in this system reflect the group structure?

In Problems 11.65 through 11.67, G_h denotes a cyclic group of order h whose elements $\mathbf{g}_k$ are m vectors of the form $\binom{0}{k}$ for $k = 1, \ldots, h - 1$ and $\mathbf{g}_h = \binom{0}{0}$.

Problem 11.65 Find all faces and vertices of the master polyhedron P^G for $G = G_2$ when $\mathbf{g}_t$ in (5) is (a)$\mathbf{g}_1$, (b) $\mathbf{g}_2$. Illustrate your results geometrically.

Problem 11.66 Find all faces and vertices of the master polyhedron P^G for $G = G_3$ when $\mathbf{g}_t$ in (5) is (a)$\mathbf{g}_1$, (b) $\mathbf{g}_2$, and (c) $\mathbf{g}_3$. Illustrate your results geometrically.

Problem 11.67 Show that the faces of the master polyhedron for $G = G_4$, when $\mathbf{g}_t = \mathbf{g}_3$, are

$$t_1 \qquad + \quad t_3 \quad = \quad 1$$
$$t_1 \ + \ 2t_2 \ + \ 3t_3 \quad = \quad 3$$
$$t_1 \geq 0, \ t_2 \geq 0, \ t_3 \geq 0.$$

Problem 11.68 Using the results appearing in Section 11.7.5, generate three valid inequalities from master polyhedra, using the row

$$x = \frac{3}{15} + \frac{24}{15}x_1 - \frac{13}{15}x_2 + \frac{12}{15}x_3$$

from a linear programming tableau in which all variables are required to be integral and $D =|\det \mathbf{B}| = 15$.

Problem 11.69 Solve Example 11.4 by any *enumeration* algorithm. Generate valid inequalities from each row of the optimal simplex tableau and use them to implicitly enumerate.

Problem 11.70 Given the mixed integer program

$$
\begin{array}{lll}
\text{maximize} & \mathbf{cx} + \mathbf{dy} \\
\text{subject to} & \mathbf{Ax} + \mathbf{Dy} \leq \mathbf{b}, \\
& \mathbf{x} \geq \mathbf{0}, \mathbf{y} \geq \mathbf{0}, \\
\text{and} & \mathbf{x} \quad \text{integer},
\end{array}
$$

where $\mathbf{A}$ is m by n, $\mathbf{D}$ is m by n', and all data is integral. Let $\mathbf{I}$ be the identity matrix corresponding to the slack variables. Select a dual feasible basis $\mathbf{B}$ from the columns of $(\mathbf{A}, \mathbf{D}, \mathbf{I})$ and express the mixed integer program as a minimization problem where the nonnegativity constraints on the basic variables have been dropped. Discuss the structure of the coefficient columns. In particular, do they form an abelian group? What algorithms might be used to solve the minimization problem?

Problem 11.71 Express the mixed integer program

$$
\begin{array}{rcrcrcrcrcl}
\text{maximize} & -x_1 & - & x_2 & - & x_3 & - & 2x_4 & - & 4y \\
\text{subject to} & x_1 & + & 2x_2 & + & x_3 & + & x_4 & + & 5y & = & 10 \\
& 3x_1 & - & 3x_2 & + & 2x_3 & - & 3x_4 & + & 3y & = & 5, \\
\end{array}
$$
$$y \geq 0, \ x_1, x_2, x_3, x_4 \geq 0 \text{ and integer},$$

as a minimization problem as explained in the previous problem. Solve the mixed integer program by solving the minimization problem. (*Hint*: This is Example 11.4 with $x_5 = y$, and the optimal linear programming solution is in Section 11.7.5.)

Problem 11.72 Discuss the corner polyhedra and master polyhedron for the convex hull of mixed integer solutions to the minimization problem discussed in Problem 11.70. Illustrate your answer with a small example.

Problem 11.73 Suppose all the variables in the integer program are required to be either 0 or 1. What simplifications occur in the dynamic programming algorithm? In the enumerative algorithm? Illustrate your answer by solving Example 11.1 with all variables required to be either 0 or 1.

Problem 11.74 Discuss the group structure when all the variables are required to be either 0 or 1. Does the binary restriction influence the corner polyhedra and master polyhedron? Can their faces be found using Theorem 11.9? explain.

The following problems relate to Appendix A and Section 11.4.

Problem 11.75 List the shortest route (or minimal chain) algorithm discussed in Appendix A. Assume that computations do not terminate until all nodes in the network are members of the tree. Draw a flow chart of the algorithm. What must be stored in computer implementation? What is a possible bookkeeping scheme which keeps track of the computations?

Problem 11.76 Show that a tree constructed by the algorithm in Appendix A for a network with D nodes has precisely $D-1$ arcs. Why does a tree contain a minimal chain from the source node N_D to every other node in the network?

Problem 11.77 Show that any subchain of a minimal chain is a minimal chain. Why does this infer that the shortest route algorithm described in Appendix A can find a minimal chain between *many* pair of nodes in the network? Is this useful when solving a group problem?

Problem 11.78 (Hu [336]) Use the algorithm described in Appendix

A and the result appearing in the previous problem to find a minimal chain between every pair of nodes for the network below (arcs without arrows can be traversed in either direction). Tabulate your results using the array.

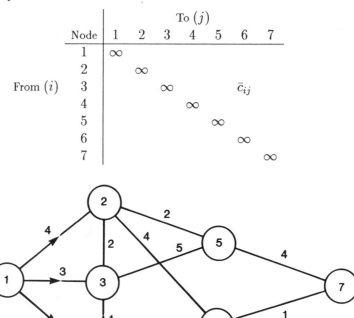

Problem 11.79   Solve Example 11.3, whose representative network appears in Figure 11.7(c), by the shortest route algorithm presented in Appendix A. Does the solution solve the integer program? (Arcs without arrows are undirected.)

Problem 11.80   Suppose a group problem is solved as a shortest route problem. If the minimal chain is a minimal solution to the group problem which does not solve the integer program, what must be done? What are various alternatives? Which seems best? Why?

Problem 11.81   Discuss various changes to the shortest route algorithm discussed in Appendix A so that it can explicitly take advantage of the special structure of a network which represents a group minimization problem. Do your alterations seem computationally worthwhile? Explain.

Problem 11.82   Suppose that a group problem has zero costs and a minimal chain has cycles. Show that this means that the integer program may have an optimal solution with some of its components arbitrarily large and integer. Illustrate, using a small example.

The following problems relate to Appendix B.

Using the algorithm appearing in Appendix B, obtain the Smith Normal Form $\hat{\mathbf{B}}$ of the following basis $\mathbf{B}$. Find the matrices $\mathbf{R}$ and $\mathbf{C}$ such that $\hat{\mathbf{B}} = \mathbf{RBC}$.

Problem 11.83

$$\mathbf{B} = \begin{pmatrix} 3 & 2 \\ -1 & 6 \end{pmatrix}$$

Problem 11.84

$$\mathbf{B} = \begin{pmatrix} 1 & 6 & 4 \\ 0 & 3 & -1 \\ 2 & -3 & 1 \end{pmatrix}$$

Problem 11.85 (Wolsey [603])

$$\mathbf{B} = \begin{pmatrix} 4 & 6 & -4 \\ 7 & 1 & -5 \\ -2 & 9 & -1 \end{pmatrix} \qquad Hint : \hat{\mathbf{B}} = \begin{pmatrix} 1 & 0 & 0 \\ 0 & 1 & 0 \\ 0 & 0 & 18 \end{pmatrix}$$

Problem 11.86   Why does the algorithm appearing in Appendix B converge? What is an upper bound on the number of loops of Step 1 through Step 6 until b_{11} will be 1 or divide the other $b_{ij}(i \neq 1, j \neq 1)$? Assume that the elements of the matrix are between 0 and $D = |\det \mathbf{B}|$.

Problem 11.87   Draw a flow chart for the algorithm appearing in

484   Foundations of Integer Programming

Appendix B. In terms of a computer program, what has to be stored?
What is a possible bookkeeping scheme which keeps track of the computations? Would you program the algorithm as it is outlined? In
particular, can Step 5 be omitted until $\mathbf{B}$ is diagonalized and we determine whether the diagonal elements divide each other? What is
involved in this strategy? Does it seem more efficient?

Problem 11.88   Consider the basis $\mathbf{B}$ with columns $\mathbf{b}_1, \ldots, \mathbf{b}_m$. In
terms of the standard identity or unit vectors $\mathbf{e}_i$ $(i = 1, \ldots, m)$, we can
write each

$$\mathbf{b}_j = \sum_{i=1}^{m} b_{ij} \mathbf{e}_i,$$

where b_{ij} is the ith component $(1 \leq i \leq m)$ of the basis vector
$\mathbf{b}_j$ $(j = 1, \ldots, m)$. Show that putting a matrix into Smith's Normal Form is equivalent to choosing a new set of basis vectors and unit
vectors. That is, the basis vectors $\mathbf{b}_j$ are transformed to basis vectors
$\mathbf{b}'_j$ $(j = 1, \ldots, m)$, which are the columns of $\hat{\mathbf{B}}$, the Smith Normal
Form of $\mathbf{B}$, and the unit vectors $\mathbf{e}_i$ are transformed to unit vectors
$\mathbf{e}'_i$ $(i = 1, \ldots, m)$ so that each

$$\mathbf{b}'_j = \sum_{i=1}^{m} b'_{ij} \mathbf{e}'_i = \epsilon_j \mathbf{e}'_j \text{ for } j = l, \ldots, m.$$

(*Hint*: Step 3 in the algorithm is equivalent to letting $\mathbf{e}'_1 = \mathbf{e}_1 + n_k \mathbf{e}_k$
and $\mathbf{e}'_i = \mathbf{e}_i$ for $i \neq 1$. Step 4 is the same as letting $\mathbf{b}_k' = \mathbf{b}_k - n_k \mathbf{b}_1$ for
$k = 2, \ldots, m$, and $\mathbf{b}'_1 = \mathbf{b}_1$.)

Problem 11.89   Find the new basis vectors and unit vectors as described in the previous problem for Problem 11.83. Are there any relationships between these vectors and the matrices $\mathbf{R}$ and $\mathbf{C}$?

Problem 11.90   For each unit vector $\mathbf{e}'_i$ $(i = 1, \ldots, m)$ defined in
Problem 11.88, define $\mathbf{t}_i = (\mathbf{B}^{-1} \mathbf{e}'_i)$ reduced modulo 1. Let $G(\mathbf{t}_i)$ be the
abelian group generated by $\mathbf{t}_i$, where addition in the group is modulo
1. Show that

$$G(\bar{\alpha}) = G(\mathbf{t}_1) \oplus G(\mathbf{t}_2) \oplus \ldots \oplus G(\mathbf{t}_m), \tag{8}$$

where $\oplus$ means that addition is as defined in $G(\bar{\alpha})$ and thus modulo 1.

(*Hint*: By a previous result, the group $G(\bar{\alpha})$ is isomorphic to the factor group $G(\hat{A})/G(\hat{B})$, where $\hat{B} = (b'_1, \ldots, b'_m)$ is the Smith Normal Form of B. As $b'_i = \epsilon_i e'_i$, each subgroup $G(t_i)$ generates precisely ϵ_i different elements in $G(\bar{\alpha})$, and since $D = |\det B| = \sum_{i=1}^m \epsilon_i$, the result follows.)

Problem 11.91 (Hu [335])

$$\text{Let } A = \begin{pmatrix} -1 & 2 & 0 & 1 & 0 & 0 \\ 2 & 0 & 0 & 0 & 1 & 0 \\ 1 & -2 & 2 & 0 & 0 & 1 \end{pmatrix} \text{ and } B = \begin{pmatrix} -1 & 2 & 0 \\ 2 & 0 & 0 \\ 1 & -2 & 2 \end{pmatrix}.$$

Using (8) in the previous problem, show that

$$G(\bar{\alpha}) = \begin{pmatrix} 0 \\ 0 \\ 4/8 \end{pmatrix} \oplus \begin{pmatrix} 4/8 \\ 2/8 \\ 0 \end{pmatrix}$$

$$\text{Hint : } B^{-1} = \begin{pmatrix} 0 & 4/8 & 0 \\ 4/8 & 2/8 & 0 \\ 4/8 & 0 & 4/8 \end{pmatrix} \text{ and } \hat{B} = \begin{pmatrix} 1 & 0 & 0 \\ 0 & 2 & 0 \\ 0 & 0 & 4 \end{pmatrix}.$$

Appendix A
A Shortest Route Algorithm

(Dijkstra[171])

We have shown (Section 11.4) that the group minimization problem may be formulated as a network problem where a shortest route is desired. In this appendix we show how to find the shortest route, and thus present another way of solving the group problem.

Recall that a *network* is a collection of *nodes* with *arcs* joining them. To restate the problem: Label the nodes, N_k $(k = 1, \ldots, D)$, and the directed arc joining nodes N_i and N_k by A_{ik}. Assign a nonnegative cost (or distance) $\bar{c}_{ik}$ to arc A_{ik}. That is, if arc A_{ik} is traversed, a cost of $\bar{c}_{ik}$ is paid. We wish to travel from an initial *source* node N_D to a final or *sink* node N_k $(1 \le k \le D - 1)$. The problem is to reach the sink node from the source node, using existing arcs, so that a minimal total cost is paid. Before discussing the algorithm we need the following definitions.

Definitions

1. A *chain* from N_i to N_k is a collection of directed arcs and nodes which emanate from N_i and end in N_k so that each node in the chain, except N_i and N_k, has a single arc entering and leaving it. Node N_i has a single emanating arc and node N_k a single incident arc.

2. A *cycle* from N_i to N_k is a chain when $N_i = N_k$.

3. A node N_j is a *neighbor* of the node N_k $(J \neq k)$ if the arc A_{kj} is in the network.

Example In the network depicted in Figure 11.15:

1. Chains from N_D to N_2 are

which are denoted by (N_D, N_1, N_2), and (N_D, N_2), respectively.
2. Cycles from N_D to N_D are

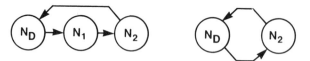

3. Nodes N_1 and N_2 are neighbors of node N_D

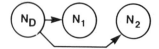

Figure 11.15: Sample Network

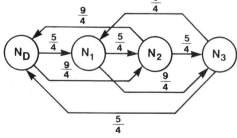

since arcs A_{D1} and A_{D2} are in the network.

Using the definitions, the problem is to obtain a minimal cost chain, or simply minimal chain, from the source node to the sink node. As the costs are nonnegative it is evident that a minimal chain need not have any cycles.[25]

The algorithm works by successively finding the first closest node to N_D, the second closest node to N_D, and so on. In particular, let N_{i1} be the closest node to N_D. As it is a neighbor of N_D, we have a minimal chain from N_D to N_{i1} (namely, (N_D, N_{i1})). Since the costs are nonnegative, there is a second closest node N_{i2} to N_D which is a neighbor of N_D or N_{i1}, and thus a minimal chain from N_D to N_{i2} is either (N_D, N_{i2}) or (N_D, N_{i1}, N_{i2}). A third closest node N_{i3} to N_D is a neighbor of N_D, N_{i1} or N_{i2}, and therefore a minimal chain from N_D to N_{i3} may be computed. The process is continued until a minimal chain from N_D to the sink is found.

In computations, suppose minimal chains from the source node N_D to some nodes N_t $(1 \le t \le D-1)$ are known. Denote the network containing the arcs and nodes which comprise the minimal chains as a *tree*, and arcs (nodes) in the tree as *tree arcs* (nodes). To find the next closest node to N_D, we wish to find a minimal chain from N_D to a neighbor of some tree node. (In Fig. 11.16, the tree is the composition of the nodes N_D, N_1, N_4, N_5 and the arcs joining them. A minimal chain is sought from N_D to either N_2, N_6, N_7, N_8, or N_9). This may be done as follows. Define C_{Dt} to be the total cost of the minimal chain from the source node N_D to the tree node N_t. For each neighboring node N_i of the tree, compute

$$C'_{Di} = \underset{t}{\text{minimum}} \ (C_{Di} + \bar{c}_{ti}), \tag{1}$$

which is the cost of a chain from N_D to N_i using a minimal chain and exactly one nontree arc, namely A_{ti}. (If arc A_{ti} does not exist, take $\bar{c}_{ti}$ as infinity in (1).) Then if

$$C'_{Dr} = (C_{Dt} + \bar{c}_{tr}) = \text{minimum } C'_{Di}, \tag{2}$$

---

[25]Cycles in a minimal chain are possible only if sum of the costs around a cycle is zero. However, we may ignore this possibility (see Problem 11.82). Also, if distances are used in place of costs, we have a shortest route problem. The two names are used interchangeably.

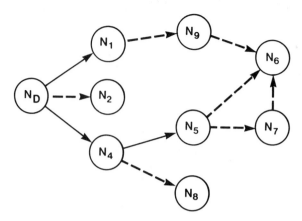

Figure 11.16: Tree nodes: N_D, N_1, N_4, N_5;
neighbors of tree nodes: N_2, N_6, N_7, N_8, N_9

the minimal chain from N_D to N_r is the minimal chain from N_D to N_t plus the arc A_{tr}. The cost of the chain C_{Dr} is C'_{Dr}. To see this, observe that $C'_{Dr} \leq C'_{Di}$ for all neighboring nodes N_i of the tree; or, any chain to N_r passing through any neighbor N_i must cost at least C'_{Dr}. Therefore, two or more nontree arcs need never be used in a minimal chain to N_r, since there will be a first arc A_{il} where N_i is a tree node and N_l is not, so that the resulting chain must cost at least C'_{Dr}. (In Fig. 11.16, if $N_r = N_6$, the cost of the chain (N_D, N_1, N_9, N_6) must be at least C'_{D6}, since by (2), $C'_{D6} \leq C'_{D9}$. Also, by (1), the cost of the chain (N_D, N_4, N_5, N_7) is at least C'_{D6}, so the cost of the chain to N_6 through N_7 must be at least C'_{D6}.)

Once the minimal chain to N_r is found, A_{tr} becomes a tree arc and N_r a tree node. As the tree has changed, (1) must be recalculated, then (2) is used again, a new tree arc and node are found, and so forth. Initially, there are no tree arcs and the source is the single tree node in which case (1) and (2) reduce to

$$C_{Dr} = \underset{i}{\text{minimum}} \ \bar{c}_{Di} \tag{3}$$

for all neighbors N_i of N_D. Therefore the algorithm starts with (3) and then uses (1) and (2). Computations may be terminated once the sink node becomes a member of the tree.

To illustrate the algorithm we find the shortest route from the source N_D to the sink N_3 for the network appearing in Fig. 11.15. The numbers adjacent to the arcs represent their cost.[26] The computations are tabulated in Table 11.2.

Remarks

1. Every time a new arc A_{tr} is added to the tree the values of C'_{Di} in (1) have to be compared with $C_{Dr} + \bar{c}_{ri}$. If it exceeds this sum, a minimal chain using one nontree arc passes through the new tree node N_r. Thus C'_{Di} should be updated to $C_{Dr} + \bar{c}_{ri}$. Otherwise it remains the same. In the example, $C'_{D2} = 9/4$ is compared with $C_{D1} + \bar{c}_{12} = 5/4 + 5/4 = 10/4$ in Step 2. However, the shortest route to N_2, using one nontree arc, is still A_{D2}, and thus C'_{D2} stays at $9/4$.

2. The tree contains the minimal chain from the source node N_D to every other tree node (Problem 11.76). Therefore, the computations may be continued, even after the sink node becomes a member of the tree, until all nodes are tree members, in which case we have the minimal chain from N_D to every other node in the network. If this is done, the tree will have precisely $D - 1$ arcs with no loops (Problem 11.76). In our illustration, the minimal chain from N_D to N_1 is A_{D1} and to N_2 is A_{D2}.

3. In context of the group minimization problem (Section 11.4), the node N_k represents the group element $\mathbf{g}_k$ $(k = 1, \ldots, D)$, and $D = |\det \mathbf{B}|$ is the order of the group. The zero element of the group is $\mathbf{g}_D$ and is the source. The sink represents the right-hand side. The solution to the example is therefore $x_{J(1)} = x_{J(2)} = 1$ (which satisfied $1\mathbf{g}_1 + 1\mathbf{g}_2 = \mathbf{g}_3$).

4. A careful examination of the procedure will indicate that the computations are proportional to D^2, where $D = |\det \mathbf{B}|$ is the number of

---

[26]An examination of Section 11.4 reveals that this network represents a group minimization problem, where the group is cyclic of order 4, the first two elements $\mathbf{g}_1 = \begin{pmatrix} 0 \\ 1 \end{pmatrix}$, $\mathbf{g}_2 = \begin{pmatrix} 0 \\ 2 \end{pmatrix}$ are columns in the problem, their costs are $5/4$ and $9/4$, respectively, and the right-hand side is $\mathbf{g}_3 = \begin{pmatrix} 0 \\ 3 \end{pmatrix}$. The zero element in the group is $\mathbf{g}_D = \begin{pmatrix} 0 \\ 4 \end{pmatrix} = \begin{pmatrix} 0 \\ 0 \end{pmatrix}$.

Step	Tree nodes	Tree arcs	Tree	Neighbors N_i	Chain cost C'_{Di}	Minimum C'_{Dr}
1	N_D	—		N_1, N_2	$C'_{D1}=\frac{5}{4}$ $C'_{D2}=\frac{9}{4}$	$C'_{D1}=\frac{5}{4}$
2	N_D, N_1	A_{D1}		N_2, N_3	$C'_{D2}=\min\{\frac{5}{4}+\frac{5}{4},\frac{9}{4}\}=\frac{9}{4}$ $C'_{D3}=\frac{5}{4}+\frac{9}{4}=\frac{14}{4}$	$C'_{D2}=\frac{9}{4}$
3	N_D, N_1, N_2	A_{D1}, A_{D2}		N_3	$C'_{D3}=\frac{9}{4}+\frac{5}{4}=\frac{14}{4}$	$C'_{D3}=\frac{14}{4}$
4	N_D, N_1, N_2, N_3	A_{D1}, A_{D2}, A_{23}		—	—	—

Table 11.2: Shortest Route Calculations

nodes in the network. Therefore, the network approach becomes very inefficient as the order of the group increases.

The shortest path algorithm is formally listed below:

Step 1 [Initialization] Initialize the origin node N_D to be the only tree-node with $C_{Dt} = 0$. All other nodes are non-tree nodes with $C'_{Di} = A_{Di}$. Go to Step 2.

Step 2 [Finding a new tree node] Find a non-tree node N_r with the minimum C'_{Di} value amongst all the non-tree nodes. This non-tree node becomes a new tree node with the shortest distance from the origin N_D equal to C'_{Dr}. Go to Step 3.

Step 3 [Termination] If all nodes become tree nodes, stop. Otherwise, go to Step 4.

Step 4 [Update C'_{Di} values for non-tree nodes] For each non-tree node i, determine if the distance from the source, C'_{Di}, can be reduced by traversing through the new tree-node N_r. That is, C'_{Di} becomes minimum $\{C'_{Di}, C_{Dr} + A_{ri}\}$. Go to Step 2.

A.1 Specialization of the Shortest Path Algorithm for the Group Minimization Problem (Chen and Zionts [129], Morito [434])

As indicated in Section 11.4, the network for the group problem is very large but is highly structured. The algorithm just presented does not take this structure into account. Morito [434] exploited the following properties of the network to modify the shortest path algorithm so as to produce an efficient procedure to solve the group minimization problem.

1. In a chain from the origin, if an arc corresponding to the variable $x_{J(j_1)}$ is used just before an arc corresponding to the variable $x_{J(j_2)}$, then to find the shortest path, without loss of optimality, we may require $j_2 \geq j_1$ (Shapiro [530]). This result can be used in Step 4 in that we need only update the distance C'_{Di} for those non-tree nodes i which may be reached through a variable $x_{J(j)}$ satisfying the ascending order.

2. Each node of the network has as many arcs of the same cost directed into it as it has leaving; and the number of arcs going out of or coming into a node is the same as the number of distinct nonzero group elements in the constraints. This property suggests generating arcs as required rather than attempting to store the often huge network in the computer.

3. Often the number of non-tree nodes which can be reached from the origin with a finite distance is very small when compared with the total number of nodes. Hence significant computational effort can be saved in Step 2 by restricting the minimization over only the non-tree nodes with finite distance.

The detailed listing of the shortest path algorithm with these modifications is left to Morito [434].

Appendix B
Diagonalizing the Basis–Smith's Normal Form

In Theorem 11.2 we said that any (m by m) basis $\mathbf{B}$ can be converted to a diagonal matrix (all elements not on the diagonal are zero)

$$
\hat{\mathbf{B}} = \begin{pmatrix} \epsilon_1 & & & \\ & & & 0 \\ & & \ddots & \\ 0 & & & \\ & & & \epsilon_m \end{pmatrix},
$$

using elementary row and column operations; where ϵ_i $(i = 1, \ldots, m)$ is a positive integer, ϵ_i divides ϵ_{i+1} $(i = 1, \ldots, m-1)$, and $D = |\det \mathbf{B}| = \prod_{i=1}^{m} \epsilon_i$. $\hat{\mathbf{B}}$ is sometimes referred to as the Smith Normal Form for $\mathbf{B}$. Recall (Section 11.5) that a row (column) operation is the permutation of two rows (columns), or the addition or subtraction of integer multiples of one row (column) from another, and is equivalent to premultiplying (postmultiplying) the matrix by an appropriate integral matrix whose determinant is plus or minus 1. We explain the diagonalization procedure and thus verify the theorem. The discussion is combined with a listing of the algorithm, and is similar to the one appearing in Hu [336]. The element in the i, j position of $\mathbf{B}$ is denoted by b_{ij}.

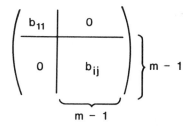

Figure 11.17: First Pass Through Algorithm

Step 1 Interchange rows and columns such that b_{11} is the element of smallest absolute value among all nonzero elements in the matrix. Go to Step 2.

The intent of this step is to eventually allow b_{11} and hence ϵ_1 to divide each element in row 1 and column 1. When this happens, multiples of column 1 (row 1) subtracted from other columns (rows) may yield a matrix as in Fig. 11.17.

In computations, **B** is successively multiplied on the left (right) by the matrix which defines each row (column) transfer.

Step 2 If b_{11} divides the remaining row 1 elements b_{1j} $(j = 2, \ldots, m)$, go to Step 3. Otherwise, if b_{11} does not divide b_{1j} for some $j = k$, it follows that $b_{1k} = n_k b_{11} + q$, where n_k is some integer and $0 < q < |b_{11}|$.[27] Subtract n_k times column 1 from column k so that b_{1k} becomes q. Return to Step 1.

This step determines whether b_{11} divides row 1. If not, an element b_{1k} is found which has a positive remainder q. The value of b_{1k} is reduced to q, and as $q < |b_{11}|$, a number smaller than $|b_{11}|$ enters the (1,1) position of the matrix in Step 1.[28] Notice that Step 3 must eventually be reached, since in the worst case, q (hence b_{11}) is reduced to 1, which divides every b_{1j}.

---

[27]This is a statement of the Euclidean algorithm (see Lang [384]).

[28]A positive integer smaller than q may appear in the matrix, and thus q may not enter the (1,1) position. For example,

$$\begin{pmatrix} 3 & 8 \\ 4 & 9 \end{pmatrix} \quad \text{becomes} \quad \begin{pmatrix} 3 & 2 \\ 4 & 1 \end{pmatrix} \quad \text{and } q = 2 > 1.$$

In computations, if b_{11} does not divide b_{1k}, the matrix is postmultiplied by the (m by m) matrix $\mathbf{C}_k = (c_{ij})$, which has the form

$$
\begin{array}{c}
\text{column } k \\
\downarrow
\end{array}
$$

$$
\begin{pmatrix}
1 & & -n_k & & \\
& \ddots & \vdots & & \\
& & 1 & & \\
& & & \ddots & \\
& & & & 1
\end{pmatrix}
\qquad
\begin{array}{l}
c_{ii} = 1, \quad i = 1, \ldots, m \\
c_{1k} = -n_k \\
c_{ij} = 0 \quad \text{otherwise}
\end{array}
$$

Step 3 If b_{11} divides the remaining column 1 elements b_{i1} $(i = 2, \ldots, m)$, go to Step 4. Otherwise, there is a row k for which $b_{k1} = n_k b_{11} + q$, where n_k is an integer and $0 < q < |b_{11}|$. Subtract n_k times row 1 from row k, so that b_{k1} becomes q. Go to Step 1.

This step checks whether b_{11} divides column 1. If not, an element b_{k1} with a positive remainder q is reduced to that value. As $q < |b_{11}|$, a number smaller than $|b_{11}|$ enters the $(1,1)$ position in Step 1. As in Step 2, we must eventually reach Step 4, since a lower bound for q is 1.

In computations, if b_{11} does not divide b_{k1}, the matrix is premultiplied by the (m by m) matrix $\mathbf{R}_k = (r_{ij})$, which has the form

$$
\text{row } k \rightarrow
\begin{pmatrix}
1 & & & & \\
& \ddots & & & \\
-n_k & \ldots & 1 & & \\
& & & \ddots & \\
& & & & 1
\end{pmatrix}
\qquad
\begin{array}{l}
r_{ii} = 1, \quad i = 1, \ldots, m \\
r_{k1} = -n_k \\
r_{ij} = 0 \quad \text{otherwise}
\end{array}
$$

Step 4 Now b_{11} divides row one and column one. So let $b_{1j} = n_j b_{11}$ and $b_{i1} = m_i b_{11}$, where m_i and n_j are integers $(i, j = 2, \ldots, m)$. From each column j $(j \geq 2)$, subtract n_j times column 1, and from each row i $(i \geq 2)$ subtract m_i times row 1. The resulting matrix appears in Fig. 11.17. Go to Step 5.

In computations, the composition of all the column subtractions is equivalent to postmultiplying the matrix by

$$\prod_{k=2}^{m} \mathbf{C}_k,$$

where $\mathbf{C}_k$ appears in Step 2. Similarly, the composition of all the row subtractions is equivalent to premultiplying the matrix by

$$\prod_{k=2}^{m} \mathbf{R}_k,$$

where $\mathbf{R}_k$ appears in Step 3.

Step 5 This step involves the following operations:

If b_{11} divides every element in the submatrix (b_{ij}) $(i,j = 2,\ldots,m)$, go to Step 6. Otherwise, there is an entry b_{ij} $(2 \leq i, j \leq m)$ for which $b_{ij} = nb_{11} + q$, where n is an integer and $0 < q < |b_{11}|$. In this case, bring q into the $(1,1)$ position of the matrix, using the following elementary operations.

$$
\begin{array}{cc}
 & \text{col. 1} \quad\ \text{col. } j \\
\begin{array}{c}\text{row 1}\\ \text{row } i\end{array} &
\begin{pmatrix} b_{11} & 0 \\ 0 & nb_{11} + q \end{pmatrix}
\end{array}
$$

$$\downarrow (a)$$

a) Add n times row 1 to row i, in which case b_{i1} becomes nb_{11} and b_{ij} remains at $nb_{11} + q$.

$$
\begin{array}{cc}
 & \text{col. 1} \quad\ \text{col. } j \\
\begin{array}{c}\text{row 1}\\ \text{row } i\end{array} &
\begin{pmatrix} b_{11} & 0 \\ nb_{11} & nb_{11} + q \end{pmatrix}
\end{array}
$$

$$\downarrow (b)$$

b) Subtract column 1 from column j so that b_{ij} becomes q.

$$
\begin{array}{cc}
 & \text{col. 1} \quad\ \text{col. } j \\
\begin{array}{c}\text{row 1}\\ \text{row } i\end{array} &
\begin{pmatrix} b_{11} & -b_{11} \\ nb_{11} & q \end{pmatrix}
\end{array}
$$

$$\downarrow (c)$$

c) Interchange column j with column 1 so that b_{i1} becomes q.

$$
\begin{array}{cc}
 & \text{col. 1} \quad\ \text{col. } j \\
\begin{array}{c}\text{row 1}\\ \text{row } i\end{array} &
\begin{pmatrix} -b_{11} & b_{11} \\ q & nb_{11} \end{pmatrix}
\end{array}
$$

$$\downarrow (d)$$

d) Interchange row i with row 1
so that q enters the (1,1) position
of the matrix.

$$\begin{array}{cc} & \text{col. 1} \quad\ \text{col. } j \\ \text{row 1} \\ \text{row } i \end{array} \begin{pmatrix} q & nb_{11} \\ -b_{11} & b_{11} \end{pmatrix}$$

Return to Step 1.

This step is necessary in order to guarantee that the diagonal elements in $\hat{\mathbf{B}}$, ϵ_i divide ϵ_{i+1} $(i = 1, \ldots, m-1)$. In particular, if b_{11} is to become the diagonal element ϵ_1, it must divide the diagonal element ϵ_2. If b_{11} now divides every element in the submatrix, it will divide every element in the submatrix even after it is submitted to elementary row and column operations. This follows because if there are no remainders then

$$b_{ij} = n_{ij}b_{11} \tag{4}$$

for every $i, j = 2, \ldots, m$ and n_{ij} is an integer. So if an element in the submatrix becomes an integer multiple of some b_{ij}, by (4), we can still write it as an integer times b_{11}; thus, b_{11} still divides it. Since the next pass through Steps 1 through 4 will yield a matrix as in Fig. 11.18, we have that b_{11} divides b_{22} so that it can serve as the first diagonal element ϵ_1 in $\hat{\mathbf{B}}$. Note that we must eventually reach Step 6, since a lower bound for the remainder in (d) is 1, which divides all elements.

The computations in (a) through (d) reduce to simply multiplying the matrix on the left or right by the appropriate m by m matrix $\mathbf{R}$ or $\mathbf{C}$, respectively. The matrix for each case is (where $r_{11} = c_{11} = 0$)

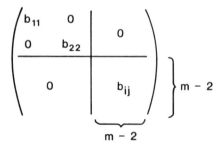

Figure 11.18: Second Pass Through Algorithm

a) **R**: $r_{tt} = 1$ $t = 1, \ldots, m$, $r_{i1} = n$, $r_{tk} = 0$ otherwise.

b) **C**: $c_{tt} = 1$ $t = 1, \ldots, m$, $c_{1j} = -1$, $c_{tk} = 0$ otherwise.

c) **C**: $c_{tt} = 1$ $t = 1, \ldots, m$, $t \neq j$, $c_{11} = 0$, $c_{1j} = 1$,
$c_{j1} = 1$, $c_{tk} = 0$ otherwise.

d) **R**: $r_{tt} = 1$ $t = 1, \ldots, m$, $t \neq i$, $r_{11} = 0$, $r_{1i} = 1$,
$r_{i1} = 1$, $r_{tk} = 0$ otherwise.

In each case (a through d), subscript $k = 2, \ldots, m$.

Step 6 If $b_{11} < 0$, multiply row 1 by minus one. Apply Steps 1 through
4 to the $m-1$ by $m-1$ submatrix appearing in Fig. 11.17. This
will produce the matrix where b_{11} divides b_{22}. Take $\epsilon_1 = b_{11}$ in
$\hat{\mathbf{B}}$ and pass to Step 5. (Row 1 and column 1 may be dropped.)
Continue this process until $\hat{\mathbf{B}}$ is obtained.

It is necessary to check the sign of b_{11}, since positive diagonal elements are required.

Multiplying row 1 by minus one is equivalent to premultiplying the
matrix by an identity matrix except that the (1,1) entry is -1 rather
than 1.

Example

Initially, $\mathbf{B} = \begin{pmatrix} 1 & 3 & 4 \\ 4 & 2 & 6 \\ 0 & 2 & 4 \end{pmatrix}$ and $|\det \mathbf{B}| = 20$.

The computations are tabulated in Table 11.3, which gives

$$\mathbf{R} = \begin{pmatrix} 1 & 0 & 0 \\ 0 & 1 & 0 \\ 0 & 5 & 1 \end{pmatrix} \begin{pmatrix} 1 & 0 & 0 \\ 0 & 0 & 1 \\ 0 & 1 & 0 \end{pmatrix} \begin{pmatrix} 1 & 0 & 0 \\ -4 & 1 & 0 \\ 0 & 0 & 1 \end{pmatrix}$$

$$= \begin{pmatrix} 1 & 0 & 0 \\ 0 & 0 & 1 \\ -4 & 1 & 5 \end{pmatrix}$$

Step	R			B			C			= Updated matrix B		
$1 \to 2 \to 3 \to 4$	1	0	0	1	3	4				1	3	4
	-4	1	0	4	2	6				0	-10	-10
	0	0	1	0	2	4				0	2	4
4				1	3	4	1	-3	0	1	0	4
				0	-10	-10	0	1	0	0	-10	-10
				0	2	4	0	0	1	0	2	4
4				1	0	4	1	0	-4	1	0	0
				0	-10	-10	0	1	0	0	-10	-10
				0	2	4	0	0	1	0	2	4
$4 \to 5 \to 6 \to 1$ ($\epsilon_1 = 1$)	1	0	0	1	0	0				1	0	0
	0	0	1	0	-10	-10				0	2	4
	0	1	0	0	2	4				0	-10	-10
$2 \to 3 \to 4$	1	0	0	1	0	0				1	0	0
	0	1	0	0	2	4				0	2	4
	0	5	1	0	-10	-10				0	0	10
4				1	0	0	1	0	0	1	0	0
				0	2	4	0	1	-2	0	2	0
				0	0	10	0	0	1	0	0	10
$5 \to 6$ ($\epsilon_2 = 2$)				1	0	0						
				0	2	0						
				0	0	10						
$1 \to 2 \to 3 \to 4$ $\to 5 \to 6$ ($\epsilon_3 = 10$)												

Table 11.3: Computations for the example

$$
C = \begin{pmatrix} 1 & -3 & 0 \\ 0 & 1 & 0 \\ 0 & 0 & 1 \end{pmatrix} \begin{pmatrix} 1 & 0 & -4 \\ 0 & 1 & 0 \\ 0 & 0 & 1 \end{pmatrix} \begin{pmatrix} 1 & 0 & 0 \\ 0 & 1 & -2 \\ 0 & 0 & 1 \end{pmatrix}
$$

$$
= \begin{pmatrix} 1 & -3 & 2 \\ 0 & 1 & -2 \\ 0 & 0 & 1 \end{pmatrix}
$$

Hence **B** is diagonalized to

$$
\hat{\mathbf{B}} = \overset{\mathbf{R}}{\begin{pmatrix} 1 & 0 & 0 \\ 0 & 0 & 1 \\ -4 & 1 & 5 \end{pmatrix}} \overset{\mathbf{B}}{\begin{pmatrix} 1 & 3 & 4 \\ 4 & 2 & 6 \\ 0 & 2 & 4 \end{pmatrix}} \overset{\mathbf{C}}{\begin{pmatrix} 1 & -3 & 2 \\ 0 & 1 & -2 \\ 0 & 0 & 1 \end{pmatrix}}
$$

$$
= \begin{pmatrix} 1 & 0 & 0 \\ 0 & 2 & 0 \\ 0 & 0 & 10 \end{pmatrix}
$$

Appendix C
Computational Experience

Some computational work has been done on solving the group minimization problem. Several of the computer programs (or codes) also have the ability to seek the best solution to the group problem which yields nonnegative values for the basic variables, so that they can always solve the integer program. We now summarize the results appearing in the literature. A more detailed analysis may be found in the references. In addition, computational experience generating faces of corner polyhedra may be found in Devine and Glover [170], and in Rubin and Links [494].

C.1 Computer Programmed Group Minimization Algorithms

C.1.1 A Dynamic Programming Algorithm (White [599])

A dynamic programming algorithm which finds the rth $(r = 1, 2, 3, \ldots)$ best solution to the group problem is described in White [599]. The procedure first solves the linear program, reduces the updated nonbasic columns and right-hand side modulo 1, and then constructs the $G(\bar{\alpha})$ (Section 11.2) group minimization problem. If the minimal solution to the group problem solves the integer program, the recursive relationships defined in [599] with $r = 1$ reduces to those appearing in Section 11.3.1. Otherwise, additional (enumerative) work is required and the

Problem					Optimal Solution			
No.	m	$m+n$	Description		D	Linear Program	Group Problem	Integer Program
1	6	11	Fixed Charge	1	183	8.79	7.00	7.00
2	6	11	Fixed Charge	2	258	9.61	8.00	8.00
3	6	11	Fixed Charge	3	320	11.81	10.00	10.00
4	6	11	Fixed Charge	4	205	9.22	8.00	8.00
5	6	11	Fixed Charge	6	> 1798	?	?	?
6	9	15	Fixed Charge	9	2000	12.00	9.00	9.00
7	16	28	Fixed Charge	10	> 1192	?	?	?
8	21	56	Machine scheduling	1	> 639	?	?	?
9	7	14	IBM test	1	32	7.50	8.00	8.00
10	7	14	IBM test	2	32	5.75	7.00	7.00
11	3	7	IBM test	3	72	179.78	182.00	187.00
12	15	30	IBM test	4	> 989	?	?	?
13	15	30	IBM test	5	> 846	?	?	?
14	12	49	IBM test	8	2856	0.00	0.00	0.00
15	50	65	IBM test	9	96	5.33	6.00	9.00

Table 11.4: Haldi [307] Test Problems

computations are equivalent to finding the cheapest route in a network for which the basic variables turn out to be nonnegative.

The algorithm has been programmed in FORTRAN IV and tested on an IBM 7094 computer. It is set up to solve problems with a maximum of 100 constraints and 200 variables (100 nonbasic variables). Some peripheral equipment is used for storage. The seven small examples described in Problem 11.2 through 11.8 were solved in about 10 seconds each. Some of the problems appearing in Haldi [307] were also tried. The results for the Haldi [307] problems are summarized in Table 11.4, where m is the number of constraints and $m+n$ is the number of variables (including slacks) in the original integer program, $D = |\det \mathbf{B}|$, $r = k$ ($k = 1, 2, \ldots$) means that the kth solution to the group problem solved the integer program, and the time is in seconds.

The algorithm was unable to solve Problems 5, 7, 8, 12, or 13. It is evident that the algorithm's efficiency depends crucially on the size of D, the order of the group. (In the next section we describe some techniques for reducing the value of the determinant.) Note that the minimal solution to the group problem solved the integer program in problems numbered 1, 2, 3, 4, 6, 9, 10, and 14.

C.1.2 Branch and Bound Algorithm (Shapiro [532], Morito [434])

The technique described by Shapiro [532] and Morito [434] first solves the linear program and diagonalizes the optimal basis $\mathbf{B}$ (using an algorithm similar to the one in Appendix B), then constructs the group $G(\boldsymbol{\lambda})$ (Section 11.5) and the associated group minimization problem. It solves the group problem, using a specialized shortest route algorithm. If some of the basic variables turn out to be negative, the branch and bound scheme discussed in Section 11.3.3 is used.

The algorithm was tried on several of the Haldi [307] problems (Table 11.4 lists the problem characteristics), using an IBM 360/65 computer by Shapiro[532], and an UNIVAC 1108 computer by Morito[434]. The results are listed in Table 11.5. All times are in seconds.

		White	Shapiro		Morito	
No.	r	Time IBM 7094	Branch and Bound used	Time IBM 360/65	Branch and Bound used	Time UNIVAC 1108
1	1	36.0	No	1.28	— —	— —
2	1	59.0	No	0.97	No	.2669
3	1	68.0	No	1.27	No	.3060
4	1	9.0	No	0.91	— —	— —
5	—	273.0	–	— —	— —	— —
6	1	239.0	No	6.59	No	3.2631
7	—	124.0	–	— —	— —	— —
8	—	73.0	–	— —	— —	— —
9	1	12.0	No	2.19	No	.1366
10	1	11.0	No	1.95	— —	— —
11	5	27.0	Yes	0.82	— —	— —
12	—	197.0	–	— —	— —	— —
13	—	88.0	–	— —	— —	— —
14	1	209.0	No	9.50	No	3.5926
15	—	—	Yes	20.20	— —	— —

Table 11.5: Summary of computational results

Even though different machines were used, it appears that these algorithms are more efficient than the one introduced by White [599]. However, note that none of the five problems the White algorithm could not solve was listed by Shapiro and Morito. Also, it is often agreed that it is worth diagonalizing the basis and constructing the group problem using $G(\boldsymbol{\lambda})$ rather than $G(\bar{\boldsymbol{\alpha}})$, since the former group contains elements which are easier to manipulate and keep track of. Both Morito [434] and Shapiro [532] used the group $G(\boldsymbol{\lambda})$, whereas White [599] used $G(\bar{\boldsymbol{\alpha}})$. Of the 10 runs appearing in the table, the branch and bound routine only had to be used twice, and in all cases the linear programming solution was near the minimal integer one.

C.1.3   Shortest Route Algorithms (Chen and Zionts [129])

In Section 11.4 we showed that the group problem can be represented by a network where a shortest or minimal cost route is desired. Any standard shortest route algorithm may be used to solve the group problem. However, the representative network has special properties which can simplify the computations. In particular, unlike the ordinary shortest route problem, a principle concern of the algorithm for the representative network is in avoiding the usually enormous number of routes of equal cost that correspond to the same values for the nonbasic variables. Thus the *number* of different types of arcs which have been traversed is important, not the *order* in which they were traversed.[29] A second computational efficiency is made possible by noting that every node in the network has an equal number of incoming and outgoing arcs. Moreover, the set of arcs emanating from one node is identical to the set emanating from every other node, both with respect to costs associated with them and with respect to the symmetry of the nodes to which they lead (see Fig. 11.7).

Chen and Zionts [129] used these properties to simplify the shortest route algorithms described in Dantzig [159], Farbey, Land, and Murchland [195], Floyd [202] and Murchland [436]. The properties allowed them to perform computations on a row of length $D = |\det \mathbf{B}|$, in place of using an entire D by D matrix (see Problem 11.78). In order to make comparisons in computational efficiency, the dynamic programming algorithms described by Gomory [278] (Section 9.3) and Shapiro [531], and the network approach appearing in Hu [335] were also programmed

---

[29]Arcs of the same type correspond to increasing the value of the the same nonbasic variable by 1 and have equal cost.

and tested. All computer programs were written in FORTRAN IV and run on a CDC 6400 computer.

Table 11.6 summarizes the results on a series of cyclic group problems generated as follows. The number of nonbasic variables n, and the order of the group D, were selected. Then each reduced cost $\bar{c}_j$ $(j = 1, \ldots, n)$ was sampled from a uniform distribution on the interval 20 to 90. The coefficient column λ_j (a generator of the group $G(\lambda)$, Section 11.5) was selected by randomly selecting λ_1 from the set $G_D = \{g_1, g_2, \ldots, g_{D-1}\}$, λ_2 from the set $G_D - \{\lambda_1\}$, λ_3, from the set $G_D - \{\lambda_1, \lambda_2\}$, and so on until λ_n was picked. The group element which was the right-hand side ranged from g_1 to g_{D-1}, and every value was used for each problem. The times in the table are in seconds and reflect the mean computation time. In this way, the effect of the right-hand side were supressed and only the parameters D and n had to be considered.

Examination of the table indicates that for a fixed value of the determinant the Hu [335], the modified Floyd [202], and the modified Farbey, Land, and Murchland [195] algorithms yield constant computation time. Other mean computation time increases as the number of variables increase. The times for the Gomory [280] algorithm are roughly proportional to n. Chen and Zionts [129] feel that the modified Farbey, Land, and Murchland algorithm is best when the ratio of n/D is moderate or large, while Shapiro's method is most efficient when the ratio is small.

C.2 Reducing the Order of the Group

As already noted, the efficiency of the algorithms heavily depends on the value of D or the order of the group. Gorry, Shapiro, and Wolsey [288] indicate that group problems of order greater than 100,000 are not very amenable to computation, and determinants less than 10,000 in magnitude are definitely preferred. Several methods for circumventing the difficulties encountered when D is large have appeared in the literature. These include scaling, relaxation, and decomposition. We briefly discuss these techniques.

C.2.1 Scaling (White [599])

As pointed out by Haldi and Isaacson [307], any initial rescaling of rows or columns in the constraint matrix could be of great help in reducing

Shortest Route Algorithm		D = 20			D = 60			D = 100		
		$n=2$	$n=10$	$n=19$	$n=6$	$n=30$	$n=59$	$n=10$	$n=50$	$n=99$
Gomory [278]	original	.004	.017	.033	.034	.155	.306	.083	.391	.791
Hu [335]		.008	.008	.008	.059	.059	.059	.163	.163	.163
Shapiro [531]		.006	.010	.015	.029	.070	.139	.064	.147	.196
Floyd [202]	modified	.014	.014	.014	.124	.124	.124	.344	.342	.344
Farbey, et al. [195]		.007	.007	.007	.053	.053	.053	.146	.146	.146
Dantzig [159]		.005	.006	.011	.055	.068	.098	.169	.175	.184

Table 11.6: Computation Times for Shortest Route Algorithms

the magnitude of the determinant since the basis is made up of columns from the constraint matrix. A first step in this direction is to ensure that no row has a common factor. A common factor in a column can also be taken out, provided we do not allow the group problem to generate noninteger solutions to the integer program. In particular, suppose a column has a common factor K, where K is a positive integer. If, at the optimal linear programming solution, this column turns out to be nonbasic, then it should be multiplied by K prior to generating the group problem. On the other hand, if the column is in the basis, then some subroutine would have to be added to the algorithm so that it guarantees that the associated basic variable is a multiple of K. In this case, D is reduced by a factor of $1/K$, but the amount of additional computations may be substantial.

C.2.2 Relaxation (Gorry, Shapiro, and Wolsey [288])

The determinant of the basis will tend to decrease in absolute value as its entries decrease. Thus it seems natural to use a basis containing many zeroes and other entries close to zero. One possibility is to seek an optimal solution to the linear program which is highly degenerate. Then, starting with an identity matrix, the positive basic variables can be pivoted in so that the resulting optimal basis will have many identity columns and hence many zeroes.

Another possibility uses the concept of a Gomory all integer cut (Chapter 6). In particular, the constraints

$$\sum_{j=1}^{n} a_{ij} x_{J(j)} \le b_i \quad (i = 1, \ldots, m) \tag{5}$$

in the integer program imply the inequalities

$$\sum_{j=1}^{n} \left\lceil \frac{a_{ij}}{\lambda_i} \right\rceil x_{J(j)} \le \left\lceil \frac{b_i}{\lambda_i} \right\rceil \quad (i = 1, \ldots, m), \tag{6}$$

where the data is initially integral, λ_i is a positive integer, and $[y]$ means the largest integer $\le y$. The procedure for showing that (5) implies (6) is equivalent to showing that a source row (5) yields the Gomory all integer cut (6), and this procedure is given in Chapter 6.

Suppose that some or all of the m inequalities (5) in the integer program are initially replaced by (6). Then every solution to the original problem is a solution to the new problem. In this sense, the problem with the scaled coefficients is termed the relaxed problem. Notice that it is possible to have solutions to the *relaxed problem* which do not satisfy the constraints (5) and therefore are not solutions to the integer program. The idea is to scale some or all of the constraints using divisors λ_i, which yield a small value for D, and yet an optimal solution to the relaxed problem (with the same objective function) also satisfies (5) and thus solves the more constrained original integer program. Heuristics for choosing λ_i are discussed in Gorry, Shapiro, and Wolsey [288]. That article also discusses other related ways for reducing D. Computational experience on a very limited number of problems indicates that the size of the group may be substantially reduced, but the difficulty is to find a good or optimal solution to the relaxed problem which does not violate the constraints (5) to the degree that it is unsatisfactory. (For example, slightly violated constraints may be discounted because of inaccuracies in the data.)

C.2.3 Decomposition (Hefley and Thomas [317])

The coefficient columns in the group minimization problem generate a cyclic group of order D. Moreover, each coefficient column generates a cyclic subgroup of this group whose order divides D (Section 11.2). If at least two coefficient columns can be selected so that they generate nonzero subgroups whose direct sum is the group, then it is possible to solve the group problem by solving a series of subproblems, each over one of the subgroups, and then by relating the results using a linking group problem. Each subproblem may be solved by any group minimization algorithm and the linking problem is solved using an enumeration. The details are in Hefley and Thomas [317].

A decomposition approach was programmed by Hefley and Thomas [317] in PL1, and test runs were made on an IBM 360/65 computer. Each problem is first solved as a linear program using a revised simplex routine. The optimal basis **B** is diagonalized (Appendix B) and the $G(\boldsymbol{\lambda})$ group (Section 11.5) is found. A heuristic is then used to decompose the group. The Gomory [280] dynamic programming algorithm solves each group subproblem, and the results are linked using another

	Problem		Group Problem				Times with (without) decomposition		Ratio
No.	m	$m+n$	D	r	t	N	with	(without)	
1	4	14	64	3	4	3	.20	(.51)	2.58
2	4	14	64	3	4	3	.12	(.46)	3.67
3	6	18	2000	3	6	3	38.66	(108.16)	1.80
4	7	21	32	4	7	3	.36	(.43)	1.16
5	14	68	156	4	12	5	.34	(8.06)	23.40
6	8	34	576	5	10	6	.28	(25.57)	90.21
7	8	34	576	5	10	6	.35	(25.13)	70.88
8	7	54	192	3	19	4	5.70	(7.26)	1.27
9	8	34	192	5	9	7	.47	(5.21)	11.07
10	8	34	288	6	8	7	.20	(8.55)	42.94
11	6	32	336	3	20	3	12.05	(13.82)	1.15
12	6	32	408	2	20	2	8.31	(13.63)	1.64

Table 11.7: Decomposition Algorithm Results

group problem. The results are tabulated in Table 11.7. For purposes of comparison, each group problem was also solved directly by the Gomory technique. In the table, m is the number of constraints and $m+n$ is the number of variables (including the slacks) for the original integer program, $D = |\det \mathbf{B}|$ is the order of the group, r is the number of nonzero (congruence) equations in the group problem (the number of diagonal terms in the Smith Normal Form for $\mathbf{B}$ which exceed 1; see Section 11.5), t is the number of variables in the group problem (the number of distinct nonzero group elements which come from nonbasic columns), N is the number of group subproblems solved, and all times are in seconds.

The rightmost column in Table 11.7 represents the time without decomposition divided by the time with decomposition. Since in every case it is larger than 1, the decomposition approach was always more efficient. It is especially worthwhile on problems 5, 6, 7, 9, and 10. These results indicate that it seems worthwhile to decompose the group whenever possible, and solve the group subproblems plus a linking problem in place of solving the group problem directly. However, further investigation on larger problems is necessary prior to drawing more specific conclusions.

Appendix D
Implementation of a Group Theoretic Algorithm

(Morito [434])

+his algorithm first constructs the Factor Group Minimization Problem (Section 11.5) using the Smith Normal Form of the basis **B** (Appendix B). It solves the group problem using a specialized shortest route algorithm. If some of the basic variables turn out to be negative, then a branch and bound scheme similar to an improved version of the one in Section 11.3.3 is used. The algorithm can be divided into the four major modules discussed in Section D.1 through Section D.4.

D.1   Linear Programming Module

Since we expect that the integer optimal solution is "close" to the LP optimal solution, it seems logical to use the optimal LP basis to construct the group problem. Unfortunately, even for small problems, the determinant D of the optimal basis may be very large resulting in an unmanageable network problem. On the other hand, we can select a dual feasible, but not optimal, basis with a smaller value for the determinant. This will result in a smaller network problem. However, since the basis is not primal feasible, there is a high probability that the solution to the group problem is infeasible to the integer problem, and hence the branch and bound part of the algorithm will have to be used. Computational experience indicates that branch and bound time increases as the basis (i.e., extreme point) used to construct the group problem moves away from the LP optimum. Morito [434] suggests the following tradeoff criterion: choose the "closest to the optimal" dual feasible LP basis whose determinant is less than a prespecified number D_{max}. Morito [434] used an LP code which is a modification of the one written by Salkin and Spielberg [519]. It contains a revised simplex method applied to the dual problem. The dual method maintains a dual feasible basis. This routine terminates with either the optimal basis or as soon as the determinant of the basis exceeds 2000 (the value of D_{max}).

D.2   Construction of FGMP

The dual feasible basis is first diagonalized by the Smith Normal Form algorithm outlined in Appendix B, and then the FGMP is constructed.

D.3 Solution of the FGMP (shortest path algorithm)

The Factor Group Minimization Problem is solved using the network approach. The shortest route algorithm used is the Dijkstra's algorithm specialized for a network corresponding to the group problem (see Appendix A). The solution to the FGMP is then converted to the solution of the original problem. If the solution is nonnegative, it is optimal. If the solution is not feasible, the branch and bound algorithm is applied.

D.4 Branch and Bound Algorithm

The basic approach is the algorithm discussed in Section 11.3.3. This scheme inspects the integer solutions to the group problem by successively adding constraints of the form $x_{J(j)} \geq k_j$, where k_j starts at 0 and is incremented by 1. To improve the efficiency of the basic algorithm, Morito [434] uses the following node selection and implicit branching and node omission rules.

D.4.1 Node Selection Rule

Whenever more than one node is available to branch from, the node selection rule used is to pick the candidate node with a minimal objective value for the associated group problem.

D.4.2 Implicit Branching Rule

We label each node as $(k_1, k_2, \ldots, k_n)$, which corresponds to the original group problem with the additional constraints $x_{J(j)} \geq k_j$, $j = 1, \ldots, n$. When a node $(k_1, \ldots, k_n)$ is selected for branching, n nodes, namely $(k_1 + 1, k_2, k_3, \ldots, k_n), (k_1, k_2 + 1, k_3, \ldots, k_n), \ldots, (k_1, k_2, k_3, \ldots, k_n + 1)$, are created at the next level. However, as illustrated in Figure 11.19 (cf. dashed lines), for an example consisting of three nonbasic variables, this branching process results in the creation of many duplicate nodes. (For example, node 6 is reached first from node 2 and latter from node 3.) Morito [434] uses the following simple rule to avoid the creation of such duplicate nodes: From the node $(k_1, k_2, \ldots, k_n)$, only create nodes from the branches corresponding to the constraints $x_{J(j)} \geq k_j + 1$, $j \geq j^*$, where $j^* = \text{maximum}\{j | k_j > 0\}$.

Note that by using this rule, in Figure 11.19, node 6 will not be created from node 3, and nodes 7 and 9 will not be created from node

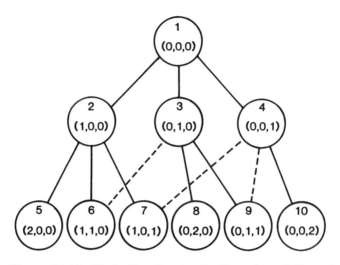

Figure 11.19: Node Duplication in Branch and Bound

4. Also, it is easy to verify that this rule will enumerate all nodes. Finally, note that this branching rule is not applicable for the first node $(0, 0, \ldots, 0)$ in the tree.

D.4.3 Node Omission Rule

Here we show that certain nodes need not be explicitly generated. Consider Figure 11.19 again. Suppose at node 1, the solution to FGMP gives $x_{J(1)} = 1, x_{J(2)} = 1$ and $x_{J(3)} = 1$. Then nodes 2, 3, 4, 6, 7, and 9 need not be created as we know that these nodes give the same optimal solution to FGMP as node 1. This observation suggests the following node omission rule: If at node $(k_1, k_2, \ldots, k_n)$, the solution to the FGMP is $x_{J(j)} = \hat{x}_{J(j)}$, $(j = 1, \ldots, n)$. Then from this node, we create n subproblems by adding the constraints $x_{J(j)} \geq \hat{x}_{J(j)} + 1$, one at a time, for $j = 1, \ldots, n$, respectively. Using this rule, in Figure 11.19, from node 1, we will create nodes 5, 8, and 10 at the next level.

D.4.4 Combining the Implicit Branching Rule and the Node Omission Rule

If either the implicit branching rule or the node omission rule is used, it results in a valid algorithm in the sense that all nodes are either implicitly or explicitly enumerated. In this section, we show that if both the rules are used simultaneously, it will result in an incorrect algorithm. Note that the combined implicit branching and node omission rule is: If at node $(k_1, k_2, \ldots, k_n)$ the resulting group problem has solution $x_{J(j)} = \hat{x}_{J(j)}$ $(j = 1, \ldots, n)$, then from this node we create $(n - j^* + 1)$ subproblems by adding constraint $x_{J(j)} \geq \hat{x}_{J(j)} + 1$, one at a time, for $j^* \leq j \leq n$, where $j^* = \text{minimum}\{j | k_j > 0\}$. To illustrate, consider the example below.

Example 11.5 (Haldi [306])

Maximize $13x_1 + 15x_2 + 14x_3 + 11x_4$
subject to
$$
\begin{array}{llllll}
4x_1 + 5x_2 + 3x_3 + 6x_4 & - x_5 & & = 96 \\
20x_1 + 21x_2 + 17x_3 + 12x_4 & & - x_6 & = 200 \\
11x_1 + 12x_2 + 12x_3 + 7x_4 & & - x_7 & = 101
\end{array}
$$
and $\quad x_j \geq 0$ and integer $(j = 1, \ldots, 7)$.

The corresponding FGMP is,

maximize $46x_{J(1)} + 34x_{J(2)} + 238x_{J(3)} + 64x_{J(4)}$
subject to
$$
\begin{array}{llll}
x_{J(1)} + & + x_{J(3)} + x_{J(4)} & \equiv 0 & (\text{mod } 2) \\
5x_{J(1)} + 35x_{J(2)} + 29x_{J(3)} + 32x_{J(4)} & \equiv 8 & (\text{mod } 36)
\end{array}
$$
and $\quad x_{J(j)} \geq 0$ and integer $(j = 1, 2, 3, 4)$.

Here, $x_{J(1)} = x_2, x_{J(2)} = x_6, x_{J(3)} = x_3$, and $x_{J(4)} = x_5$. The solution to FGMP (i.e., node $(0, 0, 0, 0)$) is $x_{J(1)} = x_{J(2)} = 2, x_{J(3)} = x_{J(4)} = 0$, which yields the infeasible basic solution $x_1 = -1$, $x_4 = 15$, and $x_7 = 17$.

If the combined branching rule is used, the first branching from node 1 $(0, 0, 0, 0)$ whose solution is $(2, 2, 0, 0)$, yields the four nodes $(3, 0, 0, 0), (0, 3, 0, 0), (0, 0, 1, 0)$, and $(0, 0, 0, 1)$ (Figure 11.20). From these nodes, using the combined branching rule, node $(0, 2, 1, 0)$ will never explicitly or implicitly be examined. In fact, many descendants of an omitted node may never be examined under this branching rule. So, if a node is omitted, all nodes which are to be created from the

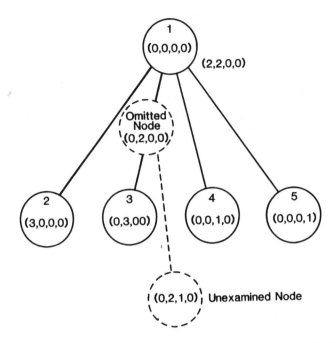

Figure 11.20: Problem with Combined Branching Rule

omitted node will never be examined and never appear in the tree, unless special care is taken to recover them. Morito uses an ingenious
scheme to recover omitted nodes "only if necessary." The details are in
[434].

Chapter 12

The Knapsack Problem

12.1 Introduction

Consider the integer program with a single constraint

$$\text{maximize} \quad \sum_{j=1}^{n} c_j x_j$$

$$\text{subject to} \quad \sum_{j=1}^{n} a_j x_j \leq b$$

$$x_j \geq 0 \text{ and integer} \quad (j = 1, 2, \ldots, n),$$

where the costs c_j, coefficients a_j, and right-hand side (number) b are integers, and each a_j $(j = 1, 2, \ldots, n)$ and b are positive.[1] Such a program is called a *knapsack problem*. The name is in reference to packing a knapsack so that its capacity is not exceeded and the total value–that is, the sum of all relative values–is maximized. The model is precisely the above problem where each variable x_j is either 0 (item j is not packed) or 1 (item j is packed).

Although the simplest integer program, the knapsack problem is worth studying because: (a) it is representative of many industrial situations such as capital budgeting, project selection and capital investment, budget control, and numerous loading problems; and (b) it appears as a subproblem that has to be solved in many integer programming algorithms (e.g., in the group minimization algorithm presented in the previous chapter). In the next section, some of the more common applications and algorithm uses are discussed.

---

[1]In most application the costs will also be positive. In this case, we could say $c_j \geq 0$ $(j = 1, 2, ..., n)$ and $b \geq 0$. However $c_j = 0$ implies $x_j = 0$ in any optimal solution, and $b = 0$ means that the only feasible solution is $x_j = 0$ $(j = 1, 2, \ldots, n)$, since $a_j > 0$. If some $a_j = 0$ with $c_j > 0$, the maximum solution is unbounded since x_j can be made arbitrarily large and integral.

Background

Algorithms (Section 12.4) which solve the knapsack problem are usually of the dynamic programming, enumerative, Lagrangian multiplier, or network type.

Early papers which discuss the knapsack problem in context of dynamic programming (Section 12.4.1) include Bellman [66], [67], [68], [69], and Dantzig [155]. Later expositions are in Bellman and Dreyfus [71] and Dantzig [159]. Computational improvements in dynamic programming algorithms are described by Greenberg [294], Dreyfus and Prather [176], Yormark [611], and Martello and Toth [414].

Enumerative algorithms (Section 12.4.2) specialized to solve the knapsack problem include the branch and bound (Chapter 8) procedure for the zero-one problem appearing in Kolesar [371] and the search algorithms (Chapter 9) described by Guignard and Spielberg [298]. Other enumerative techniques are in Balas and Zemel [46], Cabot [121], Cabot and Hurter [122], Faaland [191], Greenberg and Hegerich [295], Horowitz and Sahni [501], Nauss [442], and Martello and Toth [412],[413],[414].

Everett [190] introduced the concept of utilizing Lagrangian multipliers (Section 12.4.3) to solve discrete programming problems. A more detailed specialization appears in Brooks and Geoffrion [101], Geoffrion [226], Nemhauser and Ullman [443], and Shapiro [528]. Application of the Lagrangian approach to the knapsack problem is described by Fox and Landi [205]. The use of Lagrangian multipliers in solving nonlinear zero-one programming problems is discussed by Hammer, Rosenberg, and Rudeanu [310], and by Hammer and Rudeanu [311].

The knapsack problem can also be represented as a network in which a minimal cost or shortest route is sought (Section 12.4.4). Shapiro [527], [528] discusses the formulation and solution techniques. In addition, Shapiro and Wagner [529] solve the knapsack problem by considering a related problem which is solved by a network approach.

A classical application which is a knapsack problem is the Lorie and Savage [401] capital budgeting model (Section 12.2.1). The model, extensions, dynamic programming algorithms, and some computational results are extensively discussed in Weingartner [595], [596], and in Weingartner and Ness [597]. Lagrangian techniques (Section 12.4) applied to capital budgeting models are described by Cord [149] and by Kaplan [356]. Extensions of the model, which are not of the knapsack type, have also been considered. In particular, the case where the objective function is quadratic[2] is treated by Radhakrishnan [469], Mao and Wallingford [405], Unger [581], and Mathur, Salkin, and Morito [420]. Knapsack problems with multiple-choice constraints have been discussed by Glover and Klingman [263], Sinha and Zoltner [538], Mathur, Salkin, and Morito [419], and Zemel [618]. The use of Balas duality [20], [22] - [24], [27], in integer programming[3] to give economic interpretations and properties of optimal solutions to capital budgeting models is discussed by Radhakrishnan [469] and Unger [581]. Other articles relating mathematical programming to capital budgeting models include Baumol and Quandt [56], Byrne, Charnes, Cooper, and Kortanek [119], Naslund [441], and Woolsey [609].

Other applications involving the knapsack problem or its solution include the cutting stock problem (Section 12.2.2) extensively discussed by Gilmore and Gomory [232],[233], [236], [237], the journal selection problem appearing in Kraft and Hill [376], [377] and in Glover and Klingman [262] (Fig. 12.1), and the capital investment problem as explained by Hansmann [312].

Using a basic number theoretic result appearing in Mathews [417], it can be shown that an integer program with any finite number of constraints can be transformed to a knapsack problem (Section 12.3). Although the transformed (knapsack) problem has the same number of variables, its constraint coefficients are usually enormous. Thus, roundoff errors and overflow problems often make computer solution impossible. Articles describing the transformation process and derivations which may reduce the magnitude of the coefficients include Elmagrahby and Wig [188], Anthonisse [6], [7], Bradley [96], Bush and Carlson [118], Garfinkel and Nemhauser [217], Glover [256], Glover and Woolsey [267], Hammer and Rudeanu [311], Onyekwelu [450], and Padberg [454]. An excellent survey is in Kendell and Zionts [361].[4]

---

[2]A quadratic function is of the form $cx + \frac{1}{2}xCx$, where c (an n vector) and C (an n by n matrix) are known.

[3]Under certain circumstances it is possible to write the "dual" of an integer program and say such things as the dual of the dual is the primal, and if one has an optimal solution the other does too, and vice versa. However, the results are difficult to present and considerably weak relative to linear programming duality theory, and thus do not appear to be of substantial computational use. The details are left to the references.

[4]With regard to transforming the integer programs, Bradley [92], [93] has shown that every integer programming problem is equivalent to an infinite number of different integer programming problems. The equivalence is such that an optimal

Computational results with knapsack algorithms are in Cabot [121], Cabot and Hurter [122], Fayard and Plateau [196], Gilmore and Gomory [233], Guignard and Spielberg [298], Kolesar [372], Nauss [442], Weingartner and Ness [597], and in Martello and Toth [414].

A general survey article which condenses some of the material presented in this chapter is in Salkin and DeKluyver [511].

12.2　Applications and Uses of Knapsack and Related Problems

12.2.1　Capital Budgeting (Lorie and Savage [401], Weingartner [595])

In the simplest case the problem consists of choosing among n competing independent investment possibilities so as to maximize the total payoff from the investments subject to a constraint on available funds. To model the problem, let c_j $(j = 1, \ldots, n)$ be the payoff from including project j, a_j $(j = 1, \ldots, n)$ be the cost of project j, and b be the total available funds. Further, let x_j be 0 (1) if project j is rejected (accepted). The basic model is then

$$\text{maximize} \quad \sum_{j=1}^{n} c_j x_j$$
$$\text{subject to} \quad \sum_{j=1}^{n} a_j x_j \le b$$
$$\text{and} \quad x_j = 0 \text{ or } 1 \quad (j = 1, \ldots, n),$$

---

solution to any problem in the equivalence class yields an optimal solution to the original integer program. The equivalent programs are usually more structured than the original problem so that they may be easier to solve. The transformation process is similar to the one explained in Chapter 11, Section 11.2, where a linear programming basis was first selected from the integer program's coefficient matrix, then its Smith Normal Form was found, and the entire integer program was transformed accordingly. In this case a different representation of the linear programming basis matrix, called its Hermite Normal Form, is used. The transformation process, algorithms for the equivalent problems, extensions for transforming a mixed integer program, and other related details are discussed by Bradley [94], [95], [97], [98], Bradley and Wahi [99], Bowman [85], Faaland [192], Faaland and Hillier [193], and Hillier [327], [328].

which is a knapsack problem with zero-one variables. Notice that the parameters c_j, a_j, and b may be taken as positive and integral.

The investment situation just described often can be divided into several periods with costs and available funds broken down to period requirements. In this situation, the nonnegative cost of project j in period t is denoted by a_{tj} $(j = 1, \ldots, n, t = 1, \ldots, T)$, and the available funds in period t are b_t $(t = 1, \ldots, T)$. The *multi-period Lorie-Savage capital budgeting model* is then

$$\text{maximize} \quad \sum_{j=1}^{n} c_j x_j$$

$$\text{subject to} \quad \sum_{j=1}^{n} a_{tj} x_j \leq b_t \quad (t = 1, \ldots, T)$$

$$\text{and} \quad x_j = 0 \text{ or } 1 \quad (j = 1, \ldots, n).$$

Unless $T = 1$, the above problem is a standard zero-one integer program with a nonnegative T by n constraint matrix,[5] and may be solved by the search enumeration techniques discussed in Chapter 9 (Problem 12.3). Other solution possibilities include aggregating the T constraints into a single constraint with the same set of integer solutions (Section 12.3) and solving the resulting knapsack problem, using a Lagrangian multiplier approach to find a near optimal solution (Section 12.4.4), or possibly dynamic programming when there are only a few periods (Problem 12.2).

A second extension to the model occurs when the n investment possibilities are partitioned into disjoint subsets $n_1, \ldots, n_p$ $(p \leq n)$ and it is specified that precisely one project in each subset must be selected.[6] The additional requirements have the form

$$\sum_{j \in n_i} x_j = 1 \quad (i = 1, \ldots, p).$$

Equalities of this type are called *multiple choice constraints*, since each one specifies (when x_j is 0 or 1) that a single variable is selected to be

---

[5] Project j may not require any funds in period t, so it is possible to have $a_{tj} = 0$. We may also note that an integer program with more than one constraint and a nonnegative coefficient matrix is sometimes termed a *multidimensional* knapsack problem.

[6] For example, to obtain an equal (geographic) distribution of capital investment.

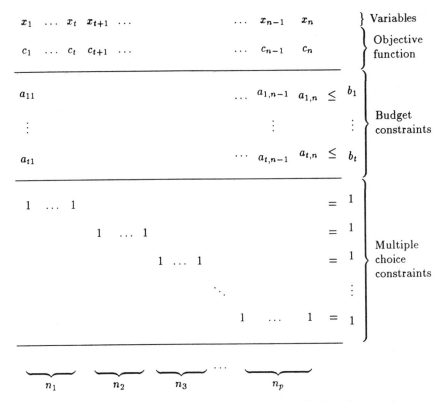

Figure 12.1: Integer Program with Multiple Choice Constraints

set to 1 among a set of variables. The structure of the multi-period model with these constraints appears in Figure 12.1.[7]

An algorithm for an integer program such as the one appearing in the figure should exploit the structure of the multiple choice constraints. For example, they dictate obvious branching rules in an enumeration and natural constraint feasibility tests, and they can be handled im-

---

[7]The model in Fig. 12.1 also represents the problem of selecting scientific journals to purchase so that their expected usage is maximized while (library) budget constraints are not exceeded; see Glover and Klingman [262]. Swart [556] has shown that multiple choice constraints can also be used in place of precedence constraints of the form $x_j \leq x_k$ for some j, k, where the variables are restricted to be 0 or 1. However, more variables are introduced and all zero-one solutions to the precedence constraints must be enumerated prior to making the conversion.

plicitly using a technique called *generalized upper bounding* (see Lasdon [385]) when linear programming is used (Problem 12.4)[8]

12.2.2   The Cutting Stock Problem (Gilmore and Gomory [232], [233], [236], [237])

There are numerous situations in which a material comes in standard lengths and must be cut up according to requirements and costs (or profit). These include goods which come in rolls (e.g., paper, sheet metals, and textiles) as well as less flexible materials (e.g., glass, metal pipe, and other tubing).

The One-Dimensional Case ([232], [233])

Consider the case in which only straight vertical cuts are allowed. That is, each standard length or roll is to be sliced into lengths l_i ($i = 1, \ldots, m$). One possible cutting pattern for a roll of length L appears in Fig. 12.2. Note that it produces two pieces of length l_1, three pieces of length l_2, three pieces of length l_3, and some waste. The cutting stock problem is to cut up rolls of the material so that the demand for the number of pieces of each length l_i ($i = 1, \ldots, m$) is satisfied while the total cost of the rolls used is minimized. Suppose that there are n possible cutting patterns and each can be pictured as in Fig. 12.2. Then the problem can be expressed as an integer program by letting N_i ($i = 1, \ldots, m$) be the number of pieces of length l_i demanded, x_j ($j = 1, \ldots, n$) be the number of times the jth cutting pattern is used (each time to cut up a roll), c_j be the cost of the roll from which the

---

[8]Healy [315] probably presented the first algorithm which takes advantage of multiple choice constraints. It uses cutting planes and does not always converge. Specialized enumerative procedures are in Brown [102], and other algorithms appear in Balintfy [53], DeMaio and Roveda [167], and Liggett [396]. Aggregating multiple choice constraints is discussed by Glover [256], and their appearance in integer programs which are solved by group theoretic algorithms (Chapter 11) is discussed by White [599]. Their use in producing cutting planes is in Glover [251]. Computational experience with integer programs having multiple choice constraints using branch and bound algorithms is discussed by Tomlin [568]. Algorithms for single knapsack problem with multiple choice constraints are discussed in Mathur, Salkin, and Morito [419], and Sinha and Zoltner [538].

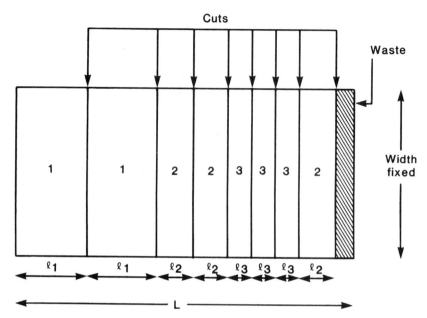

Figure 12.2: Sample Cutting Pattern (j), One-dimensional Case

jth cutting pattern is cut,[9] and a_{ij} is the number of pieces of length l_i produced each time the jth cutting pattern is used. The cutting stock problem is then

$$\text{maximize} \quad \sum_{j=1}^{n} c_j x_j$$

$$\text{subject to} \quad \sum_{j=1}^{n} a_{ij} x_j \geq N_i \qquad (i = 1, \ldots, m),$$

$$x_j \geq 0 \text{ and integer} \quad (j = 1, \ldots, n).$$

The difficulty of the problem lies in the usually enormous number of cutting patterns (columns j). For example, with a roll of length 200 in. and demand for 40 different lengths ranging from 20 in. to 80 in., the number of cutting patterns can exceed 100 million (Gilmore and Gomory [233]). Therefore, the integer program can rarely be solved directly. One possibility is to solve the problem as a *linear* program using a column generation technique. That is, the columns

---

[9]There are usually a relatively few standard stock lengths, and thus many of the costs are equal. If there is only one standard length, c_j may be taken as 1 in the integer programming model.

$$\mathbf{a}_j = \begin{pmatrix} a_{1j} \\ \vdots \\ a_{mj} \end{pmatrix}$$

are generated as needed. The optimal linear programming solution may then be rounded to produce a hopefully good set of cutting patterns.

The linear programming procedure developed by Gilmore and Gomory [232], [233] is a standard revised simplex algorithm in which the column that enters the basis is found by solving one or more knapsack problems. In particular, suppose a basic feasible solution $(\mathbf{x_B})$ to the equations

$$\sum_{j=1}^{n} a_{ij}x_j - s_i = N_i \quad (i = 1, \ldots, m)$$

is known, where s_i $(i = 1, \ldots, m)$ is a nonnegative slack variable. Then from linear programming we know the corresponding dual solution $\mathbf{w} = (w_1, \ldots, w_m)$ can be found by premultiplying the basis inverse $(\mathbf{B}^{-1})$ by the cost of the basic variables $(\mathbf{c_B})$, i.e., $\mathbf{w} = \mathbf{c_B}\mathbf{B}^{-1}$. The current basic feasible solution is optimal if for every other column (cutting pattern) $\mathbf{a}_j$, we have

$$\sum_{i=1}^{m} w_i a_{ij} - c_j \leq 0.$$

(Why?) But any set of nonnegative integers, a_{ij} $(i = 1, \ldots, m)$ is a cutting pattern provided that

$$\sum_{i=1}^{m} l_i a_{ij} \leq L_k \quad \text{for any standard length } L_k. \tag{1}$$

Thus, suppose that for each standard roll length L_k the knapsack problem

$$\begin{aligned}
\text{maximize} \quad & \sum_{i=1}^{m} w_i a_{ij} \\
\text{subject to} \quad & \sum_{i=1}^{m} l_i a_{ij} \leq L_k \\
\text{and} \quad & a_{ij} \geq 0 \text{ and integer } (i = 1, \ldots, m)
\end{aligned}$$

is solved. Let $\mathbf{a}_j^k$ be the optimal solution to each problem. Then if

$$\mathbf{wa}_j^k - c_k \leq 0 \quad \text{for each } k, \tag{2}$$

the current basic feasible solution $(\mathbf{x_B})$ is optimal. Otherwise, the column $\mathbf{a}_j^k$ which yields the largest positive value for $\mathbf{wa}_j^k - c_k$ enters the basis. Here, c_k is the cost of the roll whose length is L_k and is thus the cost of column $\mathbf{a}_j^k$. The leaving column is found after $\mathbf{B}^{-1}\mathbf{a}_j^k$ is calculated, and subsequent to the usual pivoting a new nonbasic column (cutting pattern) is found by solving the knapsack problems. The linear program is solved when (2) is satisfied.

The algorithm is crucially dependent on how many knapsack problems (and how quickly each of them) can be solved. If there is a single standard length, only one knapsack problem is solved at each iteration. Further, if the initial basic feasible solution is near (in the sense of extreme points) the optimal one, only a few knapsack problems may have to be solved before satisfying (2). (Procedures for obtaining an initial solution are in Problem 12.7 and in the references.) Computational experience indicates that a specialized enumeration algorithm (see [233]) appears to be the best way of solving the knapsack problems.

The Two-Dimensional Case ([236], [237])

Suppose now that we have the same problem as before except that straight width cuts are allowed and N_i is the demand for the number of rectangles of length l_i and width w_i. A sample cutting pattern appears in Fig. 12.3. If the problem is to select cutting patterns so that demands for rectangular pieces of various sizes are met while total cost of the rolls used is minimized, the model is precisely the integer program developed for the one dimensional case–where a_{ij} is the number of rectangles of dimensions (l_i, w_i) produced by the jth cutting pattern. However, because of the added dimension, inequality (1) does not necessarily produce a cutting pattern and the column generation scheme can no longer be used. Without explicitly enumerating all the columns, the linear program (hence the integer program) cannot be solved.

A related problem is to ignore the demand constraints and find cutting patterns which maximize the value of a roll when it is cut up into rectangles of given sizes. In particular, we wish to

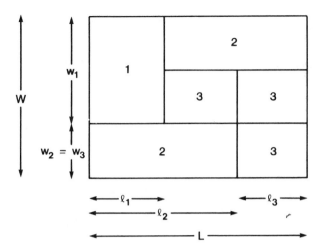

Figure 12.3: Sample Cutting Pattern (j), Two-dimensional Case

$$\text{maximize } f(L, W) = p_1 a_1 + \ldots + p_m a_m,$$

where $a_1, \ldots, a_m$ are nonnegative integers such that there exists a way of dividing a rectangle (roll) (L, W) into a_i rectangles (l_i, w_i) for $i = 1, \ldots, m$ and each rectangle a_i has a value p_i $(i = 1, \ldots, m)$. The case where the cuts are always continued to the end of the existing roll (termed "guillotine cuts" [236]) is extensively discussed by Gilmore and Gomory [236], [3237]. Algorithms for this problem which take advantage of certain properties of the function $f(L, W)$ are also presented.[10] The details are left to the references. A sample guillotine cut is in Fig. 12.4, where the cuts are numbered in the order they are made.

---

[10]The function

$$f(x) = max \left\{ \sum_{j=1}^{n} c_j x_j \,\middle|\, \sum_{j=1}^{n} a_j x_j \leq b_j, x_j \geq 0 \text{ and integer } (j = 1, \ldots, n) \right\}$$

is sometimes termed a *one-dimensional knapsack function*; $f(L, W)$ is termed a *two-dimensional knapsack function*. Properties of one- and two-dimensional functions are discussed in [236] and [237]. A periodic property for the one-dimensional function is given in Section 12.4.1, and other properties for the two-dimensional function are presented in Problem 12.8.

Also, an enumerative procedure for the integer program representing the one-dimensional cutting stock problem is presented by Pierce [461], and a heuristic technique for the two-dimensional case is given by Art [14].

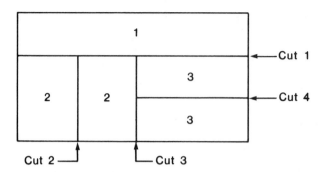

Figure 12.4: Sample Guillotine Cut, Two-dimensional Case

12.2.3 Loading Problems

The loading problem, as an alternative name to the knapsack problem, was given by Bellman [66]. However, this reference is now used for a more generalized form of the knapsack model where instead of one knapsack, there are m knapsacks available, but as before, items are indivisible (i.e., cannot be split). As an example consider a situation where we have a convoy of m trucks carrying various items. To model this scenario, we let a_j $(j = 1, \ldots, n)$ be the weight of item j and c_j $(j = 1, \ldots, n)$ its cost, and let b_i $(i = 1, \ldots, m)$ be the weight capacity of the ith $(i = 1, \ldots, m)$ truck. The model is then

$$
\begin{aligned}
\text{maximize} \quad & \sum_{j=1}^{n} \sum_{i=1}^{m} c_j x_{ij} \\
\text{subject to} \quad & \sum_{j=1}^{n} a_j x_{ij} \;\leq\; b_i \quad (i = 1, \ldots, m), \\
& \sum_{i=1}^{m} x_{ij} \;\leq\; 1 \quad (j = 1, \ldots, n), \\
\text{and} \quad & x_{ij} = 0 \text{ or } 1, \qquad (i = 1, \ldots, m; \; j = 1, \ldots, n),
\end{aligned}
$$

where x_{ij} takes the value zero (one) if the ith truck does not (does) carry the jth item.

Various authors have described several other variations of the above loading problem. For example, we may wish to minimize the number of trucks used when there is an unlimited number available.

12.2.4   Change Making Problem

A change making problem arises in the following way: Suppose there are n types of coins, where each type j has denomination w_j. A cashier wishes to make change to meet a given amount c using the least number of coins. It is assumed that the number of coins available in each denomination is unlimited and that the parameters w_j $(j = 1, \ldots, n)$ and c are expressed in the same terms (e.g., in dollars). If we let x_j $(j = 1, \ldots, n)$ be the number of coins of denomination w_j selected, the model is

$$\text{minimize} \quad \sum_{j=1}^{n} x_j$$
$$\text{subject to} \quad \sum_{j=1}^{n} w_j x_j = c,$$
$$\text{and} \quad x_j \geq 0 \text{ and integer} \quad (j = 1, \ldots, n).$$

12.2.5   Other Uses

Group Theory (Gomory [280])

In Chapter 11 we showed that if we drop the nonnegativity requirements on certain variables, any integer program can be defined as an integer program on an abelian group. Moreover, its constraints can be treated as a single congruence relationship, and thus the problem is a knapsack problem with addition as defined in the group (Section 11.2).

Enumerative Algorithms (Balas [42])

In Chapter 9 we mentioned that a surrogate constraint is a linear combination of the integer program's constraints. Thus, the problem

$$\begin{aligned}
\text{maximize} \quad & \mathbf{cx} \\
\text{subject to} \quad & (\mathbf{wA})\mathbf{x} \leq \mathbf{wb}, \\
& \mathbf{x} \geq \mathbf{0} \text{ and integer}
\end{aligned}$$

is a knapsack problem, where the surrogate constraint $\mathbf{wAx} \leq \mathbf{wb}$, with a given nonnegative weighting vector $\mathbf{w}$, has replaced the constraints

$\mathbf{Ax} \leq \mathbf{b}$.[11] Since the set of integer solutions to $\mathbf{Ax} \leq \mathbf{b}$ is contained in the set of integer solutions to $\mathbf{wAx} \leq \mathbf{wb}$, it follows that the integer program may be solved by finding the best solution to the knapsack problem which satisfies $\mathbf{Ax} \leq \mathbf{b}$. An algorithm for doing this is similar to the one presented in Chapter 11 (Section 11.3.3), where an enumeration is used if the optimal solution to the knapsack problem violates the integer program's constraints (Problem 12.10).

A similar algorithm involving multiple choice constraints is in Glover and Klingman [262]. The knapsack problem can also be used to generate relatively strong cuts and/or bounds for zero-one integer programs (see Kianfar[367], Djerdjour, Mathur, and Salkin [173]).

12.3   Reducing Integer Programs to Knapsack Problems: Aggregating Constraints

In this section we shall show that a system of linear equations with integer coefficients can usually be transformed to a single linear equation which has the same set of nonnegative integer solutions as the parent equations. This means that the constraints of an integer program can first be transformed to a single constraint and the integer program can then be solved by solving the knapsack problem. If the resulting knapsack problem is easier to solve than the integer program and its construction can be accomplished in a reasonable amount of time, the transformation is worthwhile.

Consider the m linear equations

$$\sum_{j=1}^{n} a_{ij}x_j = b_i \quad (i = 1, \ldots, m), \tag{3}$$

with every a_{ij} and b_i integer. The problem is to find weights $w_1, \ldots, w_m$ so that every nonnegative integer solution to the single equation

$$\sum_{j=1}^{n} a_j x_j = b, \tag{4}$$

---

[11]Components of the vector $\mathbf{wA}$ may be negative. To obtain positive coefficients in $(\mathbf{wA})\mathbf{x} \leq \mathbf{wb}$, complementing variables may have to be used; see Section 12.3.1 and especially Remark 1.

where $a_j = \sum_{i=1}^{m} w_i a_{ij}$ $(j = 1, \ldots, n)$ and $b = \sum_{i=1}^{m} w_i b_i$, is a solution to (3). Observe that, since the constraints (3) imply (4), every nonnegative integer solution to (3) is a solution to (4). For arbitrary weights, however, the set of nonnegative integers $\mathbf{x} = (x_1, \ldots, x_n)$ satisfying (4) is usually larger than the set satisfying (3). We give an example illustrating this and then show how to choose the weights, or, equivalently, how to aggregate the equations so that the solution set does not enlarge.

Example 12.1 (Balas [19]) [12]

The program initially has three inequality constraints with five zero-one variables. To obtain equations the nonnegative slack variables x_6, x_7, and x_8 are added, which must be integer since the coefficients are integers. The result is

$$\text{minimize} \quad 5x_1 + 7x_2 + 10x_3 + 3x_4 + x_5$$

$$\text{subject to} \quad -x_1 + 3x_2 - 5x_3 - x_4 + 4x_5 + x_6 \qquad\qquad = -2, \quad (i)$$

$$2x_1 - 6x_2 + 3x_3 + 2x_4 - 2x_5 \qquad + x_7 \qquad = 0, \quad (ii)$$

$$x_2 - 2x_3 + x_4 + x_5 \qquad\qquad + x_8 = -1, (iii)$$

$$x_j = 0 \text{ or } 1 \qquad\qquad (j = 1, 2, 3, 4, 5)$$

$$\text{and} \qquad x_j \geq 0 \text{ and integer} \qquad (j = 6, 7, 8).$$

There are $2^5 = 32$ possible zero-one values for $(x_1, \ldots, x_5)$. By complete enumeration, the only binary solutions are $(0, 1, 1, 0, 0)$ and $(1, 1, 1, 0, 0)$. If we arbitrarily choose weights $w_1 = 1, w_2 = 1$, and $w_3 = 1$, the three equations are added to produce $(i) + (ii) + (iii)$: $x_1 - 2x_2 - 4x_3 + 2x_4 + 3x_5 + x_6 + x_7 + x_8 = -3$, which has the above two zero-one solutions plus five others, namely, $(0, 0, 1, 0, 0), (0, 1, 1, 0, 1), (0, 1, 1, 1, 0), (1, 0, 1, 0, 0)$ and $(1, 1, 1, 1, 0)$.

12.3.1   An Aggregation Process (Mathews [417], Elmaghraby and Wig [188])

A theorem originating in Mathews [417] indicates how to aggregate two equations with positive coefficients so that the nonnegative integer solu-

---

[12]These computations are also in Kendell and Zionts [361].

tion set does not enlarge. In most cases this will allow us to reduce any system of m equations (3) to the single equation (4). Before explaining the transformation process, we state and prove Mathews' result.

Theorem 12.1 *(Mathews [417]) Consider a system of two linear equations*[13]

$$s_1 \equiv \sum_{j=1}^{n} a_{1j}x_j = b_1, \qquad\qquad (i)$$

$$s_2 \equiv \sum_{j=1}^{n} a_{2j}x_j = b_2, \qquad\qquad (ii)$$

with strictly positive integer coefficients a_{ij} $(i = 1, 2, \; j = 1, \ldots, n)$.

(a) If there exist nonnegative values $x_1, \ldots, x_n$ satisfying (i) and (ii), then

$$b_2 a_{1j}/a_{2j} \geq b_1 \text{ for at least one } j \quad (1 \leq j \leq n);$$

(b) If w is any positive integer such that

$$w > b_2 \; \underset{j}{maximum} \; \{a_{1j}/a_{2j}\}$$

then the solution set of (i) and (ii) in nonnegative integer variables is the same as that of the single equation

$$s_1 + ws_2 = b_1 + wb_2. \qquad\qquad (iii)$$

Proof: (a) Suppose the result is false. By hypothesis, $a_{1j} > 0$ and $a_{2j} > 0$ for every j. This with the inequality $b_2 a_{1j}/a_{2j} < b_1$ for every j means that both b_1 and b_2 cannot be 0. Thus, for any nonnegative values $x_1, \ldots, x_n$ satisfying (i) and (ii) it follows that $x_j > 0$ for at least one j $(1 \leq j \leq n)$. As $a_{1j} > 0$ and $a_{2j} > 0$, it follows that $b_2 a_{1j}x_j < b_1 a_{2j}x_j$. Summing over j yields

$$b_2 \sum_{j=1}^{n} a_{1j}x_j < b_1 \sum_{j=1}^{n} a_{2j}x_j$$

---

[13]The symbol $\equiv$ means "defined to be"; so $s_1 \equiv \sum_j a_{1j}x_j$ means that s_1 takes on the value of the summation.

for every $x_j \geq 0$ and at least one x_j strictly positive. But, by hypothesis, there is at least one nonnegative solution (x_j^0) to (i) and (ii), so that

$$\sum_{j=1}^{n} a_{1j} x_j^0 = b_1 \text{ and } \sum_{j=1}^{n} a_{2j} x_j^0 = b_2.$$

Substituting in the last inequality produces $b_2 b_1 < b_1 b_2$, which is a contradiction.

(b) For any solution to (i) and (ii), we have $s_1 = b_1$ and $s_2 = b_2$, which means that $ws_2 = wb_2$ and $s_1 + ws_2 = b_1 + wb_2$, or it is a solution to (iii). To prove the converse, consider any nonnegative integer solution (x_j^0) to (iii). If it is not an integer solution to (i) and (ii), then $s_2 \neq b_2$, because if $s_2 = b_2$, it follows that $ws_2 = wb_2$, and subtracting in (iii) yields $s_1 = b_1$. If

$$s_2 = \sum_{j=1}^{n} a_{2j} x_j^0 \neq b_2,$$

then there is a nonzero integer q such that $s_2 = b_2 + q$. Substituting $b_2 + q$ for s_2 in (iii) (with $x_j = x_j^0$) yields $s_1 + w(b_2 + q) = b_1 + wb_2$, or $s_1 = b_1 - wq$. We now show that q must be 0. Since the coefficients a_{1j} are positive,

$$s_1 = b_1 - wq \geq 0$$

for any nonnegative (integer) solution.

For $q > 0, b_1 - wq \geq 0$ implies $b_1 \geq wq > b_1 q$, or $1 > q$, which is a contradiction. ($w > b_1$ follows from the hypothesis and (a), and is used to obtain $wq > b_1 q$.)

When $q < 0$, it must be $-1, -2, -3$, etc., and thus $s_1 = b_1 - wq$ implies $s_1 \geq b_1 + w$. But, by definition,

$$w > b_2 \underset{j}{\text{maximum}} \frac{a_{1j}}{a_{2j}},$$

and thus $wa_{2j} > b_2 a_{1j}$ for $j = 1, \ldots, n$. Multiplying both sides by x_j and summing yields $ws_2 > b_2 s_1$. Since $b_1 b_2 \geq 0$, this implies $ws_2 > b_2 s_1 - b_1 b_2$. Substituting $s_1 \geq b_1 + w$ gives $ws_2 > wb_2$, or $s_2 > b_2$. But $s_2 = b_2 + q$ and $q < 0$, so $s_2 < b_2$; hence a contradiction. So q must be 0 and the theorem is proved. ∎

Mathews observed that this theorem can be applied to two equations with nonnegative (but not necessarily all positive) integer coefficients by first replacing them with the pair

$$s_1 + s_2 = b_1 + b_2 \qquad (\bar{i})$$

$$s_1 + 2s_2 = b_1 + 2b_2. \qquad (\bar{\bar{i}})$$

Equations $(\bar{i})$ and $(\bar{\bar{i}})$ have strictly positive coefficients since they are of the form $a_{1j} + a_{2j}$ in $(\bar{i})$, or $a_{1j} + a_{2j}$ in $(\bar{\bar{i}})$ and for each j $(j = 1, \ldots, n)$, $a_{ij} > 0$ for at least one i $(i = 1, 2)$. Thus, these equations can assume the role of (i) and (ii) in Mathews' theorem. Also, note that any solution to $(\bar{i})$ and $(\bar{\bar{i}})$ is a solution to (i) and (ii) (and conversely). To see this, just subtract $(\bar{i})$ from $(\bar{\bar{i}})$ to obtain $s_2 = b_2$ and then, by $(\bar{i})$, $s_1 = b_1$.

Elmaghraby and Wig [188] have further noted that Mathews' theorem can be used to aggregate a system of equations (3) with nonnegative coefficients by recursively using the construction $(\bar{i})$ and $(\bar{\bar{i}})$. That is, the first two equations in the system are replaced by $(\bar{i})$ and $(\bar{\bar{i}})$ and then aggregated. The aggregated equation becomes the first equation and replaces the first two, so that the system now has one less equation. The process is then repeated until a single equation is left.

We can also use the above pairwise aggregation process on systems of equations containing negative coefficients so long as upper bounds on the corresponding variables can be found. In particular, if $a_{1j} < 0$ and it is known that $x_j \le u_j$ (some positive integral upper bound), then substituting $u_j - \bar{x}_j$ for x_j in constraint 1, where the complementing variable $\bar{x}_j = u_j - x_j$ changes the sign of a_{1j}. Notice that this and the requirements in Mathews' theorem mean that any integer program which has a bounded linear programming feasible region with at least one integer point can be transformed to an equivalent knapsack problem.

Example (from Example 12.1)

Equation (i) is $-x_1 + 3x_2 - 5x_3 - x_4 + 4x_5 + x_6 = -2$. As the first five variables are 0 or 1, we have $u_j = 1$ $(j = 1, \ldots, 5)$. To eliminate the negative coefficients, $1 - \bar{x}_1, 1 - \bar{x}_3$, and $1 - \bar{x}_4$ replace x_1, x_3, and x_4, respectively. This gives

$$-(1 - \bar{x}_1) + 3x_2 - 5(1 - \bar{x}_3) - 1(1 - \bar{x}_4) + 4x_5 + x_6 = -2;$$

or

$$\bar{x}_1 + 3x_2 + 5\bar{x}_3 + \bar{x}_4 + 4x_5 + x_6 = 5. \tag{i}'$$

Similarly, equations (ii) and (iii) become

$$2x_1 + 6\bar{x}_2 + 3x_3 + 2x_4 + 2\bar{x}_5 + x_7 = 8 \tag{ii}'$$

$$x_2 + 2\bar{x}_3 + x_4 + x_5 + x_8 = 1 \tag{iii}'$$

So the initial system (a_{ij}) is:

Eq./Variable	x_1	$\bar{x}_1$	x_2	$\bar{x}_2$	x_3	$\bar{x}_3$	x_4	$\bar{x}_4$	x_5	$\bar{x}_5$	x_6	x_7	x_8	b
$(i)'$		1	3			5	1		4		1			5
$(ii)'$	2			6	3		2			2		1		8
$(iii)'$			1			2	1		1				1	1

So

	x_1	$\bar{x}_1$	x_2	$\bar{x}_2$	x_3	$\bar{x}_3$	x_4	$\bar{x}_4$	x_5	$\bar{x}_5$	x_6	x_7	x_8	b
$(\bar{i}) = (i)' + (ii)'$	2	1	3	6	3	5	2	1	4	2	1	1		13
$(\bar{ii}) = (i)' + 2(ii)'$	4	1	3	12	6	5	4	1	4	4	1	2		21

Now $w > 21 \max_j \left\{\frac{a_{1j}}{a_{2j}}\right\} = 21\left(\frac{1}{1}\right)$, so selecting $w = 22$ yields the system with one less equation:

	x_1	$\bar{x}_1$	x_2	$\bar{x}_2$	x_3	$\bar{x}_3$	x_4	$\bar{x}_4$	x_5	$\bar{x}_5$	x_6	x_7	x_8	b
$(i)'' = (\bar{i}) + 22(\bar{ii})$	90	23	69	270	135	115	90	23	92	90	23	45		475
$(ii)'' = (iii)'$			1			2	1		1				1	1

After forming the new equations $(\bar{i}) = (i)'' + (ii)''$, and $(\bar{\bar{ii}}) = (i)'' + 2(ii)''$ we find that $w > 477$, so a multiplier $w = 478$ is selected, which results in the aggregated equation

$$(i)''' = (\bar{\bar{i}}) + 478(\bar{\bar{ii}}) :$$

$$43110x_1 + 11017\bar{x}_1 + 34008x_2 + 129330\bar{x}_2 + 64665x_3$$
$$+ 56999\bar{x}_3 + 44067x_4 + 11017\bar{x}_4 + 45025x_5$$
$$+ 43110\bar{x}_5 + 11017x_6 + 21555x_7 + 957x_8 = 228482.$$

Substituting $1 - x_j$ $(j = 1, \ldots, 5)$ for the complemented variables $\bar{x}_j$ yields the single equation

$$32093x_1 - 95322x_2 + 7666x_3 + 33050x_4 + 1915x_5$$
$$+ 11017x_6 + 21555x_7 + 957x_8 = -22991.$$

There are only two binary solutions $(x_1, \ldots, x_5)$ to this equation and they are exactly the same as for the three original equations, namely, $(0, 1, 1, 0, 0)$ and $(1, 1, 1, 0, 0)$.

Remarks

1. The example indicates that the aggregated equation may have negative coefficients. Thus, if the integer program is to be solved as a knapsack problem with positive values for the weights, upper bounds must be available for variables having negative coefficients so that their complements can be introduced. (If the problem has a nonnegative solution, the right-hand side becomes positive subsequent to the complementation.) In the example, $-95322 < 0$ and $x_2 \leq 1$; introducing $1 - \bar{x}_2$ for x_2 in the aggregated equation and writing the integer program's objective function in maximization format (by initially changing the sign of the costs) yields the equivalent knapsack problem

$$+7 - \text{maximize} \quad -5x_1 + 7\bar{x}_2 - 10x_3 - 3x_4 - x_5$$

$$\text{subject to } 32093x_1 + 95322\bar{x}_2 + 7666x_3 + 33050x_4 + 1915x_5$$

$$+ 11017x_6 + 21555x_7 + 957x_8 = 72331$$

$$x_1 \leq 1, \bar{x}_2 \leq 1, x_3 \leq 1, x_4 \leq 1, x_5 \leq 1$$

$$x_1, x_2, x_3, \ldots, x_8 \geq 0 \text{ and integer.}$$

Note that this problem has negative values for the costs and upper bounds on certain variables. Thus, if it is to be solved as a knapsack problem, the solution technique must allow for negative cost values and upper bounds. It must also permit an equality, rather than an inequality, constraint. Such algorithms are discussed in the next section.

2. Multipliers can be found once the final equation is known. In particular, in the example

$$
\begin{aligned}
(i)''' &= (\bar{\bar{i}}) + 478(\bar{\bar{ii}}) = ((i)'' + (ii)'') + 478((i)'' + 2(ii)'') \\
&= 479(i)'' + 957(ii)'' = 479((\bar{i}) + 22(\bar{ii})) + 957(iii)' \\
&= 479((i)' + (ii)') + 10538((i)' + 2(ii)') + 957(iii)' \\
&= 11017(i)' + 21555(ii)' + 957(iii)'.
\end{aligned}
$$

Notice that the weights $w_1 = 11017, w_2 = 21555$, and $w_3 = 957$ are precisely the coefficients of the slack variables x_6, x_7, and x_8, respectively, in the final equation $(i)'''$. It is a simple matter to prove that if a slack variable appears in every equation i in the initial system (3), then its coefficient in the single equation (4) is the weight used to aggregate each equation i (Problem 12.12).

3. It is evident that recursively aggregating pairs of equations into a single equation using $(\bar{i})$ and $(\bar{ii})$ causes the coefficients to grow very rapidly. To illustrate, consider the case in which m equations have a common right-hand side value $\bar{b} = b_i$ for $i = 1, \ldots, m$. Then, as indicated in Glover and Woolsey [267], the result of recursively aggregating m equations (3) will yield an equation with a right-hand side exceeding $2^u \bar{b}^v$, where $u = 3(2^{m-2})$ and $v = 2^{m-1}$. This number is clearly astronomical for as few as 7 constraints. These huge numbers obviously cause computer roundoff errors and overflow problems. Therefore, if the aggregated problem is to be solved, a different aggregation process which yields smaller coefficients must be found. We now discuss an improved scheme.

12.3.2 An Improved Aggregation Process (Glover [256])

As already mentioned in an earlier section, several articles have appeared in the literature discussing aggregation techniques which tend to yield smaller coefficients at the expense of additional computations. We now present a result, due to Glover [256], which perhaps gives one of the best reduction procedures. It is similar to the previous one in the sense that constraints are successively pairwise aggregated. However, fewer conditions on the coefficients of the equations to be aggregated are required. Other results suggesting reduction procedures are in Problems 12.15 to 12.18.

Theorem 12.2 *(Glover [256]) Consider a system*

$$s_1 \equiv \sum_{j=1}^{n} a_{1j} x_j = b_1 \tag{i}$$

$$s_2 \equiv \sum_{j=1}^{n} a_{2j} x_j = b_2, \tag{ii}$$

where all coefficients (a_{ij}, b_i) are integers, and at least one of b_1 and b_2 is not zero. Let w_1 and w_2 be relatively prime (nonzero) integers (their greatest common divisor is plus or minus 1). If there exists at least one nonnegative integer solution to (i) and (ii), then every nonnegative integer solution to

$$w_1 s_1 + w_2 s_2 = w_1 b_1 + w_2 b_2 \tag{iii}$$

is a nonnegative integer solution to (i) and (ii), and conversely, provided that

$$w_1 a_{1j} + w_2 a_{2j} \geq |b_2 a_{1j} - b_1 a_{2j}| \tag{5}$$

for $j = 1, \ldots, n$, and (5) holds as a strict inequality for j in J, where J is any nonempty subset of $\{1, \ldots, n\}$ such that all nonnegative solutions to (iii) satisfy $x_j > 0$ for at least one j in J.

Proof: Clearly (i) and (ii) imply (iii), so that every (nonnegative integer) solution to (i) and (ii) satisfies (iii). To show the converse, select any nonnegative integer solution x_j ($j = 1, \ldots, n$) to (iii) and multiply both sides of (5) by $x_j = |x_j|$, so that

$$(w_1 a_{1j} + w_2 a_{2j}) x_j \geq |b_2 a_{1j} - b_1 a_{2j}| |x_j| = |(b_2 a_{1j} - b_1 a_{2j}) x_j|.$$

Summing the last expression over j gives

$$w_1 b_1 + w_2 b_2 > |b_2 s_1 - b_1 s_2|, \tag{6}$$

where the strict inequality follows from the definition of J.[14] Now suppose $s_1 \neq b_1$, which implies $s_2 \neq b_2$ (and vice versa). Then let α and β be any integers such that

$$s_1 = b_1 + \alpha \quad \text{and} \quad s_2 = b_2 - \beta.$$

For the current integer solution, $w_1 s_1 + w_2 s_2 = w_1 b_1 + w_2 b_2$. Substituting for s_1 and s_2 yields

$$w_1 b_1 + w_1 \alpha + w_2 b_2 - w_2 \beta = w_1 b_1 + w_2 b_2;$$

thus, $w_1 \alpha = w_2 \beta$. But, by hypothesis, w_1 and w_2 are relatively prime. Hence, for the last equality to be true $\alpha = q w_2$, where q is an integer. This follows because $\beta = (w_1/w_2)\alpha$, and w_1/w_2 is not an integer, so the only way for β to be an integer is for α to be a multiple of w_2. A similar argument can be made for β. Substituting for α and β in expressions for s_1 and s_2 gives $s_1 = b_1 + q w_2$ and $s_2 = b_2 - q w_1$. These last two equalities in (6) give

$$\begin{aligned} w_1 b_1 + w_2 b_2 &> |b_2 b_1 + b_2 q w_2 - b_1 b_2 + b_1 q w_1| \\ &= |q||w_1 b_1 + w_2 b_2|, \end{aligned}$$

which means that $|q| = 0$ or $q = 0$, and thus $s_1 = b_1$ and $s_2 = b_2$, completing the proof. ∎

The theorem implies by (6) that w_1 and w_2 be chosen so that $w_1 b_1 + w_2 b_2 > 0$. Also, by (5), the coefficients in (iii) (i.e., $w_1 a_{1j} + w_2 a_{2j}$ for $j = 1, \ldots, n$) are nonnegative. Therefore, every nonnegative integer solution to (iii) must have $x_j > 0$ for at least one j where its coefficient in (iii), $w_1 a_{1j} + w_2 a_{2j}$, is positive. Consequently, the set J can consist of those j for which $w_1 a_{1j} + w_2 a_{2j} > 0$. Condition (5) also imposes some additional restrictions on the coefficients in (i) and (ii). By (5) a sufficient, but not necessary, condition for the existence of relatively prime weights w_1 and w_2 is that coefficients a_{ij} are nonnegative, since we can then take $w_1 = 1$. This may not yield the aggregated equation

---

[14]By (5) each coefficient in the aggregated constraint is nonnegative. Thus, $x_j = 0$ $(j = 1, \ldots, n)$ is the only solution if and only if $w_1 b_1 + w_2 b_2 = 0$. We are ignoring this possibility, and hence $x_j > 0$ for some j in every solution to (iii).

with coefficients of smallest magnitude, although it does simplify the computations.

Example (from Example 12.1)

To simplify computations we initially complement those variables with negative coefficients. From the previous illustration this yields

Eq./Variable	x_1	$\bar{x}_1$	x_2	$\bar{x}_2$	x_3	$\bar{x}_3$	x_4	$\bar{x}_4$	x_5	$\bar{x}_5$	x_6	x_7	x_8	b
$(i)'$		1	3			5		1	4		1			5
$(ii)'$	2			6	3		2			2		1		8
$(iii)'$		1			2	1		1					1	1

To combine equations $(i)'$ and $(ii)'$, let $J = \{1,\ldots,7\}$ and by (5) we have

j	$\lvert b_2 a_{1j} - b_1 a_{2j}\rvert$	Equation (5)	
1	10	$2w_2 > 10$	$\Rightarrow\ w_2 > 5$
$\bar{1}$	8	$w_1 > 8$	
2	24	$3w_1 > 24$	$\Rightarrow\ w_1 > 8$
$\bar{2}$	30	$6w_2 > 30$	$\Rightarrow\ w_2 > 5$
3	15	$3w_2 > 15$	$\Rightarrow\ w_2 > 5$
$\bar{3}$	40	$5w_1 > 40$	$\Rightarrow\ w_1 > 8$
4	10	$2w_2 > 10$	$\Rightarrow\ w_2 > 5$
$\bar{4}$	8	$w_1 > 8$	
5	32	$4w_1 > 32$	$\Rightarrow\ w_1 > 8$
$\bar{5}$	10	$2w_2 > 10$	$\Rightarrow\ w_2 > 5$
6	8	$w_1 > 8$	
7	5	$w_2 > 5$	

Selecting $w_1 = 9$ means that w_2 cannot be 6 (since 9 and 6 are not relatively prime), so we set $w_2 = 7$. Therefore.

j	1	$\bar{1}$	2	$\bar{2}$	3	$\bar{3}$	4	$\bar{4}$	5	$\bar{5}$	6	7	8	b
$9(i)' + 7(ii)' = (i)''$	14	9	27	42	21	45	14	9	36	14	9	7		101
$(iii)' = (ii)''$		1			2	1		1					1	1

To combine $(i)''$ with $(ii)''$, let $J = \{1, \ldots, 8\}$, and again using (5) we obtain

j	$\lvert b_2 a_{1j} - b_1 a_{2j}\rvert$	Equation (5)
1	14	$14w_1 > 14 \quad \Rightarrow \quad w_1 > 1$
$\bar{1}$	9	$9w_1 > 9 \quad \Rightarrow \quad w_1 > 1$
2	74	$27w_1 + w_2 > 74$
$\bar{2}$	42	$42w_1 > 42 \quad \Rightarrow \quad w_1 > 1$
3	21	$21w_1 > 21 \quad \Rightarrow \quad w_1 > 1$
$\bar{3}$	157	$45w_1 + 2w_2 > 157$
4	87	$14w_1 + w_2 > 87$
$\bar{4}$	9	$9w_1 > 9 \quad \Rightarrow \quad w_1 > 1$
5	65	$36w_1 + w_2 > 65$
$\bar{5}$	14	$14w_1 > 14 \quad \Rightarrow \quad w_1 > 1$
6	9	$9w_1 > 9 \quad \Rightarrow \quad w_1 > 1$
7	7	$7w_1 > 7 \quad \Rightarrow \quad w_1 > 1$
8	101	$w_2 > 101$

The last inequality indicates that w_2 must be at least 102, and since $w_1 > 1$, taking $w_2 = 103$ and $w_1 = 2$ yields a set of relatively prime multipliers which satisfies all the inequalities. The aggregated equation is

$$2(i)'' + 103(ii)'' = (i)''':$$

$$28x_1 + 18\bar{x}_1 + 157x_2 + 84\bar{x}_2 + 42x_3 + 296\bar{x}_3 + 131x_4$$
$$+ 18\bar{x}_4 + 175x_5 + 28\bar{x}_5 + 18x_6 + 14x_7 + 103x_8 = 305.$$

After substituting $1 - x_j$ for each $\bar{x}_j$, we have the final equation

$$(i)''': \quad 10x_1 + 73x_2 - 254x_3 + 113x_4 + 147x_5$$
$$+ 18x_6 + 14x_7 + 103x_8 = -139.$$

It is evident that this aggregation scheme yields a single equation with considerably smaller weights. The weights using Glover's procedure are $w_1 = 18, w_2 = 14$, and $w_3 = 103$ (coefficients of the slack variables, see Problem 12.12). This is in contrast to the former technique which produced weights $w_1 = 11017, w_2 = 21555$, and $w_3 = 957$. Although the initial constraints did not have to be added so that the resulting system had all positive coefficients, more work is required at each step in computing the multipliers using Glover's algorithm. As a final note, Kendell and Zionts [361] present an aggregation theory (scheme) for integer programs with some zero-one variables. It allows us to use weights $w_1 = 18, w_2 = 14$, and $w_3 = 13$ on our problem so that the three constraints become

$$10x_1 - 17x_2 - 74x_3 + 23x_4 + 57x_5 + 18x_6 + 14x_7 + 13x_8 = -49.$$

This equation has uniformly smaller coefficients than equation $(i)'''$. However, more computations may be necessary in using their result.

12.4   Algorithms

An algorithm for the knapsack problem can usually be classified as a dynamic programming technique, a branch and bound enumeration, a Lagrangian multiplier method, or a network method. We now discuss these types of algorithms. For ease of reference the knapsack problem is rewritten

$$\text{maximize} \quad \sum_{j=1}^{n} c_j x_j$$
$$\text{subject to} \quad \sum_{j=1}^{n} a_j x_j \le b$$
$$x_j \ge 0 \text{ and integer} \quad (j = 1, \ldots, n),$$

where all data are integral and each a_j $(j = 1, \ldots, n)$ and b are positive.

12.4.1   Dynamic Programming Techniques (Bellman [67], Dantzig [155])

Let us define $f(k, g)$ to be the maximal value of the objective function using only the first k $(k = 1, 2, \ldots, n)$ items (variables) with weight limitation (right-hand side) g $(g = 0, 1, 2, \ldots, b)$. That is,

$$f(k, g) \quad = \quad \text{maximum} \quad \sum_{j=1}^{k} c_j x_j$$
$$\text{subject to} \quad \sum_{j=1}^{k} a_j x_j \le g$$
$$x_j \ge 0 \text{ and integer} \quad (j = 1, \ldots, k).$$

The problem is to find $f(n, b)$, which we shall do by first finding $f(k, g)$ for $k = 1, \ldots, n-1$ and $g = 0, 1, \ldots, b-1$. Notice that $f(k, 0) = 0$ for $k = 1, \ldots, n$.

Recursion I: Bounded Variables

The positive coefficients a_j $(j = 1, \ldots, n)$ and the positivity of g imply that in every integer solution making up $f(k, g)$, $x_j \leq [g/a_j]$, where $[y]$ means the largest integer $\leq y$. In certain instances, however, it may be necessary to enforce smaller upper bounds on certain variables. (For example, in a zero-one problem.) That is, for some j we have the additional constraints $x_j \leq u_j < [b/a_j]$, where u_j is a positive integer. A dynamic programming technique which allows for these upper bounds is as follows. Suppose that for a given k $(k = 2, \ldots, n)$, $f(k-1, g)$ is known for all possible right-hand sides $g = 0, 1, \ldots, b$. We wish to find $f(k, g)$ for a given g $(0 \leq g \leq b)$. To do this write

$$f(k, g) = \text{maximum} \quad c_k x_k + \sum_{j=1}^{k-1} c_j x_j$$

$$\text{subject to} \quad \sum_{j=1}^{k-1} a_j x_j \leq g - a_k x_k$$

$$x_j \geq 0 \text{ and integer} \quad (j = 1, \ldots, k-1),$$

$$\text{and} \quad x_k = 0, 1, \ldots, [g/a_k].$$

For a given integer value of x_k $(0 \leq x_k \leq [g/a_k])$ [15] $f(k, g)$ reduces to

$$c_k x_k + \left(\begin{array}{l} \text{maximum} \sum_{j=1}^{k-1} c_j x_j \\ \text{subject to} \sum_{j=1}^{k-1} a_j x_j \leq g - a_k x_k \\ \text{and} \quad x_j \geq 0 \text{ and integer} \quad (j = 1, \ldots, k-1). \end{array} \right)$$

But the bracketed term is precisely $f(k-1, g - a_k x_k)$, which is known

---

[15]We are using $x_k > 0$ and integer with the inequality constraint to obtain $x_k = 0, 1, \ldots, [g/a_k]$. If an upper bound $u_k < [g/a_k]$ exists, then it should replace $[g/a_k]$ in $f(k, g)$ so that $x_k \leq u_k$ is enforced.

since $g - a_k x_k$ is a nonnegative integer not exceeding b. (Why?) Therefore,

$$f(k,g) = \underset{x_k = 0,1,\ldots,[g/a_k]}{\text{maximum}} (c_k x_k + f(k-1, g - a_k x_k)). \qquad (7)$$

For each $k = 2,\ldots,n$, equation (7) can be used to find $f(k,g)$ for $g = 0,1,\ldots,b$. To start the computations $f(1,g)$ is first found using

$$
\begin{aligned}
f(1,g) \quad = \quad & \text{maximum} \quad c_1 x_1 \\
& \text{subject to} \quad a_1 x_1 \; \le \; g \\
& \qquad\qquad\quad x_1 \; \ge \; 0 \text{ and integer.}
\end{aligned}
$$

$$
= \underset{x_1 = 0,1,\ldots,[g/a]}{\text{maximum}} c_1 x_1
$$

$$
= \begin{cases}
0 & \text{if} \quad c_1 \le 0 \quad (\text{ with } x_1 = 0) \\
[g/a_1] c_1 & \text{if} \quad c_1 > 0 \quad (\text{ with } x_1 = [g/a_1]).
\end{cases}
$$

Note that by defining $f(0,g) = 0$ for $g = 0,1,\ldots,b$ we may use (7) for the $k = 1$ term also.

To retrieve the values of the variables making up $f(k,g)$ we list the value of x_k with $f(k,g)$. Then the value of x_{k-1} is listed with $f(k-1, g - a_k x_k)$, and so forth until x_1 is known.

Example 12.2

$$
\begin{aligned}
\text{Maximize} \quad & 3x_1 \; + \; 5x_2 \; + \; x_3 \; + \; x_4 \\
\text{subject to} \quad & 2x_1 \; + \; 4x_2 \; + \; 3x_2 \; + \; 2x_4 \; \le \; 5, \\
& x_1 \le 1, x_2 \le 1, x_3 \le 1, \\
\text{and} \quad & x_1, x_2, x_3, x_4 \ge 0 \text{ and integer.}
\end{aligned}
$$

As x_1, x_2, and x_3 cannot exceed 1, we replace $[g/a_k]$ by 1 in (7) whenever $[g/a_k] \ge 1$. (When $[g/a_k] = 0, x_k$ must be 0 in $f(k,g)$.) When $k = 4$, $[g/a_k] = [g/2]$. Note that x_4 cannot exceed $[5/2] = 2$.

For $k = 1$, $a_1 = 2$, so $f(1,g) = 0$ with $x_1 = 0$ for $g = 0,1$, since $[g/2] = 0$; when $g = 2,3,4,5$, then $f(1,g) = \text{ maximum } 3x_1 = 3$ with $x_1 = 0,1$
$x_1 = 1$.

For $k = 2, a_2 = 4$, and $f(2, g) = \underset{x_2 = 0, 1}{\text{maximum}} \ (5x_2 + f(1, g - 4x_2))$.

Whenever $g = 0, 1, 2, 3$, $[g/4] = 0$; so $x_2 = 0$ and $f(2, g) = f(1, g)$; and $f(1, g) = 0$ for $g = 0, 1$, and $f(1, g) = 3$ for $g = 2, 3$.

For $g = 4 \quad f(2, 4) = \underset{x_2 = 0, 1}{\text{maximum}} \ (5x_2 + f(1, 4 - 4x_2))$

$$= \text{maximum} \ \{0 + f(1, 4), 5 + f(1, 0)\}$$

$$= \text{maximum} \ \{3, 5\} = 5 \text{ with } x_2 = 1.$$

For $g = 5 \quad f(2, 5) = \underset{x_2 = 0, 1}{\text{maximum}} \ (5x_2 + f(1, 5 - 4x_2))$

$$= \text{maximum} \ \{0 + f(1, 5), 5 + f(1, 1)\}$$

$$= \text{maximum} \ \{3, 5\} = 5 \text{ with } x_2 = 1.$$

For $k = 3$, $a_3 = 3$, and $f(3, g) = \underset{x_3 = 0, 1}{\text{maximum}} \ (x_3 + f(2, g - 3x_3))$; when $g = 0, 1, 2$, we have $[g/3] = 0$, so $x_3 = 0$ and $f(3, g) = f(2, g)$.

For $g = 3, f(3, 3) = \underset{x_3 = 0, 1}{\text{maximum}} \ (x_3 + f(2, 3 - 3x_3)) = 3$ with $x_3 = 0$;

etc. The computations are tabulated below.

							g		
k	a_k	c_k		0	1	2	3	4	5 = b
1	2	3	$f(1, g)$	0	0	3	3	3	3
			x_1	–	0	1	1	1	1
2	4	5	$f(2, g)$	0	0	3	3	5	5
			x_2	–	0	0	0	1	1
3	3	1	$f(3, g)$	0	0	3	3	5	5
			x_3	–	0	0	0	0	0
$n = 4$	2	1	$f(4, g)$	0	0	3	3	5	5
			x_4	–	0	0	0	0	0

The optimal solution is $f(4,5) = 5$ with $x_4 = 0$ (b is reduced to $5 - 2(0) = 5$), $x_3 = 0$ (b is reduced to $f - 3(0) = 5$), $x_2 = 1$ (b is reduced to $5 - 4(1) = 1$), and $x_1 = 0$. The optimal values of the variables are boxed. Notice that the value of the slack is $1 = 5 - a_2$.

Remarks

1. Horowitz and Sahni [501], and Martello and Toth [414] have pointed out that for a problem with 0-1 variables (see Example 12.2), at each stage, there are numerous dominated states; where, the value g_1 is dominated by g_2 at stage k if $f(k, g_2) \geq f(k, g_1)$ and $g_2 < g_1$. Considerable computational effort may be saved by eliminating dominated states at each stage. Martello and Toth [414] give a dynamic programming algorithm for the $0 - 1$ knapsack problem exploiting this observation.

2. Consider following knapsack problems

$$\text{maximize} \quad \sum_{i=1}^{q} c_i x_i$$
$$\text{subject to} \quad \sum_{i=1}^{q} a_i x_i \leq g_1$$
$$x_i \geq 0 \text{ and integer},$$

and

$$\text{maximize} \quad \sum_{i=q+1}^{n} c_i x_i$$
$$\text{subject to} \quad \sum_{i=q+1}^{n} a_i x_i \leq g_2$$
$$x_i \geq 0 \text{ and integer}.$$

Let $f_1(q, g_1)$, $g_1 = 0, 1, \ldots, b$ and $f_2(n - q, g_2)$, $g_2 = 0, 1, \ldots, b$ be the respective optimal solution to the above two problems. Then, $f(n, b)$, the optimal solution to the original knapsack problem, can be found by using the relationship

$$f(n, b) = \underset{b_1 = 0, 1, \ldots, b}{\text{maximum}} \quad \{f_1(q, b_1) + f_2(n - q, b - b_1)\}$$

Horowitz and Sahni [501] present encouraging computational results using $q = [n/2]$.

When the Inequality Is an Equality To solve a knapsack problem having bounded variables and an equality constraint we simply add the initial conditions $f(0, 0) = 0$ and $f(0, g) = -\infty$ for $g = 1, \ldots, b$. This replaces $f(0, g) = 0$ and allows us to use equation (7) directly. To see that this procedure works, note that for $k = 1$, equation (7) reduces to

$$f(1, g) = \underset{x_1 = 0, 1, \ldots, [g/a_1]}{\text{maximum}} (c_1 x_1 + f(0, g - a_1 x_1)),$$

and since $f(0, g - a_1 x_1) = -\infty$ for $g - a_1 x_1 > 0$, x_1 will be selected, if possible, so that $g - a_1 x_1 = 0$ or there is no slack. Now $f(2, g)$ involves an $f(1, g - a_2 x_2)$ term, and thus the maximum in (7) guarantees that x_2 will be selected, if possible, so that the solution in $f(1, g - a_2 x_2)$ has no slack. This means that the solution in $f(2, g)$ has no slack. We can continue this argument and conclude that the solution in $f(n, b)$ has no slack. A formal proof is left to Problem 12.19.

Example (from Example 12.2)

Consider the previous problem, where we now have an equality:

$$
\begin{array}{llllllll}
\text{maximize} & 3x_1 & + & 5x_2 & + & x_3 & + & x_4 \\
\text{subject to} & 2x_1 & + & 4x_2 & + & 3x_3 & + & 2x_4 & = & 5, \\
& x_1 \leq 1, x_2 \leq 1, x_3 \leq 1, \\
& x_1, x_2, x_3, x_4 \geq 0 \text{ and integer.}
\end{array}
$$

The computations are tabulated below. Entries of $-\infty$ indicate that there is no solution making up $f(k, g)$ which satisfies the equality, and thus a value for x_k is not recorded. The optimal solution is $f(4, 5) = 4$, with $x_4 = 0, x_3 = 1, x_2 = 0$, and $x_1 = 1$. Note that $a_1 + a_3 = 5 = b$.

				0	1	2	3	4	$5 = b$
						g			
k	a_k	c_k							
1	2	3	$f(1,g)$	0	$-\infty$	3	$-\infty$	$-\infty$	$-\infty$
			x_1	$-$	$-$	$\boxed{1}$	$-$	$-$	$-$
2	4	5	$f(2,g)$	0	$-\infty$	3	$-\infty$	5	$-\infty$
			x_2	$-$	$-$	$\boxed{0}$	$-$	1	$-$
3	3	1	$f(3,g)$	0	$-\infty$	3	1	5	4
			x_3	$-$	$-$	0	1	0	$\boxed{1}$
$n = 4$	2	1	$f(4,g)$	0	$-\infty$	3	1	5	4
			x_4	$-$	$-$	0	0	0	$\boxed{0}$

The dynamic programming procedure developed in this section does not take advantage of the linearity of the constraint and objective function. It may involve considerable computation, especially when b is large and when the variables may take on many values. The technique is important, however, since it can treat problems with bounded variables.[16] When bounds are not present the technique presented below is considerably more efficient.

Recursion II: Variables Not Bounded

Again suppose for some index k $(k = 2, \ldots, n)$ we know $f(k-1, g)$ for all possible g. We wish to use this information to calculate $f(k, g)$. As the $f(k, g)$ problem differs from the $f(k-1, g)$ problem in only the variable x_k, we must decide whether it should be fixed at 0, or be at least 1. Naturally we would use x_k only if it does not violate the constraint and can yield an improved (or larger) value for the objective function. Thus, for a given g,

(i) $x_k = 0$ if $c_k + f(k, g - a_k) \leq f(k-1, g) = f(k, g)$,

(ii) $x_k \geq 1$ if $f(k, g) = c_k + f(k, g - a_k) > f(k-1, g)$
                        and $g - a_k \geq 0$.

---

[16]It can also be easily generalized for a nonlinear objective function and (or) constraint; see Problem 12.22.

As we wish a maximal solution, (i) and (ii) imply

$$f(k,g) = \text{maximum} \{ f(k-1,g), c_k + f(k, g - a_k), a_k \leq g \}. \tag{8}$$

This allows us to compute $f(k,g)$ for $k = 1, \ldots, n$ and $g = 0, 1, \ldots, b$. To start the computations, $f(0,0)$ is defined to be 0 and $f(0,g) = 0$ for $g = 1, \ldots, b$. For the knapsack problem with an equality constraint, $f(0,g) = -\infty$ for $g = 1, \ldots, b$ (see Problem 12.19).

To keep track of the solution making up $f(k,g)$, we enter a 0 under $f(k,g)$ if x_k is not used and a 1 if x_k is used at least once. To retrieve the solution we find the value under $f(n,b)$. If it is $0, x_n = 0$ and the value under $f(n-1,b)$ is examined. If it is 1, we increment x_n by 1 making $x_n = 1$, and examine the value under $f(n, b - a_n)$. (If the value under $f(n, b - a_n)$ is $1, x_n$ becomes 2.) This procedure is repeated until the value of x_1 is found.[17]

Example (from Example 12.2)

Suppose we drop the zero-one requirements on x_1, x_2, and x_3. This yields the problem

$$\begin{array}{rl} \text{maximize} & 3x_1 + 5x_2 + x_3 + x_4 \\ \text{subject to} & 2x_1 + 4x_2 + 3x_3 + 2x_4 \leq 5 \\ & x_1, \ldots, x_4 \geq 0 \text{ and integer.} \end{array}$$

$f(0,g) = 0$ for $g = 1, \ldots, 5,$

$f(1,g) = \text{maximum} \{ f(0,g), 3 + f(1, g - 2) \}$

$\quad\quad = f(0,g) = 0$ for $g = 0, 1 \quad (x_1 = 0)$

$f(1,2) = 3 \quad (x_1 \geq 1)$

$f(1,3) = 3 \quad (x_1 \geq 1)$

$f(1,4) = 6 \quad (x_1 \geq 1)$

$f(1,5) = 6 \quad (x_1 \geq 1)$

$f(2,g) = \text{maximum} \{ f(1,g), 5 + f(2, g - 4) \}$

$\quad\quad = f(1,g)$ for $g = 1, 2, 3 \quad (x_2 = 0)$

---

[17]A different bookkeeping scheme appears in Chapter 11, Section 11.3.1. The dynamic programming technique presented here is equivalent to the one presented in Section 11.3.1 for the group minimization problem.

$$f(2,4) = \text{maximum} \{f(1,4), 5 + f(2,0)\} = 6 \quad (x_2 = 0)$$

$$f(2,5) = \text{maximum} \{f(1,5), 5 + f(2,1)\} = 6 \quad (x_2 = 0); \text{etc.}$$

The computations are tabulated below. To find the solution, $f(4,5) = 6$ and the 0 indicates $x_4 = 0$. The entry below $f(3,5)$ is a 0, so $x_3 = 0$, the 0 under $f(2,5)$ indicates $x_2 = 0$, and the 1 below $f(1,5)$ means that $x_1 = 1$; the current value of the right-hand side is reduced to $5 - 1(2) = 3$, the 1 below $f(1,3)$ indicates that x_1 is incremented to 2, and the right-hand side becomes $3 - 1(2) = 1$. The dash below $f(1,1)$ means that we have the whole solution. (Why?)

				g					
k	a_k	c_k		0	1	2	3	4	$5 = b$
1	2	3	$f(1,g)$	0	0	3	3	6	6
				-	-	1	$\boxed{1}$	1	$\boxed{1}$
2	4	5	$f(2,g)$	0	0	3	3	6	6
				-	-	0	0	0	$\boxed{0}$
3	3	1	$f(3,g)$	0	0	3	3	6	6
				-	-	0	0	0	$\boxed{0}$
$n = 4$	2	1	$f(4,g)$	0	0	3	3	6	6
				-	-	0	0	0	$\boxed{0}$

12.4.2   A Periodic Property (Gilmore and Gomory [237], Hu [336])

If we drop the x integer requirements, the knapsack problem is a linear program. Both the optimal integer and linear programming solutions to the knapsack problem are dependent on the weight capacity b. Therefore, the difference in the maximal value of these objective functions may be considered as a function of b. For b sufficiently large, this function has a periodic property which is worth examining, because for certain weight limits $b' \neq b$ we can obtain the optimal integer solution to a knapsack problem with capacity b' from the optimal solution with capacity b. This will mean that for certain stages k $(1 \le k \le n)$ in a dynamic programming procedure we may only have to compute $f(k, g)$ for $g = 0, 1, \ldots, T < b$. If b is large and the integer T is relatively small, an enormous number of calculations are saved. We now proceed to define the difference function, develop the periodic property, and explain how it can be used. We first view the knapsack problem as a linear program.

The Linear Programming (LP) Solution

Dropping the x integer constraints in the knapsack problem yields the linear problem

$$\text{maximize} \quad \sum_{j=1}^{n} c_j x_j$$
$$\text{subject to} \quad \sum_{j=1}^{n} a_j x_j = b, \quad x_j \ge 0 \quad (j = 1, \ldots, n),$$

where x_n is a slack variable (with $a_n = 1$) and $c_n = 0$ when the constraint is initially an inequality. By defining $y_j = a_j x_j$ $(j = 1, \ldots, n)$, we have $y_j \ge 0$ since $a_j > 0$, and after substituting, the linear program becomes

$$\text{maximize} \quad \sum_{j=1}^{n} (c_j / x_j) y_j$$
$$\text{subject to} \quad \sum_{j=1}^{n} y_j = b, \quad y_j \ge 0 \quad (j = 1, \ldots, n).$$

To simplify the notation we shall assume that the variables have been reordered so that $c_1/a_1 \ge c_2/a_2 \ge \ldots \ge c_n/a_n$. Then the optimal linear programming solution has $y_1 = b$ and $y_j = 0$ for $j = 2, \ldots, n$, or $x_1 = b/a_1$ and $x_j = 0$ for $j = 2, \ldots, n$. If $x_1 = b/a_1$ is integer, the knapsack problem is solved; otherwise the maximal value of the objective function $f(n, b)$ (in dynamic programming notation) cannot exceed $c_1(b/a_1)$, the LP solution, and must at least be $c_1[b/a_1]$, since $x_1 = [b/a_1]$ and $x_j = 0$ $(j \ne 1)$ is an integer solution. Here, $[y]$ means the largest integer $\le y$. (Note that this means that if $c_1[b/a_1]$ is reasonably close to $c_1(b/a_1)$, then $x_1 = [b/a_1]$, and $x_j = 0$ $(j \ne 1)$ may be an "acceptable solution", and the knapsack problem has been "solved.")

Recursion III: The Difference Function

The value of the optimal linear programming solution is $c_1(b/a_1)$ and the value of the optimal integer solution is $f(n, b)$. Define $D(b)$, a nonnegative function in b, to be the difference in these values. That is,

$$D(b) = (c_1/a_1)b - f(n, b). \qquad (9)$$

We shall show that $D(b - a_1) = D(b)$ when b is sufficiently large. To do this we need the following result.

Theorem 12.3 *If the variables are ordered such that $c_j/a_j \geq c_{j+1}/a_{j+1}$ ($j = 1, \ldots, n-1$), then for b sufficiently large, $x_1 \geq 1$ is in every optimal integer solution provided that $c_1/a_1 \neq c_2/a_2$. When $c_1/a_1 = c_2/a_2$, then for b sufficiently large, $x_1 \geq 1$ in at least one optimal integer solution.*

Proof: If $x_1 \geq 1$ in an *optimal* integer solution, the objective function must be at least

$$c_1[b/a_1] > c_1(b/a_1 - a_1/a_1) = (c_1/a_1)(b - a_1),$$

where the inequality follows from $[y] > y - 1$ for any scalar y. Now, if $x_1 = 0$ in every optimal integer solution, the objective function cannot exceed $c_2(b/a_2)$, the value of the linear programming solution with $x_1 = 0$. This means that x_1 must be at least 1 in every optimal integer solution whenever

$$\underbrace{c_1[b/a_1]}_{\substack{\text{Lower bound on} \\ \text{objective function} \\ \text{with } x_1 \geq 1}} > (c_1/a_1)(b - a_1) \geq \underbrace{c_2(b/a_2)}_{\substack{\text{Upper bound on} \\ \text{objective function} \\ \text{with } x_1 = 0}}$$

Suppose $c_1/a_1 \neq c_2/a_2$, then we may solve for b in $(c_1/a_1)(b - a_1) \geq c_2(b/a_2)$ to obtain

$$b \geq \frac{c_1}{(c_1/a_1 - c_2/a_2)} = \bar{b}.$$

Therefore, for $b \geq \bar{b}$, $x_1 \geq 1$ in any optimal integer solution, provided that $c_1/a_1 \neq c_2/a_2$.

To show that $x_1 \geq 1$ for sufficiently large b when $c_1/a_1 = c_2/a_2$ we can no longer use the above argument. One way to prove that $x_1 \geq 1$ for at least one optimal integer solution is to consider an optimal linear programming solution $\mathbf{x} = (\mathbf{x_B}, \mathbf{x_N})$, where the basic variable is $\mathbf{x_B} = x_1$ and the nonbasic variables are $\mathbf{x_N} = (x_2, \ldots, x_n)$. Then the basis is $\mathbf{B} = a_1$, the basis inverse is $\mathbf{B}^{-1} = 1/a_1$, and the nonbasic columns (of single entries) are $\mathbf{N} = (a_2, \ldots, a_n)$. Proceeding as in Chapter 11, we may derive an optimal integer solution from $(\mathbf{x_B}, \mathbf{x_N})$ by finding nonnegative integer values for $(x_2, \ldots, x_n) = \mathbf{x_N}$ so that

$$\mathbf{x_B} = \mathbf{B}^{-1}\mathbf{b} - \mathbf{B}^{-1}\mathbf{N}\mathbf{x_N} \geq 0 \quad \text{and integer,}$$

and $\sum_{j=2}^{n} \bar{c}_j x_j$ is minimized, where $\bar{c}_j = \mathbf{c_B}\mathbf{B}^{-1}a_j - c_j \geq 0$ $(j = 1, \ldots, n)$ and $\mathbf{c_B} = c_1$. Substituting, we have

$$
\begin{aligned}
x_1 \; = \; \mathbf{x_B} &= \mathbf{B}^{-1}(\mathbf{b} - \mathbf{N}\mathbf{x_N}) \\
&= (1/a_1)\left(b - \sum_{j=2}^{n} a_j x_j\right) \geq (1/a_1)\left(b - a_M \sum_{j=2}^{n} x_j\right), \quad (10)
\end{aligned}
$$

where $a_M = \underset{j \geq 2}{\text{maximum}} \; a_j$. Now we appeal to a result proved in Chapter 11 (Section 11.4.2), which says that in every integer solution obtained from $(\mathbf{x_B}, \mathbf{x_N})$, the nonbasic variables $(x_2, \ldots, x_n)$ must satisfy

$$\sum_{j=2}^{n} x_j \leq |\det \mathbf{B}| - 1 = a_1 - 1. \qquad (11)$$

Substituting (11) into (10) yields

$$x_1 \geq (1/a_1)(b - a_M(a_1 - 1)).$$

As $a_1 > 0$, this means $x_1 > 0$ provided that $b > a_M(a_1 - 1) = \bar{b}$.[18] ∎

We have just shown that for b sufficiently large, $x_1 \geq 1$ in at least one optimum integer solution. By equation (8) this means

$$f(n, b) = c_1 + f(n, b - a_1) \text{ for } b > \bar{b}. \tag{12}$$

We use this to show $D(b) = (c_1/a_1)b - f(n, b) = D(b - a_1)$ as follows.

$$
\begin{aligned}
D(b - a_1) &= (c_1/a_1)(b - a_1) - f(n, b - a_1) \\
&= (c_1/a_1)(b - a_1) + c_1 - f(n, b) \quad \text{(by (12) for } b > \bar{b}) \\
&= (c_1/a_1)b - f(n, b) \\
&= D(b).
\end{aligned}
$$

Therefore, the function is periodic with period a_1 (Fig. 12.5). This result can be useful. For suppose $f(n, b)$ is known and we wish to find

---

[18]When $c_1/a_1 = c_2/a_2$, then selecting x_2 as the basic variable gives an alternate optimal linear programming solution. In this case, we may proceed, using (10) and (11) with $\mathbf{B} = a_2$ to show that $x_2 \geq 1$ in at least one optimal integer solution (namely, the one constructed from $\mathbf{x_B} = x_2$, $\mathbf{x_N} = (x_1, x_3, \ldots, x_n)$) provided that $b > a_M(a_2 - 1)$ where $a_M = \underset{j \neq 2}{\text{maximum }} a_j$ For this optimal integer solution, x_1 may, in fact, have value 0.

As an example consider

$$
\begin{array}{llllllll}
\text{maximize} & 7x_1 & + & x_2 & + & 8x_3 & = & z \\
\text{subject to} & 7x_1 & + & x_2 & + & 9x_3 & \leq & b \quad (b > 0) \\
\text{and} & & & \multicolumn{5}{l}{x_1, x_2, x_3 \geq 0, \text{ and integer.}}
\end{array}
$$

Then $c_1/a_1 = c_2/a_2 = 1$, and for $b > a_M(a_1 - 1) = 9(7 - 1) = 54$, $x_1 \geq 1$ in at least one optimal solution. For example, if $b = 60$, then optimal integer solutions have $z = 60$ and are of the form

$$x_1 = k, \ x_2 = 60 - 7k, \ x_3 = 0 \ (k = 0, 1, \ldots, 8),$$

thus the point $x_1 = 0, x_2 = 60, x_3 = 0$ is optimal. Note that for $b > a_M(a_2 - 1) = 9(1 - 1) = 0$, $x_2 \geq 1$ in at least one optimal integer solution.

When $c_1/a_1 \neq c_2/a_2$, then the only basic variable that will yield $\bar{c}_j \geq 0$ ($j = 1, \ldots, n$) is x_1. Thus, optimal integer solutions can only be constructed from a single optimal basic feasible solution, namely $\mathbf{x_B} = x_1$, $\mathbf{x_N} = (x_2, \ldots, x_n)$. In this case, $x_1 \geq 1$ in every optimal integer solution, provided that $b > a_M(a_1 - 1)$. Thus, the entire lemma could be proven using (10) and (11). Inequality (11), however, required considerable effort to prove.

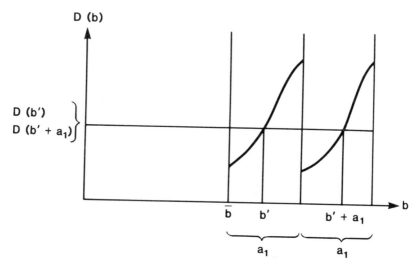

Figure 12.5: Periodic Function with Period a_1

$f(n, b + \lambda a_1)$, where λ is a positive integer and $b > \bar{b}$. As $D(b) = D(b + \lambda a_1)$ (Why?), it follows that the optimal solution in $f(n, b + \lambda a_1)$ is precisely the optimal solution in $f(n, b)$, except that x_1 in the latter solution is incremented by λ, and

$$f(n, b + \lambda a_1) = f(n, b) + c_1 \lambda. \tag{13}$$

To see this, note that

$$D(b) = (c_1/a_1)b - f(n, b) = D(b + \lambda a_1)$$

$$= (c_1/a_1)(b + \lambda a_1) - f(n, b + \lambda a_1).$$

Solving for $f(n, b + \lambda a_1)$ gives (13), and as the solution making up $f(n, b)$ is a solution to $f(n, b + \lambda a_1)$ whose value agrees with (13) when x_1 is incremented by λ, it must be optimal for the knapsack problem with weight capacity $b + \lambda a_1$.

We can develop an equation which successively finds $D(b)$ for $b = 0, 1, 2, \ldots$ To do this, note that for $b > 0$ at least one $x_j \geq 1$ in the optimal integer solution. By (8) this means that $f(n, b) = c_j + f(n, b - a_j)$ for some j, which gives

$$D(b - a_j) = (c_1/a_1)(b - a_j) - f(n, b - a_j)$$

$$= (c_1/a_1)(b - a_j) + c_j - f(n, b)$$

$$= D(b) - (c_1/a_1)a_j + c_j,$$

or

$$D(b) = D(b - a_j) + (c_1/a_1 - c_j/a_j)a_j. \tag{14}$$

As the value of the optimal integer solution will be as close as possible to the optimal linear programming solution, the $x_j \geq 1$ in the optimal integer solution will make the right-hand side in (14) as small as possible. Therefore,

$$D(b) = \min_{j} \; \{D(b - a_j) + (c_1/a_1 - c_j/a_j)a_j\}. \tag{15}$$

Expression (15) allows us to calculate $D(b)$ for all b starting with the initial conditions $D(0) = 0$ and $D(y) = +\infty$ for $y < 0$.

Example 12.3 (Hu [336])

j	1	2	3	4	5 (slack variable)
c_j	18	14	8	4	0
a_j	15	12	7	4	1
c_j/a_j	1.2	1.167	1.143	1	0
$(c_1/a_1 - c_j/a_j)a_j$	0	0.40	0.40	0.80	1.2

Using (15) we can develop the following table, which yields $D(b)$ for $b = 1, \ldots, 12$.

b	j	1	2	3	4	5	6
$D(b - 15)$	1	$+\infty$	$+\infty$	$+\infty$	$+\infty$	$+\infty$	$+\infty$
$D(b - 12) + .40$	2	$+\infty$	$+\infty$	$+\infty$	$+\infty$	$+\infty$	$+\infty$
$D(b - 7) + .41$	3	$+\infty$	$+\infty$	$+\infty$	$+\infty$	$+\infty$	$+\infty$
$D(b - 4) + .80$	4	$+\infty$	$+\infty$	$+\infty$	.8	2.0	3.2
$D(b - 1) + 1.20$	5	1.2	2.4	3.6	4.8	2.0	3.2
$D(b)$		1.2	2.4	3.6	0.8	2.0	3.2

b	j	7	8	9	10	11	12
$D(b-15)$	1	$+\infty$	$+\infty$	$+\infty$	$+\infty$	$+\infty$	$+\infty$
$D(b-12)+.40$	2	$+\infty$	$+\infty$	$+\infty$	$+\infty$	$+\infty$	0.4
$D(b-7)+.40$	3	0.4	1.6	2.8	4.0	1.2	2.4
$D(b-4)+.80$	4	4.4	1.6	2.8	4.0	1.2	2.4
$D(b-1)+1.20$	5	4.4	1.6	2.8	4.0	5.2	2.4
$D(b)$		0.4	1.6	2.8	4.0	1.2	0.4

Continuing this process reveals that for $b \geq 26$, $D(b)$ is periodic with period a_1. In particular, $D(b) = D(b+15\lambda)$ for $b \geq 26$ and

b	26 $+15\lambda$	27 $+15\lambda$	28 $+15\lambda$	29 $+15\lambda$	30 $+15\lambda$	31 $+15\lambda$	32 $+15\lambda$
$D(b)$	1.2	0.4	1.6	0.8	0	1.2	2.4

b	33 $+15\lambda$	34 $+15\lambda$	35 $+15\lambda$	36 $+15\lambda$	37 $+15\lambda$	38 $+15\lambda$	39 $+15\lambda$	40 $+15\lambda$
$D(b)$	1.6	0.8	2.0	1.2	0.4	1.6	0.8	2.0

Remarks

1. $b \geq \bar{b}$ is a sufficient but *not a necessary* condition for $D(b)$ to be periodic. In most instances the function is periodic for much smaller b. In our example,

$$\bar{b} = \frac{c_1}{(c_1/a_1 - c_2/a_2)} = 540,$$

which is considerably larger that 26. Also, if $c_1/a_1 = c_2/a_2$, it may pay to reindex the variables so that x_2 comes first, since this may mean that $\bar{b} = a_M(a_j - 1)$ is smaller.

2. Using the definition (9) we can find $f(n, b)$, using $f(n, b) = (c_1/a_1)b - D(b)$, once $D(b)$ is known. Also, if we keep track of those j in which the minimum is attained in (15), it is possible to retrieve the values of

the variables in $f(n, b)$. For example, using the first table for $b = 10$, we have $D(10) = 4.0$. As $(c_1/a_1)b = 12$, $f(5, 10) = 12 - 4 = 8$. To obtain the solution, the minimum in (15) for $b = 10$ occurs when $j = 4$ or $j = 5$. Selecting $j = 4$ means that $x_4 \geq 1$, and when $x_4 = 1$, b is reduced to $10 - a_4 = 6$. When $b = 6$, the minimum in (15) occurs when $j = 4$ or 5. Selecting $j = 4$ means that x_4 becomes 2 and b is reduced to $6 - a_4 = 2$. For $b = 2$, the minimum occurs when $j = 5$ or $x_5 = 1$; b decreases to $2 - a_5 = 1$, and x_5 is incremented to 2 with b reduced to 0. Thus an optimal integer solution when $b = 10$ is $x_1 = x_2 = x_3 = 0$, and $x_4 = x_5 = 2$. Note that $\sum_{j=1}^{5} c_j x_j = 8 = f(5, 10)$ and $x_1 = 0$, and b is not sufficiently large. Had $j = 5$ initially been selected, we would have obtained the same optimal solution. We have just shown that *equation (15) gives another dynamic programming recursion* for the knapsack problem.

3. In using (15) we find that when $b = 33$ the optimal solution is $x_1 = 1, x_2 = 0, x_3 = 2, x_4 = 1$, and $x_5 = 0$, with $f(5, 33) = 38$. So for $b = 33 + 15\lambda$ ($\lambda = 0, 1, \ldots$), the optimal solution is $x_1 = 1 + \lambda, x_2 = 0, x_3 = 2, x_4 = 1$, and $x_5 = 0$, with

$$f(5, 33 + 15\lambda) = 38 + 18\lambda.$$

This means that regardless of the size of b, for $k = 5$, we need only compute $f(k, g)$ for $g = 1, \ldots, 40$ ($40 = 26 + (a_1 - 1)$) in a dynamic programming algorithm. The number of calculations saved is enormous when b is substantially larger than 40. This suggests that it may be advantageous, especially when b is large, to use (15) to solve the knapsack problem.

12.4.3   Branch and Bound Algorithms: General Knapsack Problem

For ease of reference, the general knapsack problem is rewritten below:

$$\text{maximize} \quad \sum_{j=1}^{n} c_j x_j$$

$$\text{subject to} \quad \sum_{j=1}^{n} a_j x_j \leq b$$

$$\text{and} \quad 0 \leq x_j \leq u_j \text{ and integer,}$$

where u_j is an upper bound on variable x_j, which if not specified, can be taken as $[b/a_j]$.

In the previous section we have shown that if the variables are initially ordered so that $c_1/a_1 \geq c_2/a_2 \geq \ldots \geq c_n a_n$, then the linear programming solution to the knapsack problem is $x_1 = b/a_1$, $x_j = 0$ $(j \neq 1)$ and the objective function is $c_1(b/a_1)$. We can extend this result to the case where the variables are bounded. If we define $y_j = a_j x_j$ $(j = 1, 2, \ldots, n)$, the knapsack problem without the integrality constraint is,

$$\text{maximize} \quad \sum_{j=1}^{n}(c_j/a_j)y_j$$

$$\text{subject to} \quad \sum_{j=1}^{n} y_j = b$$

$$\text{and} \quad 0 \leq y_j \leq a_j u_j \quad (j = 1, \ldots, n).$$

Clearly, the optimal linear programming solution to the above problem is $y_1 = b, y_j = 0$ $(j \geq 2)$ when $a_1 u_1 \geq b$. If $a_1 u_1 < b$, the LP solution is $y_j = a_j u_j$ $(j = 1, \ldots, t)$, $y_{t+1} = b - \sum_{j=1}^{t} a_j u_j$, and $y_j = 0$ $(j \geq t+2)$, where t is the smallest index such that $b - \sum_{j=1}^{t+1} a_j u_j < 0$. Stated in terms of the variables, x_j the optimal LP solution is

$$\mathbf{x} = (b/a_1, 0, 0, \ldots, 0) \text{ when } a_1 u_1 \geq b,$$

and

$$\mathbf{x} = (u_1, u_2, \ldots, u_t, \underbrace{(1/a_{t+1})(b - \sum_{j=1}^{t} a_j u_j)}_{x_{t+1}}, 0, 0, \ldots, 0),$$

otherwise.

A Modified Dakin Procedure

We have just shown that a knapsack problem with or without bounded variables can be solved as a linear program by inspection. Thus natural procedures to solve knapsack problems are often a specialized form of a branch and bound procedure (Chapter 8). If we adopt Dakin's branch and bound algorithm (Section 8.5), the LP solution of the problem is initially examined, if it is integer, the knapsack problem is solved. Otherwise, if in the LP solution the variable x_k has fractional value θ, then two nodes are created, the first by introducing the constraint $x_k \leq [\theta]$, and the second by imposing the constraint $x_k \geq [\theta] + 1$. The linear program at each of these nodes is solved by inspection and the process is repeated. Nodes are thus created whenever they have potential to produce an improved integer solution; that is, when the LP solution is not integer and its value exceeds the value of the current best integer solution. Also, observe that if the LP solution is not integer, there is exactly one variable (namely, x_1 or x_{t+1}) which is fractional. The fractional variable when rounded down to the nearest integer produces a feasible solution to the knapsack problem, the corresponding objective function value can therefore be taken as a lower bound on the objective function. The basic procedure can be further illustrated using the example below.

Example 12.4 (from Example 12.3)
We solve the problem appearing in the previous example when $b = 33$. For ease of reference, we restate the problem where the slack variable x_5 has been dropped.

$$\begin{array}{llllllllll}
\text{Maximize} & 18x_1 & + & 14x_2 & + & 8x_3 & + & 4x_4 \\
\text{subject to} & 15x_1 & + & 12x_2 & + & 7x_3 & + & 4x_4 & \leq & 33 \\
& x_1, \ldots, x_4 \geq 0 \text{ and integer.}
\end{array}$$

The computations are tabulated in Table 12.1 and the associated tree appears in Fig. 12.6. We are using the node selection rule: "Choose the most recently created node with the highest linear programming solution for further branching." (Other criteria are described in Section 8.7, Chapter 8.) The integer solution $x_1 = [33/15] = 2$, and $x_2 = x_3 = x_4 = 0$ gives the initial lower bound $2(18) = 36$. Note that it is very close to 38, the value of the optimal integer solution. (Unfortunately, this does not always happen.)

The nodes in Table 12.1 and Fig. 12.6 are numbered in order of their creation. An active node naturally is one that can possibly produce an improved integer solution. As a sample calculation, at Node 13 we have the additional constraint (see Fig. 12.6) $x_1 \leq 2, x_2 \geq 1, x_1 \leq 1$, and $x_2 \geq 2$, which are equivalent to $x_1 \leq 1$ and $x_2 \geq 2$. Letting $\bar{x}_2 = x_2 - 2 \geq 0$, and substituting $\bar{x}_2 + 2$ for x_2 in the knapsack constraint, yields

Node #	Constraints Active at Node				LP Solution — Variables				Value	Value of Best Int. Sol.	Active Nodes (○ Selected)	Remarks
	x_1	x_2	x_3	x_4	x_1	x_2	x_3	x_4				
1	—	—	—	—	$\frac{32}{15}$	0	0	0	$39\frac{2}{5}$	36	①	36 from $c_1, x_1 = 18[\frac{22}{15}]$
2	≤ 2	—	—	—	2	$\frac{1}{4}$	0	0	$39\frac{1}{2}$	36	①.2	
3	≥ 3	—	—	—	Infeasible				$-\infty$	36	②	
4	≤ 2	$= 0$	—	—	2	0	$\frac{2}{7}$	0	$39\frac{2}{7}$	36	②.4	$x_2 \leq 0 \Rightarrow x_2 = 0$
5	≤ 2	≥ 1	—	—	$\frac{21}{15}$	1	0	$\frac{2}{4}$	$39\frac{1}{5}$	36	④.5	$39\frac{1}{5} < 39\frac{1}{2}$
6	≤ 2	$= 0$	$= 0$	—	2	0	0	0	39	36	④.5,6	
7	≤ 2	$= 0$	≥ 1	—	$\frac{26}{15}$	0	1	0	$39\frac{1}{5}$	36	5,6,⑦	
8	≤ 1	$= 0$	≥ 1	—	1	0	$\frac{18}{7}$	0	$38\frac{4}{7}$	36	5,6,⑦.8	
9	$= 2$	$= 0$	≥ 1	—	Infeasible				$-\infty$	36	⑤.6.8	
10	≤ 1	≥ 1	—	—	1	$\frac{1}{2}$	0	0	39	36	⑤.6,8,10	$x_1 \leq 2, x_1 \leq 1 \Rightarrow x_1 \leq 1$
11	$= 2$	≥ 1	—	—	Infeasible				$-\infty$	36	6,8,⑩	$x_1 \leq 2, x_1 \geq 2 \Rightarrow x_1 = 2$
12	≤ 1	$= 1$	—	—	1	1	$\frac{6}{7}$	0	$38\frac{6}{7}$	36	6,8,⑩.12	$x_2 \geq 1, x_2 \leq 1 \Rightarrow x_2 = 1$
13	≤ 1	≥ 2	—	—	$\frac{3}{5}$	2	0	0	$38\frac{4}{5}$	36	⑥.8,12,13	$x_2 \geq 1, x_2 \geq 2 \Rightarrow x_2 > 2$
14	≤ 2	$= 0$	$= 0$	$= 0$	2	0	0	0	36	36	⑥.8,12,13	
15	≤ 2	$= 0$	$= 0$	≥ 1	Infeasible				$-\infty$	36	8,⑫.13	
16	≤ 1	$= 1$	$= 0$	—	$\frac{14}{15}$	1	0	$\frac{3}{2}$	38	36	⑫.13,16	
17	≤ 1	$= 1$	≥ 1	—	0	1	1	0	$38\frac{4}{5}$	36	8,13,16,⑰	
18	$= 0$	$= 1$	≥ 1	—	0	1	3	0	38	38	⑰	Integer solution: nodes 8,13,16 dropped
19	$= 1$	$= 1$	≥ 1	—	Infeasible				$-\infty$	38	None	

Table 12.1: Summary of Computations

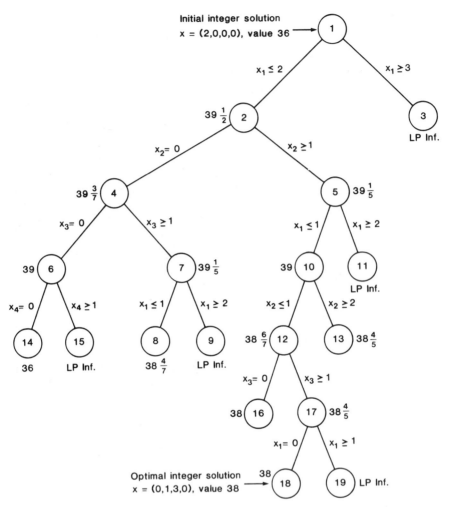

Figure 12.6: Branch and Bound Tree (Dakin Variation)

$$15x_1 + 12\bar{x}_2 + 7x_3 + 4x_4 \leq 9.$$

As $15(1) > 9$, the LP solution is

$$x_1 = 9/15, \ \bar{x}_2 = 0 \text{ or } x_2 = 2, \ x_3 = 0, \ x_4 = 0,$$

with value $\frac{9}{15}(18) + 2(14) = 38\frac{4}{5}$. At this stage, $38\frac{4}{5} > 36$, so Node 13 is added to the active list. However, note that it is dropped once the integer solution with value 38 (Node 18) is found, since $38\frac{4}{5} - 38 < 1$

and the costs are integer. That is, any integer solution found from Node 13 cannot be larger than 38 since the LP solution is $38\frac{4}{5}$.

A Modified Land and Doig Procedure

Ingargiola and Korsh [339] modified Land and Doig's algorithm (cf. Chapter 8) to solve a general knapsack problem. In particular, the LP problem is initially solved and its solution is examined. If it is integer, the knapsack problem is solved. Otherwise, if in the LP solution, variable x_k has the fractional value θ_k, nodes are created by fixing x_k successively to values $[\theta_k] + 1, [\theta_k] + 2, \ldots$. These first nodes are then searched until either the value of x_k reaches its upper bound or the LP solution at the node is less than the current best integer solution. In either case, nodes are then created and searched by fixing x_k successively to values $[\theta_k], [\theta_k] - 1, \ldots, 1, 0$. A similar branching rule is used when newly created nodes are searched. Also, the most recently created node is selected for further branching. The procedure is further illustrated by the example below

Example 12.5 (modified Example 12.4)

$$
\begin{aligned}
\text{Maximize} \quad & 18x_1 + 14x_2 + 8x_3 + 4x_4 \\
\text{subject to} \quad & 15x_1 + 12x_2 + 7x_3 + 4x_4 \leq 33 \\
& x_1 \leq 1, x_2 \leq 2, x_3 \leq 3, x_4 \leq 2 \\
\text{and} \quad & x_j \leq 0 \quad (j, 1, 2, 3, 4).
\end{aligned}
$$

The computations are tabulated in Table 12.2 and the associated tree appears in Fig. 12.7. The optimal solution is $\mathbf{x} = (0, 1, 3, 0)$ with value of the objective function equal 38.

12.4.4 Lagrangian Multiplier Methods (Everett [190], Brooks and Geoffrion [101])

A result due to Everett [190] is now presented which may allow us to solve an integer program, and hence a knapsack problem, by only enforcing the nonnegativity and integrality requirements on the variables. The constraint inequalities are included in the objective function. The result is given in terms of an integer program but can easily be extended to a more general optimization problem (Problem 12.19).

| Node | Variables Fixed | | | | LP Solution | | | | | Lower | |
No.	x_1	x_2	x_3	x_4	x_1	x_2	x_3	x_4	value	Bound	Remarks
1	--	--	--	--	1	3/2	0	0	39	32	IP Sol. $(1,1,0,0)$
2	--	2	--	--	3/5	2	0	0	$38\frac{4}{5}$	32	
3	1	2	--	--					--	32	Infeasible
4	0	2	--	--	0	2	9/7	0	$38\frac{2}{7}$	36	IP Sol. $(0,2,1,0)$
5	0	2	2	--					--	36	Infeasible
6	0	2	1	--	0	2	1	1/2	38	36	IP Sol. $(0,2,1,0)$
7	0	2	1	1					--	36	Infeasible
8	0	2	1	0	0	2	1	0	36	36	Integer Sol.
9	0	2	0	--	0	2	0	2	36	36	Integer Sol.
10	--	1	--	--	1	1	6/7	0	$38\frac{6}{7}$	36	
11	--	1	1	--	14/15	1	1	0	$38\frac{4}{5}$	36	
12	1	1	1	--					--	36	Infeasible
13	0	1	1	--	0	1	1	2	30	36	LP Sol. < 36
14	--	1	2	--	7/15	1	2	0	$38\frac{2}{5}$	36	
15	1	1	2	--					--	36	Infeasible
16	0	1	2	--	0	1	2	7/4	37	36	
17	--	1	3	--	0	1	3	0	38	38	Integer Sol.
18	--	1	0	--	1	1	0	3/2	38	38	LP Sol. ≤ 38
19	--	0	--	--	1	0	18/7	0	$38\frac{4}{7}$	38	[LP Sol.] ≤ 38

Table 12.2: Summary of Computations (Example 12.5)

Theorem 12.4 *(Everett [190]) Consider the integer program:*

$$\begin{aligned}
\text{Maximize} \quad & \mathbf{cx} \\
\text{subject to} \quad & \mathbf{Ax} \leq \mathbf{b} \\
\text{and} \quad & \mathbf{x} \geq \mathbf{0} \quad \text{and integer,}
\end{aligned}$$

where $\mathbf{A}$ is an m by n matrix. Let $\boldsymbol{\lambda}$ be an (m) column of nonnegative numbers ("multipliers"). Then if $\mathbf{x}^0$ solves:

$$\begin{aligned}
\text{L} \quad \text{Maximize} \quad & \mathbf{cx} - \boldsymbol{\lambda}\mathbf{Ax} \\
\text{subject to} \quad & \mathbf{x} \geq \mathbf{0} \quad \text{and integer,}
\end{aligned}$$

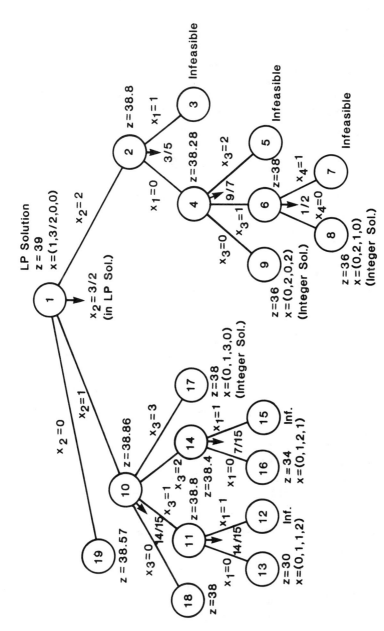

Figure 12.7: Branch and Bound Tree (Land and Doig Variation)

it also solves the integer program with **b** *replaced by* $\mathbf{A}\mathbf{x}^0$.[19] *Therefore, if* λ *is chosen so that the optimal solution* $\mathbf{x}^0$ *gives* $\mathbf{b} = \mathbf{A}\mathbf{x}^0$, *the original integer program has been solved.*

Proof: As $\mathbf{x}^0$ gives the maximum value to $\mathbf{c}\mathbf{x} - \lambda\mathbf{A}\mathbf{x}$, we have that $\mathbf{c}\mathbf{x} - \lambda\mathbf{A}\mathbf{x} \leq \mathbf{c}\mathbf{x}^0 - \lambda\mathbf{A}\mathbf{x}^0$. Therefore, for all $\mathbf{x} \geq \mathbf{0}$ and integer, $\mathbf{c}\mathbf{x}^0 + \lambda(\mathbf{A}\mathbf{x}^0 - \mathbf{A}\mathbf{x}) \leq \mathbf{c}\mathbf{x}^0$. So for all nonnegative integer solutions to $\mathbf{A}\mathbf{x} \leq \mathbf{A}\mathbf{x}^0$, the previous inequality is true. But for these solutions, $\mathbf{A}\mathbf{x}^0 - \mathbf{A}\mathbf{x} \geq \mathbf{0}$, and thus $\mathbf{c}\mathbf{x} \leq \mathbf{c}\mathbf{x}^0$, since $\lambda \geq \mathbf{0}$. Hence $\mathbf{x}^0$ solves the integer program:

$$
\begin{array}{lrcl}
\text{Maximize} & \mathbf{c}\mathbf{x} & & \\
\text{subject to} & \mathbf{A}\mathbf{x} & \leq & \mathbf{A}\mathbf{x}^0 \\
\text{and} & \mathbf{x} & \geq & \mathbf{0} \text{ and integer.}
\end{array}
$$

∎

The theorem indicates that if the multipliers λ are selected such that $\mathbf{b} = \mathbf{A}\mathbf{x}^0$, we can solve the integer program without the inequality constraints. This problem is usually considerably easier to solve.

The difficulty with the Lagrangian approach is that multipliers λ may be difficult to find such that $\mathbf{A}\mathbf{x}^0 = \mathbf{b}$. In fact, as shown by Everett [190], in certain instances the multipliers which give this equality may not even exist. However, the procedure is valuable since it may be relatively easy to find multipliers such that $\mathbf{A}\mathbf{x}^0$ is within an "acceptable" range of $\mathbf{b}$, in which case the approximate solution $\mathbf{x}^0$ is taken as optimal. This procedure is especially useful when the given data (in particular the right-hand side) has been estimated.

Lagrangian multiplier algorithms usually work by selecting the vector λ, finding $\mathbf{x}^0$, and then finding $\mathbf{A}\mathbf{x}^0$. If $\mathbf{A}\mathbf{x}^0$ is sufficiently close to $\mathbf{b}$, the process stops. Otherwise, another vector λ is selected, a new value of $\mathbf{x}^0$ and $\mathbf{A}\mathbf{x}^0$ is found, and so on. Several enumerative and/or linear programming schemes for finding the successive values for the vector λ have appeared in the literature (e.g., [101],[205],[226]). The details are left to the references.

Problem L (see Theorem 12.4) for the knapsack problem is

$$
\text{maximize} \quad \sum_{j=1}^{n} c_j x_j - \lambda \sum_{j=1}^{n} a_j x_j,
$$

$$
x_j \geq 0 \text{ and integer} \quad (j = 1, \ldots, n)
$$

---

[19] The function $L(\mathbf{x}) = \mathbf{c}\mathbf{x} - \lambda\mathbf{A}\mathbf{x}$ is termed a *Lagrangian function.*

or

$$\text{maximize} \quad \sum_{j=1}^{n}(c_j - \lambda a_j)x_j,$$

$$x_j \geq 0 \text{ and integer} \quad (j = 1, \ldots, n),$$

where λ is now a nonnegative number, since the integer program has a single constraint. The problem can be solved by inspection, namely,

$$
\begin{array}{llll}
x_j^0 = u_j & \text{if} & c_j - \lambda a_j > 0 & (u_j \text{ is an integer upper bound for } x_j) \\
x_j^0 = t & \text{if} & c_j - \lambda a_j = 0 & (t \text{ is any integer satisfying } 0 \leq t \leq u_j) \\
x_j^0 = 0 & \text{if} & c_j - \lambda a_j < 0.
\end{array}
$$

Thus the optimal solution depends on the sign not the magnitude of $c_j - \lambda a_j$. Suppose the costs are positive and the variables are ordered such that

$$c_j/a_j \geq c_{j+1}/a_{j+1} \quad (j = 1, \ldots, n-1).$$

As $\mathbf{x}^0$ changes only when $c_j - \lambda a_j = 0$, the value for $\mathbf{x}^0$ remains the same in the intervals

$$0 \leq \lambda < c_n/a_n, \; c_n/a_n < \lambda < c_{n-1}/a_{n-1}, \ldots, c_2/a_2 < \lambda < c_1/a_1.$$

(If $c_j/a_j = c_{j+1}/a_{j+1}$, one of the two terms is dropped in the interval calculation.) So for the knapsack problem the only values for λ that might be tested are $0, c_n/a_n, \ldots, c_1/a_1$.

If the upper bound u_j are not known, they can be taken as $[b/a_j]$. However, in this instance

$$\sum_{\geq 0} a_j x_j^0 \geq \sum_{> 0} a_j[b/a_j]$$

may be considerably larger than b. Here, the first summation is taken over those j such that $c_j - \lambda a_j \geq 0$ and the second summation is over those j with $c_j - \lambda a_j > 0$. For this reason, Lagrangian methods are usually applied to integer programs with known upper bounds (e.g., with zero-one variables).[20]

---

[20]Lagrangian multiplier techniques applied to the capital budgeting problem, a knapsack problem with zero-one variables, are discussed in Kaplan [356], and Weingartner and Ness [597].

Example (from Example 12.5)
Consider the integer program just solved by a branch and bound algorithm except that the variables are now required to be 0 or 1. The problem is

$$\begin{array}{ll} \text{maximize} & 18x_1 \;+\; 14x_2 \;+\; 8x_3 \;+\; 4x_4 \\ \text{subject to} & 15x_1 \;+\; 12x_2 \;+\; 7x_3 \;+\; 4x_4 \;\le\; 33 \\ & x_j = 0 \text{ or } 1 \quad (j = 1,\ldots,4) \end{array}$$

Problem L is

$$\text{maximize} \;\; (18 - 15\lambda)x_1 + (14 - 12\lambda)x_2 + (8 - 7\lambda)x_3 + (4 - 4\lambda)x_4.$$
$$x_j = 0 \text{ or } 1$$

Computations for $\lambda = 0, 4/4, 8/7, 14/12$, and $18/15$ are given in Table 12.3. If an extra unit of weight is allowed (i.e., b becomes 34), $1 \le \lambda \le 8/7$ gives the approximate solution $\mathbf{x}^0 = (1,1,1,0)$ with value 40. The graph of $b - \sum_j a_j x_j^0$ versus λ is a step function as plotted in Fig. 12.8. As we are interested in λ so that $b - \sum_j a_j x_j^0$ is near zero, the computations may be terminated once λ is selected so that $b - \sum_j a_j x_j^0 > 0$. In our case, this would be after the $\lambda = 8/7$ calculation. Note that for this problem there does not exists a λ for which $b - \sum_{j=1}^{4} a_j x_j^0 = 0$.

12.4.5   Network Approaches (Shapiro [527])

A knapsack problem with an equality constraint can be represented by a network in which a shortest route is sought. In particular, there are $b+1$ nodes representing the numbers $0,1,2,\ldots,b$. A directed arc joins two nodes N_g and N_t $(t > g)$ (representing the numbers g and t, respectively) whenever a coefficient a_j exists so that $a_j + g = t$. Traversing the arc between N_g and N_t corresponds to incrementing x_j by 1, and thus a cost $-c_j$ is assigned to the arc. (Initially, all $x_j = 0$.) The problem is then a minimal cost or shortest route problem, in which the origin node is N_0 and the destination node is N_b. It can be solved by any standard shortest route algorithm ([336], [527]) such as the one given in Chapter 11 (Appendix A). However, note that if b is large and (or) the weights a_j are relatively small there are many nodes and

λ	Sign of $c_j - \lambda a_j$				x^0				$\sum\limits_{j=1}^{4} a_j x_j^0$	$b - \sum\limits_{j=1}^{4} a_j x_j^0$	$\sum\limits_{j=1}^{4} c_j x_j^0$	Same x^0 for
	$j=1$	$j=2$	$j=3$	$j=4$	x_1^0	x_2^0	x_3^0	x_4^0				
0	$+$	$+$	$+$	$+$	1	1	1	1	38	-5	44	$0 \leq \lambda < 1$
1	$+$	$+$	$+$	0	1	1	1	0	34	-1	40	$1 \leq \lambda < \frac{8}{7}$
					1	1	1	1	38	-5	44	
$\frac{8}{7}$	$+$	$+$	0	$-$	1	1	0	0	27	$+6$	32	$\frac{8}{7} \leq \lambda < \frac{14}{12}$
					1	1	1	0	34	-1	40	
$\frac{14}{12}$	$+$	0	$-$	$-$	1	0	0	0	15	$+18$	18	$\frac{14}{12} \leq \lambda < \frac{18}{15}$
					1	1	0	0	27	$+6$	32	
$\frac{18}{15}$	0	$-$	$-$	$-$	0	0	0	0	0	$+33$	0	$\lambda > \frac{18}{15}$
					1	0	0	0	15	$+18$	18	

Table 12.3: Summary of Computations

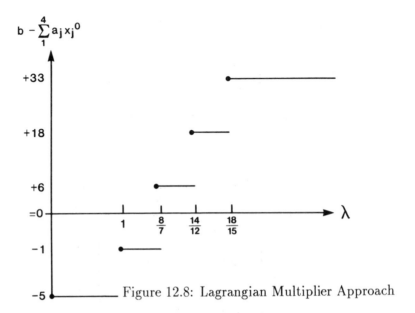

Figure 12.8: Lagrangian Multiplier Approach

(or) arcs, and the approach becomes very inefficient. Also, knapsack problems with bounded variables cannot be represented this way.

Example 12.6

$$\begin{aligned}
\text{Maximize} \quad & 3x_1 \; + \; 2x_2 \\
\text{subject to} \quad & 4x_1 \; + \; 2x_2 \; + \; x_3 \; = \; 8 \\
& x_1, x_2, x_3 \geq 0 \text{ and integer.}
\end{aligned}$$

The network is

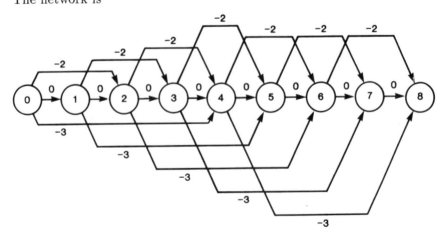

Applying the algorithm given in Appendix A of Chapter 11 gives the shortest route

which means that the optimal solution is $x_2 = 4, x_1 = x_3 = 0$, with value 8. Note the similarity between the network representation given here and the one given in Chapter 11 (Section 11.4) for the group minimization problem.

12.5 Branch and Bound Algorithms: 0-1 Knapsack Problem

Because the 0-1 knapsack problem has numerous applications, it has attracted considerable attention. In recent years, various specialized algorithms of the branch and bound type have been developed and computer implemented. We discuss a few of the more recent algorithms. For ease of reference, the 0-1 knapsack problem is rewritten below.

$$\text{Maximize} \quad \sum_{j=1}^{n} c_j x_j$$

$$\text{subject to} \quad \sum_{j=1}^{n} a_j x_j \leq b$$

$$\text{and} \quad x_j = 0 \text{ or } 1 \quad (j = 1, \ldots, n).$$

Also, without loss of generality, we assume that $c_j > 0$ $(j = 1, \ldots, n)$, $a_j > 0$ $(j = 1, \ldots, n)$, $b > 0$, and the ordering $c_1/a_1 \geq c_2/a_2 \geq \ldots \geq c_n/a_n$.

12.5.1 The Linear Programming Solution

In section 12.4.3, we derived the linear programming solution to the general knapsack problem, which when specialized for 0-1 variables (that is, the upper bound $u_j = 1$ for $j = 1, \ldots, n$) yields the optimal LP solution:

$$\mathbf{x}^* = (b/a_1, 0, 0, \ldots, 0) \qquad \text{if} \quad a_1 \leq b,$$

and $\quad \mathbf{x}^* = (1,\ldots,1,\bar{b}/a_t,0,\ldots,0) \quad$ if $\quad a_1 > b,$

where t is the smallest index $(1 < t \leq n)$ which satisfies $\sum_{j=1}^{t} a_j \geq b$ and $\bar{b} = b - \sum_{j=1}^{t-1} a_j.$

Example 12.7 (Kolesar [372])

Consider the knapsack problem below.

Maximize $\quad 60x_1 + 60x_2 + 40x_3 + 10x_4 + 20x_5 + 10x_6 + 3x_7$

subject to $\quad 30x_1 + 50x_2 + 40x_3 + 10x_4 + 40x_5 + 30x_6 + 10x_7 \leq 100$

and $\qquad x_j = 0$ or $1 \quad (j = 1,\ldots,7).$

Note that $c_1/a_1 \geq c_2/a_2 \geq \ldots \geq c_7/a_7$. Also, the smallest index t for which $\sum_{j=1}^{t} a_j \geq b$ is 3. Hence the optimal linear programming solution is $\mathbf{x}^* = (1,1,1/2,0,0,0,0).$

12.5.2   The Upper Bound Solution

The value of the objective function corresponding to the LP optimal solution is an upper bound for the value of the objective function for the 0-1 knapsack problem. Martello and Toth [414] improved the bound by using the result provided by Land and Doig (cf. Theorem 8.1, Chapter 8). In particular, the upper bound on the value of the objective function corresponding to the integer problem is the maximum of the value of the LP solutions found by fixing variable x_t to 0 and 1, respectively, where, x_t is the variable that has fractional value in the optimal LP solution.

Example (from Example 12.7)

The LP solution is $\mathbf{x} = (1,1,1/2,0,0,0,0)$ and the value of the objective function is 140. This provides an initial upper bound, denoted by UB_1, to the value of the optimal integer solution. Following the discussion in Martello and Toth, the upper bound is improved by first finding the optimal LP solution to the two knapsack problems obtained by fixing variable x_3, the fractional component in $\mathbf{x}$, to 0 and 1, respectively.

$x_3 = 0$:   The optimal LP solution is $(1,1,0,1,1/4,0,0)$ with the objective function equal to 135.

$x_3 = 1$:   The optimal LP solution is $(1,3/5,1,0,0,0,0)$ with the objective function equal to 136.

The upper bound $UB_2 = $ maximum $(135, 136) = 136$.

12.5.3   The Lower Bound Solution

The values of the variables in the optimal LP solution can be used with heuristics to construct good feasible solutions to the knapsack problem. The corresponding value of the objective function then provides a lower bound which may be used in any enumeration scheme. We present a few heuristics below.

Heuristic 1

If the optimal LP solution is not all integer, then the solution obtained by rounding down x_t to 0 is a feasible integer solution. In Example 12.6, setting x_3 to 0 yields $\mathbf{x} = (1, 1, 0, 0, 0, 0, 0)$, a feasible solution to the knapsack problem, with the objective function equal to 120.

Heuristic 2

The feasible solution obtained in Heuristic 1 may sometimes be improved by noting that when x_t is set to 0, the slack variable in the constraint is increased above 0. This may allow one or more additional variables, currently at value 0, to be increased to 1. For example, in Heuristic 1, when x_3 is fixed to 0, the value of the slack variable becomes 20. Now examining the constraint coefficients of x_4, x_5, x_6 and x_7, the variables currently at value 0, in this order reveals that x_4 can be fixed to 1 and then x_7 can be fixed to 1 without exceeding the right hand side. Hence, an improved feasible solution is $(1,1,0,1,0,0,1)$ with the objective function equal to 133.

12.5.4   Reduction Algorithm

Using the optimal LP solution, the size of the knapsack problem (i.e., the number of variables n) can sometimes be reduced through the use of a reduction procedure first suggested by Ingargiola and Korsh [339] and later improved by Nauss [442], Balas and Zemel [46], and Martello and Toth [414]. The procedure by Martello and Toth tries to identify the partitions I_0, I_1 and R of the index set $N = \{1, 2, \ldots, n\}$ such that any optimal solution of the knapsack problem satisfies $x_j = 0$, $j \in I_0$, and $x_j = 1$, $j \in I_1$. Once I_0, I_1, and R are identified, then the knapsack problem is the following reduced knapsack problem.

$$\text{Maximize} \quad \sum_{j \in I_1} c_j + \sum_{j \in R} c_j x_j$$

$$\text{subject to} \quad \sum_{j \in R} a_j x_j \leq b - \sum_{j \in I_1} a_j$$

$$\text{and} \quad x_j = 0 \text{ or } 1 \quad j \in R$$

To find subsets I_0 and I_1, let LB be a lower bound on the value of the objective function obtained using one of the heuristic algorithms previously discussed. Subsets I_0 and I_1 can then be obtained using the following steps.

(1) For each i ($i = 1, 2, \ldots, t$), compute an upper bound (cf. Section 12.5.2) UB_i obtained after fixing x_i to 0. Include i in the set I_1 if $UB_i < LB$.

(2) For each i ($i = t, t + 1, \ldots, n$), compute upper bound UB_i to the knapsack problem subsequent to fixing x_i to 1. Include i in the set I_0 if $UB_i < LB$.

Note that in Steps (1) and (2), to compute UB_i, we need to solve the *reduced problem* as defined by the current sets I_0 and I_1. Also, if during the process of finding UB_i, a better lower bound solution is identified, then the new lower bound may be used in the subsequent application of Steps (1) and (2).

Example (from Example 12.7)

In this example, $t = 3$, and via Heuristic 2, the lower bound solution is $(1, 1, 0, 1, 0, 0, 1)$ with the $LB = 133$.

Step 1 (Construct I_1)

(i) ($i = 1$) After fixing $x_1 = 0$, the Martello and Toth upper bound (cf. Section 12.5.2) is $UB_1 = 110$. Since $UB_1 < LB$, $I_1 = \{1\}$.

(ii) ($i = 2$) In the reduced problem (i.e., fix $x_1 = 1$) the Martello and Toth upper bound after fixing x_2 to 0 is, $UB_2 = \max\{116\frac{2}{3}, 110\} = 116\frac{2}{3}$. Since $UB_2 < LB$, include 2 in I_1. So $I_1 = \{1,2\}$.

(iii) ($i = 3$) Since $a_3 > b - \sum_{j \in I_1} a_j = 20$, fix x_3 to 0. Hence $I_0 = \{3\}$.

Step 2 (Construct I_0)

(i) ($i = 4$) In the reduced problem (i.e., fix x_1 and x_2 to 1, and x_3 to 0), the upper bound after fixing x_4 to 1 is, $UB_4 = 133\frac{1}{3}$. Since $UB_4 > LB$, x_4 is not fixed.

(ii) ($i = 5$) Since $a_5 > b - \sum_{j \in I_1} a_j = 20$, fix x_5 to 0, or $I_0 = \{3,5\}$.

(iii) ($i = 6$) Since $a_6 > b - \sum_{j \in I_1} a_j = 20$, fix x_6 to 0, or $I_0 = \{3,5,6\}$.

(iv) ($i = 7$) After fixing x_7 to 1, $UB_7 = 133$. Hence x_7 is not fixed.

The reduced problem is therefore:

$$
\begin{array}{lrcrcl}
\text{maximize} & 120 & + & 10x_4 & + & 3x_7 \\
\text{subject to} & & & 10x_4 & + & 10x_7 \leq 20 \\
\text{and} & & & \multicolumn{3}{l}{x_j = 0 \text{ or } 1 \quad (j = 4,7),}
\end{array}
$$

which by inspection has optimal solution $x_4 = 1$ and $x_7 = 1$. Hence $\mathbf{x} = (1,1,0,1,0,0,1)$ is the optimal solution to the original knapsack problem. In general, however, the reduced knapsack problem may be solved by a branch and bound algorithm as discussed in the next section.

12.5.5 The Branch and Bound Algorithm

One of the earliest Branch and Bound algorithms for the knapsack problem, proposed by Kolesar [372], is a straightforward specialization of the Land and Doig algorithm for the general integer programming problem. The algorithm used the node with the highest upper bound (LP solution) as the node selection rule, and the variable at fractional value in the LP solution provides the branching variable. This algorithm, however, normally requires large computer memory because of its breadth-first nature of search causing many dangling nodes. The storage requirement was greatly reduced by Greenberg and Hegerich [295] who proposed a depth-first search by replacing the node selection rule to one that selects the most recently created node. Since then various authors (Horowitz and Sahni [501], Fayard and Plateau [196], Nauss [442], Martello and Toth [413],[414], etc.) have suggested variations of the Greenberg-Hegerich scheme which exploit special characteristics of the knapsack problem to make the search computationally more efficient. In this section we discuss the algorithm given by Martello and Toth [414], which like Greenberg and Hegerich, selects the most recently created node for further branching but uses the free variable with the smallest index (assumming the ordering $c_1/a_1 \geq c_2/a_2 \geq \ldots \geq c_n/a_n$) as a branching variable. The algorithm exploits the following remarks.

1. Given a particular node with free variables $(x_1, x_2, \ldots, x_n)$ and x_t, the variable with the fractional value in the optimal LP solution, then the optimal LP solution remains unchanged at each of the $(t-1)$ nodes created by taking forward steps by successively fixing variables $x_1, x_2, \ldots, x_{t-1}$ to value 1. Hence if the free variable with the smallest index is used as a branching rule, we may take $(t-1)$ forward steps at one time by fixing $x_j = 1$ for $j = 1, 2, \ldots, (t-1)$.

2. If we define $M_i = \min\{a_j \mid i < j \leq n\}$, then given a particular node with free variables $(x_1, x_2, \ldots, x_n)$ and right hand side b, if for any i, $M_i > b$, successive forward steps may be taken by fixing all free variable x_j $(j > i)$ to value 0.

3. Suppose we use the free variable with the smallest index as the branching variable and take a forward step by first fixing it to value 1 and then to value 0. Then consider a node N at level ℓ (i.e., the first ℓ variables are fixed to value 1 or 0). As illustrated in Figure 12.9, if a backward step is taken from this node, we move up the tree until variable x_k fixed at value 1 is found. Then proceeding with our branching strategy, we set x_k to 0 creating node Q at level k. Let $R =$

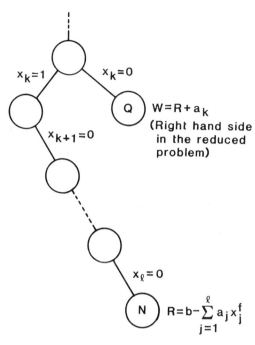

Figure 12.9: Illustration of Remark 3

$b - a_k - \sum_{j=1}^{k-1} a_j x_j^f$, where x_j^f is the value of variables x_j $(j = 1, \ldots, k-1)$ fixed at the node. Now consider the case when $R < M_k$.

(a) If $a_{k+1} - a_k = 0$, then for node Q to produce a better solution than node N, x_{k+1} cannot take value 1. This follows because $c_k/a_k \geq c_{k+1}/a_{k+1}$ and $a_{k+1} = a_k$.

(b) If $a_{k+1} - a_k > R$, then for node Q to produce a feasible solution, x_{k+1} should be equal to 0.

(c) If $a_{k+1} - a_k < 0$, then for node Q to produce a better solution, x_{k+1} should be equal to 0 whenever $R - (a_{k+1} - a_k) < M_{k+1}$. This is because if x_{k+1} is fixed to 1 at node Q, all other free variables will have to be fixed to 0, as their coefficient a_j is larger than the remaining right hand side. However, since $c_k/a_k \geq c_{k+1}/a_{k+1}$ and $a_{k+1} < a_k$, the solution at node Q cannot be better than the solution at node N.

(d) If LB is the current best value for the lower bound, then a backward step may be taken from node Q if $(\sum_{j=1}^{k-1} c_j x_j^f) + W(c_{k+1}/a_{k+1}) < LB$, where $W = R + a_k$. This follows because at node Q, $\sum_{j=1}^{k-1} c_j x_j^f$ is the contribution of the fixed variables to the value of the objective function and $W(c_{k+1}/a_{k+1})$ is an upper bound on the contribution of the free variables.

Notice that (a), (b), or (c) results in $x_{k+1} = 0$ and defines a forward step from node Q. In this case, we should reapply (a),(b), and (c) successively to $x_{k+2}, \ldots, x_n$ until a variable is not set to 0.

We now list the Martello and Toth [414] algorithm which is of the branch and bound type with depth first search, specialized for the 0-1 knapsack problem via the results previously discussed. We assume that $\sum_{j=1}^{n} a_j > b$ (otherwise $x_j = 1$, $j = 1, \ldots, n$ is the optimal solution), $a_j \leq b$ $(j = 1, \ldots, n)$ (otherwise x_j can be fixed to 0), and the ordering $c_1/a_1 \geq c_2/a_2 \geq \ldots \geq c_n/a_n$.

Step 1 (Initialization)

Find the smallest index t such that $\sum_{j=1}^{t} a_j > b$. Then the LP optimal solution is

$$
x_k = \begin{cases} 1 & k < t \\ (b - \sum_{j=1}^{t-1} a_j)/a_t & k = t \\ 0 & k > t. \end{cases}
$$

If the LP solution is all integer, it is optimal for the knapsack problem; stop. Otherwise the current best integer feasible solution $\mathbf{x}^* = (x_k^*)$ is

$$
x_k^* = \begin{cases} 1 & k < t \\ 0 & k \geq t \end{cases}
$$

with the value of the objective function equal to $z^* = \sum_{j=1}^{t-1} c_j$. Set the lower bound LB on the value of the objective function to z^*. Let M_n be arbitrarily large and $M_i = \min\{a_k | i < k \leq n\}$ for $i = 1, \ldots, n - 1$. Also, find the upper bound UB for the value of the objective function using the Martello and Toth method. Set level $\ell = t$ and create a node by fixing variables $x_j = 1$, $j = 1, \ldots, t - 1$ and $x_t = 0$. If $t = n$, stop. Otherwise, go to Step 2.

Step 2 (LP solution)

At the current node, the subproblem in the free variables can be written as

$$\sum_{j=1}^{\ell} c_j x_j^f \quad + \quad \text{maximize} \quad \sum_{j=\ell+1}^{n} c_j x_j$$

$$\text{subject to} \quad \sum_{j=\ell+1}^{n} a_j x_j \;\le\; b - \sum_{j=1}^{\ell} a_j x_j$$

$$\text{and} \quad x_j = 0 \text{ or } 1 \;\; (j = \ell+1, \ldots, n),$$

where x_j^f is the value of the variables fixed. If $a_{\ell+1} > b - \sum_{j=1}^{\ell} a_j x_j^f$, then fix $x_{\ell+1}$ to 0, set level $\ell = \ell + 1$, and repeat Step 2. Otherwise, find the LP optimal solution of the reduced problem at the current node; that is, if t is the smallest index such that $\sum_{j=\ell+1}^{t} a_j > b - \sum_{j=1}^{\ell} a_j x_j^f$, the LP solution to the reduced subproblem is

$$x_k^0 = \begin{cases} 1 & \ell < k < t \\[2mm] (b - \sum_{j=1}^{\ell} a_j x_j^f - \sum_{j=1}^{t-1} a_j)/a_t & k = t \\[2mm] 0 & k > t \end{cases}$$

with the value of the objective function equals $z^0 = \sum_{j=1}^{\ell} c_j x_j^f + \sum_{j=\ell+1}^{t} c_j x_j^0$. Consider the following possibilities:

(a) if $z^0 \le LB$, go to Step 4.

(b) if $z^0 > LB$, and if $\sum_{j=1}^{\ell} c_j x_j^f + \sum_{j=\ell+1}^{t-1} c_j > LB$, a better lower bound solution is found. Namely, set

$$x_j^* = \begin{cases} x_j^f & j \le \ell \\ 1 & \ell < j < t \\ 0 & \text{otherwise} \end{cases}$$

with $LB = \sum_{j+1}^{n} c_j x_j^*$. Otherwise, the lower bound remains the same. In either case, if $x_t^0 = 0$, that is the LP solution is all integer, go to Step 4; otherwise go to Step 3.

Step 3 (Forward step)

If $\ell = n$, go to Step 4. Otherwise, as per Remark 1, take a forward steps by successively fixing $x_{\ell+1}, \ldots, x_{t-1}$ to value 1 and x_t to 0, and go to Step 2 with level $\ell = t$.

Step 4 (Backward step)

To take a backward step (cf. Fig. 12.9), find the largest $k < \ell$ for which $x_k^f = 1$. If no such k exists, stop. Otherwise, let $R = b - \sum_{j=1}^{k} a_j x_j^f$, and $W = R + a_k$. Fix x_k to value 0 and set $\ell = k$. If $R \ge M_k$ (computed in Step 1) go to Step 2; otherwise let $h = k + 1$ and go to Step 5.

Step 5 (Using Remark 3)

If $h > n$ or the condition of case (d) (cf. Remark 3) is satisfied, that is, $LB \ge \sum_{j=1}^{\ell} c_j x_j^f + W(c_h/a_h)$, go to Step 4. Otherwise, set $D = a_h - a_k$; three possibilities now exist; namely,

(a) $D = 0$: As discussed in case (a) of Remark 3, set $x_h = 0$, $\ell = \ell + 1$, and repeat Step 5.

(b) (b) $D > 0$: If $D > R$ or $LB \ge \sum_{j=1}^{\ell} c_j x_j^f + c_h$, then due to case (b) in Remark 3, set $x_h = 0$, $\ell = \ell + 1$, and repeat Step 5. Otherwise, a better lower bound solution $x_j^* = x_j^f$ $(j = 1, \ldots, \ell)$, $x_h^* = 1$, and $x_j^* = 0$ $j > h$ is found with $LB = \sum_{j=1}^{n} x_j^*$. Set $\ell = \ell + 1$, $x_h = 0$, $R = R - D$, $k = h$, $h = h + 1$ and repeat Step 5.

(c) $D < 0$: If $R - D < M_h$, then due to case (c) in Remark

3, set $x_h = 0$, $\ell = \ell + 1$ and repeat Step 5. Otherwise, if $LB \geq \sum_{j=1}^{\ell} c_j x_j^f + c_h + (R - D)(c_{h+1}/a_{h+1})$, this node can not produce a better integer solution and hence we go to Step 4. Otherwise, set $x_h = 1$, $\ell = \ell + 1$, and go to Step 2.

The vector $\{x_j^*\}$ and LB gives the optimal integer solution and the corresponding value of the objective function when the algorithm stops. We illustrate the algorithm below.

Example 12.8 (Martello and Toth [414])
Consider the knapsack problem below.

$$\text{maximize} \quad 70x_1 + 20x_2 + 39x_3 + 37x_4 + 7x_5 + 5x_6 + 10x_7$$

$$\text{subject to} \quad 31x_1 + 10x_2 + 20x_3 + 19x_4 + 4x_5 + 3x_6 + 6x_7 \leq 50$$

$$\text{and} \qquad x_j = 0 \ \text{ or } \ 1 \ (j = 1, \ldots, 7).$$

Step 1: $t = 3$. The LP solution is $(1, 1, 9/20, 0, 0, 0, 0)$ and the knapsack solution from which the lower bound is found is $\mathbf{x}^* = (1, 1, 0, 0, 0, 0, 0)$ with $LB = 90$, $UB = 107$, and $(M_i) = (3, 3, 3, 3, 3, 6, \infty)$. Create Node 2 by fixing x_1 and x_2 to value 1 and x_3 to value 0. Set $\ell = 3$.

Step 2: $a_1 + a_2 + a_4 = 60 > 50$, hence from Node 2, $x_4 = 1$ cannot produce a feasible solution, so create node 3 by fixing x_4 to value 0, set $\ell = 4$.

Step 2: $t = 7$. The LP solution to the reduced problem (formed by setting $x_1 = x_2 = 1$, and $x_3 = x_4 = 0$) is $x_5 = x_6 = 1$, and $x_7 = 1/3$ with $z^* = 1051/3$. Also, the integer solution $(1, 1, 0, 0, 1, 1, 0)$ with objective function value 102 is a better feasible solution and thus provides a better lower bound. Thus set $LB = 102$ and $\mathbf{x}^* = (1, 1, 0, 0, 1, 1, 0)$.

Step 3: Create Node 4 by fixing x_5 and x_6 to 1 and x_7 to 0. Set $\ell = 7$. Since $\ell = n$, go to Step 4.

Step 4: $k = 6$. Create Node 5 with $x_1 = x_2 = x_5 = 1$, and $x_3 = x_5 = x_6 = 0$. Level $\ell = 6$, $R = 2$ and $W = 5$. Set $h = 7$, go to Step 5.

Step 5: $D = 3$. Since $D > R$, set x_7 to 0 and go to Step 4.

Step 4: $k = 5$. Create Node 6 with $x_1 = x_2 = 1$ and $x_3 = x_4 = x_5 = 0$. Level $\ell = 5$, $R = 5$, and $W = 9$.

Step 2: The LP optimal solution to the reduced problem is an all integer solution with $x_1 = x_2 = x_6 = x_7 = 1$, $x_3 = x_4 = x_5 = 0$, and $z^* = 105$. The new integer solution is $\mathbf{x}^* = (1,1,0,0,0,1,1)$ with the better lower bound $LB = 105$.

Step 4: $k = 2$. Create Node 7 with $x_1 = 1$ and $x_2 = 0$. $R = 9$, $W = 19$, and $\ell = 2$.

Step 2: $a_3 + a_1 > 50$, set $\ell = 3$, and create node 8 by fixing x_3 to 0.

Step 2: An integer optimal LP solution $(1,0,0,1,0,0,0)$ is found with the value of the objective function equal 107. Hence $\mathbf{x}^* = (1,0,0,1,0,0,0)$ and $LB = 107$. Since $LB = UB$, stop.

Fig. 12.10 gives the corresponding tree, where only nodes which are evaluated are numbered, and the numbering is in the order of their creation.

Problems

Problem 12.1   Consider the capital budgeting models appearing in Section 12.2.1. Suppose that the investment situation can be divided into several periods with costs and available funds broken down to period requirements so that the multiperiod Lorie-Savage model may be used. Can we transform this model to the single period one by letting $a_j = \sum_t a_{tj}$, $b_j = \sum_t b_{tj}$, and $c = \sum_t c_t$? Why or why not? What are the economic implications?

Problem 12.2   Give a set of dynamic programming recursive equations for the multiperiod Lorie-Savage capital budgeting problem (Section 12.2.1). Use these equations to solve the following two period problem with $T = 2, n = 4$.

j	1	2	3	4	
c_j	3	2	1	2	
a_{1j}	10	12	8	18	$21 = b_1$
a_{2j}	8	9	6	14	$16 = b_2$

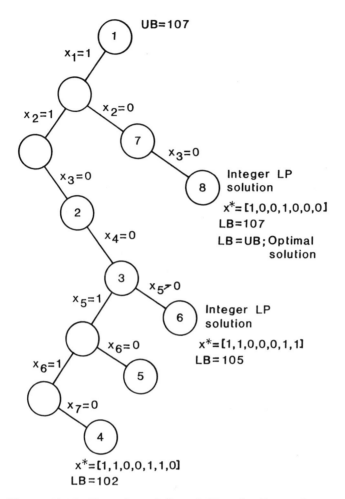

Figure 12.10: Branch and Search Tree for Example 12.8

Does the dynamic programming procedure seem efficient when T is large? When n is large?

Problem 12.3 The multiperiod capital budgeting problem (Section 12.2.1) is a zero-one integer program with a nonnegative constraint matrix. If it is to be solved by a search enumeration (Chapter 9), can the nonnegativity of the coefficients be used to reduce computational effort?

Problem 12.4 Suppose an integer programming model has a full set of multiple choice constraints as in Figure 12.1. How can these be used to reduce computational effort when the problem is to be solved by an enumerative algorithm? By a cutting plane algorithm? By a group theoretic algorithm? Illustrate your answers.

Problem 12.5 In the one dimensional cutting stock problem (Section 12.2.2) what is the largest waste that may appear on an optimal cutting pattern? Why?

Problem 12.6 Consider the following one dimensional cutting stock problem (Section 12.2.2), where 10 rolls, each of length 50, are available,

length	2	4	5	8	10
demand	10	10	15	10	5

If the cost of each roll is the same, develop the integer program which represents this problem. Give what you believe to be an optimal or near optimal solution. Explain why you selected it.

Problem 12.7 Consider the following procedure for finding an initial basic feasible solution to the linear program which solves the one dimensional cutting stock problem (Section 12.2.2):

List a sufficient number of cutting patterns (columns) until a linear programming Phase I routine gives a basic feasible solution.

a) Does this approach seem computationally efficient? Why or why not?

b) Can particular columns be listed so that a basic feasible solution appears immediately? For example, can identity columns be used; are they acceptable cutting patterns?

c) Discuss other techniques for finding a basic feasible solution. Can the costs be taken into account?

Problem 12.8 (Gilmore and Gomory [237]) Show that when the rectangles (l_i, w_i) of worths p_i $(i = 1, \ldots, m)$ are given, the two dimensional knapsack functions $f(L, W)$ defined from them (Section 12.2.2) satisfy the following three sets of inequalities:

i) $f(L, W) \geq 0$

ii) $f(L_1 + L_2, W) \geq f(L_1, W) + f(L_2, W)$
$f(L, W_1 + W_2) \geq f(L, W_1) + f(L, W_2)$

iii) $f(l_i, w_i) \geq p_i \ (i = 1, \dots, m)$.

Is f the only function that can satisfy these inequalities?

Problem 12.9   What difficulty arises when the cut in the cutting stock problem is not restricted to straight-across cuts? In particular, what if cuts of the forms below are allowed?

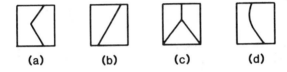

(a)   (b)   (c)   (d)

Problem 12.10   Discuss the nature of an algorithm which solves the integer program:

$$
\begin{aligned}
\text{Maximize} \quad & \mathbf{cx} \\
\text{subject to} \quad & \mathbf{Ax} \leq \mathbf{b} \\
& \mathbf{x} \geq \mathbf{0} \text{ and integer,}
\end{aligned}
$$

by solving the knapsack problem:

$$
\begin{aligned}
\text{Maximize} \quad & \mathbf{cx} \\
\text{subject to} \quad & \mathbf{wAx} \leq \mathbf{wb} \\
& \mathbf{x} \geq \mathbf{0} \text{ and integer,}
\end{aligned}
$$

where $\mathbf{w}$ is a nonnegative vector of numbers and $\mathbf{A}$ is an m by n matrix. Does the procedure seem efficient? Can anything be said about the relative size of the *linear programming* feasible regions?

Problem 12.11   When is it not possible to aggregate the constraints of an integer program so that it can be solved as a knapsack problem?

When the aggregation is not possible, can anything be said about the optimal solution to the integer program?

Problem 12.12   Suppose the system of equations $\mathbf{A}\mathbf{x} \leq \mathbf{b}$ is to be aggregated into a single equation $\mathbf{w}\mathbf{A}\mathbf{x} \leq \mathbf{w}\mathbf{b}$ which has the same set of nonnegative integer solutions (Section 12.3). Show that the multiplier w_i $(i = 1, \ldots, m)$ is the coefficient of the slack variable s_i $(i = 1, \ldots, m)$ in the aggregated equation. Here, $\mathbf{w} = (w_1, \ldots, w_m)$, $\mathbf{s} = (s_1, \ldots, s_m) \geq \mathbf{0}$, and $\mathbf{s} = \mathbf{b} - \mathbf{A}\mathbf{x}$.

Problem 12.13   (Glover [256]) Aggregate the following two equations so that the single equation has the same set of nonnegative integer solutions as the original two:

$$
\begin{array}{rcrcrcl}
7x_1 & + & 9x_2 & + & 5x_3 & = & 84 \\
6x_1 & + & 7x_2 & + & 5x_3 & = & 72.
\end{array}
$$

Problem 12.14   (Glover and Woolsey [267]) Aggregate the following system of equations, using the technique suggested by Theorem 12.2, so that the single equation has the same set of nonnegative integer solutions as the system below. What are the multipliers?

$$
\begin{array}{rcrcrcrcrcrcl}
2x_1 & + & 4x_2 & + & x_3 & + & x_4 & & & & & = & 9 \\
7x_1 & + & 2x_2 & + & 4x_3 & & & + & x_5 & & & = & 13 \\
3x_1 & + & 5x_2 & + & 6x_3 & & & & & + & x_6 & = & 17.
\end{array}
$$

Problem 12.15   (Glover and Woolsey [267]) For the system of two equations, where (i) has all nonnegative coefficients with $b_1 > 0$, and (ii) has all positive integer coefficients

$$
s_1 \equiv \sum_{j=1}^{n} a_{1j}x_j = b_1 \tag{i}
$$

$$
s_2 \equiv \sum_{j=1}^{n} a_{2j}x_j = b_2, \tag{ii}
$$

prove that

    a) part (a) of Theorem 12.1 is true.

b) part (b) of Theorem 12.2 is true for all positive integers t satisfying

$$t > b_1 - \operatorname*{minimum}_{j} \{a_{1j}\}$$

and

$$t > b_2 \operatorname*{maximum}_{j} \{(a_{1j}/a_{2j}) - (b_1/Za_{2j})\},$$

where

$$Z = \operatorname*{maximum}_{\delta = 0,1} \left\{ (\delta b_1 + b_2)/ \operatorname*{minimum}_{j} (\delta a_{1j} + a_{2j}) \right\}.$$

(*Hint*: The direction of the proof is very similar to the one for Mathews' theorem.)

Problem 12.16 For the result given in the previous problem:

a) Show that any value of Z will do as long as $Z \geq \sum_j x_j$ holds for all nonnegative integer solutions x_j satisfying the aggregated equations ((iii) in Theorem 12.1).

b) Use (a) to show that the theorem can be strengthened by specifying that

$$Z = (b_1 + b_2 t)/ \operatorname*{minimum}_{j} (a_{1j} + a_{2j})t.$$

Is it difficult in computations to determine the smallest t satisfying the conditions of the previous result for Z so defined?

c) Is the t found here always smaller than the one found using Theorem 12.1? Than the one found in Theorem 12.2?

d) What aggregation procedure is suggested by the previous problem? Does it seem more efficient than the one suggested by Theorem 12.1? By Theorem 12.2?

Problem 12.17 (Glover and Woolsey [267]) Assume that (i) and (ii) in Problem 12.15 have nonnegative integer coefficients with b_1 and b_2 positive. Let t_1 and t_2 be two positive integers satisfying the following conditions:

A) t_1 and t_2 are relatively prime,

B) t_1 does not divide b_2 and t_2 does not divide b_1,

C) $t_1 > b_2 - a_2$ and $t_2 > b_1 - a_1$, where a_i denotes the smallest of the *positive* a_{ij} $(j = 1, \ldots, n)$.

a) Show that the set of nonnegative integer solutions to (i) and (ii) is the same as the set of nonnegative solutions to $t_1 s_1 + t_2 s_2 = t_1 b_1 + t_2 b_2$ (*Hint*: Write the aggregated equation in the form $s_1 = b_1 - (s_2 - b_2) t_2 / t_1$. (A) implies $s_2 - b_2 = t_1 q$ and $s_1 = b_1 - t_2 q$ for some positive integer q. Using (B) and (C), show that $q = 0$.)

b) Discuss the relative efficiency of the aggregation process suggested here with the ones suggested by Theorem 12.1, Theorem 12.2, and by Problem 12.15.

Problem 12.18 Aggregate the two equations appearing in Problem 12.13, using:

a) the procedure indicated by Problems 12.15 and 12.16, and

b) the result appearing in the previous problem.

c) For these equations, which aggregation process is "best"? Why?

Problem 12.19 For dynamic programming recursions I and II (Section 12.4.1), prove that the solution making up $f(k, g)$ has no slack if and only if $f(k, g) \neq -\infty$, where $k = 1, \ldots, n$, $g = 1, \ldots, b$, $f(0, 0) = 0$, and $f(0, g) = -\infty$ for $g = 0, 1, \ldots, b$.

Problem 12.20 How can alternate optimal solutions to the knapsack problem be found when:

a) a dynamic programming algorithm is used (Sections 12.4.1, 12.4.2)?

b) an enumerative algorithm is used (Section 12.4.3)?

c) a network approach is used (Section 12.4.5)?

Problem 12.21   Draw a flow chart for each of the following algorithms. Discuss what has to be stored if each algorithm is computer programmed. Can anything be said about the relative amount of storage required? About the relative efficiency of the algorithm?

a) Dynamic programming recursion I (Section 12.4.1).

b) Dynamic programming recursion II (Section 12.4.1).

c) Dynamic programming recursion III (Section 12.4.2).

d) A branch and bound enumeration (Section 12.4.3).

e) A Lagrangian multiplier method (Section 12.4.4).

Problem 12.22

a) Explain how dynamic programming recursion I (Section 12.4.1) can be used to solve the following nonlinear knapsack problem:

$$\begin{aligned}
\text{Maximize} \quad & h(\mathbf{x}) \\
\text{subject to} \quad g(\mathbf{x}) \leq{} & b \\
\mathbf{x} \geq{} & \mathbf{0} \text{ and integer,}
\end{aligned}$$

where $h(\mathbf{x})$ and $g(\mathbf{x})$ are real valued functions in the vector $\mathbf{x}$, $g(\mathbf{x}) \geq 0$ for all nonnegative $\mathbf{x}$, and b is a positive number.

b) Use this procedure to solve the following problem:

$$\begin{aligned}
\text{Maximize} \quad h(\mathbf{x}) ={} & 3x_1 x_2 \\
\text{subject to} \quad g(\mathbf{x}) ={} & x_1^2 + x_2^2 \leq 5 \\
\mathbf{x} ={} & (x_1, x_2) \geq \mathbf{0} \text{ and integer.}
\end{aligned}$$

Check your solution graphically.

Solve Problems 12.23 through 12.27 using the technique indicated. All problems are of the form:

$$\text{Maximize} \quad \sum_{j=1}^{n} c_j x_j$$

$$\text{subject to} \quad \sum_{j=1}^{n} a_j x_j \leq b$$

$$x_j \geq 0 \text{ and integer} \quad (j = 1, \ldots, n).$$

Problem 12.23 $n = 4, b = 10$.

c_j	1	1	2	3
a_j	4	7	7	3

Use a dynamic programming algorithm.

Problem 12.24 $n = 5, b = 24$.

c_j	1	2	3	2	1
a_j	7	7	5	6	9

Use a branch and bound algorithm. Draw the associated tree.

Problem 12.25 $n = 8, b = 105$.

c_j	−5	−7	10	−3	−1	0	0	0
a_j	10	73	254	113	147	18	4	103

Use a Lagrangian multiplier method to get a "good" solution. (*Hint*: This is the knapsack equivalent of Example 12.1.)

Problem 12.26   As in Problem 12.24, solved by a dynamic programming algorithm.

Problem 12.27   As in Problem 12.23, solved (by inspection) as a shortest route problem.

Problem 12.28   As in Problem 12.24, but with $b = 20$ and additional constraint $x_j = 0$ or 1 $(j = 1, 2, 3, 4, 5)$, solved by branch and bound algorithm of Section 12.5.

Problem 12.29   List the branch and bound algorithm discussed in Section 12.4.3. Use as little notation as possible, and be clear and concise. Why must the algorithm converge?

Problem 12.30

a) Explain why Theorem 12.4 is applicable to the general optimization problem

$$\text{maximize} \quad h(\mathbf{x})$$
$$\text{subject to} \quad g_i(\mathbf{x}) \le b_i \quad (i = 1, \ldots, m),$$
$$\text{and} \quad \mathbf{x} \text{ in } S,$$

where $h(\mathbf{x})$ and $g_i(\mathbf{x})$ are real valued functions in the vector $\mathbf{x}$, b_i is a scalar, and S is an arbitrary set (e.g., the set of nonnegative integers).

b) Use (a) to find an approximate solution to the integer nonlinear program appearing in Problem 12.22(b).

Problem 12.31 (Dantzig [155]) Consider the knapsack problem with zero-one variables:

$$\text{Maximize} \quad \sum_{j=1}^{n} c_j x_j$$
$$\text{subject to} \quad \sum_{j=1}^{n} a_j x_j \le b,$$
$$0 \le x_j \le 1,$$
$$\text{and} \quad x_j \text{ integer} \quad (j = 1, \ldots, n),$$

where the costs and weights are positive. Show that if we plot (a_j, c_j) in two-space, the linear programming solution can be found by rotating a ray clockwise starting from the c_j axis until the sum of the weights swept out by the ray exceeds the weight limitation. If the weight limitation would first be exceeded for the jth item, the value of x_j is set to the fraction f_j that gives $a_j f_j$ exactly equal to b. With the exception of this one item, all items swept out by the ray have value 1, while those not swept have value 0. For example, in Fig. 12.11, $n = 5$, and the optimal solution is

$$x_1 = x_3 = 1, x_2 = x_4 = 0, \quad \text{and} \quad x_5 = 1/a_5(b - a_1 - a_3).$$

Problem 12.32 For $g = 0, 1, \ldots, b$ define

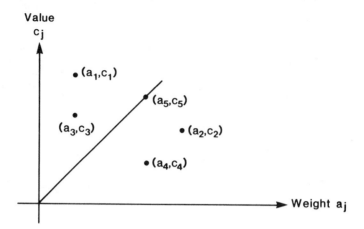

Figure 12.11: (a_j, c_j) Plot for Problem 12.31

$$f(n, g) \;=\; \text{maximum} \;\sum_{j=1}^{n} c_j x_j$$
$$\text{subject to} \;\sum_{j=1}^{n} a_j x_j \le g$$
$$x_j \ge 0 \;\text{ and integer }\; (j = 1, \ldots, n),$$

where the c_j and a_j are positive integers $(j = 1, \ldots, n)$. Consider the expression

$$f(n, g) = \text{maximum} \; \{c_j + f(n, g - a_j)\}, \tag{1}$$
$$\hspace{2cm} J$$

where $J = \{j | g - a_j \ge 0, j = 1, \ldots, n\}$.

a) Explain why (1) is a dynamic programming recursive relationship which allows us to find $f(n, b)$ and thus to solve the knapsack problem. Is it more efficient than recursions I, II, and III
   presented in Section 12.4.1? Why?

b) Can (1) be used to solve a knapsack problem with bounded
   variables?

c) Solve Problem 12.23 using (1).

d) Show that expression (14) (Section 12.4.2) can be derived using (1). (*Hint*: Using (1) substitute for $f(n, b)$ in the expression $D(b) = (c_1/a_1)b - f(n, b)$. Add and subtract $(c_1/a_1)a_j$ to the right-hand side in the resulting equation and appeal to the fact that $-\max t = \min(-t)$).

Chapter 13

The Set Covering Problem, the Set Partitioning Problem

13.1   Introduction

The *set covering problem* (SC) is the zero-one integer program

$$
\begin{array}{lll}
\text{minimize} & \mathbf{cx} & \\
\text{subject to} & \mathbf{Ex} \ \geq \ \mathbf{e} & \quad (1) \\
\text{and} & x_j \ = \ 0 \text{ or } 1 \ \ (j = 1, \ldots, n), & \quad (2)
\end{array}
$$

where $\mathbf{E} = (e_{ij})$ is an m by n matrix whose entries e_{ij} are 0 or 1, $\mathbf{c} = (c_j)$ $(j = 1, \ldots, n)$ is a cost row with positive components, $\mathbf{x} = (x_j)$ $(j = 1, \ldots, n)$ is a vector of zero-one variables, and $\mathbf{e}$ is an m vector of 1's. If the $\mathbf{Ex} \geq \mathbf{e}$ constraints are replaced by the equalities

$$
\mathbf{Ex} = \mathbf{e}, \qquad\qquad (1)'
$$

the integer program is referred to as a set *partitioning problem* (SP). Note that, in either case, the constraints (2) are equivalent to $\mathbf{x} \geq \mathbf{0}$ and integer, since in any minimal solution these conditions imply $x_j = 0$ or 1 for $j = 1, \ldots, n$.[1] Also, if we think of the columns of $\mathbf{E}$ and $\mathbf{e}$ as sets, the set covering problem is equivalent to finding a cheapest union of sets from $\mathbf{E}$ that covers every component of $\mathbf{e}$, where component i of $\mathbf{e}$

---

[1]This is evident in the set partitioning problem. For the set covering problem, suppose $\mathbf{x}$ is optimal and it has a component x_j greater than 1. Then x_j can be reduced to value 1 and the constraints (1) are still satisfied; as the costs are positive, we have an improved solution contradicting that $\mathbf{x}$ is optimal.

is covered if at least one of the selected sets (columns) from $\mathbf{E}$ has a 1 in row i. In the set partitioning problem, we seek a cheapest union of disjoint sets from $\mathbf{E}$ which equals $\mathbf{e}$. Hence, the "set" reference in the problems. [2]

As indicated in Chapter 1 (Section 1.2.2) the set covering problem and the set partitioning problem represent numerous real-world situations. In some of the applications the right hand side is a positive integer vector $\mathbf{d}$ instead of the column of ones $\mathbf{e}$, and/or the variables $\mathbf{x}$ can take on any integer value (Section 13.2.4). Also, in some instances, in addition to the constraints (1) or (1)' and (2), a third set of constraints appears. These constraints are often of the form

$$\mathbf{L} \leq \mathbf{Dx} \leq \mathbf{U}, \tag{3}$$

where $\mathbf{D}$ is an m' by n nonnegative matrix having row diagonal structure (Fig. 13.1). $\mathbf{L}$ is a nonnegative m' vector, and $\mathbf{U}$ is a positive m' vector. The inequalities (3) are sometimes referred to as *base constraints* (see Section 13.3.1) and a programming problem with (1) (or (1)'), (2), and (3) *as a base constrained set covering (or partitioning) problem*. Because of the extensive applicability of the set covering and set partitioning model, considerable effort has been devoted to developing special purpose algorithms which take into account the positive costs, the zero-one constraint matrix, the right-hand side of 1's, and, when applicable, the structure of the base constraints (Fig. 13.1).

In the next section we discuss the relationship of the set covering and set partitioning problems to certain problems defined on a network. This is followed by a description of some of the more popular situations which are modeled as either a set covering or a set partitioning problem. The following part presents results which can be used in developing algorithms. Subsequently, descriptions of two apparently efficient techniques are given. The appendix lists computational experience of some computer implemented algorithms. We begin with a brief history of the literature.

Background

Certain network problems can be modeled as a set covering or a set partitioning problem with a particular $\mathbf{E}$ matrix (Section 13.2). The

---

[2]The covering problems may be related to sets in a different way; see Baliniski [50].

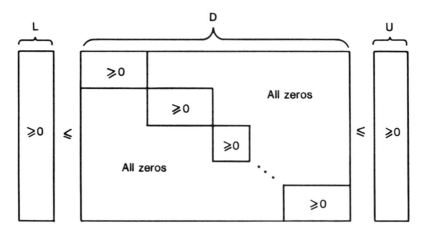

Figure 13.1: Structure of the Base Constraints

problems and formulations are in Balinski [50], Balinski and Quandt [48], Bellmore and Ratliff [77], and Garfinkel and Nemhauser [216], [217]. Efficient techniques for the network problems, which take advantage of the structure of **E**, are discussed by Edmonds [185], Edmonds and Johnson [186], and Norman and Rabin [447]. Extensions of these algorithms to any set covering or set partitioning problem are in Edmonds [184] and Ray-Chaudhuri [474].

As already mentioned, numerous situations have been modeled as set covering or a set partitioning problems. These include applications in the areas of airline crew scheduling [3] ([8], [424], [549], [561]), truck dispatching or vehicle routing ([1], [48], [137], [162], [386], [447]), political redistricting ([212], [214], [215], [589]), and information retrieval ([165]) (Section 13.3). The problems have also appeared in assembly line balancing ([520], [552]), capital investment ([583]), circuit design ([382]), map coloring ([117]), facilities location ([478], [571]), network attack and defense ([75], [76], [345]), personnel scheduling ([189]), PERT/CPM analysis ([138]), railroad crew scheduling ([127]), routing ([1], [393]), stock cutting ([462]), switching circuit scheduling ([50], [138], [332], [457], [484]), symbolic logic ([138], [483]), and other instances ([77], [214], [315], [502]).

Algorithms for the set covering and set partitioning problems are

---

[3]More sophisticated airline crew scheduling models which are not of the set covering or set partitioning type are discussed by Marshall and Tabak [408]. A non-analytical description of the problem appears in Rubin [495] and in Rothstein[488].

of the following type: enumerative, cutting plane, group theoretic, and heuristic. Enumerative techniques are described by Garfinkel and Nemhauser [214], [215] (Section 13.5), Guha [296], and Lemke, Salkin, and Spielberg [392] (Section 13.5, Originally this algorithm appeared in Salkin [502]), Marsten [409], [410], Michaud [429], Pierce [460], and Pierce and Lasky [463]. Cutting plane techniques are described by Bellmore and Ratliff [77] (Problems 13.20, 21, 22, 23), House, Nelson, and Rado [334] (Appendix A), and Salkin and Koncal [512], [513] (Problem 13.19). A group theoretic procedure is in Thirez [561], and heuristic methods are given by Ignizio and Harnett [338] and Rubin [496] (Problem 13.7). Also, a simplex type procedure, based on a result presented in Andrew, Hoffman, and Krabek [5], has been formulated by Balas and Padberg [39], [40] (Problem 13.10). Related work is in Padberg [453] and in Trubin [575].

Computational experience with computer programmed algorithms is given by Bellmore and Ratliff [76], Garfinkel and Nemhauser [214], Geoffrion [223], Lemke, Salkin, and Spielberg [392], Martin [415], Michaud [429], Pierce [460], Pierce and Lasky [463], Rao [472], Roth [486], and Salkin and Koncal [514]. The performances are summarized in Appendix A.

Survey articles describing applications, algorithms, and computations include Balinski [50], Balinski and Spielberg [51], Christofides and Korman [134], Garfinkel and Nemhauser [216], [217], and Salkin and Saha [517].

13.2   Set Covering and Networks

Certain problems defined on a network with undirected arcs [4] can be posed as either set covering or set partitioning problems, where the constraint matrix has particular structure. Recall that a *network* is a collection of nodes with arcs joining them. An arc joining the nodes r and s will be denoted by (r, s), where r and s are called the *endpoints* of the arc.

13.2.1   The Node Covering Problem

Define a *cover* to be a subset of arcs in the network such that each node

---

[4]An undirected arc is one which can be traversed in either direction.

of the network is an end point of at least one of the arcs in the subset. The *simple covering problem* is to find a cover with a minimum number of arcs.

To model the problem, number the arcs and let x_j represent the jth arc of the network, where $x_j = 1$ if the jth arc is in the cover and $x_j = 0$ otherwise. To make sure that each node i in the network is an end point of at least one arc in the cover, we must have $x_j = 1$ where the jth arc has node i as one of its end points. Thus, define $\mathbf{E} = (e_{ij})$ by letting $e_{ij} = 1$ if the ith node in the network is an endpoint of the jth arc, and $e_{ij} = 0$ otherwise. The problem of finding a minimal cover is then a set covering problem in which the costs are equal to 1. More generally, if there is a distinct cost associated with every arc, we have a *minimal cost covering problem* which is also a set covering problem.

Example 13.1 (Salkin and Saha [517]) Consider the network in Fig. 13.2. It has 5 nodes and 8 arcs. The simple covering problem is represented by the set covering problem below.

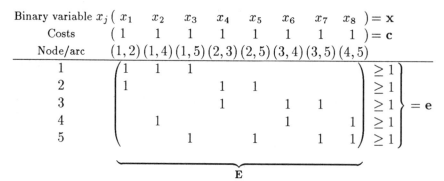

Binary variable x_j (	x_1	x_2	x_3	x_4	x_5	x_6	x_7	x_8) = x
Costs (	1	1	1	1	1	1	1	1) = c
Node/arc	$(1,2)$	$(1,4)$	$(1,5)$	$(2,3)$	$(2,5)$	$(3,4)$	$(3,5)$	$(4,5)$
1	1	1	1						≥ 1
2	1			1	1				≥ 1
3				1		1	1		≥ 1
4		1				1		1	≥ 1
5			1		1		1	1	≥ 1

$$\underbrace{}_{\mathbf{E}} \qquad \Big\} = \mathbf{e}$$

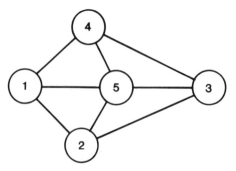

Figure 13.2: Network for Example 13.1

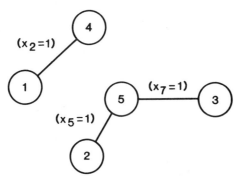

Figure 13.3: A Minimal Cover

By inspection, any three variables at value 1 and the remaining five variables at value 0 such that the $\mathbf{Ex} \geq \mathbf{e}$ constraints are satisfied is an optimal solution. One such point is $x_2 = x_5 = x_7 = 1$, and $x_1 = x_3 = x_4 = x_6 = x_8 = 0$. Thus, a minimal cover contains three arcs and the one singled out is in Fig. 13.3.

13.2.2 The Matching Problem (Balinski [50], Balinski and Quandt [48])

A *matching* for a network is a subset of the arcs in the network such that no two arcs in the subset have a common endpoint. That is, a subset of arcs is a matching if and only if each node in the network has at most one arc in the subset incident to it. The *simple matching problem* is to find a matching which has a maximum number of arcs.

To model this problem, we define the binary variables (x_j) and the $\mathbf{E}$ matrix as in the minimal covering problem. As a maximum matching is sought, the objective function is to maximize $\sum_j x_j$, and since each node is to be the endpoint of at most one arc, the constraints are $\mathbf{Ex} \leq \mathbf{e}$. Thus the simple matching problem is to

$$\text{maximize} \quad \sum_j x_j$$
$$\text{subject to} \quad \mathbf{Ex} \leq \mathbf{e}$$
$$\text{and} \quad x_j = 0 \text{ or } 1 \quad (j = 1, 2, \ldots),$$

which is a set partitioning problem with a maximization objective function [5]. If there is a weight c_j associated with each arc, the objective function becomes $\sum_j c_j x_j$ and we have a *weighted matching problem*. Note that assignment problem (cf. Chapter 3) is a special case of this problem.

Example (from Fig. 13.2)

For the network appearing in Fig. 13.2, the simple matching problem is represented by the integer program in Example 13.1 except that the $\geq$ signs are reversed.

By inspection, any two variables at value 1 and the remaining six variables at value 0 such that the $\mathbf{E}x \leq \mathbf{e}$ constraints are satisfied is an optimal solution. Two such solutions are (a) $x_2 = x_5 = 1$ and $x_1 = x_3 = x_4 = x_6 = x_7 = x_8 = 0$, and (b) $x_2 = x_7 = 1$ and $x_1 = x_3 = x_4 = x_5 = x_6 = x_8 = 0$. Thus, a maximum matching has two arcs and the two matchings singled out are in Fig. 13.4.

Remarks

1. Consider the relationship between the simple covering problem and the simple matching problem. Observe that a maximum matching has at most as many arcs as in a minimum cover. This follows since to obtain a maximum matching from a minimum cover we delete arcs from the cover so that each node has only one arc incident to it; or to obtain a minimum cover from a maximum matching we adjoin a minimum number of arcs to obtain a cover (Problems 13.4 and 13.5). To illustrate, consider the minimum cover appearing in Fig. 13.3. A maximum matching is obtained by omitting arc (5,3) (Fig. 13.4(a)) or by omitting arc (5,2) (Fig. 13.4(b)). Similarly, from the maximum matching in Fig. 13.4(a) (Fig. 13.4(b)), a minimum cover is obtained by adjoining an arc to node 3 (node 2). If arc (5,3) (arc (5,2)) is added we have the minimal cover appearing in Fig. 13.3.

2. In both problems the $\mathbf{E}$ matrix is specially structured. In particular, each column j has exactly two 1's. Moreover, the 1's occur in the

---

[5]The inequalities $\mathbf{E}x \leq \mathbf{e}$ are equivalent to $\mathbf{E}x + \mathbf{I}s = \mathbf{e}$ where $\mathbf{I}$ is an identity matrix and s is a vector of nonnegative slack variables. The zero-one requirements for x, the zero-one composition of $\mathbf{E}$, and the right-hand side of 1's imply that the slack variables are 0 or 1 in any integer solution.

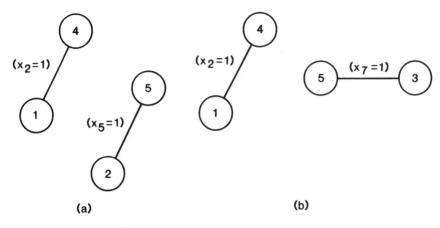

Figure 13.4: Maximum Matchings

two rows corresponding to the endpoints of the arc associated with the index j. (Such an array is often called a *node/arc incidence matrix*.) Because of this it is possible to devise relatively simple solution techniques (see Balinski [50], Edmonds [185], Edmonds and Johnson [186], and Garfinkel and Nemhauser [218]).

13.2.3 Disconnecting Paths

Given a network, define a *path* from a node s to a node t as a sequence of distinct nodes

$$s, i_1, i_2, \ldots, i_r, t,$$

where (s, i_1), (i_1, i_2), $\ldots$, (i_r, t) are arcs in the network. Suppose all the paths in a network are known, and also assume there is a cost associated with removing an arc from the network. The problem is to find a set of arcs which, if removed from the network, will disconnect all the paths from s to t with the total cost of the removed arcs minimized.

To model the problem, number the arcs and paths in the network. Define the binary variable $x_j = 1$ if arc j is to be removed, and let $x_j = 0$ otherwise. The constraint matrix is found by letting $e_{ij} = 1$ if the jth arc is in the ith path and $e_{ij} = 0$ otherwise. The situation is then represented by a set covering problem in which the **Ex** $\geq$ **e** inequalities ensure that an arc is omitted from each path.

Example (from Fig. 13.2)

Consider the network in Fig. 13.2. Suppose we wish to disconnect all paths from node 1 to node 3. The paths between these nodes are listed and numbered below.

i	paths
1	1, 4, 3
2	1, 5, 3
3	1, 2, 3
4	1, 4, 5, 3
5	1, 5, 4, 3
6	1, 2, 5, 3
7	1, 5, 2, 3
8	1, 4, 5, 2, 3
9	1, 2, 5, 4, 3

Thus, the representative set covering problem has the constraint matrix

Path/Arc	$(1,2)$ $(2,1)$	$(1,4)$ $(4,1)$	$(1,5)$ $(5,1)$	$(4,5)$ $(5,4)$	$(2,5)$ $(5,2)$	$(4,3)$ $(3,4)$	$(3,2)$ $(2,3)$	$(3,5)$ $(5,3)$
1		1				1		
2			1					1
3	1						1	
4		1		1				1
5			1	1		1		
6	1				1			1
7			1		1		1	
8		1		1	1		1	
9	1			1	1	1		

The disconnecting path problem can also be considered as what is termed a "minimum cut problem," which can be solved by the ef-

ficient labeling network algorithm developed by Ford and Fulkerson
[203]. Considerable effort is usually required to pose the disconnecting
path problem as a set covering problem and then to solve the resulting integer program, and this precludes the computational use of the
transformation given here.

13.2.4   The Maximum Flow Problem

Consider the directed network in Figure 13.5 depicting (e.g.) a pipeline
network between a refinery s and a terminal t. Each arc j represents
a pipeline segment with the arrowhead pointing in the direction of
gasoline flow and d_j is the maximum flow rate (*bbl*sec). The max flow
problem is to find the maximum flow possible from s to t through the
network without exceeding the arc capacities.

Define a directed path from node s to node t as a sequence of distint
nodes

$$s, i_1, i_2, \ldots, i_r, t$$

with a directed arc between each successive pair of nodes. The matrix
E below lists all the directed paths for this network. In the matrix,
each column corresponds to a directed path and an element e_{ji} of **E** is
1 (0) if the arc denoted by row j is on (is not on) path i.

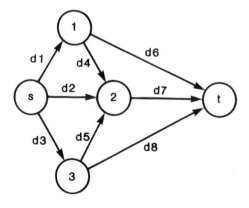

Figure 13.5: Directed Network

| | path(i) |
Arc (j)	1	2	3	4	5
1 : $(s,1)$	1	1	0	0	0
2 : $(s,2)$	0	0	1	0	0
3 : $(s,3)$	0	0	0	1	1
4 : $(1,2)$	0	1	0	0	0
5 : $(3,2)$	0	0	0	0	1
6 : $(1,t)$	1	0	0	0	0
7 : $(2,t)$	0	1	1	0	1
8 : $(3,t)$	0	0	0	1	0

If we define x_i as the amount of flow through path i, then the max flow problem is:

$$\text{Maximize} \quad \sum_{i=1}^{5} x_i$$
$$\text{subject to} \quad \mathbf{Ex} \leq \mathbf{d}$$
$$\text{and} \quad \mathbf{x} \geq \mathbf{0}, \quad \text{and integer,}$$

where $\mathbf{d} = (d_1, d_2, \ldots, d_8)$. The above is a generalized set partitioning problem where the variables can take on any non-negative integer value and the right hand side contains positive integers rather than ones.

13.3 Applications

In this section, a few specific applications are given. As indicated in Section 13.1, numerous other situations may be modeled as set covering or set partitioning problems.

13.3.1 Airline Crew Scheduling (Spitzer [549])

In this situation we have to assign a set of m flights which must be flown over a given time period to a set of N crews so that the cost of operations is minimized. Subject to time, geography, crew ability (e.g., not all pilots are licensed to fly Boeing 767 aircraft), and other limitations, the combinations of flights that each crew may take are enumerated.[6] Let $x_{k(t)}$ denote the tth way that the kth crew can make

---

[6]In practice this is done by a computer program (i.e., a matrix generator code) which takes the limitations as input. As there are usually an enormous number of combinations, those which "obviously" will not be used are discarded initially. After these omissions, a good sized problem may have several hundred constraints and, thousands of combinations (columns) for all crews.

the flights; $x_{k(t)} = 1$ if the kth crew is assigned the series of flights designated in the tth combination, and $x_{k(t)} = 0$ otherwise. Let $e_{ik(t)} = 1$ if the ith "flight leg" (e.g., Flight 347, Cleveland to Chicago) can be assigned to the kth crew in the tth combination, and let $e_{ik(t)} = 0$ otherwise. Since each flight leg must be assigned to exactly (or at least) one crew, we have the standard constraints

$$\sum_k \sum_t e_{ik(t)} x_{k(t)} = (\text{or} \geq) 1 \quad (i = 1, 2, \ldots, m), \tag{4}$$

where $x_{k(t)} = 0$ or 1 and $e_{ik(t)} = 0$ or 1. Let $c_{k(t)}$ be the cost of the tth combination of the kth crew. So the problem is to determine the value of the variables $x_{k(t)}$ such that $\sum_k \sum_t c_{k(t)} x_{k(t)}$ is minimized and the constraints (4) are satisfied.

Note that in this model we usually have the requirement that each crew be assigned to only one combination or series of flights. Hence the additional constraints

$$\sum_t x_{k(t)} \leq 1 \quad (k = 1, 2, \ldots, N) \tag{5}$$

may appear in the model (Fig. 13.6). Observe, however, that these inequalities are of the form $(\sum_t x_{k(t)} + s_k = 1)$ for each k, where s_k is a nonnegative slack variable which must be 0 or 1 whenever all the $x_{k(t)}$'s are binary. Thus, if these constraints are added and there is an equality in (4), we still have a set partitioning problem.

There are, nevertheless, situations in which additional crew base constraints of the form (3) (Section 13.1) have to be added. In particular, suppose the crews originate from one of (usually) a few home cities or crew bases. Then, because of flight time restrictions, equal distributions of flight assignments, union rules, and other factors, we may have these constraints, where the crews corresponding to the indices in the nonnegative row part of **D** (Fig. 13.1) originate from the same crew bases, and there are m' crew bases.

Example 13.2 Consider Fig. 13.6, which represents the case in which there are $N = 4$ crews, $m = 5$ flight legs, and $m' = 2$ crew bases. Crew 1 can take flights numbered 1,2, 4, and 5, or 1, 3, and 4, or 1 and 3. Crew 2 can be assigned to flights 1,2, and 3, or 1 and 2, and so on. We assume that crews 1 and 2 and crews 3 and 4 originate from the same crew base and that at most one crew from each base is to be used. Note that in this example, the crew base constraints imply the constraints (5) so that they can be dropped.

Crew (k)	1			2		3		4		
Combination (t)	1	2	3	1	2	1	2	1	2	
Crew base	1			1		2		2		
Variable $x_{k(t)}$	$x_{1(1)}$	$x_{1(2)}$	$x_{1(3)}$	$x_{2(1)}$	$x_{2(2)}$	$x_{3(1)}$	$x_{3(2)}$	$x_{4(1)}$	$x_{4(2)}$	
Flight leg (i)										
1	1	1	1	1	1	0	0	0	0	$\geq (=)1$
2	1	0	0	1	1	1	1	1	1	$\geq (=)1$
3	0	1	1	1	0	0	1	1	0	$\geq (=)1$
4	1	1	0	0	0	1	0	0	0	$\geq (=)1$
5	1	0	0	0	0	0	1	0	0	$\geq (=)1$
	1	1	1							≤ 1
				1	1					≤ 1
						1	1			≤ 1
								1	1	≤ 1
	1	1	1	1	1					≤ 1
						1	1	1	1	≤ 1

Figure 13.6: Airline Crew Scheduling Example

13.3.2 Truck Scheduling (Balinski and Quandt [48])

In this situation, there is a central warehouse with trucks, and m clients at different locations who have placed orders for goods at the warehouse. The problem is to transport the goods by trucks to the customers at a minimum cost of filling the orders. Given the m orders, the possible delivery schedules are determined on the basis of truck capabilities (e.g., size) and route constraints. For each truck, we may list all possible schedules (which may be a huge number) and assign a weight (or cost) to each. Thus, let N denote the total number of route possibilities and let $c_{k(t)}$ denote the cost of the delivery associated with the tth possibility of the kth truck. Suppose $x_{k(t)} = 1$ if the tth schedule for truck k is selected and let $x_{k(t)} = 0$ otherwise. Also, let $e_{ik(t)} = 1$ if the ith order is to be delivered in the tth schedule of truck k and let $e_{ik(t)} = 0$ otherwise; then the problem of transporting goods to m customers is a set partitioning problem where constraints of the form (5) ensure that each truck is used at most once. Further details of this formulation are in Chapter 15.

13.3.3 Political Redistricting (Garfinkel and Nemhauser[215], Wagner [589])

Political districting is a process by which an area (e.g., a state) is par-

titioned into smaller areas, each of which is assigned a single representative. Let m be the number of basic population units (counties,
census tracts, etc.). A district is a combination of these units which
meets population, compactness, and other requirements. The various
ways the population units can compose a district can be enumerated.
Let n be the number of possible districts. The problem is to select
k ($k \leq n$) districts so that every population unit belongs to exactly one
district. We define x_j to be 1 if the jth district is in the redistricting
plan and 0 otherwise. Further, let $e_{ij} = 1$ if the ith population unit
is in the jth district, and let $e_{ij} = 0$ if it is not. Then if c_j is defined
to be some measure of unacceptability of the jth district, the problem
of finding the districts which will cover all the population units at a
minimal measure of unacceptability is a set partitioning problem with
the additional constraint $\sum_{j=1}^{n} x_j = k$.

13.3.4 Information Retrieval (Day [165])

The problem is to retrieve a given set of m requests for information
from a set of n files so that the length of the search is minimized. Here
the cost c_j corresponds to the length of the jth file, x_j is equal to 1 if
the jth file is selected and 0 otherwise; also $e_{ij} = 1$ means that the ith
information requested is in the jth file, and $e_{ij} = 0$ if it is not.

13.4 Relevant Results

In this section we state certain facts relating to the set covering and the
set partitioning problems. Some of these are used in the development
of the algorithms described in the next part. Note that the facts per-
tain equally well to the subproblems that are resolved during a search
enumeration procedure (Chapter 9), since the subproblems are set covering or set partitioning problems in their own right. (Naturally, by a
subproblem we mean the resulting integer program with some of the
variables x_j fixed at 0 or 1. In the set partitioning problem we shall
see, in Facts 2 and 3, that by fixing $x_j = 1$ certain other variables must
be set to 0 and thus are deleted as well. The constraints which become
redundant are also dropped.) Recall that the set covering problem is

$$
\begin{array}{lll}
\text{SC} & \text{minimize} & \mathbf{cx} \\
& \text{subject to} & \mathbf{Ex} \geq \mathbf{e} \\
& \text{and} & x_j = 0 \text{ or } 1 \quad (j = 1, \ldots, n),
\end{array}
$$

and the set partitioning problem is

$$
\begin{array}{lll}
\text{SP} & \text{minimize} & \mathbf{cx} \\
& \text{subject to} & \mathbf{Ex} = \mathbf{e} \\
& \text{and} & x_j = 0 \text{ or } 1 \quad (j = 1, \dots, n).
\end{array}
$$

Facts

1. ([392]) If any row r of $\mathbf{E}$ has all 0's, neither problem has a solution as constraint r cannot be satisfied.

2. ([50], [216]) If in a row r of $\mathbf{E}$ there is only one 1 and it occurs in the kth column, then $x_k = 1$ in every zero-one solution, and hence the rth constraint and the other constraints satisfied by $x_k = 1$ may be deleted.

In addition, in the set partitioning problem, variables x_t, where $e_{it} = e_{ik} = 1$ $(i = 1, \dots, m)$, must be deleted. That is to say, the variables having nonzero coefficients in any constraint where x_k also has a nonzero coefficient must have value 0, since $x_k = 1$ and the variables must sum to 1. In the first array below, row 3 of $\mathbf{E}$ has only one 1, and thus to satisfy this constraint we must have $x_2 = 1$. As the second and fourth constraints are then satisfied, x_1, x_3, and x_5 must be set to 0. The problem is reduced to the second array, which indicates the only solution to the entire problem is $x_2 = x_4 = 1$ and $x_1 = x_3 = x_5 = x_6 = 0$.

	Variable							
Constraint	x_1	x_2	x_3	x_4	x_5	x_6		
1	1	0	0	1	0	0	= 1	
2	0	1	1	0	0	0	= 1	
3	0	1	0	0	0	0	= 1	← Row with one 1
4	1	1	1	0	1	0	= 1	
5	0	0	0	1	0	1	= 1	

$$\underbrace{}_{\mathbf{E}}$$

$x_2 = 1 \Rightarrow x_1 = x_3 = x_5 = 0$, so the problem becomes

	Variable			
Constraint	x_4	x_6		
1	1	0	= 1	← Row with one 1
5	1	1	= 1	

$$x_4 = 1 \Rightarrow x_6 = 0.$$

3. ([216], [392]) (Row Dominance) Suppose $\mathbf{E}^s$ and $\mathbf{E}^r$ are two rows of $\mathbf{E}$ such that $\mathbf{E}^r \geq \mathbf{E}^s$; i.e., if x_k has a nonzero coefficient in the sth constraint, it has a nonzero coefficient in the rth constraint. So for the set covering problem, a variable which satisfies the sth constraint also satisfies the rth constraint; or the rth row of $\mathbf{E}$ may be deleted, as it is dominated by row s. This is illustrated below.

$$
\begin{array}{llllllll}
 & & & x_k \\
\text{Row } r & 1 & 1 & 1 & 0 & 1 & 1 & \geq (=)1 & \leftarrow \text{dominated row} \\
\text{Row } s & 1 & 0 & 1 & 0 & 0 & 1 & \geq (=)1
\end{array}
$$

For the set partitioning problem, in addition to deleting row r, the variables that have zero coefficients in the sth constraint and a nonzero coefficient in the rth constraint, must have value 0, since both the rth and the sth constraints must be satisfied. Here, both x_h and x_l must have value 0. Note that this must be checked for prior to deleting the dominating row.

$$
\begin{array}{llllllll}
 & x_h & & & x_l \\
\text{Row } r & 1 & 1 & 1 & 0 & 1 & 1 & = 1 \\
\text{Row } s & 1 & 0 & 1 & 0 & 0 & 1 & = 1
\end{array}
$$

4. ([216], [392]) (Column Dominance) Suppose for some column $\mathbf{E}_j$ of $\mathbf{E}$ in a set covering problem there exists a set S of other columns of $\mathbf{E}$ whose sum is greater than or equal to $\mathbf{E}_j$ (in a vector sense), and the cost of x_j is greater than or equal to sum of the costs of the variables corresponding to the columns in S. Then $\mathbf{E}_j$ may be deleted. This follows because the constraints that can be satisfied by setting $x_j = 1$ can also be satisfied by setting $x_t = 1$ for $t \in S$ at a reduced cost.

For the set partitioning problem, suppose for some column $\mathbf{E}_j$ of $\mathbf{E}$ there exists a set S' of columns such that $\mathbf{E}_j$ is equal to the sum of columns of $\mathbf{E}$ in S' and the cost of x_j is greater than or equal to the sum of the costs of the variables corresponding to the columns in S'. Then $\mathbf{E}_j$ may be deleted.

Note that the last three results may enable us to reduce the row and/or column size of a set covering or set partitioning problem.

5. ([392]) Any set partitioning problem can be converted to a set covering problem, and hence an algorithm which solves SC will also solve

SP.[7] The set partitioning problem is solved if we can solve the set covering problem and obtain a minimal solution with all slack variables (i.e., $\mathbf{s} = -\mathbf{e} + \mathbf{Ex}$) equal to 0. By assigning a high positive cost, say M, to all the slacks, we will tend to get a minimal SC solution with few positive slack variables. Thus, for M large enough (e.g., at least the sum of the costs), we can either find the minimal solution to the set partitioning problem or show that it has no solution (indicated by the appearance of a slack variable with a positive value in the minimal solution to problem SC). Thus, the set partitioning is equivalent to

$$
\begin{array}{llll}
\text{SP} & \text{minimize} & \mathbf{cx} + M\mathbf{e}^T\mathbf{s} \\
& \text{subject to} & \mathbf{Ex} - \quad \mathbf{Is} = \mathbf{e}, \\
& & \quad\quad\quad \mathbf{s} \geq \mathbf{0}, \\
& \text{and} & x_j = 0 \text{ or } 1 & (j = 1, \ldots, n),
\end{array}
$$

where $\mathbf{e}^T$ is a row of 1's and M is a large positive number (whose value is at least $\mathbf{ce}$). This can readily be converted to a 0–slack cost (set covering) problem by pre-multiplying $\mathbf{Ex} - \mathbf{Is} - \mathbf{e} = \mathbf{0}$ by $M\mathbf{e}^T$ and substituting for $M\mathbf{e}^T\mathbf{s}$ in the objective function. Or, SP is equivalent to

$$
\begin{array}{llll}
\text{SP} & -M\mathbf{e}^T\mathbf{e}+ \text{minimize} & \bar{\mathbf{c}}\mathbf{x} \\
& \text{subject to} & \mathbf{Ex} - \mathbf{Is} = \mathbf{e}, \\
& & \quad\quad\quad \mathbf{s} \geq \mathbf{0}, \\
& \text{and} & x_j = 0 \text{ or } 1 & (j = 1, \ldots, n).
\end{array}
$$

Here $\bar{\mathbf{c}} = \mathbf{c} + M\mathbf{e}^T\mathbf{E}$: that is, to each cost element c_j we add M times the number of 1's in the associated column of $\mathbf{E}$.

As an example, consider the set partitioning problem

$$
\begin{array}{llcccccccc}
\text{SP} & \text{minimize} & 5x_1 & + & 4x_2 & + & 1x_3 & + & 2x_4 \\
& \text{subject to} & x_1 & & & & & + & x_4 & = 1, \\
& & & & x_2 & + & x_3 & & & = 1, \\
& & x_1 & & & + & x_3 & + & x_4 & = 1, \\
& \text{and} & \multicolumn{8}{l}{x_1, x_2, x_3, x_4 = 0 \text{ or } 1.}
\end{array}
$$

---

[7]While this result permits the solution of the set partitioning problem by means of an algorithm designed for the set covering problem, serious roundoff errors may occur during computer implementation (see Appendix A). This is especially true if the sum of the costs is large while the individual costs are relatively small and close to one another in value.

Suppose we wish to solve this problem as a set covering problem. The set partitioning problem is equivalent to (where $M = \mathbf{ce} = 12$)

$$\text{SP} \quad \text{minimize} \quad 5x_1 + 4x_2 + 1x_3 + 2x_4 + 12s_1 + 12s_2 + 12s_3$$

$$\text{subject to (i)} \quad x_1 + \qquad\qquad x_4 - \quad s_1 \qquad\qquad = 1,$$

$$\text{(ii)} \qquad x_2 + x_3 \qquad\qquad - \quad s_2 \qquad = 1,$$

$$\text{(iii)} \quad x_1 \qquad + x_3 + x_4 \qquad\qquad - \quad s_3 = 1,$$

$$\text{and} \qquad x_1, x_2, x_3, x_4 = 0 \text{ or } 1, s_1 \geq 0, s_2 \geq 0, s_3 \geq 0.$$

Using equalities (i), (ii), and (iii) to substitute for s_1, s_2, and s_3 in the objective function yields the equivalent set covering problem

$$\text{SC} \quad \text{minimize} \quad 29x_1 + 16x_2 + 25x_3 + 26x_4 - 36$$

$$\text{subject to} \quad x_1 \qquad\qquad + \quad x_4 \qquad \geq 1,$$

$$x_2 + \quad x_3 \qquad\qquad \geq 1,$$

$$x_1 \qquad + \quad x_3 + \quad x_4 \qquad \geq 1,$$

$$\text{and} \qquad x_1, x_2, x_3, x_4 = 0 \text{ or } 1.$$

The minimal $\mathbf{x}$ is $(0, 1, 0, 1)$ and $\mathbf{cx} - 36 = 6$ with $\mathbf{s} = (0, 0, 0)$.

By replacing the $x_j = 0, 1$ requirements by $x_j \geq 0$ $(j = 1, \ldots, n)$ in problems SC and SP we have the associated linear programs LPC and LPP, respectively. The following facts relate to these problems.

6. ([392]) SC has a binary solution if and only if LPC has a feasible solution. Clearly, if SC has an integer solution, LPC has a feasible solution because a solution for SC satisfies the constraints of LPC. On the other hand, if we have a feasible solution x_j^0 $(j = 1, \ldots, n)$ to LPC, we can construct a binary solution to SC as follows:

(i) set $x_j = 0$ for $x_j^0 = 0$

(ii) set $x_j = 1$ for $x_j^0 > 0$.

Note that a feasible solution to LPP does not necessarily imply an integer solution to the set partitioning problem. For example, a feasible solution with $\mathbf{E}$ equal to

$$\begin{pmatrix} 1 & 0 & 1 \\ 1 & 1 & 0 \\ 0 & 1 & 1 \end{pmatrix}$$

is 1/2, 1/2, 1/2, and yet there is no zero-one solution to SP.

7. ([392]) If $\mathbf{c} > \mathbf{0}$, the dual linear programs of LPC and LPP initially reveal basic feasible solutions. This follows since the dual of LPC is:

$$
\begin{aligned}
\text{Minimize} \quad & \mathbf{we} \\
\text{subject to} \quad & \mathbf{wE} \leq \mathbf{c}, \\
& \mathbf{w} \geq \mathbf{0}, \\
\text{and} \quad & \mathbf{c} > \mathbf{0}.
\end{aligned}
$$

So by setting the slack variables $\mathbf{s} = \mathbf{c} - \mathbf{wE}$ equal to $\mathbf{c}$, and $\mathbf{w} = \mathbf{0}$, we have the basic feasible solution $(\mathbf{w}, \mathbf{s}) = (\mathbf{0}, \mathbf{c})$. The dual of LPP is the same except that the variables $(\mathbf{w})$ are unrestricted in sign. The result suggests that if linear programming is to be used in an algorithm we might use the dual simplex method.

8. ([392]) LPC and LPP have the same optimal solutions as the respective problems we obtain with the additional constraints $x_j \leq 1$ $(j = 1, 2, \ldots, n)$, added to each. Moreover:

9. ([392]) At any LPC or LPP extreme point, $0 \leq x_j \leq 1$ $(j = 1, \ldots, n)$ is actually satisfied. In LPP the constraints $\mathbf{Ex} = \mathbf{e}$ evidently guarantee that every extreme point has $x_j \leq 1$ $(j = 1, \ldots, n)$. For LPC, let $\mathbf{x}$ denote any extreme point. Form the basis matrix $\mathbf{B}$ as follows: take the columns of $\mathbf{E}$ corresponding to the positive x_j's, the columns of $-\mathbf{I}$ corresponding to positive slacks s_j's, and additional columns of $-\mathbf{I}$ (if necessary), so as to make $\mathbf{B}$ a nonsingular matrix. After permuting rows ($\mathbf{x}, \mathbf{s}$ becomes $\bar{\mathbf{x}}, \bar{\mathbf{s}}$), we have

$$
\mathbf{Ex} - \mathbf{Is} = \mathbf{B} \begin{pmatrix} \bar{\mathbf{x}} \\ \bar{\mathbf{s}} \end{pmatrix} = \begin{pmatrix} \mathbf{B}_{11} & \mathbf{0} \\ \mathbf{B}_{12} & -\mathbf{I} \end{pmatrix} \begin{pmatrix} \bar{\mathbf{x}} \\ \bar{\mathbf{s}} \end{pmatrix} = \begin{pmatrix} \mathbf{e} \\ \mathbf{e} \end{pmatrix},
$$

where $\bar{\mathbf{x}} > \mathbf{0}$. Then, since $\mathbf{B}$ is nonsingular, so is $\mathbf{B}_{11}$; hence, each column of $\mathbf{B}_{11}$ has at least one 1. But $\mathbf{B}_{11}$ and $\mathbf{e}$ contain only 0's and 1's, with $\mathbf{x}$ containing all positive components. Therefore, since $\mathbf{B}_{11}\bar{\mathbf{x}} = \mathbf{e}$, no positive component of $\mathbf{x}$ may exceed 1.

In any solution for the set partitioning problem, the positive variables must correspond to a set of zero-one columns which sum to a column of 1's. Thus, the columns in the set are linearly independent. We may therefore conclude that any solution to SP is an extreme point of its linear programming feasible region.[8] The analagous result for the set covering problem is now presented.

10. ([392]) If an integer solution to SC is not an extreme point of its linear programming feasible region, it is possible to reduce this solution to another SC integer solution which is an extreme point of the linear program's feasible region and the value of the objective function at this new solution is lower.

To show this, let $\bar{x} = (\bar{x}_j)$ be any SC integer solution that is not an extreme point for LPC. Since $\mathbf{Ex} - \mathbf{Is} = \mathbf{e}$, some of the slacks must be positive (otherwise $\mathbf{Ex} = \mathbf{e}$, which would mean $\bar{x}$ is an extreme point for LPC, since the columns of $\mathbf{E}$ corresponding to $\bar{x}_j = 1$ will be linearly independent). Permuting rows to get positive slacks last, we obtain

$$\mathbf{E}\bar{x} - \mathbf{I}\bar{s} = \begin{pmatrix} \mathbf{E}_1 \\ \mathbf{E}_2 \end{pmatrix} \bar{x} - \begin{pmatrix} \mathbf{0} \\ \mathbf{I} \end{pmatrix} \bar{s} = \begin{pmatrix} \mathbf{e} \\ \mathbf{e} \end{pmatrix}, \quad \text{with } \bar{s} \geq \mathbf{e}$$

(each component $\bar{s}_j$ of $\bar{s}$ is positive and integral). Then, a permutation of columns to get positive $\bar{x}$'s first, leads to

$$\begin{pmatrix} \mathbf{E}_1 \\ \mathbf{E}_2 \end{pmatrix} \bar{x} = \begin{pmatrix} \mathbf{E}_{11} & \mathbf{E}_{12} \\ \mathbf{E}_{21} & \mathbf{E}_{22} \end{pmatrix} \begin{pmatrix} \mathbf{e} \\ \mathbf{0} \end{pmatrix} = \begin{pmatrix} \mathbf{E}_{11} \\ \mathbf{E}_{21} \end{pmatrix} \mathbf{e};$$

so that,

$$\mathbf{E}\bar{x} - \mathbf{Is} = \begin{pmatrix} \mathbf{E}_{11} \\ \mathbf{E}_{21} \end{pmatrix} \mathbf{e} + \begin{pmatrix} \mathbf{0} \\ -\mathbf{I} \end{pmatrix} \bar{s} = \begin{pmatrix} \mathbf{e} \\ \mathbf{e} \end{pmatrix}, \quad \bar{s} \geq \mathbf{e} \tag{6}$$

The last expression means that $\mathbf{E}_{21}\mathbf{e} - \bar{s} = \mathbf{e}$ (with $\bar{s} \geq \mathbf{e}$), i.e., each row of $\mathbf{E}_{21}$ contains at least two 1's, and $\mathbf{E}_{11}\mathbf{e} = \mathbf{e}$; that is, each row of $\mathbf{E}_{11}$ contains exactly one 1. Thus, permuting rows and columns further (if necessary), we may exhibit

$$\mathbf{E}_{11} = \begin{pmatrix} \mathbf{I} & \mathbf{0} \\ \mathbf{E}_{11}^1 & \mathbf{0} \end{pmatrix}, \quad \text{and} \quad \begin{pmatrix} \mathbf{E}_{11} \\ \mathbf{E}_{21} \end{pmatrix} = \begin{pmatrix} \mathbf{I} & \mathbf{0} \\ \mathbf{E}_{11}^1 & \mathbf{0} \\ \mathbf{E}_{21}^1 & \mathbf{E}_{21}^2 \end{pmatrix}.$$

---

[8]We are using a well–known result from linear programming which says that $\hat{x} = (\hat{x}_j)$ is an extreme point of the region defined by $\mathbf{Ax} = \mathbf{b}, \mathbf{x} \geq \mathbf{0}$ ($\mathbf{A}$ is an m by n matrix and $\mathbf{b}$ is an m column) if and only if the columns of $\mathbf{A}$ corresponding to $\hat{x}_j > 0$ are linearly independent.

With the above, equation (6) may be rewritten as:

$$\begin{pmatrix} \mathbf{I} & \mathbf{0} \\ \mathbf{E}_{11}^1 & \mathbf{0} \\ \mathbf{E}_{21}^1 & \mathbf{E}_{21}^2 \end{pmatrix} \begin{pmatrix} \mathbf{e} \\ \mathbf{e} \\ \mathbf{e} \end{pmatrix} - \begin{pmatrix} \mathbf{0} \\ \mathbf{0} \\ \mathbf{I} \end{pmatrix} \bar{\mathbf{s}} = \begin{pmatrix} \mathbf{e} \\ \mathbf{e} \\ \mathbf{e} \end{pmatrix}.$$

Now, the columns of

$$\begin{pmatrix} \mathbf{I} & \mathbf{0} \\ \mathbf{E}_{11}^1 & \mathbf{0} \\ \mathbf{E}_{21}^1 & -\mathbf{I} \end{pmatrix}$$

clearly form a linearly independent set, whereas those of

$$\begin{pmatrix} \mathbf{I} & \mathbf{0} & \mathbf{0} \\ \mathbf{E}_{11}^1 & \mathbf{0} & \mathbf{0} \\ \mathbf{E}_{21}^1 & -\mathbf{I} & \mathbf{E}_{21}^2 \end{pmatrix}$$

are linearly dependent by the hypothesis ($\bar{\mathbf{x}}$ is not an extreme point).

Therefore, $\mathbf{E}_{21}^2$ has at least one column, say column $\mathbf{E}_j^2$. Then, $\bar{\mathbf{s}} \geq \mathbf{e} \geq \mathbf{E}_j^2$ demonstrates that: (i) deleting the $\mathbf{E}_j^2$ column, and (ii) replacing $\bar{\mathbf{s}}$ by $\bar{\mathbf{s}}' = \bar{\mathbf{s}} - \mathbf{E}_j^2$, yields a new feasible integer solution, with cost reduced by the positive cost of the deleted column.

Now suppose we repeat the above procedure with $\bar{\mathbf{x}}$ set to $\bar{\mathbf{x}}'$ and $\bar{\mathbf{s}}$ set to $\bar{\mathbf{s}}'$. Then either we do not have an extreme point and may obtain another cost reduction, or an extreme point is discovered by virtue of either $\bar{\mathbf{s}} = \mathbf{0}$ or $\mathbf{E}_{21}^2 = \mathbf{0}$.

As an example, consider $\mathbf{c} = (1, 2, 1, 1, 2, 3, 1)$ and

$$\mathbf{E} = \begin{pmatrix} 1 & 0 & 1 & 0 & 0 & 0 & 1 \\ 0 & 1 & 1 & 0 & 0 & 1 & 0 \\ 1 & 0 & 0 & 0 & 0 & 0 & 1 \\ 1 & 0 & 0 & 1 & 1 & 1 & 0 \\ 0 & 1 & 0 & 1 & 0 & 0 & 0 \end{pmatrix}$$

Then $\bar{\mathbf{x}} = (1, 1, 0, 1, 1, 0, 0)$ is a solution to SC with $\mathbf{c}\bar{\mathbf{x}} = 6$ and $\mathbf{s} = (0, 0, 0, 2, 1)$. Hence,

$$\mathbf{E} = \begin{pmatrix} \mathbf{E}_1 \\ \mathbf{E}_2 \end{pmatrix}, \quad \text{since } \mathbf{s} = \begin{pmatrix} \mathbf{0} \\ \bar{\mathbf{s}} \end{pmatrix} = \begin{pmatrix} 0 \\ 0 \\ 0 \\ 2 \\ 1 \end{pmatrix}.$$

$$\begin{pmatrix} \mathbf{E}_{11} & \mathbf{E}_{12} \\ \mathbf{E}_{21} & \mathbf{E}_{22} \end{pmatrix} = \begin{array}{ccccccc} 1 & 2 & 4 & 5 & 3 & 6 & 7 \\ \begin{pmatrix} 1 & 0 & 0 & 0 & 1 & 0 & 1 \\ 0 & 1 & 0 & 0 & 1 & 1 & 0 \\ 1 & 0 & 0 & 0 & 0 & 0 & 1 \\ \hline 1 & 0 & 1 & 1 & 0 & 1 & 0 \\ 0 & 1 & 1 & 0 & 0 & 0 & 0 \end{pmatrix} \end{array}$$

$$\mathbf{E}_{11} = \begin{pmatrix} \mathbf{I} & \mathbf{0} \\ \mathbf{E}_{11}^1 & \mathbf{0} \end{pmatrix} = \begin{pmatrix} 1 & 0 & 0 & 0 \\ 0 & 1 & 0 & 0 \\ \hline 1 & 0 & 0 & 0 \end{pmatrix}$$

$$\begin{pmatrix} \mathbf{I} & \mathbf{0} \\ \mathbf{E}_{11}^1 & \mathbf{0} \\ \mathbf{E}_{21}^1 & \mathbf{E}_{21}^2 \end{pmatrix} = \begin{array}{cccc} 1 & 2 & 4 & 5 \\ \begin{pmatrix} 1 & 0 & 0 & 0 \\ 0 & 1 & 0 & 0 \\ \hline 1 & 0 & 0 & 0 \\ \hline 1 & 0 & 1 & 1 \\ 0 & 1 & 1 & 0 \end{pmatrix} \end{array}$$

Therefore, set $\bar{x}_4 = 0$; then $\begin{pmatrix} \bar{s}_4' \\ \bar{s}_5' \end{pmatrix} = \begin{pmatrix} 2 \\ 1 \end{pmatrix} - \begin{pmatrix} 1 \\ 1 \end{pmatrix} = \begin{pmatrix} 1 \\ 0 \end{pmatrix}$, or $\mathbf{s}' = (0, 0, 0, 1, 0)$.

$$\begin{pmatrix} \mathbf{I} & \mathbf{0} \\ \mathbf{E}_{11}^1 & \mathbf{0} \\ \mathbf{E}_{21}^1 & \mathbf{E}_{21}^2 \end{pmatrix} = \begin{array}{c} 1 \\ 2 \\ 3 \\ 5 \\ 4 \end{array} \begin{array}{ccc} 1 & 2 & 5 \\ \begin{pmatrix} 1 & 0 & 0 \\ 0 & 1 & 0 \\ \hline 1 & 0 & 0 \\ 0 & 1 & 0 \\ \hline 1 & 0 & 1 \end{pmatrix} \end{array}$$

Finally, set $\bar{x}'_5 = 0$; then $s''_4 = 1 - 1 = 0$ or $\mathbf{s}'' = (0,0,0,0,0)$, and $\bar{\mathbf{x}}'' = (1,1,0,0,0)$ is an extreme point for LPC and $\mathbf{c}\bar{\mathbf{x}}'' = 3 < \mathbf{c}\bar{\mathbf{x}} = 6$.

The next four facts suggest a technique for extracting a "good" solution for the set covering problem from the associated optimal linear programming solution. The discussion is *not* valid for the set partitioning problem.

11. Let x^0_j ($j = 1,\ldots,n$) be any feasible solution to LPC. Then the roundup solution ($x_j = 1$ whenever $x^0_j > 0, x_j = 0$ otherwise) is a solution to SC. Further:

12. ([392]) A roundup solution obtained from a nonintegral extreme point can always be reduced to another integer solution with a lower cost. To show this, let $\mathbf{x}$ be an extreme point with not all integer values. Then we have $\mathbf{Ex} \geq \mathbf{e}$, and $\mathbf{0} \leq \mathbf{x} \leq \mathbf{e}$. Consider the columns of $\mathbf{E}$ associated with the positive nonintegral x_j variables, and the rows of $\mathbf{E}$ corresponding to the constraints that these variables explicitly satisfy. Identify the positive part of $\mathbf{x}$ and the corresponding matrix (selected as above) by a superscript *. Then we have $\mathbf{E}^*\mathbf{x}^* \geq \mathbf{e}$ and $\mathbf{0} < \mathbf{x}^* < \mathbf{e}$ (where $\mathbf{e}$ is a column of 1's of appropriate size.); that is, every row of $\mathbf{E}^*$ has at least two 1's. Therefore, setting some of the x^*_j variables to 1 maintains $\mathbf{E}^*\mathbf{x}^* \geq \mathbf{e}$, whereas the roundup solution calls for setting all the x^*_j variables to 1.

Properties 10, 11, and 12 suggest a procedure for obtaining improved SC integer solutions from any (nonintegral) extreme point of LPC. In particular:

$i)$ Change the LPC extreme point to an SC integer solution.

$ii)$ By considering the structure of the submatrix $\mathbf{E}^*$ of $\mathbf{E}$, reduce the roundup solution to another SC integer solution with a lower cost. $\qquad$ (7)

$iii)$ Reduce the integer solution found in (ii) to an LPC extreme point integer solution.

Clearly, (i) and (iii) are well defined. To aid in the development of rules for (ii), we explore the structure of $\mathbf{E}^*$ for an LPC extreme point associated with the minimal objective function, say $\bar{\mathbf{x}}$. We may assume that $\bar{\mathbf{x}}$ contains positive nonintegral entries (otherwise, LPC solves SC).

The submatrix $\mathbf{E}^*$ of $\mathbf{E}$ is obtained from the columns of $\mathbf{E}$ associated with the positive nonintegral $\bar{x}_j$ variables, which makes up $\mathbf{x}^*$, and the rows of $\mathbf{E}$ corresponding to the constraints these variables satisfy in the optimal LPC solution.

13. ([392]) At least one of the constraints satisfied by the positive nonintegral $\bar{x}_j$ variables holds with strict equality. For, $\mathbf{E}^*\mathbf{x}^* \geq \mathbf{e}$ or $\mathbf{E}^*\mathbf{x}^* - \mathbf{Is} = \mathbf{e}$ and $\mathbf{s} \geq \mathbf{0}$. Thus, with

$$\mathbf{x}^{**} = \mathbf{x}^* - \underset{i}{\text{minimum}}\ [s_i/(\mathbf{E}^*\mathbf{e})_i]\mathbf{e} \geq \mathbf{0}$$

(observe that $(\mathbf{E}^*\mathbf{e})_i$ is the sum of row i, that $s_i/(\mathbf{E}^*\mathbf{e})_i$ is the value by which each of the variables x_j^* can be reduced while retaining feasibility in row i, and that equality holds at the rows yielding the minimum ratio), we have

$$\mathbf{E}^*\mathbf{x}^{**} = \mathbf{E}^*\mathbf{x}^* - \underset{i}{\text{minimum}}\ [s_i/(\mathbf{E}^*\mathbf{e})_i]\mathbf{E}^*\mathbf{e} \geq \mathbf{E}^*\mathbf{x}^* - \mathbf{Is} = \mathbf{e},$$

or $\mathbf{x}^{**}$ is a feasible solution to LPC and $\mathbf{x}^{**} \leq \mathbf{x}^*$. But $\mathbf{x}^*$ is extracted from the minimal $\mathbf{x}$ (that is, $\bar{\mathbf{x}}$). Therefore, $\mathbf{x}^* = \mathbf{x}^{**}$, or $\underset{i}{\min}\ [s_i/(\mathbf{E}^*\mathbf{e})_i] = 0$, which means $\underset{i}{\min}\ s_i = 0$, proving the assertion.

14. ([392]) If a constraint satisfied by the positive nonintegral minimal LPC variables (that is, $\mathbf{x}^*$) has a slack variable at a positive value (or holds as a strict inequality), then every variable corresponding to a 1 in that row must contribute to the strict equality of another constraint satisfied by these variables. Otherwise we could produce an $\mathbf{x}^{**} < \mathbf{x}^*$ in a manner suggested by Fact 13, contradicting the hypothesis that $\mathbf{x}^*$ is extracted from the minimal $\mathbf{x}$ (Problem 13.11).

As an example, let us obtain a possible $\mathbf{x}^*$ based on Facts 13 and 14 for

$$\mathbf{E}^* = \begin{pmatrix} 1 & 0 & 0 & 1 & 0 & 1 \\ 0 & 0 & 1 & 1 & 0 & 1 \\ 1 & 1 & 0 & 0 & 1 & 0 \\ 0 & 0 & 1 & 1 & 0 & 0 \end{pmatrix} = (e_{ij}^*).$$

Initially suppose

$$\mathbf{x}^* = (1/3, 1/2, 2/3, 1/2, 1/3, 1/6).$$

Then,

$$\mathbf{s} = (0, 1/3, 1/6, 1/6),$$

so that property (13) is satisfied. But $s_2 > 0$ with $e_{23}^* = e_{43}^* = 1$ and $s_4 > 0$, contradicting (14). Therefore, set $x_3^* = 2/3 - \min(1/3, 1/6) = 1/2$, and then,

$$\mathbf{s} = (0, 1/6, 1/6, 0) \quad \text{with} \quad \mathbf{x}^* = (1/3, 1/2, 1/2, 1/2, 1/3, 1/6).$$

Again $s_2 > 0$, only now $e_{23}^* = e_{43}^* = 1$, $s_4 = 0$; $e_{24}^* = e_{44}^* = 1$, $s_4 = 0$; and $e_{26}^* = e_{16}^* = 1$, $s_1 = 0$. But $s_3 > 0$, and $e_{32}^* = 1$ with column 3 not contributing to the equality of any of the three other constraints. Hence, reset $x_2^* = 1/2 - \min(1/6) = 1/3$ and

$$\mathbf{s} = (0, 1/6, 0, 0) \quad \text{with} \quad \mathbf{x}^* = (1/3, 1/3, 1/2, 1/2, 1/3, 1/6).$$

This last $\mathbf{x}^*$ satisfies properties 13 and 14.

Facts 13 and 14 and the example suggested that columns of $\mathbf{E}^*$ will tend to have many 1 entries. Thus it is reasonable to suppose that only a relatively few columns of $\mathbf{E}^*$ will be needed to satisfy $\mathbf{E}^* \mathbf{x}^* \geq \mathbf{e}$ constraints. In this example, only two columns of $\mathbf{E}^*$ are needed to produce a set covering solution (i.e. we set the corresponding two components to 1, as well as those which are already integral in $\bar{\mathbf{x}}$), whereas the roundup solution included all the $\mathbf{E}^*$ columns. (The algorithm presented by Lemke, Salkin, and Spielberg[392] verifies numerically that the set covering solution extracted via (i), (ii), and (iii) in (7) is substantially better than the roundup solution, and, in fact, is often very near the value of the optimal binary solution; see Appendix A.)

13.5   Algorithms

We now discuss two techniques for solving the set covering and set partitioning problem.[9] Both are search algorithms (Chapter 9) and have performed reasonably well in computations (see Appendix A). Recall that a search algorithm is a basic approach (Section 9.2) coupled with a point or node algorithm (Section 9.3).

---

[9]Other procedures are suggested in the problems.

The basic approach is a bookkeeping scheme which keeps track of the enumeration and ensures that the 2^n possible binary values for the vector $\mathbf{x} = (x_1, \ldots, x_n)$ are either explicitly or implicitly examined. In terms of the description given in Chapter 9, we assume that the enumeration starts from the point $\mathbf{x}^0$ with all variables free and that the point $\mathbf{x}^l$ is the point $\mathbf{x}$ with l variables fixed at 1. Then the basic approach is:

i) Fix a free variable x_k from $\mathbf{x}^l$ (initially $\mathbf{x}^l = \mathbf{x}^0$) at value 1 (forward step).

ii) Resolve the subproblem in the remaining free variables, then

iii) Fix x_k at value 0 (or cancel x_k at level l, backward step), and

iv) Repeat this process for the problem with x_k fixed at 0.

The subproblem associated with $\mathbf{x}^l$ is resolved when all branches from $\mathbf{x}^l$ have been canceled, in which case (a) x_k is canceled at level $l - 1$; (b) all variables canceled at level l revert to the status free; and (c) the process reverts to $\mathbf{x}^{l-1}$ at level $l - 1$. The search is complete when all variables are canceled at level 0. Reverting to $\mathbf{x}^{l-1}$ automatically defines the backward step from $\mathbf{x}^l$.

The node algorithm contains the implicit enumeration tests and branching rules which are developed by taking the problem's structure into account. It is used at each subproblem to reduce the number of free variables and to select the next variable to fix at 1.

The search algorithm, which is a composite of the basic approach and point algorithm, can be related to a tree composed of nodes (or points) and branches. A node corresponds to a point $\mathbf{x}^l$, and a branch joining two nodes $\mathbf{x}^l$ and $\mathbf{x}^{l+1}$ implies that the point $\mathbf{x}^{l+1}$ is defined by setting some free variable in $\mathbf{x}^l$ to 1 (Chapter 9, Section 9.2). As the procedures described in the next two sections are search algorithms with the same basic approach (i) through (iv), we center our discussion only about the different point algorithms.

13.5.1 A Search Algorithm for the Set Partitioning Problem (Garfinkel and Nemhauser [214])

A vector $\mathbf{x} = (x_1, \ldots, x_n)$ is an integer solution to the set partitioning problem only if the binary columns $\mathbf{E}_j = (e_{ij})$ $(i = 1, \ldots, m)$ of $\mathbf{E}$ which correspond to $x_j = 1$ $(j = 1, \ldots, n)$ sum to a column of 1's. Thus, if during an enumeration, a variable x_j is set to 1, then for it to be a part of an integer solution vector $\mathbf{x}$, all variables x_h $(1 \leq h \leq n)$ which have $e_{ih} = e_{ij} = 1$ $(j \neq h)$ for any row i $(i = 1, \ldots, m)$ must be fixed at value 0. Also for x_j to yield an improved solution, its cost added to those already fixed at 1 must be less than the value of the current best integer solution z^* (initially set very large). These observations allow us to develop a point algorithm. To expedite the cancellations, the variables (indices j) are initially placed in lists, where j is in list i $(i = 1, \ldots, m)$ if the first 1 in column $\mathbf{E}_j$ occurs in row i.

The cancellations (implicit enumeration tests) and branching process are explained together. The list corresponding to the first unsatisfied constraint is selected. If all the constraints are satisfied, an improved solution is found and a backward step is taken. Otherwise, suppose constraint $\bar{i}$ is not satisfied (initially $\bar{i} = 1$ since originally no constraints are satisfied). All variables in the list which cannot produce an improved integer solution are canceled (at the current level). If no free variables remain in the list, constraint $\bar{i}$ cannot be satisfied with the variables currently fixed at 1 and a backward step is taken. Otherwise, let column j correspond to the free variable in list $\bar{i}$ with the least cost. That is, column $\mathbf{E}_j$ does not have a 1 in any row corresponding to a satisfied constraint, and c_j plus the costs of the variables at 1 is less than z^*. Then x_j is set to 1, defining a forward step. (Note that constraint $\bar{i}$ and possibly others become satisfied.) This process continues until all variables in list 1 have been canceled, in which case the current best solution is optimal.

Example 13.3   Consider the four-constraint and six-variable set partitioning problem with the cost vector and coefficient matrix below.

		x_1	x_2	x_3	x_4	x_5	x_6	
		1	2	3	4	5	6	
	1	1	0	1	0	0	1	$= 1$
	2	0	1	1	0	1	1	$= 1$
i	3	0	0	1	1	1	0	$= 1$
	4	1	1	0	1	0	1	$= 1$
	c_j	3	2	1	1	3	2	

j

Step no.	Level l	Variables at 1	Constraints satisfied	List selected $\bar{\imath}$	Variables free in $\bar{\imath}$	Step type	Variable selected	Cost of var. at 1	z^*
1	0	none	none	1	1,3,6	forward	3	1	∞
2	1	3	1,2,3	4	none	backward	--	--	∞
3	0	none	none	1	1,6	forward	6	2	∞
4	1	6	1,2,4	3	none	backward	--	--	∞
5	0	none	none	1	1	forward	1	3	∞
6	1	1	1,4	2	5	forward	5	6	∞
7	2	1,5	1,2,3,4	solution	--	backward	--	3	6
	1	1	1,4	2	none	forward	--	--	6
8	0	none	none	1	none	backward	--	--	6
9	-1	termination							

Table 13.1: Computations for Example 13.3

Initially the indices are put in the lists

list	column
1	1, 3, 6
2	2, 5
3	4
4	--

The computations are summarized in Table 13.1, and are related to the tree appearing in Fig. 13.7. Keep in mind that the free variable with the least cost in list $\bar{\imath}$ is selected to branch on.

Remarks

1. It is evident that the efficiency of the algorithm crucially depends on the density of 1's in the coefficient matrix. (Why?) The algorithm becomes more effective as the number of 1's increases.

2. The branching process is constructed so that once a list $\bar{\imath}$ is selected and a forward step (at level l) is taken, all free variables on lists $i \leq \bar{\imath}$ should be canceled (at level $l-1$). This follows because the first 1 in these columns occurs in row i and $\bar{\imath}$ is the first unsatisfied constraint. Note that since the branching procedure only considers free variables

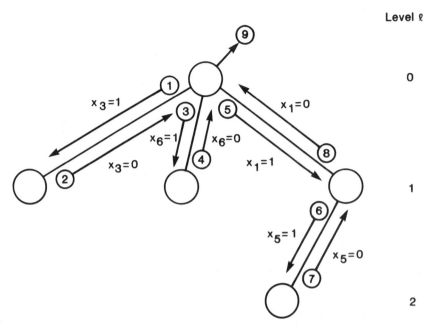

Figure 13.7: Enumeration Tree for Example 13.3

in the *current* list $\bar{i}$ as branch candidates, we are, in fact, canceling all variables on lists preceding $\bar{i}$.

3. Prior to performing the computations, it may be worth checking the **E** matrix for rows with one 1, and possible column and row dominance. Using the results (1 through 4) in the preceding section this may allow us reduce the size of the problem. Also, to simplify the bookkeeping it probably pays to reorder the variables within each list in order of increasing cost.

13.5.2   A Search Algorithm for the Set Covering Problem
  (Lemke, Salkin, and Spielberg [392], Salkin [502])

We have shown (Section 13.4, Facts 10 through 12) that integer solutions to the set covering problem can be constructed from feasible solutions to the associated linear program. The search technique presented here contains a point algorithm which is developed about solving the linear program and extracting integer solutions. We discuss the composition of the point algorithm. The criteria are described in the order in which they will be used.

1. Check for a Row of Zeroes

The current set covering subproblem is defined over the free variables and unsatisfied constraints. If the resulting coefficient matrix does not have a 1 in each row, the subproblem has no solution and a backward step may be taken.

2. Solve the Linear Program

By replacing the integrality requirements on the free variables with non-negativity constraints we have the associated linear program LPC. The linear program, when solved, yields the following useful information:

i) If the optimal solution to LPC is integer, this solution is optimal for the subproblem. Thus, it is recorded and a backward step is allowed.[10]

ii) If LPC has no feasible solution, then there is no zero-one solution to the subproblem; or a backward step may be taken.

iii) The optimal value of the objective function z^L for the linear program is a lower bound on the value z^I for *any* zero-one solution to the subproblem; that is, $z^L \leq z^I$.

Suppose LPC is solved by the dual simplex algorithm. (Recall that a dual feasible solution is initially present; see Section 13.4, Fact 7.) Then the value of the objective function is non-decreasing, and we may use (iii) to derive a ceiling test which may signal for a backward step prior to terminating the simplex computations. That is, we have the ceiling test:

iv) If at any dual simplex iteration the value of the linear program's

---

[10]Note that because $\mathbf{E}$ contains only 0's and 1's, it seems reasonable to conjecture that when $\mathbf{E}$ contains mostly 0's the optimal solution to LPC will often be all-integer. This is because the elements in the updated simplex tableaux tend to remain at 1; thus the pivot may always equal 1 and/or the determinant of the optimal linear programming basis tends to be small, and often 1 or −1.

objective function z (which includes the costs of the variables at 1) exceeds the current best integer solution z^*, a backward step is allowed.

If LPC does not signal for a backward step, we can proceed as suggested in (7) to extract an integer solution from the optimal linear programming solution. To do this, we have to define the submatrix $\mathbf{E}^*$ which contains the columns corresponding to the variables at fractional values (denoted by $\mathbf{x}^*$) and the rows that these variables explicitly satisfy. Then an integer solution to the reduced system $\mathbf{E}^*\mathbf{x}^* \geq \mathbf{e}$ is sought, where $\mathbf{e}$ is a vector of 1's of appropriate dimension. To obtain a single zero-one solution to these inequalities, variables are set to 1 (e.g., using the branching rule discussed shortly) until all the constraints corresponding to $\mathbf{E}^*$ are satisfied. The zero-one solution to the set covering problem is then defined by:

$x_j = 1$ if the variable is at 1 in the optimal solution to LPC
or if it is at a fractional value and set to 1 so that the
$\mathbf{E}^*\mathbf{x}^* \geq \mathbf{e}$ constraints are satisfied.

$x_j = 0$ otherwise.

If this solution yields an improved value for the objective function, z^* is updated and we have a more effective ceiling test (iv). Otherwise, no new information is obtained. However, note that after solving the *first* (i.e., level 0) linear program this procedure gives an immediate integer solution. In fact, in computations (see Example 13.4 and Appendix A), the value of the first integer solution produced this way has proved to be either optimal or very close to being optimal. (This suggests a heuristic technique for getting good solutions to massive set covering problems; namely, that we should solve LPC, extract the first integer solution, and terminate.)

3. The Preferred Set

To reduce the number of branch candidates it seems natural to find the row $\bar{i}$ of the current subproblem containing the smallest number of 1's. Then if P corresponds to the column indices j such that $e_{\bar{i}j} = 1$, where x_j is a free variable, only entries in P have to be considered as candidates for a forward step from the current node. The subproblem is resolved when all $j \in P$ have been canceled. In the context of Chapter 9 (Section 9.4.1), P is the *preferred set*.

4. Branching Process

On reaching a node for the first time (i.e., after a forward step), the preferred set is found and a variable is to be selected from it to define the next forward step. After revisiting a node (i.e., following a backward step) the preferred set is known and another branch is to be selected from P. In any case, a natural branching rule is to choose the $j \in P$ such that

  i) x_j satisfies the largest number of unsatisfied constraints at the cheapest per unit cost, and

  ii) the cost of c_j plus the cost of the variables at 1 (denoted by z) does not exceed z^*.

That is, if $N(l, j)$ is the number of constraints of the level l subproblem that variable j can satisfy, the variable $x_{\bar{j}}$ to be set to 1 satisfies

$$\left. \begin{array}{ll} i) & \dfrac{c_{\bar{j}}}{N(l,\bar{j})} = \underset{j \in P}{\text{minimum}} \ \dfrac{c_j}{N(l,j)} \\[4mm] ii) & c_{\bar{j}} + z < z^* \end{array} \right\} \tag{8}$$

The $N(l, j)$ array can also be used to keep track of the enumeration. In particular, an l by n matrix is kept where l is the current level of the search. After a forward step a new row $N(l+1, j)$ $(j = 1, \ldots, n)$ is added to the matrix by updating the $N(l, j)$ $(j = 1, \ldots, n)$ values to take into account those constraints satisfied by $x_{\bar{j}}$ and those variables canceled at level l. After a backward step to the level $l-1$ subproblem, row $N(l, j)$ $(j = 1, \ldots, n)$ is erased and the $N(l-1, j)$ values are used; also, $N(l-1, \bar{\bar{j}})$ is set to 0, where $x_{\bar{j}}$ defined the level l node from the level $l-1$ node. Initially, $N(0, j) = \sum_{i=1}^{m} e_{ij}$ for all j. The enumeration may then be terminated when $\sum_{j=1}^{n} N(0, j)$ becomes $< m$. (Why?)

Example (from Example 13.3)

To illustrate the bookkeeping scheme, the preferred set, and the branching process, we solve the previous example as a set covering problem.[11] From that problem

$$
\mathbf{E} = \begin{pmatrix}
1 & 0 & 1 & 0 & 0 & 1 \\
0 & 1 & 1 & 0 & 1 & 1 \\
0 & 0 & 1 & 1 & 1 & 0 \\
1 & 1 & 0 & 1 & 0 & 1
\end{pmatrix}
$$

Step 1 (level 0)

j	1	2	3	4	5	6
c_j	3	2	1	1	3	2
$N(0,j)$	2	2	3	2	2	3
$c_j/N(0,j)$	3/2	1	1/3	1/2	3/2	2/3
	↑		↑			↑

(Preferred set elements are indicated by arrows.)

Rows 1 and 3 have three 1's. We arbitrarily choose row 1. In this case, the preferred set $P = \{1, 3, 6\}$ since row 1 has a 1 in columns 1, 3, and 6. As $1/3 = \min\{3/2, 1/3, 2/3\}$, x_3 is set to 1.

Step 2 (level 1) As constraints 1, 2, and 3 become satisfied, the preferred set $P = \{1, 2, 4, 6\}$ is derived from row 4 and $N(1,j)$ is below.

j	1	2	3	4	5	6
$N(1,j)$	1	1	0	1	0	1
$c_j/N(1,j)$	3	2	–	1	–	2
	↑	↑		↑		↑

(Preferred set elements are indicated by arrows.)

So x_4 is set to 1.

---

[11]The use of the linear program and extracting integer solutions is given in the next example.

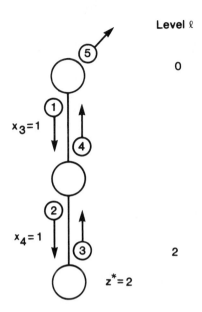

Figure 13.8: Search Enumeration Tree for Set Covering Problem

Step 3 (level 2) As all the constraints are satisfied, $N(2, j) = 0$ for $j = 1, \ldots, 6$, and the solution $x_3 = x_4 = 1, x_1 = x_2 = x_5 = x_6 = 0$ is recorded; z^* is set to $c_3 + c_4 = 2$, and the algorithm reverts back to the previous node.

Step 4 (level 1) The preferred set is $P = \{1, 2, 6\}$, but setting any of these variables to 1 cannot yield an improved solution since $z^* = 2$, and $c_3 = 1$. So they are all canceled and another backward step is taken.

Step 5 (level 0) The preferred set is $P = \{1, 6\}$. But $c_1 = 3, c_6 = 2$, and $z^* = 2$, so an improved integer solution is not possible; P becomes null and the procedure terminates. Figure 13.8 relates the computations to the tree.

Example 13.4 (Lemke, Salkin, and Spielberg [392])

To illustrate the potential efficacy of solving the linear program and extracting an integer solution, consider the 15 by 32 set covering problem appearing in Fig. 13.9. The optimal solution to the associated linear program is $x_4 = .6, x_6 = 1., x_7 = .4, x_8 = .8, x_{10} = .4, x_{11} = .6, x_{18} =$

Column	1	2	3	4	5	6	7	8	9	10	11	12	13	14	15	16	17	18	19	20	21	22	23	24	25	26	27	28	29	30	31	32
Costs c	1	1	1	1	1	1	1	1	2	2	2	2	2	2	2	3	3	3	3	4	4	4	4	5	5	5	6	6	6	7	8	9
E matrix Row																																
1	1			1													1				1			1			1				1	
2	1												1	1				1			1	1						1	1		1	1
3		1			1													1						1	1				1		1	
4		1																			1			1			1				1	
5		1									1				1						1	1				1					1	
6		1																		1				1	1				1		1	
7			1	1									1				1								1						1	1
8			1														1			1				1	1			1			1	
9			1											1	1												1			1	1	1
10						1															1					1					1	1
11														1	1			1						1	1			1		1	1	
12																	1	1			1	1		1				1			1	
13						1												1						1			1		1		1	
14			1													1		1			1										1	
15				1														1									1	1	1		1	1

Figure 13.9: **E** matrix for Example 13.4 (a blank means a 0)

.8, $x_{19} = .6$, $x_{23} = .2$, $x_{25} = .4$, $x_{31} = .2$, and $x_j = 0$ otherwise. Its value is 13.4. To extract the $\mathbf{E^*x^*} \geq \mathbf{e}$ constraints, we note that only variable $x_6 = 1$ and it satisfies equation 6. So $\mathbf{E^*}$ contains every row of $\mathbf{E}$ except row 6, and it includes columns $4, 7, 8, 10, 11, 18, 19, 23, 25$, and 31 (Fig. 13.10). Using the branching rule (8) (i), where P is replaced by all indices in $\mathbf{x^*}$, we obtain the first integer solution (the variables are ordered as they are set to 1) $x_8 = x_{19} = x_4 = x_7 = x_{11} = x_{18} = x_{10} = 1$, and $x_{23} = x_{25} = x_{31} = 0$ with cost 13. So a solution for the set covering problem is

$$x_j = 1 \quad \text{for} \quad j = 4, 6, 7, 8, 10, 11, 18, 19$$
$$x_j = 0 \quad \text{otherwise,}$$

and $z^* = 13 + c_6 = 14$. As $14 - 13.4 = .6 < 1$, and all the costs are integral, the above solution is optimal.

The search procedure discussed here contains a point algorithm outlined in Fig. 13.11.[12] Note that the conversion explained in Fact 5 (Section 13.4) can be used so that the technique can be applied to a set partitioning problem. However, as already indicated, computer round-off errors may become substantial and the linear programs naturally

---

[12] A more sophisticated version of this algorithm appears in Lemke, Salkin, and Spielberg [392].

				Columns of E						
Rows of E	4	7	8	10	11	18	19	23	25	31
1						1		1		
2						1	1			1
3			1			1				
4	1							1		
5				1			1			1
7		1						1	1	
8			1				1			
9			1							1
10				1				1	1	
11						1		1		
12						1	1			1
13				1				1		
14						1	1			1
15						1				1
$c^* = $ (1	1	1	2	2	3	3	4	5	8)	

Figure 13.10: The array E_v^* and cost row c_v^* for Example 13.4

become harder to solve because the value of the costs become almost equal. Also, the dominance criteria used to develop the first algorithm are not, and probably should be used. (These are discussed as Facts 2, 3, and 4 in Section 13.4.)

Problems

Problem 13.1 Describe several applications which can be modeled as minimum cover or matching problems. In each situation given an estimate on the number of nodes and arcs in the network.

Problem 13.2 Given the network in Figure 13.12:

a) set up the set covering problem which when solved gives a minimal cost cover (the cost of using an arc is adjacent to each arc).

b) solve the set covering problem found in (a) (by inspection) and give a minimum cost cover.

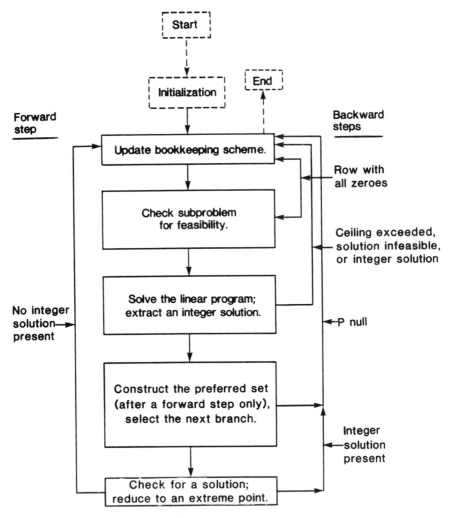

Figure 13.11: The General Structure of the Point Algorithm

Problem 13.3 For the network given in the previous problem, do (a) and (b) where a maximal weighted matching is sought (the costs become the weights).

Problem 13.4 Prove that a simple (i.e., all arc weights are equal) maximum matching can be obtained from a simple minimum cover by deleting the smallest number of arcs from the cover so that after the

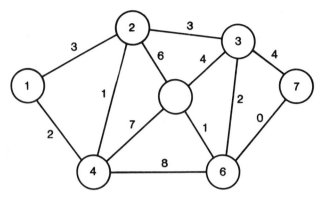

Figure 13.12: Network for Problem 13.2

removals each node is the end point of at most one arc. Is this result true when the arcs have *unequal* weights? Why?

Problem 13.5 Prove that a simple minimum cover can be obtained from a simple maximal matching by adding a single arc to each node which has no arcs incident to it in the matching. Is this result true when the arcs have unequal weights? Why?

Problem 13.6 Given the network with undirected arcs shown in Figure 13.13:

a) set up the set covering problem which then solved gives a set of arcs that disconnects all paths from node 1 to node 5 at a minimal total cost (the cost of each arc is adjacent to it).

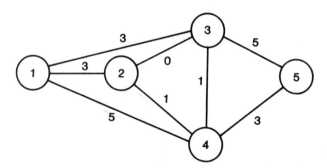

Figure 13.13: Network for Problem 13.6

b) solve the set covering problem found in (a) (by inspection) and indicate the arcs to be removed so that all paths from node 1 to node 5 are disconnected at a minimal total cost.

c) discuss situations that may produce this type of problem.

Problem 13.7 (Rubin [496] Suppose you are given an enormous set covering problem so that computer storage limitations prohibit finding an optimal solution. Discuss why the following procedure may yield "good" solutions. What are its drawbacks? Why?

i) "Intelligently" omit some of the columns.

ii) Select the first i rows (constraints) not satisfied so that the set covering subproblem in these constraints and the remaining columns can be solved efficiently. Solve the subproblem.

iii) If the solution found in (ii) does not satisfy all the constraints, record the variables at value 1, cross off the constraints they satisfy, and repeat (ii). If the solution found in (ii) satisfies all the constraints, record the solution and either stop or repeat (ii) after interchanging some columns omitted in (i) with some at value 0 in the current solution. Naturally, the second option is only exercised a finite number of times.

Problem 13.8 Consider the set covering problem below with four constraints and eight variables.

$$\mathbf{c} \; = \; (\quad 3 \quad 1 \quad 1 \quad 6 \quad 4 \quad 2 \quad 2 \quad 4 \quad)$$

$$\mathbf{E} \; = \; \begin{pmatrix} 1 & 0 & 1 & 1 & 0 & 0 & 0 & 1 \\ 0 & 1 & 1 & 0 & 0 & 1 & 0 & 1 \\ 0 & 1 & 0 & 0 & 1 & 1 & 0 & 1 \\ 1 & 1 & 1 & 0 & 0 & 0 & 1 & 0 \end{pmatrix}$$

a) Find an optimal solution (by inspection). Which constraints have positive slack variables?

b) Change the cost vector (Fact 5, Section 13.4) so that the optimal solution to the set covering problem is the optimal solution to the set partitioning problem whose cost vector c and constraint matrix E are above. What is the price paid for not having positive slack variables in the optimal solution?

Problem 13.9   List all binary solutions to the set covering problem appearing in the previous problem. Single out those which are extreme points of the associated linear program. Knowing the integer extreme points, find the optimal solution to the set covering problem. In general, why can't this procedure be used to solve any set covering problem?

Problem 13.10 (Balas and Padberg [39]) Consider the following result for the set partitioning problem (SP). Keep in mind that every integer solution to SP is an extreme point of the linear programming feasible region and thus may be associated with a basis.

Result: Let x^1 be an integer, but not optimal, solution to SP with basis B_1. If x^2 is an optimal solution to SP, then there exists a sequence of adjacent bases (i.e., the extreme points corresponding to the basic feasible solutions are adjacent) $B_{10}, B_{11}, B_{12}, \ldots, B_{1p}$, such that $B_{10} = B_1, B_{1p} = B_2$, is a basis associated with x^2, and (i) the associated basic feasible solutions $x^1 = x^{10}, x^{11}, x^{12}, \ldots, x^{1p} = x^2$ are all feasible and integer, (ii) $cx^{10} \geq cx^{11} \geq \cdots \geq cx^{1p}$, and (iii) p is the number of basic variables in x^1 that have value 1 in x^2.

Discuss an algorithm which is suggested by the result. What information does your technique require? Is it available? Does the theorem actually define an algorithm for the set partitioning problem? Why?

Problem 13.11   Prove Fact 14 given in Section 13.4 (*Hint:* The argument is similar to the one given in Fact 13.)

Solve the set problem below for the conditions given in Problems 13.12 through 13.14. In each case, relate the computations to a tree.

$$c = (\,1 \ \ 2 \ \ 3 \ \ 4 \ \ 5 \ \ 4 \ \ 3 \ \ 2 \ \ 1 \ \ 4\,)$$

$$E = \begin{pmatrix} 1 & 0 & 1 & 0 & 1 & 0 & 0 & 0 & 0 & 0 \\ 0 & 1 & 1 & 0 & 0 & 0 & 1 & 0 & 1 & 0 \\ 0 & 1 & 0 & 1 & 0 & 0 & 1 & 0 & 0 & 0 \\ 0 & 0 & 0 & 0 & 1 & 0 & 1 & 1 & 1 & 0 \\ 1 & 0 & 1 & 1 & 0 & 1 & 0 & 1 & 0 & 0 \\ 1 & 1 & 1 & 0 & 0 & 0 & 0 & 1 & 1 & 1 \end{pmatrix}$$

Problem 13.12 Use the algorithm presented in Section 13.5.1, where we have a set partitioning problem.

Problem 13.13 Use the algorithm and bookkeeping scheme (to keep track of the enumeration) presented in Section 13.5.2, where we have a set covering problem.

Problem 13.14 Use the algorithm and bookkeeping scheme presented in Section 13.5.2, where we have a set partitioning problem. (*Hint*: Use the conversion explained in Section 13.4.)

Problem 13.15 Discuss the differences and similarities between the two algorithms described in Section 13.5. Is the preferred set used in the technique given in Section 13.5.1?

Problem 13.16 In terms of computer implementation, discuss what has to be stored for each of the algorithms given in Section 13.5. Draw a flow chart for each method. Which seems simpler to computer program? Why?

Problem 13.17 Show that the structure of the submatrix $\mathbf{E}^*$ found in Example 13.4 agrees with Facts 13 and 14 (Section 13.4).

Problem 13.18 Why does it seem best to use the algorithm presented in Section

    a) 13.5.1 for the set partitioning problem when $\mathbf{E}$ has mostly 1 entries?

    b) 13.5.2 for the set covering or set partitioning problem when $\mathbf{E}$ has mostly 0 entries?

What if $\mathbf{E}$ has about the same number of 0's and 1's? Which technique would you try? Why?

Problem 13.19 (Salkin and Koncal [512], [513]) Consider the following all-integer cutting plane algorithm. Using the conversion explained in Section 13.4, it is valid for the set covering and set partitioning problem.

    i) Set the problem up in a lexicographic dual feasible all integer simplex tableau.

ii) Perform dual simplex pivot steps whenever the pivot is -1, otherwise add a Gomory all integer cut (Chapter 6) and use it as a pivot row (the pivot is then -1).

    a) What are the advantages of this technique in comparison to the ones explained in Section 13.5? What are the disadvantages?

    b) Under what circumstances will the technique work efficiently? Why?

Problem 13.20 (Bellmore and Ratliff [76]) Let $\mathbf{x}$ be an integer solution to the set covering problem which is an extreme point of the linear programming feasible region. Construct an associated basis $\mathbf{B}$ as follows:

i) take the columns corresponding to the $\mathbf{x}$ variables at value 1.

ii) take the columns corresponding to the nonzero slack variables.

iii) for each nonzero $\mathbf{x}$ variable consider the constraints which it contributes to a strict equality. If this number is more than 1, include in the basis those columns corresponding to the slack variables of all but one of these constraints. (Naturally, these variables are at value 0 in the associated basic feasible solution.)

    a) Show that each $\mathbf{x}$ variable at 1 must contribute to the strict equality of at least one constraint so that step (iii) is well defined.

    b) Show that the above method of filling out the basis will select exactly M columns which are linearly independent.

    c) Show that by permuting the rows the basis can be partitioned such that

$$\mathbf{B} = \begin{pmatrix} \mathbf{I}_1 & \mathbf{0} \\ \mathbf{B}_1 & -\mathbf{I}_2 \end{pmatrix}$$

    which means $\mathbf{B} = \mathbf{B}^{-1}$. ($\mathbf{I}_1$ and $\mathbf{I}_2$ are identity matrices).

    d) Illustrate the technique using the data preceding Problem 13.12. Select any extreme point solution.

Problem 13.21 (Bellmore and Ratliff [77])   The following cutting plane algorithm for the set covering problem is based on the result given in the previous problem.[13]

i) Find an integer solution to the set covering problem which is an extreme point of the linear programming feasible region.

ii) Construct an associated basis $\mathbf{B}$ as in the previous problem so that $\mathbf{B} = \mathbf{B}^{-1}$.

iii) Find the updated cost factors $\mathbf{c_B}\mathbf{B}^{-1}\mathbf{E}_j - c_j = \mathbf{c_B}\mathbf{E}_j - c_j$ for each column $\mathbf{E}_j$ of $(\mathbf{E}, -\mathbf{I})$, where $\mathbf{c_B}$ are the costs associated with the basic feasible solution corresponding to the basis $\mathbf{B}$. (The constraints of the set covering problem are $\mathbf{E}\mathbf{x} - \mathbf{I}\mathbf{s} = \mathbf{e}$, where $\mathbf{s}$ are slack variables which are constrained to be 0 or 1.)

iv) If all the updated cost factors are nonpositive, the current basic feasible solution solves the set covering problem. Otherwise, let $Q = \{j \mid \mathbf{c_B}\mathbf{B}^{-1}\mathbf{E}_j - c_j > 0\}$ and introduce the inequality

$$\sum_{j \in Q} x_j \geq 1 \tag{1}$$

to the constraint set so that we have a new set covering problem. Go to (i).

a) Show that Q can never contain indices corresponding to slack variables, which means that the number of variables in the set covering problem found in (iv) is always the same as in the original problem.

b) Show that inequality (1) does not eliminate any integer solution to the original problem and that it is implied by its constraints. (Hence, it is a valid cut.)

c) Suppose SC_0 is the original problem and that $SC_1, \ldots, SC_t$ are the problems generated in Step (iv). Let $\mathbf{x}^0, \mathbf{x}^1, \ldots, \mathbf{x}^t$ be the integer extreme point solutions to the corresponding set covering problems. If the process terminated at SC_t, why is the optimal solution to the original problem $\mathbf{x}^r$, where $\mathbf{x}^r$ satisfies

---

[13]A somewhat similar procedure is given by House, Nelson, and Rao [334]; it is also explained in Appendix A and in Salkin and Saha [517].

$$\mathbf{cx}^r = \begin{array}{c} \text{minimum} \\ 1 \le i \le r \end{array} (\mathbf{cx}^i)?$$

(*Hint:* The constraint region defined by problem SC_i becomes more constrained as i increases so that a solution to problem SC_k is also a solution to problem SC_i for $1 \le i \le k - 1$.)

d) Prove that the computations must eventually terminate with a basic feasible solution whose updated cost factors are nonpositive. (*Hint:* Use the limit given in (c), and appeal to the finite number of integer solutions, and let it be that each time a new problem is created at least one of the integer solutions to the original problem is excluded.)

Problem 13.22   For the algorithm given in the previous problem:

a) What does its efficiency depend on? Why?

b) How might extreme point solutions be found in (i)?

c) What is an upper bound on the number of constraints (1) generated?

d) Prove that the computations must eventually terminate with a basic feasible solution whose updated cost factors are nonpositive. (*Hint:* Use the limit given in (c), and appeal to the finite number of infinite solutions, and let it be that each time a new problem is created at least one of the integer solutions to the original problem is excluded.)

e) Use it to solve the set covering problem whose data is given in Problem 13.12.

Appendix A
Computational Experience

A.1   Cutting Plane Algorithms

(Bellmore and Ratliff [77]) The algorithm proposed by Bellmore and Ratliff (Problem 13.21) was coded in FORTRAN II for an IBM 7094 computer. Set covering problems with 30 rows and variables ranging from 30 to 90 were solved within 1.12 to 7.64 seconds with unit cost coefficients. They were also solved within .98 to 7.64 seconds with random cost coefficients.

(House, Nelson and Rao [334]) Their algorithm is similar to the previous one. It starts with an integer solution of problem SC which is also an extreme point of the corresponding linear program's feasible region. If this solution does not have all the variables x_j at value one, a new SC problem is created by adding a constraint which does not eliminate the optimal binary solution and guarantees that another component x_j, not at 1, will have value 1. (Note that if all x_j's are at 1 and $\mathbf{x}$ corresponds to an extreme point, then it must be optimal since the columns are linearly independent.) The new problem has the same number of variables but the additional constraint. An integer solution to the new problem is obtained so that it is an extreme point of the corresponding linear programming feasible region. The process is repeated. That is, if this solution does not contain all the x_j's at a nonzero value, another constraint is added and a new problem is created, and so on. The algorithm terminates when an integer extreme point solution is obtained for the corresponding SC problem such that it contains all the x_j's at a positive value. The authors indicate than an 81 by 499 problem was reduced to an 81 by 180 problem by "precomputational analysis" and then solved in 1.261 minutes using an IBM 7094 computer.

(G.L. Martin ([415]) Martin uses a variation of Gomory's cutting plane algorithm (Chapter 4) in solving large set covering problems. His Accelerated Euclidean Algorithm basically consists of solving the associated linear program, returning the optimal simplex tableau to a (not necessarily optimal) tableau with an integer solution, resolving the linear program, and so on. The process terminates when an all integer solution in an optimal linear programming tableau is reached or when reoptimization is not possible. Transforming the optimal tableau to one with an all integer solution in an optimal linear programming tableau is reached or when reoptimization is not possible. Transforming the optimal tableau to one with an all integer solution is accomplished by "composite Gomory cuts," which is equivalent to applying a series of classical Gomory cuts derived from the same row. The details of the algorithm are in Chapter 4 (Appendix B). The algorithm appears to be successful. Problems with several hundred rows and a few thousand variables were reported to have been solved in a relatively short time, after adding a few cuts.

(Salkin and Koncal [514]) Using their all integer algorithm (Problem 13.19), computational experience on a series of smaller problems was excellent. Each of 14 smaller problems, with a maximum of 30 rows and 90 columns, was solved twice. Once as a set covering problem and once as a set partitioning problem. Total computation time on a UNIVAC 1101 was 53 seconds. In addition, a 104 by 133 problem was solved in 1.4 seconds as a set covering problem and in 5.7 seconds as a set partitioning problem. However, because of computer storage limitations a revised simplex method had to be adopted for larger problems. As a result, the bookkeeping, and thus the computation time, substantially increased. Performance was erratic and is tabulated below.

Problem Size		Time	
		Inequality	Equality
m	n	problem	problem
200	300	15.9	+
200	413	26.7	+
50	450	++	++
36	455	18.6	++
104	498	11.0	++
46	683	++	++
26	777	++	++
50	905	++	++

+ Storage exceeded.
++ Time limit of 120 seconds was exceeded.

A.2 Enumerative Algorithms

(Garfinkel and Nemhauser [214]) The algorithm (Section 13.5.1) has been programmed in FORTRAN IV and MAP (FORTRAN IV assembly language for the IBM 7094) subroutines for an IBM 7094 machine. Computational experience is encouraging. A randomly generated problem with 100 rows and 1400 variables (the density of $E = 0.15$) was solved in 895 seconds. A 37 row and 1790 variable problem of political redistricting type was solved in 49 seconds. The density of E was 0.25.

A crew scheduling problem of 104 rows and 133 variables was solved in 2 seconds. This algorithm, because of the reliance on dominance criteria, tends to work well on problems which have $\mathbf{E}$ matrices with a relatively high density of 1's.

(Lemke, Salkin, and Spielberg [392]) The method of Lemke, Salkin, and Spielberg appears to be reasonably successful in solving set covering and set partitioning problems. Fourteen 30-row problems with variable size ranging from 30 to 90 were solved in a total of 5.3 minutes as covering problems and also as partitioning problems. The one density of $\mathbf{E}$ for all these problems was 0.07. Computational experiences with some of the larger problems are given below.

Problem	m	n	density of 1's	Time (minutes)
1	200	300	0.020	7.689
2	200	413	0.020	10.431
3	104	498	0.024	2.631
4	200	500	0.019	13.392
5	46	683	0.088	0.948
6	50	905	0.064	11.166
7	134	1642	0.023	24.800

The algorithm was coded in FORTRAN for an IBM 360 model 50 computer. In addition, the authors report that for the set covering problem, the integer solution found from the first linear program, using the procedure outlined in Section 13.4 (expression 7), always proved to be less than 1% away from the minimal one. (This suggests aborting the enumeration and accepting the extracted solution as "optimal.")

(Marsten [410]) Marsten presents an enumerative algorithm for the set partitioning problem which uses linear programming and dominance procedures. A special class of finite mappings is enumerated rather than the possible zero-one vectors. However, the underlying method appears to be a search enumeration (Chapter 9). It can also accommodate crew base constraints (Fig. 13.1). Computational experience is summarized below.

Problem	m	n	density of 1's	(Time)
1	90	303	.07	40.64
2	63	1,641	.11	188.76
3	111	4,826	.05	479.77
4	200	2,362	.015	418.86
5	419	21,585	.02	> 3600.00

All times are in seconds and an IBM 360/91 computer was used. Although these results are quite good, we should note that in each problem the optimal integer solution is near the optimal linear programming one. This tends to curtail the length of the enumeration since the implicit enumeration criteria derived from the linear program become more effective.

(Michaud [429]) Michaud presents an enumerative algorithm for the set partitioning problem. The associated linear program is first solved. Then the basic variables (x_B) are solved for in terms of the nonbasic variables (x_N), and the objective function is written in terms of the reduced costs ($\bar{c}$) and nonbasic variables (x_N). An algorithm similar to the one explained in Chapter 11 (Section 11.3.3), which finds the smallest value of $\bar{c}x_N$ so that (x_B, x_N) is a feasible integer solution, is then employed. Computational experience on some larger problems, using an IBM 360/75 computer, is tabulated below.

Problem	m	n	Time(seconds)
1	42	249	6.5
2	67	536	20.0
3	12	538	5.0
4	18	273	5.0

(Pierce [460]) Pierce discusses an enumerative procedure which does not use linear programming. A summary of the algorithm may be found in Balinski and Spielberg [51]. In this method, the possible binary vectors are considered in a certain lexicographic order. At each node attempts are made to discard some of the potential solutions. The actual devices are similar to those used in standard enumerative techniques (Chapter 9, Section 9.3). The procedure takes advantage of the structure of the matrix by pre-ordering the columns so that the 1's in the matrix form a staircase pattern. Also, lower bounds on the objective function are obtained by solving either a continuous or an integer knapsack problem (Chapter 12) with the constraint obtained by summing over those not yet satisfied rows. A number of randomly generated problems ranging in size between 10 by 60 and 35 by 100 were solved in less than .3 second using an IBM 7094 computer. A number of problems representing truck dispatching situations of sizes 12 by 538, 12 by 793, and 15 by 575 were solved in usually less than a minute. It appears that the larger of these problems were solved much faster with the knapsack subproblems used to generate lower bounds than without it. For randomly generated problems, however, the use of the knapsack subproblem increased computation time.

(Rao [472]) Rao presents a search enumerative algorithm for the "multiple set covering problem" – a set covering problem where the 1's in the right-hand side are replaced by positive integers. Computational results on a series of 11 smaller problems ranging in size from 5 constraints and 10 variables to 50 constraints and 100 variables were excellent. Using an IBM 360/65 computer the total computation time was 1.88 minutes.

A.3 Summary

The efficiency of the algorithms, as in most integer programming procedures, is largely data dependent. For the set covering or set partitioning problem, this principally relates to the structure and density of **E**. In particular, if **E** has relatively few 1's, then linear programming solutions will often be integer. Thus, algorithms which solve or use linear programs (e.g., [392], [410], [415], [514]) should work relatively well for these problems. On the other hand, for problems whose coefficient matrices are dense in 1's, algorithms which rely on dominance procedures (e.g., [214]) and not on linear programming should work best. Of course, for those techniques which rely on linear programming and dominance criteria and for those problems whose coefficient matrices have "in-between densities," it is not clear which method and problem should be matched.

Chapter 14

The Fixed Charge Problems: The Plant Location Problem and Fixed Charge Transportation Problem

14.1 Introduction

14.1.1 The Plant Location Problem

The *capacitated plant location problem* is the mixed integer program

$$(FCP) \quad \text{minimize} \quad \sum_{i=1}^{m}\sum_{j=1}^{n} g_{ij}z_{ij} + \sum_{i=1}^{m} f_i x_i = z \tag{1}$$

$$\text{subject to} \quad \sum_{i=1}^{m} z_{ij} = d_j \quad (j=1,\ldots,n), \tag{2}$$

$$\sum_{j=1}^{n} z_{ij} \leq M_i x_i \quad (i=1,\ldots,m), \tag{3}$$

$$z_{ij} \geq 0 \quad (\text{all } i,j), \tag{4}$$

$$\text{and} \quad x_i = 0 \text{ or } 1 \quad (i=1,\ldots,m), \tag{5}$$

where, as indicated in Chapter 1 (Example 1.5), the model represents the situation in which there are m plants that produce a single commodity for n customers. Each plant i can produce at most M_i units and each customer j requires d_j units, f_i is a positive fixed cost associated with plant i, and g_{ij} is a positive per unit shipping cost from plant i to customer j. The variables are z_{ij} and x_i, which represent the amount shipped from i to j and whether a plant is open ($x_i = 1$) or closed

($x_i = 0$), respectively. The structure of the problem suggests several useful results.

Remarks

1. If $x_i = 0$, constraints (3) and (4) imply $z_{ij} = 0$ for $j = 1, \ldots, n;$[1] when $x_i = 1$, inequality (3) reduces to

$$\sum_{j=1}^{n} z_{ij} \leq M_i. \tag{6}$$

Thus, for a *fixed* binary vector $\mathbf{x} = (x_1, \ldots, x_m)$ it follows that problem FCP is a transportation problem (which may be easier to solve than a general linear program). As an example, consider Fig. 14.1 which represents a 3-plant, 4-customer problem where the constraints (3) are written first. If $x_1 = x_3 = 1$ and $x_2 = 0$ (i.e., plants 1 and 3 are open, and plant 2 is closed), the second constraint and the part of the matrix corresponding to plant 2 are dropped. The result is a 2-origin (plant), 4-destination (customer) transportation problem with origin supplies M_1 and M_3, and destination requirements d_1, d_2, d_3, d_4. Note that if the slack variable in (6) is positive implies that there is storage at the origin.

2. As every extreme point solution of a transportation problem is integer, it follows that an optimal solution to problem FCP has z_{ij} integer (for all i, j) and occurs at an extreme point of the region defined by the inequalities (2), (4), and (6) (where $i = 1, \ldots, m$). (Why?) Problem FCP is often referred to as a fixed charge problem.[2]

---

[1]That is, shipments cannot be made from a closed plant.

[2]In general, any mixed integer program of the form:

$$\begin{aligned} \text{Minimize} \quad & \mathbf{cy} + \mathbf{dx}, \\ \text{subject to} \quad & \mathbf{y} \in Y = \{\mathbf{y} | \mathbf{Ay} = \mathbf{b}, \mathbf{y} \geq 0\} \end{aligned}$$

is termed a fixed charge problem, where $\mathbf{A}$ is a coefficient matrix, $\mathbf{x} = (x_j), \mathbf{y} = (y_i)$, and $x_j = 0$ if $y_j = 0$ and $x_j = 1$ if $y_j > 0$.

Hirsch and Dantzig [329] have shown that an optimal solution to the general fixed charge problem is an extreme point of the region Y, and that under certain circumstances (see Problem 14.13) the fixed charge problem can be solved as the linear program: Minimize $\mathbf{cy}$, where $\mathbf{y} \in Y$. Heuristic techniques for the general problem, which seek good solutions by examining extreme points of the set Y, are described by Balinski [47], Baumol and Wolfe [57], Cooper and Drebes [146], Denzler [168], Murty [437], and Steinberg [551]. Related results are discussed in Problems 14.2, 13, 14, and 15.

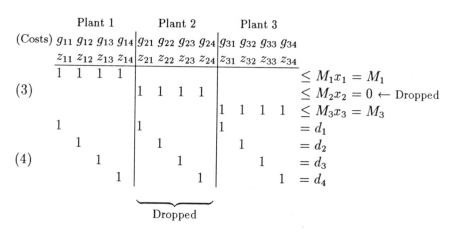

Figure 14.1: A three-plant, four-customer problem (with plants 1 and 3 open, 2 closed)

3. Suppose $M_i \geq \sum_{j=1}^{n} d_j$ for $i = 1, \ldots, m$; that is, each plant is capable of satisfying all customer demands. Then constraint (3) will always be satisfied when $x_i = 1$. Thus, for a *fixed* binary vector $\mathbf{x}$, the origin constraints (3) are either redundant ($x_i = 1$) or indicate that certain variables must have value 0 ($x_i = 0$). This means that the resulting transportation problem may be solved by inspection. In particular, the total shipping cost is minimized if the demand of each customer j is met from the open plant i with minimal cost g_{ij}. In Fig. 14.1, an optimal solution is $z_{p_j j} = d_j$ ($j = 1, 2, 3, 4$), where $p_j \in \{1, 3\}$ is selected such that

$$g_{p_j j} = \text{minimum} \ \{g_{1j}, g_{3j}\}$$

for $j = 1, 2, 3, 4$. When the plants are uncapacitated, problem FCP is often referred to as a simple plant location problem or, more conveniently, as a *plant location problem* (PLP).

14.1.2 The Fixed Charge Transportation Problem

The fixed charge transportation problem is the mixed integer program

$$(FTP) \quad \text{minimize} \quad \sum_{i=1}^{m}\sum_{j=1}^{n}(g_{ij}z_{ij} + f_{ij}y_{ij}) \tag{7}$$

$$\text{subject to} \quad \sum_{i=1}^{m} z_{ij} = d_j \qquad (j = 1,\ldots,n) \tag{8}$$

$$\sum_{j=1}^{n} z_{ij} \leq M_i \qquad (i = 1,\ldots,m) \tag{9}$$

$$z_{ij} \leq u_{ij}y_{ij} \quad \text{(all } i,j) \tag{10}$$

$$z_{ij} \geq 0 \quad \text{(all } i,j) \tag{11}$$

$$\text{and} \qquad y_{ij} = 0 \text{ or } 1 \quad \text{(all } i,j); \tag{12}$$

where $u_{ij} = \text{minimum }(M_i, d_j)$. The model represents the situation in which there are m plants that produce a single commodity for n customers. Each plant i can produce at most M_i units and each customer j requires d_j units. The per unit shipping cost from i to j is g_{ij} and a fixed cost f_{ij} is also incurred whenever there is a shipment from i to j. That is, letting the variable z_{ij} be the number of units shipped from plant i to customer j, the total shipping cost from i to j, denoted by $h_{ij}(z_{ij})$, is given by

$$h_{ij}(z_{ij}) = \begin{cases} f_{ij} + g_{ij}z_{ij} & \text{if } z_{ij} > 0 \\ 0 & \text{if } z_{ij} = 0. \end{cases}$$

The variable y_{ij} represents whether any shipment from plant i to customer j is made ($y_{ij} = 1$), or not made ($y_{ij} = 0$). Note that if $y_{ij} = 0$, constraint (10) implies $z_{ij} = 0$, and there is no fixed or variable shipping cost from plant i to customer j. Also, as in the case of the plant location problem, for a fixed binary matrix $\{y_{ij}\}$, problem FTP is a classical transportation problem.

The previous discussion suggests that the problem just described may be solved by an enumerative technique which solves linear programming subproblems. Before presenting the algorithms, we discuss the development in the literature.

Background

In 1954 Hirsch and Dantzig [329] formulated the general fixed charge problem. Other formulations include the fixed cost transportation problem by Balinski [47], and the plant location problem by Balinski [49], and (as it appears here) by Efroymson and Ray [187].

Algorithms for these and related problems include the enumerative techniques appearing in Brown [102], Davis and Ray [163], Efroymson and Ray [187], Gray [292], [293], Jones and Soland [355], Khumawala [366], Marsten [411], Pinkus, Gross, and Soland [464], Sá [499], Spielberg [542], [544], [545], Steinberg [551], Kennington and Unger [365], and Barr, Glover, and Klingman [55]. A partitioning algorithm is given by Balinski and Wolfe [52] (see Chapter 9, Appendix A), group theoretic techniques are described by Kennington and Unger [364], and by Tompkins [569], and Lagrangian multiplier approaches (Chapter 12) are given by Geoffrion [227]. Heuristic procedures are in Armour and Buffa [9], Balinski [47], [49], Baumol and Wolfe [57], Cooper [142], Cooper and Drebes [146], Denzler [168], Drysdale and Sandiford [178], Dwyer [179], Feldman, Lehrer, and Ray [197], Kuehn and Hamburger [379], Kuhn and Baumol [381], Manne [404], Sá [499], Shannon and Ignizio [524], Steinberg [551], and Walker [591].

Computational experience (Appendix A) is reported by Cooper and Drebes [146], Denzler [168], Efroymson and Ray [187], Jain [342], Jones and Soland [355], Marsten [411], Sá [499], Spielberg [544], [545], Steinberg [551], Kennington and Unger [365], Barr, Glover, and Klingman [55], and Walker [591]. Applications and case studies are discussed by Beale and Tomlin [63], Ciochetto, Swanson, Lee and Woolsey [136], Popolopas [467], Revelle, Marks, and Liebman [478], Robers [479], and Wanty [592]. A survey is given by Bair and Hefley [17].

Related work includes the network problems described by Hakimi [304], [305], Levy [395], Marazana [406], Meihle [425], and Teitz and Bart [560]. Other articles that may be consulted are Almogy and Levin [4], Beckman and Marschak [65], Bellman [72], Bowman [83], Cooper [141], [143], [145], Deighton [166], Francis [206], Goldstone [273], Hirsch and Hoffman [330], Koopmans and Beckmann [374], Shycon and Maffei [537], Trehan [574], and Vietorisz and Manne [586].

14.2 Algorithms for the Plant Location Problem

In this section we discuss a branch and bound algorithm (Chapter 8) for the fixed charge problem and for the plant location problem. The procedure begins at the node (or point) with all x_i variables free. In general, as x_i can be 0 or 1, the enumeration proceeds by creating two lower nodes from a higher one whenever the linear programming solution (in the free x_i and the z_{ij} variables) at the higher node indicates that an improved solution is possible and x_i has a fractional value. One node is created by setting x_i to 0 and the second by fixing x_i at 1. The algorithm terminates when there are no nodes left which can possibly produce an improved solution, in which case the current best mixed integer solution is optimal. In the terminology of Chapter 8, a node which may produce an improved solution is a *dangling* node.

14.2.1   A Branch and Bound Algorithm for the Fixed Charge Problem

One technique for solving problem FCP is by a straightforward branch and bound enumeration. We briefly discuss its contents. The algorithm itself is not listed or illustrated.

The Linear Programs

The linear programs are *not* transportation problems since they have free x_i variables which can vary between 0 and 1, and thus they have to be solved by a standard simplex method.[3] If optimal tableaux are kept, successive problems may be solved by simple post-optimality procedures (see, e.g., Chapter 8, Appendix A).

Node and Branch Selection Strategies

Criteria which may be used in selecting the dangling node to create nodes from are discussed in Chapter 8 (Section 8.7). Two obvious strategies are (i) pick the dangling node with the lowest linear programming solution, and (ii) select the most recently created dangling node with the lowest linear programming solution. Rule (i) is intended to obtain the optimal mixed integer solution as quickly as possible; criterion (ii) allows for a simpler bookkeeping scheme to keep track of the

---

[3]This does not contradict what was said earlier. In particular, if *all* the x_i variables are fixed, then the linear program is a transportation problem. Also, Davis and Ray [163] have apparently been successful with a linear programming decomposition technique (see [166]) applied to the dual linear program, where the fixed charge problem is first formulated as suggested in Problem 14.3. The details are left to the references.

enumeration and usually requires less computer storage. It yields a tree as in Fig. 8.13 (Chapter 8).

Once the dangling node has been selected, a variable which is at a fractional value in its linear programming solution must be selected to set to 0 and 1 so that two new nodes are created. Besides the strategies described in Chapter 8, one rule is to select the eligible variable x_i whose plant has the largest capacity M_i. The intent is to obtain a mixed integer solution quickly. As the fixed cost for high capacity plants may be very large, another criterion is to pick x_i so that it yields the smallest value of f_i/M_i; this ratio may be thought of as the per unit production cost at plant i. Note that criteria involving the capacity M_i lose meaning when solving the plant location problem.

Heuristics for Obtaining Mixed Integer Solutions

If a good solution to the mixed integer program can be found early, it will naturally curtail the length of the enumeration. Note that when the x_i variables which are at fractional values in the linear programming solution are rounded to 1, we have a solution to the fixed charge (or plant location) problem since the right-hand side in the constraint (10) (or (3) for plant location) increases. Also, observe that an integer solution to problem FCP exists whenever a subset $S \subset \{1, \ldots, m\}$ of the plants are open (that is, $x_i = 1$ for $i \in S$) which have capacity exceeding the total demand,[4] that is $\sum_{i \in S} M_i \geq \sum_{j=1}^{n} d_j$. Thus one way to obtain a solution is to open a minimum number of plants (S) so that all the demands can be met and the total fixed cost (or the total value of f_i/M_i) for the plants opened is minimized. The values of the shipping variables z_{ij} can then be found by solving the resulting transportation problem (where $x_i = 1$ for $i \in S$ and $x_i = 0$ for $i \notin S$). To illustrate this procedure we present an example. Other heuristic techniques for obtaining a solution are left to the problems.

Example 14.1
Consider the $m = 4$ plant and $n = 5$ customer fixed charge problem below.

---

[4]This assumes that every open plant can service all the customers. If this is not the case, more plants than those in S may have to be opened; see Problem 14.4.

i \\ j	1	2	g_{ij} 3	4	$5 = n$	f_i	M_i	f_i/M_i
1	4	10	5	8	9	110	190	.58
2	6	1	18	2	11	130	350	.37
3	1	2	11	21	15	140	360	.39
$m = 4$	1	13	2	11	21	160	200	.80
d_j	100	90	110	120	50			

As $\sum_{j=1}^{5} d_j = 470$, we must open at least two plants. If we wish to minimize $\sum_i f_i$, plants 1 and 2 are opened. In this case, we have the standard transportation tableau below, where an optimal solution: $z_{11} = 80, z_{13} = 110, z_{21} = 20, z_{22} = 90, z_{24} = 120, z_{25} = 50$, and all other $z_{ij} = 0$, is found by inspection.

Plant \\ Customer	1	2	3	4	5	slack	Supplies
1	4 ⟨80⟩	10	5 ⟨110⟩	8	9	0	190
2	6 ⟨20⟩	1 ⟨90⟩	18	2 ⟨120⟩	11 ⟨50⟩	0 ⟨70⟩	350
Demands	100	90	110	120	50	70	540

The total value of the objective function including the fixed cost is $1870 + 240 = 2110$. Had we initially wanted to minimize $\sum_i f_i/M_i$, plants 2 and 3 would be opened. Then an optimal solution has $z_{22} = 90, z_{24} = 120, z_{25} = 50, z_{31} = 100, z_{33} = 110$, and $z_{ij} = 0$ otherwise. Its total cost is $2190 + 270 = 2460$. An optimal solution to the fixed charge problem has plants 1, 2, and 4 open with shipping variables $z_{13} = 10, z_{15} = 50, z_{22} = 90, z_{24} = 120, z_{41} = 100, z_{43} = 100$, and all other $z_{ij} = 0$. The total value of the objective function is $1130 + 400 = 1530$. Note that the best heuristic solution we found incurs about a 25% error. (Why?)

14.2.2 A Branch and Bound Algorithm for the Plant Location Problem (Efroymson and Ray [187])

The efficiency of a branch and bound algorithm largely depends on how quickly the linear programs can be solved. When each plant can satisfy all customer demands, it is possible to reformulate the plant location problem (1) through (5) so that the linear programs can be solved *by inspection*. To transform the problem, define y_{ij} to be the fraction of customer j's demand satisfied by plant i; that is,

$$y_{ij} = z_{ij}/d_j \tag{13}$$

for all i, j. Before making direct use of (13), we first note that when $M_i \geq \sum_{j=1}^{n} d_j$ $(i = 1, \ldots, m)$, the inequality (3) is only necessary to ensure that shipments are not allowed from a closed plant. This allows us to replace constraint (3) by

$$\sum_{j=1}^{n} y_{ij} \leq n x_i \quad (i = 1, \ldots, m), \tag{3'}$$

since $x_i = 0$ implies $y_{ij} = 0$ $(i = 1, \ldots, m)$ and thus no demand will be satisfied by plant i. Now, substituting $y_{ij} d_j$ for z_{ij} in expressions (1), (2), and (4), and letting $c_{ij} = g_{ij} d_j$ yields the equivalent plant location problem

$$\text{(PLP)} \quad \text{minimize} \quad \sum_{i=1}^{m} \sum_{j=1}^{n} c_{ij} y_{ij} + \sum_{i=1}^{m} f_i x_i \tag{1'}$$

$$\text{subject to} \quad \sum_{i=1}^{m} y_{ij} = 1 \quad (j = 1, \ldots, n), \tag{2'}$$

$$\sum_{j=1}^{n} y_{ij} \leq n x_i \quad (i = 1, \ldots, m), \tag{3'}$$

$$y_{ij} \geq 0 \quad (\text{all } i, j), \tag{4'}$$

$$\text{and} \quad x_i = 0 \text{ or } 1 \quad (i = 1, \ldots, m). \tag{5'}$$

Problem PLP with $x_i = 0$ or 1 replaced by $0 \leq x_i \leq 1$ $(i = 1, \ldots, m)$ may be solved by inspection. In particular, every optimal linear programming solution will satisfy the constraints (2)', and, as $f_i > 0$, will have the smallest nonnegative values for x_i $(i = 1, \ldots, m)$ so that the constraints (3)' are satisfied. Therefore, at any linear programming optimum

$$\sum_{j=1}^{n} y_{ij} = nx_i \quad (i = 1, \ldots, m),$$

or

$$\frac{1}{n} \sum_{j=1}^{n} y_{ij} = x_i.$$

Substituting for x_i in the linear program yields

$$\text{LP}^0 \quad \text{minimize} \quad \sum_{i=1}^{m} \sum_{j=1}^{n} (c_{ij} + f_i/n) y_{ij} \tag{1$'$}$$

$$\text{subject to} \quad \sum_{i=1}^{m} y_{ij} = 1 \quad (j = 1, \ldots, n) \tag{2$'$}$$

$$\text{and} \quad y_{ij} \geq 0 \quad (\text{all } i, j), \tag{4$'$}$$

An optimal solution to this problem (see Fig. 14.2) is

$$\left. \begin{array}{l} y_{ij} = \begin{cases} 1 & \text{if } c_{ij} + f_i/n = \displaystyle\min_{k=1,\ldots,m} (c_{kj} + f_k/n) \\ 0 & \text{otherwise} \end{cases} \\[2em] \text{for } j = 1, \ldots, n, \text{ and} \\[2em] x_i = \dfrac{1}{n} \sum_{j=1}^{n} y_{ij} \quad (i = 1, \ldots, m). \end{array} \right\} \tag{14}$$

i	1			2				m			
j	1	2 ... n		1	2 ... n		...	1	2 ... n		
Variables	y_{11} y_{12}	$\cdots$ y_{1n}		y_{21} y_{22}	$\cdots$ y_{2n}			y_{m1} y_{m2}	$\cdots$ y_{mn}	(≥ 0)	
Costs*	$\bar{c}_{11}$ $\bar{c}_{12}$	$\cdots$ $\bar{c}_{1n}$		$\bar{c}_{21}$ $\bar{c}_{22}$	$\cdots$ $\bar{c}_{2n}$			$\bar{c}_{m1}$ $\bar{c}_{m2}$	$\cdots$ $\bar{c}_{mn}$	(> 0)	
	1	1 ... 1								$= 1$	
constraints				1	1 ... 1					$= 1$	
(2)$'$							$\ddots$				
								1	1 ... 1	$= 1$	

$$^*\bar{c}_{kj} = c_{kj} + f_k/n$$

Figure 14.2: Linear program LP^0

Expression (14) can be generalized so that the linear programs which appear during a branch and bound enumeration may also be solved by inspection. At any node, let I_0, I_1, and I_f be the set of indices i such that x_i is fixed at 0, fixed at 1, and free, respectively. Therefore, the linear program LP^0 becomes

$$LP^k \sum_{i \in I_1} f_i + \text{ minimize } \left(\sum_{i \in I_f} \sum_{j=1}^{n} (c_{ij} + f_i/n) y_{ij} + \sum_{i \in I_1} \sum_{j=1}^{n} c_{ij} y_{ij} \right)$$

$$\text{subject to } \sum_{i \in I_1 \cup I_f} y_{ij} = 1 \quad (j = 1, \ldots, n)$$

$$\text{and } \quad y_{ij} \geq 0 \quad \quad (\text{all } i, j; i \notin I_0).$$

An optimal solution to this problem may be written as

$$
\left.
\begin{array}{l}
y_{ij} = \begin{cases} 1 & \text{if } c_{ij} + l_i/n = \displaystyle\operatorname*{minimum}_{k \in I_1 \cup I_f} (c_{kj} + l_k/n) \\ 0 & \text{otherwise} \end{cases} \\[2em]
\text{for } j = 1, \ldots, n, \text{ and} \\[1em]
x_i = \dfrac{1}{n} \displaystyle\sum_{j=1}^{n} y_{ij} \quad (i \in I_f).
\end{array}
\right\} \quad (15)
$$

where

$$l_k = \begin{cases} f_k & \text{if } x_k \text{ is free} & (k \in I_f) \\ 0 & \text{if } x_k \text{ is fixed at 1} & (k \in I_1). \end{cases}$$

Using (15), the linear programs are solved by inspection. The other ingredients in the branch and bound enumeration (e.g., node and branch selection strategies) are essentially the same as in the previous section. To illustrate the algorithm we solve Example 14.1, where it is now assumed that each plant can produce all that is needed. Notice that solutions to the plant location problem may be found by partitioning the set of plants $\{1, \ldots, m\}$ into the subsets I_1 and I_0 and using (15) with I_f null.[5] It is not clear how to choose I_1 (which must have at least one element), so that this heuristic produces a good solution (Problem 14.9). Other criteria which may allow implicit enumeration are in

---

[5] An examination of expression (9) with I_f null verifies what was said in Remark 3 (Section 14.1).

Node Description						Linear Programming Solution (from (9))												
Node no.	Node selected	x_1	x_2	x_3	x_4	(A) $\sum_{i \in I_l} f_i$	i such that $y_{ij}=1$					(B) $\sum_{i,j} \bar{c}_{ij}$	x_i				(A)+(B) value	Dangling nodes	Solution z^*
							$j=1$	2	3	4	5		x_1	x_2	x_3	x_4			
1	—	≥0	≥0	≥0	≥0	0	3	2	4	2	1	1234	1/5	2/3	1/5	1/5	1234	1	—
2	1	=0	≥0	≥0	≥0	0	3	2	4	2	2	1338	0	3/5	1/5	1/5	1338	1,2	—
3	1	=1	≥0	≥0	≥0	110	3	2	4	2	1	1212	1	2/5	1/5	1/5	1322	2,③	—
4	3	=1	=0	≥0	≥0	110	3	3	4	1	1	1998	1	0	2/5	1/5	2108	2,3,4	—
5	3	=1	=1	≥0	≥0	240	3	2	4	2	1	1160	1	1	1/5	1/5	1400	2,4,⑤	—
6	5	=1	=1	=0	≥0	240	4	2	4	2	1	1164	1	1	0	2/5	1404	2,4,5,6	—
7	5	=1	=1	=1	≥0	380	3	2	4	2	1	1132	1	1	1	0	1512	2,4,⑥,7	—
8	6	=1	=1	=0	=0	240	1	2	1	2	1	1730	1	1	0	0	1970*	2,6,7	1970
9	6	=1	=1	=0	=1	400	4	2	4	2	1	1100	1	1	0	1	1500*	②	1500
10	2	=0	=1	≥0	≥0	0	3	3	4	4	3	2718	0	0	2/5	2/5	2718	②	1500
11	2	=0	=1	≥0	≥0	130	3	2	4	2	2	1260	0	1	1/5	2/5	1390	⑪	1500
12	11	=0	=1	=0	≥0	130	4	2	4	2	2	1264	0	1	0	2/5	1394	11,12	1500
13	11	=0	=1	=1	≥0	270	3	2	4	2	2	1232	0	1	1	1/5	1502	⑫	1500
14	12	=0	=1	=0	=0	130	2	2	2	2	2	3460	0	1	0	0	3590	12	1500
15	12	=0	=1	=0	=1	290	4	2	4	2	2	1200	0	1	0	1	1490*	—	1490

*Indicates improved integer solution found.

Table 14.1: Summary of Computations

Problem 14.10. A partitioning algorithm which uses some of the results given here is in Appendix A of Chapter 10.

Example (from Example 14.1)

Consider the previous problem, where we now assume that $M_i \geq \sum_{j=1}^{n} d_j$ $= 470$ $(i = 1, 2, 3, 4)$. Nodes are selected according to the smallest linear programming solution at the most recently created pair of nodes, and branches correspond to the minimum value of f_i. The computations are tabulated in Table 14.1 (where $\bar{c}_{ij} = c_{ij} + f_i/n$) and the corresponding tree is in Fig. 14.3. Remember that $c_{ij} = g_{ij}d_j$ for all i, j, so that the data are now:

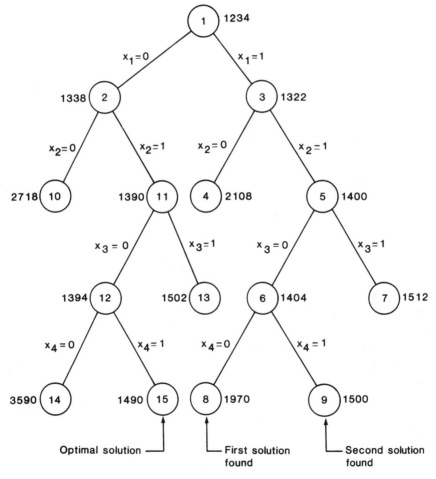

Figure 14.3: Branch and bound tree

j			c_{ij}				
i	1	2	3	4	$5 = n$	f_i	f_i/n
1	400	900	550	960	450	110	22
2	600	90	1980	240	550	130	26
3	100	180	1210	2520	750	140	28
$m = 4$	100	1170	220	1320	1050	160	32

and

j		$\bar{c}_{ij} = c_{ij} + f_i/n$			
i	1	2	3	4	5
1	422	922	572	982	472
2	626	116	2006	266	576
3	128	208	1238	2548	778
4	132	1202	252	1352	1082

14.3 Branch and Bound Algorithm for the Fixed Charge Transportation Problem

The fixed charge transportation problem can also be solved by the branch and bound procedure described in Section 14.2 where the variables y_{ij} are used as the branching variables. In the plant location problem we had to solve a general linear programming problem in order to find the lower bound at each dangling node. However, at each dangling node the linear programming relaxed problem (that is, $y_{ij} = 0$ or 1 is replaced by $0 \leq y_{ij} \leq 1$) corresponding to the fixed charge transportation problem can be converted to a transportation problem using the following theorem.

Theorem 14.1 *If (z_{ij}^*, y_{ij}^*), is the optimal solution to the linear programming relaxation of FTP, then constraint (10) satisfies $z_{ij}^* = u_{ij} y_{ij}^*$.*

The proof of the above result follows directly from the fact that f_{ij} is positive, and if $z_{ij}^* < u_{ij} y_{ij}^*$, y_{ij}^* could be reduced to value z_{ij}^*/u_{ij} without violating any other constraint. This would yield a lower value for the objective function, contradicting the optimality of (z_{ij}^*, y_{ij}^*), and thus the result.

Using Theorem 14.1, we can simplify the linear programming relaxation of FTP. In particular, using the strict equality of constraint (10), we may eliminate y_{ij} by substituting $y_{ij} = (z_{ij}/u_{ij})$. The problem reduces to a transportation problem in the variables z_{ij} as indicated below.

$$\text{Minimize} \quad \sum_{i=1}^{m} \sum_{j=1}^{n} (g_{ij} + \frac{f_{ij}}{u_{ij}}) z_{ij}$$

$$\text{subject to} \quad \sum_{i=1}^{m} z_{ij} = d_j \quad (j = 1, \dots, n)$$

$$\sum_{j=1}^{n} z_{ij} \leq M_i \quad (i = 1, \dots, m)$$

$$\text{and} \quad z_{ij} \geq 0 \quad (\text{all } i, j),$$

where as defined before, $u_{ij} = \text{minimum } (d_j, M_i)$.

Remarks

1. If z_{ij}^* is the optimal solution to the transportation problem, then $z_{ij}^*, y_{ij}^* = z_{ij}^*/u_{ij}$ is the optimal solution to the linear programming relaxation of FTP.

2. z_{ij}, y_{ij} obtained by

$$z_{ij} = x_{ij}^*$$
,

$$y_{ij} = \begin{cases} 1 & \text{if } z_{ij}^* > 0 \\ 0 & \text{if } z_{ij}^* = 0 \end{cases}$$

is a feasible solution to FTP with cost $z = \sum_{i=1}^{m} \sum_{j=1}^{n} (g_{ij} z_{ij} + f_{ij} y_{ij})$. Balinski [47] has shown that in many cases this solution is very close to the optimal one. Therefore, in a branch and bound algorithm, it makes sense to use this solution as an upper bound. This is illustrated via an example.

Example 14.2 (Balinski [47])
Consider the fixed charge transportation problem with 4 plants and 3 customers given below. Each entry (i, j) in the matrix lists (g_{ij}, f_{ij}).

i \ j	Customers			Capacity
	1	2	3	(M_i)
Plants 1	$(2, 10)$	$(3, 30)$	$(4, 20)$	10
2	$(3, 10)$	$(2, 30)$	$(1, 20)$	30
3	$(1, 10)$	$(4, 30)$	$(3, 20)$	40
4	$(4, 10)$	$(5, 30)$	$(2, 20)$	20
Demand (d_j)	20	50	30	

1. To solve the linear programming relaxation of FTP, we construct the equivalent transportation problem with cost $\bar{g}_{ij}$ given by

$$\bar{g}_{ij} = g_{ij} + \frac{f_{ij}}{u_{ij}} = g_{ij} + \frac{f_{ij}}{\min\,(M_i, d_j)}.$$

Using this equation we obtain the cost matrix $(\bar{g}_{ij})$ below.

$(\bar{g}_{ij})$

i \ j	1	2	3	M_i
1	3	6	6	10
2	3.5	3	1.67	30
3	1.5	4.75	3.67	40
4	4.5	6.5	3	20
d_j	20	50	30	

2. The optimal solution to the transportation problem, found by any standard transportation method, gives $z = 321.7$ and is listed below.

$$(z_{ij}^*)$$

$$j$$

		1	2	3
	1	0	10	0
i	2	0	20	10
	3	20	20	0
	4	0	0	20

3. From (z_{ij}^*) we can derive a solution to FTP using

$$z_{ij} = z_{ij}^*$$

$$\text{and } y_{ij} = \begin{cases} 1 & \text{if } z_{ij}^* > 0 \\ 0 & \text{if } z_{ij}^* = 0. \end{cases}$$

This gives the solution listed below which has $z = 360$.

$$(z_{ij}, y_{ij})$$

$$j$$

		1	2	3
	1	$(0,0)$	$(10,1)$	$(0,0)$
i	2	$(0,0)$	$(20,1)$	$(10,1)$
	3	$(20,1)$	$(20,1)$	$(0,0)$
	4	$(0,0)$	$(0,0)$	$(20,1)$

Hence 321.7 and 360 are the lower bound and upper bounds, respectively, on the objective function z to FTP.

Kennington and Unger [365], and Barr, Glover, and Klingman [55] recently developed efficient branch and bound algorithms using these results and a node selection rule based on penalties which is a specialization of the penalty calculations first suggested by Tomlin [567] in a branch and bound algorithm for the general mixed integer programming problem. An efficient heuristic algorithm is also suggested by Walker [591]. Some computational results with these algorithms are in Appendix A.

Problems

Unless otherwise stated, the fixed charge problem (FCP) refers to the capacitated plant location problem and the plant location problem (PLP) means the simple (or uncapacitated) plant location problem.

Problem 14.1 Consider the following mixed integer program, referred to as a *fixed charge transportation problem.*

$$\text{(FCTP)} \quad \text{minimize} \quad \sum_{i=1}^{m}\sum_{j=1}^{n}(g_{ij}z_{ij} + f_{ij}x_{ij})$$

$$\text{subject to} \quad \sum_{i=1}^{m}z_{ij} = d_j \quad (j = 1,\ldots,n), \tag{1}$$

$$\sum_{j=1}^{n}z_{ij} \le M_i \quad (i = 1,\ldots,m), \tag{2}$$

$$z_{ij} \ge 0 \qquad \text{(all } i,j), \tag{3}$$

$$z_{ij} \le m_{ij}x_{ij} \quad \text{(all } i,j),$$

$$\text{and} \qquad x_{ij} = 0 \text{ or } 1 \quad \text{(all } i,j),$$

where g_{ij} is the per unit cost of transporting goods from source i to customer j, f_{ij} is a positive fixed charge if route i,j is open (i.e., if the amount shipped $z_{ij} > 0$), d_j is the demand of customer j, M_i is the production capacity at source i, and $m_{ij} \ge \min(d_j, M_i)$.

a) Discuss situations in which problem FCTP would appear.

b) Show that problem FCTP is a special case of the fixed charge problem FCP (Section 14.1).

c) Describe what seems to be an efficient algorithm for this problem. Explain why it should be efficient. Can expression (15) be used?

Problem 14.2 Given the fixed charge transportation problem described in the preceding problem:

a) Prove that an optimal solution occurs at an extreme point of the region defined by the inequalities (1), (2), and (3). (*Hint:* Show that the function

$$h_{ij}(z_{ij}) = g_{ij}z_{ij} + h_{ij}x_{ij}$$

is concave for $z_{ij} \geq 0$, where $h(t)$ is concave by definition if

$$h(\lambda t_1 + (1 - \lambda)t_2) \geq \lambda h(t_1) + (1 - \lambda)h(t_2), \text{ for } 0 \leq \lambda \leq 1.$$

Then use the fact that the sum of concave functions is concave and appeal to the well-known result that for a closed convex set, bounded from below, the global minimum of a concave function is taken on at one or more of the extreme points of the set.)

b) Can an algorithm for problem FCTP be developed from this result? What difficulties might arise?

Problem 14.3   Explain why the constraints $\sum_{j=1}^{n} z_{ij} \leq M_i$ $(i = 1, \ldots, m)$ and $z_{ij} \leq M_i x_i$ (all i, j) can be used in place of $\sum_{j=1}^{n} z_{ij} \leq M_i x_i$ $(i = 1, \ldots, m)$ in the fixed charge problem. What analogous replacement is available in the plant location problem? If either substitution is made, does the resulting problem seem easier to solve when an enumerative technique is used? When a cutting plane algorithm is used? Why?

Problem 14.4   Suppose, in the fixed charge problem or in the plant location problem, that plant i can only service some (but not all) of the customers and that customer j can receive the goods from only some (but not all) of the plants.

a) Reformulate both problems so that these constraints are taken into account.

b) Are the branch and bound algorithms described in Section 14.2 valid for the mixed integer program found in (a)? What modifications (if any) must be made? Can these additional requirements be used to develop branching rules?

c) Under what conditions can we say that either problem found in (a) has an integer solution?

Problem 14.5 (Ciochetto, et al. [136]) Show that the following situation, referred to as the *lock box problem*, can be modeled as a plant location problem: A company which operates over a wide geographic area has to decide where to locate one or more "lock boxes"; a lock box ·

is a post office box-local bank combination which receives and processes customer remittance. A set of possible lock box locations is known. The cost of each lock box (which includes the post office box rental and the bank charges for processing checks), as well as the time required for each to process the checks, can be determined. The company would like to know (i) how many lock boxes it should have, (ii) where they should be located, and (iii) which customers should remit to which lock boxes; where (i), (ii), and (iii) are answered while some weighted average of total processing time and lock box cost is minimized.

Problem 14.6   In Section 14.2.1 heuristic techniques for obtaining solutions to the fixed charge problems are given. The plants opened are determined on the basis of the fixed costs. Give analogous heuristics which open plants that tend to minimize (i) total shipping costs, and (ii) shipping and fixed costs. Apply your heuristics to Example 14.1. Under what circumstances will the heuristics given in Section 14.2.1 produce better (worse) solutions than the ones you developed? Why?

Problem 14.7   Solve the fixed charge problem below by the algorithm suggested in Section 14.2.1.

i \ j	1	2	g_{ij} 3	$4 = n$	f_i	M_i
1	2	4	7	6	450	350
2	1	3	8	7	400	300
$m = 3$	3	2	5	10	500	400
d_j	100	50	150	200		

Problem 14.8   Why can't expression (15) be used to solve linear programs associated with the fixed charge problem? Illustrate your answer.

Problem 14.9   A solution to the plant location problem exists whenever one or more plants are open. How might this be used to develop a technique which seeks good (but not necessarily optimal) solutions? In particular, what criteria would you use to open and close plants? Illustrate, by applying your scheme to the data appearing in Example 14.1 (with $M_i \geq \sum_{j=1}^{4} d_j$ for $i = 1, 2, 3, 4$).

Problem 14.10   (Efroymson and Ray [187]) At any node in a branch and bound algorithm for the plant location problem, let I_0, I_1, and I_f

be the set of indices i such that x_i is fixed at 0, fixed at 1, and free, respectively. Explain why each of the following results is true. Illustrate how they may implicitly enumerate by resolving the example appearing in Section 14.2.2. Relate your computations to a tree and compare it to Fig. 14.3. Can these criteria be used to develop branching rules? Explain.

a) Let

$$\Delta_{ij} = \min_{k \in I_1 \cup I_f} \{\max (c_{kj} - c_{ij}, 0)\} \quad \text{for } i \in I_f.$$

Then if $\sum_{j=1}^n \Delta_{ij} - f_i > 0, x_i = 1$ in any improved integer solution found from the node.

b) If for any $i \in I_f$

$$\min_{k \in I_1} (c_{kj} - c_{ij}) < 0 \text{ for some } j, \tag{4}$$

then $y_{ij} = 0$. If (4) holds for $j = 1, \ldots, n$, then $x_i = 0$ in any improved integer solution found from the node.

c) Let

$$\bar{\Delta}_{ij} = \min_{k \in I_1} \{\max (c_{kj} - c_{ij}, 0)\} \quad \text{for } i \in I_f.$$

Then if $\sum_{j=1}^n \bar{\Delta}_{ij} - f_i < 0, x_i = 0$ in all improved integer solutions produced from the node.

Problem 14.11 Consider the following plant location problem:

i \ j	1	2	c_{ij} 3	$4 = n$	f_i
1	100	150	80	100	800
2	110	140	70	90	600
3	90	160	100	80	700
$m = 4$	120	130	110	60	500

a) Solve it using the algorithm discussed in Section 14.2.2.

b) Repeat part(a) where it is now assumed that plants 2 and 4 cannot service customers 3 or 4.

In each case give the associated tree.

Problem 14.12

a) Treat the plant location problem as a zero-one integer program. Outline a search enumeration (Chapter 9) for the integer program. Does the enumeration proceed over the x_i variables, the y_{ij} variables, or both? Why? Describe the contents of a point algorithm. Does the procedure seem more efficient than the branch and bound algorithm given in Section 14.2.2? Why?

b) Repeat (a), where the problem is thought of as a mixed integer program with integer variables x_i.

c) Does the algorithm given in (a) or in (b) seem best? Why?

Problem 14.13   (Hirsch and Dantzig [329]) Given the general fixed charge problem:

$$\begin{aligned}
&\text{Minimize} \quad \mathbf{cy} + \mathbf{dx}\\
&\text{subject to} \quad \mathbf{y} \in Y = \{\mathbf{y} | \mathbf{Ay} = \mathbf{b}, \mathbf{y} \geq \mathbf{0}\},
\end{aligned}$$

where $\mathbf{A}$ is an m by n coefficient matrix with rank m, $\mathbf{x} = (x_j)$, $\mathbf{y} = (y_j)$, and $x_j = 0$ if $y_j = 0$ and $x_j = 1$ if $y_j > 0$. Show that if every extreme point of Y corresponds to a nondegenerate basic feasible solution and all fixed costs (i.e., elements of $\mathbf{d}$) are equal, then any optimal basic solution to the linear program

$$\begin{aligned}
&\text{minimize} \quad \mathbf{cy}\\
&\text{subject to} \quad \mathbf{y} \in Y
\end{aligned}$$

solves the fixed charge problem. (*Hint*: Any optimal solution to the fixed charge problem has $my_j > 0$ and uses the fact that minimum $\mathbf{cy} + \mathbf{dx} \geq$ minimum $\mathbf{cy}$ for $\mathbf{y} \in Y$.)

Problem 14.14   (Steinberg [551]) Hirsch and Dantzig [329] have shown that an optimal solution to the general fixed charge problem (given in the previous problem) occurs at an extreme point of the region $Y = \{y | Ay = b, y \geq 0\}$. Consider the following technique which seeks adjacent extreme points that hopefully yield improved solutions; $y_B = (y_{Bi})$ is a basic feasible solution to $Ay = b$, c_B are the corresponding costs, $B = (b_i)$ is a basis matrix from A with inverse B^{-1}, a_j is the jth column from A, $y_j = (y_{ij}) = B^{-1}a_j$, $z_j = c_B y_j$, $c = (c_j)$ are the shipping costs, and $d = (d_i)$ are the fixed costs.

Step 1 Find a basic feasible solution y_B to $Ay = b$. Go to step 2.

Step 2 For each nonbasic column a_j, compute

$$\frac{x_{Br}}{y_{rj}} = \underset{k}{\text{minimum}} \left\{ \frac{x_{Bk}}{y_{kj}} \text{ for } y_{kj} > 0 \right\}; \tag{5}$$

then find the column a_j which satisfies

$$e_j = \underset{k}{\text{minimum}} \{ e_k \text{ for } y_{rk} > 0 \}, \tag{6}$$

where

$$e_k = \frac{x_{Br}}{y_{rk}}(c_k - z_k) + (d_{Br} - d_k).$$

If $e_j \geq 0$, record the value of the variables and solution; go to step 4. Otherwise $(e_j < 0)$, go to step 3.

Step 3 (New extreme point) Bring a_j into the basis and remove b_r, by pivoting on y_{rj}. Go to step 2.

Step 4 If the number of passes through this step has exceeded a predetermined upper bound, terminate. Otherwise, find a_j so that

$$e_j = \underset{k}{\text{minimum}} \, e_k \tag{7}$$

and go to step 3. If the value of $c_B y_B$ decreases subsequent to the. *second* pivot (or return to Step 3), continue. If not, return to Step 4 (from Step 3).

a) Show that if $e_j < 0$, an improved (basic feasible) solution to the fixed charge problem is found. What is the motivation for using (5) and (6)?

b) Show that $e_j \geq 0$ does not necessarily imply that $\mathbf{y_B}$ is an optimal solution and therefore, this technique does not always solve the fixed charge problem. (*Hint*: see Problem 14.15).

c) The intent of Step 4 is to find a new extreme point which is not adjacent to the current one so that an improved solution is found. Why is this done? Can expression (7) be changed to (for example) $e_j = \underset{k}{\text{maximum}}\ e_k$?.

d) Explain why the algorithm breaks down when many of the basic feasible solutions $\mathbf{y_B}$ are degenerate.

Problem 14.15 (Kuhn and Baumol [381]) The data below represent a fixed charge transportation problem (see Problem 14.1).

i \ j	g_{ij}			m_{ij}			
	1	2	$3 = n$	1	2	3	M_i
1	6	5	3	0	0	2	9
2	3	2	1	3	1	0	7
d_j	8	5	3				

The extreme points of the region Y (defined in Problem 14.13), written

$$\begin{pmatrix} z_{11} & z_{12} & z_{13} \\ z_{21} & z_{22} & z_{23} \end{pmatrix}$$

are

$$\mathbf{a} = \begin{pmatrix} 6 & 0 & 3 \\ 2 & 5 & 0 \end{pmatrix}$$

$$\mathbf{b} = \begin{pmatrix} 8 & 0 & 1 \\ 0 & 5 & 2 \end{pmatrix}$$

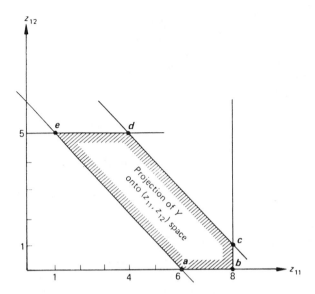

Figure 14.4: Projection of Y for Problem 14.15

$$\mathbf{c} = \begin{pmatrix} 8 & 1 & 0 \\ 0 & 4 & 3 \end{pmatrix}$$

$$\mathbf{d} = \begin{pmatrix} 4 & 5 & 0 \\ 4 & 0 & 3 \end{pmatrix}$$

$$\mathbf{e} = \begin{pmatrix} 1 & 5 & 3 \\ 7 & 0 & 0 \end{pmatrix}$$

Their total costs are

Extreme point	a	b	c	d	e
Total cost	67	66	65	67	66

Starting with the extreme point $\mathbf{e}$, apply the algorithm given in the preceding problem until the optimal extreme point $\mathbf{c}$ is found. Notice that both neighbors of $\mathbf{e}$ have higher total cost and yet $\mathbf{c}$ represents the optimal solution (see Fig. 14.4).

Problem 14.16 Find the optimal solution to the fixed charge transportation problem given in example 14.2 using a branch and bound algorithm. Relate the computations to a tree. List and explain the motivation for your node and branch selection rules.

Appendix A
Computational Experience

A.1 The General Fixed Charge Problem

As previously indicated (cf. Problem 14.13), the general fixed charge problem has the form:

$$\text{Minimize} \quad \mathbf{cy} + \mathbf{dx}$$
$$\text{subject to} \quad \mathbf{y} \in Y = \{\mathbf{y} | \mathbf{Ay} = \mathbf{b}, \mathbf{y} \geq \mathbf{0}\},$$

where $\mathbf{A} = (a_{ij})$ is an m by n matrix, $\mathbf{c} = (c_j), \mathbf{d} = (d_j), \mathbf{y} = (y_j), \mathbf{x} = (x_j)$, and $x_j = 0$ if $y_j = 0$ and $x_j = 1$ if $y_j > 0$ $(j = 1, \ldots, n)$. Computational experience with heuristic methods which seek extreme points of Y that are improved solutions to the fixed charge problem, similar to the algorithm discussed in Problem 14.14, is reported by Cooper and Drebes [146], Denzler [168], Steinberg [551], and Walker [591]. These articles list results for problems generated by Cooper and Drebes which have the following data ranges: $|a_{ij}| \leq 2, 1 \leq c_j \leq 20, 1 \leq d_j \leq 999$, and the vector $\mathbf{b}$ is selected so that Y is not empty. Using a random problem generator computer program, 290 (5 by 10) (that is, m by n) and 11 (7 by 15) problems were constructed. In addition, a series of 15 by 30, 30 by 60, and 50 by 100 problems were devised, each by aggregating three, six, and ten, respectively, 5 by 10 problems. That is, the coefficient matrix $\mathbf{A}$ for each of these problems has block diagonal structure with 5 by 10 submatrices along the diagonal. The computations are summarized in Table 14.2. Row 1 of the table may be explained as follows. Cooper and Drebes [146] applied their heuristic technique to 253 (5 by 10) problems. In about 245 cases, or 97% of the time, the best solution found turned out to be optimal. (The optimal solution to each of the 5 by 10 and 7 by 15 problems was found by enumerating all the extreme points of Y. The solution to each of the larger problems is then obtained by aggregating the optimal solutions to the 5 by 10 problems which compose it.) Of the remaining eight problems, the best solution found was, on the average, within

Source	Problem	no.	% of solutions optimal*	Average % error	Average time per problem	Computer
Cooper and	5 × 10	253	97	6.7	20	IBM 7072
Drebes [146]	7 × 15	97	96	1.9	96	IBM 7072
(1967)	15 × 30	70	90	1.8	900	IBM 7072
Denzler [168]	5 × 10	200	94	0.5	7.8	IBM 7072
(1969)	7 × 15	66	89	0.9	13.7	IBM 7072
	15 × 30	22	71	0.8	75.0	IBM 7072
Steinberg [551]	5 × 10	250	94	5.2	0.7	IBM 360/50
(1970)	15 × 30	90	83	2.6	2.2	IBM 360/50
	30 × 60	36	84	1.7	7.6	IBM 360/50
	50 × 100	9	35	1.6	54.7	IBM 360/50
Walker [591]	5 × 10	52	100	n.a.	4	CDC 1604
SWIFT-2 alg.	15 × 30	5	100	n.a.	18	CDC 1604

Table 14.2: Summary of Computational Experience: The General Fixed Charge Problem

6.7% of the optimal one. The computation time, using an IBM 7072 computer, averaged about 20 seconds per problem. Walker [591] also reports computational results for his SWIFT-2 algorithm and its other variations.

Besides the results listed in the table, Steinberg [551] applied a branch and bound algorithm to 15 (5 by 10) and 10 (15 by 30) problems. The average number of nodes examined was 32 for each 5 by 10 problem and 1208 for each 15 by 30 problem. Average computation time per problem was 9.6 seconds and 1266 seconds, respectively, using an IBM 7072 computer.

A general fixed charge problem, where the fixed charge term $\mathbf{dx}$ is replaced by the piecewise linear function

$$h(y) = \sum_{j=1}^{k} h_j(y_j), \quad \text{where} \quad k < n$$

is discussed by Jones and Soland [255]; here $h_j(y_j) = f_{ij} + \alpha_{ij} y_j$, where f_{ij} and α_{ij} are known and depend on the interval i. A branch and bound algorithm is developed in which the branching process selects a particular interval so that the objective function becomes linear. It has been tested on a series of seven problems which range in size from a 1 by 5 to a 35 by 43, with k between 5 and 15, inclusive. All problems were solved and at most 276 nodes had to be examined. No times are given.

A.2 The Fixed Charge (or Capacitated Plant Location) Problem

The branch and bound algorithm for the fixed charge problem (FCP, Section 14.1) developed by Davis and Ray [163] has been implemented on a Burroughs B-5500 computer. It has been tested on several problems possessing between 20 and 50 integer variables (plants) and a maximum of 150 customers. No problem required more than eight branches[6] and the longest computer running time was 25 minutes.

Sá [499] reports on computations with problems ranging in size from 12 plants and 12 customers to 24 plants and 50 customers. Using an IBM 360/65 computer, his branch and bound algorithm was able to solve the 12 by 12 problem and six 15 by 20 problems. Times ranged between 0.77 minute and 7.62 minutes. The algorithm was unable to solve a 15 by 20 and three 24 by 50 problems after 15 minutes. A heuristic technique (which seeks improved solutions on an iterative basis in the spirit of the discussion appearing in Section 14.1) was able to produce the optimal solution for every problem the branch and bound algorithm solved. It also generated what appear to be excellent solutions for a series of other problems. Times ranged between 0.71 minute for a 10 by 12 problem to 9 minutes for a 24 by 50 problem.

A.3 The (Uncapacitated) Plant Location Problem

Efroymson and Ray [187] report that their branch and bound algorithm (Section 14.2.2) was able to solve a number of 50 plant and 200 customer problems. The average computer time was about 10 minutes on an IBM 7094 computer. In addition, Jain [342] has successfully solved several 39-plant and 63-customer problems using Efroymson and Ray's method. Solution times ranged from 1 to 45 minutes on an IBM 360/44 computer.

A search enumeration technique (Chapter 9) which has a generalized origin/restart capability (Section 9.5) and a very sophisticated point algorithm was developed by Spielberg [544], [545]. Computational experience is reported on numerous problems ranging in size from 20 plants and 35 customers to 100 plants and 150 customers. All problems appear to have been solved with times ranging between 1 minute on an IBM 360/40 computer for a 20 by 35 problem to 60 minutes on an IBM 7094 computer for a 100 by 150 problem.[7]

Ciochetto, Swanson, Lee, and Woolsey [136] report that their enumerative algorithm solved ten problems whose size ranged from 5 plants (or potential lock box sites, see Problem 14.5) and 72 customers to 50 plants and 50 customers. Times, using a PDP 10 computer, ranged between 1.75 seconds for a 10 by 10 problem to 23 seconds for a 20 by 300 problem.

The enumerative algorithm developed by Marsten [411] was applied to ten 100-customer problems with 8 to 16 plants. Except for the 16 by 100 problem, which the algorithm failed to solve, the heuristic used to obtain an initial solution produced an average error of 2% and, on the average, required 2.2 seconds to obtain. The average solution time for the nine solved problems was 12.7 seconds using a CDC 6400 computer.

A branch and bound algorithm for a plant location type problem where the demand constraints (2)′ are replaced by

$$\sum_{j=1}^{n} a_{ij} y_{ij} \leq b_i x_i \quad (i = 1, \ldots, m) \tag{1}$$

is given by Brown [102]. In inequality (1), a_{ij} and b_i are nonnegative known constants, the y_{ij} are zero-one variables, and the variables x_i may take on *any* nonnegative integer value. Using a UNIVAC 1108 computer, five problems ranging in size from 10 x_i variables and 100 y_{ij} variables to 19 x_i variables and 950 y_{ij} variables were solved in about 36 seconds each. A heuristic used to produce the first solution gave an average error of 12.2%.

A.4 Summary

It is apparent that heuristics are essential to produce initial solutions and hence bounds for enumerative algorithms. Moreover, unless the initial bound is very low and the linear programming solutions ap-

---

[6]This suggests that the optimal solutions to the fixed charge problems were near their optimal linear programming solutions.

[7]An earlier version of this program which does not have a generalized origin/restart capability is discussed by Spielberg in [545]. The computer program, SPLT1, is available as an IBM Type III program (no. 360D-15.6.001).

proach and exceed the bound relatively quickly, fixed charge problems of only modest size can be solved. This is especially true when the coefficient matrix does not have special structure. When structure is present (for example, in the plant location problem,), the chances of solving the problem increase substantially. Capacitated plant location problems with up to 100 plants and several hundred customers may be solved if the constraints are very binding. These sizes increase when the capacity constraints are erased since the linear programs can then be solved rapidly by inspection. Finally, we should mention that although heuristics generally produced good solutions, they are naturally very data dependent and will not work uniformly well.

Chapter 15

The Traveling Salesman Problem

15.1 Introduction

Consider a salesman who wants to start from his home city, visit each of n cities exactly once, and return home while traveling the least possible distance. The situation may be represented by a network of nodes and branches, where each node corresponds to a city and the directed arc (i, j) represents the direct route from city i to city j with distance d_{ij}. The Traveling Salesman Problem (TSP) is to find the shortest tour (or cycle) which begins and ends at the salesman's home city and visits each of the other cities (nodes) exactly once.

Background

Though the origin of the term "traveling salesman problem" is not clear, the first significant attempt to solve this problem was due to Dantzig, Fulkerson, and Johnson [160] and utilized a branch and bound framework. The approach was later formalized by Little, Murty, Sweeney and Karel [400] in their branch and bound algorithm for the TSP. Since then, various authors have suggested numerous types of enumerative algorithms using assignment problems and/or Lagrangian relaxation subproblems. A partial list of researchers includes Shapiro [525], Bellmore and Malone [73], Murty [438], Garfinkel [213], Smith, Srinivasan and Thompson [539], Carpaneto and Toth [124], Christofides [133], Helbig, Hansen, and Krarup [318], Smith and Thompson [540], Gavish and Srikanth [220], and Balas and Christofides [37]. A more complete list of references is in Lawler, et al. [389].

It is well known that the TSP belongs to a general class of NP-hard problems.[1] Therefore, enumerative algorithms are computationally inefficient for large problems, and the most commonly used methods to solve large TSP are of the heuristic or approximation type. A partial list of these includes methods proposed by Croes [150], Lin [397], Lin and Kernighan [398], Norback and Love [446], Stewart [553], Rosenkrantz [485], Golden [271], and Golden, Bodin, Doyle and Stewart [272]. A more comprehensive list is in [389].

15.2   Mathematical Formulation

In this section, we develop a mathematical formulation of the TSP, which is a slight variation of the one given in Chapter 1. In particular, there are n cities, the distance traveled from city i to city j is d_{ij}, and we label the home city as 1 and the other $n-1$ cities as city $2, 3, \ldots, n$. We also define the zero-one variable x_{ij} $(i = 1, \ldots, n,\ j = 1, \ldots, n)$, where $x_{ij} = 1$ if the salesman travels from city i to j, and $x_{ij} = 0$, otherwise. Moreover, to force variable x_{ii} $(i = 1, \ldots, n)$ to take value 0, we will assume that d_{ii} is a large positive distance. Since we wish to minimize the total distance traveled, the objective function is

$$\text{minimize} \quad \sum_{i=1}^{n} \sum_{j=1}^{n} d_{ij} x_{ij}. \tag{1}$$

To guarantee that each city j $(j = 1, 2, \ldots, n)$ is entered exactly once (cf. Figure 15.1 (a)) and each city i $(i = 1, 2, \ldots, n)$ is left exactly once (cf. Figure 15.1. (b)), we have the constraints (2) and (3), respectively, as below.

$$\sum_{i=1}^{n} x_{ij} = 1 \quad (j = 1, \ldots, n) \tag{2}$$

$$\text{and} \quad \sum_{j=1}^{n} x_{ij} = 1 \quad (i = 1, \ldots, n). \tag{3}$$

---

[1]A problem is NP-hard if computational effort required to solve it grows exponentially as a function of its size (e.g., the number of cities n in the TSP). For a formal defination of NP-hard, see Garey, M. R., and D. S. Johnson : *Computers and intractibility: A Guide to the Theory of NP-Completeness*, Freeman, (1979).

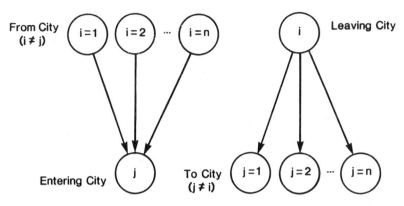

Figure 15.1: The TSP Constraints

Note that the variables x_{jj} and x_{ii} appear in constraints (2) and (3), respectively. This is contrary to Figure 15.1. However, they are forced to have value 0 due to the corresponding large positive value d_{ii} in the objective function (1).

The integer programming problem (1)-(3), together with the integrality constraints $x_{ij} = 0$ or 1, is the well-known Assignment Problem (cf. Chapter 3.) which can be solved efficiently by the Hungarian Method (cf. Appendix A). Unfortunately, the constraints (2)-(3) do not eliminate the possibility of subtours or "loops" as indicated in Figure 15.2., where the solution to (2) and (3) is $x_{12} = x_{23} = x_{31} = 1$, $x_{45} = x_{56} = x_{64} = 1$ and $x_{ij} = 0$ otherwise. A subtour Q is characterized by the property that there is no arc (i, j) connecting any node i of Q with any node j of the complement $\bar{Q} = V - Q$, where $V = \{1, 2, \ldots, n\}$. We may therefore eliminate the possibility of subtours by adding the additional constraints that for each proper subset Q of V, there is at least one arc (i, j) on the tour (that is, $x_{ij} = 1$) with $i \in Q$ and $j \in \bar{Q}$. In Figure 15.2 for example, $V = \{1, 2, 3, 4, 5, 6\}$, choosing $Q = \{1, 2, 3\}$ and $\bar{Q} = \{4, 5, 6\}$ violates this requirement. Hence to eliminate subtours, we require that

$$\sum_{i \in Q} \sum_{j \in \bar{Q}} x_{ij} \geq 1 \quad \text{for all} \quad Q \subset V. \tag{4}$$

The traveling salesman problem is thus

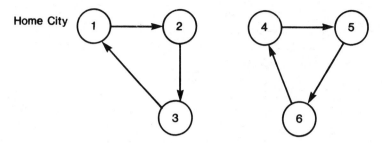

Figure 15.2: Subtours in the TSP

$$\text{minimize} \quad \sum_{i=1}^{n}\sum_{j=1}^{n} d_{ij}x_{ij} = z \tag{5}$$

$$\text{subject to} \quad \sum_{i=1}^{n} x_{ij} \;\; = \;\; 1 \quad (j = 1,\ldots,n) \tag{6}$$

$$\sum_{j=1}^{n} x_{ij} \;\; = \;\; 1 \quad (i = 1,\ldots,n) \tag{7}$$

$$\sum_{i\in Q}\sum_{j\in\bar{Q}} x_{ij} \;\; \geq \;\; 1 \quad \text{for all } Q \subset V \tag{8}$$

$$\text{and} \quad x_{ij} = 0 \text{ or } 1; \tag{9}$$

where, whenever $i = j$ it is assumed that d_{ii} in (5) is large enough to force $x_{ii} = 0$ in any optimal solution. For example,

$$d_{ii} = n \times \;\; \underset{i,j\ (i\neq j)}{\text{maximum}}\ d_{ij} \quad (i = 1,\ldots,n).$$

An alternative to the subtour breaking constraint (8) is to require that in any *proper* subset Q of V there must be fewer arcs than nodes. This requirement is violated in Figure 15.2 by both subtours $1 \to 2 \to 3 \to 1$ and $4 \to 5 \to 6 \to 4$, each of which has 3 nodes and 3 arcs. Therefore, to disallow loops, if $|Q|$ denotes the number of cities (nodes) in Q, we require that there be at most $|Q| - 1$ arcs in Q. Hence we have the alternative constraints

$$\sum_{i\in Q}\sum_{j\in Q} x_{ij} \leq |Q| - 1 \quad \text{for all } Q \subset V. \tag{10}$$

Referring to Figure 15.2, $V = \{1,2,3,4,5,6\}$ and taking $Q = \{1,2,3\}$

(or $Q = \{4,5,6\}$), the left hand side of (10) is 3 while $|Q| - 1 = 2$, violating the inequality.

15.3   Algorithms

Due to the large number of subtour breaking constraints ((8) or (10)), it is not practical to solve the TSP directly, and a specialized algorithm is required. As discussed earlier, the TSP without subtour breaking constraints is an assignment problem which is relatively easy to solve. Also, as the assignment problem is a relaxation of the corresponding TSP, its solution will be optimal for the TSP provided there are no subtours. Otherwise, the assignment solution will yield a lower bound on z, the value of the objective function of the TSP. Therefore, a branch and bound algorithm, with assignment subproblems, is a natural way to solve the TSP. To solidify the discussion and suggest a particular branch and bound procedure we examine two examples.

Example 15.1   Consider a 6 city traveling salesman problem with distance matrix given in Table 15.1. To force $x_{ii} = 0$, we have set $d_{ii} = n \times$ maximum $d_{ij} = 6 \times 30 = 180$.
$$i, j \ (i \neq j)$$
The optimal solution to the assignment problem (see Appendix A) is $x_{14} = x_{21} = x_{32} = x_{46} = x_{53} = x_{65} = 1$, and $x_{ij} = 0$ otherwise, with $z = 59$. This solution corresponds to the cycle $1 \rightarrow 4 \rightarrow 6 \rightarrow 5 \rightarrow 3 \rightarrow 2 \rightarrow 1$ which has no subtours, (see Figure 15.3.), and thus solves the traveling salesman problem.

		(City j)					
		1	2	3	4	5	6
	1	180	6	11	6	16	19
	2	5	180	16	13	21	9
(City i)	3	8	12	180	22	18	11
	4	20	18	27	180	22	15
	5	10	22	11	25	180	18
	6	16	10	17	30	10	180

Table 15.1: Distance Matrix for Example 15.1

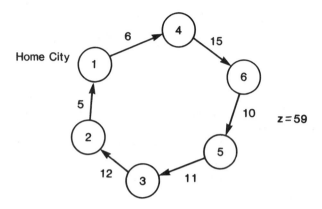

Figure 15.3: Optimal Solution to the TSP (Example 15.1)

Example 15.2 (Balas in [389]) Consider the 8 city TSP whose distance matrix is shown in Table 15.2. The diagonal values d_{ii} are computed as in Example 15.1. Solving the assignment relaxation gives the optimal solution $x_{12} = x_{23} = x_{31} = x_{45} = x_{56} = x_{64} = x_{78} = x_{87} = 1$, and $x_{ij} = 0$ otherwise, with $z = 17$. The optimal assignment solution contains three subtours and hence is not a solution to the TSP (see Figure 15.4). However, the optimal value of the assignment solution, $z = 17$, is a lower bound on any solution to the TSP.

If the optimal assignment solution is not a solution to the TSP we use a branch and bound procedure to solve the TSP. At each node of the

		(City j)						
	1	2	3	4	5	6	7	8
1	96	2	11	10	8	7	6	5
2	6	96	1	8	8	4	6	7
(City i) 3	5	12	96	11	8	12	3	11
4	11	9	10	96	1	9	8	10
5	11	11	9	4	96	2	10	9
6	12	8	5	2	11	96	11	9
7	10	11	12	10	9	12	96	3
8	7	10	10	10	6	3	1	96

Table 15.2: Distance Matrix for Example 15.2

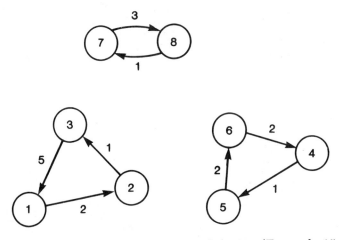

Figure 15.4: Optimal Assignment Solution (Example 15.2)

branch and bound tree, we may solve a relaxed subproblem (that is, the corresponding assignment problem) and apply various bounding rules to decide if a node is a candidate for further branching. If so, it enters the list of dangling nodes. A dangling node is selected for examination using some node selection rule (cf. Chapter 8), and new nodes are created from the dangling node selected based on particular branching rules (cf. Chapter 8). Branching and bounding rules specialized to the TSP are discussed below.

15.3.1 Bounding Rules

1. At a dangling node, an assignment subproblem is created by dropping the constraints (8) or (10). If Z_A is the value of the objective function corresponding to the optimal assignment solution, then Z_A is a lower bound on any solution to the TSP that can be found from this node. Hence, if Z^0 is the value of the current best solution to the TSP, then the current dangling node need not be examined further if

$$Z_A \geq Z^0.$$

2. The lower bound can sometimes be increased above Z_A by using subtour breaking constraints. In particular, the solution to the

(City j)

		1	2	3	4	5	6	7	8	u_i
	1	92	0	9	8	6	5	4	3	2
	2	3	95	0	7	7	3	5	6	1
	3	0	9	93	8	5	9	0	8	3
	4	8	8	9	95	0	8	7	9	1
(City i)	5	7	9	7	2	94	0	8	7	2
	6	8	6	3	0	9	94	9	7	2
	7	5	8	9	7	6	9	93	0	3
	8	4	9	9	9	5	2	0	95	1
	v_j	2	0	0	0	0	0	0	0	

Table 15.3: Matrix of Reduced Cost for Example 15.2

assignment problem (cf. Appendix A) yields the reduced costs
$\{\bar{d}_{ij}\}$ for $i, j = 1, \ldots, n$. (cf. Table 15.3. for Example 15.2.),
and we may write the objective function as

$$z = Z_A + \sum_{i=1}^{n} \sum_{j=1}^{n} \bar{d}_{ij} x_{ij}. \tag{11}$$

Define the graph G_0 to include all n nodes and only those arcs
(i, j) corresponding to $\bar{d}_{ij} = 0$. (For instance, in Example 15.2,
we use Table 15.3 to produce the graph G_0 in Figure 15.5.)

Let $K_1, K_2, \ldots, K_p$ be the components of G_0, where a nonempty
set of nodes K is defined as a component if there does not exist
an arc (i, j) with $i \in K$ and $j \in \bar{K}$, or $i \in \bar{K}$ and $j \in K$, and no
proper subset of K satisfies this condition. The subtour break-
ing constraints (8) or (10) require that the optimal solution
must contain at least one arc going out of each K_l $(l = 1, \ldots p)$.
Using (11), this means that the value of any solution to the TSP
must be at least

$$Z_A + \sum_{l=1}^{p} \underset{i \in K_l, j \in \bar{K}_l}{\text{minimum}} \{\bar{d}_{ij}\}. \tag{12}$$

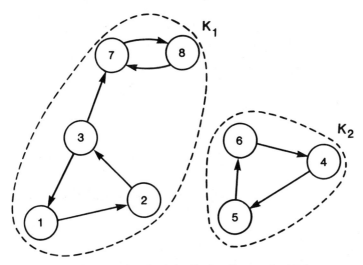

Figure 15.5: The Graph G_0 for Example 15.2

Example (from Example 15.2)

The graph G_0 in Figure 15.5 is constructed from Table 15.3. It has two components, $K_1 = \{1, 2, 3, 7, 8\}$ and $K_2 = \{4, 5, 6\}$. As

$$\underset{i \in K_1, j \in \bar{K}_1}{\text{minimum}} \{\bar{d}_{ij}\} = \bar{d}_{8,6} = 2$$

and

$$\underset{i \in K_2, j \in \bar{K}_2}{\text{minimum}} \{\bar{d}_{ij}\} = \bar{d}_{6,3} = 3,$$

we have a lower bound for any solution to the TSP equal to $Z_A + \bar{d}_{8,6} + \bar{d}_{6,3} = 17 + 2 + 3 = 22$.

15.3.2 Branching Rules

Any node k in the branch and bound tree may be represented by two sets E and I, where E consists of all arcs (i, j) excluded from (i.e., all x_{ij}'s fixed to zero) and I consists of all arcs to be included in (that is, x_{ij}'s fixed to 1) the TSP tour. Using this notation, the assignment subproblem at node k may be written as

$$\text{minimize} \quad \sum_{(i,j)\in I} d_{ij} + \sum_{i\in S}\sum_{j\in T} d'_{ij}x_{ij} \tag{13}$$

$$\text{subject to} \quad \sum_{i\in S} x_{ij} = 1 \qquad \forall\, j \in T \tag{14}$$

$$\sum_{j\in T} x_{ij} = 1 \qquad \forall\, i \in S \tag{15}$$

$$\text{and} \qquad x_{ij} = 0 \text{ or } 1 \;\; \forall\,(i,j) \ni i \in S, j \in T, \tag{16}$$

where $S = \{i|(i,j) \notin I \text{ for any } j\}$ and $T = \{j|(i,j) \notin I \text{ for any } i\}$, and

$$d'_{ij} = \begin{cases} d_{ij} & (i,j) \notin E \\ M \text{ (a large positive number)} & (i,j) \in E. \end{cases}$$

Note that assigning a large positive distance M to d'_{ij} for $(i,j) \in E$ in (13) is equivalent to fixing the corresponding x_{ij} to value 0.

The subproblem may be solved using the Hungarian Method (see Appendix A). A branching rule is then to select a variable x_{rs}, where $r \in S$ and $s \in T$, and create two nodes by fixing x_{rs} to value 1 or 0, respectively. Little, Murty, Sweeney, and Karel [400] suggest selecting variable x_{rs} for branching which has the maximum potential to increase the subproblem's objective function when fixed to 0.[2] To do this, let $\{\bar{d}_{ij}\}$, where $i \in S$ and $j \in T$, be the reduced costs at the subproblem's optimal solution. Then for each arc (i,j), $i \in S, j \in T$, with $\bar{d}_{ij} = 0$, we compute the penalty

$$P_{ij} = \min\,\{\bar{d}_{ih}|h \in T - \{j\}\} + \min\,\{\bar{d}_{hj}|h \in S - \{i\}\}. \tag{17}$$

P_{ij} is therefore the minimum amount by which the value of the optimal assignment subproblem will increase if variable x_{ij} is fixed to 0. We choose x_{rs} such that

$$P_{rs} = \text{maximum}\{P_{ij} \mid i \in S, j \in T, \text{and } \bar{d}_{ij} = 0\}. \tag{18}$$

Note that whenever a variable x_{rs} is fixed to value 1 to create a new

---

[2]The argument for this is similar to the one described in Section 8.7.2

node, other variables may be fixed to value 0 depending on the following conditions.

(i) If $x_{rs} = 1$, then to avoid a loop between nodes r and s, variable x_{sr} may be fixed to value 0.

(ii) If $x_{i_1 i_2} = x_{i_2 i_3} = \ldots = x_{i_{q-1} i_q} = x_{i_q r} = x_{rs} = 1$, where $(i_1, i_2, \ldots, i_q, r, s)$ is a proper subset of V, then to avoid a subtour between the nodes $(i_1, i_2, \ldots, i_k, r, s)$, the variable x_{si_j} can be fixed to value 0 for $j = 1, \ldots, q$.

We continue our illustration by solving Example 15.2.

Example (from Example 15.2)

The complete branch and bound tree is in Figure 15.6. For ease of exposition, we relist the reduced costs $\bar{d}_{ij}$ corresponding to the optimal assignment solution at node 1 below.

		(City j)							
		1	2	3	4	5	6	7	8
	1	92	0	9	8	6	5	4	3
	2	3	95	0	7	7	3	5	6
	3	0	9	93	8	5	9	0	8
	4	8	8	9	95	0	8	7	9
(City i)	5	7	9	7	2	94	0	8	7
	6	8	6	3	0	9	94	9	7
	7	5	8	9	7	6	9	93	0
	8	4	9	9	9	5	2	0	95

As discussed before, $Z_A = 17$ and the lower bound on the solution to the TSP is $L = 22$. Since the assignment solution has subtours (cf. Figure 15.4), we select a variable x_{rs} for branching. The candidates according to the rule of Little et al. [400] just described, are the variables x_{ij} with $\bar{d}_{ij} = 0$; or, $x_{12}, x_{23}, x_{31}, x_{37}, x_{45}, x_{56}, x_{64}, x_{78}, x_{87}$ and the penalties for these variables are computed below:

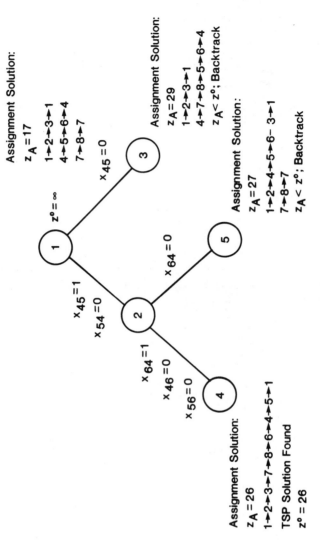

Figure 15.6: Branch and Bound Tree for Example 15.2

$$P_{12} = 3+6 = 9,$$
$$P_{23} = 3+3 = 6,$$
$$P_{31} = 0+3 = 3,$$
$$P_{37} = 0+0 = 0,$$
$$P_{45} = 7+5 = 12,$$
$$P_{56} = 2+2 = 4,$$
$$P_{64} = 3+2 = 5,$$
$$P_{78} = 5+3 = 8,$$
$$P_{87} = 2+0 = 2.$$

Hence x_{45}, with the maximum penalty is selected for branching. By fixing x_{45} to value 1 and 0 we create Nodes 2 and 3, respectively. At node 2, since $x_{45} = 1$, x_{54} cannot be 1 and is thus fixed to 0. However, the previous assignment solution is still optimal with the reduced costs listed below. For the variables fixed to 0, the reduced cost is a large positive number and is denoted by a ∗.

(City j)

		1	2	3	4	6	7	8
	1	92	0	9	8	5	4	3
	2	3	95	0	7	3	5	6
	3	0	9	93	8	9	0	8
(City i)	5	7	9	7	∗	0	8	7
	6	8	6	3	0	94	9	7
	7	5	8	9	7	9	93	0
	8	4	9	9	9	2	0	95

The penalties for x_{ij} with $\bar{d}_{ij} = 0$ are recomputed below.

$$P_{12} = 3+6 = 9,$$
$$P_{23} = 3+3 = 6,$$
$$P_{31} = 0+3 = 3,$$
$$P_{37} = 0+0 = 0,$$
$$P_{56} = 7+2 = 9,$$
$$P_{64} = 3+7 = 10,$$
$$P_{78} = 5+3 = 8,$$
$$P_{87} = 2+0 = 2.$$

As P_{64} is the maximum penalty, we create Nodes 4 and 5 from Node 2 by fixing x_{64} to value 1 or 0, respectively. At Node 4, since $x_{45} = 1$ and $x_{64} = 1$, to avoid the subtour $6 \to 4 \to 5 \to 6$, we fix the variable x_{56} to value 0. The optimal solution to the assignment problem at Node 4 is the TSP solution $1 \to 2 \to 3 \to 7 \to 8 \to 6 \to 4 \to 5 \to 1$ with $Z_A = 26$. Hence, set Z^0, the value of the current best solution, to 26. We now examine Node 5, where x_{45} is fixed to value 1 and x_{54} and x_{64} to value 0. The optimal solution to the assignment problem here has two subtours $1 \to 2 \to 4 \to 5 \to 6 \to 3 \to 1$ and $7 \to 8 \to 7$, with $Z_A = 27$. Since $Z_A > Z^0 = 26$, Node 5 cannot produce a better TSP solution and is omitted. At Node 3, only the variable x_{45} is fixed to 0. The optimal solution to the assignment problem at this node has two subtours $1 \to 2 \to 3 \to 1$ and $4 \to 7 \to 8 \to 5 \to 6 \to 4$, with $Z_A = 29$. Since $Z_A > Z^0$, we backtrack, and as there are no other dangling nodes, the TSP solution found at Node 4 is optimal. Figure 15.6 gives the complete branch and bound tree.

Various authors have suggested alternative branching rules and different ways to strengthen the lower bounds computed at each node. Some of these are discussed in the problems.

15.4   Approximate Algorithms

The branch and bound algorithm discussed in the previous section is computationally unwieldy for problems with more than a few hundred nodes or cities. Therefore, for larger problems, approximate or heuristic algorithms are necessary. The procedures discussed here, can usually be classified as methods which either construct tours or improve upon a given tour. Composite algorithms that both construct tours and improve them have also been devised.

15.4.1   Tour Construction Procedures

Tour construction algorithms normally consist of the following three steps.

Step 1 Find an initial subtour. That is, choose a starting subtour consisting of 2 or more cities.

Step 2 Based on some node selection rule, select a node (city) to be added to the subtour.

Step 3 Based on some insertion criterion, insert the node selected in Step 2 (somewhere) in the current subtour.

Step 2 and Step 3 are repeated until all nodes or cities have been inserted; that is, until a TSP solution is found. Particular choices for the node selection rule (Step 2) and/or the insertion criterion (Step 3) yield different tour construction algorithms. Some of these are explained below, while others are left to the references.

The Cheapest Insertion Algorithm

This algorithm starts with an initial tour T consisting of two cities i_1 and i_2, such that

$$d_{i_1 i_2} + d_{i_2 i_1} = \underset{i \in V, j \in V \, (i \neq j)}{\text{minimum}} (d_{ij} + d_{ji}).$$

That is, the algorithm commences with a shortest two city subtour. The node selection and insertion criteria are combined by first finding the cheapest insertion cost c_k for each node k not on the subtour. Note that if node k is added between nodes i and j on the subtour T, the distance (i.e., cost) added to the tour is $d_{ik} + d_{kj} - d_{ij}$. Hence, we have

$$c_k = \underset{(i,j) \in T}{\text{minimum}} (d_{ik} + d_{kj} - d_{ij}).$$

Now insert node k^* with the minimum insertion cost, c_{k^*}, between the nodes corresponding to the computation of c_{k^*}.

Example 15.3 (from Example 15.2.)

For ease of exposition, the distance matrix for the eight city problem is rewritten below.

	1	2	3	4	5	6	7	8
1	–	2	11	10	8	7	6	5
2	6	–	1	8	8	4	6	7
3	5	12	–	11	8	12	3	11
4	11	9	10	–	1	9	8	10
5	11	11	9	4	–	2	10	9
6	12	8	5	2	11	–	11	9
7	10	11	12	10	9	12	–	3
8	7	10	10	10	6	3	1	–

Starting Subtour: The matrix below lists, for each pair of nodes (i,j), the subtour distance $d_{ij} + d_{ji}$. Since the matrix is symmetric, only the upper triangular part is given

	1	2	3	4	5	6	7	8
1		8	16	21	19	19	16	12
2			13	17	19	12	17	17
3				21	17	17	15	21
4					5	11	18	20
5						13	19	15
6							23	12
7								4
8								

As $7 \to 8 \to 7$ is the shortest 2 city subtour, with total distance of 4, it is selected as a starting subtour.

Iteration 1

Compute the insertion cost for node k $(k = 1, 2, \ldots, 6)$; that is

$$c_k = \text{minimum}\{d_{7k} + d_{k8} - d_{78}, d_{8k} + d_{k7} - d_{87}\}.$$

The table below illustrates the necessary computations. In the table i_k and j_k correspond to the nodes which result in the cheapest insertion cost c_k for node k.

k	$d_{7k} + d_{k8} - d_{78}$	$d_{8k} + d_{k7} - d_{87}$	c_k	i_k	j_k
1	$10+ 5-3=12$	$7+ 6-1=12$	12	7 / 8	8 / 7
2	$11+ 7-3=15$	$10+ 6-1=15$	15	7 / 8	8 / 7
3	$12+11-3=20$	$10+ 3-1=12$	12	8	7
4	$10+10-3=17$	$10+ 8-1=17$	17	7 / 8	8 / 7
5	$9+ 9-3=15$	$6+10-1=15$	15	7 / 8	8 / 7
6	$12+ 9-3=18$	$3+11-1=13$	13	8	7

The node with the least insertion cost is either Node 1 or Node 3 with insertion cost of 12. Breaking ties arbitrarily, we select Node 1, and insert it between Nodes 8 and 7.

Table 15.4 summarizes the remaining iterations. Note that the heuristic solution for this particular problem is the same as the optimal solution found using the branch and bound algorithm (cf. Section 15.3). Naturally, this is not guaranteed.

Insertion Heuristics Based on a Convex Hull of the Node Set as the Starting Subtour

Heuristics here are designed for the *Euclidean* TSP, where a TSP is called Euclidean if the locations to be visited all lie in the same geometric plane and the cost of traveling between any pair of nodes is the Euclidean distance between them. We state a few important results for the Euclidean TSP.

Theorem 15.1 *For a Euclidean TSP, the optimal route cannot have intersecting paths.*

#	Subtour	k	c_k	i_k	j_k	k^*	c_{k^*}	i_{k^*}	j_{k^*}
2		2	2	1	7	2	2	1	7
		3	8	$\begin{cases}1\\8\end{cases}$	$\begin{cases}7\\1\end{cases}$				
		4	12	1	7				
		5	10	8	1				
		6	8	8	1				
3		3	-2	2	7	3	-2	2	7
		4	10	2	7				
		5	10	8	1				
		6	8	8	1				
4		4	14	8	1	6	8	8	1
		5	10	8	1				
		6	8	$\begin{cases}8\\2\end{cases}$	$\begin{cases}1\\3\end{cases}$				
5		4	1	6	1	4	1	6	1
		5	5	8	6				
6		5	1	4	1	5	1	4	1

TSP tour: $7 \rightarrow 8 \rightarrow 6 \rightarrow 4 \rightarrow 5 \rightarrow 1 \rightarrow 2 \rightarrow 3 \rightarrow 7$

Table 15.4: Cheapest Insertion Algorithm: Computations for Example 15.3

Proof: Suppose an optimal route does have intersecting paths, as illustrated in Figure 15.7(a).

Consider the alternative route shown in Figure 15.7(b). To construct this route from the one in Figure 15.7(a), we have removed arcs

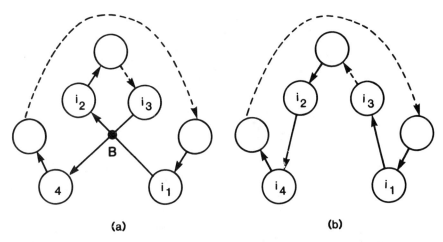

Figure 15.7: The Euclidean TSP and Intersecting Paths

(i_1, i_2) and (i_3, i_4), added arcs (i_1, i_3) and (i_2, i_4), and reversed the direction of the arc from i_2 to i_3. Since the distances in a Euclidean TSP are symmetric, the savings in the new route (cf. Figure 15.7(b)) when compared to the old one (cf. Figure 15.7(a)) is

$$\Delta = d(i_1, i_2) + d(i_3, i_4) - d(i_1, i_3) - d(i_2, i_4),$$

where $d(i, j)$ is the Euclidean distance between points i and j. Now, if we let B be the intersection point of the paths (i_1, i_2) and (i_3, i_4) in Figure 15.7(a), then we have savings

$$\Delta = d(i_1, B) + d(B, i_2) + d(i_3, B) + d(B, i_4) - d(i_1, i_3) - d(i_2, i_4)$$

or

$$\Delta = \{(d(i_1, B) + d(i_3, B) - d(i_1, i_3)\} + \{(d(B, i_2) + d(B, i_4) - d(i_2, i_4)\}.$$

Using the triangle inequality, we have $d(i_1, i_3) < d(i_1, B) + d(i_3, B)$ and $d(i_2, i_4) < d(B, i_2) + d(B, i_4)$, and hence $\Delta > 0$. This, however, contradicts the assumption that the route in Figure 15.7(a) is an optimal route to the TSP. ∎

If $V = \{1, 2, \ldots, n\}$, are n points (cities) in Euclidean space, the convex

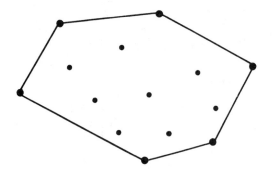

Convex Hull of the Set V

Figure 15.8: The Convex Hull of the Set V

hull of the set V is the smallest convex set that includes V. This is geometrically illustrated in Figure 15.8.

Theorem 15.2 *Every Euclidean TSP has an optimal solution in which the cities which lie on the boundary of the convex hull of the set $V = \{1, \ldots, n\}$ are visited in the same order as they appear on the boundary of the convex hull*

The proof of this result can be obtained geometrically by showing that an optimal route which does not satisfy the theorem will have at least one intersecting path, contradicting the previous theorem. The proof is left to Problem 15.8.

Theorem 15.2 suggests that for a Euclidean TSP, an obvious starting subtour for a tour construction method is the convex hull. In this regard, we state the heuristic algorithm suggested by Norback and Love [446]. Other variations to the procedure may be found in [389].

Step 1 Find the convex hull of the set V of the n cities. Use the cities on the boundary of the convex hull as the starting subtour.

Step 2 Choose a city k not on the subtour and an edge (i_k, j_k) in the subtour such that the angle formed by the edges (i_k, k) and (k, j_k) is maximum.

Step 3 Insert city k between i_k and j_k.

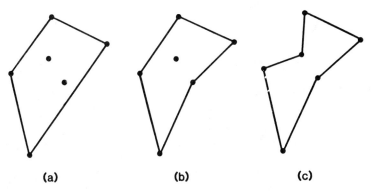

(a)   (b)   (c)

Figure 15.9: Convex Hull Heuristic

Steps 2 and 3 are repeated until a complete tour is obtained. Note that Step 2 tends to insert cities into the subtour with minimum insertion cost. Figure 15.9 gives a geometric illustration of this algorithm for a six city Euclidean TSP.

15.4.2   Tour Improvement Algorithms

Tour improvement algorithms begin with a feasible TSP solution obtained for example, from one of the tour construction heuristics just discussed. An attempt is then made to improve the solution by a sequence of edge interchanges which delete k $(k \geq 2)$ edges from the current tour and add k new edges to form a new tour. As the computations tend to work best for the symmetric TSP $(d_{ij} = d_{ji})$, we restrict our discussion to this situation.

The simplest tour improvement algorithm is the two edge interchange algorithm listed below and subsequently illustrated.

Step 1 Find a starting TSP solution.

Step 2 Evaluate all possible two edge exchanges as illustrated in Figure 15.10.

Step 3 If the exchange which yields the most improvement, reduces the length of the tour, make the exchange and go to Step 2; otherwise, stop.

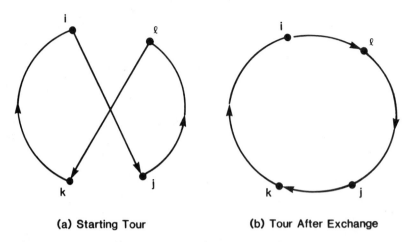

(a) Starting Tour   (b) Tour After Exchange

Figure 15.10: Two Edge Exchange

Example 15.4   Consider the six city symmetric traveling salesman problem with the distances below. Since $d_{ij} = d_{ji}$, only the upper half of the matrix is listed.

		(City j)					
		1	2	3	4	5	6
	1	–	5	7	8	11	10
	2		–	6	3	5	9
(City i)	3			–	4	11	8
	4				–	7	3
	5					–	2
	6						–

Step 1 We start with an arbitrary solution (say) $1 \rightarrow 2 \rightarrow 3 \rightarrow 4 \rightarrow 5 \rightarrow 6 \rightarrow 1$ with total length 34.

Step 2 Table 15.5 evaluates all two edge exchanges

Step 3 Table 15.5 indicates that the maximum saving of 3 is achieved when edges (4,5) and (6,1) are interchanged with edges (4,6) and (5, 1), resulting in the new solution $1 \rightarrow 2 \rightarrow 3 \rightarrow 4 \rightarrow 6 \rightarrow 5 \rightarrow 1$ with a length of 31 (Figure 15.11).

i	j	l	k	$(d_{ij}$	$+$	$d_{lk})$	$-$	$(d_{il}$	$+$	$d_{kj})$	$=$	Savings
1	2	3	4	(5	+	4)	−	(7	+	3)	=	−1
		4	5	(5	+	7)	−	(8	+	5)	=	−1
		5	6	(5	+	2)	−	(11	+	9)	=	−13
2	3	4	5	(6	+	7)	−	(3	+	11)	=	−1
		5	6	(6	+	2)	−	(5	+	8)	=	−5
		6	1	(6	+	10)	−	(9	+	7)	=	0
3	4	5	6	(4	+	2)	−	(11	+	3)	=	−8
		6	1	(4	+	10)	−	(8	+	8)	=	−2
4	5	6	1	(7	+	10)	−	(3	+	11)	=	3

Table 15.5: Computations for Two Edge Exchange

Steps 2 and 3 are repeated until no pairwise exchange offers any positive savings (Problem 15.13).

Some authors (see, e.g., [389]) have suggested evaluating three or more edge exchanges, which may result in a better solution than the two

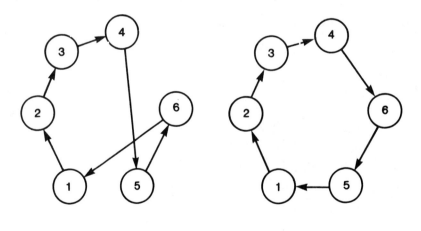

(a) Starting Solution   (b) New Solution

Figure 15.11: Two Edge Interchange for Example 15.4

edge exchange procedure. However, considerably more computational work is required. A detailed computational study comparing various heuristic algorithms is in [389].

Problems

Problem 15.1   List in detail the steps of the branch and bound algorithm for the TSP discussed in Section 15.3. Draw a simple flow chart describing the procedure.

Using the branch and bound procedure outlined in Problem 15.1, find the optimal traveling salesman tour for the problems whose distance matrices are given in Problems 15.2 through 15.6.

Problem 15.2

		(City j)				
		1	2	3	4	5
	1	— —	31	90	48	24
	2	31	— —	59	26	35
(City i)	3	90	59	— —	32	67
	4	48	26	32	— —	25
	5	24	35	67	25	— —

Problem 15.3

		(City j)					
		1	2	3	4	5	6
	1	— —	26	32	96	110	96
	2	40	— —	47	61	79	73
(City i)	3	51	43	— —	91	97	72
	4	96	61	91	— —	14	41
	5	110	79	97	23	— —	35
	6	96	73	72	46	28	— —

Problem 15.4

		(City j)					
		1	2	3	4	5	6
	1	— —	2	11	10	8	7
	2	6	— —	1	8	8	4
(City i)	3	5	12	— —	11	11	12
	4	11	9	10	— —	1	9
	5	11	11	9	4	— —	2
	6	12	8	5	2	2	— —

Problem 15.5

		(City j)						
		1	2	3	4	5	6	7
	1	— —	15	51	48	21	63	76
	2	15	— —	35	44	35	49	71
	3	51	35	— —	29	43	11	42
(City i)	4	48	44	29	— —	28	21	22
	5	21	35	43	28	— —	50	53
	6	63	49	11	21	50	— —	30
	7	76	71	42	22	53	30	— —

Problem 15.6

		(City j)							
		1	2	3	4	5	6	7	8
	1	— —	2	5	1	5	7	7	8
	2	3	— —	3	2	5	4	5	9
	3	4	5	— —	3	3	2	4	8
(City i)	4	2	2	3	— —	3	6	6	7
	5	5	7	4	3	— —	5	5	4
	6	7	4	5	6	5	— —	1	5
	7	7	5	5	4	5	1	— —	4
	8	9	10	8	7	4	4	5	— —

Problem 15.7 An alternate branching rule to the one discussed in Section 15.3 is as follows: let k be any dangling node defined by sets E_k and I_k consisting of arcs which are excluded and included, respectively. Also, in the assignment solution at this node, $A_k = \{(i_1, i_2), (i_2, i_3), \ldots, (i_t, i_1)\}$ is a subtour which has the fewest number of arcs (ties are broken arbitrarily). If we define $A_k - I_k$, the set of free arcs in the subtour as $\{(j_1, j_2), (j_3, j_4), (j_5, j_6), \ldots, (j_{2s-1}, j_{2s})\}$, then generate s successors to the node k as follows:

$$E_{k_1} = E_k \cup \{(j_1, j_2)\}$$
$$I_{k_r} = I_k$$

$$E_{k_r} = E_k \cup \{(j_3, j_4)\}$$
$$I_{k_r} = I_k \cup \{(j_1, j_2)\}$$

and

$$\left. \begin{array}{rl} E_{k_r} &= E_k \cup \{(j_{2r-1}, j_{2r})\} \\ I_{k_r} &= I_k \cup \{(j_1, j_2), (j_3, j_4), \ldots, (j_{2r-3}, j_{2r-1})\} \end{array} \right\} r = 3, 4, \ldots, s.$$

a) Show that the above procedure is a valid branching rule, that is, the resulting tree either explicitly or implicitly enumerates all possible TSP tours.

b) Solve Problem 15.2 by the branch and bound algorithm which uses this branching rule. Does this procedure seem to be more efficient? Why or why not?

In Problems 15.8 through 15.12, find the TSP tour constructed by the cheapest insertion algorithm discussed in Section 15.4.1. Compare the approximate solution obtained to the optimal solution found using the branch and bound algorithm.

Problem 15.8 The TSP given in Problem 15.2.

Problem 15.9 The TSP given in Problem 15.3.

Problem 15.10 The TSP given in Problem 15.4.

Problem 15.11 The TSP given in Problem 15.5.

Problem 15.12   The TSP given in Problem 15.6.

Problem 15.13   Use the two edge interchange algorithm given in Section 15.4.2 for the problem given in Example 15.4.

Problem 15.14   Solve the TSP given in Problem 15.5 using the two edge interchange algorithm. Is the solution obtained an optimal one? Why or why not?

Problem 15.15   Consider a network consisting of n nodes with distance matrix $D = \{d_{ij}\}$. Suppose there are m salesman located at a common city (for example city 1). Each salesman must travel along a subtour of nodes, which includes city 1, in such a way that each node in the network is visited exactly once (excluding city 1) by exactly one of the salesmen. Formulate this problem as a mathematical programming problem which minimizes the combined distance travelled by all of the m salesmen.

Problem 15.16 (Vehicle Routing Problem) Suppose we have m trucks located at a central depot (node 1) and $n - 1$ demand points (nodes numbered 2 through n) each with a demand of d_i units. Let c_{ij} be the cost of traveling from node i to node j $(i = 1, 2, \ldots, n; \; j = 1, 2, \ldots, n)$. Furthermore, let W_k be the capacity of vehicle k $(k = 1, 2, \ldots, m)$.

(a) Assuming that each customer can be served by only one truck, develop a mathematical programming model which minimizes the total distance travelled by all of the m trucks, while satisfying the demands of each of the $n - 1$ customers so that no vehicle capacity is exceeded.

(b) Develop a branch and bound algorithm anologous to the one developed for the traveling salesman problem to solve this problem.

Appendix A
The Assignment Algorithm

Consider the following assignment problem.

$$\text{Minimize} \quad \sum_{i=1}^{n}\sum_{j=1}^{n} d_{ij}x_{ij} \tag{1}$$

$$\text{subject to} \quad \sum_{j=1}^{n} x_{ij} = 1 \quad (i=1,\dots,n) \tag{2}$$

$$\sum_{i=1}^{n} x_{ij} = 1 \quad (j=1,\dots,n) \tag{3}$$

$$\text{and} \quad x_{ij} = 0 \text{ or } 1 \quad (i,j=1,\dots,n) \tag{4}$$

As discussed in Chapter 3, the coefficient matrix of this problem is unimodular and any basic feasible solution of (2)-(3) will be integer. Hence we can view the assignment problem as a linear program where constraint (4) is replaced by $x_{ij} \geq 0$.

If we define u_i $(i = 1,\dots,n)$ and v_j $(j = 1,\dots,n)$ as dual multipliers corresponding to constraints (2) and (3) respectively, then the dual of the assignment problem is,

$$\text{maximize} \quad \sum_{i=1}^{n} u_i + \sum_{j=1}^{n} v_j$$

$$\text{subject to} \quad u_i + v_j \leq d_{ij} \quad (i,j=1,\dots,n)$$

$$\text{and} \quad u_i, v_j \text{ unrestricted} \quad (i,j=1,\dots,n)$$

If for this primal-dual pair,

i) $\{x_{ij}\}$ is primal feasible,

ii) $(\{u_i\}, \{v_j\})$ is dual feasible, and

iii) they satisfy the complementary slackness conditions,

$$x_{ij}(d_{ij} - u_i - v_j) = 0 \quad (i,j=1,\dots,n), \tag{5}$$

then $\{x_{ij}\}$ is the optimal assignment solution.

The method described in this appendix, called the "Hungarian Method", is a primal-dual algorithm which at each iteration finds a dual feasible solution and then tries to generate a primal feasible solution which satisfies the complemetary slackness conditions (5). If this is not possible, a new dual feasible solution is derived. The process continues until a dual feasible solution is obtained from which a primal feasible solution satisfying complementary slackness conditions can be generated. We describe the steps of the algorithm using the $\{d_{ij}\}$ matrix of Example 15.1. For ease of reference, the $\{d_{ij}\}$ matrix is reproduced below.

		(City j)					
		1	2	3	4	5	6
	1	180	6	11	6	16	19
	2	5	180	16	13	21	9
(City i)	3	8	12	180	22	18	11
	4	20	18	27	180	22	15
	5	10	22	11	25	180	18
	6	16	10	17	30	10	180

Step 1 (Initial dual feasible solution)

It is easy to verify that

$$v_j = \underset{i}{\text{minimum }} d_{ij} \qquad (j = 1, \ldots, n)$$

$$\text{and} \quad u_i = \underset{j}{\text{minimum }} (d_{ij} - v_j) \quad (i = 1, \ldots, n)$$

is a feasible solution to the dual problem. Tableau # 1 lists the values of u_i, v_j, and $\bar{d}_{ij} = d_{ij} - u_i - v_j$ for this example. Note that the dual feasible solution generated in this step will produce a matrix $\bar{d}_{ij}$ (called the "reduced cost matrix") in which each row and column will have at least one 0 entry. (Why ?)

Step 2 (Generate primal solution)

Using only the cells (i, j) in the reduced cost tableau with $\bar{d}_{ij} = 0$ as possible candidates for $x_{ij} = 1$, try to find a complete assignment solution. That is, identify a set of n cells with $\bar{d}_{ij} = 0$ such that exactly one cell is picked from each row and column. If such a set is found, then a primal solution with $x_{ij} = 1$ whenever cell (i, j) is in this set, $x_{ij} = 0$ otherwise, is feasible and satisfies the complementary slackness conditions; hence it is optimal, stop. Otherwise go to Step 3.

In Tableau # 1, such a solution is not available as only 5 assignments (indicated by the circled entries) can be made. Hence we go to step 3.

Step 3 (Modify Dual Solution)

Cross out a minimum number of rows and columns of the reduced cost matrix such that all the 0 elements are crossed out. Note that total number of rows and columns crossed out will be less than n. Define,

$$\delta = \text{minimum } \bar{d}_{ij} \text{ among all uncrossed rows}$$

Now a new dual feasible solution can be derived as below.

$$u'_i = \begin{cases} u_i & \text{if row } i \text{ is uncrossed} \\ \\ u_i - \delta & \text{if row } i \text{ is crossed} \end{cases}$$

$$v'_j = \begin{cases} v_j & \text{if column } j \text{ is crossed} \\ \\ v_j + \delta & \text{if column } j \text{ is uncrossed} \end{cases}$$

It is easy to show that this solution is dual feasible with a higher value of the objective function. Using this solution, find the new reduced cost matrix. One can derive the new reduced cost matrix from the old one by simply subtracting δ from all uncrossed cells, adding δ to all cells whose row and column are both crossed, and keeping all other cell values same. With this modified dual solution, we go to step 2.

In Tableau # 1, $\delta = \bar{d}_{31} = 1$. The updated values of the dual variables and the reduced cost matrix is given in Tableau #2. In this tableau also, a primal feasible solution in step 2 is not available. In step 3, $\delta = \bar{d}_{32} = 3$. Tableau # 3 gives the updated reduced cost matrix. In this tableau, $x_{14} = x_{21} = x_{32} = x_{46} = x_{53} = x_{65} = 1$ and all other $x_{ij} = 0$ is an assignment solution which satisfies the complementary slackness conditions and hence is optimal with objective function value $z = 59$.

j

#1	1	2	3	4	5	6	u_i
1	175	(0)	0	0	6	10	0
2	(0)	174	5	7	11	0	0
3	1	4	167	14	6	(0)	2
4	9	6	10	168	6	(0)	6
5	5	16	(0)	19	170	9	0
6	11	4	6	24	(0)	171	0

i (left label)

v_j :   5   6   11   6   10   9

j

#2	1	2	3	4	5	6	u_i
1	175	(0)	0	0	6	11	-1
2	0	174	5	7	11	1	-1
3	(0)	3	166	13	5	0	2
4	8	5	9	167	5	(0)	6
5	5	16	(0)	19	170	10	-1
6	11	4	6	24	(0)	172	-1

i (left label)

v_j :   6   7   12   7   11   9

j

#3	1	2	3	4	5	6	u_i
1	178	0	0	(0)	6	14	-4
2	(0)	171	2	4	8	1	-1
3	0	(0)	163	10	2	0	2
4	8	2	6	164	2	(0)	6
5	8	16	(0)	19	170	13	-4
6	14	4	6	24	(0)	175	-4

i (left label)

v_j :   6   10   15   10   14   9

Assuming that all d_{ij}'s are finite and integer, this algorithm will terminate with an optimal solution in a finite number of iterations.

Appendix B
Some Counterexamples to Heuristic Algorithms for the Traveling Salesman Problem

(Contributed by Dr. Robert Haas, Case Western Reserve University, Cleveland, Ohio.)

Examples for which the heuristic algorithms do not produce an optimal solution are useful to mark out the boundaries of the methods and suggest future improvements. Here such examples will be presented for several heuristic algorithms used to solve the traveling salesman problem (TSP).

B.1 Cheapest Insertion Algorithm (Nonsymmetric Case)

The cheapest insertion algorithm constructs a path for the traveling salesman by stepwise enlargement of smaller subtours, beginning with one that visits only two cities. A commonly used starting tour is the shortest two-city tour. A weakness of the method lies in the fact that while only one arc of the initial two-city subtour will become part of the later paths, the other arc is given equal weight in selection of the initial subtour. To construct a counterexample, one may therefore trick the algorithm into accepting a poor starting subtour.

Consider thus the 4-city TSP given by the following network and distance matrix:

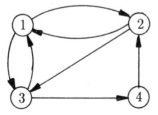

	1	2	3	4
1	—	b	a	M
2	b	—	a	M
3	2b	M	—	a
4	M	a	M	—

Here $a < b$, and arcs not drawn in the network have the large length M. By inspection, the shortest route is $1 \rightarrow 3 \rightarrow 4 \rightarrow 2 \rightarrow 1$, with length $3a + b$. It will now be shown, however, that the algorithm yields route $1 \rightarrow 2 \rightarrow 3 \rightarrow 4 \rightarrow 1$ of length $2a + b + M$, which by choice of M can be made arbitrarily larger in magnitude or proportion than $3a + b$.

For the initial two-city tour, all pairs except $(1, 2)$ and $(1, 3)$ involve a large term M and can be excluded. Because of the high "return" cost $2b$ for the cycle $1 \rightarrow 3 \rightarrow 1$, the algorithm makes the bad choice of tour $1 \rightarrow 2 \rightarrow 1$. The subsequent enlargements of the tour are calculated in the following table:

Iteration	Subtour	Length
2	* $1 \rightarrow 2 \rightarrow 3 \rightarrow 1$	$a + 3b$
	$1 \rightarrow 3 \rightarrow 2 \rightarrow 1$	$a + b + M$
	$1 \rightarrow 2 \rightarrow 4 \rightarrow 1$	$a + 2M$
	$1 \rightarrow 4 \rightarrow 2 \rightarrow 1$	$a + b + M$
3	$1 \rightarrow 4 \rightarrow 2 \rightarrow 3 \rightarrow 1$	$2a + 2b + M$
	$1 \rightarrow 2 \rightarrow 4 \rightarrow 3 \rightarrow 1$	$3b + 2M$
	* $1 \rightarrow 2 \rightarrow 3 \rightarrow 4 \rightarrow 1$	$2a + b + M$

* denotes path of minimal length at the given iteration

B.2 Cheapest Insertion Algorithm (Symmetric Case)

The previous counterexample made heavy use of non-symmetric distances. By enlarging the problem to 5 cities, however, one can give a counterexample even using a symmetric distance matrix. Consider the symmetric TSP given by the following network and symmetric distance matrix:

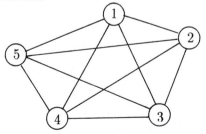

	1	2	3	4	5
1	–	1	2	2	3
2		–	2	3	M
3			–	2	2
4				–	M
5					–

The shortest tour is $1 \to 2 \to 4 \to 3 \to 5 \to 1$, of length 11. Because paths $2 \to 5$ and $4 \to 5$ have high cost M, there is only a single low-cost path available through city 5, namely $1 \to 5 \to 3$. As shown in the table below, however, the algorithm is misled by short-term gains at iterations 2 and 3 to insert cities 2 and 4 on the path between 1 and 3, so that there is no good path left at the last step.

Iteration		Subtour	Length
1		$1 \to 2 \to 1$	2
2	*	$1 \to 2 \to 3 \to 1$	5
	*	$1 \to 3 \to 2 \to 1$	5
		$1 \to 4 \to 2 \to 1$	6
		$1 \to 2 \to 4 \to 1$	6
		$1 \to 5 \to 2 \to 1$	$> M$
		$1 \to 2 \to 5 \to 1$	$> M$
3		$1 \to 4 \to 2 \to 3 \to 1$	9
		$1 \to 2 \to 4 \to 3 \to 1$	8
	*	$1 \to 2 \to 3 \to 4 \to 1$	7
		$1 \to 5 \to 2 \to 3 \to 1$	$> M$
		$1 \to 2 \to 5 \to 3 \to 1$	$> M$
		$1 \to 2 \to 3 \to 5 \to 1$	8
4		$1 \to 5 \to 2 \to 3 \to 4 \to 1$	$> M$
		$1 \to 2 \to 5 \to 3 \to 4 \to 1$	$> M$
		$1 \to 2 \to 3 \to 5 \to 4 \to 1$	$> M$
		$1 \to 2 \to 3 \to 4 \to 5 \to 1$	$> M$

* denotes path of minimal length at the given iteration

B.3 Convex Hull Heuristic

This heuristic for Euclidean TSP's uses the convex hull around the cities as its starting subtour, as illustrated in the 5-city TSP of Fig. 15.12(a). In the version of the algorithm discussed in Section 15.4.1, the cities within the initial region are then inserted into the tour sequentially, selecting city i to be inserted between j and k in such a way that angle jik is maximal. For this example, Fig. 15.12(c) gives the TSP

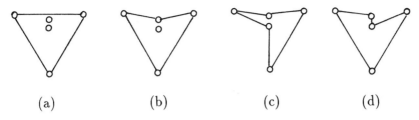

Figure 15.12: The Convex Hull Heuristic

tour, however, a more efficient tour here would be the one given in Fig. 15.12(d).

B.4   Two Edge Interchange (2-opt) Algorithm

Suppose a 7-city symmetric TSP is given by the following network and distance matrix:

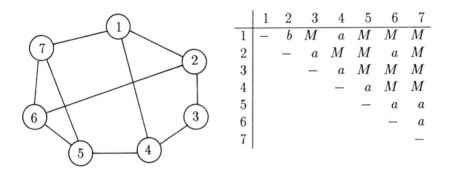

	1	2	3	4	5	6	7
1	–	b	M	a	M	M	M
2		–	a	M	M	a	M
3			–	a	M	M	M
4				–	a	M	M
5					–	a	a
6						–	a
7							–

Here the edges indicated in the network all have length a, except for edge 1-2 of distance b ($b > a$), and all edges not drawn in are of large length M. Consider now an initial path $1 \to 2 \to 3 \to 4 \to 5 \to 6 \to 7 \to 1$ of length $6a + b$ and an the optimal route $1 \to 4 \to 3 \to 2 \to 6 \to 5 \to 7 \to 1$ of length $7a$. (By appropriate choice of b, the initial tour can be made arbitrarily larger in length or proportion than the optimal one.) It is not possible, however, to reach the optimal route from the initial one by using the two-edge interchange algorithm. For a two-edge interchange, say of 1-2 and 4-5 with 1-4 and 2-5 to yield tour $1 \to 4 \to 3 \to 2 \to 5 \to 6 \to 7 \to 1$, will add the large length M of arc 2-5. Inspection shows that every two-edge exchange will add atleast one segment of length M in this way, so no pairwise exchange offers any savings and the exchange algorithm will not proceed past the initial solution.

Bibliography

[1] Agarwal, Y., K. Mathur, and H. M. Salkin; "A Set Partitioning Based Exact Algorithm for The Vehicle Routing Problem," To appear in *Networks* (1989)

[2] Agin, N.; "Optimum Seeking with Branch and Bound," *Management Science* **13**(4), 176-185 (1966).

[3] Alcaly, R., and A. Klevorick; "A Note on the Dual Prices of Integer Programs," *Econometrica* **34**(1), 206-214 (1966).

[4] Almogy, Y., and O. Levin; "The Fractional Fixed-Charge Problem," *Naval Research Logistics Quarterly* **18**(3), 307-315 (1971).

[5] Andrew G., T. Hoffman, and C. Krabek; "On the Generalized Set Covering Problem," Control Data Corporation, Data Centers Division, Minneapolis. (Paper presented at the ORSA meeting in San Francisco, 1968.)

[6] Anthonisse, J.; "A Note on Reducing a System to a Single Equation," Stichting Mathematisch Centrum preliminary report, BN 1/70 Amsterdam, December 1970.

[7] _____; "A Note on Equivalent Systems of Linear Diophantine Equations," Stichting Mathematisch Centrum preliminary report, BW 12/71 Amsterdam, July 1971.

[8] Arabeyre, T., J. Fearnley, F. Steiger, and W. Teather; "The Airline Crew Scheduling Problem: A Survey," *Transportation Science* **3**(2), 140-163 (1969).

[9] Armour, G., and E. Buffa; "A Heuristic Algorithm and Simulation Approach to Relative Location of Facilities," *Management Science* **9**(2), 294-309 (1965).

[10] Arnoff, E., and S. Sengupta; "The Traveling Salesman Problem," in *Progress in Operations Research* Vol. I (ed.: Ackoff), New York, John Wiley & Sons, 1961.

[11] Arnold, L., and M. Bellmore; "Iteration Skipping in Primal Integer Programming," *Operations Research* **22**(1), 129-136 (1974).

[12] _____; "A Bounding Minimization Problem for Primal Integer Programming," *Operations Research* **22**(2), 383-392 (1974).

[13] _____; "A Generated Cut for Primal Integer Programming," *Operations Research* **22**(1), 137-143 (1974).

[14] Art, R.; "An Approach to the Two Dimensional Irregular Cutting Stock Problem," IBM Cambridge Scientific Center Report No. 320-2006, September 1966.

[15] Ashour, S.; "An Experimental Investigation and Computational Evaluation of Flow Shop Scheduling Problems," *Operations Research* **18**(3), 541-548 (1970).

[16] Ashour, S., and A. Char; "Computational Experience on Zero-One Programming Approach to Various Combinatorial Problems," *Journal of the Operations Research Society of Japan* **13**(2), 78-108 (1970).

[17] Bair, B., and G. Hefley; "Fixed Charge Transportation Problem: A Survey," Center for Urban Transportation Studies Technical Report No. 19, the University of Iowa at Iowa City, May 1973.

[18] Balas, E.; "Extension de l'algorithme additif á la programmation en nombres entiers et á la programmation non linéaire," *C. R. Acad. Sc. Paris*, May 1964.

[19] ———————; "An Additive Algorithm for Solving Linear Programs with Zero-One Variables," *Operations Research* **13**(4), 517-548 (1965).

[20] ———————; "Duality in Discrete Programming," Department of Operations Research Technical Report No. 67-5, Stanford University, July 1967.

[21] ———————; "A Note on the Branch and Bound Principle," *Operations Research* **16**(2), 442-445 (1968).

[22] ———————; "Duality in Discrete Programming II: The Quadratic Case," *Management Science* **16**(1), 14-32 (1969).

[23] ———————; "Duality in Discrete Programming III: Nonlinear Objective Function and Constraints," Management Science Research Report No. 145, Carnegie-Mellon University, February 1968.

[24] ———————; "Duality in Discrete Programming IV: Applications," Management Science Research Report No. 145, Carnegie-Mellon University, October 1968.

[25] ———————; "Machine Sequencing via Disjunctive Graphs: An Implicit Enumeration Algorithm," *Operations Research* **17**(6), 941-957 (1969).

[26] ———————; "The Intersection Cut–A New Cutting Plane for Integer Programming," Management Sciences Research Report No. 187, Carnegie-Mellon University, October 1969.

[27] ———————; "Minimax and Duality for Linear and Nonlinear Mixed-Integer Programming," in *Integer and Nonlinear Programming* (ed. Abadie), North Holland 1970.

[28] ———————; "Alternative Strategies for Using Intersection Cuts in Integer Programming," Management Sciences Research Report No. 209, Carnegie-Mellon University, June 1970.

[29] ———————; "Intersection Cuts from Maximal Convex Extensions of the Ball and the Octahedron," Management Sciences Research Report No. 214, Carnegie-Mellon University, August 1970. To appear in *Mathematical Programming*.

[30] —————————; "Intersection Cuts–A New Type of Cutting Plane for Integer Programming," *Operations Research* **19**(1), 19-39 (1971).

[31] —————————; "Ranking the Facets of the Octahedron," Management Sciences Research Report No. 252, Carnegie-Mellon University, May 1971.

[32] —————————; "Integer Programming and Convex Analysis: Intersection Cuts from Outer Polars," *Mathematical Programming* **2**(3), 330-382 (1972).

[33] —————————; "A Constraint-Activating Outer Polar Method for Solving Pure or Mixed Integer 0-1 Programs," Management Sciences Research Report No. 275, Carnegie-Mellon University, June 1972.

[34] —————————; "Nonconvex Quadratic Programming via Generalized Polars," Management Sciences Research Report No. 275, Carnegie-Mellon University, July 1972.

[35] —————————; "A Note on Outer Polar Cuts from 0-1 Points," Management Sciences Research Report No. 298, Carnegie-Mellon University, November 1972.

[36] —————————, V. Bowman, F. Glover, and D. Sommer; "An Intersection Cut from the Dual of the Unit Hypercube," *Operations Research* **19**(1), 40-44 (1971).

[37] —————————, and N. Christofides; "A Restricted Lagrangean Approach to The Traveling Salesman Problem," *Math. Programming* **21**, 19-46 (1981).

[38] —————————, and R. Jeroslow; "Canonical Cuts on the Unit Hypercube," *SIAM Journal on Applied Mathematics* **23**(1), 61-69 (1972).

[39] —————————, and M. Padberg; "On the Set Covering Problem," *Operations Research* **20**(6), 1152-1161 (1972).

[40] —————————; "On the Set Covering Problem–II, An Algorithm," Management Sciences Research Report No. 295, Carnegie-Mellon University, May 1972.

[41] —————————; "Linear Programming with Zero-One Variables" (in Rumanian), *Proceedings of the Third Scientific Session on Statistics*, Bucharest, December 5-7, 1963.

[42] —————————; "Discrete Programming by the Filter Method," *Operations Research* **15**(5), 915-957 (1967).

[43] —————————; "A Note on the Group Theoretic Approach to Integer Programming and the 0-1 Case," *Operations Research* **21**(1), 321-322 (1973).

[44] Balas, E. and M. W. Padberg; "Set Partitioning: A Survey," *SIAM Rev.* **18**, 710-780 (1981).

[45] Balas, E. and P. Toth; "Branch and Bound Methods," in *The Traveling Salesman Problem: A Guided Tour of Combinatorial Optimization* (ed. E. L. Lawler et al.), John Wiley & Sons, (1985)

[46] Balas, E. and E. Zemel; "An Algorithm for Large Zero-one Knapsack Problems," *Operations Research* **28**, 1130-1154 (1980).

[47] Balinski, M.; "Fixed Cost Transportation Problem," *Naval Research Logistics Quarterly* **8**(1), 41-54 (1961).

[48] _____, and R. Quandt; "On an Integer Program for a Delivery Problem," *Operations Research* **12**(2), 300-304 (1964).

[49] _____; "On Finding Integer Solutions to Linear Programs," *Mathematica*, Princeton, New Jersey, May 1964.

[50] _____; "Integer Programming: Methods, Uses, Computation," *Management Science* **12**(3), 253-313 (1965).

[51] _____, and K. Spielberg; "Methods for Integer Programming: Algebraic, Combinatorial, and Enumerative," in *Progress in Operations Research* Vol. III (ed.: J. Arnofsky), 195-292, New York, John Wiley & Sons, 1968.

[52] _____, and P. Wolfe; "On Benders Decomposition and a Plant Location Problem," Mathematica Working Paper ARO-27, 1963.

[53] Balintfy, J., and C. Blackburn, II; "General Purpose Multiple Choice Programming," Graduate School of Business Administration Report, Tulane University, November 1969.

[54] Barachet, L.; "Graphic Solution of the Traveling-Salesman Problem," *Operations Research* **5**(6), 841-845 (1957).

[55] Barr, R. S., F. Glover, and D. Klingman; "A New Optimization Method for Large-Scale Fixed Charge Transportation Problems," *Operations Research*, **29**, 448-463 (1981)

[56] Baumol, W., and R. Quandt; "Investment and Discount Rates under Capital Rationing-A Programming Approach," *The Economic Journal* **75**, 317-329 (1965).

[57] _____, and P. Wolfe; "A Warehouse Location Problem," *Operations Research* **6**(2), 252-263 (1968).

[58] Baxter, W.; "A Group Theoretic Branch and Bound Algorithm for the Traveling Salesman Problem," Masters's Thesis, Massachusetts Institute of Technology, 1968.

[59] Bazaraa, M. S. and J. J. Jarvis; *Linear Programming and Network Flows*, Wiley, (1977)

[60] Beale, E.; "A Method of Solving Linear Programming Problems When Some But Not All of the Variables Must Take Integral Values," Statistical Techniques Research Group, Technical Report No. 19, Princeton University, July 1958.

[61] _____; *Mathematical Programming in Practice*, London, Pitmans, 1968.

[62] _____, and R. Small; "Mixed Integer Programming by a Branch

and Bound Technique," *Proceedings of the Third IFIP Congress* **2**, 450-451 (1965).

[63] —————————, and J. Tomlin; "An Integer Programming Approach to a Class of Combinatorial Problems," *Mathematical Programming* **3**(2), 339-344 (1972).

[64] Beasley, J. E.; "An Exact Two-Dimensional Non-Guillotine Cutting Search Procedure," *Operations Research* **33**, 49-64 (1985)

[65] Beckman, M., and T. Marschak; "An Activity Analysis Approach to Location Theory," *Proceedings of the Second Symposium on Linear Programming* **1**, 331-379 (Washington, D.C.), 1955.

[66] Bellman, R.; "Some Applications of the Theory of Dynamic Programming-A Review," *Operations Research* **2**(2), 275-288 (1954).

[67] —————————; "Notes on the Theory of Dynamic Programming IV–Maximization Over Discrete Sets," *Naval Research Logistics Quarterly* **3**(1), 67- 70 (1956).

[68] —————————; "Comment on Dantzig's Paper on Discrete Variable Extremum Problems," *Operations Research* **5**(5), 723-724 (1957).

[69] —————————; *Dynamic Programming*, Princeton University Press, 1957.

[70] —————————;" Dynamic Programming Treatment of the Traveling Salesman Problem," *Journal of the Association for Computing Machinery* **9**(1), 61-63 (1962).

[71] —————————, and S. Dreyfus; *Applied Dynamic Programming*, Princeton University Press, 1962.

[72] —————————; "An Application of Dynamic Programming to Location-Allocation Problems," *SIAM Review* **7**(1), 126-128 (1965).

[73] Bellmore, M., and J. C. Malone; "Pathology of Traveling Salesman Subtour elimination Algorithms," *Operations Research* **19**, 278-307 (1971).

[74] Bellmore, M., and G. Nemhauser; "The Traveling Salesman Problem: A Survey," *Operations Research* **16**(3), 538-558 (1968).

[75] —————————, H. Greenberg, and J. Jarvis; "Multi-Commodity Disconnecting Sets," *Management Science* **16**(6), 427-433 (1970).

[76] —————————, and H. Ratliff; "Optimal Defense of Multi-Commodity Networks," *Management Science* **18**(4), 174-185 (1971).

[77] —————————; "Set Covering and Involutory Bases," *Management Science* **18**(3), 194-206 (1971).

[78] —————————, and J. Malone; "Pathology of Traveling-Salesman Subtour- Elimination Algorithms," *Operations Research* **19**(2), 278-307 (1971).

[79] Benders, J.; "Partitioning Procedures for Solving Mixed-Variables Programming Problems," *Numerische Mathematik* **4**, 238-252 (1962).

[80] Benichou, M., J. Gauthier, P. Girodet, G. Hentges, G. Ribiere, and D. Vin-

cent; "Experiments in Mixed Integer Linear Programming," *Mathematical Programming* **1**(1), 76-94 (1971).

[81] Ben-Israel, A., and A. Charnes; "On Some Problems of Diophantine Programming," *Cahiers du Centre d'Etudes de Recherche Operationnelle (Bruxelles)* **4**, 215-280 (1962).

[82] Bertier, P., and B. Roy; "Une Procédure de Résolution pour une Classe de Problèmes pouvant avoir un Caractère Combinatoire," *Cahiers du Centre D'Etudes de Recherche Opèrationnelle* **6**, 202-208 (1964). Translated by W. Jewell, "A Solution Procedure for a Class of Problems having Combinatorial Structure," Operations Research Center, University of California at Berkeley, September 1967.

[83] Bowman, E.; "Scale of Operations–An Empirical Study," *Operations Research* **6**(3), 320-328 (1958).

[84] —————; "The Schedule-Sequencing Problem," *Operations Research* **7**(5), 621-624 (1959).

[85] Bowman, V.; "The Structure of Integer Programs under the Hermitian Normal Form," Graduate School of Industrial Administration Working Paper No. 76-70-1, Carnegie-Mellon University, May 1971.

[86] —————; "Sensitivity Analysis in Linear Integer Programming," available from the Graduate School of Industrial Administration, Carnegie-Mellon University. (Paper presented at the August 1972 AIIE Conference.)

[87] —————; "0-1 Programming for Column-Chained Matrices under Vector Partial Ordering," Graduate School of Industrial Administration Research Report No. 288, Carnegie-Mellon University, May 1972.

[88] —————; "Constraint Classification on the Unit Hypercube," Management Sciences Research Report No. 287, Carnegie-Mellon University, March 1972.

[89] —————, and F. Glover; "A Note on Zero-One Integer and Concave Programming," *Operations Research* **20**(1), 182-183 (1972).

[90] Bowman, V., and G. Nemhauser; "A Finiteness Proof for Modified Dantzig Cuts in Integer Programming," *Naval Research Logistics Quarterly* **17**(3), 309-313 (1970).

[91] —————; "Deep Cuts in Integer Programming," *OPSEARCH* **8**, 89-111 (1971).

[92] Bradley, G.; "Equivalent Integer Programs," in the *Proceedings of the Fifth International Conference of Operational Research* (ed. Lawrence), London, Travistock Publications, 1970.

[93] —————; "Equivalent Integer Programs and Canonical Problems," *Management Science* **17**(5), 354-366 (1971).

[94] —————; "Heuristic Solution Methods and Transformed Integer Linear Programming Problems," Administrative Sciences Report No. 43, Yale University, March 1971. Also in *Proceedings of the Fifth Annual Princeton Conference on Information Sciences and Systems*.

[95] —————————; "Algorithms for Hermite and Smith Normal Matrices and Linear Diophantine Equations," *Mathematics of Computation* **25**(116), 897- 898 (1971).

[96] —————————; "Transformation of Integer Programs to Knapsack Problems," *Discrete Mathematics* **1**, 29-45 (1971).

[97] —————————; "Modulo Optimization Problems and Integer Linear Programming," in *Applications of Number Theory to Numerical Analysis* (ed.: Zaremba), New York, Academic Press, 1972.

[98] —————————; "Equivalent Mixed Integer Programming Problems," *Operations Research* **21**(1), 323-326 (1973).

[99] —————————, and P. Wahi; "An Algorithm for Integer Linear Programming: A Combined Algebraic and Enumeration Approach," *Operations Research* **21**(1), 45-60 (1973).

[100] —————————, P. Hammer, and L. Wolsey; "Coefficient Reduction for Inequalities in 0-1 Variables," Department of Combinatorics and Optimization Research Report CORR 73-6, the University of Waterloo (Canada), March 1973.

[101] Brooks, R., and A. Geoffrion; "Finding Everett's Lagrange Multipliers by Linear Programming," *Operations Research* **14**(16), 1149-1153 (1966).

[102] Brown, W.; "A Generalized Plant Location Problem," Ph.D. Thesis, Department of Operations Research, Case Western Reserve University, January 1973.

[103] Burdet, C.; "The Method of Enumerative Cuts," IBM T.J. Watson Research Report RC 3174, 1970.

[104] —————————; "A Class of Cuts and Related Algorithms in Integer Programming," Management Sciences Research Report No. 220, Carnegie-Mellon University, September 1970.

[105] —————————; "A Note on Minimal Integer Conditions and Fractional Intersection Cuts," GSIA Working Paper #117-71-2, Carnegie-Mellon University, 1971.

[106] —————————; "Enumerative Inequalities in Integer Programming," *Mathematical Programming* **2**(1), 32-64 (1972).

[107] —————————; "Polaroids: A New Tool in Non-convex and in Integer Programming," Management Sciences Research Report No. 284, Carnegie-Mellon University, February 1972.

[108] —————————; "On the Diamond Cuts," Graduate School of Industrial Administration, Carnegie-Mellon University, April 1972. (Notes for presentation at the 41st National ORSA meeting at New Orleans.)

[109] —————————; "Enumerative Cuts II," GSIA Working Paper #18-71-2, Carnegie-Mellon University, April 1972.

[110] —————————; "Polar Programming," GSIA Working Paper #98-71-2, Carnegie-Mellon University, 1972.

[111] ——————————; "Generating All the Faces of a Polyhedron," Management Sciences Research Report No. 271, Carnegie-Mellon University, 1972.

[112] ——————————; "On Polaroid Intersections," Management Sciences Research Report No. 279, Carnegie-Mellon University, April 1972.

[113] ——————————; "Simple Polaroids for Non-convex Quadratic Programming," Management Sciences Research Report No. 283, Carnegie-Mellon University, May 1972.

[114] ——————————; "The Facial Decomposition Method," Management Sciences Research Report No. 280, Carnegie-Mellon University, May 1972.

[115] ——————————; "On the Algebra and Geometry of Integer Programming Cuts," Management Sciences Research Report No. 291, Carnegie-Mellon University, October 1972.

[116] ——————————; "Enumerative Cuts I," Operations Research 21(1), 61-89 (1973).

[117] Busacker, R., and T. Saaty; *Finite Graphs and Networks*, New York, McGraw-Hill, 1965.

[118] Bush, R., and R. Carlson; "Equivalent Knapsack Problem Forms for Specially Structured Integer Linear Programming Problems," Department of Industrial Engineering, Stanford University. (Paper presented at the ORSA/TIMS/AIIE Joint National Meeting, November 1972.)

[119] Byrne, R., A. Charnes, W. Cooper, and K. Kortanek; "A Chance Constrained Approach to Capital Budgeting with Portfolio Type Payback and Liquidity Constraints and Horizon Posture Controls," *Journal of Financial and Quantitative Analysis* 11, 339-364 (1967).

[120] Cabot, A.; "Variations on a Cutting Plane Method for Solving Concave Minimization Problems with Linear Constraints," Graduate School of Business paper, Indiana University at Bloomington, January 1972.

[121] Cabot, V.; "An Enumeration Algorithm for Knapsack Problems," *Operations Research* 18(2), 306-311 (1970).

[122] ——————————, and A. Hurter, Jr.; "The Application of an Approach to Zero-One Programming to Knapsack Problems," presented at the joint ORSA/TIMS Meeting in San Francisco (1968).

[123] Camion, P.; "Characterization of Totally Unimodular Matrices," *Proceedings of the American Mathematical Society* 16, 1068-1073 (1965).

[124] Carpaneto, G. and P. Toth; "Some new Branching and Bounding criteria for the Asymmetric Traveling Salesman Problem," *Management Science* 26, 736-743 (1980).

[125] Chandrasekaran, R.; "Total Unimodularity of Matrices," *SIAM Journal of Applied Mathematics* 17(6), 1032-1034 (1969).

[126] Charnes, A., and W. Cooper; *Management Models and Industrial Applications of Linear Programming* Vols. I and II, New York, John Wiley & Sons, 1961.

[127] _____, and M. Miller; "A Model for Optimal Programming of Railway Freight Train Movements," *Management Science* **3**(1), 74-92 (1956).

[128] Chen, D.; "A Group Theoretic Algorithm for Solving Integer Linear Programming Problems," Ph.D. Thesis, Department of Industrial Engineering, State University of New York at Buffalo, September 1970.

[129] _____, and S. Zionts; "An Exposition of the Group Theoretic Approach to Integer Linear Programming," School of Management Working Paper No. 78, State University of New York at Buffalo, April 1970.

[130] _____; "On the Use of Shortest Route Solution Methods to Solve the Group Theoretic Integer Programming Problem," School of Management Working Paper No. 140, State University of New York at Buffalo, April 1972.

[131] Christofides, N.; "The Shortest Hamiltonian Chain of a Graph," *SIAM Journal on Applied Mathematics* **19**, 689-696 (1970).

[132] _____; "Bounds for the Traveling-Salesman Problem," *Operations Research* **20**(5), 1044-1056 (1972).

[133] _____; "The Traveling Salesman Problem," in *Combinatorial Optimization* (ed. N. Christofides, et al.), Wiley, 131-149 (1979).

[134] _____, and S. Korman; "A Computational Survey of Methods for the Set Covering Problem," Department of Management Science Report 73/2, Imperial College of Technology (London), April 1973.

[135] Chvatal V.; "A Greedy Heuristic for The Set Covering Problem," *Math. of Operations Research* **4**, 233-235 (1979).

[136] Ciochetto, F., H. Swanson, J. Lee, and R. Woolsey; "The Lockbox Problem and Some Startling But True Computational Results for Large Scale Systems," available from R.E.D. Woolsey, Colorado School of Mines, Golden, Colorado. (Paper presented at the 41st national ORSA meeting in New Orleans, April 1972).

[137] Clarke, G., and S. Wright; "Scheduling of Vehicles from a Central Depot to a Number of Deliver Points," *Operations Research* **12**(4), 568-581 (1964).

[138] Cobham, A., R. Fridshal, and J. North; "An Application of Linear Programming to the Minimization of Boolean Functions," IBM Research Center Report RC 472, Yorktown Heights, New York, June 1961.

[139] Colmin, J., and K. Spielberg; "Branch and Bound Schemes for Mixed Integer Programming," IBM New York Scientific Center Report No. 320-2972, May 1969. Available from IBM Philadelphia Scientific Center, 3401 Market Street.

[140] Colville, A.; "Mathematical Programming Codes: Programs Available from IBM Program Information Department," IBM Philadelphia Scientific Center Report No. 320-2925, January 1968.

[141] Cooper, L.; "Location-Allocation Problems," *Operations Research* **11**(3), 331-343 (1963).

[142] _____; "Heuristic Methods for Location-Allocation Problems," *SIAM Review* **6**(1), 37-52 (1964).

[143] _____ __; "Solutions of Generalized Locational Equilibrium Models," *Journal of Regional Sciences* **7**(1), 1-18 (1967).

[144] _____; "An Extension of the Generalized Weber Problem," *Journal of Regional Sciences* **8**(2), 181-197 (1968).

[145] _____; "The Transportation-Location Problem," *Operations Research* **20**(1), 94-108 (1972).

[146] _____, and C. Drebes; "Investigations in Integer Linear Programming by Direct Search Methods," Department of Applied Mathematics and Computer Science Report No. C00-1493-10, Washington University (St. Louis), 1967.

[147] _____; "An Approximate Solution Method for the Fixed Charge Problem," *Naval Research Logistics Quarterly* **14**, 101-113 (1967).

[148] _____, and R. Echols; "The Solution of Integer Linear Programming Problems by Direct Search," Department of Applied Mathematics and Computer Science Report No. AM-66-1, Washington University (St. Louis), September 1966.

[149] Cord, J.; "A Method for Allocating Funds to Investment Projects when Returns are Subject to Uncertainty," *Management Science* **10**(2), 335-341 (1964).

[150] Croes, G.; "A Method for Solving Traveling-Salesman Problems," *Operations Research* **6**(8), 791-812 (1958).

[151] Crowder, H., E. L. Johnson, and M. W. Padberg; "Solving Large-Scale Zero-One Linear Programming Problems," *Operations Research* **31**, 803-834 (1983).

[152] Dacey, M.; "Selection of an Initial Solution for the Traveling Salesman Problem," *Operations Research* **8**(1), 133-134 (1960).

[153] Dakin, R.; "A Tree Search Algorithm for Mixed Integer Programming Problems," *Computer Journal* **8**, 250-255 (1965).

[154] Dantzig, G.; "An Application of the Simplex Method to a Transportation Problem," in *Activity Analysis of Production and Allocation* (ed.: T. C. Koopmans), Cowles Commission Monograph No. 13, New York, John Wiley & Sons, 1951.

[155] _____; "Discrete-Variable Extremum Problems," *Operations Research* **5**(2), 266-277 (1957).

[156] _____; "Note on Solving Linear Programs in Integers," *Naval Research Logistics Quarterly* **6**, 75-76 (1959).

[157] _____; "A Machine-Job Scheduling Model," *Management Science* **6**(1), 191-196 (1960).

[158] _____; "On the Significance of Solving Linear Programming Problems with Some Integer Variables," *Econometrica* **28**(1), 30-44 (1960).

[159] ——————; *Linear Programming and Extensions*, Princeton University Press, 1963.

[160] ——————, D. Fulkerson, and S. Johnson; "Solution of a Large Scale Traveling Salesman Problem," *Operations Research* **2**(4), 393-410 (1954).

[161] ——————, D. Fulkerson, and S. Johnson; "On a Linear Programming, Combinatorial Approach to the Traveling Salesman Problem," *Operations Research* **7**(1), 58-66 (1959).

[162] ——————, and J. Ramser; "The Truck Dispatching Problem," *Management Science* **6**(1), 80-91 (1960).

[163] Davis, P., and T. Ray; "A Branch-Bound Algorithm for the Capacitated Facilities Location Problem," *Naval Research Logistics Quarterly* **16**, 331-344 (1969).

[164] Davis, R., D. Kendrick, and M. Weitzman; "A Branch-and-Bound Algorithm for Zero-One Mixed Integer Programming Problems," *Operations Research* **19**(4), 1036-1044 (1971).

[165] Day, R.; "On Optimal Extracting From a Multiple File Data Storage System: An Application of Integer Programming," *Operations Research* **13**(3), 482-494 (1965).

[166] Deighton, D.; "A Comment on Location Models," *Management Science* **18**(1), 113-115 (1971).

[167] DeMaio, A., and C. Roveda; "An All Zero-One Algorithm for a Certain Class of Transportation Problems," *Operations Research* **19**(6), 1406-1418 (1971).

[168] Denzler, D.; "An Approximate Algorithm for the Fixed Charge Problem," *Naval Research Logistics Quarterly* **16**, 411-416 (1969).

[169] Derman, C., and M. Klein; "Surveillance of Multicomponent Systems: A Stochastic Traveling Salesman's Problem," *Naval Research Logistics Quarterly* **13**, 103-112 (1966).

[170] Devine, M., and F. Glover; "Generating the Nested Faces of the Gomory Polyhedron," Research Report R-69-1, School of Industrial Engineering, the University of Oklahoma, December 1969.

[171] Dijkstra, E.; "A Note on Two Problems in Connection with Graphs," *Numerische Mathematik* **1**, 269-271 (1959).

[172] Dixon, J.; "A Simple Estimate for the Number of Steps in the Euclidean Algorithm," *American Mathematical Monthly*, 374-376 (April 1971).

[173] Djerdjour, M., K. Mathur, and H. M. Salkin; "A Surrogate Relaxation based algorithm for a General Quadratic Multi-dimensional Knapsack Problem," *Operations Research Letters* **7**, 253-258 (1988)

[174] Dragan, I.; "An Improvement of the Lexicographical Algorithm for Solving Linear Discrete Programming Problems," CORE Discussion Paper No. 7029, Louvain, Belgium, August 1970.

[175] Dreyfus, S.; "An Appraisal of Some Shortest-Path Algorithms," *Operations Research* **17**(3), 395-412 (1969).

[176] —————, and K. Prather; "Improved Algorithms for Knapsack Problems," Operations Research Center Report ORC 70-30, the University of California at Berkeley, 1970.

[177] Driebeek, N.; "An Algorithm for the Solution of Mixed Integer Programming Problems," *Management Science* **12**(7), 576-587 (1966).

[178] Drysdale, J., and P. Sandiford; "Heuristic Warehouse Location-A Case History Using a New Method," *Canadian Operational Research* **7**(1), 45-61 (1969).

[179] Dwyer, P.; "The Direct Solution of the Transportation Problem with Reduced Matrices," *Management Science* **13**(1), 77-98 (1966).

[180] Dyckhoff, H.; "A New Linear Programming Approach to the Cutting Stock Problem," *Operations Research* **29**, 1092-1104 (1981)

[181] Dyer, M. E.; "An O(n) algorithm for The Multiple-choice Knapsack Linear Program," *Mathematical Programming* **29**, 57- (1984)

[182] Eastman, W.; "Linear Programming with Pattern Constraints," Ph.D. Thesis, Harvard University, 1958.

[183] Echols, R.; "The Solution of Integer Linear Programming by Direct Search," M.S. Thesis, Washington University (St. Louis), 1966.

[184] Edmonds, J.; "Covers and Packing in a Family of Sets," *Bulletin of the American Mathematical Society* **68**, 494-499 (1962).

[185] —————; "Paths, Trees, and Flowers," *Canadian Journal of Mathematics* **17**, 848-856 (1969).

[186] —————, and E. Johnson; "Matching, Eulers Tour and The Chinese Postman," *Mathematical Programming* **5**(1), 88-124 (1973).

[187] Efroymson, M., and T. Ray; "A Branch-Bound Algorithm for Plant Location," *Operations Research* **14**(3), 361-368 (1966).

[188] Elmaghraby, S., and M. Wig; "On the Treatment of Stock Cutting Problems as Diophantine Programs," Operations Research Center Report No. 61, North Carolina State University at Raleigh, May 11, 1970.

[189] Ernst, E., L. Lasdon, L. Ostrander, and S. Divell; "Anesthesiologist Scheduling Using a Set Partitioning Algorithm," available from Edward A. Ernst, M.D., University Hospitals, Case Western Reserve University, Cleveland, Ohio 44106, October 1972.

[190] Everett III, H.; "Generalized Lagrange Multiplier Method for Solving Problems of Optimum Allocation of Resources," *Operations Research* **11**(3), 399-471 (1963).

[191] Faaland, B.; "Solution of the Value-Independent Knapsack Problem by Partitioning," *Operations Research* **21**(1), 332-336 (1973).

[192] —————; "Estimates and Bounds on Computational Effort in the

Accelerated Bound-and-Scan Algorithm," Department of Operations Research Technical Report No. 35, Stanford University, May 15, 1972.

[193] —————————, and F. Hillier; "The Accelerated Bound-and-Scan Algorithm for Integer Programming," Department of Operations Research Technical Report No. 34, Stanford University, May 15, 1972.

[194] Faden, B. (ed.); *Computer Program Directory*, published by CCM Information Corporation, New York, for the Joint Users Group of the Association for Computing Machinery, 1971.

[195] Farbey, B., A. Land, and J. Murchalnd; "The Cascade Algorithm for Finding All Shortest Distances in a Directed Graph," *Management Science* **14**(1), 19-28 (1967).

[196] Fayard, D., and G. Plateau; "Resolution of the 0-1 Knapsack Problem: Comparison of Methods," *Mathematical programming* **8**, 272-307 (1975).

[197] Feldman, E., F. Lehrer, and T. Ray; "Warehouse Location under Continuous Economies of Scale," *Management Science* **12**(9), 670-684 (1966).

[198] Fiorot, J.; "Generation of All Integer Points for Given Sets of Linear Inequalities," *Mathematical Programming* **3**(3), 276-295 (1972).

[199] Fisher, M. L.; "The Lagrangean Relaxation Method for Solving Integer Programming Problems," *Management Science* **27**(1), 1-18 (1981)

[200] Fleischman, B.; "Computational Experience with the Algorithm of Balas," *Operations Research* **15**(1), 153-155 (1967).

[201] Flood, M.; "The Traveling Salesman Problem," *Operations Research* **4**(1), 61-75 (1956).

[202] Floyd, R.; "Algorithm 97: Shortest Path," *Communications of the ACM* **5**(6), 345 (1962).

[203] Ford, L., and D. Fulkerson; *Flows in Networks*, Princeton University Press, 1962.

[204] Forrest, J., J. Hirst, and J. Tomlin; "Practical Solution of Large and Complex Integer Programming Problems with Umpire," *Management Science* **20** (1973).

[205] Fox, B., and D. Landi; "Searching for the Multiplier in One Constraint Optimization Problems," *Operations Research* **18**(2), 253-262 (1970).

[206] Francis, R.; "Some Aspects of a Minimax Location Problem," *Operations Research* **15**(6), 1163-1169 (1967).

[207] Freeman, R.; "Computational Experience with the Balas Integer Programming Algorithm," *Operations Research* **14**(6), 935-940 (1966).

[208] —————————, D. Gogerty, G. Graves, and R. Brooks; "A Mathematical Model of Supply Support for Space Operations," *Operations Research* **14**(1), 1-15 (1966).

[209] Fricks, R.; "Nonconvex Linear Programming," Ph.D. Thesis, Department of Operations Research, Case Western Reserve University, January 1971.

[210] Fuchs, L.; *Abelian Groups*, New York, Pergamon Press, 1960.

[211] Gass, S. I.; "Linear Programming: Methods and Applications", McGraw-Hill Book Co., (1964)

[212] Garfinkel, R.; "Optimal Political Districting," Ph.D. Thesis, Johns Hopkins University, 1968.

[213] —————; "On Partitioning the Feasible Set in a Branch-and-Bound Algorithm for the Asymmetric Traveling-Salesman Problem," *Operations Research* **2**(1), 340-342 (1973).

[214] —————, and G. Nemhauser; "The Set Partitioning Problem: Set Covering with Equality Constraints," *Operations Research* **17**(5), 848-856 (1969).

[215] —————; "Optimal Political Districting by Implicit Enumeration Techniques," *Management Science* **16**(8), 495-508 (1970).

[216] —————; "Optimal Set Covering: A Survey," in *Perspectives in Optimization: A Collection of Expository Articles* (ed.: Geoffrion), Reading, Mass., Addison-Wesley, 1972.

[217] —————; "A Survey of Integer Programming Emphasizing Computation and Relation Among Models," Department of Operations Research Technical Report No. 156, Cornell University, August 1972.

[218] —————; *Integer Programming*, New York, John Wiley & Sons, 1972.

[219] Garfinkel, R. S.; "On Partitioning the Feasible Set in a Branch and Bound Algorithm for The Asymmetric Traveling Salesman Problem," *Management Science* **26**, 736-743 (1980).

[220] Gavish, B., and K. N. Srikanth; "Efficient Branch and Bound Code for Solving Large Scale Traveling Salesman Problem to Optimality," *Working Paper QM8329*, Graduate School of Management, University of Rochester, (1983).

[221] Gearing, C., W. Swart, and T. Var; "A Decision Structure for Touristic Investment Allocations," *Tourist Review* **1**, January-March 1972.

[222] Geoffrion, A.; "Integer Programming by Implicit Enumeration and Balas' Method," *SIAM Review* **9**(2), 178-190 (1967).

[223] —————; "An Improved Implicit Enumeration Approach for Integer Programming," *Operations Research* **17**(3), 437-454 (1969).

[224] —————; "Duality in Nonlinear Programming," *SIAM Review* **13**(1), 1-38 (1971).

[225] —————; "Generalized Benders Decomposition," *Journal of Optimization Theory and its Applications* **10**(4), 237-260 (1972).

[226] —————; "Lagrangian Relaxation and Its Uses in Integer Programming," Western Management Science Institute Working Paper No. 195, the University of California at Los Angeles, December 1972.

[227] —————; "The Capacitated Facility Location Problem with Additional Constraints," Operations Research Study Center Discussion paper, the University of California at Los Angeles, February 1973.

[228] _____, and G. W. Graves; "Multicommodity Distribution System Design By Benders Decomposition", *Management Science* **20**(5), 822-844 (1974).

[229] _____, and R. Marsten; "Integer Programming: A Framework and State-of-the-Art Survey," *Management Science* **18**(9), 465-491 (1972). Also in *Perspectives in Optimization: A Collection of Expository Articles* (ed.: Geoffrion), Reading, Mass., Addison-Wesley, 1972.

[230] _____, and A. Nelson; "Users Instructions for 0-1 Integer Linear Programming Code RIP30C," RAND Report RM-5627-PR, May 1968.

[231] Giglio, R., and H. Wagner; "Approximate Solutions to the Three-Machine-Scheduling Problem," *Operations Research* **12**(2), 306- 324 (1964).

[232] Gilmore, P., and R. Gomory; "A Linear Programming Approach to the Cutting Stock Problem I," *Operations Research* **9**(6), 849-858 (1961).

[233] _____; "A Linear Programming Approach to the Cutting Stock Problem II," *Operations Research* **11**(6), 863-888 (1963).

[234] _____; "A Solvable Case of the Traveling Salesman Problem," *Proceedings of the National Academy of Science* **51**, 178-181 (1964).

[235] _____; "Sequencing a One State-Variable Machine: A Solvable Case of the Traveling Salesman Problem," *Operations Research* **12**(5), 655-679 (1964).

[236] _____; "Multi-stage Cutting Stock Problems of Two and More Dimensions," *Operations Research* **13**(1), 94-120 (1965).

[237] _____; "The Theory and Computation of Knapsack Functions," *Operations Research* **14**(6), 1045-1074 (1966).

[238] Glover, F.; "A Bound Escalation Method for the Solution of Integer Linear Programs," Graduate School of Industrial Administration Reprint No. 175, Carnegie-Mellon University, 1964.

[239] _____; "A Multiphase-Dual Algorithm for the Zero-One Integer Programming Problem," *Operations Research* **13**(6), 879-929 (1965).

[240] _____; "Generalized Cuts in Diophantine Programming," *Management Science* **13**(3), 254-268 (1966).

[241] _____; "An Algorithm for Solving the Linear Integer Programming Problem Over a Finite Additive Group, with Extensions to Solving General and Certain Nonlinear Integer Programs," Operations Research Center Report 66-29, the University of California at Berkeley, September 1966.

[242] _____; "Stronger Cuts in Integer Programming," *Operations Research* **15**(6), 1174-1177 (1967).

[243] _____; "A Pseudo Primal-Dual Integer Programming Algorithm," *Journal of Research of the National Bureau of Standards* **71B**(4), 187-195 (1967).

[244] _____; "A New Foundation for a Simplified Primal Integer Programming Algorithm," *Operations Research* **16**(4), 727-748 (1968).

[245] ⎯⎯⎯⎯⎯⎯⎯; "A Note on Linear Programming and Integer Feasibility," Management Sciences Research Report No. 127, Carnegie-Mellon University, August 1968.

[246] ⎯⎯⎯⎯⎯⎯⎯; "Surrogate Constraints," *Operations Research* **16**(4), 741- 749 (1968).

[247] ⎯⎯⎯⎯⎯⎯⎯; "Integer Programming Over a Finite Additive Group," AMM-4, School of Business, the University of Texas at Austin, 1968.

[248] ⎯⎯⎯⎯⎯⎯⎯; "Faces of an Integer Polyhedron for Cyclic Groups," Management Science Research Report No. 135, Carnegie-Mellon University, July 1968.

[249] ⎯⎯⎯⎯⎯⎯⎯; "Faces of an Integer Polyhedron for Additive Groups," AMM-11, School of Business, the University of Texas at Austin.

[250] ⎯⎯⎯⎯⎯⎯⎯; "Face of the Gomory Polyhedron," in *Integer and Nonlinear Programming* (ed.: Abadie), North Holland, 1970.

[251] ⎯⎯⎯⎯⎯⎯⎯; "Convexity Cuts for Multiple Choice Problems," Management Science Series Report No. 71-1, the University of Colorado at Boulder, January 1971.

[252] ⎯⎯⎯⎯⎯⎯⎯; "Accelerating the Determination of the First Facets of the Octahedron," Management Science Series Report No. 71-6, the University of Colorado at Boulder, September 1971.

[253] ⎯⎯⎯⎯⎯⎯⎯; "A Note on Extreme Point Solutions and a Paper by Lemke, Salkin, and Spielberg," *Operations Research* **19**(4), 1023-1025 (1971).

[254] ⎯⎯⎯⎯⎯⎯⎯; "Improved Linear Representations of Discrete Mathematical Programs," Management Science Report No. 72-8, the University of Colorado at Boulder, February 1972.

[255] ⎯⎯⎯⎯⎯⎯⎯; "Cut Search Methods in Integer Programming," *Mathematical Programming* **3**(1), 86-100 (1972).

[256] ⎯⎯⎯⎯⎯⎯⎯; "New Results for Reducing Linear Programs to Knapsack Problems," Management Science Report Series 72-7, the University of Colorado at Boulder, April 1972.

[257] ⎯⎯⎯⎯⎯⎯⎯; "Convexity Cuts and Cut Search," *Operations Research* **21** (1), 123-134 (1973).

[258] ⎯⎯⎯⎯⎯⎯⎯, and D. Klingman; "Concave Programming Applied to a Special Class of Zero-One Integer Programs," Graduate School of Business Report No. AMM-11, the University of Texas at Austin, January 1969.

[259] ⎯⎯⎯⎯⎯⎯⎯; "Concave Programming Applied to a Special Class of 0-1 Integer Programs," *Operations Research* **21**(1), 135-140 (1973).

[260] ⎯⎯⎯⎯⎯⎯⎯; "The Generalized Lattice Point Problem," Management Science Report Series 71-3, the University of Colorado at Boulder, August 1971.

[261] ⎯⎯⎯⎯⎯⎯⎯; "The Generalized Lattice Point Problem," *Operations Research* **21**(1), 141-155 (1973).

[262] ——————; "Mathematical Programming Models and Methods for the Journal Selection Problem," *Operations Research* **21**(1), 141-155 (1973).

[263] Glover, F. and D. Klingman; "An O(n *log* n) Algorithm for LP Knapsacks with GUB Constraints," *Mathematical Programming* **17**, 345-361 (1979).

[264] Glover, F., D. Klingman, and J. Stutz; "The Disjunctive Facet Problem: Formulation and Solution Techniques," Management Science Series Report No. 72-10, June 1972.

[265] Glover, F., and L. Litzler; "Extension of an Asymptotic Integer Programming Algorithm to the General Integer Programming Problem," School of Business Report, the University of Texas at Austin, October 1969.

[266] Glover, F., and D. Sommer; "Pitfalls of Rounding in Discrete Management Decision Problems," Management Science Report No. 72-2, the University of Colorado at Boulder, February 1972.

[267] Glover, F., and R. Woolsey; "Aggregating Diophantine Equations," Management Science Report Series 70-4, the University of Colorado at Boulder, October 1970.

[268] ——————; "Further Reduction of Zero-One Polynomial Programming Problems to Zero-One Linear Programming Problems," *Operations Research* **21**(1), 156-161 (1973).

[269] ——————; "Converting the 0-1 Polynomial Programming Problem to a 0-1 Linear Program," *Operations Research* **21**(1), 156-161 (1973).

[270] Glover, F., and S. Zionts; "A Note on the Additive Algorithm of Balas," *Operations Research* **13**(4), 546-549 (1965).

[271] Golden, B. L.; "A Statistical Approach to the TSP", *Networks*, **7**, 209-225 (1977)

[272] Golden, B. L., L. D. Bodin, T. Doyle, and W. Stewart Jr.; "Approximate Traveling Salesman Algorithms", *Operations Research*, **28**, 694-711 (1980)

[273] Goldstone, L.; "A Further Note on Warehouse Location," *Management Science* **15**(4), 132-133 (1968).

[274] Golomb, S., and L. Baumert; "Backtrack Programming," *Journal of the Association for Computing Machinery* **12**(4), 516-524 (1965).

[275] Gomory, R.; "An Algorithm for the Mixed Integer Problem," RM-2597 Rand Corporation, July 1960.

[276] ——————; "An Algorithm for Integer Solutions to Linear Programs," Princeton IBM Mathematical Research Report, November 1958; also in *Recent Advances in Mathematical Programming* (eds.: Graves and Wolfe), New York, McGraw-Hill, 1963.

[277] ——————; "All-Integer Integer Programming Algorithm," IBM Research Center Report RC-189, January 1960; also in *Industrial Scheduling* (eds.: Muth and Thompson), Englewood Cliffs, NJ., Prentice-Hall, 1963.

[278] ——————; "On the Relation Between Integer and Noninteger Solutions to Linear Programs," *Proceedings of the National Academy of Science* **53**(2), 260-265 (1965).

[279] ⸺; "Faces of an Integer Polyhedron," *Proceedings of the National Academy of Science* **57**(1), 16-18 (1967).

[280] ⸺; "Some Polyhedra Related to Combinatorial Problems," *Journal of Linear Algebra and Its Applications* **2**(4), 451-558 (1969).

[281] ⸺; "Properties of a Class of Integer Polyhedra," in *Integer and Nonlinear Programming* (ed.: Abadie), North Holland, 1970.

[282] ⸺, and W. Baumol; "Integer Programming and Pricing," *Econometrica* **28**(3), 521-550 (1960).

[283] ⸺, and A. Hoffman; "On the Convergence of an Integer Programming Process," *Naval Research Logistics Quarterly* **10**(1), 121-123 (1963).

[284] ⸺, and E. Johnson; "Some Continuous Functions Related to Corner Polyhedra," *Mathematical Programming* **3**(1), 23-85 (1972).

[285] ⸺; "Some Continuous Functions Related to Corner Polyhedra II," *Mathematical Programming* **3**(3), 359-389 (1972).

[286] Gonzales, R.; "Solution to the Traveling Salesman Problem by Dynamic Programming on the Hypercube," Operations Research Center Technical Report No. 18, Massachusetts Institute of Technology, 1962.

[287] Gorry, A., and J. Shapiro; "An Adaptive Group Theoretic Algorithm for Integer Programming Problems," *Management Science* **17**(5), 285-306 (1971).

[288] Gorry, A., J. Shapiro, and L. Wolsey; "Relaxation Methods for Pure and Mixed Integer Programming Problems," *Management Science* **18**(5), 229-239 (1972).

[289] Gorry, G., W. Northup, and J. Shapiro; "Computational Experience with a Group Theoretic Integer Programming Algorithm," *Mathematical Programming* **4**(2), 171-192 (1973).

[290] Granot, F., and P. Hammer; "On the Role of Generalized Covering Problems," Center of Cybernetic Studies Research Report CS96, the University of Texas at Austin, September 1972.

[291] Graves, G., and A. Whinston; "A New Approach to Discrete Mathematical Programming," Institute for Research in the Behavioral, Economic, and Management Sciences Paper No. 180, Purdue University, July 1967.

[292] Gray, P.; "Mixed Integer Programming Algorithms for Site Selection and Other Fixed Charge Problems Having Capacity Constraints," Ph.D. Thesis, Stanford University, November 1967. (Available as Stanford Research Institute Report No. 101, 1967.)

[293] ⸺; "Exact Solution of the Fixed Charge Transportation Problem," *Operations Research* **19**(6), 1529-1538 (1971).

[294] Greenberg, H.; "An Algorithm for the Computation of Knapsack Functions," *Journal of Mathematical Analysis and Applications* **26**, 159-162 (1969).

[295] _____, and R. Hegerich; "A Branch Search Algorithm for the Knapsack Problem," *Management Science* **16**(5), 327-332 (1970).

[296] Guha, D.; "The Set-Covering Problem with Equality Constraints," *Operations Research* **21**(1), 348-351 (1973).

[297] Guignard, M., and K. Spielberg; "Search Techniques with Adaptive Features for Certain Integer and Mixed Integer Programming Problems," *Proceedings of the Fifth IFIPS Congress* **1**, 141-146, August 1968.

[298] _____; "The State Enumeration Method for Mixed Zero-One Programming," IBM Philadelphia Scientific Center Report No. 320-3000, February 1971.

[299] _____; "A Realization of the State Enumeration Procedure," IBM Philadelphia Scientific Center Working Paper, 1971.

[300] _____; "Mixed Integer Algorithms for the Zero-One Knapsack Problem," *IBM Journal of Research and Development* **16**(4), 424-430 (1972).

[301] Gupta, J.; "The Generalized n-Job, M-Machine Scheduling Problem," *OPSEARCH* **8**(3), 171-185 (1971).

[302] Haas, R., K. Mathur, and H. M. Salkin; "An Improved Counterexample to the Rudimentary Primal Algorithm", Tech. Memo, Department of Operations Research, Case Western Reserve University, (1989)

[303] Hadley, G.; *Linear Programming*, Reading, Mass., Addison-Wesley, 1962.

[304] Hakimi, S.; "Optimum Location of Switching Centers and the Absolute Centers and Medians of a Graph," *Operations Research* **12**(3), 450-459 (1964).

[305] _____; "Optimal Distribution of Switching Centers in a Communication Network and Some Related Graph Theoretic Problems," *Operations Research* **17**(1), 85-111 (1969).

[306] Haldi, J.; "25 Integer Programming Test Problems," Working Paper No. 43, Graduate School of Business, Stanford University, December 1964.

[307] _____, and L. Issacson; "A Computer Code for Integer Solutions to Linear Programs," *Operations Research* **13**(6), 946-959 (1965).

[308] Halfin, S.; "Arbitrarily Complex Corner Polyhedra are Dense in R^n", Bell Telephone Laboratories paper, Holmdel, New Jersey, 1972.

[309] Hammer, P., and S. Nguyen; "A Partial Order in the Solution Space of Bivalent Programs," Centre de Recherches Mathematiques Paper No. C.R.M.-163, the University of Montreal (Canada), April 1972.

[310] Hammer, P., I. Rosenberg, and S. Rudeanu; "On the Determination of the Minima of Pseudo-Boolean Functions," (in Romanian), *Stud. Cerc. Mat.* **14**, 359-364 (1963).

[311] Hammer, P., and S. Rudeanu; *Boolean Methods in Operations Research and Related Areas*, New York, Springer-Verlag, 1968.

[312] Hansmann, F.; "Operations Research in the National Planning of Under-developed Countries," *Operations Research* 9(1), 203-248 (1961).

[313] Hardgrave, W., and G. Nemhauser; "On the Relation Between the Traveling Salesman and the Longest Path Problem," *Operations Research* 10(5), 647-657 (1962).

[314] Hatfield, D., and J. Pierce; "Production Sequencing by Combinatorial Programming," in *Operations Research and the Design of Management Information Systems* (ed.: Pierce), Technical Association of the Pulp and Paper Industry, New York, 1967.

[315] Healy Jr., W.; "Multiple Choice Programming," *Operations Research* 12(1), 122-138 (1964).

[316] Hefley, G.; "Group Programming Decomposition in Integer Programming," Systems Research Center THEMIS Report No. 52, Department of Industrial and Systems Engineering, the University of Florida at Gainesville, 1971.

[317] _____, and M. Thomas; "Decomposition of the Group Problem in Integer Programming," Report No. IME 72-101, Department of Industrial and Management Engineering, the University of Iowa at Iowa City, 1972.

[318] Helbig Hansen K., and J. Krarup; "Improvements of the Held-Karp Algorithm for the Symmetric Traveling Salesman Problem," *Math. Programming*, 7, 87-96 (1974).

[319] Held, M., and R. Karp; "A Dynamic Programming Approach to Sequencing Problems," *Journal of the Society for Industrial and Applied Mathematics* 10, 196-210 (1962).

[320] _____; "The Traveling-Salesman Problem and Minimum Spanning Trees," *Operations Research* 18(6), 1138-1162 (1970).

[321] _____; "The Traveling Salesman Problem and Minimum Spanning Trees-Part II," *Mathematical Programming* 1(1), 6-25 (1971).

[322] Heller, I.; "On the Traveling Salesman's Problem," *Proceedings of the Second Symposium in Linear Programming* (National Bureau of Standards, Washington, D.C.), 1955.

[323] _____; "On Linear Systems with Integral Valued Solutions," *Pacific Journal of Mathematics* 7, 1351-1364 (1957).

[324] _____; "On a Class of Equivalent Systems of Linear Inequalities," *Pacific Journal of Mathematics* 13, 1209-1227 (1963).

[325] _____; "On Unimodular Sets of Vectors," in *Recent Advances in Mathematical Programming* (eds.: Graves and Wolfe), New York, McGraw-Hill, 1963.

[326] _____, and A. Hoffman; "On Unimodular Matrices," *Pacific Journal of Mathematics* 12, 1321-1327 (1962).

[327] Hillier, F.; "Efficient Heuristics Procedures for Integer Programming with an Interior," *Operations Research* **17**(4), 600-637 (1969).

[328] —————————-; "A Bound-and-Scan Algorithm for Pure Integer Linear Programming with General Variables," *Operations Research* **17**(4), 638-679 (1969).

[329] Hirsch, W., and G. Dantzig; "The Fixed Charge Problem," *Naval Research Logistics Quarterly* **15**(3), 413-424 (1968). (Originally appeared as a "Notes on Linear Programming: Part XIX, The Fixed Charge Problem," RAND Research Memorandum No. 1383, December 1954.)

[330] Hirsch, W., and A. Hoffman; "Extreme Varieties, Concave Functions, and the Fixed Charge Problem," *Communications on Pure and Applied Mathematics* **14**(3), 355-369 (1961).

[331] Hoffman, A., and J. Kruskal; "Integral Boundary Points of Convex Polyhedra," in *Linear Inequalities and Related Systems* (eds.: Kuhn and Tucker), Princeton University Press, 1956.

[332] Hohn, F.; "Some Mathematical Aspects of Switching," *American Math. Monthly* **71**, 75-90 (1955).

[333] Hong, S.; "A Linear Programming Approach to the Traveling Salesman Problem," Ph.D. Thesis, Johns Hopkins University, 1971.

[334] House, R., L. Nelson, and T. Rado; "Computer Studies of a Certain Class of Linear Integer Programs," in *Recent Advances in Optimization Techniques* (eds.: Lavi and Vogl), New York, John Wiley & Sons, 1966.

[335] Hu, T. C.; "On the Asymptotic Integer Algorithm," *Journal of Linear Algebra and Its Applications* **3**(2), 279-294 (1970).

[336] —————————; *Integer Programming and Network Flows*, Reading Mass., Addison-Wesley, 1969.

[337] Hwang, C., L. Fan, F. Tillman, and S. Kumar; "Optimization of Life Support Systems Reliability by an Integer Programming Method," *AIIE Transactions* **3**(3), 229-238 (1971).

[338] Ignizio, J., and R. Harnett; "Heuristically Aided Set Covering Algorithms," available from the University of Alabama at Huntsville, December 1972.

[339] Ingargiola, G. P. and J. F. Korsh; "A General Algorithm for One-Dimensional Knapsack Problems," *Operations Research* **25**(5), 752-759 (1977).

[340] Isaacson, L.; SC LIPI, SHARE General Program Library #3335, February 15, 1965.

[341] Jaikumar, R.; "An Effective Constraint Algorithm for Large 0-1 Programming Problems," available from Booth Fisheries Division, Consolidated Foods Corporation, Two N, Riverside Plaza, Chicago. (Paper presented at the joint ORSA/TIMS/AIIE meeting held at Atlantic City, November 1972).

[342] Jain, A.; Private Communication (at Indian Explosives Ltd., Calcutta, India), June 6, 1972.

[343] Jambekar, A., and D. Steinberg; "Computational Experience with a New Algorithm for 0-1 Integer Programming," School of Business and Engineering Administration paper, Michigan Technological University, Houghton, Michigan, May 1973.

[344] _____; "An Implicit Enumeration Algorithm for the General Integer Programming Problem," School of Business and Engineering Administration paper, Michigan Technological University, Houghton, Michigan, May 1973.

[345] Jarvis, J.; "Optimal Attack and Defense of a Command and Control Communications Network," Ph.D. Thesis, Johns Hopkins University, 1968.

[346] Jensen, R.; "Sensitivity Analysis and Integer Linear Programming," *Accounting Review* **43**(3), 425-446 (1968).

[347] Jeroslow, R.; "On the Unlimited Number of Faces in Integer Hulls of Linear Problems with Two Constraints," Department of Operations Research Technical Report No. 67, Cornell University, April 1969.

[348] _____; "Comments on Integer Hulls of Two Linear Constraints," *Operations Research* **19**(4), 1061-1069 (1971).

[349] _____, and K. Kortanek; "Dense Sets of Two Variable Integer Programs Requiring Arbitrarily Many Cuts by Fractional Algorithms," Management Sciences Research Report 174, Carnegie-Mellon University, 1969.

[350] _____; "On an Algorithm of Gomory," *SIAM Journal of Applied Mathematics* **21**(1), 55-60 (1971).

[351] Johnson, E.; "Cyclic Groups, Cutting Planes, and Shortest Paths," IBM T. J. Watson Research Center Report RC 4010, Yorktown Heights, New York, August 1972.

[352] _____; "On the Group Problem for Mixed Integer Programming," IBM T. J. Watson Research Center Report, Yorktown Heights, New York, January 1973.

[353] _____, H. Crowder, and M. W. Padberg; "Solving Large-Scale Zero-One Linear Programming Problems," *Operations Research* **31**, 803-834 (1983).

[354] _____, and K. Spielberg; "Inequalities in Branch and Bound Programming," IBM Thomas J. Watson Research Center Report RC 3649, Yorktown Heights, New York, December 13, 1971.

[355] Jones, A., and R. Soland; "A Branch and Bound Algorithm for Multilevel Fixed Charge Problems," *Management Science* **16**(1), 67-76 (1969).

[356] Kaplan, S.; "Solution of the Lorie-Savage and Similar Integer Programming Problems by the Generalized Lagrange Multiplier Method," *Operations Research* **14**(6), 1130-1136 (1966).

[357] Karg, L., and G. Thompson; "A Heuristic Approach to Solving Traveling Salesman Problems," *Management Science* **10**(2), 225-248 (1964).

[358] Karp, R.; "Reducibility Among Combinatorial Problems," available from the University of California at Berkeley. (Paper presented at the Symposium on Mathematical Programming at the University of Wisconsin at Madison, September 1972.)

[359] Karwan, M. H., and R. Rardin; "Surrogate Dual Multiplier Search Procedures in Integer Programming," *Operations Research* **32**, 52-71 (1984)

[360] Kelley, J.; "The Cutting-Plane Method for Solving Convex Programs," *J. Soc. Indust. Appl. Math.* **8**(4), 703-712 (1960).

[361] Kendell, K., and S. Zionts; "Solving Integer Programming Problems by Aggregating Constraints," School of Management Working Paper No. 155, S.U.N.Y. at Buffalo, November 1972.

[362] Kennington, J.; "The Fixed Charge Transportation Problem: A Computational Study with a Branch and Bound Code," *IEEE Transactions* **8**, (1976).

[363] Kennington, J., and D. Fyffe; "A Note on the Quadratic Programming Approach to the Solution of the 0-1 Linear Integer Programming Problem," School of Industrial and Systems Engineering Report, Georgia Institute of Technology, 1972.

[364] Kennington, J., and V. Unger; "The Group Theoretic Structure in the Fixed- Charge Transportation Problem," *Operations Research* **21**(5), 1142-1153 (1973).

[365] Kennington, J., and V. Unger; "A New Branch and Bound Algorithm for the Fixed Charge Transportation Problem," *Management Science* **22**(10), 1116-1126 (1976).

[366] Khumawala, B.; "An Efficient Branch and Bound Algorithm for the Warehouse Location Problem," *Management Science* **18**(12), 718-731 (1972).

[367] Kianfar, F.; "Stronger Inequalities for 0,1 Integer Programming Using Knapsack Functions," *Operations Research* **19**(6), 1373-1392 (1971).

[368] Kirby, M., H. Love, and K. Swarup; "Extreme Point Mathematical Programming," *Management Science* **18**(9), 540-549 (1972).

[369] Kirby, M., and P. Scobey; "Production Scheduling of N Identical Machines," *Canadian Operational Research* **8**(1), 14-27 (1970).

[370] Kohler, W., and K. Steiglitz; "Characterization and Theoretical Comparison of Branch-and-Bound Algorithms for Permutation Problems," Department of Electrical and Computer Engineering paper, the University of Massachusetts at Amherst, 1972.

[371] Kolesar, P.; "Assignment of Optimal Redundancy in Systems Subject to Failure," Operations Research Group Technical Report, Columbia University, 1966.

[372] ⎯⎯⎯⎯⎯⎯⎯⎯; "A Branch and Bound Algorithm for the Knapsack Problem," *Management Science* **13**(9), 723-735 (1967).

[373] Kolner, T.; "Some Highlights of a Scheduling Matrix Generator System," paper presented at the Sixth AGIFORS Symposium, September 1966.

[374] Koopmans, T., and M. Beckmann; "Assignment Problems and the Location of Economic Activities," *Econometrica* **25**(1), 53-76 (1957).

[375] Kortanek, K., and R. Jeroslow; "An Exposition on the Constructive Decomposition of the Group of Gomory's Round-Off Algorithm," Department of Operations Research Technical Report No. 39, Cornell University, January 1968.

[376] Kraft, D., and T. Hill; "The Journal Selection Problem in a University Library System," *Management Science* **19**(6), 613-626 (1973).

[377] _____; "A Lagrangian Formulation of the Journal Selection Model," School of Industrial Engineering Technical Report, Purdue University, 1971.

[378] Kruskal, J.; "On the Shortest Spanning Subtree of a Graph and the Traveling Salesman Problem," *Proceedings of the American Mathematical Society* **1**, 48-50 (1956).

[379] Kuehn, A., and M. Hamburger; "A Heuristic Program for Locating Warehouses," *Management Science* **10**(4), 643-444 (1963).

[380] Kuehn, A., and R. Kuenne; "An Efficient Algorithm for the Numerical Solutions of the Generalized Weber Problem in Spatial Economics," *Journal of Regional Science* **4**(1), 21-28 (1962).

[381] Kuhn, H., and W. Baumol; "An Approximate Algorithm for the Fixed Charge Transportation Problem," *Naval Research Logistics Quarterly* **9**(1), 1-16 (1962).

[382] Labordere, H.; "Partitioning Algorithm in Mixed and Pseudo Mixed Integer Programming," Ph.D. Thesis, Rensselaer Polytechnic Institute, January 1969.

[383] Land, A., and A. Doig; "An Automatic Method of Solving Discrete Programming Problems," *Econometrica* **28**(3), 497-520 (1960).

[384] Lang, S.; *Algebraic Structures*, Reading, Mass., Addison-Wesley, 1967.

[385] Lasdon, L.; *Optimization Theory for Large Systems*, New York, MacMillan, 1970.

[386] Lasky, J.; "Optimal Scheduling of Freight Trucking," M.S. Thesis, Massachusetts Institute of Technology, June 1969.

[387] Lawler, E.; "Covering Problems: Duality Relations and a New Method of Solution," *SIAM Journal of Applied Mathematics* **14**, 1115-1132 (1966).

[388] _____, and M. Bell; "A Method for Solving Discrete Optimization Problems," *Operations Research* **14**(6), 1098-1112 (1966).

[389] Lawler, E. L, J. K. Lenstra, A. H. G. Rinnooy Kan, and D. B. Shmoys; "*The Traveling Salesman Problem: A Guided Tour of Combinatorial Optimization*", John Wiley & Sons, (1985)

[390] Lawler, E., and D. Wood; "Branch and Bound Methods: A Survey," *Operations Research* **14**(4), 699-719 (1966).

[391] Lemke, C., and K. Spielberg; "Direct Search Algorithm for Zero-One and Mixed Integer Programming," *Operations Research* 15(5), 892-914 (1967).

[392] Lemke, C., H. Salkin, and K. Spielberg; "Set Covering by Single Branch Enumeration with Linear Programming Subproblems," *Operations Research* 19(4), 998-1022 (1971).

[393] Levin, A.; "Fleet Routing and Scheduling Problems for Air Transportation Systems," Ph.D. Thesis, Massachusetts Institute of Technology, January 1969.

[394] Levy, F.; "An Application of Heuristic Problem Solving to Accounts Receivable Management," *Management Science* 12(1), 236-244 (1966).

[395] Levy, J.; "An Extended Theorem for Location on a Network," *Operational Research Quarterly* 18(4), 433-442 (1967).

[396] Liggett, J.; "A General Multiple Choice Formulation of an Algorithm by Balintfy," Computer Science Center Technical Report CP-68004, Southern Methodist University, March 1968.

[397] Lin, S.; "Computer solutions of The Traveling Salesman Problem", *Bell System Tech. J.*, **44**, 2245-2269 (1965)

[398] Lin, S. and B. W. Kernighan; "An Effective Heuristic Algorithm for The Traveling Salesman Problem", *Operations Research*, **21**(2), 498-516 (1973)

[399] Lind, M.; "Application of Mixed Integer Programming to Letter Processing Systems," Mitre Corporation Report M71-102, October 1971.

[400] Little, J., K. Murty, D. Sweeney, and C. Karel; "An Algorithm for the Traveling Salesman Problem," *Operations Research* 11(5), 972-989 (1963).

[401] Lorie, J., and L. Savage; "Three Problems in Capital Rationing," *Journal of Business* 28, 229-239 (1955).

[402] MacDuffee, C.; *An Introduction to Abstract Algebra*, New York, John Wiley & Sons, 1940.

[403] Manne, A.; "On the Job-Shop Scheduling Problem," *Operations Research* 8(2), 219-223 (1960).

[404] —————; "Plant Location under Economics of Scale: Decentralization and Computation," *Management Science* 11(2), 213-235 (1964).

[405] Mao, J., and B. Wallingford; "An Extension of Lawler and Bell's Method of Discrete Optimization with Examples from Capital Budgeting," *Management Science* 15(2), 51-60 (1968).

[406] Marazana, F.; "On the Location of Supply Points to Minimize Transport Costs," *Operational Research Quarterly* 15, 261-270 (1964).

[407] Markowitz, H., and A. Manne; "On the Solution of Discrete Programming Problems," *Econometrica* 25(1), 84-110 (1957).

[408] Marshall, L., and D. Tabak; "Airline Route Allocation by Mixed Integer Programming Using OPHELIE II," Pratt and Whitney Aircraft, East Hartford, Connecticut. (Paper presented at the joint ORSA/TIMS/AIEE Atlantic City Meeting, November 1972.)

[409] Marsten, R.; "An Implicit Enumeration Algorithm for the Set Partitioning Algorithm with Side Constraints," Ph.D. Thesis, the University of California at Los Angeles, October 1971.

[410] —————; "An Algorithm for Large Set Partitioning Problems," Center for Mathematical Studies in Economics and Management Science, Discussion Paper No. 8, Northwestern University, July 14, 1972.

[411] —————; "Approximate and Exact Solutions for a Class of Fixed Charge Problems," Center for Mathematical Studies on Economics and Management Science Discussion Paper No. 10, Northwestern University, August 21, 1972.

[412] Martello, S. and P. Toth; "An Upper Bound for the Zero-One Knapsack Problem and a Branch and Bound Algorithm," *European Journal of Operations Research* **1**, 169-175 (1977).

[413] Martello, S. and P. Toth; "Algorithm 37, Algorithm for the solution of the 0-1 Single Knapsack Problem," *Computing* **21**, 81-86 (1978).

[414] Martello, S. and P. Toth; "The 0-1 Knapsack Problem," (in *Combinatorial Optimization*, ed. N. Christofides, et al.), Wiley, (1979)

[415] Martin, G.; "An Accelerated Euclidean Algorithm for Integer Linear Programming," in *Recent Advances in Mathematical Programming* (eds.: Graves and Wolfe), New York, McGraw-Hill, 1963.

[416] —————; "Solving the Traveling Salesman Problem by Integer Linear Programming," CEIR, New York, 1966.

[417] Mathews, G.; "On the Partition of Numbers," *Proceedings of the London Mathematical Society* **28**, 486-490 (1897).

[418] Mathis, J.; "A Counterexample to the Rudimentary Primal Integer Programming Algorithm," *Operations Research* **19**(6), 1518-1522 (1971).

[419] Mathur, K., H. M. Salkin, and S. Morito; "An Efficient Algorithm for the General Multiple-choice Knapsack Problem (GMKP)," *Annals of Operations Research* **4**, 253-283 (1985).

[420] Mathur, K., H. M. Salkin, and S. Morito; "A Branch and Search Algorithm for a Class of Nonlinear Knapsack Problem," *Operations Research Letters* **2**, 155-160 (1983).

[421] Mathur, K., H. M. Salkin, and B. Mohanty; "A Note on a General Nonlinear Knapsack Problem," *Operations Research Letters* **5**, 79-81 (1986).

[422] Mathur, K., H. M. Salkin, K. Nishimura, and S. Morito; "Application of Integer Programming in Radio Frequency Management," *Management Science* **31**(7), 829-839 (1985).

[423] Mathur, K., H. M. Salkin, K. Nishimura, and S. Morito; "The Design of an Interactive Computer Software System for the Frequency-Assignment Problem," *IEEE Transaction on Electromagnetic Compatibility* **26**, 207-212 (1984).

[424] McCloskey, J., and F. Hanssmann; "An Analysis of Stewardess Require-

ments and Scheduling for a Major Airline," *Naval Research Logistics Quarterly* 4, 183-192 (1957).

[425] Meihle, W.; "Link-Length Minimization in Networks," *Operations Research* 6(2), 232-243 (1958).

[426] Mensch, G.; "Single-Stage Linear Programming Zero-One Solutions to Some Traveling Salesman Type Problems," *INFOR* (Canada) 9(3), 282-292 (1971).

[427] Merville, L.; "An Investment Decision Model for the Multinational Firm: A Chance-Constrained Programming Approach," Ph.D. Thesis, University of Texas, May 1971.

[428] Mezei, J.; "Structure in the Traveling Salesman's Problem, The Shklar Algorithm," IBM Philadelphia Scientific Center Report No. 320-3003, September 1971.

[429] Michaud, P.; "Exact Implicit Enumeration Method for Solving the Set Partitioning Problem," *IBM Journal of Research and Development* 16(6), 573-578 (1972).

[430] Miller, C., A. Tucker, and R. Zemlin; "Integer Programming Formulations and Traveling Salesman Problems," *Journal of the Association for Computing Machinery* 7, 326-329 (1960).

[431] Mitten, L.; "Branch-and-Bound Methods: General Formulation and Properties," *Operations Research* 18(1), 24-34 (1970).

[432] Moder, J., and C. Phillips; *Project Management with CPM and PERT*, New York, Reinhold, 1964.

[433] Morito, S., H. M. Salkin, and K. Mathur; "Computational Experience with a Dual Backtrack Algorithm for Identifying Frequencies Likly to Create Intermodulation Problems," *IEEE Transaction on Electromagnetic Compatibility* 23, 32-36 (1981)

[434] Morito, S; *"Integer Programming by Group Theory"* Ph.D. Thesis, Case Western Reserve University, 1976.

[435] Mostow, G., J. Sompson, and J. Meyer; *Fundamental Structures of Algebra*, New York, McGraw-Hill, 1963.

[436] Murchland, J.; "A New Method for Finding All Elementary Paths in a Complete Directed Graph," Transport Network Theory Unit Report LSE-TNT-22, London School of Economics, October 1965.

[437] Murty, K.; "Solving the Fixed Charge Transportation Problem by Ranking the Extreme Points," *Operations Research* 16(2), 268-279 (1968).

[438] Murty, K. G.; "On the Tours of a Traveling Salesman Problem," *SIAM J. Control Optimization* 7, 122-131 (1969).

[439] Murty, K. G.; *Linear and Combinatorial Programming*, Wiley, (1976)

[440] _____; "A Fundamental Problem in Linear Inequalities with Application to the Traveling Salesman Problem," *Mathematical Programming* 2(3), 296-308 (1972).

[441] Naslund, B.; "A Model of Capital Budgeting under Risk," *Journal of Business* **39**,(2), 257-271 (1966).

[442] Nauss, R. M.; "An Efficient Algorithm for the 0-1 Knapsack Problem," *Management Science* **23**(1), 27-31 (1976).

[443] Nemhauser, G., and Z. Ullmann; "A Note on the Generalized Lagrange Multiplier Solution to an Integer Programming Problem," *Operations Research* **16**(2), 450-452 (1968).

[444] ———————; "Discrete Dynamic Programming and Capital Allocation," *Management Science* **15**(9), 494-505 (1969).

[445] Netto, E.; *Lehrbuch der Combinatorik*, New York, Chelsea.

[446] Norback, J. P. and R. F. Love; "Geometric Approaches to Solving The Traveling Salesman Problem", *Management Science*, **23**(11), 1208-1223 (1977)

[447] Norman, R., and M. Rabin; "An Algorithm for the Minimum Cover of a Graph," *Proceedings of the American Mathematical Society* **10**, 315-319 (1969).

[448] Obruca, A.; "Spanning-Tree Manipulation and the Traveling-Salesman Problem," *Computer Journal* **10**, 374-377 (1968).

[449] O'Neil, P.; "Hyperplane Cuts of an n-Cube," *Discrete Mathematics* **1**(2), 193-195 (1971).

[450] Onyekwelu, D. C.; "Computational Viability of a Constraint Aggregation Scheme for Integer Linear Programming Problems," *Operations Research* **31**, 795-801 (1983).

[451] "Operations Research," Issue No. 4, Control Data Corp., Minneapolis, Minnesota, December 1, 1968.

[452] OPHELE User's Manual, Control Data Corp., Minneapolis, Minnesota, 1972.

[453] Padberg, M.; "On the Facial Structure of Set Covering Problems," IIM Reprint I/72-13, International Institute of Management, West Berlin, Germany, April 1972.

[454] ———————; "Equivalent Knapsack Type Formulations of Bounded Integer Linear Programs," Management Science Research Report No. 227, Carnegie-Mellon University, September 1970.

[455] ———————, and M. Rao; "The Traveling Salesman Problem and A Class of Polyhedra of Diameter Two," International Institute of Management Preprint I/73-5, D-1000 Berlin 33, Griegstrasse 5, January 1973.

[456] Panwalkar, S., and W. Iskander; "Comparison of Various Branching Methods," available from Texas Tech University, Lubbock, Texas. (Paper presented at the joint AIIE/TIMS/ORSA meeting held at Atlantic City, November 1972.)

[457] Paul, M., and S. Unger; "Minimizing the Number of States in Incompletely

Specified Sequential Functions," *IRE Transactions on Electronic Computers* **EC-8**, 356-367 (1959).

[458] Peterson, C.; "Computational Experience with Variants of the Balas Algorithm Applied to the Selection of R&D Projects," *Management Science* **13**(9), 736-750 (1967).

[459] _____; "A Note on Transforming the Product of Variables to Linear Form in Linear Programs," Working Paper, Purdue University, 1971.

[460] Pierce, J,; "Application of Combinatorial Programming to a Class of All Zero- One Integer Programming Problems," *Management Science* **15**(1), 191-209 (1968).

[461] _____; "On the Solution of Integer Programming Cutting Stock Problems by Combinatorial Programming–Part I," IBM Cambridge Scientific Center Report No. 320-2001, May 1966.

[462] _____; "Pattern Sequencing and Matching in Stock Cutting Operations," *TAPPI* (Journal of the Technical Association of the Pulp and Paper Industry) **53**(4), 668-678 (1970).

[463] _____, and J. Lasky; "Improved Combinatorial Programming Algorithms for a Class of All Zero-One Integer Programming Problems," *Management Science* **19**(5), 528-543 (1973).

[464] Pinkus, C., D. Gross, and R. Soland; "Optimal Design of Multiactivity Multifacility Systems by Branch and Bound," *Operations Research* **21**(1), 270-283 (1973).

[465] Piper, C., and A. Zoltners; "A Pragmatic Approach to 0-1 Decision Making," Graduate School of Industrial Administration report, Carnegie-Mellon University, 1972.

[466] Pollatschek, M., and B. Avi-Itzhak; "A Class of Deep-Cut Procedures for Integer or Mixed Integer Linear Programs," Operations Research, Statistics and Economics Mimeograph Series No. 85, Technion, Haifa, Statistics and Economics Mimeograph Series No. 85, Technion, Haifa, Israel, June 1971.

[467] Popolopas, L.; "Optimum Plant Numbers and Locations for Multiple Produce Processing," *Journal of Farm Economics* **49**(2), 287-295 (1965).

[468] "Proposal for Continuing Research at the NBER Computer Research Center from February 1974 through January 1975, Volume I: Project Descriptions," National Bureau of Economic Research, Inc., 575 Technology Square, Cambridge, Mass., August 1973.

[469] Radhakrishnan, S.; "Capital Budgeting and Mixed Zero-One Integer Quadratic Programming," Division of Systems and Computer Services Technical Report, Medical College of Georgia at Augusta, November 1972.

[470] Raghavachari, M.; "On Connections Between Zero-One Integer Programming and Concave Programming Under Linear Constraints," *Operations Research* **17**(4), 680-684 (1969).

[471] _____; "Supplement," *Operations Research* **18**(3), 564-565 (1970).

[472] Rao, A.; "The Multiple Set Covering Problem: A Side Stepping Algorithm," Operations Research and Statistics Center Research Paper No. 37-71-P6, Rensselaer Polytechnic Institute, September 1971.

[473] Rardin, R., and V. Unger; "A Surrogate Constraint for 0-1 Mixed-Integer Programs," School of Industrial and Systems Engineering report, Georgia Institute of Technology at Atlanta. (Paper presented at the 43rd National ORSA meeting, May 1973.)

[474] Ray-Chaudhuri, D.; "An Algorithm for a Maximum Cover of Abstract Complex," *Canadian Journal of Mathematics* **15**, 11-24 (1963).

[475] Raymond, T.; "New Heuristic Algorithms for the Traveling-Salesman Problem," M.S. Thesis, Syracuse University, 1968.

[476] _____; "Heuristic Algorithm for the Traveling Salesman Problem," *IBM Journal of Research and Development* **13**(4), 400-407 (1969).

[477] Reiter, S., and G. Sherman; "Discrete Optimizing," *Journal of the Society for Industrial and Applied Mathematics* **13**, 864-889 (1965).

[478] Revelle, C., D. Marks, and J. Liebman; "An Analysis of Private and Public Sector Location Models," *Management Science* **16**(11), 692-707 (1970).

[479] Robers, P.; "Some Comments Concerning Revelle, Marks, and Liebman's Article on Facility Location," *Management Science* **18**(1), 109-111 (1971).

[480] Roberts, S., and B. Flores; "An Engineering Approach to the Traveling Salesman Problem," *Management Science* **13**(3), 269-288 (1966).

[481] Rodriquez, E., H. Stern, and M. Utter; "The *M*-Traveling Salesman Problem with Ordered Visits," Operations Research and Statistics Research Paper No. 37-71-P5, Rensselaer Polytechnic Institute (Troy, New York), September 1971.

[482] Roodman, G.; "Postoptimality Analysis in Zero-One Programming by Implicit Enumeration," Amos Tuck School of Business Administration working paper, Dartmouth College, January 1972.

[483] Root, J.; "An Application of Symbolic Logic to a Selection Problem," *Operations Research* **12**(4), 519-526 (1964).

[484] Roth, J.; "Algebraic Topological Methods for the Synthesis of Switching Systems–I," *Transactions of the American Mathematical Society* **88**, 301-326 (1950).

[485] Rosenkrantz, D. J., R. E. Stearns, and P. M. Lewis; "An Analysis of Several Heuristics for the Travelling Salesman Problem", *SIAM J. of Computing*, **6**, 563-581 (1977)

[486] Roth, R.; "Computer Solutions to Minimum Cover Problems," *Operations Research* **17**(3), 455-466 (1969).

[487] Rothkopf, M.; "The Traveling Salesman Problem: On the Reduction of

Certain Large Problems to Smaller Ones," *Operations Research* **14**(3), 532-533 (1966).

[488] Rothstein, M.; "Operations Research in the Airlines," Bureau of Business Research and Services, the University of Connecticut, Storrs, Connecticut. (Paper presented at the joint ORSA/TIMS/AIEE Atlantic City meeting, November 1972.)

[489] Rubin, D.; "The Neighboring Vertex Cut and Other Cuts Derived with Gomory's Asymptotic Algorithm," Ph.D. Thesis, University of Chicago, June 1970.

[490] ——————; "On the Unlimited Number of Faces in Integer Hulls of Linear Programs with a Single Constraint," *Operations Research* **18**(5), 940-946 (1970).

[491] ——————; "Redundant Constraints and Extraneous Variables in Integer Programs," *Management Science* **18**(7), 423-427 (1972).

[492] ——————; "A Hybrid Cutting Plane-Enumeration Algorithm for Integer Programming," Operations Research and Systems Analysis Report, the University of North Carolina at Chapel Hill, May 1973.

[493] ——————, and R. Graves; "Strengthened Dantzig Cuts for Integer Programming," *Operations Research* **20**(1), 178-182 (1972).

[494] Rubin D., and G. Links; "The Asymptotic Algorithm as Cut Generator: Some Computational Results," Operations Research and Systems Analysis Report, the University of North Carolina at Chapel Hill, April 1972.

[495] Rubin, J.; "Airline Crew Scheduling–the Non Mathematical Problem," IBM Philadelphia Scientific Center Report No. 320-3006, September 1971.

[496] ——————; "A Technique for the Solution of Massive Set Covering Problems, with Application to Airline Crew Scheduling," *Transportation Science* **7**(1), 34-48 (1973).

[497] Rudeanu, S., and P. Hammer; *Boolean Methods in Operations Research and Related Areas*, Berlin, Springer-Verlag, 1968.

[498] Rutten, D.; "Structured Integer Programs," School of Business working paper, Indiana University at Bloomington. (Paper presented at the 43rd National ORSA meeting, May 1973.)

[499] Sá, G.; "Branch and Bound and Approximate Solutions to the Capacitated Plant Location Problem," *Operations Research* **17**(6), 1005-1016 (1969).

[500] Saaty, T.; *Optimization in Integers and Related Extremal Problems*, New York, McGraw-Hill, 1970.

[501] Sahni, S. and E. Horowitz; "Computing Partitions with Applications to the Knapsack Problem," *Journal of the ACM*, **21**, (1974)

[502] Salkin, H.; "Enumerative Algorithms for Integer and Mixed Integer Programs," Ph.D. Thesis, Rensselaer Polytechnic Institute, 1969; reprinted as Technical Memorandum No. 162, Department of Operations Research, Case Western Reserve University, August 1969.

[503] —————————; "On the Merit of the Generalized Origin and Restarts in Implicit Enumeration," *Operations Research* **18**(3), 549-555 (1970).

[504] —————————; "A Note on Gomory Fractional Cuts," *Operations Research* **19**(6), 1538-1541 (1971).

[505] —————————; "A Note Comparing Glover's and Young's Simplified Primal Algorithms," *Naval Research Logistics Quarterly* **19**(2), 399-402 (1972).

[506] —————————; "Binding Inequalities in Benders' Partitioning Algorithm," Department of Operations Research Technical Memorandum No. 266, Case Western Reserve University, February 1972.

[507] —————————; "A Brief Survey of Integer Programming," *OPSEARCH* **10**(2), 81-123 (1973).

[508] —————————; "A Brief Survey of Algorithms and Recent Results in Integer Programming," *OPSEARCH* **10**(2), 81-123, 1973.

[509] —————————, and W. Balinski; "Integer Programming Models and Codes in the Urban Environment," *Socio Economic Planning Sciences* **7**, 739-753 (1973).

[510] Salkin, H., and P. Breining; "Integer Points on the Gomory Fractional Cut (Hyperplane)," *Naval Research Logistics Quarterly* **18**(4), 491- 496 (1971).

[511] Salkin, H., and C. DeKluyver; "The Knapsack Problem: A Survey," Department of Operations Research Technical Report No. 281, Case Western Reserve University, December 1972.

[512] Salkin, H., and R. Koncal; "A Pseudo Dual All-Integer Algorithm for the Set Covering Problem," Department of Operations Research Technical Memorandum No. 204, Case Western Reserve University, November 1970.

[513] —————————; "A Dual All-Integer Algorithm (in Revised Simplex Form) for the Set Covering Problem," Department of Operations Research Technical Memorandum No. 250, Case Western Reserve University, August 1971.

[514] —————————; "Set Covering by an All-Integer Algorithm: Computational Experience," *Journal for the Association of Computing Machinery* **20**(2), 189-193 (1973).

[515] Salkin, H., and C. Lin; "Computational Experience with an Enumerative Algorithm for the Set Covering Problem with Base Constraints," Department of Operations Research Technical Memorandum No. 313, Case Western Reserve University, July 1973.

[516] Salkin, H. M., K. Mathur, and C. Leach; "Satellite Constellation Design: An Integer Programming Approach," *Applications of Management Science* **3**, 249-259 (1983).

[517] Salkin, H., and J. Saha; "Set Covering: Uses, Algorithms, Results," Department of Operations Research Technical Memorandum No. 272, Case Western Reserve University, July 1973.

[518] Salkin, H., P. Shroff, and S. Mehta; "All-Integer Integer Programming Algorithms Applied to Tableaux with Rational Coefficients," Technical Memorandum No. 298, Department of Operations Research, Case Western Reserve University, April 1973.

[519] Salkin, H., and K. Spielberg; "Adaptive Binary Programming," IBM New York Scientific Center Report No. 320-2951, April 1968. Available from IBM Philadelphia Scientific Center, 3401 Market Street.

[520] Salverson, M.; "The Assembly Line Balancing Problem," *Journal of Industrial Engineering* **6**(1), 18-25 (1955).

[521] Schrijver, A.; "Theory of Linear and Integer Programming", John Wiley & Sons, (1986).

[522] Senju, S., and Y. Toyoda; "An Approach to Linear Programming with 0-1 Variables," *Management Science* **15**(4), 196-207 (1968).

[523] Seymore, G.; "Interval Integer Programming," Department of Statistics Technical Report 22, Oregon State University at Corvallis, May 1971.

[524] Shannon, R., and J. Ignizio; "A Heuristic Programming Algorithm for Warehouse Location," *AIIE Transactions* **2**, 361-364 (1970).

[525] Shapiro, D.; "Algorithms for the Solution of the Optimal Cost Traveling Salesman Problem," Sc.D. Thesis, Washington University (St. Louis), 1966.

[526] Shapiro, G.; "Shortest Route Methods for Finite State Space Deterministic Dynamic Programming Problems," *SIAM Journal of Applied Mathematics* **16**(6), 1232-1250 (1968).

[527] ——————; "Dynamic Programming Algorithms for the Integer Programming Problem I: The Integer Programming Problem Viewed as a Knapsack Type Problem," *Operations Research* **16**(1), 103-121 (1968).

[528] ——————; "Generalized Lagrange Multipliers in Integer Programming," *Operations Research* **19**(1), 68-77 (1971).

[529] ——————, and H. Wagner; "A Finite Renewal Algorithm for the Knapsack and Turnpike Models," *Operations Research* **15**(2), 319-341 (1967).

[530] Shapiro, J.; "A Group Theoretic Branch and Bound Algorithm for the Zero-One Integer Programming Problem," Sloan School of Management Working Paper 302-367, Massachusetts Institute of Technology, December 1967.

[531] ——————; "Dynamic Programming Algorithms for the Integer Programming Problem-I: The Integer Programming Problem Viewed as a Knapsack Type Problem," *Operations Research* **16**(1), 103-121 (1968).

[532] ——————; "Group Theoretic Algorithms for the Integer Programming Problem-II: Extension to a General Algorithm," *Operations Research* **16**(5), 928- 947 (1968).

[533] ——————; "Turnpike Theorems for Integer Programming Problems," *Operations Research* **18**(3), 432-440 (1970).

[534] Shareshian, R.; "A Modification of the Mixed Integer Algorithm of N. Driebeek," IBM Philadelphia Scientific Center Report No. 2, July 1966.

[535] —————————; BBMIP, S/360 General Program Library #360D-15.2005, April 1967.

[536] —————————, and K. Spielberg; "The Mixed Integer Algorithm of N. Driebeek," IBM Philadelphia Scientific Center Report No. 320-2931, March 1968.

[537] Shycon, H., and R. Maffei; "Simulation–Tool for Better Distribution," *Harvard Business Review* **38**(6), 65-75 (1960).

[538] Sinha, P. and A. A. Zoltner; "The Multiple-choice Knapsack Problem," *Operations Research* **27**, 503- (1979).

[539] Smith, T. H. C., V. Srinivasan, and G. L. Thompson; "Computational Performance of Three Subtour Elimination Algorithm for Solving Asymmetric Traveling Salesman Problems," *Ann. Discrete Math.*, **1**, 495-506 (1977).

[540] Smith, T. H. C., and G. L. Thompson; "A LIFO Implicit Enumeration Search Algorithm for The Symmetric Traveling Salesman Problems using Held and Karp's 1-Tree Relaxation," *Ann. Discrete Math.*, **1**, 479-493 (1977).

[541] Solow, D.; *"Linear Programming: An Introduction to Finite Improvement Algorithms"*, North-Holland, 1984

[542] Spielberg, K.; "On the Fixed Charge Transportation Problem," *Proceedings of the 19th National ACM Conference*, 1964.

[543] —————————; "Enumerative Methods for Integer and Mixed Integer Programming," IBM New York Scientific Center Report No. 320-2938, March 1968. This article is a part of: M. Balinski, and K. Spielberg; "Methods for Integer Programming: Algebraic, Combinatorial, and Enumerative," in *Progress in Operations Research* Vol. III (ed.: Aronofsky), New York, John Wiley & Sons, 1969.

[544] —————————; "Plant Location with Generalized Search Origin," *Management Science* **16**(3), 165-178 (1969).

[545] —————————; "Algorithms for the Simple Plant-Location Problem with Some Side Constraints," *Operations Research* **17**(1), 85-111 (1969).

[546] —————————; "On Solving Plant Location Problems," available from IBM Philadelphia Scientific Center, 3401 Market St. (Paper presented at the NATO conference on "Applications of Mathematical Programming Techniques," Cambridge, England, June 1968.)

[547] —————————; "Minimal Preferred Variable Reduction for Zero-One Programming," IBM Philadelphia Scientific Center Reprot No. 320-3013, July 1972.

[548] —————————; "A Minimal-Inequality Branch-Bound Method," IBM Philadelphia Scientific Center working paper, October 1972.

[549] Spitzer, M.; "Solution to the Crew Scheduling Problem," presented at the first AGIFORS symposium, October 1961.

[550] Steckhan, H.; "A Theorem on Symmetric Traveling Salesman Problems," *Operations Research* **18**(6), 1163-1167 (1970).

[551] Steinberg, D.; "The Fixed Charge Problem," *Naval Research Logistics Quarterly* **17**, 217-236 (1970).

[552] Steinmann, H., and R. Schwinn; "Computational Experience with a Zero-One Programming Problem," *Operations Research* **17**(5), 917-920 (1969).

[553] Stewart, W. R.; "A Computationally Efficient Heuristic for The Travelling Salesman Problem", *Proc. of 13th Annual Meeting of S. E. TIMS*, 75-85 (1977).

[554] Story, A., and H. Wagner; "Computational Experience with Integer Programming for Job Shop Scheduling," in *Industrial Scheduling* (eds.: Muth and Thompson), Englewood Cliffs, N.J., Prentice-Hall, 1963.

[555] Svestka, J., and V. Huckfeldt; "Computational Experience with an M-Salesman Traveling Salesman Algorithm," *Management Science* **19**(7), 790-799 (1973).

[556] Swart, W.; "A Multi-Compartment Knapsack Problem," West Virginia College of Graduate Studies report, Institute, West Virginia, November 1972.

[557] —————, and H. Spriggs; "A Decision Structure for Regional Planning of Pollution Abatement Activity," *Proceedings of the Eighth Annual Conference of Southeastern Chapters of TIMS*, Knoxville, Tennessee, October 1972.

[558] Sweeney, D.; "The Exploration of a New Algorithm for Solving the Traveling Salesman Problem," M.S. Thesis, Massachusetts Institute of Technology, 1963.

[559] Taha, H.; "A Balasian-Based Algorithm for 0-1 Polynomial Programming," Research Report No. 70-2, the University of Arkansas, May 1970.

[560] Teitz, M., and P. Bart; "Heuristic Methods for Estimating Generalized Vertex Median of a Weighted Graph," *Operations Research* **16**(5), 955-961 (1968).

[561] Thirez, H.; "Airline Crew Scheduling: A Group Theoretic Approach," Ph.D. Thesis, Massachusetts Institute of Technology, October 1969.

[562] Thompson, G., F. Tonge, and S. Zionts; "Techniques for Removing Constraints and Extraneous Variables from Linear Programming Problems," *Management Science* **12**(7), 588-608 (1966).

[563] Thompson, G.; "The Stopped Simplex Method: I.Basic Theory for Mixed Integer Programming; Integer Program," *Revue Française de Recherche Opérationnelle*, 155-182, 1964. Also available as Graduate School of Industrial Administration Reprint No. 170, Carnegie-Mellon University.

[564] Thompson, R.; "Unimodular Group Matrices with Rational Integers as Elements," *Pacific Journal of Mathematics* **14**(2), 719-726 (1964).

[565] Tillman, F., and R. Hering; "A Study of a Look-Ahead Procedure for

Solving the Multiterminal Delivery Problem," *Transportation Research* **5**, 225-229 (1971).

[566] Tomlin, J.; "Branch and Bound Methods for Integer and Non-Convex Programming," in *Integer and Nonlinear Programming* (ed.: Abadie), North Holland, 1970.

[567] ——————; "an Improved Branch-and-Bound Method for Integer Progarmming," *Operations Research* **19**(4), 1070-1075 (1971).

[568] ——————; "Survey of Computational Methods for Solving Large Scale Systems," Department of Operations Research Technical Report 72-25, Stanford University, October 1972.

[569] Tompkins, C.; "Group Theoretic Structures in the Fixed Charge Transportation Problem," Ph.D. Thesis, Georgia Institute of Technology, July 1971.

[570] ——————, and V. Unger; "Group Theoretic Structures in the Fixed Charge Transportation Problem," Graduate School of Business Administration Report, the University of Virginia at Charlottesville, April 1972.

[571] Toregas, C., R. Swain, C. Revelle, and L. Bergman; "The Location of Emergency Service Facilities," *Operations Research* **19**(6), 1363-1373 (1971).

[572] Trauth, C., Jr., and R. Woolsey; "Mesa, A Heuristic Integer Linear Programming Technique," Sandia Corporation Research Report SC-RR-68-299, Albuquerque, New Mexico, July 1968.

[573] ——————; "Integer Linear Programming: A Study in Computational Efficiency," *Management Science* **15**(7), 481-483 (1969).

[574] Trehan, S.; "A Survey of Facility Location Models," M.S. Thesis, Department of Applied Mathematics and Statistics, State University of New York at Stony Brook, October 1971.

[575] Trubin, V.; "On a Method of Solution of Integer Linear Programming Problems of a Special Kind," *Soviet Math. Dokl.* **10**, 1544-1546 (1969).

[576] Tuan, N.; "A Flexible Tree-Search Method for Integer Programming Problems," *Operations Research* **19**(1), 115-119 (1971).

[577] Tucker, A.; "Dual Systems of Homogeneous Linear Relations," in *Linear Inequalities and Related Systems (Annals of Mathematical Studies 38)*, Princeton University Press, 1956.

[578] ——————; "On Directed Graphs and Integer Programs," IBM Mathematical Research Project Technical Report, Princeton University, 1960.

[579] Tui, H.; "Concave Programming Under Linear Constraints," *Soviet Mathematics*, 1437-1440 (1964).

[580] Tutte, W.; "On Hamiltonian Circuits," *London Mathematical Society Journal* **21**, Part 2, No. 82, 98-101 (1946).

[581] Unger, Jr., V.; "Capital Budgeting and Mixed Zero-One Integer Programming," *AIIE Transactions* **11**(1), 28-36 (1970).

[582] ——————; "Duality Results for Discrete Budgeting Problems," to appear in the *Engineering Economist* (1973).

[583] Valenta, J.; "Capital Equipment Decisions: A Model for Optimal Systems Interfacing," M.S. Thesis, Massachusetts Institute of Technology, June 1969.

[584] Veinott, A., and G. Dantzig; "Integral Extreme Points," *SIAM Review* **10**(3), 371-372 (1968).

[585] Verblunsky, S.; "On the Shortest Path Through a Number of Points," *Proceedings of the American Mathematical Society* **2**, 6, December 1951.

[586] Vietorisz, T., and A. Manne; "Chemical Processes, Plant Location, and Economies of Scale," in *Studies in Process Analysis* (eds.: Manne and Markowitz), New York, John Wiley & Sons, 1963.

[587] Von Lanzenauer, C., and R. Himes; "A Linear Programming Solution to the General Sequencing Problem," *Canadian Operational Research* **8**(2), 129-134 (1970).

[588] Wagner, H.; "An Integer Linear Programming Model for Machine Scheduling," *Naval Research Logistics Quarterly* **6**, 131-140 (1959).

[589] ——————; "Legislative Redistricting as a 0-1 Integer Programming Set Covering Problem," CROND Inc., Wilmington, Delaware. (Paper presented at the 34th national ORSA meeting, November 1968.)

[590] ——————, R. Giglio, and R. Glaser; "Preventive Maintenance Scheduling by Mathematical Programming," *Management Science* **10**(2), 316-334 (1964).

[591] Walker, W. E.; "A Heuristic Adjacent Extreme Point Algorithm for the Fixed Charge Problem," *Management Science*, **22**(5), 587-596 (1976).

[592] Wanty, J.; "A Practical Application of the Warehousing Problem: Location of Depots in a Petroleum Company," *Operational Research Quarterly* **9**(2), 194-253 (1958).

[593] Watters, L.; "Reduction of Integer Polynomial Programming Problems to Zero-One Linear Programming Problems," *Operations Research* **15**(6), 1171-1174 (1967).

[594] Weber, A.; *Uber den Standort der Industrien*, T'ubingen, 1909; in *Theory of Location of Industries* (ed. and translator: C. J. Friedrich), Chicago University Press, 1929.

[595] Weingartner, H.; *Mathematical Programming and the Analysis of Capital Budgeting Problems*, Englewood Cliffs, N.J., Prentice Hall, 1963.

[596] ——————; "Capital Budgeting and Interrelated Projects: Survey and Synthesis," *Management Science* **12**(7), 485-516 (1968).

[597] ——————, and D. Ness; "Methods for the Solution of 0-1 Knapsack Problems," *Operations Research* **15**(1), 83-103 (1967).

[598] White, W.; "On Gomory's Mixed Integer Algorithm," Senior Thesis, Princeton University, May 1961.

[599] —————; "On a Group Theoretic Approach to Linear Integer Programming," Operations Research Center Report 66-27, the University of California at Berkeley, 1966.

[600] —————; "On the Computational Status of Mathematical Programming," IBM Philadelphia Scientific Center Report No. 320-2990, June 1970.

[601] Whiting, P., and J. Hillier; "A Method for Finding the Shortest Route through a Road Network," *Operational Research Quarterly* 11, 37-40 (1960).

[602] Wilson, R.; "Stronger Cuts in Gomory's All-Integer Programming Algorithm," *Operations Research* 15(1), 155-157 (1967).

[603] Wolsey, L.; "Mixed Integer Programming: Discretization and the Group Theoretic Approach," Operations Research Center Technical Report No. 42, Massachusetts Institute of Technology, June 1969.

[604] —————; "Extensions of the Group Theoretic Approach in Integer Programming," *Management Science* 18(1), 74-83 (1971).

[605] —————; "Group Theoretic Results in Mixed Integer Programming," *Operations Research* 19(7), 1691-1697 (1971).

[606] Woolsey, R.; "IPSC, A Machine-Independent Integer Linear Program," Sandia Corporation Research Report, SC-RR-66433, July 1966.

[607] —————; "Comments of Briskin's Note," *Management Science* 17(7), 500-501 (1971).

[608] —————; "A Candle to Saint Jude, or Four Real World Application of Integer Programming," available from the Colorado School of Mines, Golden, Colorado, 1971.

[609] —————; "Quick and Dirty Methods in Time-Shared Capital Budgeting," Department of Combinatories and Optimization Research Report CORR 72-8, University of Waterloo (Canada), August 1971.

[610] —————, and H. Swanson; *The Operations Research Quick & Dirty Manual*, New York, Harper and Row, 1973.

[611] Yormark, J.; "Accelerating Greenberg's Method for the Computation of Knapsack Functions," Jet Propulsion Laboratory Report, California Institute of Technology, Pasadena, California, November 1972.

[612] Young, R.; "A Primal (All-Integer) Integer Programming Algorithm," *Journal of Research of the National Bureau of Standards* 69B(3), 213-249 (1965).

[613] —————; "New Cuts for a Special Class of 0-1 Integer Programs," Department of Economics and Mathematical Sciences research report, Rice University, October 1968.

[614] ——————; "A Simplified Primal (All-Integer) Integer Programming Algorithm," *Operations Research* **16**(4), 750-782 (1971).

[615] ——————; "Hypercylindrically Deduced Cuts," *Operations Research* **19**(6), 1393-1405 (1971).

[616] ——————; "Finite Pure Integer Programming Algorithms Employing only Hyperspherically Deduced Cuts," Department of Economics and Mathematical Sciences working paper, Rice University, May 3, 1971.

[617] Zangwill, W.; "Media Selection by Decision Programming," *Journal of Advertising Research* **5**(3), (1965).

[618] Zemel E.; "The Linear Multiple-choice Knapsack Problem," *Operations Research* **28**, 1412- (1980).

Author Index

Agin, N., 246
Almogy, Y., 643
Andrew, G., 593
Anthonisse, J., 515
Armour, G., 643
Arnold, L., 210
Art, R., 523
Ashour, S., 345

Balas, E., 8, 131, 142, 246, 298, 311-312, 317, 327-328, 332-333, 347, 381, 514, 515, 525, 527, 570, 593, 629, 669, 674
Balinski, M., 3, 90, 152, 174, 246, 264, 299, 347, 370, 374, 397, 591-593, 595, 602, 604, 637, 640, 642, 643, 653
Balintfy, J., 519
Barr, R. S., 643, 655
Bart, P., 643
Baumert, L., 335
Baumol, W., 515, 640, 643, 662
Baxter, W., 382
Bazaraa, M. S., 34
Beale, E., 246, 272, 275, 643
Beckman, M., 643
Bellman, R., 396, 514, 524, 538, 643
Bellmore, M., 210, 592-593, 632-633, 669
Benders, J., 79, 346-347,
Benichou, M., 277
Ben-Israel, A., 197
Bertier, P., 246
Bodin, L. D., 670
Bowman, E., 643
Bowman, V., 8, 132, 516
Blackburn, C., 519
Bradley, G., 298, 515-516
Breining, P., 108
Brooks, R., 514, 559
Brown, W., 519, 643
Buffa, E., 643

Burdet, C., 382
Bush, R., 515
Byrne, R., 515

Cabot, V., 514, 516
Carpaneto, G., 669
Char, A., 345
Charnes, A., 197, 515
Chen, D., 381-382, 422, 473, 491, 502-503
Christofides, N., 593, 669
Ciochetto, F., 643, 657, 667
Colmin, J., 246, 271
Cooper, L., 640, 643-644, 655
Cooper, W., 515
Cord, J., 514
Croes, G., 670
Crowder, H., 246, 299, 343

Dakin, R., 246, 264, 265, 410
Dantzig, G., 3, 34, 86, 132, 502, 514, 538, 587, 642, 660, 669
Davis, P., 643-644
Davis, R., 246, 271
Day, R., 603
Deighton, D., 643-644
DeKluyver, K., 516
DeMaio, A., 519
Denzler, D., 640, 643
Devine, M., 381, 499
Dijkstra, E., 485
Dixon, J., 114
Djerdjour, M., 312, 526
Doig, A., 246, 248, 257, 261, 298
Doyle, T., 670
Drebes, C., 640, 643, 655
Dreyfus, S., 396, 420, 514
Driebeek, N., 246, 298, 310, 336
Drysdale, P., 643
Dwyer, P., 643

Edmonds, J., 592

Efroymson, M., 642-643, 646, 658
Elmagrahby, S., 515, 527, 530
Everett, H., III, 514, 559, 560

Faaland, B., 514, 516
Fan, L., 345
Farbey, B., 502, 503,
Fayard, D., 516, 572
Feldman, E., 643
Fisher, M. L., 298
Fleischman, B., 299, 343
Floyd, R., 502-503
Forrest, J., 277
Fox, B., 514
Francis, R., 643
Freeman, R., 299, 343
Fricks, R., 4
Fuchs, L., 431
Fulkerson, D., 669

Gass, S. I., 34
Gauthier, J., 277
Garfinkel, R., 8, 515, 592-593, 602,
    604-605, 615, 635, 669
Geoffrion, A., 11, 246, 298-299, 311,
    334, 343, 347, 367, 379, 514,
    559, 593, 643
Gilmore, P., 515-516, 519-521, 523,
    546, 580
Girodet, P., 277
Glover, F., 8, 77, 194, 197-199, 210,
    215-216, 226, 311-312, 328,
    343, 381, 499, 515, 518-519,
    526, 533-534, 582, 584, 643,
    645
Golden, B. L., 670
Goldstone, L., 643
Golomb, S., 335
Gomory, R., 79, 80, 87, 108, 128, 132,
    150, 173, 176, 188, 197, 380-
    382, 410, 457, 459, 461, 463,
    467, 470, 479, 502-503, 506,
    515, 519, 521, 525
Gorry, A., 381, 503, 505-506
Granot, F., 8
Graves, G. W., 11, 347, 379
Gray, P., 643
Greenberg, H., 514, 572

Gross, D., 643
Guha, D., 593
Guignard, M., 299, 330-331, 381, 466,
    514, 516

Haas, R., 203, 240, 700
Hadley, G., 34, 37-38, 60
Hakimi, S., 643
Haldi, J., 343, 500-501, 503
Halfin, S., 86
Hamburger, M., 643
Hammer, P., 8, 298, 307, 514-515
Hansmann, F., 515
Healey, W., Jr., 519
Hefley, G., 381-382, 506
Helbig, H. K., 669
Hegerich, R., 514, 572
Hentges, G., 277
Hill, T., 515
Hillier, F., 516
Hirsch, W., 642-643, 660
Hirst, J., 277
Hoffman, A., 87, 132, 593, 643
Horowitz, E., 514, 542, 543, 572
House, R., 593, 632, 634
Hu, T. C., 381, 396, 400, 481, 485,
    502-503, 546
Hurter, A., Jr., 514, 516
Hwang, C., 345

Ingargiola, G. P., 559, 570
Ignizio, J., 643
Issacson, L., 500-501, 503

Jain, A., 643, 666
Jambekar, A., 299, 345
Jarvis, J., 34
Jeroslow, R., 86, 246, 298-299, 381-
    382, 467
Johnson, E., 80, 343, 381-382, 461,
    463, 467, 592
Johnson, S., 669
Jones, A., 643

Kaplan, S., 514, 563
Karel, C., 246, 669, 678
Kendell, K., 515, 527, 538
Kendrick, D., 246, 271
Kennington, J., 8, 382, 643, 655

Kernighan, B. W., 670
Khumawala, B., 643
Kianfar, F., 526
Klingman, D., 515, 643, 645
Kolesar, P., 514, 516, 568, 571
Koncal, R., 593, 630
Koopmans, T., 643
Korman, S., 593
Korsh, J. F., 559, 570
Kortanek, K., 86, 382, 515
Krabek, C., 593
Kraft, D., 515
Kuehn, A., 643
Kuhn, H., 643, 662
Kumar, S., 345

Land, A., 246, 248, 257, 261, 298, 502-503
Landi, D., 514
Lang, S., 99, 431, 493
Lasdon, L., 347, 519
Lasky, J., 593
Lawler, E., 246, 299, 335, 669-670,
Leach, C., 25
Lee, J., 643, 657, 667
Lehrer, F., 643
Lemke, C., 298-299, 314, 316, 322,
    593, 604-605, 607-609, 612-
    613, 618, 623-624, 636
Lenstra, J. K., 669-670
Levy, J., 643
Leibman, J., 643
Liggett, J., 519
Links, G., 381, 499
Lin, S., 670
Little, J., 246, 678, 669
Litzler, L., 381
Lorie, J., 514, 516
Love, R. F., 670, 688

Maffei, R., 643
Malone, J. C., 669
Manne, A., 643
Mao, J., 515
Marazana, F., 643
Marks, D., 643
Marschak, T., 643
Marshall, L., 592

Marsten, R., 246, 299, 593, 636, 643
Martello, S., 514, 516, 542, 568, 570,
    572, 577
Martin, G., 135, 137, 593, 643
Mathews, G., 515, 527, 528
Mathis, J., 203, 240
Mathur, K., 18-19, 25, 203, 240, 246,
    276, 312, 382, 515, 519, 526
Mehta, S., 193, 211, 235
Meihle, W., 643
Michaud, P., 593, 637
Mitten, L., 246
Morito, S., 18, 19, 246, 276, 381-382,
    412, 422, 491-492, 501-502,
    515, 519
Murchland, J., 502-503
Murty, K., 34, 60, 246, 640, 669, 678

Naslund, B., 515
Nauss, R. M., 514, 516, 570, 572
Nelson, L., 593, 632, 634
Nemhauser, G., 132, 514, 593, 615,
    635
Ness, D., 514, 516, 563
Netto, E., 335
Nishimura, K., 18, 19, 276
Norback, J. P., 670, 688
Norman, R., 592

Onyekwelu, D. C., 515

Padberg, M., 246, 299, 343, 515, 593,
    629
Petersen, C., 8, 299, 343
Pierce, J., 523, 593, 637
Pinkus, C., 643
Plateau, G., 516, 572
Popolopas, L., 643
Prather, K., 514

Quandt, R., 515, 592, 595, 602

Rabin, M., 592
Radhakrishnan, S., 515
Rado, T., 593, 632, 634
Raghavachari, M., 8, 21
Rao, A., 593, 638
Rardin, R., 347
Ratliff, H., 593, 631
Ray, T., 643-644, 646, 658

Ray-Chaudhuri, D., 592
Revelle, C., 643
Ribiere, G., 277
Rinnooy Kan, A. H. G., 669-670
Robers, P., 643
Rosenberg, I., 514
Rosenkrantz, D. J., 670
Roth, R., 593
Rothstein, M., 592
Roveda, C., 519
Roy, B., 246
Rubin, D., 86, 381, 499
Rubin, J., 592-593, 628
Rudeanu, S., 8, 514-515
Rutten, D., 347

Sa, G., 643
Saaty, T., 112, 114
Saha, J., 593, 594, 632
Sahni, S., 514, 542-543, 572
Salkin, H., 8, 18, 19, 25, 94, 108, 159,
        193, 203, 208, 211, 235, 240,
        246, 276, 298-299, 312, 318,
        320, 339-341, 376, 382, 515,
        516, 519, 526, 593-594, 604-
        605, 607-609, 612-613, 618,
        623-624, 630, 632, 635, 636
Sandiford, P., 643
Savage, L., 514, 516
Schwinn, R., 410
Shannon, R., 643
Shapiro, D., 669
Shapiro, G., 514, 564
Shapiro, J., 381-382, 491, 501-502, 503,
        505-506
Shareshian, R., 298, 310
Shroff, P., 193, 211, 235
Shycon, H., 643
Shmoys, D. B., 669-670
Sinha, P., 515, 519
Small, R., 246, 272, 275
Smith, T. H. C., 669
Soland, R., 643
Solow, D., 34, 38, 59-60, 64
Sommer, D., 77
Spielberg, K., 3, 246, 269, 271, 298-
        299, 310, 314, 316, 318, 322,
        339, 341, 381, 467, 593, 604-

        605, 607-609, 612-613, 618,
        623-624, 636, 643, 666
Spitzer, M., 600
Srinivasan, V., 669
Steinberg, D., 299, 345, 640, 643, 660,
        665
Steinmann, H., 410
Stewart, W. R., 670
Swanson, H., 643, 657, 667
Swart, W., 518
Sweeney, D., 246, 669, 678

Tabak, D., 592
Taha, H., 8
Teitz, M., 643
Thirez, H., 593
Thomas, M., 381-382, 506
Thompson, G., 246, 285, 669
Tillman, F., 345
Tomlin, J., 246, 272, 277, 284, 519,
        643, 655
Tompkins, C., 382, 643
Toth, P., 514, 516, 542, 568, 570, 572,
        577
Trehan, S., 643
Trubin, V., 593
Tuan, N., 299
Tucker, A., 18, 367
Tui, H., 142

Ullman, Z., 514
Unger, V., 347, 382, 515, 643, 655

Vietorisz, 643
Vincent, D., 277

Wahi, P., 516
Wagner, H., 514, 602
Walker, W. E., 643, 655, 665
Wallingford, B., 515
Wanty, J., 643
Watters, L., 8
Weingartner, H., 514, 516, 563
Weitzman, M., 246, 271
White, W., 151, 167, 381-382, 410,
        413, 420, 468-469, 499, 502-
        503, 519
Wig, M., 515, 527, 530
Wilson, R., 184

Wolfe, P., 347, 370, 640, 643
Wolsey, L., 298, 381-382, 503, 505-506
Wood, D., 246
Woolsey, R., 515, 533, 582, 584

Yormark, J., 514
Young, R., 142, 197-199, 201, 206, 208-209, 211, 215, 221, 235

Zangwill, W., 8
Zemel, E., 514, 515, 570
Zionts, S., 298-299, 311, 328, 343, 381, 382, 422, 473, 491, 502-503, 515, 527, 538
Zoltner, A. A., 515, 519

Subject Index

Accelerated Euclidean Algorithm, 129, 135-142, 634

Acceptable source row selection rules (see Simplified Primal Algorithm)

Additive algorithm, 298-299, 343

Aggregating constraints, (see also transformations) 526-538

Airline crew scheduling, 23, 592, 600-602

Assembly line balancing, 592

Assignment problem, 74-76, 286, 671

Assignment algorithm, 695

Asymptotic problem, (see also Groups and group theory) 408, 471-472

Basic feasible solution, 33-34

Beale tableau, 62-67

Bound Escalation Algorithm, 194-196

Branch and bound, 245-297, 298-299, 381-382, 410-413, 467, 472, 501-502, 508-512, 514, 519, 538, 554-559, 586
    algorithm, 261-263
    the Dakin variation, 264-267
    enumeration tree, 255-257
    geometric interpretation, 248-255
    illustrations, 257-260, 263-265, 267
    implicit enumeration result, 248-255
    specialized for the plant location problem, 643-655, 666-668
    specialized for the zero-one problem, 267-269, 567-578
    (see also Enumeration)

Canonical form, 36

Capital budgeting, 513-519, 563, 578-579

Capital investment, 513, 592

Circuit design, 592

Complementing variables, 318-319, 331, 413, 526, 530-533, 556

Complementary slackness, 56-57

Complete reduction, 316, 330

Composite algorithms, 186-187, 192, 334-335

Composite Gomory cut, 135-136, 138-139, 643

Computational results, 80
    with algorithms for the fixed charge and plant location problems, 664-668
    with algorithms for the knapsack problem, 516, 519, 522
    with algorithms for the set covering and set partitioning problems, 593, 633-638
    with branch and bound algorithms, 277
    with group theoretic algorithms, 382, 466-467, 499-507
    with partitioning algorithms, 379
    with search algorithms, 343-345

Concave function, 248, 656-657

Cone, 405-407

Congruence, 99-100, 384

Convex hull, 685, 702

Convexity, 29-30, 34

Convexity cut (see Geometrically derived cut)

Corner polyhedra, 443-445, 446-449, 475
    faces of, 446, 450-451
    (see also Master polyhedra)

Cosets, 430-432, 475

Cover, 593-594

Cutting plane algorithm (dual all integer), 170-196
    algorithm, 170-172
    convergence, 182, 188-190
    illustrations, 173-179

properties, 184-186

tableau and notation, 171-173

Cutting plane algorithm (dual fractional integer programming),
84-149

algorithm, 84-85

convergence, 121-123

illustrations, 90-99

properties, 102-116

strategies, 116-118

tableau and notations, 87-90

Cutting plane algorithm (dual fractional mixed integer programming), 150-169

algorithm, 150-151

convergence, 151, 163-166, 167-
169

illustration, 152-155

properties, 154-155

tableau and notation, 151-152

Cutting plane algorithm (primal all
integer), 197-244

algorithms, 200-201, 211-212

convergence, 228-233

cycling, 240-244

illustrations, 201-203, 216-225, 240-
244

properties, 213-215, 226-228

rudimentary primal algorithm, 199-
200

simplified primal algorithm, 204-
213

tableau, 198-199

Cutting planes, (see also Geometrically derived cuts, Gomory
all integer cut, Gomory fractional cut) 79, 380, 466, 526,
593, 630-633, 633-635

Cutting stock problem, 515, 519-522,
580-581, 592

two dimensional, 522-523, 580-581

Dantzig cut, 132-135

Direct algorithm, 197

Disconnecting paths, 597-599, 627-628

Dominance, 605

Dual problem, 53-54

Dual lexicographic simplex method,
60, 64-66, 87-99

basis relationships, 104-108

illustrations, 90-94

notation, 87-89

properties, 94-99, 121-122, 170,
181

Duality, 53-61

Duality in integer programming, 515

Dual simplex algorithm, 57-61, 69

Dynamic programming, 381, 395-405,
420, 471, 499-502, 514, 517,
538-554, 578-579, 585-586

Elementary row (column) operations
(see Smith normal form)

Enumeration, 79, 298, 347, 365, 381,
410-415, 420, 466-467, 514,
562, 593, 635-639, 643, 667
(see also Branch and bound,
Search, Group minimization
problem (algorithms for))

Euclidean algorithm, 114, 493

Euclidean distance, 405

Extreme half line, 33-34, 68

Extreme homogeneous solution, 33

Extreme point, 29-31, 34

Factor group, 426, 432-434 (see also
Cosets)

Farkas' lemma, 366

Filter algorithm, 312, 328-329

Fixed charge problem, 640, 644-646,
657-658

general fixed charge problem, 664-
666

as the uncapacitated plant location problem, 666-667
(see Plant location problem)

Fixed cost transportation problem, 641-
643, 652-656

Four color problem, 24, 592

Generalized upper bounding, 518-519

Geometrically derived cut, 142-149, 382

algorithm use, 146-147

degeneracy situations, 149

illustration, 143-146

strength, 149

Gomory all integer cut, 170-196, 505,
593, 630-631

derivation, 179-181
form of, 172-173
relation to Gomory fractional cut, 186
Gomory fractional cut (integer program), 84-149, 387-388, 447, 470-471
　analytical derivation, 99-102
　form of, 89-90
　geometric derivation, 118-121
　illustration, 102
　properties, 102-116, 118-121
　(see also Composite Gomory cut)
Gomory fractional cut (mixed integer program), 150-169
　derivation, 155-159
　form of, 151-152
　illustration, 152-155
　use in integer programming algorithm, 159-163
　use in penalty calculation, 275-276
　weak form of, 166
Graphical solution, 28-31, 68, 70-73, 80-81
Graph theory, 246
Groups and group theory, 79-80, 380-512
　abelian group, 380
　cyclic group, 380
　(see also Group minimization problem)
Group minimization problem, 380-393, 513, 519, 525, 545, 580, 593, 643
　algorithms for, 394-415
　computational experience, 497-507
　derivation, 383-385
　geometric interpretations, 443-467
　illustration, 386-387
　network equivalence, 415-420
　representing mixed integer program, 382, 467, 481
　representing zero-one integer program, 381-382, 407, 481
　(see also Cosets, Factor group, Groups and group theory)

Hermite normal form, 516

Heuristics, 76-78, 83, 523
　applied to the fixed charge and plant location problem, 640-643, 645, 658, 664-667
　applied to the knapsack problem, 569
　applied to the set covering and set partitioning problem, 593, 620, 628
　(see also Lagrangian multiplier methods)
Homogeneous solution, 33,68
Homomorphism, 430

Implicit enumeration, 79, 160, 298-299, 303, 322, 412, 616, 658-659 (see also Branch and bound, search)
Information retrieval, 592, 603
Intersection cut (see Geometrically derived cut)
Irreducible point, 476
Isomorphic group, 426, 430, 434-436, 474-475

Journal selection problem, 515, 518

Knapsack function, 523, 580-581
Knapsack problem, 15, 23-24, 311-312, 329, 381, 513-589, 638
　algorithms for, 538-567
　applications of, 516-526
　multidimensional, 517
　periodic property, 549-550
　quadratic objective function, 514-515
　solved as a linear program, 447
　uses, 526-538

Lagrangian multiplier methods, 514, 517, 538, 559-564, 643
Lexicographic order, 64
Linear programming,
　algorithm, 28-29, 35-42, 42-47, 47-53, 57-61, 67-69
　alternate optimal solutions, 30
　canonical form, 36
　cycling, 38
　degeneracy, 33, 60

dual problem, 53-54
duality, 53-61
dual lexicographic simplex method,
    60, 64-66
dual simplex algorithm, 57-61, 69
graphical solution, 28-31, 67
primal problem, 54
reduced costs, 37
revised simplex algorithm, 47-53
simplex multipliers, 48, 57
weak duality, 54-55
Loading problems, 524
Lock box problem, 657-658, 667

Mappings between groups, 426, 430,
    474-475
Master polyhedra, 453-461, 477
Matching problem, 595-597, 626
Maximum flow problem, 599-600
Minimum cost network flow, 74
Minimum cut problem, 598-599
Multibranch-branch and bound, 271
Multidimensional knapsack problem,
    517
Multiple choice constraints, 517-518,
    526, 580

Network, 3, 74, 643
    use in group theory, 381-382, 415-
        422, 473, 481-483, 485-492
    use in knapsack problem, 514, 538,
        564-567
    use in set covering problem, 591-
        600
    (see also Disconnecting paths, Match-
        ing problem, Node-arc inci-
        dence matrix, Node covering
        problem)
Node-arc incidence matrix, 597
Node covering problem, 595-597, 625
Nonlinear integer programming, 336-
    337, 365-367, 515, 544, 560,
    585, 587

Parametric linear programming, 250-
    255
Partitioning, 79, 346-379, 643, 651
    algorithm, 352-356

convergence, 360-363
properties, 356-359
transformation to integer program,
    347-352
Penalties, 272-277, 309-310, 318, 327,
    334, 472 (see also Branch and
    bound)
Piecewise linear function, 6-8, 665
Plant location problem, 9-11, 382, 639-
    667
    algorithms for, 646-652
    computational results, 666-667
    solved using a partitioning algo-
        rithm, 347, 368-379
Point or node algorithm, 302-304, 340-
    341, 660, 666-667
    specialized for the set covering prob-
        lem, 615-616, 618-625
    (see also Search)
Political redistricting, 592, 602-603
Polynomial programming problem, 8
Precedence constraints, 518
Preferred sets, 314-317
Preferred variable, 316, 620-621
Projection, 248-249
Project selection, 513

Quadratic programming problem, 8,
    21-22, 514-515
Quadratic knapsack problem, 514-515

Recursive relationship, 396, 400, 420-
    421, 539-540, 542-545, 550,
    585, 587-589 (see also Dynamic
    programming)
Reduced costs, 37
Redundant cutting planes, 94, 191, 356-
    359, 458-460
Reference equation (see Simplified pri-
    mal algorithm)
Revised simplex algorithm, 47-53
Routing, 592
Rudimentary primal algorithm, 199-
    201 (see also Cutting plane
    algorithm (primal all integer))

Scheduling, 3, 592

Search, 298, 345, 381, 514, 517, 579,
        603, 660, 666-667
    algorithm, 299-304
    computational results, 299, 318-
        319, 343-345
    generalized origin, 318-321
    implementation, 339
    implicit enumeration criteria, 304-
        314, 339-342
    linear programming uses, 308-309
    specialized for the set covering prob-
        lem, 614-625
    tree, 300-304
    (see also Enumeration, Point or
        node algorithm)
Sequencing, 3
Set covering problem, 13-15, 329-330,
        590-637
    algorithms for, 614-625
    application, 600-603
    computational results, 633-638
    network equivalence, 593-600
    properties, 603-614
    solved as a linear program, 607-
        614
Set partitioning problem, 13-15, 590-
        637 (see also Set covering prob-
        lem)
Shortest route, 481-482, 485-491, 514,
        564-567
Simplex algorithm, 31, 35-42, 68-69
Simplex multipliers, 48, 57
Simplified primal algorithm, 204-213
        (see also Cutting plane algo-
        rithm (primal all integer))
Single branch-branch and bound, 271
Smith normal form, 423-424
Stationary cycle, 209, 235-236

Stopped simplex method, 276
Surrogate constraint, 216, 310-314, 323-
        325, 334, 525-526
Symbolic logic, 592

Transformations, 2-4, 8, 20-22, 347-
        352, 515, 526-538, 581-582,
        605-607
Transition cycle, 209, 235-236
Transportation problem, 74, 81-82, 359
    use in solving the plant location
        problem, 640
Transshipment problem, 82
Traveling salesman problem, 16-18, 24,
        146, 275, 286-287, 382, 669-
        705
    algorithms, 673-691
    approximate algorithms, 682
    branch and bound algorithm, 673
    branching rules, 677
    bounding rules, 675
    cheapest insertion, 683
    convex hull, 685, 702
    edge exchange, 703
    mathematical formulation, 670
    network, 669
    subtours
    tour improvement, 689
    two-edge exchange, 703
Tree, 556-558, 615 (see also Branch
        and bound, Search, Shortest
        route)
Truck dispatching, 592, 602

Unimodular, 74
Unimodularity, 74-76, 81-82

Vehicle routing problem, 695